OXFORD ENGINEERING SCIENCE SERIES

THE OXFORD ENGINEERING SCIENCE SERIES

---

1 D. R. RHODES: *Synthesis of planar antenna sources*
2 L. C. WOODS: *Thermodynamics of fluid systems*
3 R. N. FRANKLIN: *Plasma phenomena in gas discharges*
4 G. A. BIRD: *Molecular gas dynamics*
5 H.-G. UNGER: *Planar optical waveguides and fibres*
6 J. HEYMAN: *Equilibrium of shell structures*
7 K. H. HUNT: *Kinematic geometry of mechanisms*
8 D. S. JONES: *Methods in electromagnetic wave propagation* (Two volumes)
9 W. D. MARSH: *Economics of electric utility power generation*
10 P. HAGEDORN: *Non-linear oscillations* (Second edition)
11 R. HILL: *Mathematical theory of plasticity*
12 D. J. DAWE: *Matrix and finite element displacement analysis of structures*
13 N. W. MURRAY: *Introduction to the theory of thin-walled structures*
14 R. I. TANNER: *Engineering rheology*
15 M. F. KANNINEN and C. H. POPELAR: *Advanced fracture mechanics*
16 R. H. T. BATES and M. J. McDONNELL: *Image restoration and reconstruction*
18 K. HUSEYIN: *Multiple-parameter stability theory and its applications*
19 R. N. BRACEWELL: *The Hartley transform*
20 J. WESSON: *Tokamaks*
21 P. B. WHALLEY: *Boiling, condensation, and gas-liquid flow*
22 C. SAMSON, M. LeBORGNE, and B. ESPIAU: *Robot control: the task function approach*
23 H. J. RAMM: *Fluid dynamics for the study of transonic flow*
24 R. R. A. SYMS: *Practical volume holography*
25 W. D. McCOMB: *The physics of fluid turbulence*
26 Z. P. BAZANT and L. CEDOLIN: *Stability of structures: principles of elastic, inelastic, and damage theories*
27 J. D. THORNTON: *Science and practice of liquid-liquid extraction* (Two volumes)
28 J. VAN BLADEL: *Singular electromagnetic fields and sources*
29 M. O. TOKHI and R. R. LEITCH: *Active noise control*
30 I. V. LINDELL: *Methods for electromagnetic field analysis*
31 J. A. C. KENTFIELD: *Nonsteady, one-dimensional, internal, compressible flows*
32 W. F. HOSFORD: *Mechanics of crystals and polycrystals*
33 G. S. H. LOCK: *The tubular thermosyphon*
34 A. LINÁN and F. A. WILLIAMS: *Fundamental aspects of combustion*
35 N. FACHE, D. DE ZUTTER, and F. OLYSLAGER: *Electromagnetic and circuit modeling of multiconductor transmission lines*
36 A. N. BERIS and B. J. EDWARDS: *Thermodynamics of flowing systems: with internal microstructure*

Thermodynamics of Flowing Systems

with internal microstructure

ANTONY N. BERIS

Department of Chemical Engineering
University of Delaware

BRIAN J. EDWARDS

Department of Chemical Engineering
University of Illinois
Urbana-Champaign

New York Oxford
OXFORD UNIVERSITY PRESS
1994

Oxford University Press

Oxford New York Toronto
Delhi Bombay Calcutta Madras Karachi
Kuala Lumpur Singapore Hong Kong Tokyo
Nairobi Dar es Salaam Cape Town
Melbourne Auckland Madrid

and associated companies in
Berlin Ibadan

Published by Oxford University Press, Inc.,
200 Madison Avenue, New York, New York 10016

Library of Congress Cataloging-in-Publication Data
Beris, Antony N., 1957–
Thermodynamics of flowing systems : with internal microstructure /
Antony N. Beris, Brian J. Edwards.
p. cm.—(Oxford engineering science series : 36)
Includes bibliographical references and index.
ISBN 0-19-507694-X
1. Transport theory—Mathematical models.
2. Microstructures—Mathematical models.
3. Thermodynamcs—Mathematical models.
4. Hamiltonian systems. 5. Viscoelasticity. 6. Poisson brackets.
I. Edwards, Brian J., 1964– . II. Title. III. Series.
QC175.2.B47 1993
530.4'25—dc20 93-20886

2 4 6 8 9 7 5 3 1

Printed in the United States of America
on acid-free paper

To Deborah and Gretchen

Preface

I believe in keeping things as simple as possible, but not simpler.
—Albert Einstein

As the number of applications for complex materials continues to increase, from tennis rackets to hypersonic, transatmospheric vehicles, so does the desire for a better theoretical description of their behavior during processing. Complex materials are substances with an internal microstructure, such as advanced composites, polymer blends, liquid crystals, and electrorheological materials. They accordingly exhibit highly unusual properties with respect to conventional materials, which can prove to be beneficial in a great number of applications. With the properties of these materials becoming the limiting barrier to new technological applications, the interest in a theoretical understanding of the structure/property relationship and, more importantly, of how this structure is created and altered during processing, is of immediate concern. This book aims to discuss, in accordance with the preceding motto, the simplest possible way to describe transport phenomena within structured media without unduly sacrificing the capability of the theoretical representation to describe the underlying physics. Of course, when one says that something is the "simplest possible," he does not necessarily imply that it is simple. The description of certain physical phenomena can be inherently complex, at least to an uneducated eye.

The advent in the late 1970s of high-performance supercomputers, together with the spectacular successes of numerical simulations in the realm of Computational (Newtonian) Fluid Dynamics (CFD), led to a belief that similar breakthroughs in materials-processing design and optimization were not far away. However, one crucial difference between the Newtonian fluid dynamics and materials processing is that the complexity of the phenomena taking place in the latter is significantly greater. One no longer has a microscopically "structureless" medium which can safely be abstracted as a continuum described by a small number of macroscopic parameters (as, for example, the velocity and the pressure), which are functions of time and space only. Instead, one needs to recognize the presence of an internal microstructure, the time evolution of which is tightly coupled with the manufacturing process. Clearly, any progress in the area of macroscopic simulations of the processing of complex, "structured" media relies heavily on the successful modeling of the thermodynamics and transport phenomena from a continuum perspective.

The underlying physics of the thermodynamics and transport phenomena within structured media is quite complex and is the subject of active and extensive investigation. To a large degree, the methods used in these investigations (e.g., statistical mechanics, kinetic theory, renormalization theory, etc.) depend upon the nature of the medium (i.e., solid, fluid, plasma, etc.) under consideration. It is impossible for any work of the size of the present one to comprehensibly cover this very large subject and, consequently, this is not the objective of this book. Instead, this book will attempt to provide a mechanism by which the necessary information from the above-mentioned microscopic investigations can be transferred into a consistent description of dynamical phenomena from a macroscopic, continuum perspective. Despite the wealth of information that is now becoming available on the microscopic description of phenomena through, for example, the specification of the distribution function of several characteristic parameters of the problem (such as the end-to-end distance of polymer chains, or the correlation function of particle aggregations), it is only the reduction of this information to a few macroscopic parameters which might allow any progress in the simulation of processes within complex, multi-dimensional geometries by preserving the lower dimensionality of the problem.

This reduction is accomplished, for example, in the thermodynamic description of phase equilibria, leading to models valid for large time and space scales. When flow is involved, the equilibrium thermodynamic description needs to be extended to accommodate changes taking place within shorter length and time scales. The hope is that there exist experimentally meaningful minimum length and time scales above which a "local thermodynamic description" is valid, thus allowing the description of the system through only a few macroscopic variables characterizing both the internal structure and the flow kinematics. This situation is similar to the quantum orbital calculations of complex chemical structures which, if done properly, require the solution of the Schrödinger equation for the wave function of the electrons of dimensionality, which is equal to the number of electrons present multiplied by three! Instead, several simplifications are used, the most popular of which requires only the solution of the electron-density function in a three-dimensional space.

This reduction of dimensionality is necessary because of the significant increases in computational requirements accompanying any increase in the dimensionality of the problem. Characteristically, most of the CFD simulations in the literature refer to the solution of a two-dimensional problem (planar or axisymmetric flows). It is only within the previous decade that fully three-dimensional calculations have become possible through advances both in computational power (supercomputers like Cray in the Gigaflop range of performance) and computational algorithms (spectral methods, conjugate gradients, multigrids, etc.). Even the computational power of the forthcoming parallel computers (offering another factor of 1000—Teraflop performance!) is sufficient to accommodate only, perhaps, one additional dimension in the numerical calculations. However, this additional computational power can allow, together with the evaluation of the traditional macroscopic flow variables (such as pressure, velocity, etc.), for the solution of several internal variables, describing the microstructure of the medium under investigation. This is the motivation for undertaking the study for which the results are presented in this book.

System theory tells us that the dynamical behavior of a system can be described in general following either one of two alternative approaches. In the first one, the output (i.e., measurable quantities) at a specific time is modeled as a memory functional over the time history of all the variables (both input and output) of the system. This is mathematically expressed through general but complex, nested integrals expressions. In the second approach, the time evolution of the output variables is expressed in terms of partial differential equations (in general) which involve, in addition to the input and output variables of the system, internal parameters characterizing the state of the system. Additional equations are then, in turn, provided which dictate the time evolution of these new variables. It is interesting to note here that, from an abstract theoretical point of view, the two approaches are exactly equivalent (see §1.2); i.e., the same information (knowledge) about the system can be presented interchangeably using either one of the above approaches.

In the present work, the internal variables approach is followed, primarily for the following reasons: first, because excellent treatises are already available where the memory-functional description is explained in detail (see §1.1); second, because the use of memory functionals defeats our stated objective of providing a mathematical modeling approach that would result in equations which are easily solvable numerically. This is especially true when multidimensional, time-dependent phenomena are under investigation where the memory-functional approach leads to integro-differential equations which are, at least for the foreseeable future, computationally intractable. Finally, we believe that in most cases a macroscopic, internal-variable description can most easily be connected with the underlying physics due to a natural association of the internal variables with some measure of the internal microstructure. Thus, this book proceeds down the internal-variable path, although in a non-traditional way: the internal variables which are used here are directly associated with the microstructure itself; they do not necessarily involve the transport fluxes, as sometimes required by other versions of extended (irreversible) thermodynamics.

The traditional engineering approach to the modeling and simulation of transport phenomena in continua involves the solution of the conservation equations (mass, momentum, and energy) together with constitutive equations for the fluxes. The latter are algebraic or differential equations which are either phenomenologically postulated or derived from the truncation of an infinite (in general) series of equations, resulting from a statistical description of the underlying microstructure in terms of suitable distribution functions. There are several disadvantages associated with either one of the above approaches. If phenomenologically postulated, the resulting equations usually necessitate redundant parameters to describe the effect of the same factor on several evolution equations or, even worse, can fail to describe the coupling between different transport processes altogether. If microscopically derived, the resulting truncated equations might lead to a thermodynamically aphysical behavior due to the fact that the effect of the truncation is, in general, unpredictable. Although this last problem can be rectified, in principle, by an *a posteriori* calculation of extended thermodynamic quantities, such as the internal entropy production, this action is seldom followed in current practice because of the inherent mathematical complexity of the calculations and an ambiguity in the definitions.

In this book, we directly incorporate in the mathematical formalism such fundamental thermodynamic concepts as the Hamiltonian—an extended free energy—and the rate of entropy production. It is actually through the definition of the above quantities as functionals of the state variables over the system's volume that the governing equations are subsequently derived following a systematic and mathematically straightforward procedure. The only other pieces of information that are necessary are the definitions of an appropriate Poisson and dissipation bracket.

Since the early 1800s, the Poisson bracket formalism of dynamical equations of motion has been used very successfully for the description of dynamics through the Hamiltonian for either discrete or continuum conservative (non-dissipative) systems. It has distinct advantages in terms of its generality: it applies to almost all areas of physics, from classical mechanics and electrodynamics to quantum mechanics and the new theories of subatomic particles, and it is applicable to nonlinear stability analyses. However, its restriction to conservative media has limited its impact to only a few applications of engineering interest. This limited range of applications, coupled with the sophisticated mathematical language of the principal work on the subject, has discouraged the use of the Poisson bracket techniques by the majority of engineers and applied scientists working with applications of transport phenomena within real (dissipative) materials.

The principal objective of this book is to present a framework for modeling transport phenomena in complex systems, a framework which is based on a thermodynamically consistent generalization of the Poisson bracket formalism and which is able to describe dissipative processes. The foundations of the formalism are laid in the first part of the book, whereas specific applications are discussed in detail in the second. As seen in Part I, this new description is based on a consistent combination of the non-canonical Poisson bracket of dissipationless continuum systems and the local-thermodynamic-equilibrium, internal-variable representation of extended irreversible thermodynamics. These two theories actually can be recovered as subcases of the more general description, valid in the absence of dissipation or flow, respectively. The applications presented in Part II include the modeling of viscoelastic transport phenomena, the description of relaxation phenomena associated with heat and mass transfer, the modeling of flow-induced phase transitions discussed in general as well as within the specific context of transport phenomena in liquid-crystalline systems, the description of transport and reaction phenomena in multi-fluid systems, and others. Although these applications are discussed with the bracket formalism in mind, an effort has been made to simultaneously present a comprehensive review of, in the authors' opinion, the most significant progress in modeling in these areas today. The material in any one of the chapters of the second part is independent from the others; thus, any one of them can be studied individually in any order. We believe the modeling overview can profit the researcher interested in the corresponding field irrespective of whether or not he is interested in the bracket formalism. Moreover, despite the fact that the applications presented in this monograph are by necessity limited to the areas of expertise of the authors, we believe that the formalism has a much wider applicability.

For each system investigated in this book, application of the bracket formalism leads to a set of governing equations which contain, in general, a number of parameters. Simultaneously, a set of constraint inequalities is derived, the satisfaction of which guarantees the thermodynamic admissibility of the equations. Although the parameters can, in principle, be determined phenomenologically through fitting of experimental data, the power of the formalism is best harnessed when it is used in conjunction with a microscopic model for the underlying dynamics. In this fashion, the macroscopic (continuum) equations are expressed in terms of a few physically meaningful parameters, with their values chosen so that the thermodynamic consistency of the resulting macroscopic model is guaranteed. This approach is best illustrated in the applications to viscoelastic and liquid-crystalline media discussed in Part II. As demonstrated there, the formalism does not replace the need of an understanding of the dynamic phenomena at the microscopic level in order to describe the macroscopic behavior, as it might appear to a casual user of the formalism. In contrast, it makes the undertaking of an investigation of the physical phenomena a necessary step if an *a priori* predictive capability of complex, dynamic behavior is sought. The authors of this book hope that the formalism will be the medium through which information is effectively transferred from the microscopic to the macroscopic level. This transfer of information from one level of description to the next is of course related to the ideas on phase-space reduction associated with projection operators in statistical mechanics. Although there are a few comments dispersed throughout the book on how a reduction in the number of variables can lead to changes in the dissipation, this very important theoretical issue is not explored further here, since it is felt that it lies outside the practical scope set forth for the present monograph. However, it is anticipated that investigation of this issue might lead to potential new discoveries in the future.

By necessity, for didactic purposes, many of the examples discussed in Part I, as well as some which are referred to in Part II, employ the bracket formalism to develop equations already existing in the literature. However, several of the model systems presented in Part II would be nearly impossible to obtain with the previously existing methodology. Even in cases where the application of the bracket formalism does not lead to new results, it very often leads to new insight, based on a clearer understanding of the thermodynamic nature of each one of the terms which are incorporated into the models. In addition, with the basic information provided in this book, it is anticipated that the reader will be able to apply the exposed theoretical framework to other areas of applications. These might include the description of transport phenomena in suspensions (which are only mentioned briefly as an example in chapter 7), free surface problems, composite materials, etc.

Putting this book together has been a great learning experience for us, as we have been forced to become fluent in a wide range of scientific arenas. In writing the basics of chapters 2, 3, and 6, we have relied heavily upon the works of Guillemin and Sternberg [1984], Lanczos [1970], and Woods [1975], respectively, to provide us with the insight to accomplish the purposes of these chapters. If imitation is the most sincere form of flattery, then we have flattered these authors a great deal indeed.

The first part of this book resulted from an outgrowth of the notes prepared for a course on the same subject which was jointly taught by the authors in the Department of Chemical Engineering at the University of Delaware during the spring semesters of 1990 and 1991. The help and constant encouragement received by the then chairman of the department, T.W. Fraser Russell in setting up this course is most gratefully appreciated. The second part of the book relies heavily on a series of papers published in various scientific journals as well as from our students' term projects prepared within the context of the above-mentioned course. We thank V. Mavrantzas, N. Kalospiros, M. Avgousti, A. Souvaliotis, and J. Gustafson for their input during the course and for their help later in the preparation of selected material for the book, as more specifically acknowledged in the corresponding places. We would also like to thank Stevan Wilson for his help in the preparation of some of the graphs and figures, and Gretchen Edwards for her valuable help in the proofreading. Furthermore, we thank Prof. G. Marrucci and Dr. K. Wissbrun for their helpful comments on a draft of chapter 11, and Prof. A. McHugh for his constructive comments on chapter 5. We would also like to acknowledge Prof. M. Grmela for introducing us to this subject of research. In addition, we thank the Center for Composite Materials at the University of Delaware for providing partial financial support for this work in an era in which fundamental research is consciously neglected. Moreover, the encouragement received at various stages of this work from the director of CCM, Prof. R. McCullough, and from Prof. A. Metzner is gratefully acknowledged. ANB would also like to acknowledge additional funding support received from NSF, division of hydraulics (S. Traugott, director), and a gift from Schlumberger Foundation, as well as the moral support received at various stages during this long project by Profs. M. Crochet and R. Tanner and Dr. A. Pearson. Last, but not least, the very significant help received during the long hours of the typesetting and editing of the final manuscript by Deborah Beris and Vlasis Mavrantzas is immeasurably acknowledged.

Newark, DE *A.N.B.*
Urbana,IL *B.J.E.*
January, 1994

Contents

p. xvii: the page number for §8.2.3 should be 321

Part I

Theory

The whole question of imagination in science is often misunderstood by people in other disciplines. They try to test our imagination in the following way. They say, "Here is a picture of some people in a situation. What do you imagine will happen next?" When we say, "I can't imagine," they may think we have a weak imagination. They overlook the fact that whatever we are allowed to imagine in science must be *consistent with everything else we know*; that the electric fields and the waves we talk about are not just some happy thoughts which we are free to make as we wish, but ideas which must be consistent with all the laws of physics we know. We can't allow ourselves to seriously imagine things which are obviously in contradiction to the known laws of nature. And so our kind of imagination is quite a difficult game. One has to have the imagination to think of something that has never been seen before, never been heard of before. At the same time the thoughts are restricted in a straightjacket, so to speak, limited by the conditions that come from our knowledge of the way nature really is. The problem of creating something which is new, but which is consistent with everything which has been seen before, is one of extreme difficulty.

—*Richard Phillips Feynman*

1
Introduction

It seems that if one is working from the point of view of getting beauty in one's equations, and if one has really a sound insight, one is on a sure line of progress.

—*Paul Adrien Maurice Dirac*

1.1 Overview

The investigation of dynamical phenomena in gases, liquids, and solids has attracted the interest of physicists, chemists, and engineers from the very beginning of the modern science. The early work on transport phenomena focussed on the description of ideal flow behavior as a natural extension to the dynamical behavior of a collection of discrete particles, which dominated so much of the classical mechanics of the last century. As far back as 1809, the mathematical techniques which later came to be known as Hamiltonian mechanics began to emerge, as well as an appreciation of the inherent symmetry and structure of the mathematical forms embodied by the Poisson bracket. It was in this year that S. D. Poisson introduced this celebrated bracket [Poisson, 1809, p. 281], and in succeeding years that such famous scholars as Hamilton, Jacobi, and Poincaré laid the foundation for classical mechanics upon the earlier bedrock of Euler, Lagrange, and d'Alembert.

This surge of interest in Hamiltonian mechanics continues well into the waning years of the twentieth century, where scholars are just beginning to realize the wealth of information to be gained through the use of such powerful analytic tools as the Hamiltonian/Poisson formalism and the development of symplectic methods on differential manifolds. Specifically, the study of the dynamics of ideal continua, which is analogous to the discrete particle dynamics studied by Hamilton, Jacobi, and Poisson, has recently benefited significantly by the adaptation of the equations of motion into Hamiltonian form. The inherent structure and symmetry of this form of the equations is particularly well suited for many mathematical analyses which are extremely difficult when conducted in terms of the standard forms of the dynamical equations, for instance, stability and perturbation analyses of ideal fluid flows. Thus, classical mechanics and its outgrowth, continuum mechanics, seem to be on the verge of some major developments. Yet, further progress in this area was hindered by the fact that the traditional form of the Hamiltonian structure can only describe conservative systems, thus placing a severe constraint on the applicability of these mathematically elegant and computationally powerful techniques to real systems.

Parallel to the skeletal sequence of events recorded above, irreversible thermodynamics was developed in order to describe non-ideal, i.e., dissipative, system behavior, largely during the present century. To be sure, the origins of the study of irreversible phenomena can be traced back as far as Newton, who introduced the concept of viscosity long before the emergence of thermodynamics as a science (which was concerned with equilibrium phenomena within simple systems). With the rise of classical thermodynamics came the realization of a new concept, entropy, which had the unusual property of being created, but not destroyed. Thus, irreversibility came upon the scene, embodied initially as the "non-compensated heat" of Clausius. The early investigations in dissipative transport processes were phenomenological in nature and occurred quite separately from the more traditional equilibrium thermodynamics. W. Thomson is the first to be credited with the application of thermodynamic knowledge to an irreversible process, as long ago as 1854. Later, non-equilibrium transport phenomena were linked to thermodynamics through the study of transitions to equilibrium of states close to equilibrium in processes involving perfect gases through kinetic theory and the emergence of statistical mechanics. Many years afterwards, in 1931, Onsager established the initial form of his famous "reciprocal relations," which were reformulated in 1945 into their present form by Casimir. Finally, in 1962, a compilation of work in the area of irreversible (non-equilibrium) thermodynamics appeared by de Groot and Mazur, which remains to this day the seminal text in the field.[1]

Despite the fact that there exists a consistent link between the macroscopic, phenomenological form of irreversible thermodynamics and the formal theory of non-equilibrium statistical mechanics (Boltzmann equation, Chapman/Enskog theory, kinetic theory, etc.) through which dissipative processes can be studied at a deeper level,[2] the complexity of the involved calculations is so large that this link is rigorously established only for the simplest, homogeneous media. For the scientist or the

[1]For a brief overview of the history of further theoretical developments since Onsager, within the context of nonequilibrium statistical mechanics, see Balescu [1975, pp. 567-70].

[2]However, one has still to keep in mind that the formal description of dissipative phenomena from the true "first principles" (i.e., the dynamics of elementary particles and the description of the four fundamental forces in nature) has never been accomplished. All of the available approaches through non-equilibrium statistical mechanics involve in one way or another the introduction of "randomness" or "loss of information" which, ultimately, leads to the principle of entropy production.

engineer who wants to study complex transport phenomena taking place in macroscopic systems, no other solution exists but to use a more-or-less phenomenological model, based on intuition and following a specific set of assumptions with respect to the dominant part of the physics involved. Then, if the model is a microscopic one, more approximations usually have to be made in order to deduce a final set of partial differential equations which can be used to describe the dynamic behavior of the system under investigation from a continuum perspective.

During the overall modeling process, the degrees of freedom (i.e., the number of variables used to describe the system under investigation) are necessarily drastically reduced. This is similar to what happens in the transition from a microscopic state description in equilibrium statistical mechanics to a macroscopic variables description in classical equilibrium thermodynamics. Thus, non-equilibrium (irreversible) thermodynamics emerges as the phenomenological (macroscopic) theory which describes non-equilibrium irreversible phenomena in the same way equilibrium thermodynamics is used to describe equilibrium states. Granted, as an approximative theory it has to be realized that it has limitations (that it is applied for states close to equilibrium is one of them); however, it is the only macroscopic theory we have that can handle complex phenomena such as flow-induced phase transitions.

Within the context of irreversible thermodynamics, constitutive relations for the various "fluxes" are postulated in terms of various "affinities;" however, this purely phenomenological approach fails to work in practice as the complexity of the system under investigation increases due to a proliferation of phenomenological, "adjustable" parameters, which are very difficult, if not impossible, to be evaluated uniquely through experiments. Characteristically, when the transition from an isotropic to an anisotropic viscous fluid medium is made, not only the number of phenomenological parameters describing the flow behavior increases, but it also becomes necessary to develop additional "evolution" equations for variables describing the anisotropic structure which evolves according to the kinematics. In fact, these structural variables are often somewhat arbitrary in themselves, having a tensorial character which depends upon the nature of the system in consideration. Naturally, a higher-rank tensor variable will contain more information than a lower-rank one, but it also results in more complex evolution equations. As an example, in describing an anisotropic, viscous fluid, the number of independent parameters increases by an order of magnitude as one moves from a first-rank tensor (vector) to a second-rank tensor representation.[3]

[3]The anisotropic fluid is discussed in detail in Part II.

This dramatic increase in the number of the parameters has been largely responsible for the most severe criticisms against irreversible thermodynamics. Clearly, a mechanism is needed in order to deduce more constraints in the phenomenological parameters, and thus reduce their number to a practical sum. One solution involves the direct connection between the continuum model and a detailed microscopic model which, however, is difficult to achieve through the traditional formulation of irreversible thermodynamics. This use of irreversible thermodynamics as the framework through which microscopic models reduce to continuum ones is highly desirable, though, since it allows one to ascertain the "thermodynamic consistency" (for instance, the non-negative rate of entropy production) of the final result.

Moreover, the traditional description of dissipative transport through irreversible thermodynamics, though somewhat appealing to the intuition, lacks the powerful mathematical formalism and symmetry embodied in the Hamiltonian mechanics of ideal systems, as discussed in the opening paragraphs. It describes the phenomena at the level of fluxes and affinities, which is one level lower in generality than the Hamiltonian formulation of the equations in classical mechanics, which is now accepted as the norm for the description of conservative systems. As a result, the description of complex systems is quite difficult and remains an art known only to the few experts in the field. Thus, it is beneficial to attempt to restructure irreversible thermodynamics into a higher-level, more symmetric form suitable for complex applications, as well as to simplify subsequent mathematical analyses. Can the descriptions of ideal and dissipative systems be combined into a unified whole of Hamiltonian mechanics? Hopefully, a positive answer to this question is conveyed by this book.[4]

The complex applications mentioned in the preceding paragraph are the main emphasis of this book. Although the strength of the Hamiltonian formulation has its usefulness in mathematical analyses and numerical schemes, the symmetry and structure inherent to this formalism provide a more efficient and less confusing means of constructing complex transport equations. Specifically, the focus of the complex applications discussed in this book is on materials with internal microstructure, such materials as polymeric fluids, liquid crystals, colloidal suspensions, etc. To be sure, much work has been done since the 1940s concerning media with internal microstructure from the continuum point of view. Irreversible thermodynamics has been applied to these materials in this setting through the works of Truesdell,

[4]Ours is, of course, neither the only nor the first effort to unify the description of dynamics and thermodynamics; see, for example, Prigogine *et al.* [1973] for an alternative approach through nonequilibrium statistical thermodynamics.

Coleman, Noll, and others, who put in a consistent mathematical form the governing equations for the constitutive behavior of complex, anisotropic (mainly elastic) media.[5] Despite this progress, the proper modeling of transport processes still involves more art than science and requires a lot of experience, skill, and intuition in efficiently using the limited help offered by irreversible thermodynamics towards building thermodynamically consistent dynamical equations. As a result, it seems that today fewer researchers use any thermodynamic information in their modeling, with dramatic implications thus arising as to the soundness of the proposed models in describing physically realistic material behavior, which is sometimes reflected in the failure of the numerical solution of the corresponding equations or the prediction of aphysical results. Is there a simpler method of constructing thermodynamically consistent constitutive models? Can this method also introduce some of the advantages of Hamiltonian mechanics into the formulation? Hopefully, affirmative answers to both of these questions are conveyed by this book.

1.2 The Challenge of Multiple Time and Length Scales

The dynamics of complex systems, consisting of large numbers of particles (typically on the order of 6.023×10^{23} particles, the Avogadro number), is formally following the solution of Schrödinger's equation in a very large dimensional space (or, in the classical limit, the solution to the equations of motion for the position variables and momenta of the particles). These equations are conservative and can easily be recast into a Hamiltonian formulation (see chapter 3) which makes the conservation of the total energy and the symplectic structure of the equations readily apparent. No doubt, if an exact solution to the problem is sought, these are the equations to be solved, and this is practiced (to some extent—approximations are still used most of the time) in molecular dynamics simulations, which have increased at a rapid pace in the last decade. However, the major problem facing the molecular dynamics simulations is the limitation in the scales of length and time that can be realistically handled, even with the best of the available supercomputers. The biggest simulations involve, at most, a few thousand particles and can be used for times comparable to the fastest of the relaxation times of the system under investigation (e.g., for a protein this will correspond to a few picoseconds). How the information gathered in this way can be used for the study of significantly greater assemblages of molecules at significantly larger time scales is one fundamental problem that still

[5]For a recent exposition of this post-war work, see Truesdell [1984].

awaits a satisfactory solution. One thing that appears to be certain is that approximations have to be made in order to reduce the enormous amount of information gathered at the molecular level to the level necessary for the macroscopic description of the system. Even the best theoretical techniques available, through renormalization and cluster theories, cannot avoid introducing a certain amount of "filtering" of the microscopic information.

Once the fact that approximations are inevitable is accepted, then one must introduce the concept of "dissipation" as the framework within which all "redundant" information due to neglected degrees of freedom affects the behavior of the remaining ones. In the formal mathematical theory, dissipation is introduced in a system of Hamiltonian (or symplectic) equations through the "projection operation" of the original degrees of freedom to the final ones. Whether or not this is feasible to do with realistic systems (so far, it has only been used in idealized systems) is irrelevant to the discussion at hand. The important fact is that in the procedure of projection operation, where a large system of equations involving a large number of variables is reduced to a smaller set involving a few variables, one has a transition of type (from a symplectic or Hamiltonian structure to a dissipative one) and of length and time scales.

If the projection procedure is carried over to infinite scales of length and time, one of course obtains the framework of equilibrium thermodynamics, which, in practice, is used for length and time scales large with respect to human standards (i.e., with respect to the patience of the graduate student doing the experiments, on the order of meters and days), but certainly not infinite (which would have reduced the applicability of equilibrium thermodynamics to the study of isothermal radiation and a soup of electrons/positrons—the thermal death of the universe). Thus, the ideas of proper length and time scales are already imbedded into the use of classical or equilibrium thermodynamics, although perhaps not recognizably so. The underlying assumption is that for a given range of length and time scales, and for a given system, a particular set of variables (called the "dynamic" and "frozen" variables) can describe the macroscopic characteristics of the system with all the other statistical variables (called "relaxed" variables) having reached a "local equilibrium." Which ones of the variables can be considered as "frozen," "dynamic," or "relaxed" is solely determined by the length and time scales imposed by the observation window of the experimenter.

Let us consider a system whose state can be completely and unequivocally described by a number of different variables, v_i, each associated with a characteristic time, t_i, and ordered according to increasing values of these characteristic times. However, this system is being observed by an experimenter whose observation window is, through no fault of his/her own, unfortunately small, say between times $t_{\min}$ and $t_{\max}$. Using the terminology of the preceding paragraph, all the variables associated with a characteristic time larger than t_{max} can be considered as frozen, for all practical purposes. This implies that these variables change so slowly that they are fixed in the observation window of the experimenter, and thus must be considered as parameters of the problem, with their values specified *a priori*. Examples of frozen variables are the mass of a thermodynamic system for small enough times such that nuclear changes may be neglected, and the viscosity of a glass for times smaller than the glass relaxation time (typically, years).

On the other hand, all the variables with characteristic times smaller than t_{min} are relaxed variables, which quickly respond to changes in the system, approaching a local equilibrium which is completely determined by the values of the frozen and dynamic variables. Since these variables achieve their equilibria so quickly, one sees only their average equilibrium value at sufficiently large times. As an example, consider the fraction of particles, in a larger system, with an energy that falls within a specific energy interval. At times long enough for thermal equilibrium to be established, this fraction is completely determined by the system temperature.

Finally, the remaining variables with time constants between t_{min} and t_{max} are the dynamic variables for the system. These are the variables whose time evolution must be monitored for a successful experiment, which is dictated by a set of coupled evolution equations and the frozen variables of the system. As typical examples of the dynamic variables for an atmospheric system, consider the velocity and the pressure of the atmosphere for time scales ranging from seconds to days. If the time scale of interest decreases significantly below a second, then the velocity and the pressure values can be considered as constant (frozen) whereas if the time scale increases above a few days, then the velocity can be assumed zero and the pressure can be considered to have assumed the equilibrium value corresponding to the altitude and ambient temperature of the air. These ideas are represented schematically in Table 1.1.

Within the framework represented in Table 1.1, it is interesting to examine certain characteristic limits. First, for observation windows which do not involve any variables (i.e., there are no dynamic variables) equilibrium thermodynamics is applicable according to which the values of the relaxed variables can be determined as functions of the values of the frozen ones. Second, for very large values of time, the only frozen variables are the ones corresponding to conserved quantities, such as the total energy, electric charge, linear and angular momentum. Third, for very small time scales the problem reduces to the study of the behavior of elementary particles.

It is important to notice that as the time scale increases from the one corresponding to the characteristic times of the elementary particles to

Relevant theory	Statistical mechanics	Present theory	Equilibrium thermodynamics
Number	Many	Hopefully few	Few
Characterization	Relaxed variables	Dynamic variables	Frozen variables
Variables	$v_1, v_2, \ldots, v_n$	$v_{n+1}, v_{n+2}, \ldots, v_m$	$v_{m+1}, \ldots, v_l$
Characteristic times	$t_1 < t_2 < \ldots < t_n$ $<$	$t_{n+1} < t_{n+2} < \ldots < t_m$ $<$	$t_{m+1} < \ldots < t_l$
Comments	f(dynamic, frozen)	Subject to evolution equations	Parameters
Examples	Particle distributions	Velocity, pressure	Mass, viscosity

Table 1.1: Characterization of state variables.

infinity, the "macroscopic description" of the system reflects an increasing number of corresponding "microstates" which are, from the perspective of the observer, in some sense equivalent. This loss of information has as a consequence, as discussed in preceding paragraphs, the generation of the idea of "dissipation" into the mathematical description of the system evolution. It is best reflected through a key variable in the approach described in this book, the "entropy." The entropy is seen here as the variable which characterizes the degree of "randomness," "lack of information," or "multiplicity of corresponding microscopic states" of the system. This picture of the entropy is similar to its definition from statistical mechanics. In that theory, a general property of the entropy, which is carried over into the present work, can be described as the monotonic increase of its value as a function of time. An additional property that we postulate here (for a justification see Woods [1975, Part I]) is that there are several mathematical descriptions for the entropy, each one dependent on the time scale of the observation window, as a function of the dynamic and frozen variables. However, all of these descriptions are consistent with each other, providing the same answer in their common time interval of applicability. Furthermore, in accordance with the monotonically increasing in time value of the entropy property, as the observation window moves to longer time scales we require that the entropy similarly increase (or, more precisely, not decrease) in value.

Let us consider a simple example where we can come to grips with these concepts called dissipation and entropy. Consider an isolated box, containing two elementary particles, one red, one blue, but otherwise indistinguishable. If these particles are given some arbitrary momentum at time t_0, they will begin to bounce around the box, experiencing, let us say, perfectly elastic collisions with each other and with the walls of the box. It is easy to imagine, at some time t', that the two particles will experience a head-on collision, in which each particle will exactly reverse its momentum. In this situation, each particle will exactly reverse its path and time history until, after another increment of time equal to $t'-t_0$ has occurred, the two particles will be exactly back to where they started. If a system-dependent clock can somehow be inserted into the box, which an outside observer can see but not affect, then the clock will run forward during the first half of the experiment to t', and then backward to t_0 during the second half. Thus, the system is time reversible.

If we let the number of particles in the box become very large, it becomes hard to accept that each particle will exactly reverse its motion at the same instant in time. In fact, one way that this can happen is if all particles meet at the same location in space at the same time. Therefore, the system will always move forward in time,[6] and though they may again conceivably attain the configuration which they had at time t_0, they will not have done so reversibly. If we define some quantity, A, which depends upon the state or configuration of the particles in the system, the larger the number of particles, then the greater the amount of time in which A will remain at its average value, rather than in a "fluctuation," i.e., slightly perturbed from its average value.

Now consider that we are observing not one system, but the flow of many such systems through space and time. Let us say that our length and time scales of observation are such that we can observe each system, but not its inhabitants. Hence we work in terms of the variables A_i, one for each system, taking A_i to be a parameter which we specify *a priori*, maybe through some other experiment of a single system on much shorter time and length scales. It is obvious, however, that this procedure entails a loss of information in going from the microscopic level to the macroscopic level, and to account for this loss of information, we insert the concepts of dissipation and entropy into our system description.

In statistical mechanics, the entropy is defined in a purely probabilistic fashion, i.e., a system moves in such a way that the randomness, or entropy, is maximized, since this represents the most probable occurrence. As the number of particles approaches infinity, the average state becomes more probable statistically and the fluctuations smaller and of shorter duration. In this theory, however, we do not use such a well-defined quantity for the entropy. Rather, we view it in more of the sense envisioned by Gibbs: a state variable of the system which, together with the other state variables of the system, is enough to describe the physical state of the system.

It is important to place here the limits within which the present theory is going to be useful. These involve the cases where the number of dynamic variables within the observation window is small but non-zero. This is typically the case for most processes which correspond to organized movement of matter, ranging from Newtonian flow to the large scale (non-linear) deformation of solids and the flow behavior of polymers, plasmas, and suspensions. The area where the present formulation might be especially useful is that which corresponds to the flow and transport phenomena within media having an internal microstructure which is described by a few dynamic variables.

A last item that needs justification for its use in the present theory is the concept of the "internal" or "structural" variables. It is implicitly assumed in the present theory that the state of the system at a given time

[6]However, if we wait an infinite amount of time, it is conceivable that this situation could occur. One might wonder what the outcome would be if every particle in the universe reversed its momentum at exactly the same time.

can be completely characterized by the values of a selected set of variables *at the same time*. In general, this requires the specification of a number of internal parameters in addition to the ones easily associated with macroscopic experimentation (such as velocity, pressure, etc.). Although the internal parameters used in this book are usually associated with some characteristic property of the structure of the medium under investigation, in general they do not have to have such an explicit interpretation for the theory to be valid.

An alternative approach for the description of complex media (which is not followed in this work) is the use of integral constitutive relationships for the characterization of the system under investigation. In this approach, fewer variables are used at the expense of the much more computationally intensive evaluation of integral equations over the past history of the system. This duality in the description of material systems from a continuum point of view is very similar to the dual description of systems in information-processing systems theory, from a discrete point of view. There, the output variables $\mathbf{o}\equiv\{o_1,o_2,\ldots,o_l\}^T$ of a system at a specific time step, t_n, can be described either as functions of the input variables $\mathbf{i}\equiv\{i_1,i_2,\ldots,i_m\}^T$ at the same and all previous time steps $t_1,t_2,\ldots,t_n$,

$$\mathbf{o}(t_n) = \mathbf{f}(\mathbf{i}(t_1),\mathbf{i}(t_2),\ldots,\mathbf{i}(t_n)) \ , \tag{1.2-1}$$

(corresponding to the integral approach) or as a function of the input variables and an additional k internal variables $\mathbf{r}\equiv\{r_1(t),r_2(t),\ldots, r_k(t)\}^T$, all of which are evaluated at the current time step, t_n:

$$\mathbf{o}(t_n) = \mathbf{g}(\mathbf{i}(t_n),\mathbf{r}(t_n)) \ , \tag{1.2-2}$$

where the values of the internal variables are determined at the current time step n as a function of the input parameters at the current time step, t_n, and of their values at the previous time step, $n-1$,

$$\mathbf{r}(t_n) = \mathbf{h}(\mathbf{i}(t_n),\mathbf{r}(t_{n-1})) \ , \tag{1.2-3}$$

(corresponding to the internal variables approach). The additional variables follow special evolution equations in time, Eq. (1.2-3), reflecting the system response. It is interesting to note that information-processing theory can show the complete equivalence of the two approaches, via the *Realization Theorem* [Casti, 1989, pp. 127-28].

1.3 The Energy as the Fundamental Quantity

A quantity which may be defined in all possible descriptions of a system is the energy. In contrast to the entropy, the value of the energy remains the same regardless of the level of description one is working on, or the establishment (or not) of any sort of dynamic equilibrium. Energy is the only quantity we know of which is conserved for all systems; though other quantities (mass, momentum, etc.) may be conserved in certain limits. This conservation of energy seems to be absolute, irrespective of the number and nature of the interactions within the system. At the most fundamental description of matter, i.e., in quantum mechanics, the energy naturally arises as the quantity of primary interest defined by the eigenvalues of the Hamiltonian operator, which is used to describe the elementary particle dynamics. The energy is naturally conserved as the transition from a microscopic to a macroscopic (continuum) description is made, even though the observation window has become somewhat restricted to large length and time scales. Thus, the energy can serve to characterize a system at static equilibrium (equilibrium thermodynamics), as well as a system away from equilibrium (irreversible thermodynamics). Therefore, it offers an ideal starting point for a unified description of thermodynamics and dissipative phenomena.

In conjunction with the internal-variables description of non-equilibrium phenomena in macroscopic (continuum) systems, we need to clarify the assumptions made in connection with the preceding discussion on the different time and length scales of the system variables. A key assumption for the internal variable theory described in this work is that of the local thermodynamic equilibrium; i.e., we assume that for the time and length scales of observation, the internal energy of the system has had the time to become equilibrated among the non-resolved degrees of freedom, and that the velocity distribution of the particles around the average has had time to achieve a Maxwellian shape characterized by the temperature. In this case, it is useful to distinguish between the energy associated with the non-equilibrated degrees of freedom present in the formulation (for example, kinetic energy associated with the macroscopic motion, or potential energy due to an external field or intermolecular interactions expressed by a structural parameter) and the energy corresponding to the non-resolved degrees of freedom (internal energy) represented through an equation of state in terms of macroscopic variables like density, temperature, etc. The increase in entropy with increasing time is generally accompanied by a "degradation" of the energy "quality" of a system; i.e., in the absence of external fields, the total energy of an isolated system is eventually going to be transformed at equilibrium (long times) to internal (thermodynamic) energy. In addition, the local change of internal energy density per unit change in entropy density defines the thermodynamic temperature, a relationship that is of obvious utility in heat transfer applications.

In the applications which are dealt with in this book, the total energy (Hamiltonian) of the system is expressed in terms of a kinetic energy, a potential energy due to external fields, and an internal energy, described above. The internal energy represents a thermodynamic equation of state valid for a system close to equilibrium. It is generally assumed to be a function (or functional) of the dynamic variables of the system, with the frozen variables playing the roles of parameters. As we shall see, the form of the internal energy is highly system dependent, and most of the physics of complex applications will go into the proper specification of this term.

Let us conclude this section with a quote from R. P. Feynman [1965, pp. 70-71], who left us with these thoughts on energy:

> What we have discovered about energy is that we have a scheme with a sequence of rules. From each different set of rules we can calculate a number for each different kind of energy. When we add all the numbers together, from all the different forms of energy, it always gives the same total. But as far as we know there are no real units, no little ballbearings. It is abstract, purely mathematical, that there is a number such that whenever you calculate it does not change. I cannot interpret it any better than that.
>
> This energy has all kinds of forms... There is energy due to motion called kinetic energy, energy due to gravitational interaction (gravitational potential energy, it is called), thermal energy, electrical energy, light energy, elastic energy in springs and so on, chemical energy, nuclear energy—and there is also an energy that a particle has from its mere existence, an energy that depends directly on its mass...

1.4 The Generalized Bracket Approach

> *One of the principal objects of practical research...is to find the point of view from which the subject appears in its greatest simplicity.*
>
> —*Josiah Willard Gibbs*

There are often several different methods for solving a particular problem: some mathematical, some intuitive, some computational, and some purely logical. Each is in itself uniquely valuable, for there is no way to tell what it may induce us to think about as we try to extend our knowledge towards solving new problems. What is important to understand is that even new ways of solving old problems are useful, since these may lead us to thinking in new ways about outstanding problems for which the old methods were getting us nowhere. As we hope to convey to the reader, this seems to be the case for the approach advocated in this book. In the words of James Clerk Maxwell [Thomson, 1931, p. 31]:

> For the sake of persons of different types of mind scientific truth should be presented in different forms and should be regarded as equally scientific whether it appears in the robust form and vivid colouring of a physical illustration or in the tenuity and paleness of a symbolic expression.

Traditionally, the formulation of equations governing the dynamic behavior of a continuous medium follows one of two approaches. The first approach is macroscopic, encompassing the consideration of conservation principles (mass, momentum, and energy) in conjunction with constitutive equations for the transport fluxes. This approach works well when the constitutive equations for the fluxes are simple enough to be represented in terms of a small number of parameters. Then, a "generalized entropy" may be defined and the rate of entropy production may be evaluated and used to provide constraints upon the continuum parameters. Alternatively, one can derive dynamical equations for a given system based upon an action principle; i.e., by calculating the critical point of a quantity, called the action, which for some uncertain reason is known to produce the required result. This approach involving the action principle is usually restricted to non-dissipative systems, where one need not assume any constitutive relations for fluxes. Whether or not dissipative transport processes can be described through an equivalent action principle still remains an open question.

In the generalized bracket formalism, we consider the total energy of the system, termed the Hamiltonian for historical significance, as the fundamental quantity of interest. Indeed, for conservative systems at least, it is evident that the Hamiltonian is just the Legendre transformation of the action described above (see subsequent chapters). For a given application, the Hamiltonian of the system must be defined as a function(al) in terms of the independent dynamic variables which are assumed to uniquely determine the macroscopic (in general, non-equilibrium) state of the system.

The main idea (call it a postulate, if you prefer) of the generalized bracket is that the dynamical equation for an arbitrary function(al) of the system dynamic variables, F, can be expressed through the master equation

$$\frac{\mathrm{d}F}{\mathrm{d}t} = \{[F,H]\}, \tag{1.4-1}$$

where H is the Hamiltonian of the system and $\{(\cdot,\cdot)\}$ is the generalized bracket. At this point, the structure and symmetry of the generalized bracket leads directly to the coupled time evolution equations for the same dynamic variables through a direct comparison between the master equation, (1.4-1), and the expression that results through differentiation by parts of $\mathrm{d}F/\mathrm{d}t$ (see §1.5 below).

The generalized bracket has to satisfy certain properties in order for the results to be consistent with a few additional principles that are assumed to be valid together with the master equation. In the special circumstance that $F=H$, it is required that

$$\frac{\mathrm{d}H}{\mathrm{d}t} = \{[H,H]\} = 0 \ , \tag{1.4-2}$$

since the total energy of the system must be conserved. The generalized bracket must describe both the conservative and the non-conservative process effects, and is thus broken into two subbrackets

$$\{[F,H]\} \equiv \{F,H\} + [F,H] \ , \tag{1.4-3}$$

$\{\cdot,\cdot\}$ describing the conservative effects through the traditional Poisson bracket and $[\cdot,\cdot]$ accounting for dissipative processes. Although the former bracket has been around for nearly two hundred years, only recently have researchers begun describing non-conservative phenomena through a dissipation bracket.

Each of the brackets in (1.4-3) must satisfy certain properties consistent with its physical and/or mathematical interpretation. The Poisson bracket must be bilinear and antisymmetric, i.e.,

$$\{F,H\} = -\{H,F\} \ , \tag{1.4-4}$$

which guarantees that for a conservative system (i.e., one for which the dissipative bracket is zero) the energy is conserved, $\mathrm{d}H/\mathrm{d}t=\{H,H\}=0$, as it should. Also, the Poisson bracket must satisfy a slightly more complex property, the Jacobi identity, which guarantees that the bracket fulfills the conditions of a Lie algebra. (See chapters 2 and 3, as well as Appendix A, for a deeper discussion.) By necessity, the dissipation bracket must be a linear function of F. It may be also a linear function of H (and in the lowest-order approximation—close to equilibrium—it is assumed to be), in which case it can be symmetric with respect to both F and H, or, in general, it might be a nonlinear function of H. In the special case of Eq. (1.4-1) where $F=S$, the entropy function(al), it is required that $\mathrm{d}S/\mathrm{d}t\geq 0$, which translates into $\{S,H\}=0$ and $[S,H]\geq 0$.

Alternatively, if one is interested in describing a system where the heat effects are considered unimportant (as, for example, in the consideration of incompressible fluids, see §5.4), the generalized bracket may also be applied to the mechanical energy contribution to the total, H_m, to formulate equations of motion. In this case, the degradation of the mechanical energy to heat requires that $dH_m/dt=[H_m,H_m]\leq 0$.

It is noteworthy that a symmetric structure of the dissipation bracket, evaluated close to equilibrium, is directly indicative of the Onsager/Casimir relationships between the transport coefficients relating the various fluxes to the affinities. The question of what are the proper affinities needs not be answered in the present framework since the affinities appear naturally as the derivatives of the Hamiltonian (or the gradients thereof) with respect to the dynamic variables of the system. A distinct feature of the generalized bracket formalism is that the appropriate expressions for the fluxes arise automatically from the formulation, depending only on the dissipation bracket, and are never explicitly specified.

1.5 A Simple Application: The Damped Oscillator

As a simple illustrative example[7] of the applicability of the generalized bracket, let us consider the discrete, dynamic system composed of a linear elastic spring and a viscous dashpot, as shown in Figure 1.1. Let m be the mass of a point particle connected to a stationary wall at position $x=0$ by a Hookean spring, of force constant K, and a dashpot, with damping coefficient μ, set up in a parallel configuration. The position of the particle, as given at any time t by x, will vary in time about the equilibrium position, x_o, for a given initial condition (other than the equilibrium condition, of course), oscillating back and forth until the system is equilibrated due to the damping action of the dashpot.

The equation of motion for the particle can be written in terms of Newton's laws of motion as (*mass*)×(*acceleration*)=(*force*), which leads directly to a second-order equation for x:

$$m\frac{d^2x}{dt^2} = -\mu\frac{dx}{dt} - K(x-x_o). \tag{1.5-1}$$

[7]The damped harmonic oscillator is a famous problem which has been used many times in the past, and continues to be used, in order to test various theories (both classical and quantum) for the description of dissipative phenomena in terms of first principles—see Dekker [1981] for a comprehensive review of the (already voluminous) work on this subject before 1981.

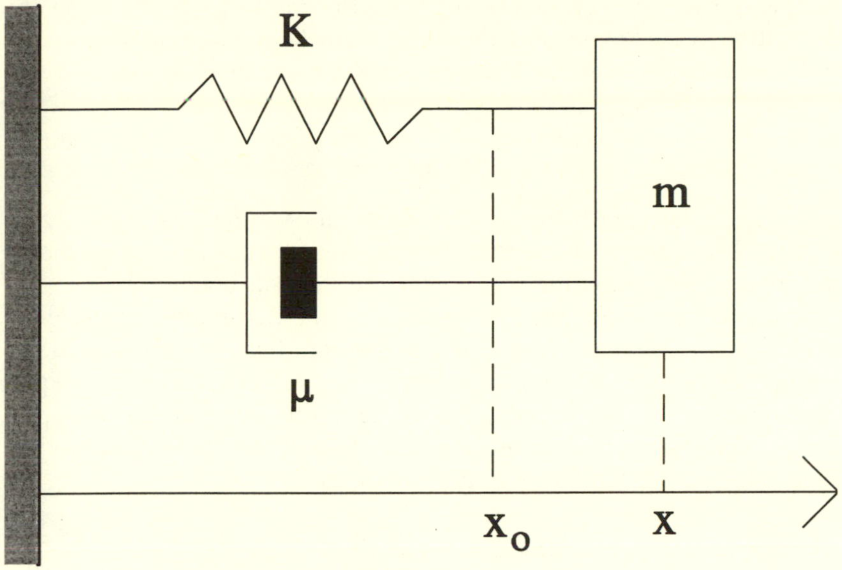

Figure 1.1: Schematic diagram of a mass connected to a spring and dashpot in parallel. The linear spring constant is K and the dashpot damping coefficient is μ. The equilibrium position of the mass is given by x_0.

This is the traditional damped oscillator equation giving oscillating solutions of diminishing amplitude in time, derived in the traditional manner by consideration of Newton's laws.

The same equation may also be derived through the bracket formalism, where the starting point is the specification of the Hamiltonian of the system. In this case, we are not interested in describing the heat effects in the system (i.e., the temperature rise due to viscous dissipation in the dashpot), so we only concern ourselves with the mechanical contribution to the total system energy, H_m. As such, the dynamic variables of the system are the particle position, x, and the particle momentum, p, defined as $p \equiv m(\mathrm{d}x/\mathrm{d}t)$. The Hamiltonian is expressed in terms of these dynamic variables as

$$H_m(x,p) = \frac{1}{2m}p^2 + \frac{1}{2}K(x - x_o)^2, \tag{1.5-2}$$

where the first term on the right-hand side is the kinetic energy of the particle and the second term is its potential energy. This is the traditional mechanical energy expression for a point particle of mass m.

Now the generalized bracket of (1.4-3) needs to be defined, via the specification of the Poisson and dissipation brackets. For a single-particle, one-dimensional system, the Poisson bracket is given as

$$\{F,G\} = \frac{\partial F}{\partial x}\frac{\partial G}{\partial p} - \frac{\partial G}{\partial x}\frac{\partial F}{\partial p}, \tag{1.5-3}$$

where F and G are two arbitrary (but continuous) functions of the dynamic variables, x and p. Note that this bracket satisfies the requirements mentioned in the preceding section for the Poisson bracket (bilinearity, antisymmetry, etc.). For the specification of the dissipation bracket, the rate at which mechanical energy dissipates into heat (via friction in the dashpot) needs to be taken into account. In this case, this rate is equal to the product of the velocity, $\mathrm{d}x/\mathrm{d}t$, and the force at the dashpot, $\mu(\mathrm{d}x/\mathrm{d}t)$. Therefore, an appropriate definition of the dissipation bracket (which satisfies the symmetry requirement) is

$$[F,G] = -\mu\frac{\partial F}{\partial p}\frac{\partial G}{\partial p} . \tag{1.5-4}$$

With this choice of the generalized bracket, $\{[H_m,H_m]\}=[H_m,H_m]\leq 0$ provided that the damping coefficient is non-negative.

The equation of motion for the point particle can now be found by consideration of the master equation expressing the dynamics of the arbitrary function F,

$$\frac{\mathrm{d}F}{\mathrm{d}t} = \{[F,H]\} = \{F,H\} + [F,H] , \tag{1.5-5}$$

via a direct comparison with the expression obtained by differentiating $F=F(x,p)$ by parts:

$$\frac{\mathrm{d}F}{\mathrm{d}t} = \frac{\partial F}{\partial x}\frac{\partial x}{\partial t} + \frac{\partial F}{\partial p}\frac{\partial p}{\partial t} . \tag{1.5-6}$$

Therefore, since

$$\begin{aligned}\{[F,H]\} &= \frac{\partial F}{\partial x}\frac{\partial H_m}{\partial p} - \frac{\partial H_m}{\partial x}\frac{\partial F}{\partial p} - \mu\frac{\partial F}{\partial p}\frac{\partial H_m}{\partial p} \\ &= \frac{\partial F}{\partial x}\frac{p}{m} + \frac{\partial F}{\partial p}[-K(x-x_o) - \mu\frac{p}{m}] ,\end{aligned} \tag{1.5-7}$$

we obtain directly the definition of the momentum, p, and the equation of motion, (1.5-1).

Hopefully, this simple example illustrates the principles of the bracket formalism; however, the usefulness of this procedure is to be drawn from much more complex applications. The body of this book is devoted to the systematic presentation of this thesis through an analysis of the fundamental principles and a validation of the approach, together with a description of the generalized bracket for general continua (Part I); followed by a number of applications in the fields of transport phenomena and fluid mechanics of such systems as polymer solutions and melts, liquid crystals, multi-fluid systems, and others (Part II).

2
Symplectic Geometry in Optics

Exact observation of reality shows that spatial relations—the phenomena of symmetry—lie at the basis of all phenomena we have studied.
—Vladimir Ivanovich Vernadskii

2.1 Introduction

The scope of this book is to address the fundamental problem of modeling transport processes within complex systems, i.e., systems with internal microstructure. The classical engineering approach involves the modeling of the systems as structured continua and the subsequent use of the models in order to derive (if possible) analytical results, exact or approximate. The advent of powerful computers and the promise through parallel processing of even more substantial computational gains in the near future have introduced yet another paragon to the established engineering practice: that of the numerical simulation.

Numerical simulation has emerged as a viable alternative to experiments (contrast Computational Fluid Dynamics (CFD) simulations versus wind tunnel experiments); however, the key limitation to a wider application of numerical simulations in engineering practice lies in the reliability of the models (as well as in their simplicity). CFD applications are successful since the Navier/Stokes equations which they employ are quite capable of describing accurately enough the hydrodynamics of air and water. However, as we move our emphasis to materials of such internal complexity as polymer melts, liquid crystals, suspensions, etc., the development of reliable continuum models becomes an increasingly arduous task.

The main objective of this treatise is to investigate a more systematic approach through which continuum models may be developed and analyzed. The key issue that the modeler has to cope with is how to construct models which describe more of the underlying physics without, at the same time, becoming excessively complex so that they either require a prohibitively large, experimentally determined number of adjustable parameters (such as current phenomenological theories) or a prohibitively large computational time (such as required for a detailed "brute force" description of the molecular dynamics).

It is the thesis of the present work that a lot of effort can be saved if the appropriate formulation is used in deriving model equations, a formulation which is capable of exploiting to a maximum degree the inherent symmetry and consistency of the collective phenomena exhibited by a large number of internal degrees of freedom. One leg of this formulation emanates from a description of the dissipative approach of a large number (almost all) of degrees of freedom to near-equilibrium (irreversible thermodynamics), and it is the subject of later chapters. The other leg is concerned with the conservative evolution of a few key degrees of freedom (Hamiltonian mechanics), and it is the subject of the current and subsequent chapters. Specifically, chapters 2 and 3 will present in turn the two classic examples of symplectic structure: optics and particle dynamics. It is truly amazing to see the same mathematical structure inherent to two such apparently diverse phenomena. Hopefully, a sense of the power gained through working in terms of these methods will be conveyed to the reader.

Although the Hamiltonian description of dynamics was developed almost two hundred years ago through the pioneering work of Poisson, Lagrange, Hamilton, and others, and although it had a grand impact by the beginning of the twentieth century, most notably in the development of quantum mechanics, it has been very slow in being introduced into the description of continua, where its use is still not widespread even today. The major reason for such slow progress is, of course, the fact that it employs mathematical concepts, such as symplectic structure, which to the outsider seem to lack the intuition that the more widely used conservation equations enjoy.

It is the aim of this chapter to help build this missing intuition by providing a series of examples where the underlying symplectic structure of the governing equations naturally emerges as the common mathematical link, which is the only thing to survive as different scales of length and time are crossed and successive approximations are made. The examples are drawn out of the description of electromagnetism and optics. Although this subject may at first appear irrelevant to the study of the dynamics of continua, actually it is very much related because of the intimate wave/particle duality of the elemental particles. This close connection of optics with mechanics was most notably observed by Hamilton (the optical/mechanical analogy) and eventually led to the development of quantum mechanics. It is presented here as the perfect example to demonstrate the perseverance of a mathematical characteristic after several degrees of approximation—all the way from quantum electrodynamics to Gaussian optics! Invoking again the optical/ mechanical analogy, we may postulate that the same symplectic structure also survives the several approximations necessary to lead to the description of conservative continua. Indeed, this can be proven to be the case, as we shall see in subsequent chapters.

2.2 Theories of Optics

Currently, the most complete theory of light is quantum electrodynamics (QED), which governs the interaction of light with charged particles down to quantum scales of length and time, and which takes into account the wave/particle duality of electromagnetic phenomena. The QED theory, originally developed in 1929, has since been tested more and more accurately over a wider range of conditions without any significant difference ever being found between experiment and theoretical predictions [Feynman, 1985, p. 7]. This theory has a truly phenomenal accuracy: for example, the value of the magnetic moment of the electron calculated from QED is $1.00115965246 \pm 20\times10^{-11}$ (in appropriate units) as compared to the best available experimental value of $1.00115965293 \pm 10\times10^{-11}$. According to Feynman, "This kind of accuracy could determine the distance between New York and Los Angeles to within the width of a human hair" [Penrose, 1989, p. 154]. However, the theory is somewhat untidy, being plagued by infinities in the answer which can be avoided only after elaborate renormalization procedures.

If one ignores quantum effects and phenomena taking place at the atomic or subatomic level by working on a length scale which is large compared with that of the elementary particles, then one can use the theory of electrodynamics developed by James Clerk Maxwell (1831-1879). Maxwell, with remarkable insight, described all electric and magnetic phenomena (including the generation and propagation of electromagnetic waves) with just four equations:

$$\frac{1}{c^2}\frac{\partial \mathbf{E}}{\partial t} = \operatorname{curl}\mathbf{B} - 4\pi\mathbf{j}\ , \qquad \frac{\partial \mathbf{B}}{\partial t} = -\operatorname{curl}\mathbf{E}\ ,$$
$$\operatorname{div}\mathbf{E} = 4\pi\rho\ , \qquad \operatorname{div}\mathbf{B} = 0\ , \tag{2.2-1}$$

where **E**, **B** are the intensities of the electric field and the magnetic induction, respectively, and ρ, **j** are the electric charge density and the electric current per unit area, respectively.

Even before Maxwell, however, there existed Fresnel's fairly well-developed wave theory of light that dealt rather successfully with the propagation of light in various media, on length scales comparable to the wavelength of light [Guillemin and Sternberg, 1984, p. 4]. Fresnel's theory correctly accounted for the wave properties of light, such as diffraction, interference, and polarization, but had nothing to say about the ways in which light is generated. Still another approximation, of Fresnel optics this time, can be developed if the wave characteristics of light are ignored: geometrical optics deals with phenomena involving light propagation when the length scale of description is taken to be much larger than the wavelength of light. Linear optics is yet one more approximation (of geometrical optics this time) valid when, on top of the necessary assumptions for the validity of geometrical optics, one adds the requirement of small angles between the light rays and the principal axis of light propagation. As the length scale continues to increase, at some point this assumption is bound to be valid. In addition, the properties of the media in consideration are assumed to change in a discontinuous fashion at the interface between different media. Finally, Gaussian optics adds the further approximation of axisymmetry of all the surfaces involved. The most important of its applications is mirror and lens theory. This amazing cascade of successive approximations pertaining to the theory of light can be better appreciated by viewing Figure 2.1 (reprinted with permission from Guillemin and Sternberg [1984, p. 6]), where all of the above-mentioned theories and their respective regions of validity are represented schematically.

It is worthwhile to mention after bringing to attention this wonderful collection of available theories for light that, although they have historically been developed at various times (almost in reverse of the order in which we present them here, going from the simplest but most restrictive to the most complex but also most rigorous), all of them harmoniously coexist today, each one applicable in its respective domain of validity. But the most important feature that these theories share, and the reason why they have been mentioned here, is an underlying mathematical structural characteristic (albeit in different forms)—the "symplectic structure." In order to understand more of what a symplectic structure is, let us pause for a moment in our discussion of optics to present some mathematical definitions and properties associated with a symplectic structure.

2.3 Symplectic Structure

As Guillemin and Sternberg [1984, p. 1] remark, the relationship between mathematics and science is a very special one: "On the one hand, mathematics is created to solve specific problems arising in physics, and, on the other hand, it provides the very language in which the laws of physics are formulated." In particular, this is true for group theory and the symmetries characterizing the mathematical structures that are represented there. A particular structure that underlies the dynamics of Hamiltonian systems and, in particular, all the above-mentioned theories for optics is the *symplectic structure.*

2.3.1 The symplectic vector space

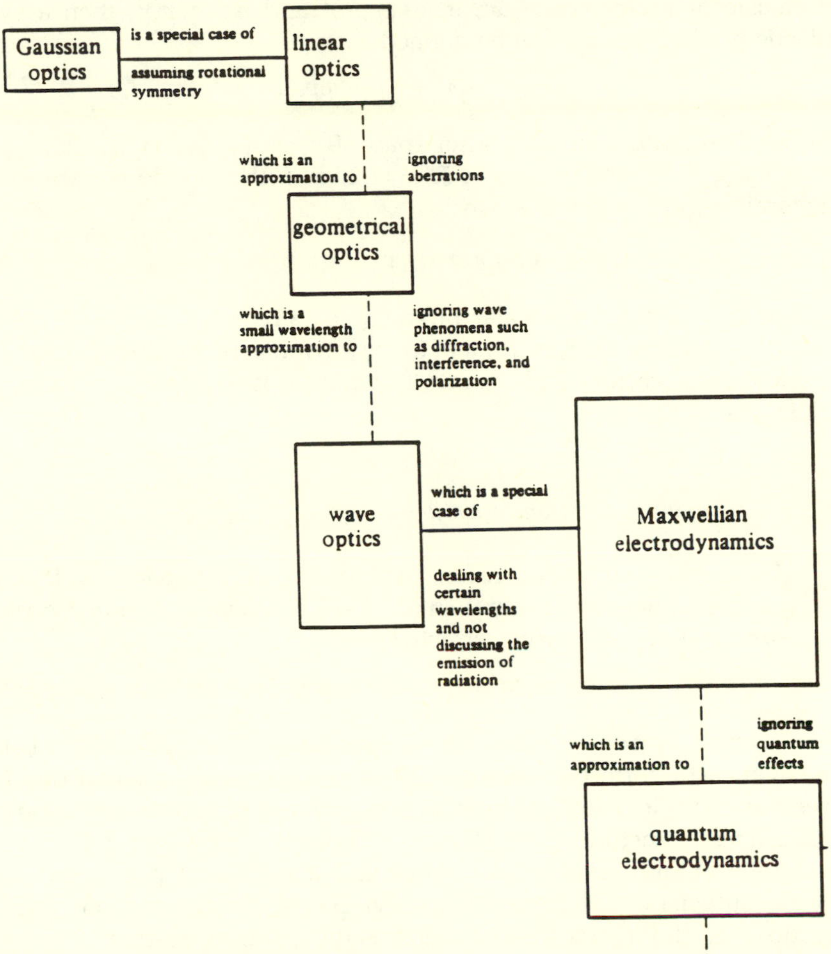

Figure 2.1: Theories of optics. Reprinted from Guillemin and Sternberg [1984, Fig.1.4] with permission of Cambridge University Press.

A symplectic vector space is a real vector space, V, of even dimensions, equipped with a non-degenerate (single-valued), antisymmetric, bilinear form, «·,·»:

$$
\begin{aligned}
«\mathbf{u},\mathbf{v}» &= 0 \;\; \forall\, \mathbf{v} \in V \text{ iff } \mathbf{u} = \mathbf{0} \;, && \text{(uniqueness)} \\
«\mathbf{u},\mathbf{v}» &= -«\mathbf{v},\mathbf{u}» \;, && \text{(antisymmetry)} \\
«\alpha\mathbf{u}+\beta\mathbf{v},\mathbf{w}» &= \alpha«\mathbf{u},\mathbf{w}» + \beta«\mathbf{v},\mathbf{w}» \;, && \text{(bilinearity)}
\end{aligned}
\tag{2.3-1}
$$

where **u**, **v**, **w** and α, β are arbitrary elements of the vector space, V, and the number field ($\mathbb{C}$ or $\mathbb{R}$), respectively. For example, in the real, two-dimensional vector space, $\mathbb{R}^2$, if $\mathbf{u}_1 \equiv (q_1, p_1)^T$ and $\mathbf{u}_2 \equiv (q_2, p_2)^T$, then a symplectic product «$\mathbf{u}_1, \mathbf{u}_2$» can be defined as

$$«\mathbf{u}_1, \mathbf{u}_2» \equiv q_1 p_2 - q_2 p_1 \ . \tag{2.3-2}$$

In the $2n$-dimensional real vector space, $\mathbb{R}^{2n}$, $\mathbf{u}_1 = (q_1^1, \ldots, q_1^n, p_1^1, \ldots, p_1^n) \equiv (\mathbf{q}_1, \mathbf{p}_1)^T$ and $\mathbf{u}_2 \equiv (\mathbf{q}_2, \mathbf{p}_2)^T$, where $\mathbf{q}_i$, $\mathbf{p}_i \in \mathbb{R}^n$, i=1,2, and the previous expression generalizes, in the obvious way, as

$$«\mathbf{u}_1, \mathbf{u}_2» \equiv \mathbf{q}_1 \cdot \mathbf{p}_2 - \mathbf{q}_2 \cdot \mathbf{p}_1 \ , \tag{2.3-3}$$

where the $\cdot$ denotes the scalar product; i.e., $\mathbf{a} \cdot \mathbf{b} \equiv a_1 b_1 + \ldots + a_n b_n$. It can be shown [Fomenko, 1988, p. 18] that for any symplectic product there exists a set of basis vectors with respect to which it is expressed in the form indicated by (2.3-3).

2.3.2 *The symplectic transformation*

Let A and B be two $2n$-dimensional, symplectic vector spaces, so that we may define a linear transformation $\mathbf{M}$: $A \to B$ as a *symplectic transformation* if it preserves the symplectic structure; i.e.,

$$«\mathbf{M} \cdot \mathbf{u}, \mathbf{M} \cdot \mathbf{v}» = «\mathbf{u}, \mathbf{v}» \quad \forall \, \mathbf{u}, \mathbf{v} \in A \ . \tag{2.3-4}$$

For example, if n=1 (i.e., in a two-dimensional vector space), and using the symplectic form provided by (2.3-2), it is easy to show that det$\mathbf{M}$=1 (see the example below), is both a necessary and sufficient condition for a 2×2 linear transformation, $\mathbf{M}$, to be symplectic. However, for n>1, the condition of unit determinant, while still a necessary one, is not any longer sufficient. Furthermore, the symplectic transformations form a group, with the matrix multiplication as the group operation:

$$\mathbf{M} \circ \mathbf{N} \equiv \mathbf{M} \cdot \mathbf{N} \ . \tag{2.3-5}$$

Therefore, if $\check{\mathrm{S}}$ is the group of symplectic transformations in a $2n$-dimensional vector space and $\check{\mathrm{U}}$ is the set of the $2n \times 2n$ matrices with unit determinant, then

$$\check{\mathrm{S}} \subseteq \check{\mathrm{U}}, \tag{2.3-6}$$

with the equality being valid only when n=1.

EXAMPLE 2.1: Let $\mathbf{u}_1 = (q_1, p_1)^T$ and $\mathbf{u}_2 = (q_2, p_2)^T \in A$ and $\mathbf{v}_1 = \mathbf{M} \cdot \mathbf{u}_1$ and $\mathbf{v}_2 = \mathbf{M} \cdot \mathbf{u}_2 \in B$, where $\mathbf{M}$ is an arbitrary transformation. Show that for $\mathbf{M}$ to be symplectic, det$\mathbf{M}$=1.

According to (2.3-4), «$\mathbf{v}_1,\mathbf{v}_2$» must be equal to «$\mathbf{u}_1,\mathbf{u}_2$» if $\mathbf{M}$ is symplectic. We know via $\mathbf{M}$ that the components of $\mathbf{v}_1$ and $\mathbf{v}_2$ are

$$\mathbf{v}_1 = (M_{11}q_1 + M_{12}p_1, M_{21}q_1 + M_{22}p_1)^{\mathrm{T}} ,$$

$$\mathbf{v}_2 = (M_{11}q_2 + M_{12}p_2, M_{21}q_2 + M_{22}p_2)^{\mathrm{T}} ,$$

which implies via (2.3-2) that

$$\begin{aligned} «\mathbf{v}_1,\mathbf{v}_2» &= (M_{11}q_1 + M_{12}p_1)(M_{21}q_2 + M_{22}p_2) \\ &\quad - (M_{11}q_2 + M_{12}p_2)(M_{21}q_1 + M_{22}p_1) \\ &= M_{11}M_{22}(q_1p_2 - q_2p_1) + M_{12}M_{21}(p_1q_2 - p_2q_1) \\ &= (M_{11}M_{22} - M_{12}M_{21})(q_1p_2 - q_2p_1) \\ &\equiv \det\mathbf{M}\,(q_1p_2 - q_2p_1) . \end{aligned}$$

Hence the above requirement is true for arbitrary vectors $\mathbf{u}_1$ and $\mathbf{u}_2$ if and only if det$\mathbf{M}$=1. ■

2.4 Gaussian and Linear Optics

2.4.1 Gaussian optics

Gaussian optics is particularly useful for the study of axisymmetric mirrors, lenses, etc. This theory deals with the propagation of light rays into homogeneous and isotropic media separated by sharp, axisymmetric interfaces. The rays are assumed to propagate within a single medium in straight lines, which never deviate too much from the axis of axisymmetry called, say, z. The rays are therefore completely characterized by the distance q of their intercept with a plane normal to the principal axis at some specific location and by their slope, determined by the angle θ between the ray and the principal axis (see Figure 2.2).

The fundamental problem in Gaussian optics can then be posed as follows. Given the intercept q_1 and the angle θ_1 of a light ray at a particular location z_1 on the principal axis, find the parameters q_2 and θ_2 corresponding to another specified location z_2. Alternatively, in mathematical language, given the coordinate locations z_1 and z_2, as well as the structure of the optical medium in between, what is the transformation (mapping) $\mathbf{M}$ between (q_2,θ_2) and (q_1,θ_1). In order to find $\mathbf{M}$, one can use empirical laws, such as the law of straight propagation of light rays through a homogeneous and isotropic medium and Snell's law of refraction for rays transmitted from one optical medium to another.

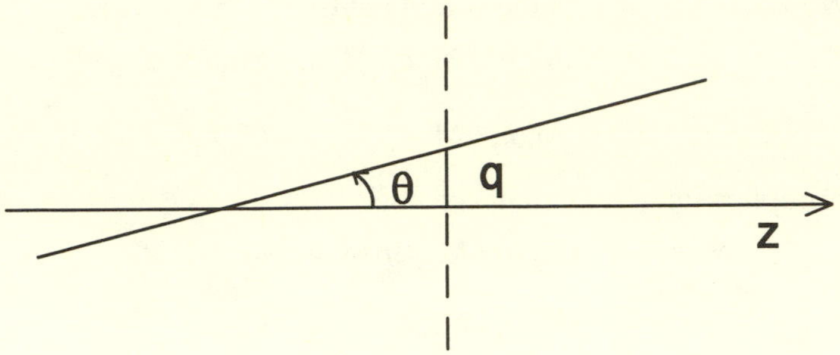

Figure 2.2: Parameters characterizing a light ray in Gaussian optics.

Based on these simple empirical laws, and on a number of mathematical simplifications made possible by the assumption of small angles θ, one can show that, if the intercept q along with the product of the angle θ and the index of refraction n, $p \equiv \theta n$, are used as principal coordinates, rather than q and θ, then the following theorem is valid [Guillemin and Sternberg, 1984, p. 17].

THEOREM 2.1: (a) The mapping **M** between (q_2,p_2) and (q_1,p_1) induced by the path of a light ray is a *linear symplectic transformation*, and (b) every symplectic transformation can be realized by the travel of a light ray through a selection of the appropriate medium between the points of interest. ■

The proof for (b) relies on the mathematical properties of symplectic 2×2 groups and the interested reader is referred to Guillemin and Sternberg [1984, pp. 9, 15]. The proof for (a), however, depends on the physics and it is instructive to reproduce it here so that the emergence of the symplectic structure can be more easily appreciated.

First, it is important to realize that an optical medium, handled within the framework of Gaussian optics, is assumed to consist of homogeneous isotropic optical media separated by axisymmetric interfaces. Therefore, any transformation between two arbitrary points z_1 and z_2 can be obtained as the composition of two elemental transformations. The first represents a straight line traveling through a homogeneous medium. The second corresponds to the case when a ray is transmitted from one optical medium to another. Since the symplectic transformations form a group and therefore the composition of two transformations is a closed operation (i.e., the composition of two symplectic transformations is also symplectic), it suffices to show that each one of the above-mentioned elemental transformations is symplectic in order to prove that every transformation is a symplectic one.

Let us first consider a light ray traveling through a piecewise-homogeneous, isotropic, optical medium with index of refraction n. Let $(q_1,p_1)\equiv(q_1,n\theta_1)$ and (q_2,p_2) be the coordinates characterizing this light ray at distances z_1 and z_2 along the principal axis, respectively (see also Figure 2.3). Then, from the geometrical triangle ABC, we can obtain the relations

$$\begin{aligned} p_2 &= n\theta_2 = n\theta_1 = p_1 , \\ q_2 &= q_1 + \Delta z\tan(\theta_1) \approx q_1 + \Delta z\theta_1 = q_1 + \frac{\Delta z}{n}p_1 , \end{aligned} \tag{2.4-1}$$

or, in matrix form,

$$\begin{pmatrix} q_2 \\ p_2 \end{pmatrix} = \begin{pmatrix} 1 & \dfrac{\Delta z}{n} \\ 0 & 1 \end{pmatrix} \cdot \begin{pmatrix} q_1 \\ p_1 \end{pmatrix} . \tag{2.4-2}$$

As we can clearly see, the transformation is linear with the transformation matrix being symplectic (detM=1). Second, let us consider a light ray entering a medium with index of refraction n_2 originating from a medium with index of refraction n_1, with the two media being separated by an axisymmetric surface which, for small enough angles of approach, can be approximated by the parabola

$$z = z_1 + \tfrac{1}{2}k_c q^2 . \tag{2.4-3}$$

Furthermore, let i_1 and i_2 be the angles between the light ray and the normal to the surface, before and after the ray enters the second optical medium, respectively (see also Figure 2.4). Then, we have

$$\frac{\pi}{2}-\psi \approx \tan(\frac{\pi}{2}-\psi) = \frac{dz}{dq} = k_c q \approx k_c q_1 , \tag{2.4-4}$$

while, from the triangle ABC in Figure 2.4,

$$(\pi-\psi) + \theta_1 + (\frac{\pi}{2}-i_1) = \pi \;\Rightarrow\; \psi = \frac{\pi}{2} + \theta_1 - i_1 . \tag{2.4-5}$$

From a combination of the previous two equations, we have

$$i_1 = \theta_1 + k_c q . \tag{2.4-6}$$

Similarly, from the other side of the surface, we have

$$i_2 = \theta_2 + k_c q , \tag{2.4-7}$$

where

$$q \approx q_1 \approx q_2 . \tag{2.4-8}$$

If Snell's law is now invoked, then

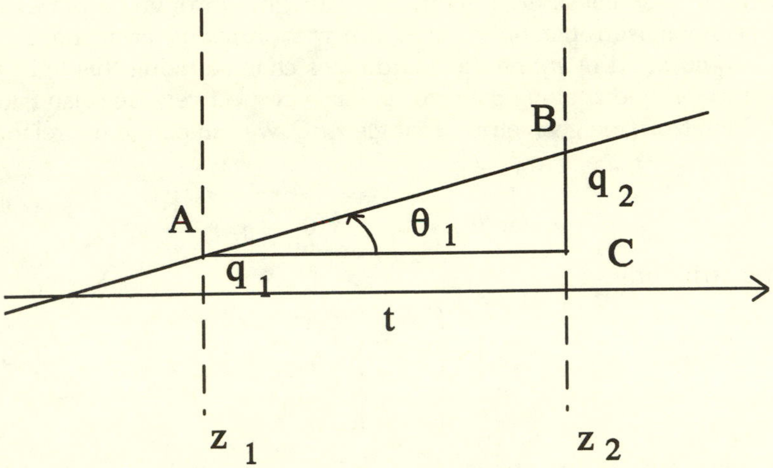

Figure 2.3: Light ray propagating within a homogeneous and isotropic optical medium.

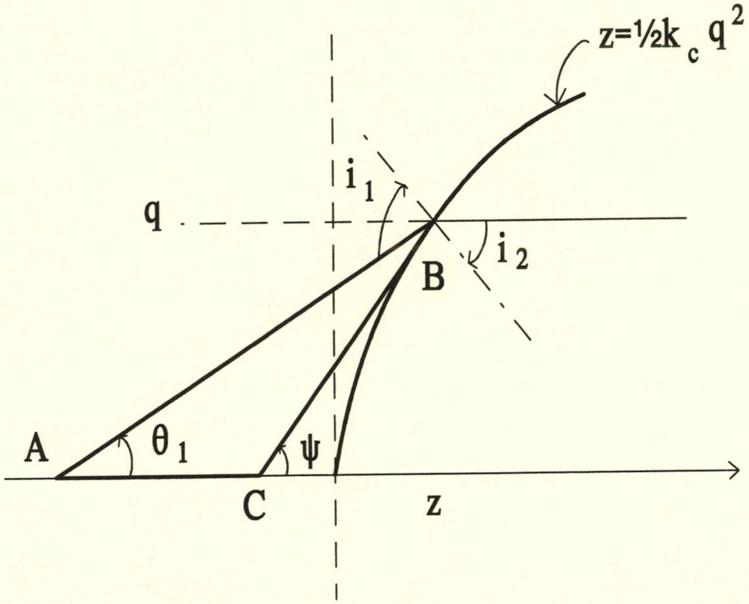

Figure 2.4: The refraction of a light ray between two homogeneous optical media.

$$n_1 i_1 = n_2 i_2 \,, \tag{2.4-9}$$

which, in conjunction with (2.4-6) and (2.4-7), implies that

$$p_1 + (n_1 k_c) q_1 = p_2 + (n_2 k_c) q_2 \,, \tag{2.4-10}$$

or, in matrix form

$$\begin{pmatrix} q_2 \\ p_2 \end{pmatrix} = \begin{pmatrix} 1 & 0 \\ -P_R & 1 \end{pmatrix} \cdot \begin{pmatrix} q_1 \\ p_1 \end{pmatrix} , \tag{2.4-11}$$

where

$$P_R = (n_2 - n_1) k_c \,. \tag{2.4-12}$$

Notice that, again, the transformation is linear and symplectic (det**M**=1). It is straightforward then to see that we can describe any axisymmetric, piecewise-continuous optical medium, within the small-angle limitation, as a combination of two or more of the elemental transformations described by Eqs. (2.4-2) and (2.4-11) corresponding to propagation and refraction, respectively.

EXAMPLE 2.2: A thin mirror can be described by a combination of two refractions

$$\begin{aligned} \mathbf{M} &= \mathbf{M}_{R_1} \cdot \mathbf{M}_{R_2} \\ &= \begin{pmatrix} 1 & 0 \\ -\dfrac{(n_2 - n_1)}{R_1} & 1 \end{pmatrix} \cdot \begin{pmatrix} 1 & 0 \\ -\dfrac{(n_1 - n_2)}{-R_2} & 1 \end{pmatrix} \\ &= \begin{pmatrix} 1 & 0 \\ -\dfrac{1}{f} & 1 \end{pmatrix} , \end{aligned} \tag{2.4-13}$$

where

$$\frac{1}{f} \equiv (n_2 - n_1) \left(\frac{1}{R_1} + \frac{1}{R_2} \right) , \tag{2.4-14}$$

i.e., f is the focal length of the mirror (see also Figure 2.5). ■

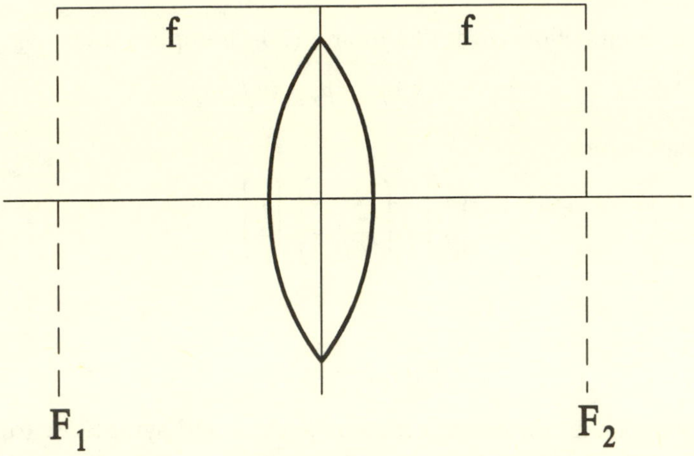

Figure 2.5: An example of a thin mirror. Locations F_1 and F_2 correspond to the focal planes.

The symplectic formulation is not the only mathematical description that is available for Gaussian optics. In fact, by 1828, Hamilton's method in Gaussian optics was developed. Hamilton's idea originated from the observation that a general symplectic transformation, $(q_1,p_1)\rightarrow(q_2,p_2)$, such as

$$\begin{pmatrix} q_2 \\ p_2 \end{pmatrix} = \begin{pmatrix} A & B \\ C & D \end{pmatrix} \cdot \begin{pmatrix} q_1 \\ p_1 \end{pmatrix} , \tag{2.4-15}$$

can be used (if $B \neq 0$) to solve for p_1 and p_2 in terms of q_1 and q_2. Physically, this means that, except when $B = 0$, for every set of positions q_1 and q_2, there is a unique light ray that goes through, corresponding to a unique set of tangents, p_1/n_1 and p_2/n_2, respectively:

$$p_1 = \frac{1}{B}(q_2 - Aq_1) \quad \text{and} \quad p_2 = \frac{1}{B}(Dq_2 - q_1) , \tag{2.4-16}$$

where the symplectic transformation condition, det**M**=1, has been used to simplify the above formulae. In the case B=0, all the rays emanating from q_1 end up at the same location $q_2 \equiv Aq_1$ (conjugate planes).

Hamilton's contribution was the realization that Eqs. (2.4-16) can also be derived from a scalar quantity W, the "point characteristic" or "eikonal" of the system, through a set of first-order, partial-differential equations,

$$p_1 = -\frac{\partial W}{\partial q_1} \quad \text{and} \quad p_2 = \frac{\partial W}{\partial q_2} . \tag{2.4-17}$$

Note that the eikonal is considered to be a function of the end-point position variables, q_1 and q_2, i.e., $W=W(q_1, q_2)$. For the transformation defined by Eqs. (2.4-15), it is straightforward to show the equivalence between Eqs. (2.4-16) and (2.4-17) if W is defined as

$$W(q_1,q_2) = \frac{1}{2B}(Aq_1^2 + Dq_2^2 - 2q_1q_2) + K , \tag{2.4-18}$$

where K is an arbitrary constant and $B \neq 0$. Alternatively, other generating functions can be derived, expressed as functions of different combinations of q_1 or p_1 with q_2 or p_2, by taking Legendre transforms of the eikonal. In exactly the same fashion, different equations of state can be derived in thermodynamics (such as the enthalpy, free energy, etc.) through Legendre transforms of the internal energy (see chapter 4).

It is also of interest to point out some of the properties of W, as, for example, additivity: the eikonal function W_{31} between two positions q_1 and q_3 is the same as the sum of the eikonal functions W_{21} and W_{32} between q_1 and q_2, and q_2 and q_3, respectively. Alternatively, in mathematical form,

$$W_{31}(q_1,q_3) = W_{21}(q_1,q_2) + W_{32}(q_2,q_3) . \tag{2.4-19}$$

Note that on the right-hand side of the above relationship, the intermediate point q_2 can, in principle, be expressed in terms of the end points, q_1 and q_3, by solving the equation

$$p_2 = \frac{\partial W_{21}(q_1,q_2)}{\partial q_2} = -\frac{\partial W_{32}(q_2,q_3)}{\partial q_2} . \tag{2.4-20}$$

Actually, we can attribute a physical meaning to the eikonal by identifying it with the "optical length," $\mathcal{L}$, as

$$W = \mathcal{L} \equiv \sum_{i=1}^{N} n_i l_i , \tag{2.4-21}$$

where N is the number of different homogeneous optical media traveled by the light ray and n_i and l_i are the index of refraction and the distance traveled within medium i, respectively. Then it is straightforward to show that $\mathcal{L}_{21}$, associated with the light ray joining q_1 and q_2, can be expressed as

$$\mathcal{L}_{21} = \mathcal{L}_{axis} + \frac{1}{2}(p_2q_2 - p_1q_1) , \tag{2.4-22}$$

where $\mathcal{L}_{\text{axis}} \equiv \int n dz$ is the optical length associated with the z coordinate axis and p_1 and p_2 are expressed in terms of q_1 and q_2 using Eq. (2.4-16).

From a physical point of view, the optical length is proportional to the time of travel associated with a particular light ray, as can easily be seen from its definition, (2.4-21), simply by expressing the index of refraction within the i-th medium, n_i, as the ratio of the speed of light in a vacuum, c, to the speed of light within the medium under consideration, c_i.

The above observation, together with Eqs. (2.4-17-19, 21), leads directly to *Fermat's principle*: among all possible light paths connecting two given points, q_1 and q_3, the actual light path is the one for which the optical path is an extremum with respect to all permissible infinitesimal variations, $\delta(\cdot)$. Alternately, in mathematical terms, any intermediate q_2 is selected so as to make $\mathfrak{L}_{31}$ an extremum:

$$\delta\mathfrak{L} = 0 \quad \Rightarrow \quad \frac{\partial\mathfrak{L}_{31}}{\partial q_2}(q_1,q_2,q_3) = 0 \ . \tag{2.4-23}$$

This is equivalent to the variational principles discussed for particle dynamics based on the minimization of the "action integral" (see chapter 3), which plays an equivalent role to the optical path.

2.4.2 *Linear optics*

The analysis of linear optics closely follows that of Gaussian optics. The only difference is that in linear optics a two-dimensional positional parameter, $\mathbf{q} \equiv (q_x,q_y)^T$ and a two-dimensional angular parameter, $\mathbf{p} \equiv (p_x,p_y)^T$, are necessary in order to completely specify an arbitrary light ray at a specific axial coordinate position, z—the axisymmetry no longer being valid. Therefore, the influence of the optical medium is modeled by symplectic transformations of four-dimensional vectors, $\mathbf{v} \equiv (\mathbf{q},\mathbf{p})^T$, represented by 4×4 symplectic matrices, $\mathbf{M}$. The most general symplectic transformation is similar in form to the one encountered in Gaussian optics represented by (2.4-15); the only difference is that the scalar parameters q_i, p_i and A, B, C, D are now two-dimensional vectors, and 2×2 square matrices, respectively. Of course, at this time, the necessary conditions for the transformation to be symplectic are not only limited to the unit determinant requirement—see also the comment in §2.3.2—but they also involve an additional set of algebraic constraints on the coefficients of the matrix $\mathbf{M}$, which can be easily recovered from (2.3-4) [Guillemin and Sternberg, 1984, pp. 26-27].

Hamilton's method and Fermat's principle are also valid in linear optics, with their corresponding mathematical expressions obtained from the ones applicable in Gaussian optics, extended in the obvious way. Essentially, no new physics is introduced, the major limitation of the small angles of Gaussian optics is still present, and Gaussian optics is recovered in the case when axisymmetry is present.

2.5 Geometrical Optics

In order to discuss the symplectic structure of geometrical optics,[1] some minimal knowledge of calculus on differential manifolds is necessary. In the last decade, several textbooks have appeared on the subject [Arnold, 1978; Burke, 1985; Olver, 1986; Abraham *et al.*, 1988; Bamberg and Sternberg, 1988; Conlon, 1993] which offer quite lengthy discussions of this subject. For the benefit of the interested reader, we offer in Appendix A a basic introduction to the fundamentals of this subject which we consider adequate for the understanding of the remainder of this chapter. Note, however, that this mathematically involved material is not necessary for the understanding of any other part of this book. In the following, the notation for differentiable manifolds introduced in Appendix A is used.

2.5.1. The symplectic structure of geometrical optics

There is a striking parallelism between the description of the theory of geometrical optics and that presented in the previous sections, §2.3 for Gaussian optics and §2.4 for linear optics. The subject of investigation continues to be in geometrical optics the same as in linear optics, i.e., the evaluation of the pathway of light rays within optically transparent media. However, this time, the properties of the medium (like the index of refraction, n) are assumed to depend continuously on the geometrical coordinates in space, x, y, and z; i.e., $n=n(x,y,z)$ and no particular assumptions are made concerning axisymmetry or small angles with respect to the z-axis. Therefore, mathematically, we are interested in evaluating, based on the properties of the medium and the initial conditions, the continuous path of a light ray. This is mathematically described as a general one-dimensional curve within a three-dimensional space $(x,y,z)=(x(\lambda),y(\lambda),z(\lambda))$, where λ is an arbitrary parametrization of the curve. Locally, we can always assume that at least one of the coordinates (say, z) is a monotonically-increasing function of the parameter λ, in which case the analysis can be somewhat simplified (without loss of generality) if we take the coordinate z for the light curve parametrization, i.e., $\lambda=z$. Then, the problem reduces to the evaluation of the functions $x(z)$ and $y(z)$.

[1]The description of geometrical optics in this section closely follows the analysis of Guillemin and Sternberg [1984, ch. 1]. For an alternative but equivalent approach which extends the development of Hamilton's equations to accommodate discontinuities in the refractive index see Sekigushi and Wolf [1987].

In order to preserve the symplectic structure, it turns out that the appropriate generalization of the parameters used in linear optics, **q** and **p**, is given as [Guillemin and Sternberg, 1984, p. 34]

$$p_x = \frac{\dot{x}n(x,y,z)}{\sqrt{1+\dot{x}^2+\dot{y}^2}}\,, \quad p_y = \frac{\dot{y}n(x,y,z)}{\sqrt{1+\dot{x}^2+\dot{y}^2}}\,, \qquad (2.5\text{-}1)$$
$$q_x = x\,, \qquad q_y = y\,,$$

where the dot over x and y denotes differentiation with respect to z. Note that in the limit of small angles, $\dot{x}\ll 1$ and $\dot{y}\ll 1$, the definition introduced by (2.5-1) reduces to the one used in linear optics in §2.4, as it should.

In the general case, (2.5-1) can be used to parametrically describe the path of a light ray within a four-dimensional manifold, $\mathfrak{I}$, isomorphic to $\mathbb{R}^4$, with respect to the axial coordinate z. This parametric description of the light curve is achieved by specifying the coordinates $(\mathbf{q},\mathbf{p})$ as functions of z: $\mathbf{q}=\mathbf{q}(z)$ and $\mathbf{p}(z)$. Thus, the curve generated by the light path passing through a given point of the manifold $(\mathbf{q}_0,\mathbf{p}_0)$ at $z=z_0$ can be considered as also generating a sequence of mappings $\gamma(z)\circ(\mathbf{q}_0,\mathbf{p}_0)\equiv(\mathbf{q}(\mathbf{q}_0,\mathbf{p}_0;z),\mathbf{p}(\mathbf{q}_0,\mathbf{p}_0;z))$ as well as a sequence of mappings between the tangent vectors, $\mathbf{M}(z)$: $T\mathfrak{I}_0\to T\mathfrak{I}(z)$ such that

$$\mathbf{M}\circ\mathbf{v} = \begin{pmatrix} \dfrac{\partial\mathbf{q}}{\partial\mathbf{q}_0} & \dfrac{\partial\mathbf{q}}{\partial\mathbf{p}_0} \\ \dfrac{\partial\mathbf{p}}{\partial\mathbf{q}_0} & \dfrac{\partial\mathbf{p}}{\partial\mathbf{p}_0} \end{pmatrix}\cdot\mathbf{v}\,, \qquad (2.5\text{-}2)$$

where $\partial\mathbf{q}/\partial\mathbf{q}_0$ is a 2×2 square matrix whose entries are the partial derivatives $\partial q_i/\partial q_{0j}$, etc. Similarly, the inverse mapping $\mathbf{M}^{-1}$ induces a transformation between the cotangent spaces $T^*\mathfrak{I}(z)\to T^*\mathfrak{I}_0$, the Cartesian product $\mathbf{M}^{-1}\times\mathbf{M}^{-1}$ a transformation between 2-forms, etc.

The manifold $\mathfrak{I}$ can acquire a symplectic structure by specifying a symplectic 2-form field ω. As it turns out, the appropriate symplectic form for the description of the geometrical optics is the canonical one in terms of the conjugate coordinates **q** and **p**:

$$\omega = \sum_{i=1}^{2} dp_i\wedge dq_i\,. \qquad (2.5\text{-}3)$$

Using the symplectic canonical form field ω on the manifold, together with the induced mappings from the light paths, the symplectic structure of geometrical optics can be expressed by the statement that the linear transformations induced from the light paths, as defined by (2.5-2), preserve the symplectic 2-form ω. Alternatively, in a mathematical form that has a direct correspondence with (2.3-4),

$$\omega(z)\circ(\mathbf{M}\cdot\mathbf{v})(\mathbf{M}\cdot\mathbf{u}) = \omega(z_0)\circ\mathbf{v}\mathbf{u}\ ,\quad \mathbf{v},\mathbf{u}\in T\Im_0\ . \tag{2.5-4}$$

Eq. (2.5-4) implies that the 2-form Ω

$$\Omega = \sum_{i=1}^{2} \mathrm{d}p_{0i}\wedge\mathrm{d}q_{0i} - \sum_{i=1}^{2} \mathrm{d}p_i\wedge\mathrm{d}q_i\ , \tag{2.5-5}$$

where $(\mathbf{q}_0,\mathbf{p}_0)$ are the canonical coordinates of the origin, z_0, of the light path, and $(\mathbf{q},\mathbf{p})$ those at the destination, z, is identically equal to zero in the extended manifold $\Im\times\gamma\circ\Im$ (i.e., the space generated by the original manifold and its image under the action of the mapping γ induced by the light ray). This in turn implies that the extended manifold $\Im\times\gamma\circ\Im$ is a Lagrangian manifold of the space $(\mathbf{q}_0,\mathbf{q},\mathbf{p}_0,-\mathbf{p})\in\mathbb{R}^8$ since, with this mapping, the 2-form Ω is a canonical symplectic form in $\mathbb{R}^8$ and at the same time Ω is identically equal to zero when restricted to the $\Im\times\gamma\circ\Im$ submanifold. Therefore, according to the theorem of Lagrangian manifolds (see the discussion in §A.6 of Appendix A), this implies that there is a function $S(\mathbf{q}_0,\mathbf{q},\mathbf{p}_0,-\mathbf{p})$ such that

$$\mathbf{p}_0 = \frac{\partial S}{\partial \mathbf{q}_0}\ ,\quad \mathbf{p} = -\frac{\partial S}{\partial \mathbf{q}}\ , \tag{2.5-6}$$

i.e., we recognize S as the eikonal function of Gaussian optics.

Alternatively, the symplectic nature of the transformation can be mathematically represented as

$$\omega(z_0) = \mathbf{M}^{-1}\times\mathbf{M}^{-1}\circ\omega(z)\ . \tag{2.5-7}$$

When z is infinitesimally close to z_o, (2.5-7) gives

$$\frac{\mathrm{d}\omega}{\mathrm{d}z} = 0\ . \tag{2.5-8}$$

The last relationship implies (see Theorem A.1 in Appendix A) that there is a Hamiltonian vector field $\mathbf{u}\equiv\mathbf{sgrad}(H)$ from which the light curve arises as an integral curve, i.e.,

$$\frac{\mathrm{d}\mathbf{q}}{\mathrm{d}z} = \frac{\partial H}{\partial \mathbf{p}}\ ,\quad \frac{\mathrm{d}\mathbf{p}}{\mathrm{d}z} = -\frac{\partial H}{\partial \mathbf{q}}\ . \tag{2.5-9}$$

Using differentiation by parts, it is easy to show that (2.5-9) is equivalent to the statement that the dynamic behavior of every arbitrary function f (i.e., its rate of change along a light ray) is governed by

$$\frac{\mathrm{d}f}{\mathrm{d}z} = \{f,H\}\ , \tag{2.5-10}$$

where $\{\cdot,\cdot\}$ is the Poisson bracket, defined as a skew-symmetric bilinear form such that for two arbitrary functions f and g, $\{f,g\}$ is a function such that

$$\{f,g\} \equiv \frac{\partial f}{\partial \mathbf{p}} \cdot \frac{\partial g}{\partial \mathbf{q}} - \frac{\partial g}{\partial \mathbf{p}} \cdot \frac{\partial f}{\partial \mathbf{q}} \ . \tag{2.5-11}$$

Note that the Poisson bracket defined by (2.5-11)—which is a particular case of the most general definition provided by Eq. (A-32) in Appendix A—satisfies all the properties assigned to a Poisson bracket, Eqs. (A-33) through (A-35).

2.5.2 *Fermat's principle*

In this subsection, we shall see how the symplectic structure described in the previous subsection is realized in geometrical optics through the prevalent optical theory, namely, Fermat's principle. We start with the definition of the optical path $\gamma(z)$,

$$\gamma(z) = \begin{pmatrix} x(z) \\ y(z) \\ z \end{pmatrix} , \tag{2.5-12}$$

where we have used the axial coordinate z, as mentioned before, for the parameterization of the curve defined by the path of the light ray into consideration. Then, the appropriate expression for the optical length, following the definition of (2.4-1), is

$$\begin{aligned} \mathcal{L}_\gamma &= \int n \sqrt{(1 + \dot{x}^2 + \dot{y}^2)} \, \mathrm{d}z \\ &= \int \mathcal{L}(x,y,\dot{x},\dot{y};z) \, \mathrm{d}z \ , \end{aligned} \tag{2.5-13}$$

where the function $\mathcal{L}(x,y,\dot{x},\dot{y};z)$ is the optical length density

$$\mathcal{L}(x,y,\dot{x},\dot{y};z) = n(x,y,z) \sqrt{(1 + \dot{x}^2 + \dot{y}^2)} \ . \tag{2.5-14}$$

Fermat's principle can now be expressed as the requirement that the optical path realized between two fixed end points corresponds to an extremum of the optical path $\mathcal{L}$. Moreover, note that the canonical coordinates defined by (2.5-1) are related to the coordinates $(x,y,\dot{x},\dot{y})$ through the Legendre transformations (see Appendix B)

$$\begin{aligned} q_x &= x \ , \qquad q_y = y \ , \\ p_x &= \frac{\partial \mathcal{L}}{\partial \dot{x}} \ , \quad p_y = \frac{\partial \mathcal{L}}{\partial \dot{y}} \ . \end{aligned} \tag{2.5-15}$$

Now let us define a function $\mathcal{H}$ of the canonical coordinates $\mathbf{q}$ and $\mathbf{p}$ as

$$\mathcal{H}(q_x,q_y,p_x,p_y;z) \equiv p_x \dot{x} + p_y \dot{y} - \mathcal{L} \ , \tag{2.5-16}$$

where on the right-hand side of (2.5-16) we consider $\dot{x}$ and $\dot{y}$ as functions of the canonical coordinates $\mathbf{q}$ and $\mathbf{p}$ through the inverse transformation defined by (2.5-15). Then, Fermat's principle can be reinstated by the requirement that the integral of the function $\mathcal{H}$ along the optical path in the canonical coordinates is an extremum [Guillemin and Sternberg, 1984, p. 42]. This implies that

$$\dot{q}_x = \frac{\partial \mathcal{H}}{\partial p_x}, \quad \dot{q}_y = \frac{\partial \mathcal{H}}{\partial p_y},$$
$$\dot{p}_x = -\frac{\partial \mathcal{H}}{\partial q_x}, \quad \dot{p}_y = -\frac{\partial \mathcal{H}}{\partial q_y}, \tag{2.5-17}$$

which implies that $\mathcal{H}$ is a Hamiltonian function with respect to the canonical coordinates, $\mathbf{q}$ and $\mathbf{p}$, as defined in Appendix A.

2.6 An Overview of Wave Optics and Electromagnetism

The symplectic structure continues to permeate through successive refinements in the theories of optics, namely wave optics and electromagnetism. The interested reader can find a full account of this subject by Guillemin and Sternberg [1984, pp. 60-150]. Rather than duplicating their discussion here, it is instructive to reproduce their final results stressing only the formalism and its associated physical meaning.

In wave optics, as in Gaussian or geometrical optics, the objective is to relate information with respect to a light path at a particular coordinate cross section, z_2, with information related to the light path at a previous cross section, z_1. However, in the wave/optical theory, the information is given in terms of a complex function, $c = c(\mathbf{q};z)$, rather than a set of discrete variables $\mathbf{q}$ and $\mathbf{p}$. The correlation is therefore obtained through an integral expression which, in the equivalent of the Gaussian optics approximation, is

$$c(x) = \exp\left(\frac{\pm\pi i}{4}\right)|B\lambda|^{-\frac{1}{2}} \int_{-\infty}^{\infty} \exp[2\pi i \frac{\mathcal{L}(x,y)}{\lambda}] c(y)\,\mathrm{d}y\,, \tag{2.6-1}$$

where x and y are the position coordinates at the locations z_2 and z_1, respectively, $\mathcal{L}(x,y)$ is the corresponding optical length, λ is the wavelength of light, B is the coefficient of $\mathbf{M}$ (in the Gaussian description of the optical properties of the medium—see Eq. (2.4-15)), and where the ± sign refers to the sign of B.

Note that since the Lagrangian function (optical length) is also determined by the optical matrix $\mathbf{M}$, the coefficients M_{ij} are involved in determining the optical response of the medium in both the Gaussian and the wave-theory approach. It is this correspondence that allows, in the limit of a piecewise-constant medium and for small angles of light propagation, the one-to-one correlation (actually, two-to-one) between the integral operators of the wave/optical theory and the matrices of Gaussian optics, thus embedding the symplectic structure of Gaussian (similarly, linear and geometric) optics into wave optics.

The physical meaning of the complex function c is that of determining the amplitude of the light, A, at a corresponding location:

$$A = \bar{c}\,c = |c|^2 , \tag{2.6-2}$$

where the overbar denotes the complex conjugate. Not surprisingly, wave optics is intimately connected with quantum mechanics and the complex function c with the wave function ψ. Actually, it turns out that (2.6-1) is formally the solution of the corresponding Schrödinger equation. The integral operators in (2.6-1) define path families whose tangent operators correspond to the Hamiltonian operators of quantum mechanics. Hence the symplectic structure of wave optics is now transferred to quantum mechanics and through its classical limit to particle dynamics. (See also the discussion on the optical/mechanical analogy in §3.6.)[2]

[2]The direct analogy between quantum and classical mechanics which is established when they are formulated in terms of the commutator and the Poisson bracket, respectively, has first been put forward by Dirac [1925, p. 648]. In fact, the commutator is equal to the Poisson bracket times $ih/(2\pi)$. This property is of such fundamental importance that it is the main route that it is used in order to obtain the quantum-mechanical description of a classical system: The quantum-mechanical description of a classical system is simply obtained from the Poisson-bracket description of the latter by simply substituting the Poisson bracket with the commutator of the corresponding quantum mechanical operators divided by $ih/(2\pi)$. Thus, it is of no wonder that it continues to attract the attention of scientific papers more than sixty-five years after its original development—the relevant literature is too extensive to be able to mention it all here, as a small example see [Bloore, 1972; Sherry, 1990; Hojman and Shepley, 1991]. Moreover, the last one of these three references goes as far as to suggest that unless a Hamiltonian formalism is available for a classical system (or a Lagrangian; the two are equivalent for non-dissipative systems [Hojman and Shepley, p. 144]) there is no hope to quantize that system! (No Lagrangian? No quantization! [Hojman and Shepley, 1991]). Independently, the absence of a (natural) Hamiltonian description of general relativity can be considered as one of the principal reasons that explains the lack of a theory for quantum gravity [Rovelli, 1991]. However, in the same work, Rovelli notes:

> We do not regard the lack of a (natural) Hamiltonian description of general relativity as a failing of the formalism. Instead, we take it as a profound indication that the absolute time t is not a physical quantity, and that only evolution with respect to clocks is observable.
>
> The suggestion of this paper ([Rovelli, 1991]) is that this indication has to be taken seriously. If so, it has to be extended also to quantum mechanics.

Elsewhere in the same paper, Rovelli suggests a "time-free" pre-symplectic formalism as an alternative to the Poisson-bracket based one—see also Hajicek [1991]. Either way, the symplectic structure has had and continues to have an important role in the investigation of the fundamental laws of nature.

The next step up in rigor from wave optics involves the description of the interaction of charged particles with electromagnetic fields. This is described by the famous Maxwell equations in conjunction with the expressions for the electromagnetic force. It is interesting to note that although the description of electromagnetic forces on charged particles can be developed in a straightforward manner through the Hamiltonian formulation of the equations of motion, the Maxwellian field equations cannot because the particular tensorial character of the principal quantities involved is different.[3] These quantities are the electric field strength, **E**, which through its definition from the potential field can be identified as a 1-form, and the electric displacement, D, which through its definition from the enclosed total charge can be identified as a 2-form. It is obvious that for any constitutive relation between these two quantities to be possible, one of them has to be transferred to the form of the other one. This can only be accomplished through the use of a metric. The particular form of Maxwell equations can only admit a Lorentzian metric in a four-dimensional space/time frame as the appropriate one, in which framework Maxwell's field equations become simply closure constraints on two specific 2-forms, $F=B+E\wedge \mathrm{d}t$ and $G=D-H\wedge \mathrm{d}t$, namely

$$\mathrm{d}F = 0 \quad \text{and} \quad \mathrm{d}G = 4\pi J' , \tag{2.6-3}$$

where B is the magnetic induction 2-form (not to be confused with the B coefficient of $\mathbf{M}$), H is the magnetic field strength, and J' is given as

$$J' = \rho \mathrm{d}x\wedge \mathrm{d}y\wedge \mathrm{d}z - j\wedge \mathrm{d}t , \tag{2.6-4}$$

with ρ being the electric charge density (0-form) and j the electric current density (2-form).

Still, the evolution equation of a charged, spinning particle can be expressed in the framework of the symplectic theory, outlined in

[3]This is related to the Lorentz invariance of the Maxwell equations. Similar problems, as mentioned in the previous footnote, arise with the (natural) Hamiltonian description of Einstein's theory of relativity. Instead, a Hamiltonian description in a higher (four) dimensionality space where both (natural) space and time are taken on an equal basis as the independent coordinates is possible (in the presence of a gravitational field [Stueckelberg, 1942, p. 25]). The evolution is then described in terms of an independent parameter which is typically taken to coincide with the time of inertial clocks. This theory was first developed by Stueckelberger [1942] and then, independently by Feynman [1950]; see also Horwitz *et al.* [1988]. More recently, Barut [1986, pp. 106-07] offered such a Hamiltonian description for a relativistic Dirac (spinning) electron particle interacting with an eletromagnetic field. Even more recently, Ashtekar offered a new Hamiltonian formulation for general relativity in terms of new variables with respect to which the constraints required in the theory for the fields acquire simple polynomial structures [Ashtekar, 1991; Husain, 1993; Maluf, 1993]. Thus, there are several ways according to which the symplectic structure can be preserved.

Appendix A, through the following selection of Hamiltonian, $\mathcal{H}$, and the symplectic form:

$$\mathcal{H}(\mathbf{q},\mathbf{p},\mathbf{u}) = \frac{1}{2m}|\mathbf{p}|^2 + \phi(\mathbf{q}) + \mu s \mathbf{u}\cdot\mathbf{B} \ , \tag{2.6-5}$$

and

$$\omega = \sum_{i=1}^{3} \mathrm{d}p_i \wedge \mathrm{d}q_i + eB + s\Omega \ , \tag{2.6-6}$$

where $\mathbf{q}$ and $\mathbf{p}$ are the position and momentum vectors of the particle, $\phi(\mathbf{q})$ is the electrostatic potential, μ is the magnetic moment, s is the spin, $\mathbf{u}$ is the axis of the magnetic moment of the particle, e is the electric charge, and Ω the standard volume 2-form on the unit sphere (acting on the spin orientation $\mathbf{u}$) [Guillemin and Sternberg, 1984, pp. 134-36]. Note that in the Hamiltonian the magnetic intensity B appears as a vector, whereas in the definition of the symplectic form it appears as a 2-form. This is consistent with a three-dimensional space since, in that case, both vectors and 2-forms are elements of the three-dimensional space. However, as mentioned before, a transformation between one and the other is formally necessary. It is in this transformation that the isotropic or anisotropic properties of the medium are expressed.

A further refinement on the Maxwell equations and wave optics leads us to quantum electrodynamics, developed by Dirac in the late 1920s. This theory uses a refinement of the integral operators described above for wave optics expressed in a statistical fashion for all possible events between two points. An extensive discussion is outside the scope of this book, but the interested reader is referred to the classical references in the field (for example, Feynman and Hibbs [1965, ch. 9]; Feynman [1985]). For an alternative description using field quantization, see the recent textbook of Cohen-Tannoudji *et al.* [1989].

This concludes our excursion into the theories of optics which brought us all the way from macroscopic to subatomic phenomena. The amazing fact remains that the same mathematical structure permeates through six levels of successive approximation practically unchanged. It is this robustness of the symplectic structure of the equations that motivates its use in continuum systems.

3
Hamiltonian Mechanics of Discrete Particle Systems

If one way be better than another, that you may be sure is Nature's way.

—*Aristotle*

In the previous chapter, we discussed briefly the fundamental nature of the symplectic structure of theories in optics in order to illustrate the underlying uniformity, physical consistency, and mathematical simplicity inherent to a symplectic mathematical formulation of the governing equations. Hence the main emphasis of chapter 2 was to "discover" the symplectic structure in the physical theories of optics and to see how this structure is interconnected with and implies fundamental theorems in optics, such as Fermat's principle and Hamilton's equations.

In the present chapter, we continue our efforts to present a coherent description of symplectic transformations and their applications to physical systems; however, here we switch our emphasis from the underlying symplectic structure of the dynamical equations to the physical integrity of the Poisson bracket and the canonical equations which find their roots in Hamilton's principle of least action and the calculus of variations. Hence we intend to cover ground in this chapter which we neglected in the previous one, and, in so doing, to gradually begin to move towards the applications of the extended bracket formalism at which this book is aimed.

3.1 The Calculus of Variations

In order to apply Hamilton's principle of least action, we first need to study a simple problem of the calculus of variations, following Bedford [1985, §1.1]. Let x be a real variable ($x \in \mathbb{R}$) on the closed interval $x_1 \leq x \leq x_2$, denoted $[x_1,x_2]$. The function $y(x)$ is defined as C^N on $[x_1,x_2]$ if the N-th derivative of $y(x)$ exists and is continuous on $[x_1,x_2]$. Now let $y(x)$ be a C^1 function on $[x_1,x_2]$ such that $y(x_1)=y_1$ and $y(x_2)=y_2$ so that $y(x)$ describes a smooth curve which joins two points in the x-y plane, (x_1,y_1) and (x_2,y_2). Henceforth, any C^1 function joining these two points will be called an *admissible function*.

Now we shall consider the integral

$$I \equiv \int_{x_1}^{x_2} f(x,y,y')\ \mathrm{d}x\ , \tag{3.1-1}$$

where $y' \equiv \mathrm{d}y/\mathrm{d}x$ and f is a given function with prescribed values y_1 and y_2 at positions x_1 and x_2. The value of this integral obviously depends on $y(x)$, and the question we need to answer is whether a particular $y(x)$ can be found such that the value of the integral is a minimum.

In order to find an admissible function $y(x)$ for which the integral is a minimum, we need to define an *admissible comparison function* by

$$y^*(x,\varepsilon) \equiv y(x) + \varepsilon\,\eta(x)\ , \tag{3.1-2}$$

where ε is an arbitrary parameter and $\eta(x)$ is an arbitrary C^1 function on $[x_1,x_2]$. $\eta(x)$ must be subject to the constraints $\eta(x_1)=\eta(x_2)=0$, since $y^*(x)$ must be an admissible function, i.e., $y^*(x_1)=y_1$ and $y^*(x_2)=y_2$. Substitution of the comparison function into the integral (3.1-1) in place of $y(x)$ yields the integral

$$I^*(\varepsilon) \equiv \int_{x_1}^{x_2} f(x,y^*,y'^*)\ \mathrm{d}x \equiv \int_{x_1}^{x_2} f^*\ \mathrm{d}x\ . \tag{3.1-3}$$

Now we shall assume that the value of the integral (3.1-3) is a minimum when the parameter $\varepsilon \to 0$, so that

$$\left(\frac{\mathrm{d}I^*(\varepsilon)}{\mathrm{d}\varepsilon}\right)_{\varepsilon\to 0} = 0\ . \tag{3.1-4}$$

Taking the derivative of the integral (3.1-3) with respect to ε gives

$$\begin{aligned}\frac{\mathrm{d}I^*(\varepsilon)}{\mathrm{d}\varepsilon} &= \int_{x_1}^{x_2}\left[\frac{\partial f^*}{\partial y^*}\frac{\partial y^*}{\partial\varepsilon} + \frac{\partial f^*}{\partial y'^*}\frac{\partial y'^*}{\partial\varepsilon}\right]\mathrm{d}x \\ &= \int_{x_1}^{x_2}\left[\frac{\partial f^*}{\partial y^*}\eta + \frac{\partial f^*}{\partial y'^*}\eta'\right]\mathrm{d}x\ ,\end{aligned} \tag{3.1-5}$$

so that the condition (3.1-4) implies

$$\int_{x_1}^{x_2}\left[\frac{\partial f}{\partial y}\eta + \frac{\partial f}{\partial y'}\eta'\right]\mathrm{d}x = 0\ , \tag{3.1-6}$$

since $f^* \to f$ as $\varepsilon \to 0$. The second term in the above integral can be integrated by parts to give

$$\int_{x_1}^{x_2} \frac{\partial f}{\partial y'} \eta' \, dx = \left(\frac{\partial f}{\partial y'} \eta \right)_{x_1}^{x_2} - \int_{x_1}^{x_2} \frac{d}{dx} \left(\frac{\partial f}{\partial y'} \right) \eta \, dx \; ; \tag{3.1-7}$$

however, since $\eta(x_1)=\eta(x_2)=0$, the integral (3.1-6) can be rewritten as

$$\int_{x_1}^{x_2} \left[\frac{\partial f}{\partial y} - \frac{d}{dx} \left(\frac{\partial f}{\partial y'} \right) \right] \eta \, dx = 0 \, . \tag{3.1-8}$$

Remember that the function $\eta(x)$ is arbitrary, other than the conditions noted above, which implies that the bracketed quantity in the integral (3.1-8) must vanish on $[x_1, x_2]$, since otherwise a function could be chosen such that equality (3.1-8) would be violated. Thus we have

$$\frac{\partial f}{\partial y} - \frac{d}{dx} \left(\frac{\partial f}{\partial y'} \right) = 0 \, , \quad \text{on } [x_1, x_2] \, , \tag{3.1-9}$$

which is called the *Euler/Lagrange equation.* It provides a differential equation which can be used to determine the function $y(x)$.

It should be noted here, however, that the condition (3.1-4) is only a necessary condition and is not sufficient to guarantee that the value of the integral is a minimum when $\varepsilon=0$. This condition is also satisfied when the value of the integral is a maximum or has an inflection point. In other words, the condition (3.1-4) is satisfied when the function $y(x)$ is *stationary relative to admissible comparison functions.*

The mathematical problem studied in this section was that of minimizing the value of an integral. Using the calculus of variations, the final result can be obtained without accounting explicitly for an infinite number of admissible comparison functions, but by restricting the comparison functions to be indefinitely close to the actual function. A comparison function which differs from the actual function by an infinitesimal degree is called a *variation* of the actual function.

3.2 Hamilton's Principle of Least Action

Let N discrete particles be described by a set of N Cartesian position vectors $\mathbf{x}_i$, $i=1,2,...,N$. Also, let $\mathbf{F}_i$ be the conservative force felt by the i-th particle

$$\mathbf{F}_i \equiv - \frac{\partial e_p^v}{\partial \mathbf{x}_i} \, , \quad i=1,2,...,N \, , \tag{3.2-1}$$

where $e_p^v = e_p^v(\mathbf{x}_1, \mathbf{x}_2, ..., \mathbf{x}_N)$ is the corresponding generating force potential. Then the dynamics of the N particles can be described exactly, provided the initial positions $\mathbf{x}_i$ and velocities $\dot{\mathbf{x}}_i$ are known, by N applications of Newton's law (*mass* × *acceleration* = *force*)

$$m_i \frac{\partial^2 \mathbf{x}_i}{\partial t^2} \equiv - \frac{\partial e_p^v}{\partial \mathbf{x}_i} \ , \quad i=1,2,...,N \ , \tag{3.2-2}$$

in which m_i is the mass of the i-th particle and $\partial e_p^v/\partial \mathbf{x}_i \equiv (\partial e_p^v/\partial x_{i1}, \partial e_p^v/\partial x_{i2}, \partial e_p^v/\partial x_{i3})^T$, where x_{i1}, x_{i2}, and x_{i3} are the Cartesian components of the position vector $\mathbf{x}_i$.

As the simplest example of the potential e_p^v, consider a system of N point particles experiencing perfectly elastic collisions. In this case, the potential is zero provided that $\mathbf{x}_i \neq \mathbf{x}_j$, and unity if $\mathbf{x}_i = \mathbf{x}_j$. Consequently, when $\mathbf{x}_i$ and $\mathbf{x}_j$ become extremely close to each other, the force on each particle becomes infinite. This potential becomes more complex when the particles are experiencing inter-particle attractive or repulsive forces, or the effects of external fields.

Alternatively, the same system of equations, (3.2-2), can be derived through the application of Hamilton's principle. Let t_1 and t_2 be fixed times, with $t_1 < t_2$, such that the system configuration at times t_1 and t_2 is known. An admissible motion of the system is specified as the set of functions $\mathbf{x}_i = \mathbf{x}_i(t)$, $i=1,2,...,N$, which both fulfill the required configurations at t_1 and t_2, and have continuous second derivatives on the time interval $[t_1, t_2]$ [Bedford, 1985, p. 5].

Hamilton's principle for a system of particles states that, among admissible motions, the actual system motion is such that the integral

$$I = \int_{t_1}^{t_2} L \mathrm{d}t \ , \tag{3.2-3}$$

is stationary with respect to neighboring admissible motions. The term "stationary" refers to the minimization (actually, locating the extremum) of the integral (3.2-3). Here, L is the system Lagrangian defined as

$$L(\mathbf{x}_1,...,\mathbf{x}_N, \dot{\mathbf{x}}_1,...,\dot{\mathbf{x}}_N) \equiv e_k - e_p^v = \sum_{i=1}^{N} \frac{1}{2} m_i \dot{\mathbf{x}}_i \cdot \dot{\mathbf{x}}_i - e_p^v(\mathbf{x}_1,...,\mathbf{x}_N) \ , \tag{3.2-4}$$

with e_k being the kinetic energy of the system. In the following, we only consider Lagrangians (and Hamiltonians) which are not explicit functions of time, these being the most widely applicable.

Physically, Hamilton's principle states that the motion of the system will be such that the time integral of the difference in the kinetic and potential energies (called the *action*) is minimized. In order to see why, let us consider a system which is at a point A at time t_1, and also assume that the system will be at point B at a time t_2 [Lanczos, 1970, p. xxiii]. Now let us connect these two points by any arbitrary C^2 function. Statistically speaking, the function which we have chosen is not likely to be the actual function which nature has chosen for this system. However, we can correct this arbitrary function and arrive at a function which can be designated as the actual path of motion. The value of the action will vary from function to function (path to path), and mathematically, we can try every possible function, but there must exist one function for which the action assumes a minimum value. The principle of least action states that the particular function chosen by nature as the actual system motion minimizes the action.

If the set of functions $\mathbf{x}_i(t)$ describes the actual motion of the system, then an admissible comparison motion for this system is

$$\mathbf{x}_i^*(t,\varepsilon) = \mathbf{x}_i(t) + \varepsilon\boldsymbol{\eta}_i(t) \ , \ i=1,2,\ldots,N \ , \tag{3.2-5}$$

where ε is a small parameter and $\boldsymbol{\eta}_i(t)$ is a set of arbitrary C^2 functions such that $\boldsymbol{\eta}_i(t_1)=\boldsymbol{\eta}_i(t_2)=\mathbf{0}$. Substituting (3.2-5) into (3.2-3) yields the integral

$$I^*(\varepsilon) \equiv \int_{t_1}^{t_2} L^*\mathrm{d}t \ , \tag{3.2-6}$$

with $L^*=L(\mathbf{x}_1^*,\ldots,\mathbf{x}_N^*,\dot{\mathbf{x}}_1^*,\ldots,\dot{\mathbf{x}}_N^*)$. Since $\mathbf{x}_i(t)$ represents the actual system motion, the integral (3.2-6) is stationary when $\mathbf{x}_i^*(t,\varepsilon)=\mathbf{x}_i(t)$, and so

$$\left(\frac{\mathrm{d}I^*(\varepsilon)}{\mathrm{d}\varepsilon}\right)_{\varepsilon\to 0} = 0 \ . \tag{3.2-7}$$

Taking the derivative of (3.2-6) with respect to ε gives

$$\begin{aligned}\frac{\mathrm{d}I^*(\varepsilon)}{\mathrm{d}\varepsilon} &= \int_{t_1}^{t_2}\sum_{i=1}^{N}\left[\frac{\partial e_k^*}{\partial \mathbf{x}_i^*}\cdot\frac{\partial \mathbf{x}_i^*}{\partial\varepsilon} + \frac{\partial e_k^*}{\partial \dot{\mathbf{x}}_i^*}\cdot\frac{\partial \dot{\mathbf{x}}_i^*}{\partial\varepsilon} - \frac{\partial e_p^{v*}}{\partial \mathbf{x}_i^*}\cdot\frac{\partial \mathbf{x}_i^*}{\partial\varepsilon}\right]\mathrm{d}t \\ &= \int_{t_1}^{t_2}\sum_{i=1}^{N}\left[\frac{\partial e_k^*}{\partial \mathbf{x}_i^*}\cdot\boldsymbol{\eta}_i + \frac{\partial e_k^*}{\partial \dot{\mathbf{x}}_i^*}\cdot\dot{\boldsymbol{\eta}}_i - \frac{\partial e_p^{v*}}{\partial \mathbf{x}_i^*}\cdot\boldsymbol{\eta}_i\right]\mathrm{d}t \ ,\end{aligned} \tag{3.2-8}$$

and thus the condition (3.2-7) becomes

$$\int_{t_1}^{t_2}\sum_{i=1}^{N}\left[\frac{\partial e_k}{\partial \mathbf{x}_i}\cdot\boldsymbol{\eta}_i + \frac{\partial e_k}{\partial \dot{\mathbf{x}}_i}\cdot\dot{\boldsymbol{\eta}}_i - \frac{\partial e_p^{v}}{\partial \mathbf{x}_i}\cdot\boldsymbol{\eta}_i\right]\mathrm{d}t = 0 \ . \tag{3.2-9}$$

When the second term in the integrand of (3.2-9) is integrated by parts (remembering that $\boldsymbol{\eta}_i(t_1)=\boldsymbol{\eta}_i(t_2)=\mathbf{0}$), the integral (3.2-9) is rewritten as

$$\int_{t_1}^{t_2}\sum_{i=1}^{N}\left[\frac{\partial L}{\partial \mathbf{x}_i} - \frac{\mathrm{d}}{\mathrm{d}t}\left(\frac{\partial L}{\partial \dot{\mathbf{x}}_i}\right)\right]\cdot\boldsymbol{\eta}_i\,\mathrm{d}t = 0 \ . \tag{3.2-10}$$

Since $\boldsymbol{\eta}_i(t)$ is an arbitrary function on $[t_1,t_2]$, Eq. (3.2-10) can only be valid provided that

$$\frac{\partial L}{\partial \mathbf{x}_i} - \frac{\mathrm{d}}{\mathrm{d}t}\left(\frac{\partial L}{\partial \dot{\mathbf{x}}_i}\right) = \mathbf{0} \ , \quad i=1,2,\ldots,N \ , \quad \text{on } [t_1,t_2] \ . \tag{3.2-11}$$

Eq. (3.2-11) is often called *Lagrange's equation of motion.* Once it is realized that $\partial L/\partial \dot{\mathbf{x}}_i = m_i \dot{\mathbf{x}}_i$ and $\partial L/\partial \mathbf{x}_i = -\partial e_p^v/\partial \mathbf{x}_i$, it is apparent that Eq. (3.2-2) is recovered. Thus, Eq. (3.2-11) represents a generalization of (3.2-2) to arbitrary coordinate systems.

A notation which is often used with respect to Hamilton's principle is

$$\delta(\cdot) \equiv \left(\frac{\partial(\cdot)^*}{\partial \varepsilon}\right)_{\varepsilon \to 0} \ , \tag{3.2-12}$$

where the symbol $\delta(\cdot)$ is termed the *variation* of $(\cdot)$. As a consequence, the variation of $\mathbf{x}_i$ from (3.2-5) is

$$\delta \mathbf{x}_i(t) = \boldsymbol{\eta}_i(t) \ , \tag{3.2-13}$$

and Eqs. (3.2-6) and (3.2-7) can be rewritten as

$$\delta\left(\int_{t_1}^{t_2} L \, \mathrm{d}t\right) = 0 \ . \tag{3.2-14}$$

Another equivalent formulation of the equations governing the dynamics of the N discrete particles is obtained using Hamilton's principle in modified form [Goldstein, 1980, p. 341]:

$$\delta\left(\int_{t_1}^{t_2} \left[\sum_{i=1}^{N} \mathbf{p}_i \cdot \dot{\mathbf{x}}_i - H\right] \mathrm{d}t\right) = 0 \ , \tag{3.2-15}$$

where $\mathbf{p}_i$ is the conjugate momentum of the i-th particle defined as

$$\mathbf{p}_i \equiv \frac{\partial L}{\partial \dot{\mathbf{x}}_i} = m_i \dot{\mathbf{x}}_i \ , \quad i=1,2,\ldots,N \ , \tag{3.2-16}$$

and H is the system Hamiltonian, given by

$$\begin{aligned} H(\mathbf{p}_1,\ldots,\mathbf{p}_N,\mathbf{x}_1,\ldots,\mathbf{x}_N) &= \sum_{i=1}^{N} \mathbf{p}_i \cdot \dot{\mathbf{x}}_i - L\left(\mathbf{x}_i, \frac{\mathbf{p}_i}{m_i}\right) \\ &= \sum_{i=1}^{N} \frac{1}{2} \frac{\mathbf{p}_i \cdot \mathbf{p}_i}{m_i} + e_v^p \ . \end{aligned} \tag{3.2-17}$$

The transformation implied by the first equality in Eq. (3.2-17) is termed a *Legendre transformation* [Modell and Reid, 1983, §5.4]. (For a more thorough discussion of the Legendre transformation, see Appendix B.)

Thus, while the Lagrangian, L, is a function of the coordinates and their time derivatives, the Hamiltonian, H, is a function of the coordinates and their conjugate momenta. The Lagrangian as defined in (3.2-15) (the quantity inside the rectangular brackets) is often called the *canonical integrand*.

In a similar manner, the evolution equations for the $\mathbf{x}_i$ and $\mathbf{p}_i$ can then be obtained as

$$\dot{\mathbf{x}}_i = \frac{\partial H}{\partial \mathbf{p}_i} = \frac{\mathbf{p}_i}{m_i} \ ; \ \dot{\mathbf{p}}_i = -\frac{\partial H}{\partial \mathbf{x}_i} = -\frac{\partial e_p^v}{\partial \mathbf{x}_i} \ ; \ i=1,2,\ldots,N \ , \tag{3.2-18}$$

which are called *Hamilton's canonical equations of motion*, meaning that the equations are invariant under all canonical transformations. Any transformation which leaves the canonical integral invariant also leaves the canonical equations invariant. In §3.4, we shall take a closer look at the canonical transformations in relation to the Poisson bracket. The effect of this procedure is to transfer from the Lagrangian system of $3N$ second-order differential equations to a system of $6N$ first-order differential equations.

At this point, it is possible to verify the nature of the Hamiltonian. From the definitions of the conjugate momenta in (3.2-16) and the Hamiltonian in (3.2-17), H may be rewritten as

$$H = \sum_{i=1}^{N} \frac{\partial L}{\partial \dot{\mathbf{x}}_i} \cdot \dot{\mathbf{x}}_i - L = \sum_{i=1}^{N} \frac{\partial e_k}{\partial \dot{\mathbf{x}}_i} \cdot \dot{\mathbf{x}}_i - L \ . \tag{3.2-19}$$

In this expression, e_k is a second-order, homogeneous function of the $\dot{\mathbf{x}}_i$, and so Euler's theorem (in rudimentary form, $y[\partial y^n/\partial y]=nyy^{-1}=ny^n$) transforms the first term on the right-hand side of (3.2-19) into

$$\sum_{i=1}^{N} \frac{\partial e_k}{\partial \dot{\mathbf{x}}_i} \cdot \dot{\mathbf{x}}_i = 2e_k \ . \tag{3.2-20}$$

Consequently,

$$H = 2e_k - L = e_k + e_p^v \ , \tag{3.2-21}$$

and thus the Hamiltonian is recognized as the total energy of the system. Differentiation of $H=H(\mathbf{x}_1,\ldots,\mathbf{x}_N,\mathbf{p}_1,\ldots\mathbf{p}_N)$ with respect to time gives

$$\frac{\mathrm{d}H}{\mathrm{d}t} = \sum_{i=1}^{N} \left[\frac{\partial H}{\partial \mathbf{x}_i} \cdot \frac{\partial \mathbf{x}_i}{\partial t} + \frac{\partial H}{\partial \mathbf{p}_i} \cdot \frac{\partial \mathbf{p}_i}{\partial t} \right] , \tag{3.2-22}$$

but the canonical equations, (3.2-18), immediately reveal that

$$\mathrm{d}H/\mathrm{d}t = 0 \ \Rightarrow \ H = \text{constant} \ . \tag{3.2-23}$$

Hence we observe that the total energy of the system is conserved.

3.3 Poisson Bracket Description of Hamilton's Equations of Motion

Now that Hamilton's equations of motion have been derived, it is possible to demonstrate the underlying symmetry of these equations. Let $F(\mathbf{p}_1,\ldots,\mathbf{p}_N,\mathbf{x}_1,\ldots,\mathbf{x}_N)$ be an arbitrary function of the canonical variables $\mathbf{p}_i$ and $\mathbf{x}_i$, i=1,2,...,N. The total derivative of F with respect to time is then obtained by the rule of implicit differentiation as

$$\frac{\mathrm{d}}{\mathrm{d}t}F(\mathbf{p}_1,\ldots,\mathbf{p}_N,\mathbf{x}_1,\ldots,\mathbf{x}_N) = \sum_{i=1}^{N}\left[\frac{\partial F}{\partial \mathbf{x}_i}\cdot\frac{\partial \mathbf{x}_i}{\partial t} + \frac{\partial F}{\partial \mathbf{p}_i}\cdot\frac{\partial \mathbf{p}_i}{\partial t}\right] . \tag{3.3-1}$$

(Note that F is not an explicit function of time, and hence $\partial F/\partial t$=0.) Substitution of Hamilton's equations, (3.2-18), into (3.3-1) yields

$$\frac{\mathrm{d}}{\mathrm{d}t}F(\mathbf{p}_1,\ldots,\mathbf{p}_N,\mathbf{x}_1,\ldots,\mathbf{x}_N) = \sum_{i=1}^{N}\left[\frac{\partial F}{\partial \mathbf{x}_i}\cdot\frac{\partial H}{\partial \mathbf{p}_i} - \frac{\partial H}{\partial \mathbf{x}_i}\cdot\frac{\partial F}{\partial \mathbf{p}_i}\right] . \tag{3.3-2}$$

The expression on the right-hand side is termed the *Poisson bracket* (after S.D. Poisson, who introduced it in 1809), which, for two arbitrary functions $F(\mathbf{p}_1,\ldots,\mathbf{p}_N,\mathbf{x}_1,\ldots,\mathbf{x}_N)$ and $G(\mathbf{p}_1,\ldots,\mathbf{p}_N,\mathbf{x}_1,\ldots,\mathbf{x}_N)$, is expressed as

$$\{F,G\} \equiv \sum_{i=1}^{N}\left[\frac{\partial F}{\partial \mathbf{x}_i}\cdot\frac{\partial G}{\partial \mathbf{p}_i} - \frac{\partial G}{\partial \mathbf{x}_i}\cdot\frac{\partial F}{\partial \mathbf{p}_i}\right] . \tag{3.3-3}$$

Thus, Hamilton's equations of motion can be expressed as

$$\frac{\mathrm{d}}{\mathrm{d}t}F(\mathbf{p}_1,\ldots,\mathbf{p}_N,\mathbf{x}_1,\ldots,\mathbf{x}_N) = \{F,H\} , \tag{3.3-4}$$

since comparison of Eqs. (3.3-4) and (3.3-1) must then yield the equations of motion directly.

EXAMPLE 3.1: For the bracket of (3.3-3), find the symplectic 2-form $\boldsymbol{\omega}$ applying the theory developed in Appendix A. In addition, identify as well a Lagrangian manifold, the skew-gradient, and the Hamiltonian vector field.

In order to find $\boldsymbol{\omega}$, we rewrite the Poisson bracket of (A-35) as

$$\{F,G\} = \sum_{i,j=1}^{2N} \boldsymbol{\omega}^{ij} : \frac{\partial F}{\partial \mathbf{z}_i}\frac{\partial G}{\partial \mathbf{z}_j} ,$$

where $\mathbf{z}_i \equiv \mathbf{x}_i$ when $i \leq N$ and $\mathbf{z}_i \equiv \mathbf{p}_i$ when $i \geq N$. It is now an easy matter to show that

$$\boldsymbol{\omega}^{ij} = \boldsymbol{\delta}^{i+N,j} - \boldsymbol{\delta}^{i-N,j} .$$

A Lagrangian manifold can be specified as any N-dimensional manifold which is defined solely in terms of the coordinates $\mathbf{x}_i$, e.g., $\mathbf{p}_i \equiv 0$ for all i is an acceptable Lagrangian manifold. The Hamiltonian vector field corresponding to (3.2-17) is identified as

$$\mathbf{sgrad}(H) = \left(\frac{\mathbf{p}_1}{m_1}, \ldots, \frac{\mathbf{p}_N}{m_N}, -\frac{\partial e_p^v}{\partial \mathbf{x}_1}, \ldots, -\frac{\partial e_p^v}{\partial \mathbf{x}_N} \right)^{\mathrm{T}} . \quad \blacksquare$$

3.4 Properties of the Poisson Bracket

From the form of (3.3-3), we can immediately deduce a number of important properties of the Poisson bracket for arbitrary functions $f=f(\mathbf{p}_1,\ldots,\mathbf{p}_N,\mathbf{x}_1,\ldots,\mathbf{x}_N)$, $g=g(\mathbf{p}_1,\ldots,\mathbf{p}_N,\mathbf{x}_1,\ldots,\mathbf{x}_N)$, and $h=h(\mathbf{p}_1,\ldots,\mathbf{p}_N,\mathbf{x}_1,\ldots,\mathbf{x}_N)$. First, and most obviously, we have the antisymmetry property:

$$\{f,g\} = -\{g,f\} , \tag{3.4-1}$$

which ensures that $dH/dt=\{H,H\}=0$ (the total energy of the system is conserved). In addition, it is very simple to prove the following properties as well:

$$\{h,f+g\} = \{h,f\} + \{h,g\} , \tag{3.4-2}$$

$$\{hf,g\} = h\{f,g\} + \{h,g\}f , \tag{3.4-3}$$

$$\frac{\partial}{\partial a_{i\alpha}}\{f,g\} = \{\frac{\partial f}{\partial a_{i\alpha}},g\} + \{f,\frac{\partial g}{\partial a_{i\alpha}}\} , \quad a_{i\alpha} = x_{i\alpha} \text{ or } p_{i\alpha} , \tag{3.4-4}$$

and

$$\{x_{i\alpha},g\} = \frac{\partial g}{\partial p_{i\alpha}} ; \quad \{p_{i\alpha},g\} = -\frac{\partial g}{\partial x_{i\alpha}} , \tag{3.4-5}$$

where the subscript α denotes the α-component of the appropriate vector. From property (3.4-5),

$$\begin{aligned} \{f,\{x_{i\alpha},g\}\} - \{\{f,x_{i\alpha}\},g\} &= \{f,\frac{\partial g}{\partial p_{i\alpha}}\} + \{\frac{\partial f}{\partial p_{i\alpha}},g\} \\ &= \frac{\partial}{\partial p_{i\alpha}}\{f,g\} = \{x_{i\alpha},\{f,g\}\} , \end{aligned} \tag{3.4-6}$$

and hence

$$\{f,\{x_{i\alpha},g\}\} + \{x_{i\alpha},\{g,f\}\} + \{g,\{f,x_{i\alpha}\}\} = 0 , \tag{3.4-7}$$

for any $x_{i\alpha}$.

The above properties are interesting in their own right, but the real significance of the Poisson bracket was discovered by Jacobi through the study of canonical transformations, i.e., transformations which left the canonical equations invariant. Before we inspect the general canonical transformation, let us first look at the simpler point transformation of Lagrangian mechanics, following Lanczos [1970, ch. 7].

In Lagrangian mechanics, the system variables are the $\mathbf{x}_i$ alone, which allows us to specify a point transformation to the new variables $\mathbf{X}_i$ of the form:

$$\mathbf{x}_i = \mathbf{f}_i(X_1, X_2, \ldots, X_N) , \quad i=1,2,\ldots N . \tag{3.4-8}$$

Therefore, from the canonical integral, (3.2-15), it is obvious that the canonical equations will be preserved if we transfer the $\mathbf{p}_i$, i=1,2,...,N, by requiring the invariance of the differential form $\Sigma\mathbf{p}_i\cdot d\mathbf{x}_i$ (remember, $d\mathbf{x}_i=[d\mathbf{x}_i/dt]dt$):

$$\sum_{i=1}^{N} \mathbf{p}_i \cdot d\mathbf{x}_i = \sum_{i=1}^{N} \mathbf{P}_i \cdot d\mathbf{X}_i . \tag{3.4-9}$$

If the above condition holds for arbitrary infinitesimal variations in $\mathbf{x}_i$, then we can replace the canonical integral

$$I = \int_{t_1}^{t_2} \left[\sum_{i=1}^{N} \mathbf{p}_i \cdot d\mathbf{x}_i - H dt \right] , \tag{3.4-10}$$

with the corresponding integral

$$I = \int_{t_1}^{t_2} \left[\sum_{i=1}^{N} \mathbf{P}_i \cdot d\mathbf{X}_i - H' dt \right] , \tag{3.4-11}$$

where H' is that Hamiltonian function of $\mathbf{X}_i$, $\mathbf{P}_i$, i=1,...,N which is necessary so that the same evolution equations for $\mathbf{X}_i$, i=1,...,N are derived from it as those derived from the corresponding evolution equations for $\mathbf{x}_i$, i=1,...,N, through the application of the transformation Eq. (3.4-8). From Eqs. (3.4-9) and (3.4-10), we can see that the canonical equations are preserved and that $H'(\mathbf{X}_1,\ldots,\mathbf{X}_N,\mathbf{P}_1,\ldots,\mathbf{P}_N)=H(\mathbf{x}_1,\ldots,\mathbf{x}_N,\mathbf{p}_1,\ldots,\mathbf{p}_N)$. Thus, the Hamiltonian is an invariant of the point transformation (3.4-9).

There is a more extended group of transformations than (3.4-9), called the *general canonical transformation*, which leaves the canonical equations invariant:

$$\sum_{i=1}^{N} \mathbf{p}_i \cdot \mathrm{d}\mathbf{x}_i = \sum_{i=1}^{N} \mathbf{P}_i \cdot \mathrm{d}\mathbf{X}_i + \mathrm{d}Z \ , \tag{3.4-12}$$

where dZ is the total differential of a function $Z=Z(\mathbf{x}_1,\ldots,\mathbf{x}_N,\mathbf{X}_1,\ldots,\mathbf{X}_N)$. In this case, the canonical integral becomes

$$I = \int_{t_1}^{t_2} \left[\sum_{i=1}^{N} \mathbf{P}_i \cdot \mathrm{d}\mathbf{X}_i - H\mathrm{d}t \right] + \int_{t_1}^{t_2} \mathrm{d}Z \ , \tag{3.4-13}$$

where the last term is a boundary term which therefore has no influence upon the variation (Z does not depend explicitly on time), which is between definite limits. Thus, once again, the canonical equations are invariant under the transformation (3.4-12).

Let us now take a closer look at the transformation (3.4-12) by taking two independent differentiations d′ and d″ [Lanczos, 1972, p. 171]:

$$\sum_{i=1}^{N} \left[\mathrm{d}''(\mathbf{p}_i \cdot \mathrm{d}'\mathbf{x}_i - \mathbf{P}_i \cdot \mathrm{d}'\mathbf{X}_i) \right] = \mathrm{d}''\mathrm{d}'Z \ , \tag{3.4-14a}$$

or, alternatively,

$$\sum_{i=1}^{N} \left[\mathrm{d}''\mathbf{p}_i \cdot \mathrm{d}'\mathbf{x}_i - \mathrm{d}''\mathbf{P}_i \cdot \mathrm{d}'\mathbf{X}_i + \mathbf{p}_i \cdot \mathrm{d}''\mathrm{d}'\mathbf{x}_i - \mathbf{P}_i \cdot \mathrm{d}''\mathrm{d}'\mathbf{X}_i \right] = \mathrm{d}''\mathrm{d}'Z . \tag{3.4-14b}$$

Exchanging the sequence of the two differentiations and taking the difference (realizing that d″d′Z=d′d″Z) yields

$$\sum_{i=1}^{N} \left[\mathrm{d}''\mathbf{p}_i \cdot \mathrm{d}'\mathbf{x}_i - \mathrm{d}'\mathbf{p}_i \cdot \mathrm{d}''\mathbf{x}_i \right] = \sum_{i=1}^{N} \left[\mathrm{d}''\mathbf{P}_i \cdot \mathrm{d}'\mathbf{X}_i - \mathrm{d}'\mathbf{P}_i \cdot \mathrm{d}''\mathbf{X}_i \right], \tag{3.4-15}$$

which shows that this bilinear differential form is invariant as well, and associated with two independent directions (d′ and d″).

For an arbitrary function of two variables, $G=G(v,w)$, $\mathrm{d}'G$ is the change in G due to v, keeping w constant:

$$\mathrm{d}'G = \left(\frac{\partial G}{\partial v}\right) \mathrm{d}v \ , \tag{3.4-16}$$

and vice versa for d″G:

$$\mathrm{d}''G = \left(\frac{\partial G}{\partial w}\right) \mathrm{d}w \ . \tag{3.4-17}$$

Thus, instead of considering f and g as functions of the $\mathbf{x}_i$ and $\mathbf{p}_i$, we consider $\mathbf{x}_i=\mathbf{x}_i(f,g)$ and $\mathbf{p}_i=\mathbf{p}_i(f,g)$, $i=1,2,\ldots,N$, so that the invariant differential form can be expressed as

$$\sum_{i=1}^{N}\left[\mathrm{d}'\mathbf{p}_i\cdot\mathrm{d}''\mathbf{x}_i-\mathrm{d}''\mathbf{p}_i\cdot\mathrm{d}'\mathbf{x}_i\right]=\sum_{i=1}^{N}\left[\frac{\partial\mathbf{x}_i}{\partial f}\cdot\frac{\partial\mathbf{p}_i}{\partial g}-\frac{\partial\mathbf{p}_i}{\partial f}\cdot\frac{\partial\mathbf{x}_i}{\partial g}\right]\mathrm{d}f\,\mathrm{d}g \qquad (3.4\text{-}18)$$
$$\equiv (f,g)_L\,\mathrm{d}f\,\mathrm{d}g\ ,$$

where $(f,g)_L$ is defined as the *Lagrangian bracket*. The Lagrangian bracket is thus an invariant, bilinear differential form [Lanczos, 1970, p. 212].

Now let us transform the variables $\mathbf{x}_i$ and $\mathbf{p}_i$, $i=1,2,\ldots,N$, to the arbitrary new variables $\mathbf{X}_i$ and $\mathbf{P}_i$, $i=1,2,\ldots,N$:

$$\begin{aligned}\mathbf{x}_i &= \phi_i(\mathbf{X}_1,\ldots,\mathbf{X}_N,\mathbf{P}_1,\ldots,\mathbf{P}_N)\ , \quad i=1,2,\ldots,N\ ,\\ \mathbf{p}_i &= \psi_i(\mathbf{X}_1,\ldots,\mathbf{X}_N,\mathbf{P}_1,\ldots,\mathbf{P}_N)\ , \quad i=1,2,\ldots,N\ .\end{aligned} \qquad (3.4\text{-}19)$$

We can now pick any pair of new variables and treat them as the parameters f and g in the Lagrangian bracket, considering the others as constants. Doing the same in the new coordinate system, we obtain (from the definition of the Lagrangian bracket) the conditions:

$$(X_{i\alpha},X_{k\beta})_L=0\ ;\quad (P_{i\alpha},P_{k\beta})_L=0\ ;\quad (X_{i\alpha},P_{k\beta})_L=\delta_{ik}\delta_{\alpha\beta}\ . \qquad (3.4\text{-}20)$$

Now we can assume that we have $6N$ independent scalar functions $h_1,h_2,\ldots,h_{6N}$ of the $2N$ variables $\mathbf{x}_i$ and $\mathbf{p}_i$, $i=1,2,\ldots,N$:

$$\begin{aligned}h_1 &= h_1(\mathbf{x}_1,\ldots,\mathbf{x}_N,\mathbf{p}_1,\ldots,\mathbf{p}_N)\\ &\ \ \vdots\\ h_{6N} &= h_{6N}(\mathbf{x}_1,\ldots,\mathbf{x}_N,\mathbf{p}_1,\ldots,\mathbf{p}_N)\ .\end{aligned} \qquad (3.4\text{-}21)$$

Inversely, we also have

$$\begin{aligned}\mathbf{x}_1 &= \mathbf{x}_1(h_1,h_2,\ldots,h_{6N})\\ &\ \ \vdots\\ \mathbf{x}_N &= \mathbf{x}_N(h_1,h_2,\ldots,h_{6N})\\ \mathbf{p}_1 &= \mathbf{p}_1(h_1,h_2,\ldots,h_{6N})\\ &\ \ \vdots\\ \mathbf{p}_N &= \mathbf{p}_N(h_1,h_2,\ldots,h_{6N})\ .\end{aligned} \qquad (3.4\text{-}22)$$

Consequently, it is easy to show, using the rule of implicit differentiation, i.e.,

$$\frac{\partial h_k}{\partial h_i}=\sum_{m=1}^{N}\left[\frac{\partial h_k}{\partial\mathbf{x}_m}\cdot\frac{\partial\mathbf{x}_m}{\partial h_i}+\frac{\partial h_k}{\partial\mathbf{p}_m}\cdot\frac{\partial\mathbf{p}_m}{\partial h_i}\right], \qquad (3.4\text{-}23)$$

that

$$\sum_{j=1}^{2N} (h_i,h_j)_L \{h_k,h_j\} = \delta_{ik} , \qquad (3.4\text{-}24)$$

which implies that the Poisson bracket is the inverse transpose of the Lagrangian bracket. Thus, the relations (3.4-20) can be reformulated in terms of the Poisson bracket:

$$\{X_{i\alpha},X_{k\beta}\} = 0 \ ; \ \{P_{i\alpha},P_{k\beta}\} = 0 \ ; \ \{X_{i\alpha},P_{k\beta}\} = \delta_{ik}\delta_{\alpha\beta} \ . \qquad (3.4\text{-}25)$$

Now, making the transformation from the Poisson bracket of the old system, $\{f,g\}$, into the Poisson bracket of the new system, $\{f,g\}'$, where

$$\{f,g\}' \equiv \sum_{i=1}^{N} \left[\frac{\partial f}{\partial \mathbf{X}_i} \cdot \frac{\partial g}{\partial \mathbf{P}_i} - \frac{\partial g}{\partial \mathbf{X}_i} \cdot \frac{\partial f}{\partial \mathbf{P}_i} \right] , \qquad (3.4\text{-}26)$$

we obtain (using the Einstein summation convention of summing over repeated spatial indices $\alpha,\beta,...$)

$$\begin{aligned} \{f,g\}' &= \sum_{i=1}^{N} \left[\sum_{j=1}^{N} \left(\frac{\partial f}{\partial x_{j\alpha}} \frac{\partial x_{j\alpha}}{\partial X_{i\beta}} + \frac{\partial f}{\partial p_{j\alpha}} \frac{\partial p_{j\alpha}}{\partial X_{i\beta}} \right) \right] \times \left[\sum_{k=1}^{N} \left(\frac{\partial g}{\partial x_{k\gamma}} \frac{\partial x_{k\gamma}}{\partial P_{i\beta}} + \frac{\partial g}{\partial p_{k\gamma}} \frac{\partial p_{k\gamma}}{\partial P_{i\beta}} \right) \right] \\ &\quad - \sum_{i=1}^{N} \left[\sum_{j=1}^{N} \left(\frac{\partial g}{\partial x_{j\alpha}} \frac{\partial x_{j\alpha}}{\partial X_{i\beta}} + \frac{\partial g}{\partial p_{j\alpha}} \frac{\partial p_{j\alpha}}{\partial X_{i\beta}} \right) \right] \times \left[\sum_{k=1}^{N} \left(\frac{\partial f}{\partial x_{k\gamma}} \frac{\partial x_{k\gamma}}{\partial P_{i\beta}} + \frac{\partial f}{\partial p_{k\gamma}} \frac{\partial p_{k\gamma}}{\partial P_{i\beta}} \right) \right] \\ &= \sum_{j,k=1}^{N} \left[\frac{\partial f}{\partial x_{j\alpha}} \frac{\partial g}{\partial x_{k\gamma}} \{x_{j\alpha},x_{k\gamma}\}' + \frac{\partial f}{\partial p_{j\alpha}} \frac{\partial g}{\partial p_{k\gamma}} \{p_{j\alpha},p_{k\gamma}\}' \right] \\ &\quad + \sum_{j,k=1}^{N} \left[\frac{\partial f}{\partial x_{j\alpha}} \frac{\partial g}{\partial p_{k\gamma}} - \frac{\partial g}{\partial x_{j\alpha}} \frac{\partial f}{\partial p_{k\gamma}} \right] \{x_{j\alpha},p_{k\gamma}\}' \ . \end{aligned} \qquad (3.4\text{-}27)$$

However, since the transformation is canonical, we have the conditions (3.4-25), and so

$$\{f,g\}' = \{f,g\} \ . \qquad (3.4\text{-}28)$$

Thus we have the most interesting property of the Poisson bracket, i.e., *the Poisson bracket is invariant of any arbitrary canonical transformation.* It is now obvious that (3.4-7) can be generalized to give the well-known Jacobi identity:

$$\{f,\{g,h\}\} + \{g,\{h,f\}\} + \{h,\{f,g\}\} = 0 , \qquad (3.4\text{-}29)$$

where we transform $x_{i\alpha} \rightarrow h$ by an arbitrary canonical transformation. This identity can be verified for arbitrary f, g, and h straight from the definition of the Poisson bracket, (3.3-3).

Due to the invariant nature of the Poisson bracket, we now understand that the motion in time of a dynamical system can be viewed as a succession of infinitesimal canonical transformations. These transformations cannot change the value of $\{f,g\}$, which implies that $\{f,g\}$ remains constant in time. Note the similarity between the above observations and, in particular, Eq. (3.4-28) and Theorem A.1 and Eq. (A-31) originating from the interconnection of the Poisson bracket and the symplectic form.

3.5 The Liouville Equation

If we now consider a large ensemble of non-interacting particle systems and assume that they differ from each other only in their initial conditions, then the state of the ensemble can be described by a set of points in phase space, i.e., by a set of *phase points*, $(\boldsymbol{\chi},\boldsymbol{\wp})=(\mathbf{x}_1,\ldots,\mathbf{x}_N,\mathbf{p}_1,\ldots,\mathbf{p}_N)$. If the number of systems in the ensemble is sufficiently large, the set of points can be described by a continuous distribution function, $\rho=\rho(\boldsymbol{\chi},\boldsymbol{\wp},t)$, which is termed the *density in phase space*. Consequently, $\rho\,d\boldsymbol{\chi}d\boldsymbol{\wp}$ is the number of phase points occupying the volume element $d\boldsymbol{\chi}d\boldsymbol{\wp}$ in phase space at time t. The probability that a given system lies in $d\boldsymbol{\chi}d\boldsymbol{\wp}$ is then defined as

$$\xi(\boldsymbol{\chi},\boldsymbol{\wp},t)\,d\boldsymbol{\chi}d\boldsymbol{\wp} \equiv \frac{\rho(\boldsymbol{\chi},\boldsymbol{\wp},t)\,d\boldsymbol{\chi}d\boldsymbol{\wp}}{\int \rho(\boldsymbol{\chi},\boldsymbol{\wp},t)\,d\boldsymbol{\chi}d\boldsymbol{\wp}} \,. \tag{3.5-1}$$

Though this density function is arbitrary at the initial time, it is fixed at all subsequent times by the equations of motion. This density must satisfy a continuity equation:

$$\frac{\partial\rho}{\partial t} + \sum_{i=1}^{N}\left[\frac{\partial}{\partial\mathbf{x}_i}\cdot(\dot{\mathbf{x}}_i\rho) + \frac{\partial}{\partial\mathbf{p}_i}\cdot(\dot{\mathbf{p}}_i\rho)\right] = 0 \,. \tag{3.5-2}$$

Then by using Hamilton's canonical equations of motion, this expression becomes

$$\frac{\partial\rho}{\partial t} + \sum_{i=1}^{N}\left[\frac{\partial\rho}{\partial\mathbf{x}_i}\cdot\frac{\partial H}{\partial\mathbf{p}_i} - \frac{\partial\rho}{\partial\mathbf{p}_i}\cdot\frac{\partial H}{\partial\mathbf{x}_i}\right] \equiv \frac{\partial\rho}{\partial t} + \{\rho,H\} = 0 \,, \tag{3.5-3}$$

since, for all i,

$$\frac{\partial}{\partial\mathbf{x}_i}\cdot\dot{\mathbf{x}}_i + \frac{\partial}{\partial\mathbf{p}_i}\cdot\dot{\mathbf{p}}_i = \frac{\partial}{\partial\mathbf{x}_i}\cdot\frac{\partial H}{\partial\mathbf{p}_i} - \frac{\partial}{\partial\mathbf{p}_i}\cdot\frac{\partial H}{\partial\mathbf{x}_i} = 0 \,. \tag{3.5-4}$$

Eq. (3.5-3) is termed the *Liouville equation*.

We can rewrite Liouville's equation by defining a *phase-space velocity vector* as

$$\mathbf{V} \equiv (\dot{\chi}, \dot{\wp})^{\mathrm{T}} , \tag{3.5-5}$$

i.e., with components $\dot{\chi}$ and $\dot{\wp}$, thus giving an alternative form to (3.5-3):

$$\frac{\partial \rho}{\partial t} + \mathbf{V} \cdot \boldsymbol{\nabla} \rho = 0 , \tag{3.5-6}$$

where the operator $\boldsymbol{\nabla}$ is defined as

$$\boldsymbol{\nabla} \equiv \left(\frac{\partial}{\partial \chi}, \frac{\partial}{\partial \wp} \right)^{\mathrm{T}} . \tag{3.5-7}$$

Thus, we can easily visualize *Liouville's theorem*, which states that phase points move through the phase space as if they were contained in an incompressible fluid. In other words, the system moves through phase space via a continuous point transformation, which transforms each infinitesimal region at time t_0 to another at time t_1 with the same volume.

3.6 The Optical/Mechanical Analogy

As we have seen in this chapter and the preceding one, the mathematical analysis of the theories of optics and mechanics has revealed an inherent underlying symplectic structure which is common to both. Thus Hamilton's optical/mechanical analogy was born, which later led to the birth of wave mechanics through the work of de Broglie, Schrödinger, and Dirac during the quantum revolution in the 1920s. (For a detailed discussion of the history of the optical/mechanical analogy, see Hankins [1980, ch. 14].) In both theories, one has to determine the stationary points of a particular integral (in optics the "time" or "optical path" and in mechanics the "action") in terms of a certain generating function (the eikonal or the Lagrangian). In Table 3.1 we illustrate the main analogies between the optical and mechanical theories.

3.7 A Historical Aside on the Principle of Least Action

In the above presentation of the main principles and concepts of classical mechanics, we have fallen into some old traps set by many decades of inconsistent lip service to the great masters of the previous three centuries. Although no one could contest the authenticity of the physical principles and mathematical analyses which have been handed down to us through the ages via textbooks, lectures, etc., the origins and terminology used to designate the same seem to have vanished from reality

	Optics	Mechanics
Hamiltonian	$\mathcal{H} \equiv \mathbf{p}\cdot\dot{\mathbf{q}} - \mathfrak{L}$	$H \equiv \sum_{i=1}^{N} \mathbf{p}_i \cdot \dot{\mathbf{x}}_i - L$
Canonical integral	$\int_{z_1}^{z_2} [\mathbf{p}\cdot d\mathbf{q} - \mathcal{H} dz]$	$\int_{t_1}^{t_2} [\sum_{i=1}^{N} \mathbf{p}_i \cdot d\mathbf{x}_i - H dt]$
Canonical equations	$\dot{\mathbf{q}} = \frac{\partial \mathcal{H}}{\partial \mathbf{p}}, \quad \dot{\mathbf{p}} = -\frac{\partial \mathcal{H}}{\partial \mathbf{q}}$	$\dot{\mathbf{x}}_i = \frac{\partial H}{\partial \mathbf{p}_i}, \quad \dot{\mathbf{p}}_i = -\frac{\partial H}{\partial \mathbf{x}_i}, \quad i = 1,2,\ldots,N$
Poisson bracket $\{F,G\}$	$\frac{\partial F}{\partial \mathbf{q}}\cdot\frac{\partial G}{\partial \mathbf{p}} - \frac{\partial G}{\partial \mathbf{q}}\cdot\frac{\partial F}{\partial \mathbf{p}}$	$\{F,G\} \equiv \sum_{i=1}^{N}\left[\frac{\partial F}{\partial \mathbf{x}_i}\cdot\frac{\partial G}{\partial \mathbf{p}_i} - \frac{\partial G}{\partial \mathbf{x}_i}\cdot\frac{\partial F}{\partial \mathbf{p}_i}\right]$

Table 3.1: The optical/mechanical analogy.

into that misty realm of mythology which continuously changes as it passes hands through the generations. In writing this chapter and the preceding one, we have, for clarity, used the terminology for these principles that is largely prevalent today, whether or not it would have offended the original creators. We are by no means justified in doing so, however, and so in this section we wish to briefly set in a historical perspective some of the ideas discussed in these chapters, with reference to the elaborate history of mechanics of Dugas [1988].[1]

Our guilty consciences will be placated on this issue if we are able, at least, to instill in the reader a sense of the often vexing attitudes and prejudices (not to mention politics) of the ancient ones, which undoubtedly has contributed to the posthumous confusion. Indeed, much of the history of mechanics (that which is most intriguing, at any rate) lies not in the discovery of new physical principles, but in the squabbling which took place over priorities and metaphysics, often in the public domain.

We shall concern ourselves here solely with the extremum principles of mechanics, which during the second half of the seventeenth century and well into the eighteenth were creating a dramatic (maybe even a theatric) upheaval in the general philosophy of the older doctrines. Extremum principles were appearing *ad infinitum* as "universal laws of Nature" during this period: to name only a few, the principles of least constraint, shortest path, shortest time, least resistance, and rest. Bitter disputes over their soundness (and—Lo!—their authorship) were as prevalent as the principles themselves.

3.7.1 *Fermat's reception*

It was in February of 1662 that Fermat wrote his classical paper on the properties of light refraction using his principle [Dugas, 1988, p. 254] "that Nature always acts in the shortest ways." Descartes had earlier established a proportion for refractions based upon the "proven" supposition that light travels through dense bodies more easily than through rare ones. In a letter written by Fermat at about this time, he states [Dugas, 1988, p. 257], "I found that my principle gave exactly the same proportion of the refractions that M. Descartes has established." Unfortunately for Fermat, his demonstration assumed correctly that light travels more easily through rare media than dense ones. Consequently, he was resoundly rebuked from all sides by those who swore by Descartes's supposition. As an example, in a letter to Fermat from

[1]Most of the meanderings in this section are derived from Part III, Chapter V of Dugas's history.

Clerselier of May, 1662, Clerselier states that Fermat's principle is [Dugas, 1988, p. 258] "a principle which is moral and in no way physical; which is not, and which cannot be, the cause of any effect of Nature." One must wonder what was meant by a "moral principle," provided that it carried any assignable meaning at all. Furthermore, Clerselier wrote, "M. Descartes...proves and does not simply suppose, that light moves more easily through dense bodies than through rare ones."

Of all the principles that were originally proposed to describe universally the passage of light, only Fermat's principle is still quoted today, in much the same form as the original statement. Therefore, it is completely ironical that, shortly after receipt of Clerselier's letter, Fermat wrote the following response [Dugas, 1988, p. 259]:

> I have often said to M. de la Chambre and yourself that I do not claim and that I have never claimed, to be in the private confidence of Nature. She has obscure and hidden ways that I have never had the initiative to penetrate; I have merely offered her a small geometric assistance in the matter of refraction, supposing that she has need of it. But since you, Sir, assure me that she can conduct her affairs without this, and that she is satisfied with the order that M. Descartes has prescribed for her, I willingly relinquish my pretended conquest of physics and shall be content if you will leave me with a geometric problem, quite pure and *in abstracto*, by means of which there can be found the path of a particle which travels through two different media and seeks to accomplish its motion as quickly as it can.

3.7.2 Maupertuis's contribution

This was not the end of the diatribe against Fermat, however, as the scholars of the day were not content to leave him with his "geometrical problem," even after his death. Still, this did not prevent others from also invoking extremum principles to explain the propagation of light; for instance, Leibnitz in 1682 proposed the principle of least resistance. The incessant bickering eventually set the stage for the next principal character, Maupertuis, who stated the principle of least action for the first time in a paper dated April 15, 1744.

Maupertuis's paper bore the rather smug title *The agreement between the different laws of Nature that had, until now, seemed incompatible*, but ultimately this title would be viewed as more or less appropriate. Although it was still generally agreed upon at this time that light traveled more quickly through dense media than through rare ones, Maupertuis's arguments (unlike Fermat's) never approached this issue, and thus he was able to avoid the deluge of wrath that otherwise would have drowned him. We lift the following disconnected passages from Maupertuis's paper, as quoted by Dugas [1988, pp. 262-64], which together should give, in the author's own words (actually, a translation), a nice summation of the main concepts of the work as well as the coinage of the words "least" and "action."

> In meditating deeply on this matter, I thought that, since light has already forsaken the shortest path when it goes from one medium to another—the path which is a straight line—it could just as well not follow that of the shortest time. Indeed, what preference can there be in this matter for time or distance? Light cannot travel along the shortest path and along that of the shortest time—why should it go by one of these paths rather than by the other? Further, why should it follow either of these two? It chooses a path which has a very real advantage—the path which it takes is that by which the quantity of *action* is the *least*.
>
> It must now be explained what I mean by the quantity of action. When a body is carried from one point to another a certain action is necessary. This action depends on the velocity that the body has and the distance that it travels, but it is neither the velocity nor the distance taken separately. The quantity of action is the greater as the velocity is the greater and the path which it travels is longer. It is proportional to the sum of the distances, each one multiplied by the velocity with which the body travels along it.
>
> It is the quantity of action which is Nature's true storehouse, and which it economizes as much as possible in the motion of light.
>
> All the phenomena of refraction now agree with the great principle that Nature in the production of her works, always acts in the most simple ways.
>
> In both these circumstances (reflections and linear propagation), the least action reduces to the shortest path and the shortest time. And it is this *consequence* that Fermat took as a principle.
>
> The error which men like Fermat and those that followed him, have committed, only shows that, too often, their use is dangerous. It can be said, however, that it is not the principle which has betrayed them, but rather, the haste with which they have taken for the principle what is merely one of the consequences of it.

Despite the grandiose title of his work, Maupertuis was also subjected to a swelling of objections, especially from those whose philosophy was bent towards the "rational" plane; i.e., those who approached mechanics from a purely material perspective rather than through "final causes." He also became embroiled in a priority dispute as a certain professor at The Hague, Koenig, hence proclaimed that Leibnitz, in a letter to Herman some forty years earlier, had inferred the principle of least action. Although no letter of the above nature was ever produced, it still became a public issue as even Voltaire became involved in the controversy. As Maupertuis wrote [Dugas, 1988, p. 271], "The strangest thing was to see as an auxiliary in this dispute a man who had no claim to take part. Not satisfied with deciding at random on this matter—which demanded much knowledge that he lacked—he took this opportunity to hurl the grossest insults at me..." Still, Maupertuis was declared the winner in this instance, essentially through the vindication of Euler. Maupertuis stated afterwards, "My only fault was that of having discovered a principle that created something of a sensation."

3.7.3 Hamilton on the principle of least action

By and large, until the beginning of the nineteenth century, mechanics was still primarily a geometric science. Only during the early years of this century did scholars such as Lagrange, Hamilton, Jacobi, Poisson, and others elevate it to the level of analytical mathematics. Our discussion here shall concern Hamilton, since it is he whose name is most often associated with the principle of least action today.

Hamilton's ideas concerning dynamics were inseparable from his ideas concerning optics, and vice versa. He was surely the most eloquent expositor on the optical/mechanical analogy, and certainly garnered much practical usefulness out of the principle of least action through the calculus of variations and subsequent analyses. Indeed, part of Hamilton's lasting contribution to the science of mechanics was his philosophy that placed emphasis on deductive reasoning, in contrast to the metaphysical ramblings of many of his contemporaries. In his own words [Dugas, 1988, p. 391], "We must gather and group appearances, until the scientific imagination discerns their hidden laws, and unity arises from variety; and then from unity we must reduce variety, and force the discovered law to utter its revelations of the future."

In particular, Hamilton was diametrically opposed to the term "least" being associated with the principle currently bearing discussion, and it is rather ironical that this principle bears his name. Surely, however, he is worthy of the accolade for his efforts in elevating the principle from the geometric plane via the calculus of variations. We thus conclude this mild polemic with Hamilton's own thoughts on the designation of this principle [Dugas, 1988, p. 391]:

> But although the law of least action has thus attained a rank among the highest theorems of physics, yet its pretensions to a cosmological necessity, on the ground of economy in the universe, are now generally rejected. And the rejection appears just, for this, among other reasons, that the quantity pretended to be economized is in fact often lavishly expended.
>
> We cannot, therefore, suppose the economy of this quantity to have been designed in the divine idea of the universe: though a simplicity of some high kind may be believed to be included in that idea.

4
Equilibrium Thermodynamics

Ones ideas must be as broad as Nature if they are to interpret Nature.
—Sir Arthur Conan Doyle: A Study in Scarlet

The ideas which we shall present in the remainder of this book are intimately connected with thermodynamics. In order to describe the various transport processes in structured continua, one must first build a solid foundation of equilibrium thermodynamics upon which to base further development. In approaching transport phenomena from an energetic viewpoint, one must first define what is meant by various thermodynamic variables as, for example, the temperature and pressure, in terms of the primitive variables used to characterize the system under investigation.

In this chapter, we present a brief overview of equilibrium thermodynamics tailored to the needs of this book. Specifically, we want to define explicitly the thermodynamic quantities which are used in subsequent chapters. For clarity and completeness, we shall re-derive, rather than merely state, some of the standard thermodynamic relationships. Of course, the experienced reader may proceed directly to the next chapter and use this chapter as a reference to notation as the need arises.

4.1 The Fundamental Equation of Thermodynamics

The starting point for our discussion of equilibrium thermodynamics is the axiomatic foundation of the description of macroscopic equilibria on certain fundamental principles. First, the macroscopic equilibrium of a closed system[1] is completely described through the specification of a number of extensive (i.e., proportional to the total mass or volume of the system, and additive between systems) or intensive (i.e., independent of the total mass or volume of the system) parameters. This is a very important point which is usually overlooked in the traditional thermodynamic development. The extensive nature of the primary variables of

[1]Here we define a "closed system" in the sense that the system may not exchange matter with its environment, although it may exchange energy.

the system introduces an additional relationship which acts on the allowed variations of the differentials, which, as we shall see, is tantamount to the Gibbs/Duhem relation. This implies, as we shall demonstrate in §4.3, that the density formalism, where every extensive quantity is reported on a unit volume basis, is a much more natural framework for describing the system in that it avoids a number of pitfalls of the traditional formalism.

Another principle is that there exists an extensive function of the state of the system, U, called the internal energy, such that, for an infinitesimal transition between two equilibrium states where $\mathrm{d}Q$ and $\mathrm{d}W$ are the infinitesimal heat and work exchanged with the environment,[2] respectively, the total infinitesimal change in U is given by

$$\mathrm{d}U = \mathrm{d}Q + \mathrm{d}W \ . \tag{4.1-1}$$

Additionally, the differential of the internal energy satisfies the *fundamental equation of thermodynamics*, often called the *Gibbs relation*, which applies to reversible changes in the closed system Ω. This is the usual starting point in elementary thermodynamics texts. This relation is given by the expression

$$\mathrm{d}U = T\mathrm{d}S - p\mathrm{d}V \ , \tag{4.1-2}$$

in which p is the pressure and the uppercase symbols denote absolute system quantities: S is the entropy, T is the temperature, and V is the volume. Here, S and V are extensive variables, while p and T are intensive. Comparing Eqs. (4.1-1) and (4.1-2), we can specify the infinitesimal heat transfer as $\mathrm{d}Q = T\mathrm{d}S$, and the work performed on the system as $\mathrm{d}W = -p\mathrm{d}V$. For the purposes of this book, it is sufficient to consider the expression of (4.1-1) as the mathematical statement of the *first law of thermodynamics*.

In this chapter, we shall consider only processes (e.g., say, an infinitesimal increase or decrease in volume, $\mathrm{d}V$) with a characteristic time scale, say $\mathrm{d}t$, of the same order as the time scale of the experimenter, τ, which is furthermore assumed within that time scale to have a complete knowledge of the system. Processes of this type are called *equilibrium processes* [Woods, 1975, p. 10]. Furthermore, we restrict the processes of this chapter to be *reversible processes*; i.e., processes which can be completely reversed upon reversal of the externally imposed changes. In the above example, this would correspond to an infinitesimal deformation of $-\mathrm{d}V$. In chapter 6, we shall look in depth at irreversible

[2]Note that $\mathrm{d}Q$ and $\mathrm{d}W$ are not differentials in the usual sense, but are rather infinitesimal changes in these quantities. Differentials refer to state functions, like U, and the dynamic variables; yet we retain the above notation for convenience since no confusion should arise.

and non-equilibrium processes, i.e., processes for which changes occur rapidly with respect to the experimenter ($dt \ll \tau$).

The fundamental equation, (4.1-2), as well as its alternate forms which will pop up shortly, plays a major role in the thermodynamic foundation for the subject of this book, and it is therefore beneficial to understand exactly what it means. The term $-p\mathrm{d}V$ represents the reversible work done on the system, with both p and V being primitive variables which may be taken directly from measurements on the system, Ω, at any instant. U represents the internal energy of the system and, as such, is not a primitive variable; yet it can be expressed as a function of such variables. Thus, its value may be inferred from measurements taken on the system. The term $T\mathrm{d}S$ is dissimilar to the previous two in that the quantity S cannot be measured for a given system, and its value can only be obtained as defined through the Gibbs relation, (4.1-2).

The fundamental equation shows us how simply U changes when S and V are varied, and thus it is sensible to view it as a function of these two quantities, $U=U(S,V)$. Note that for a one-component, closed system, one implicitly assumes in $U=U(S,V)$ a fixed number of molecules in the system, and therefore the discussion is exactly the same as is written in terms of specific quantities, for example $\hat{u}=U/N$, without any problems. Of course, we could also consider U as a function of any other pair of variables, say V and T, since they are all interrelated, but then not all of the information in terms of arbitrary derivatives of U with respect to S and V can be extracted. Thus, the information in the fundamental equation reduces to that of an equation of state. In order to preserve the same amount of information, the function to be represented needs to be changed through Legendre transformations [Callen, 1960, §5.3; Modell and Reid, 1983, §5.4].

Alternate forms of the Gibbs relation can be derived by considering different state functions (or thermodynamic potentials, as they are often called) of other pairs of the four variables p, V, T, and S as independent variables in (4.1-2). All of these forms may be obtained from (4.1-2) through the Legendre transformation, as described in Appendix B. The first of these alternate potentials is the enthalpy, E, defined as

$$E \equiv U + pV \ . \tag{4.1-3}$$

By considering the exact differential of E,

$$\mathrm{d}E = \mathrm{d}U + V\mathrm{d}p + p\mathrm{d}V \ , \tag{4.1-4}$$

and substituting $\mathrm{d}U$ into this expression from (4.1-2), we arrive immediately at

$$\mathrm{d}E = T\mathrm{d}S + V\mathrm{d}p \ . \tag{4.1-5}$$

This expression indicates that the natural variables for E are S and p, i.e., $E=E(S,p)$. Other relations are the Helmholtz free energy, defined as

$$A \equiv U - TS \,, \tag{4.1-6}$$

such that

$$dA = -SdT - pdV \,, \quad A = A(T,V) \,, \tag{4.1-7}$$

and the Gibbs free energy,

$$B \equiv A + pV = U - TS + pV \,, \tag{4.1-8}$$

such that

$$dB = -SdT + Vdp \,, \quad B = B(T,p) \,. \tag{4.1-9}$$

Therefore, we have four alternate forms of the fundamental equation, each expressing the same information in terms of different pairs of thermodynamic variables.

EXAMPLE 4.1: The simplest possible thermodynamic system is the ideal gas, which corresponds to the continuum realized by a large number of colliding spheres with no interactions at a distance. The simplicity of the interactions in the system (expressed through delta functions), allows the previously defined thermodynamic potentials to be obtained as explicit functions of the corresponding independent variables (see, for example, Reichl [1980, pp. 52-4]). Here, we include another independent variable in the state space, N, which characterizes the number of moles of the ideal gas in the system. (We shall look at this variable in more detail in §4.3.) For a gas following the relation $pV=NRT$, we find the following relations through statistical mechanics. Each may be derived from the others by the appropriate Legendre transformation. They are:

$$U(S,V,N) = U_0 + \alpha NRT_0\left[\left(\frac{V_0}{V}\right)^{\frac{1}{\alpha}}\exp\left(\frac{S-S_0}{\alpha NR}\right) - 1\right] ,$$

$$E(S,p,N) = E_0 - (\alpha+1)NRT_0\left[1 - \left(\frac{p}{p_0}\right)^{\frac{1}{\alpha+1}}\exp\left(\frac{S-S_0}{(\alpha+1)NR}\right)\right] ,$$

$$A(T,V,N) = A_0 + (\alpha NR - S_0)(T-T_0) - NRT\ln\left[\left(\frac{T}{T_0}\right)^{\alpha}\frac{V}{V_0}\right] ,$$

$$B(T,p,N) = B_0 + [(\alpha+1)NR - S_0](T-T_0) - NRT\ln\left[\left(\frac{T}{T_0}\right)^{\alpha+1}\frac{p_0}{p}\right] ,$$

where α is a parameter related to the internal degrees of freedom of the particles (α=3/2 for monatomic gases, α=5/2 for diatomic gases, and α=3 for polyatomic gases), and R is the ideal gas constant. ■

4.2 Other Fundamental Relationships of Thermodynamics

The mathematics of thermodynamics is largely an exercise in calculus, and one needs to have a fair knowledge of certain basic principles in order to derive various fundamental relationships. One of these principles is the exact differential, which is what the thermodynamic potentials are known to be. Consider a continuous function, $\phi=\phi(w,x,y,...)$, of the variables inside the parentheses. The exact differential of ϕ is given as

$$d\phi = \left.\frac{\partial\phi}{\partial w}\right|_{x,y,...} dw + \left.\frac{\partial\phi}{\partial x}\right|_{w,y,...} dx + \left.\frac{\partial\phi}{\partial y}\right|_{w,x,...} dy + \dots , \tag{4.2-1}$$

where the $|_{\alpha,\beta,..}$ denotes a partial derivative keeping the variables $\alpha,\beta,...$ constant. When dealing with functions of two or more variables in the remaining chapters of this book, we shall drop the $|_{\alpha,\beta,..}$ on the partial derivatives, and only use it occasionally when we want to call special attention to a constant variable (to avoid any possible confusion).

If we now treat the thermodynamic potentials as exact differentials, we can find a number of interesting relationships. For example, if we consider the exact differential of $U=U(S,V)$, we have

$$dU = \left.\frac{\partial U}{\partial S}\right|_V dS + \left.\frac{\partial U}{\partial V}\right|_S dV , \tag{4.2-2}$$

and by comparing with the Gibbs relation, (4.1-2), we find immediately that

$$\left.\frac{\partial U}{\partial S}\right|_V = T \quad \text{and} \quad \left.\frac{\partial U}{\partial V}\right|_S = -p . \tag{4.2-3}$$

These two relations will play a major role in succeeding chapters. (The relations (4.2-3) now make it immediately obvious how one transforms $U(S,V)$ into $A(T,V)$, $E(S,p)$, and $B(T,p)$ via the Legendre transformation,[3] as alluded to in §4.1.) Taking the exact differentials of the other three

[3]Note, however, that this transformation of the thermodynamic state functions is defined as the negative of the (Lagrangian to Hamiltonian) transformation of Eq. (B-2), which is completely equivalent mathematically. Therefore, although $b_i=\partial f/\partial a_i$, we have instead of (B-6) the relation $\partial g/\partial b_i = -a_i$. The duality of the transformation is thus restricted under the definition of negative (B-2); i.e., we could term the transformation of (B-2) a "symmetric transformation" and that of negative (B-2) an "antisymmetric transformation."

potentials, $E=E(S,p)$, $A=A(T,V)$ and $B=B(T,p)$, and then comparing with Eqs. (4.1-5,7,9), respectively, gives more relations:

$$\left.\frac{\partial E}{\partial S}\right|_p = T\ , \quad \left.\frac{\partial E}{\partial p}\right|_S = V\ , \tag{4.2-4a}$$

$$\left.\frac{\partial A}{\partial T}\right|_V = -S\ , \quad \left.\frac{\partial A}{\partial V}\right|_T = -p\ , \tag{4.2-4b}$$

$$\left.\frac{\partial B}{\partial T}\right|_p = -S\ , \quad \left.\frac{\partial B}{\partial p}\right|_T = V\ . \tag{4.2-4c}$$

The relation (4.2-4b) also will pop up occasionally later.

By studying the forms of exact differentials, it is possible to write down several useful identities, which we shall use throughout the book. Let us consider three functions, $\psi=\psi(\xi,\zeta)$, $\xi=\xi(\psi,\zeta)$, and $\phi=\phi(\xi,\zeta)$, which depend on the variables within the parentheses. The exact differentials of these functions are easily defined through (4.2-1) as

$$\mathrm{d}\psi = \left.\frac{\partial \psi}{\partial \xi}\right|_\zeta \mathrm{d}\xi + \left.\frac{\partial \psi}{\partial \zeta}\right|_\xi \mathrm{d}\zeta\ , \tag{4.2-5a}$$

$$\mathrm{d}\xi = \left.\frac{\partial \xi}{\partial \psi}\right|_\zeta \mathrm{d}\psi + \left.\frac{\partial \xi}{\partial \zeta}\right|_\psi \mathrm{d}\zeta\ , \tag{4.2-5b}$$

$$\mathrm{d}\phi = \left.\frac{\partial \phi}{\partial \xi}\right|_\zeta \mathrm{d}\xi + \left.\frac{\partial \phi}{\partial \zeta}\right|_\xi \mathrm{d}\zeta\ . \tag{4.2-5c}$$

In eliminating $\mathrm{d}\xi$ from (4.2-5a) using (4.2-5b), we obtain

$$\mathrm{d}\psi = \left.\frac{\partial \psi}{\partial \xi}\right|_\zeta \left[\left.\frac{\partial \xi}{\partial \psi}\right|_\zeta \mathrm{d}\psi + \left.\frac{\partial \xi}{\partial \zeta}\right|_\psi \mathrm{d}\zeta \right] + \left.\frac{\partial \psi}{\partial \zeta}\right|_\xi \mathrm{d}\zeta\ , \tag{4.2-6a}$$

or, rewriting,

$$\mathrm{d}\psi \left[1 - \left.\frac{\partial \psi}{\partial \xi}\right|_\zeta \left.\frac{\partial \xi}{\partial \psi}\right|_\zeta \right] = \mathrm{d}\zeta \left[\left.\frac{\partial \psi}{\partial \xi}\right|_\zeta \left.\frac{\partial \xi}{\partial \zeta}\right|_\psi + \left.\frac{\partial \psi}{\partial \zeta}\right|_\xi \right] . \tag{4.2-6b}$$

The differentials $\mathrm{d}\psi$ and $\mathrm{d}\zeta$ are independent, so that if we arbitrarily set $\mathrm{d}\zeta=0$, then, if $\mathrm{d}\psi$ is to be non-trivial, we obtain the identity

$$\left.\frac{\partial \psi}{\partial \xi}\right|_\zeta = \left(\left.\frac{\partial \xi}{\partial \psi}\right|_\zeta \right)^{-1} . \tag{4.2-7}$$

Alternatively, we can set $d\psi=0$, so that

$$\left.\frac{\partial\psi}{\partial\xi}\right|_{\zeta} = -\left.\frac{\partial\psi}{\partial\zeta}\right|_{\xi}\left(\left.\frac{\partial\xi}{\partial\zeta}\right|_{\psi}\right)^{-1} , \qquad (4.2\text{-}8)$$

which, via (4.2-7), gives the identity

$$\left.\frac{\partial\psi}{\partial\xi}\right|_{\zeta} = -\left.\frac{\partial\zeta}{\partial\xi}\right|_{\psi}\left(\left.\frac{\partial\zeta}{\partial\psi}\right|_{\xi}\right)^{-1} . \qquad (4.2\text{-}9)$$

Now, eliminating $d\xi$ from (4.2-5c) using (4.2-5b), gives

$$d\phi = \left.\frac{\partial\phi}{\partial\xi}\right|_{\zeta}\left[\left.\frac{\partial\xi}{\partial\psi}\right|_{\zeta} d\psi + \left.\frac{\partial\xi}{\partial\zeta}\right|_{\psi} d\zeta\right] + \left.\frac{\partial\phi}{\partial\zeta}\right|_{\xi} d\zeta , \qquad (4.2\text{-}10)$$

from which, when $d\zeta=0$, we recognize that the chain rule of differentiation manifests:

$$\left.\frac{\partial\phi}{\partial\psi}\right|_{\zeta} = \left.\frac{\partial\phi}{\partial\xi}\right|_{\zeta}\left.\frac{\partial\xi}{\partial\psi}\right|_{\zeta} . \qquad (4.2\text{-}11)$$

Applying (4.2-7) to this expression, we obtain our third identity,

$$\left.\frac{\partial\psi}{\partial\xi}\right|_{\zeta} = \left.\frac{\partial\psi}{\partial\phi}\right|_{\zeta}\left(\left.\frac{\partial\xi}{\partial\phi}\right|_{\zeta}\right)^{-1} . \qquad (4.2\text{-}12)$$

Also, remembering that for $\phi=\phi(\xi,\zeta)$, the order of differentiation is immaterial, we have

$$\left.\frac{\partial}{\partial\zeta}\left(\left.\frac{\partial\phi}{\partial\xi}\right|_{\zeta}\right)\right|_{\xi} = \left.\frac{\partial}{\partial\xi}\left(\left.\frac{\partial\phi}{\partial\zeta}\right|_{\xi}\right)\right|_{\zeta} , \qquad (4.2\text{-}13)$$

which implies that for $d\phi=w\,d\xi+y\,d\zeta$,

$$\left.\frac{\partial w}{\partial\zeta}\right|_{\xi} = \left.\frac{\partial y}{\partial\xi}\right|_{\zeta} . \qquad (4.2\text{-}14)$$

With the above identities, we may now derive additional thermodynamic relations. For instance, using the Gibbs relation, (4.1-2), and the identity (4.2-14), it is obvious that

$$\left.\frac{\partial T}{\partial V}\right|_{S} = -\left.\frac{\partial p}{\partial S}\right|_{V} . \qquad (4.2\text{-}15a)$$

Similarly, through use of the other thermodynamic potentials, we obtain

$$\left.\frac{\partial T}{\partial p}\right|_{S} = \left.\frac{\partial V}{\partial S}\right|_{p} , \qquad (4.2\text{-}15b)$$

$$\left.\frac{\partial S}{\partial V}\right|_T = \left.\frac{\partial p}{\partial T}\right|_V , \tag{4.2-15c}$$

$$\left.\frac{\partial S}{\partial p}\right|_T = -\left.\frac{\partial V}{\partial T}\right|_p . \tag{4.2-15d}$$

The above expressions, (4.2-15), are called the *Maxwell relations*, and represent well-established thermodynamic identities.

One further derivation needs to be performed at this point, that of calculating the heat capacities and compressibilities. From Eqs. (4.1-2) and (4.1-5), we have

$$T\mathrm{d}S = \mathrm{d}U + p\mathrm{d}V , \tag{4.2-16a}$$

and

$$T\mathrm{d}S = \mathrm{d}E - V\mathrm{d}p . \tag{4.2-16b}$$

Now we define $\mathrm{d}Q \equiv T\mathrm{d}S$ as the heat reversibly applied to Ω, corresponding to a temperature increase of $\mathrm{d}T$, with either V or p held fixed throughout the process. Thus we can define the heat capacities at constant volume, C_v, and at constant pressure, C_p, through the expressions

$$\mathrm{d}Q \equiv C_v\mathrm{d}T \quad \text{and} \quad \mathrm{d}Q \equiv C_p\mathrm{d}T , \tag{4.2-17}$$

respectively. Hence, from (4.2-16), keeping volume and pressure constant, respectively, we obtain

$$\mathrm{d}U = C_v\mathrm{d}T \quad \text{and} \quad \mathrm{d}E = C_p\mathrm{d}T , \tag{4.2-18}$$

so that, via (4.2-1), it is obvious that

$$C_v = \left.\frac{\partial U}{\partial T}\right|_V \quad \text{and} \quad C_p = \left.\frac{\partial E}{\partial T}\right|_p , \tag{4.2-19}$$

or, using $\mathrm{d}Q = T\mathrm{d}S$,

$$\mathrm{d}S = \frac{C_v}{T}\mathrm{d}T \quad \text{and} \quad \mathrm{d}S = \frac{C_p}{T}\mathrm{d}T , \tag{4.2-20a}$$

so that considering $S=S(T,V)$ and $S=S(T,p)$, respectively, implies that

$$C_v = T\left.\frac{\partial S}{\partial T}\right|_V \quad \text{and} \quad C_p = T\left.\frac{\partial S}{\partial T}\right|_p . \tag{4.2-20b}$$

Now suppose that instead of applying heat reversibly to Ω we apply a reversible amount of compression/expansion work, defined as $\mathrm{d}W \equiv -p\mathrm{d}V$, corresponding to a pressure increase $\mathrm{d}p$, with either S or T held fixed throughout the process. Hence we may define a couple of compressibility factors as

$$dW \equiv \kappa_S' \, dp \quad \text{and} \quad dW \equiv \kappa_T' \, dp \,, \tag{4.2-21}$$

and from (4.2-16a) and (4.1-7), we see that

$$dU \equiv \kappa_S' \, dp \quad \text{and} \quad dA \equiv \kappa_T' \, dp \,. \tag{4.2-22}$$

Consequently, using arguments analogous to those above concerning the heat capacity, we define the compressibilities as

$$\kappa_S' = \left.\frac{\partial U}{\partial p}\right|_S = -p\left.\frac{\partial V}{\partial p}\right|_S \quad \text{and} \quad \kappa_T' = \left.\frac{\partial A}{\partial p}\right|_T = -p\left.\frac{\partial V}{\partial p}\right|_T \,; \tag{4.2-23}$$

yet these are not the definitions which have been historically ascribed to the compressibility factors. By tradition, the isentropic and isothermal compressibilities are defined as

$$\kappa_S \equiv -\frac{1}{V}\left.\frac{\partial V}{\partial p}\right|_S \quad \text{and} \quad \kappa_T \equiv -\frac{1}{V}\left.\frac{\partial V}{\partial p}\right|_T \,, \tag{4.2-24}$$

which obviously arise through the relations $-V\kappa_a dp = dV$, where a is S or T. These, however, do not correspond to a reversible application of compression/expansion work upon the system, as do the analogous relations for κ_S' and κ_T'. We know of no explanation for why κ_S and κ_T are defined as (4.2-24) historically, yet we employ (4.2-24) rather than (4.2-23) henceforth, in keeping with tradition. In any case, the two sets of definitions are directly relatable through measurable quantities.

Of the four quantities defined above, C_v, C_p, κ_S, and κ_T, it turns out that only three of them are independent. Using our knowledge of the manipulations of calculus described previously, we see that

$$\begin{aligned} \kappa_S &= -\frac{1}{V}\left.\frac{\partial V}{\partial p}\right|_S = \frac{1}{V}\left.\frac{\partial S}{\partial p}\right|_V \left(\left.\frac{\partial S}{\partial V}\right|_p\right)^{-1} = \frac{1}{V}\left.\frac{\partial S}{\partial T}\right|_V \left.\frac{\partial T}{\partial p}\right|_V \left(\left.\frac{\partial S}{\partial T}\right|_p \left.\frac{\partial T}{\partial V}\right|_p\right)^{-1} \\ &= \frac{C_v}{C_p}\frac{1}{V}\left.\frac{\partial T}{\partial p}\right|_V \left(\left.\frac{\partial T}{\partial V}\right|_p\right)^{-1} = -\frac{C_v}{C_p}\frac{1}{V}\left.\frac{\partial V}{\partial p}\right|_T = \kappa_T \frac{C_v}{C_p} \,. \end{aligned} \tag{4.2-25}$$

Also, by taking another combination of the variables C_p, C_v, and κ_T, through manipulations of the type in (4.2-25), we may define an additional quantity through the relation [Woods, 1975, p. 34]

$$C_p - C_v = TV\frac{\beta^2}{\kappa_T} \,, \tag{4.2-26}$$

with the coefficient of thermal expansion defined through tradition as

$$\beta \equiv \frac{1}{V}\left.\frac{\partial V}{\partial T}\right|_p \,. \tag{4.2-27}$$

We shall make use of the relations (4.2-25,26) in chapter 6 when we discuss stability.

The above derivations of the Maxwell relations and other relationships, although quite standard, serve to illustrate some of the simple mathematical manipulations which we shall use in succeeding chapters. Hopefully, by reminding the reader of these calculations on a well-known subject (equilibrium thermodynamics) at this stage of the book, we shall avoid any confusion when we deal with topics which are not very common.

4.3 The Fundamental Equation for a Multicomponent System

In this section, we look at some of the consequences of the generalization of the fundamental equation to an open, multicomponent fluid system, again denoted by Ω, composed of ν different species, each with N_i (i=1,2,...,ν) number of moles present in the mixture. Thus, the total number of moles in Ω, N, is given by a simple summation over the ν components:

$$\sum_{i=1}^{\nu} N_i = N \ . \tag{4.3-1}$$

For this system, we want to define certain quantities for later use, but more importantly, we want to determine how certain thermodynamic variables (in particular, the pressure) change when we deal with either specific (per mole or per mass) or volumetric (density) quantities. The latter is necessary since usually the key thermodynamic properties are worked out in terms of specific quantities, whereas the natural variables for the generalized bracket theory are volumetric quantities, as described in succeeding chapters. As we shall see here, working in terms of volumetric quantities reveals some interesting consequences.

4.3.1 Extensive variable formulation

In our multicomponent system, Ω, we must now consider the internal energy as a function of not only V and S, but of the amounts of each component as well, in this case, N_i, i=1,2,...,ν. Although we choose to work here in terms of the moles of the ν components, we could very well work in terms of their masses, as discussed a little later. Consequently, $U=U(S,V,N_1,N_2,...,N_\nu)$, such that the exact differential of U becomes[4]

[4]Note that this relation neglects the fundamental assumption that U (and S, V, N_i) is an extensive parameter; therefore only ν+1 of the terms in (4.3-2) are independent. This oversight will result in the Gibbs/Duhem relation in the following paragraphs.

$$dU = \left.\frac{\partial U}{\partial S}\right|_{V,N_i} dS + \left.\frac{\partial U}{\partial V}\right|_{S,N_i} dV + \sum_{i=1}^{\nu} \left.\frac{\partial U}{\partial N_i}\right|_{S,V,N_j} dN_i\,, \tag{4.3-2}$$

with the subscript N_j denoting that the amounts of all components are kept constant other than component i. Defining the chemical potential of component i, μ_i^+, as

$$\mu_i^+ \equiv \left.\frac{\partial U}{\partial N_i}\right|_{S,V,N_j}\,, \tag{4.3-3}$$

and using the relations (4.2-3), we obtain the fundamental equation for a multicomponent system:

$$dU = TdS - pdV + \sum_{i=1}^{\nu} \mu_i^+ dN_i\,. \tag{4.3-4a}$$

Similarly, by considering the other three thermodynamic potentials as functions of the N_i also, we obtain the generalizations of Eqs. (4.1-5,7,9), respectively:

$$dE = TdS + Vdp + \sum_{i=1}^{\nu} \mu_i^+ dN_i\,, \tag{4.3-4b}$$

$$dA = -SdT - pdV + \sum_{i=1}^{\nu} \mu_i^+ dN_i\,, \tag{4.3-4c}$$

$$dB = -SdT + Vdp + \sum_{i=1}^{\nu} \mu_i^+ dN_i\,. \tag{4.3-4d}$$

In the above equations, it follows from (4.3-3) and (4.3-4a) that

$$\mu_i^+ = \left.\frac{\partial U}{\partial N_i}\right|_{S,V,N_j} = \left.\frac{\partial E}{\partial N_i}\right|_{S,p,N_j} = \left.\frac{\partial A}{\partial N_i}\right|_{T,V,N_j} = \left.\frac{\partial B}{\partial N_i}\right|_{T,p,N_j}\,, \tag{4.3-5}$$

so that there are four expressions for the μ_i^+, each of which contains the derivative of a potential with respect to the amount of component i in the system.

We can find an explicit expression for $U=U(S,V,N_1,\ldots,N_\nu)$ through the extensive property of the internal energy; i.e., for an arbitrary λ, we have that $\lambda U(S,V,N_1,\ldots,N_\nu)=U(\lambda S,\lambda V,\lambda N_1,\ldots,\lambda N_\nu)$. Differentiating this expression with respect to λ thus results in the equation

$$U = TS - pV + \sum_{i=1}^{\nu} \mu_i^+ N_i\,, \tag{4.3-6}$$

since the temperature, pressure, and chemical potential are intensive quantities; i.e., since $I(X)=I(\lambda X)$, with I being an arbitrary intensive parameter and X an arbitrary extensive parameter. Eq. (4.3-6) can be viewed as a specification of U as a function of temperature, pressure, and the system composition, yet not all of these variables are independent. Taking the differential of this expression, we have

$$dU = T dS + S dT - p dV - V dp + \sum_{i=1}^{\nu} \mu_i^+ dN_i + \sum_{i=1}^{\nu} N_i d\mu_i^+ , \qquad (4.3\text{-}7)$$

which implies, after subtracting (4.3-4a), that

$$S dT - V dp + \sum_{i=1}^{\nu} N_i d\mu_i^+ = 0 . \qquad (4.3\text{-}8)$$

This equation is known as the *Gibbs/Duhem equation*, and for constant T and p, it becomes

$$\sum_{i=1}^{\nu} N_i d\mu_i^+ \Big|_{p,T} = 0 , \qquad (4.3\text{-}9)$$

which shows that not all ν components can be considered as independent in this case. Thus, of the $\nu+2$ system variables, only $\nu+1$ can be varied independently, and therefore $U=U(S,V,N_1,...N_{\nu-1})$. The Gibbs/Duhem equation represents a well-established thermodynamic principle for multiple component systems.

4.3.2 *Specific variable formulation*

Now that we have studied this system via the absolute quantities, let us next look at the same system in terms of specific quantities, i.e., the per mole quantities, $\hat{u}=U/N$, $\hat{v}=V/N$, $\hat{s}=S/N$, and $\hat{n}_i=N_i/N$. By dividing (4.3-6) by N, we find that

$$\hat{u} = T\hat{s} - p\hat{v} + \sum_{i=1}^{\nu} \mu_i^+ \hat{n}_i . \qquad (4.3\text{-}10)$$

Furthermore, we know that

$$d\hat{u} = d\left(\frac{U}{N}\right) = \frac{1}{N} dU - \frac{U}{N^2} dN , \qquad (4.3\text{-}11a)$$

or, via (4.3-1),

$$d\hat{u} = \frac{1}{N} dU - \frac{U}{N^2} \sum_{i=1}^{\nu} dN_i . \qquad (4.3\text{-}11b)$$

Hence, substituting (4.3-4a) into (4.3-11b), we have

$$\mathrm{d}\hat{u} = \frac{T}{N}\mathrm{d}S - \frac{p}{N}\mathrm{d}V + \sum_{i=1}^{\nu} (\mu_i^+ - \hat{u})\frac{1}{N}\mathrm{d}N_i \ , \tag{4.3-12}$$

and if we use expressions similar to (4.3-11) for $\hat{s}$, $\hat{v}$, and $\hat{n}_i$, we can rewrite (4.3-12) as

$$\begin{aligned} \mathrm{d}\hat{u} &= T\mathrm{d}\hat{s} - p\mathrm{d}\hat{v} + \sum_{i=1}^{\nu} \mu_i^+\mathrm{d}\hat{n}_i \\ &+ \sum_{i=1}^{\nu}\left[-\hat{u} - p\hat{v} + T\hat{s} + \sum_{j=1}^{\nu}\mu_j^+\hat{n}_j\right]\frac{1}{N}\mathrm{d}N_i \ . \end{aligned} \tag{4.3-13}$$

However, the fourth term on the right-hand side of (4.3-13) vanishes, via (4.3-10), so that

$$\mathrm{d}\hat{u} = T\mathrm{d}\hat{s} - p\mathrm{d}\hat{v} + \sum_{i=1}^{\nu} \mu_i^+\mathrm{d}\hat{n}_i \ , \tag{4.3-14}$$

which implies that $\hat{u}=\hat{u}(\hat{s},\hat{v},\hat{n}_1,\ldots,\hat{n}_\nu)$ and

$$\left.\frac{\partial\hat{u}}{\partial\hat{s}}\right|_{\hat{v},\hat{n}_i} = T \ , \quad \left.\frac{\partial\hat{u}}{\partial\hat{v}}\right|_{\hat{s},\hat{n}_i} = -p \ , \quad \text{and} \quad \left.\frac{\partial\hat{u}}{\partial\hat{n}_i}\right|_{\hat{s},\hat{v},\hat{n}_j} = \mu_i^+ \ . \tag{4.3-15}$$

Thus we have the thermodynamic relationships corresponding to those involving the absolute quantities.

Of course, once again, not all of the $\hat{n}_i$ are independent, since

$$\sum_{i=1}^{\nu} \hat{n}_i = 1 \ \Rightarrow \ \sum_{i=1}^{\nu} \mathrm{d}\hat{n}_i = 0 \ , \quad \text{or} \quad \mathrm{d}\hat{n}_\nu = -\sum_{i=1}^{\nu-1} \mathrm{d}\hat{n}_i \ . \tag{4.3-16}$$

Substituting (4.3-16) into (4.3-14) gives

$$\mathrm{d}\hat{u} = T\mathrm{d}\hat{s} - p\mathrm{d}\hat{v} + \sum_{i=1}^{\nu-1} (\mu_i^+ - \mu_\nu^+)\mathrm{d}\hat{n}_i \ , \tag{4.3-17a}$$

and also from (4.3-10)

$$\hat{u} = \mu_\nu^+ + T\hat{s} - p\hat{v} + \sum_{i=1}^{\nu-1} (\mu_i^+ - \mu_\nu^+)\hat{n}_i \ , \tag{4.3-17b}$$

whereby

$$\hat{s}\mathrm{d}T - \hat{v}\mathrm{d}p + \sum_{i=1}^{\nu} \hat{n}_i\mathrm{d}\mu_i^+ = 0 \ . \tag{4.3-17c}$$

Hence the system variables could be listed as $p,T,\mu_1^+,\ldots,\mu_{\nu-1}^+$, and we now see that $\hat{u}$ is a function of the variables $\hat{s},\hat{v},\hat{n}_1,\ldots,\hat{n}_{\nu-1}$.

4.3.3 Density variable formulation

Now that we have looked at the multicomponent system in terms of both absolute and specific properties, let us explore the consequences of working in terms of volumetric quantities. We shall now consider as the variables of interest the internal energy density, $u=U/V$, the entropy density, $s=S/V$, and the mole density (concentration) of component i, $n_i=N_i/V$.

As before, let us start by dividing (4.3-6) by V, to find that

$$u = Ts - p + \sum_{i=1}^{\nu} \mu_i^{+} n_i \; . \qquad (4.3\text{-}18)$$

Also, we know that

$$du = d\left(\frac{U}{V}\right) = \frac{1}{V}dU - \frac{U}{V^2}dV \; , \qquad (4.3\text{-}19)$$

and upon substitution of (4.3-4a) into (4.3-19), we find that

$$du = \frac{1}{V}(TdS - pdV + \sum_{i=1}^{\nu} \mu_i^{+} dN_i) - \frac{u}{V}dV \; . \qquad (4.3\text{-}20)$$

Using identities similar to (4.3-19) for s and n_i, (4.3-20) can be rewritten as

$$du = Tds + \sum_{i=1}^{\nu} \mu_i^{+} dn_i + \frac{1}{V}(-u - p + Ts + \sum_{i=1}^{\nu} \mu_i^{+} n_i)dV \; , \qquad (4.3\text{-}21)$$

and from (4.3-18), it is obvious that (4.3-21) reduces to

$$du = Tds + \sum_{i=1}^{\nu} \mu_i^{+} dn_i \; . \qquad (4.3\text{-}22)$$

Thus, we see that the internal energy density can be considered as a function of the $\nu+1$ variables $s,n_1,\ldots,n_\nu$, and that

$$\left.\frac{\partial u}{\partial s}\right|_{n_i} = T \quad \text{and} \quad \left.\frac{\partial u}{\partial n_i}\right|_{s,n_j} = \mu_i^{+} \; . \qquad (4.3\text{-}23)$$

We also notice that p, no longer being a problem variable, is given by a thermodynamic constitutive relation, via (4.3-18), as

$$p = -u + \left.\frac{\partial u}{\partial s}\right|_{n_i} s + \sum_{i=1}^{\nu} \left.\frac{\partial u}{\partial n_i}\right|_{s,n_j} n_i \; . \qquad (4.3\text{-}24)$$

This relation thus replaces the Gibbs/Duhem equation when working in terms of the density variables, and we shall see it pop up very often in the following chapters.

All of the above arguments hold regardless of whether we deal with moles or masses of the ν system components. By considering the molecular weight of the *i*-th component, M_w^i, it is easy to see that

$$\mu_i = \frac{\mu_i^+}{M_w^i} \quad \text{and} \quad \rho_i = M_w^i n_i \,, \quad \text{(no summation)} \tag{4.3-25}$$

where μ_i is the chemical potential on a mass basis, and ρ_i is the mass density, both of component *i*. Thus all of the preceding thermodynamic relationships transfer over directly to a problem formulation in terms of mass variables.

One more quantity needs to be defined at this point for later use: the partial specific entropy. Using (4.1-6), we have the Helmholtz free energy density given as

$$a = u - Ts \,, \tag{4.3-26}$$

from which we may deduce that

$$\mathrm{d}a = \mathrm{d}u - T\mathrm{d}s - s\mathrm{d}T \,. \tag{4.3-27}$$

Substituting (4.3-22) into this expression yields

$$\mathrm{d}a = -\,s\mathrm{d}T + \sum_{i=1}^{\nu} \mu_i^+ \mathrm{d}n_i \,, \tag{4.3-28}$$

which, via (4.2-14), yields the partial molar and specific entropies, $M_w^i \overline{S}_i$ and $\overline{S}_i$, defined as

$$M_w^i \overline{S}_i \equiv \left.\frac{\partial s}{\partial n_i}\right|_{T,n_j} = -\left.\frac{\partial \mu_i^+}{\partial T}\right|_{n_i} \,, \tag{4.3-29}$$

and thus

$$\overline{S}_i = \left.\frac{\partial s}{\partial \rho_i}\right|_{T,\rho_j} = -\left.\frac{\partial \mu_i}{\partial T}\right|_{\rho_i} \,. \tag{4.3-30}$$

Other partial quantities may be defined in a similar manner.

4.4 Equilibrium Thermodynamics of a Material with Internal Microstructure

Now we wish to consider a microstructured, multicomponent, open system; i.e., a material with an internal energy which depends not only upon the entropy and volume, but on the amounts of the various components present as well as their structure (or distribution) within the system. We have just seen how the thermodynamics of the system is affected by the incorporation of multicomponents into the Gibbs relation; now we shall see how the structure complicates the situation. We begin by postulating that the internal energy is a function of the entropy, volume, moles of each component, and the distribution of each component within the system. As a suitable macroscopic measure of the distribution of each component, we declare a second-rank tensor for each component which quantifies the necessary structural information. (There is no need to define these tensors explicitly at this point.) In reality, a second-rank tensor might not always be the best parameter for determining the structure, but we employ it here because of the important role which the second-rank conformation tensor will play later on in the book. Hence we define $U=U(S,V,N_1,\ldots,N_\nu,\mathbf{Z}_1,\ldots,\mathbf{Z}_\nu)$, where $\mathbf{Z}_i$ is the symmetric ($Z_{i\alpha\beta}=Z_{i\beta\alpha}$) structural tensor of the i-th component which is taken to be an extensive quantity.

Now we can proceed analogously to the preceding section by taking the exact differential of U,

$$dU = \left.\frac{\partial U}{\partial S}\right|_{V,N_i,\mathbf{Z}_i} dS + \left.\frac{\partial U}{\partial V}\right|_{S,N_i,\mathbf{Z}_i} dV + \sum_{i=1}^{\nu} \left.\frac{\partial U}{\partial N_i}\right|_{S,V,N_j,\mathbf{Z}_i} dN_i + \sum_{i=1}^{\nu} \left.\frac{\partial U}{\partial \mathbf{Z}_i}\right|_{S,V,N_i,\mathbf{Z}_j} : d\mathbf{Z}_i \ , \tag{4.4-1}$$

and upon use of the Euler relation,

$$U = TS - pV + \sum_{i=1}^{\nu} \mu_i^+ N_i + \sum_{i=1}^{\nu} \mathbf{z}_i : \mathbf{Z}_i \ . \tag{4.4-2}$$

$\mathbf{z}_i$ represents a "structural potential," analogous to the chemical potential, given by

$$\mathbf{z}_i \equiv \left.\frac{\partial U}{\partial \mathbf{Z}_i}\right|_{S,V,N_i,\mathbf{Z}_j} \ , \quad i=1,2,\ldots\nu \ , \tag{4.4-3}$$

which may be written in terms of the other three thermodynamic potentials through relations similar to (4.3-5). As a consequence of the symmetry of the $\mathbf{Z}_i$, one can determine that the $\mathbf{z}_i$ are symmetric tensors, and that $z_{i\alpha\gamma}Z_{i\gamma\beta}=z_{i\beta\gamma}Z_{i\gamma\alpha}$ as well. We may now take the differential of (4.4-2), subtract from it (4.4-1), and thereby obtain a generalized Gibbs/Duhem relation for the structured medium:

$$S dT - V dp + \sum_{i=1}^{\nu} N_i d\mu_i^+ + \sum_{i=1}^{\nu} \mathbf{Z}_i : d\mathbf{z}_i = 0 \ . \tag{4.4-4}$$

Later on, we shall be working in terms of volumetric quantities, and therefore we need to describe the thermodynamics of this system in terms of per-volume variables. This is easy to do, however, following the résumé of the preceding section, with the additional variable $\mathbf{C}_i \equiv \mathbf{Z}_i/V$. Hence we find that

$$du = T ds + \sum_{i=1}^{\nu} \mu_i^+ dn_i + \sum_{i=1}^{\nu} \mathbf{z}_i : d\mathbf{C}_i , \tag{4.4-5}$$

with the constitutive equation for the pressure given as

$$p = -u + \left.\frac{\partial u}{\partial s}\right|_{n_i,\mathbf{C}_i} s + \sum_{i=1}^{\nu} \left.\frac{\partial u}{\partial n_i}\right|_{s,n_j,\mathbf{C}_i} n_i + \sum_{i=1}^{\nu} \left.\frac{\partial u}{\partial \mathbf{C}_i}\right|_{s,n_i,\mathbf{C}_j} : \mathbf{C}_i . \tag{4.4-6}$$

One may also now use (4.2-14) to obtain any partial quantities so desired.

As in §4.2, we may also define any other thermodynamic relationships of interest. For example, through (4.4-1), we may define the additional Maxwell relations to (4.2-15a) as

$$\frac{\partial T}{\partial N_i} = \frac{\partial \mu_i^+}{\partial S} ; \quad \frac{\partial T}{\partial \mathbf{Z}_i} = \frac{\partial \mathbf{z}_i}{\partial S} ; \quad \frac{\partial \mu_i^+}{\partial V} = -\frac{\partial p}{\partial N_i} ;$$

$$\frac{\partial \mathbf{z}_i}{\partial V} = -\frac{\partial p}{\partial \mathbf{Z}_i} ; \quad \frac{\partial \mu_i^+}{\partial N_j} = \frac{\partial \mu_j^+}{\partial N_i} ; \tag{4.4-7}$$

$$\frac{\partial z_{i\alpha\beta}}{\partial Z_{j\gamma\varepsilon}} = -\frac{\partial z_{j\gamma\varepsilon}}{\partial Z_{i\alpha\beta}} ; \quad \frac{\partial \mu_i^+}{\partial \mathbf{Z}_j} = \frac{\partial \mathbf{z}_j}{\partial N_i} ,$$

where we have suppressed the $|_a$ on the partial derivatives for notational convenience. The above equalities represent just a sampling of the many available for such a complex system. Using the other thermodynamic potentials, defined through the Legendre transformation, we may construct any further relationships as required.

4.5 Additivity in Compound Systems

Now instead of considering only a single system, let us consider a compound system, Ω, composed of ν subsystems, Ω_i, $i=1,2,\ldots,\nu$, such that $\Omega \subset \{\Omega_i\}$. These Ω_i are in contact with each other, and allow transfer of both heat and work (and mass in certain situations) between the subsystems. As Woods [1975, p. 44] points out, we may consider two types of compound systems: the first consisting of spatially distinct subsystems, called *phases*, and the second consisting of spatially overlapping subsystems, called *components*. Hence the phases can be thought of as distinct compartments of specified volume (in the usual thermodynamic sense); while the components occupy the same volume, yet may still retain characteristic thermodynamic variables, for example, T_i as the temperature of component Ω_i. In order for this to be the case, however, we have to assume that heat exchanges between the subsystems are much slower than the thermal equilibration within each subsystem.

The importance of the compound system will become apparent in chapter 5. For now, we wish to assert what properties such a seemingly complex thermodynamic system must fulfill. The first principle which we assert is that the total energy of the system Ω is an additive function of the energies of all the subsystems Ω_i:

$$U = \sum_{i=1}^{\nu} U_i \quad \Rightarrow \quad dU = \sum_{i=1}^{\nu} dU_i \ . \tag{4.5-1}$$

The volume is also obviously an additive quantity. Furthermore, via the Gibbs relation, we know that

$$dU_i = T_i dS_i - p_i dV_i \ , \quad i = 1,2,\ldots,\nu \ , \tag{4.5-2}$$

which, along with (4.5-1), implies that the entropy is also an additive function of the entropies of the Ω_i provided that *all of the* Ω_i *are at the same temperature*.[5] Still, it is a firmly entrenched practice in thermodynamics to *postulate* that the entropy of Ω can be expressed in terms of subsystems with different temperatures as

$$S = \sum_{i=1}^{\nu} S_i \quad \Rightarrow \quad dS = \sum_{i=1}^{\nu} dS_i = \sum_{i=1}^{\nu} (dU_i + p_i dV_i)/T_i \ . \tag{4.5-3}$$

This postulate is necessary when extending equilibrium thermodynamics to non-equilibrium processes. With expressions (4.5-1,3), it is now evident that we could have described the open, multicomponent system of §4.3 (and §4.4) as a compound system of the component type. Doing so would yield the same results as before.

EXAMPLE 4.2: Evaluate the entropy of a system which consists of two subsystems of equal volume V, each initially containing the same number of moles, N, of an ideal gas at different temperatures, T_1 and T_2, respectively. Demonstrate the additivity of the entropy by comparing

[5]This is because of the thermodynamic definition of the temperature, $T_i \equiv (\partial U_i / \partial S_i)$ at constant V_i, and is also partly a result of irreversibilities in the system.

this expression when $T_1=T_2=T$ with the one corresponding to a single system with $2N$ moles, volume $2V$, and temperature T.

The expression for the entropy of an ideal gas can be derived as a function of T, V, and N by taking the negative partial derivative of the Helmholtz free energy of Example 4.1 with respect to T. Doing so gives

$$S(T,V,N) = S_0 + NR\ln\left[\left(\frac{T}{T_0}\right)^{\alpha}\frac{V}{V_0}\right]; \qquad (4.5\text{-}4)$$

however, in order to add the individual contributions of the entropy and still obtain a meaningful result, we need to specify the integration constant S_0. Classical thermodynamics is of no help here because all of the extensive thermodynamic potentials are only defined up to within a constant. Still, (quantum) statistical mechanics can provide us with the correct answer. Indeed, using the Maxwell/Boltzmann formulation of an ideal monatomic gas ($\alpha=3/2$), we have

$$S(T,V,N) = \frac{5RN}{2} - NR\ln\left[\frac{\lambda_T^3}{V}\frac{N}{N_A}\right], \qquad (4.5\text{-}5)$$

where

$$\lambda_T \equiv \sqrt{\frac{2\pi\hbar^2}{mk_BT}}\,, \qquad (4.5\text{-}6)$$

$\hbar$ is the Plank's constant, N_A is Avogadro's number, and m is the molecular mass. Eq. (4.5-5) is called the Sackur/Tetrode equation. If (4.5-5) is now used for S_0, and for given V_0 and T_0, S_0 can readily be obtained as a constant $\hat{s}$ times N, as we could have anticipated from the extensive character of S. In this case, the entropy of the combined system is

$$\begin{aligned} S(T_1,T_2,V,N) &= 2NR\hat{S}_0 + NR\ln\left[\left(\frac{T_1}{T_0}\right)^{\alpha}\frac{V}{V_0}\right] + NR\ln\left[\left(\frac{T_2}{T_0}\right)^{\alpha}\frac{V}{V_0}\right] \\ &= 2N\hat{s}_0 + 2NR\ln\left[\left(\frac{T_1T_2}{T_0T_0}\right)^{\frac{\alpha}{2}}\frac{V}{V_0}\right], \end{aligned} \qquad (4.5\text{-}7)$$

which for $T_1=T_2=T$ collapses to the expression for a single system of $2N$ moles with temperature T and volume V. ■

The above sampling of the theory of equilibrium thermodynamics was specifically designed to suit the needs of this book; however, it does not represent the entire amount of thermodynamic knowledge which we shall make use of in later chapters. Consequently, it will become necessary in chapter 6 to return to a discussion of the theory of thermodynamics, but geared toward the irreversible processes encountered in engineering practice. Hence, at that time, we shall extend the theory presented in this chapter so that more complex applications may be covered.

5
Poisson Brackets in Continuous Media

I have yet to see any problem, however complicated, which, when you looked at it in the right way, did not become still more complicated.
—*Poul Anderson*

Now that we have defined the necessary thermodynamic quantities in chapter 4, we can turn back to the consideration of the dynamics of various physical systems. In order to apply a Poisson bracket to macroscopic transport phenomena, it is first necessary to rewrite the bracket (3.3-3) in a form which is suitable for continuum-mechanical considerations. As the number of particles increases to infinity, the transition is made from the specification of a very large number of discrete particle trajectories, $\mathbf{x}_i(t)$, $i=1,2,\ldots,N$, $N\rightarrow\infty$, to the determination of a single, continuous, vector function, $\mathbf{Y}(\mathbf{r},t)$, indicating the position of a fluid particle at time t which at a reference time $t=0$ was at position $\mathbf{r}$, i.e., $\mathbf{Y}(\mathbf{r},0)=\mathbf{r}$. This is called a *Lagrangian* or *material description*. Alternatively, an *Eulerian* or *spatial description* can be used according to which the flow kinematics are completely specified through the determination of the velocity vector field, $\mathbf{v}(\mathbf{x},t)$, indicating the velocity of a fluid particle at a fixed spatial position, $\mathbf{x}$, and time, t. (Truesdell [1966, p. 17] notes that the Lagrangian/Eulerian terminology is erroneous, however.) In this chapter, we shall use both descriptions to tackle the problem of ideal (inviscid) fluid flow and to arrive at a Poisson bracket for each case. The dissipative system will be considered in chapter 7.

Once again, the concept of time and length scales is very important in determining when the experimenter views the system in consideration as a continuum entity. In chapter 3, we studied the dynamics of a system of discrete particles bouncing around with our time scale implicitly set on the order of the mean free time of the particles between collisions, ζ, and the length scale on the order of the mean free path, λ. As the number of particles approaches infinity, however, we note that certain averages, such as the velocity and the energy of the system, are practically constant on such a small time scale. Since the number of particles is so large, it is almost impossible to get any detailed information about the system as a whole by looking at individual particles because the number of the degrees of freedom is horrendous. Still, if we increase our time and length scales to values much greater than ζ and λ, respectively, then we can determine the dynamics of the average quantities which are much fewer than the original number of degrees of freedom. In so doing, we have lost some of the information with respect to the more detailed microscopic picture; however, as the number of particles increases so does the accuracy, since the averages are better indicators of the system behavior for larger systems. When we deal with length scales, d^3r, and time scales, dt, which are large on a microscopic scale (ζ and λ) but small on a macroscopic scale, then we have entered the realm of continuum mechanics. Here, smooth functions, such as the velocity, may be defined depending on the position, $\mathbf{r}$, and the time, t, whose evolution can be expected to give a fairly accurate picture of the system dynamics.

Much of the following work on the formulation of Poisson brackets for continuous media has been done within the framework of differential geometry (see, for example, Arnold [1966a], Holm *et al.* [1985], Olver [1986, ch. 6],[1] Abraham *et al.* [1988, ch. 8] and references therein), most notably due to Marsden and his co-workers. Still others used a commutator approach [Dzyaloshinskii and Volovick, 1980], or an enlarged space through the introduction of Clebsch variables [Kuznetsov and Mikhailov, 1980; Holm and Kupershmidt, 1983]. In contrast, we opt here to present this material in terms of the more conventional tensorial Cartesian notation. Granted, in so doing we have restricted some of the power and utility of these techniques, yet, for the purposes of this text, we shall not find this simplification unduly restrictive. The main limitation imposed is that we consider only rectangular Cartesian coordinate systems in Euclidean space. However, as engineers and physicists, all of the experiments which we conduct are limited to the spatial description, i.e., a fixed, laboratory reference frame. Indeed, it is in this framework that dissipative phenomena have been studied extensively within the guidelines of irreversible thermodynamics. All of the following principles, however, transfer over directly through the formalism of differential geometry when all reference frames (in Riemannian space) are put on an equal footing. In particular, we call attention to Marsden *et al.* [1984] and Simo *et al.* [1989] as probably the most advanced discourses to date concerning transformations of Poisson brackets between various reference systems.

5.1 The Material Description of Ideal Fluid Flow

In order to derive the equations of motion for a continuous system (in this case, an ideal fluid), it is first necessary to recognize the transforma-

[1]The reader interested in the history of the development of the Poisson bracket can find a concise but lucid account in the last two pages of Olver's chapter 6.

tions of the required properties as $N \to \infty$. As was just alluded to in the introductory section, a certain fluid particle of the continuous medium is labeled by a position (reference) vector at time t=0, $\mathbf{r}$, with a coordinate function $\mathbf{Y}(\mathbf{r},t) \in \mathbb{R}^3$ specifying the trajectory of this particle in space and time. (For the sake of simplicity, we require both coordinates $\mathbf{r}$ and $\mathbf{Y}$ to be rectangular Cartesian.) At time t=0, the fluid occupies a region, Ω, with boundary $\partial\Omega$, and the initial condition on the fluid particle under consideration is obviously $\mathbf{Y}(\mathbf{r},0)=\mathbf{r}$. At the later time t, the entire body of the fluid system may have moved and deformed, and hence the fluid occupies the region Ω', with boundary $\partial\Omega'$, and the fluid particle is at position $\mathbf{Y}(\mathbf{r},t)$. Therefore $\mathbf{Y}$ is a one-to-one mapping which gives the location at time t of the particle which was at $\mathbf{r}$ at t=0. (See Figure 5.1.)

For simplicity in the following analysis, we consider the boundary of the fluid to be fixed, i.e., $\partial\Omega'=\partial\Omega$ for all times. Later on in §5.6, we shall discuss the implications of removing this restriction, but for the time being, this allows us the freedom to neglect the boundary effects upon the system. In this chapter, we treat the system in terms of Ω and Ω' (it being understood that the two quantities are identical) so that most of the mathematics will carry over directly to the free-boundary problem.[2] This simplification also allows us to study the problem using a simple formalism similar to that of Abarbanel *et al.* [1988] rather than the fully general methodology of Marsden *et al.* [1984]. In the case of a fixed boundary, all of the necessary boundary conditions can be easily specified.

Associated with this fluid particle is a volume element at t=0, $d^3r \equiv dr_1 dr_2 dr_3$, where the vector $\mathbf{r}$ has components r_1, r_2, and r_3. Though the mass of the fluid particle must remain constant, its volume element is allowed to vary with time and space due to the compressible nature of the fluid. The volume element of the particle at any time t can be related to its volume element at t=0 through the *Euler relation*,

$$d^3Y = J\, d^3r \,, \tag{5.1-1}$$

where J is the Jacobian which defines the mapping of the fluid particle from $\mathbf{r}$ at time t=0 to $\mathbf{Y}(\mathbf{r},t)$ at arbitrary time t. This Jacobian is given by the determinant of the *deformation gradient* tensor field, $\mathbf{F}$, which, as the name suggests, measures the relative deformation undergone by the fluid element compared to the reference configuration:

$$J \equiv \det \mathbf{F} \,, \quad F_{\alpha\beta} \equiv \frac{\partial Y_\alpha}{\partial r_\beta} \,. \tag{5.1-2}$$

[2]Unfortunately, size constraints do not allow us to discuss free-boundary problems in this volume.

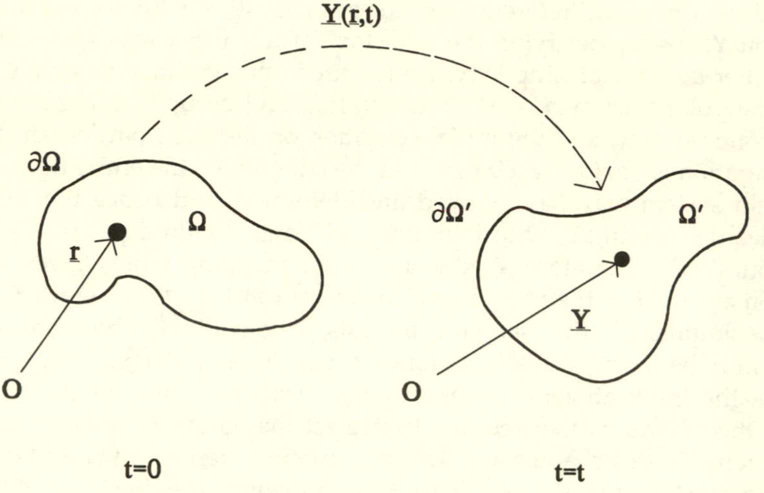

Figure 5.1: The material description of fluid flow.

Given the definitions (5.1-2), it is a simple task to prove the *Boussinesq identity* (see, for example, Herivel [1955, p. 347]), which states that

$$\frac{\partial J}{\partial F_{\alpha\beta}} = J \frac{\partial r_\beta}{\partial Y_\alpha} , \tag{5.1-3}$$

as well as the identities,

$$\frac{\partial}{\partial r_\beta}\left(\frac{\partial J}{\partial F_{\alpha\beta}}\right) = 0 \quad \text{and} \quad \frac{\partial}{\partial Y_\alpha}\left(\frac{1}{J}\frac{\partial Y_\alpha}{\partial r_\beta}\right) = 0 . \tag{5.1-4}$$

The distribution of the system mass at t=0 can be described by a density function, $\rho_o=\rho_o(\mathbf{r})$. Since the mass of the fluid particle is conserved, the mass density at any $\mathbf{Y}(\mathbf{r},t)$, denoted ρ, must satisfy the conservation equation

$$\rho\, \mathrm{d}^3 Y = \rho_o(\mathbf{r})\, \mathrm{d}^3 r . \tag{5.1-5}$$

Thus the density can be written in terms of the Jacobian of Eq. (5.1-1) as

$$\rho = \rho_o(\mathbf{r}) / J . \tag{5.1-6}$$

Hence we see that ρ depends on $\mathbf{Y}$ through the Jacobian or, more specifically, through the deformation of $\mathbf{Y}$ relative to the reference configuration, $\mathbf{F}$. Therefore we write the mass density as $\rho=\rho(\mathbf{F},t)$.

Essentially, the functions $\rho_o(\mathbf{r})$ and $\rho(\mathbf{F},t)$ are elements of $\mathbb{R}^+$, where $\mathbb{R}^+$ is the space of real, positive numbers.

In the material description of ideal fluid flow, the dynamical system variables are

$$\mathbf{Y}(\mathbf{r},t) \quad \text{and} \quad \left.\frac{\partial \mathbf{Y}(\mathbf{r},t)}{\partial t}\right|_{\mathbf{r}} = \dot{\mathbf{Y}}(\mathbf{r},t) \;, \tag{5.1-7}$$

(similar to the variables of §3.2) where $|_\alpha$ indicates that the derivative is evaluated at constant α. The operating space, P, for this problem is thus

$$P \equiv \begin{cases} \mathbf{Y}(\mathbf{r},t) \in \mathbb{R}^3, & \mathbf{Y}(\mathbf{r},0) = \mathbf{r} \text{ in } \Omega \\ \dot{\mathbf{Y}}(\mathbf{r},t) \in \mathbb{R}^3, & \dot{\mathbf{Y}}(\mathbf{r},0) = \dfrac{\partial \mathbf{Y}(\mathbf{r},0)}{\partial t} \text{ in } \Omega \end{cases} . \tag{5.1-8}$$

The Lagrangian for this system can be expressed in terms of these variables through an integral over every fluid particle as [Seliger and Whitham, 1968, p. 4]

$$\begin{aligned} L &= \int_{\Omega'} \rho(\mathbf{F},t)\,\{\tfrac{1}{2}\dot{\mathbf{Y}}\cdot\dot{\mathbf{Y}} - e_p^v(\mathbf{Y}) - \hat{U}[\rho(\mathbf{F},t), s_o(\mathbf{r})]\}\, d^3Y \\ &= \int_{\Omega} \rho_o(\mathbf{r})\,\{\tfrac{1}{2}\dot{\mathbf{Y}}\cdot\dot{\mathbf{Y}} - e_p^v(\mathbf{Y}) - \hat{U}[\rho(\mathbf{F},t), s_o(\mathbf{r})]\}\, d^3r \;. \end{aligned} \tag{5.1-9}$$

As before, the Lagrangian represents the difference in the kinetic and potential energies. Here, the potential from chapter 3 has been split into two terms: an external field potential, $e_p^v = e_p^v(\mathbf{Y})$, and the internal energy per unit mass, $\hat{U}$, which represents the attractive/repulsive intermolecular forces, etc., and which is a function of the mass density and the initial entropy density, $s_o(\mathbf{r}) = \rho_o\hat{S}$, $s_o \in \mathbb{R}$, with $\hat{S}$ constant for the fluid element (adiabatic flow—see §5.6).

With this Lagrangian, we can write the action integral for this system in typical fashion as (following Seliger and Whitham [1968, p. 4])

$$I = \int_{t_1}^{t_2} L\, dt \;, \tag{5.1-10}$$

so that we can apply the calculus of variations to the integral

$$I^* = \int_{t_1}^{t_2}\int_{\Omega} \rho_o\,[\tfrac{1}{2}\dot{\mathbf{Y}}^*\cdot\dot{\mathbf{Y}}^* - e_p^v(\mathbf{Y}^*) - \hat{U}[\rho(\mathbf{F}^*,t), s_o(\mathbf{r})]\}\, d^3r\, dt \;, \tag{5.1-11}$$

where

$$\mathbf{Y}^*(\mathbf{r},t,\varepsilon) = \mathbf{Y}(\mathbf{r},t) + \varepsilon\,\boldsymbol{\eta}(\mathbf{r},t) \,. \tag{5.1-12}$$

Doing so, holding **r** fixed and remembering that ρ depends on **Y** through its dependence on **F**, yields the variational equation

$$\delta I = \int_{t_1}^{t_2}\int_{\Omega} \rho_o\left[\dot{Y}_\alpha \frac{\partial(\delta Y_\alpha)}{\partial t} - \frac{\partial e_p^v}{\partial Y_\alpha}\delta Y_\alpha - \frac{\partial \hat{U}}{\partial \rho}\frac{\partial \rho}{\partial F_{\alpha\beta}}\delta F_{\alpha\beta}\right] d^3r\, dt \,, \tag{5.1-13}$$

and integrating the first term by parts with respect to t, keeping **r** fixed, gives

$$\delta I = -\int_{t_1}^{t_2}\int_{\Omega} \rho_o\left[\frac{\partial^2 Y_\alpha}{\partial t^2}\delta Y_\alpha + \frac{\partial e_p^v}{\partial Y_\alpha}\delta Y_\alpha + \frac{\partial \hat{U}}{\partial \rho}\frac{\partial \rho}{\partial F_{\alpha\beta}}\delta F_{\alpha\beta}\right] d^3r\, dt \,. \tag{5.1-14}$$

If we switch the order of differentiation in $\delta F_{\alpha\beta}$, i.e.,

$$\delta F_{\alpha\beta} = \delta\left(\frac{\partial Y_\alpha}{\partial r_\beta}\right) = \frac{\partial(\delta Y_\alpha)}{\partial r_\beta} \,, \tag{5.1-15}$$

then we can integrate the third term in the integrand of (5.1-14) by parts with respect to **r**, keeping t fixed, to give

$$\delta I = -\int_{t_1}^{t_2}\int_{\Omega} \left[\rho_o\frac{\partial^2 Y_\alpha}{\partial t^2} + \rho_o\frac{\partial e_p^v}{\partial Y_\alpha} - \frac{\partial}{\partial r_\beta}\left(\rho_o\frac{\partial \hat{U}}{\partial \rho}\frac{\partial \rho}{\partial F_{\alpha\beta}}\right)\right]\delta Y_\alpha\, d^3r\, dt \,. \tag{5.1-16}$$

Hence the variational equation for the ideal fluid is found to be

$$\rho_o\frac{\partial^2 Y_\alpha}{\partial t^2} = -\rho_o\frac{\partial e_p^v}{\partial Y_\alpha} + \frac{\partial}{\partial r_\beta}\left(\rho_o\frac{\partial \hat{U}}{\partial \rho}\frac{\partial \rho}{\partial F_{\alpha\beta}}\right) . \tag{5.1-17}$$

Although the expression (5.1-17) represents the equation of motion in the material description of an ideal fluid, it is not yet in a form which is immediately recognizable to a student of hydrodynamics. Therefore we have to rewrite this equation using some of the identities which have already been introduced. First we invoke the thermodynamic information of the preceding chapter, where the pressure was defined as

$$p = -\left.\frac{\partial \hat{u}}{\partial \hat{v}}\right|_{\hat{s}} . \tag{5.1-18}$$

This expression is on a unit mole basis, however, and the problem formulation at present is on a unit mass basis. Consequently, the above expression must be modified by

$$p = -\frac{M_w}{M_w}\frac{\partial \hat{u}}{\partial \hat{v}}\bigg|_{\hat{s}} = -\frac{\partial \hat{U}}{\partial \hat{V}}\bigg|_{\hat{S}} , \tag{5.1-19}$$

where M_w is the molecular weight of the ideal fluid. The volume per unit mass, $\hat{V}$, is easily recognized as the reciprocal of the mass density, $1/\rho$, so that (5.1-19) can be rewritten as

$$p = -\frac{\partial \hat{U}}{\partial (1/\rho)}\bigg|_{\hat{S}} = -\frac{\partial \hat{U}}{\partial \rho}\bigg|_{\hat{S}}\frac{\partial \rho}{\partial (1/\rho)} = \rho^2\frac{\partial \hat{U}}{\partial \rho}\bigg|_{\hat{S}} . \tag{5.1-20}$$

Substituting this expression into the equation of motion, (5.1-17), we see that

$$\rho_o\frac{\partial^2 Y_\alpha}{\partial t^2} = -\rho_o\frac{\partial e_p^v}{\partial Y_\alpha} + \frac{\partial}{\partial r_\beta}\left(p\frac{\rho_o}{\rho^2}\frac{\partial \rho}{\partial F_{\alpha\beta}}\right) , \tag{5.1-21}$$

where $p=p(\mathbf{r},t)$.

Now we may rewrite $\partial\rho/\partial F_{\alpha\beta}$ using (5.1-6) as

$$\frac{\partial \rho}{\partial F_{\alpha\beta}} = \frac{\partial(\rho_o/J)}{\partial F_{\alpha\beta}} = \rho_o\frac{\partial(1/J)}{\partial F_{\alpha\beta}} = -\frac{\rho_o}{J^2}\frac{\partial J}{\partial F_{\alpha\beta}} , \tag{5.1-22}$$

and then using the Boussinesq identity gives

$$\frac{\partial \rho}{\partial F_{\alpha\beta}} = -\frac{\rho^2}{\rho_o}\frac{\partial}{\partial F_{\alpha\beta}}\left(\frac{\partial J}{\partial F_{\gamma\varepsilon}}\frac{\partial Y_\gamma}{\partial r_\varepsilon}\right) = -\frac{\rho^2}{\rho_o}\frac{\partial J}{\partial F_{\alpha\beta}} , \tag{5.1-23}$$

whereupon (5.1-21) becomes

$$\rho_o\frac{\partial^2 Y_\alpha}{\partial t^2} = -\rho_o\frac{\partial e_p^v}{\partial Y_\alpha} - \frac{\partial}{\partial r_\beta}\left(p\frac{\partial J}{\partial F_{\alpha\beta}}\right) . \tag{5.1-24}$$

Applying the identity $(5.1\text{-}4)_1$ to this expression then yields

$$\rho_o\frac{\partial^2 Y_\alpha}{\partial t^2} = -\rho_o\frac{\partial e_p^v}{\partial Y_\alpha} - \frac{\partial J}{\partial F_{\alpha\beta}}\frac{\partial p}{\partial r_\beta} . \tag{5.1-25}$$

Using (5.1-6) again, we obtain

$$\frac{\partial^2 Y_\alpha}{\partial t^2} = -\frac{\partial e_p^v}{\partial Y_\alpha} - \frac{1}{\rho J}\frac{\partial J}{\partial F_{\alpha\beta}}\frac{\partial p}{\partial r_\beta} , \tag{5.1-26}$$

which, given the Boussinesq identity, gives

$$\frac{\partial^2 Y_\alpha}{\partial t^2} = -\frac{\partial e_p^v}{\partial Y_\alpha} - \frac{1}{\rho}\frac{\partial r_\beta}{\partial Y_\alpha}\frac{\partial p}{\partial r_\beta} , \tag{5.1-27}$$

from which we can assume that $p=p(\mathbf{Y},t)$ (i.e., that $\mathbf{r}=\mathbf{r}(\mathbf{Y},t)$ since $J\neq 0$) and arrive at

$$\rho \frac{\partial^2 Y_\alpha}{\partial t^2} = -\rho \frac{\partial e_p^v}{\partial Y_\alpha} - \frac{\partial p}{\partial Y_\alpha} . \tag{5.1-28}$$

Eq. (5.1-28) is the culmination of this section, i.e., the continuum equivalent of the Lagrangian equations of motion for a system of discrete particles, Eqs. (3.2-11). It represents the equation of motion for an ideal fluid in a material description of the fluid flow. This equation is usually derived in standard hydrodynamics textbooks using a macroscopic force balance; however, we have arrived at the same expression in this section using the techniques of variational calculus.

5.2 The Canonical Poisson Bracket for Ideal Fluid Flow

Now that we have determined the Lagrangian (5.1-9) for our system, as well as the equations of motion, we are in a position to define the canonical Poisson bracket for ideal fluid flow. In order to do this, however, it is first necessary to review some aspects of the analysis of functionals.

5.2.1 *The calculus of functionals*

Let us define an arbitrary functional, $F=F[a,b,\ldots]$, where $a,b,\ldots\in P$ (P being the operating space of the problem in consideration) are the dynamical variables of the system of interest (which depend upon the particular coordinate system, $\mathbf{x}$, and the time t), through the equation

$$F[a,b,\ldots] \equiv \int_\Omega f(a,b,\ldots)\, \mathrm{d}^3x . \tag{5.2-1}$$

In this expression, f is a scalar function of the system variables with suitable continuity, Ω is the domain of interest with boundary $\partial\Omega$, and d^3x represents the appropriate volume element for the integration. With the above definition for F, we can also define the *Volterra functional derivatives* (for unconstrained systems) as

$$\frac{\delta F}{\delta a} \equiv \frac{\partial f}{\partial a} , \quad \frac{\delta F}{\delta b} \equiv \frac{\partial f}{\partial b} , \quad \ldots \in P . \tag{5.2-2}$$

Although the variables are treated as scalars here, for simplicity, they can be tensors of arbitrary order, in general, in which case the derivatives are with respect to the components of the tensor.

The situation is slightly more complex when one considers a functional of not only a, but implicitly on $\boldsymbol{\nabla} a$, where $\boldsymbol{\nabla} \equiv \partial/\partial \mathbf{x}$, i.e., where

$$F[a] \equiv \int_\Omega f(a, \boldsymbol{\nabla} a)\, \mathrm{d}^3x \,. \tag{5.2-3}$$

In this case, taking the variation of F with respect to a gives

$$\delta F = \int_\Omega \left[\frac{\partial f}{\partial a} \delta a + \frac{\partial f}{\partial(\boldsymbol{\nabla} a)} \cdot \delta(\boldsymbol{\nabla} a) \right] \mathrm{d}^3x \,, \tag{5.2-4}$$

whereupon switching the order of differentiation and integrating by parts the second term on the right-hand side yields

$$\delta F = \int_\Omega \left[\frac{\partial f}{\partial a} - \boldsymbol{\nabla} \cdot \frac{\partial f}{\partial(\boldsymbol{\nabla} a)} \right] \delta a\, \mathrm{d}^3x \,. \tag{5.2-5}$$

Therefore, the Volterra derivative of the functional F with respect to a becomes

$$\frac{\delta F}{\delta a} \equiv \frac{\partial f}{\partial a} - \boldsymbol{\nabla} \cdot \frac{\partial f}{\partial(\boldsymbol{\nabla} a)} \in P \,. \tag{5.2-6}$$

Thus we see the intimate connection between the above functional derivatives and the calculus of variations, i.e., the functional derivative $\delta F/\delta a$ is just a notation indicating the Euler/Lagrange equation resulting from the variation of the functional F with respect to the variable a. It is straightforward to define the functional derivative for a functional which depends on higher-order spatial derivatives of a as well.

Since a and b are functions of time as well as space, we can write the total time derivative of F as

$$\begin{aligned}
\frac{\mathrm{d}F}{\mathrm{d}t} &= \frac{\mathrm{d}}{\mathrm{d}t}\left[\int_\Omega f(a, \boldsymbol{\nabla} a, b)\, \mathrm{d}^3x \right] = \int_\Omega \frac{\mathrm{d}}{\mathrm{d}t} f(a, \boldsymbol{\nabla} a, b)\, \mathrm{d}^3x \\
&= \int_\Omega \left[\frac{\partial f}{\partial a}\frac{\partial a}{\partial t} + \frac{\partial f}{\partial(\boldsymbol{\nabla} a)} \cdot \frac{\partial(\boldsymbol{\nabla} a)}{\partial t} + \frac{\partial f}{\partial b}\frac{\partial b}{\partial t} \right] \mathrm{d}^3x \\
&= \int_\Omega \left[\frac{\partial f}{\partial a}\frac{\partial a}{\partial t} - \boldsymbol{\nabla} \cdot \frac{\partial f}{\partial(\boldsymbol{\nabla} a)} \frac{\partial a}{\partial t} + \frac{\partial f}{\partial b}\frac{\partial b}{\partial t} \right] \mathrm{d}^3x \\
&= \int_\Omega \left[\frac{\delta F}{\delta a}\frac{\partial a}{\partial t} + \frac{\delta F}{\delta b}\frac{\partial b}{\partial t} \right] \mathrm{d}^3x \,.
\end{aligned} \tag{5.2-7}$$

This expression will be used throughout the book.

5.2.2 The continuum Poisson bracket

Now it is possible to define the conjugate momentum vector field in the usual manner (see Eq. (3.2-16) and Goldstein [1980, p. 563]) as

$$\Pi(\mathbf{r},t) \equiv \delta L/\delta\dot{\mathbf{Y}}(\mathbf{r},t) = \rho_o(\mathbf{r})\,\dot{\mathbf{Y}}(\mathbf{r},t) \ . \tag{5.2-8}$$

Then, as before, we can determine the Hamiltonian through the Legendre transformation of the integrand (see Appendix B) as

$$H[\mathbf{Y},\Pi] \equiv \int_\Omega \left[\frac{1}{2\rho_o}\Pi\cdot\Pi + \rho_o e_p^v(\mathbf{Y}) + \rho_o \hat{U}[\rho(\mathbf{Y},t),s_o(\mathbf{r})]\right] \mathrm{d}^3r, \tag{5.2-9a}$$

by analogy with Eqs. (3.2-17), or alternatively,

$$H = \int_\Omega [e_k + e_p + u]\,\mathrm{d}^3r \ , \tag{5.2-9b}$$

where e_k, e_p, and u are the kinetic, potential, and internal energy densities, respectively. Thus, once again, we recognize the Hamiltonian as the total system energy, where each fluid particle has a certain kinetic, potential, and internal energy density.

Now we can easily show that the equations of motion for an ideal fluid can be written in an analogous form to Eqs. (3.2-18):

$$\dot{\mathbf{Y}}(\mathbf{r},t) = \frac{\delta H[\mathbf{Y},\Pi]}{\delta\Pi(\mathbf{r},t)} = \frac{\Pi(\mathbf{r},t)}{\rho_o(\mathbf{r})} \ , \tag{5.2-10a}$$

$$\dot{\Pi}(\mathbf{r},t) = -\frac{\delta H[\mathbf{Y},\Pi]}{\delta\mathbf{Y}(\mathbf{r},t)} = -\rho_o\frac{\partial e_p^v}{\partial\mathbf{Y}} - \frac{\rho_o}{\rho}\frac{\partial p}{\partial\mathbf{Y}} \ . \tag{5.2-10b}$$

Consequently, by using the chain rule of differentiation for a functional, F, given by (5.2-7),

$$\frac{\mathrm{d}F[\mathbf{Y},\Pi]}{\mathrm{d}t} = \int_\Omega \left[\frac{\delta F}{\delta\mathbf{Y}}\cdot\frac{\partial\mathbf{Y}}{\partial t} + \frac{\delta F}{\delta\Pi}\cdot\frac{\partial\Pi}{\partial t}\right]\mathrm{d}^3r \ , \tag{5.2-11}$$

we can substitute into this expression Eqs. (5.2-10) and arrive at the dynamical equation for the arbitrary functional:

$$\frac{\mathrm{d}F}{\mathrm{d}t} = \{F,H\}_L \equiv \int_\Omega \left[\frac{\delta F}{\delta\mathbf{Y}}\cdot\frac{\delta H}{\delta\Pi} - \frac{\delta F}{\delta\Pi}\cdot\frac{\delta H}{\delta\mathbf{Y}}\right]\mathrm{d}^3r \ . \tag{5.2-12}$$

Because of the antisymmetric nature of this bracket, we have immediately that $\mathrm{d}H/\mathrm{d}t=\{H,H\}_L=0$, which indicates that the total energy of the system is again conserved. Thus we can define the continuum version of the Poisson bracket (3.3-3) for two arbitrary functionals F and G as

$$\{F,G\}_L \equiv \int_\Omega \left[\frac{\delta F}{\delta \mathbf{Y}} \cdot \frac{\delta G}{\delta \mathbf{\Pi}} - \frac{\delta F}{\delta \mathbf{\Pi}} \cdot \frac{\delta G}{\delta \mathbf{Y}}\right] d^3 r \ . \tag{5.2-13}$$

In concluding this section, it has become apparent that the mathematical symmetries which we have noted in the discrete particle case can be transferred over directly to the continuum approach. This is not a coincidence, but a reflection of the underlying physical structure of the Hamiltonian equations which is inherent to the Poisson bracket.

5.3 The Spatial Description of Ideal Fluid Flow

Now that we have derived the canonical form of the Poisson bracket from the material description, let us see how this quantity can be transformed into its noncanonical, spatial counterpart. In the spatial description, we consider a fixed coordinate point in space (again assumed to be rectangular Cartesian for convenience), $\mathbf{x}$, for which the (constant) volume element is given by d^3x. At this point, we sit and observe the properties of the fluid passing by us. For the transformation from material to spatial coordinates, we need to identify the fixed point $\mathbf{x}$ with the position function $\mathbf{Y}(\mathbf{r},t)$, which coincide at the instant t. At this time, the fluid occupies the region Ω' with boundary $\partial\Omega'$. This procedure specifies a new function for all later times $\tilde{t}$, $\mathbf{R}(\mathbf{x},\tilde{t})$, which maps $\mathbf{Y}$ back to the fixed position $\mathbf{x}$:

$$\mathbf{x} = \mathbf{Y}[\mathbf{R}(\mathbf{x},\tilde{t}\,),\tilde{t}\,] \ . \tag{5.3-1}$$

Thus this function $\mathbf{R}(\mathbf{x},\tilde{t})$ serves as a material label for the fluid particle which has position $\mathbf{x}$ at time t, which is a determined function of the spatial field. Of course, $\mathbf{R}=\mathbf{r}$ at time t. Figure 5.2 illustrates these concepts.

In the spatial description, we can now determine the dynamical variables for inviscid fluid flow as functions of $\mathbf{x}$ and t, in the domain, Ω', with fixed boundary, $\partial\Omega'$. Here the variables are the mass density, $\rho(\mathbf{x},t)$, the momentum density, $\mathbf{M}(\mathbf{x},t)=\rho(\mathbf{x},t)\mathbf{v}(\mathbf{x},t)$ (where $\mathbf{v}(\mathbf{x},t)$ is the velocity vector field), and the entropy density, $s(\mathbf{x},t)=\rho(\mathbf{x},t)\hat{S}(\mathbf{x},t)$. Thus the oper-ating space, P, for the spatial description is

$$P \equiv \begin{cases} \rho(\mathbf{x},t) \in \mathbb{R}^+ \\ \mathbf{M}(\mathbf{x},t) \in \mathbb{R}^3, \quad \mathbf{n}\cdot\mathbf{M} = 0 \ \text{ on } \ \partial\Omega' \ , \\ s(\mathbf{x},t) \in \mathbb{R} \end{cases} \tag{5.3-2}$$

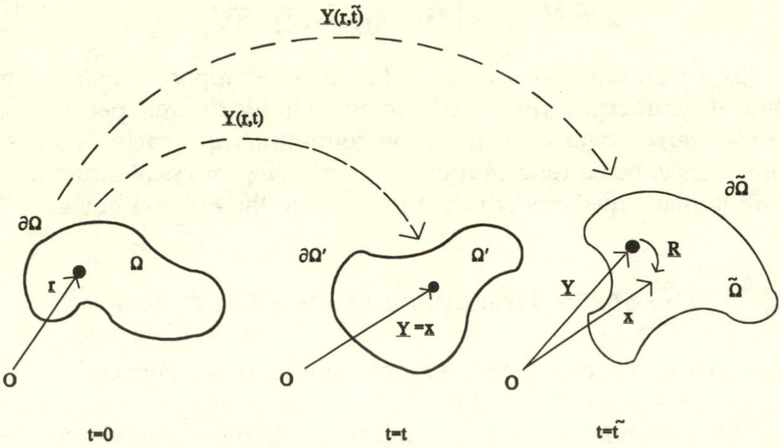

Figure 5.2: The spatial description of fluid flow.

which also includes unspecified initial and boundary conditions on ρ, **M**, and *s* Here, **n**·**M**=0 (**n** being the outwardly-directed unit vector normal to ∂Ω') ensures that the fluid does not penetrate the boundary. Thus we see that the specification of a fixed domain (for the time being), Ω'=Ω, with both initial and boundary conditions is implicit to the spatial description, i.e., in a description based on **x**. The dynamical variables can now be written solely in terms of their material counterparts through the definition

$$\int_{\Omega'} g(\mathbf{b})\, \delta^3[\mathbf{b}-\mathbf{a}]\, d^3b \equiv \begin{cases} g(\mathbf{a}) \; ; & \mathbf{b}=\mathbf{a} \\ 0 \; ; & \mathbf{b}\neq\mathbf{a} \end{cases} , \tag{5.3-3}$$

where **a** and **b** are arbitrary vectors and g is an arbitrary function with suitable continuity. Using (5.3-3), the mass density can be expressed as

$$\rho(\mathbf{x},\tilde{t}) = \rho(\mathbf{F}[\mathbf{R}(\mathbf{x},\tilde{t}),\tilde{t}],\tilde{t}) ,$$

or given that at time $\tilde{t}=t$

$$\rho(\mathbf{x},t) = \rho(\mathbf{F}(\mathbf{r},t),t) \Rightarrow$$

$$\begin{aligned} \rho(\mathbf{x},t) &= \int_{\Omega'} \rho(\mathbf{F}(\mathbf{r},t),t)\, \delta^3[\mathbf{Y}(\mathbf{r},t)-\mathbf{x}]\, d^3Y \\ &= \int_{\Omega} \rho_o(\mathbf{r})\, \delta^3[\mathbf{Y}(\mathbf{r},t)-\mathbf{x}]\, d^3r , \end{aligned} \tag{5.3-4}$$

where we have made use of (5.1-5). Similarly,

$$\mathbf{M}(\mathbf{x},t) = \int_{\Omega'} \rho(\mathbf{F}(\mathbf{r},t),t)\, \dot{\mathbf{Y}}(\mathbf{r},t)\, \delta^3[\mathbf{Y}(\mathbf{r},t) - \mathbf{x}]\, \mathrm{d}^3Y$$

$$= \int_{\Omega} \rho_o(\mathbf{r})\, \dot{\mathbf{Y}}(\mathbf{r},t)\, \delta^3[\mathbf{Y}(\mathbf{r},t) - \mathbf{x}]\, \mathrm{d}^3r \tag{5.3-5}$$

$$= \int_{\Omega} \Pi(\mathbf{r},t)\, \delta^3[\mathbf{Y}(\mathbf{r},t) - \mathbf{x}]\, \mathrm{d}^3r \ ,$$

$$s(\mathbf{x},t) = \int_{\Omega} s_o(\mathbf{r})\, \delta^3[\mathbf{Y}(\mathbf{r},t) - \mathbf{x}]\, \mathrm{d}^3r \ , \tag{5.3-6}$$

where we have multiplied both sides of (5.1-5) by $\hat{S}$:

$$s(\mathbf{F},t)\, \mathrm{d}^3Y = s_o(\mathbf{r})\, \mathrm{d}^3r \ . \tag{5.3-7}$$

Thus the spatial variables ρ, $\mathbf{M}$, and s are now expressed solely in terms of the material variables and it is apparent why we have chosen these as the natural variables, i.e., the density variables are necessary for the transformation from material to spatial coordinates through the Jacobian (see Eqs. (5.1-1,5,6)). Note that the entropy density cannot be considered constant in either a material or a spatial coordinate system.

Now we can follow a reduction procedure, originally due to Marsden *et al.* [1984], but through the simpler form of Abarbanel *et al.* [1988], to derive the noncanonical, spatial equivalent of the Poisson bracket. Therefore, we need to define the functional derivatives of (5.2-13) through the chain rule of differentiation—see (5.2-7). For an arbitrary functional, $F[\rho(\mathbf{x},t),\mathbf{M}(\mathbf{x},t),s(\mathbf{x},t)]$, these are

$$\frac{\delta F}{\delta Y_\alpha} = \int_{\Omega'} \left[\frac{\delta F}{\delta \rho(\mathbf{x},t)} \frac{\delta \rho(\mathbf{x},t)}{\delta Y_\alpha(\mathbf{r},t)} + \frac{\delta F}{\delta M_\beta(\mathbf{x},t)} \frac{\delta M_\beta(\mathbf{x},t)}{\delta Y_\alpha(\mathbf{r},t)} + \frac{\delta F}{\delta s(\mathbf{x},t)} \frac{\delta s(\mathbf{x},t)}{\delta Y_\alpha(\mathbf{r},t)} \right] \mathrm{d}^3x \ , \tag{5.3-8}$$

$$\frac{\delta F}{\delta \Pi_\alpha} = \int_{\Omega'} \left[\frac{\delta F}{\delta \rho(\mathbf{x},t)} \frac{\delta \rho(\mathbf{x},t)}{\delta \Pi_\alpha(\mathbf{r},t)} + \frac{\delta F}{\delta M_\beta(\mathbf{x},t)} \frac{\delta M_\beta(\mathbf{x},t)}{\delta \Pi_\alpha(\mathbf{r},t)} + \frac{\delta F}{\delta s(\mathbf{x},t)} \frac{\delta s(\mathbf{x},t)}{\delta \Pi_\alpha(\mathbf{r},t)} \right] \mathrm{d}^3x \ . \tag{5.3-9}$$

In order for the integrals (5.3-8,9) to be proper inner products, the derivatives must all belong to the same operating space, in this case, P,

(5.3-2). Thus, $\mathbf{n}\cdot\delta F/\delta\mathbf{M}=0$ on $\partial\Omega'$. Substitution of Eqs. (5.3-8,9) into the Poisson bracket of (5.2-13) yields the spatial bracket for two arbitrary functionals $F[\rho(\mathbf{x},t),\mathbf{M}(\mathbf{x},t),s(\mathbf{x},t)]$ and $G[\rho(\mathbf{z},t),\mathbf{M}(\mathbf{z},t),s(\mathbf{z},t)]$ (using, for notational convenience, an additional coordinate, $\mathbf{z}$, which has the same properties as $\mathbf{x}$):

$$
\begin{aligned}
\{F,G\}_E = & \int_{\Omega'}\int_{\Omega'}\left[\frac{\delta F}{\delta\rho(\mathbf{x},t)}\frac{\delta G}{\delta M_\beta(\mathbf{z},t)} - \frac{\delta G}{\delta\rho(\mathbf{x},t)}\frac{\delta F}{\delta M_\beta(\mathbf{z},t)}\right] \\
& \times \{\rho(\mathbf{x},t),M_\beta(\mathbf{z},t)\}_L \, d^3z\, d^3x \\
& + \int_{\Omega'}\int_{\Omega'}\left[\frac{\delta F}{\delta s(\mathbf{x},t)}\frac{\delta G}{\delta M_\beta(\mathbf{z},t)} - \frac{\delta G}{\delta s(\mathbf{x},t)}\frac{\delta F}{\delta M_\beta(\mathbf{z},t)}\right] \\
& \times \{s(\mathbf{x},t),M_\beta(\mathbf{z},t)\}_L \, d^3z\, d^3x \\
& + \int_{\Omega'}\int_{\Omega'}\frac{\delta F}{\delta M_\gamma(\mathbf{x},t)}\frac{\delta G}{\delta M_\beta(\mathbf{z},t)}\{M_\gamma(\mathbf{x},t),M_\beta(\mathbf{z},t)\}_L \, d^3z\, d^3x \ ,
\end{aligned}
\tag{5.3-10}
$$

where $\{\cdot,\cdot\}_L$ indicates the material bracket of (5.2-13). This expression was obtained through a lengthy (but straightforward) calculation, using antisymmetry identities of the type

$$
\{\rho(\mathbf{x},t),M_\beta(\mathbf{z},t)\} = -\{M_\beta(\mathbf{z},t),\rho(\mathbf{x},t)\} \ , \tag{5.3-11}
$$

and anticipating Eqs. (5.3-12) through (5.3-14). (Note that it is allowable to interchange the coordinates $\mathbf{x}$ and $\mathbf{z}$ under the integration.)

The three brackets $\{\cdot,\cdot\}_L$ in (5.3-10) now need to be evaluated. From Eqs. (5.3-4,5,6), we note that the necessary functional derivatives are

$$
\frac{\delta\rho(\mathbf{x},t)}{\delta Y_\alpha(\mathbf{r},t)} = \rho_o(\mathbf{r})\frac{\partial\delta^3[\mathbf{Y}(\mathbf{r},t)-\mathbf{x}]}{\partial Y_\alpha(\mathbf{r},t)} \ , \quad \frac{\delta\rho(\mathbf{x},t)}{\delta\Pi_\alpha(\mathbf{r},t)} = 0 \ , \tag{5.3-12}
$$

$$
\frac{\delta s(\mathbf{x},t)}{\delta Y_\alpha(\mathbf{r},t)} = s_o(\mathbf{r})\frac{\partial\delta^3[\mathbf{Y}(\mathbf{r},t)-\mathbf{x}]}{\partial Y_\alpha(\mathbf{r},t)} \ , \quad \frac{\delta s(\mathbf{x},t)}{\delta\Pi_\alpha(\mathbf{r},t)} = 0 \ , \tag{5.3-13}
$$

$$
\begin{aligned}
\frac{\delta M_\beta(\mathbf{x},t)}{\delta Y_\alpha(\mathbf{r},t)} &= \Pi_\beta(\mathbf{r},t)\frac{\partial\delta^3[\mathbf{Y}(\mathbf{r},t)-\mathbf{x}]}{\partial Y_\alpha(\mathbf{r},t)} \ , \\
\frac{\delta M_\beta(\mathbf{x},t)}{\delta\Pi_\alpha(\mathbf{r},t)} &= \delta_{\alpha\beta}\,\delta^3[\mathbf{Y}(\mathbf{r},t)-\mathbf{x}] \ ,
\end{aligned}
\tag{5.3-14}
$$

where $\delta_{\alpha\beta}$ is a component of the identity tensor. Consequently, we have

$$\{\rho(\mathbf{x},t), M_\beta(\mathbf{z},t)\}_L =$$

$$\int_\Omega \left[\frac{\delta \rho(\mathbf{x},t)}{\delta Y_\alpha(\mathbf{r},t)} \frac{\delta M_\beta(\mathbf{z},t)}{\delta \Pi_\alpha(\mathbf{r},t)} - \frac{\delta M_\beta(\mathbf{z},t)}{\delta Y_\alpha(\mathbf{r},t)} \frac{\delta \rho(\mathbf{x},t)}{\delta \Pi_\alpha(\mathbf{r},t)} \right] \mathrm{d}^3 r$$

$$= \int_\Omega \rho_o(\mathbf{r}) \frac{\partial \delta^3[\mathbf{Y}(\mathbf{r},t) - \mathbf{x}]}{\partial Y_\alpha(\mathbf{r},t)} \delta_{\alpha\beta}\, \delta^3[\mathbf{Y}(\mathbf{r},t) - \mathbf{z}]\, \mathrm{d}^3 r \tag{5.3-15}$$

$$= \int_{\Omega'} \rho(\mathbf{F},t) \frac{\partial \delta^3[\mathbf{Y}(\mathbf{r},t) - \mathbf{x}]}{\partial Y_\beta(\mathbf{r},t)} \delta^3[\mathbf{Y}(\mathbf{r},t) - \mathbf{z}]\, \mathrm{d}^3 Y$$

$$= \rho(\mathbf{z},t) \frac{\partial \delta^3[\mathbf{z} - \mathbf{x}]}{\partial z_\beta} ,$$

$$\{s(\mathbf{x},t), M_\beta(\mathbf{z},t)\}_L = s(\mathbf{z},t) \frac{\partial \delta^3[\mathbf{z} - \mathbf{x}]}{\partial z_\beta} , \tag{5.3-16}$$

$$\{M_\gamma(\mathbf{x},t), M_\beta(\mathbf{z},t)\}_L =$$

$$\int_\Omega \left[\frac{\delta M_\gamma(\mathbf{x},t)}{\delta Y_\alpha(\mathbf{r},t)} \frac{\delta M_\beta(\mathbf{z},t)}{\delta \Pi_\alpha(\mathbf{r},t)} - \frac{\delta M_\beta(\mathbf{z},t)}{\delta Y_\alpha(\mathbf{r},t)} \frac{\delta M_\gamma(\mathbf{x},t)}{\delta \Pi_\alpha(\mathbf{r},t)} \right] \mathrm{d}^3 r$$

$$= \int_\Omega \left[\Pi_\gamma(\mathbf{r},t) \frac{\partial \delta^3[\mathbf{Y}(\mathbf{r},t) - \mathbf{x}]}{\partial Y_\alpha(\mathbf{r},t)} \delta_{\alpha\beta}\, \delta^3[\mathbf{Y}(\mathbf{r},t) - \mathbf{z}] \right. \tag{5.3-17}$$

$$\left. - \Pi_\beta(\mathbf{r},t) \frac{\partial \delta^3[\mathbf{Y}(\mathbf{r},t) - \mathbf{z}]}{\partial Y_\alpha(\mathbf{r},t)} \delta_{\alpha\gamma}\, \delta^3[\mathbf{Y}(\mathbf{r},t) - \mathbf{x}] \right] \mathrm{d}^3 r$$

$$= M_\gamma(\mathbf{z},t) \frac{\partial \delta^3[\mathbf{z} - \mathbf{x}]}{\partial z_\beta} - M_\beta(\mathbf{x},t) \frac{\partial \delta^3[\mathbf{x} - \mathbf{z}]}{\partial x_\gamma} .$$

The brackets (5.3-15,16,17) can be substituted back into (5.3-10), which we rewrite now as

$$\{F,G\}_E = \{F,G\}_E^\rho + \{F,G\}_E^s + \{F,G\}_E^M , \tag{5.3-18}$$

where the subbrackets $\{F,G\}_E^a$ have the obvious interpretation from (5.3-10). Let us look at each subbracket in turn. The first subbracket, $\{F,G\}_E^\rho$, can be written as

$$\{F,G\}_E^\rho = \int_{\Omega'}\int_{\Omega'}\left[\frac{\delta F}{\delta\rho(\mathbf{x},t)}\frac{\delta G}{\delta M_\beta(\mathbf{z},t)} - \frac{\delta G}{\delta\rho(\mathbf{x},t)}\frac{\delta F}{\delta M_\beta(\mathbf{z},t)}\right] \times \rho(\mathbf{z},t)\frac{\partial\delta^3[\mathbf{z}-\mathbf{x}]}{\partial z_\beta}\,d^3z\,d^3x\ . \tag{5.3-19}$$

Integrating this expression by parts with respect to **z** yields two terms:

$$\begin{aligned}\{F,G\}_E^\rho = &\int_{\Omega'}\int_{\partial\Omega'}\left[\frac{\delta F}{\delta\rho(\mathbf{x},t)}\frac{\delta G}{\delta M_\beta(\mathbf{z},t)} - \frac{\delta G}{\delta\rho(\mathbf{x},t)}\frac{\delta F}{\delta M_\beta(\mathbf{z},t)}\right] \times n_\beta\,\delta^3[\mathbf{z}-\mathbf{x}]\,d^3z\,d^3x \\ &- \int_{\Omega'}\int_{\Omega'}\left[\frac{\delta F}{\delta\rho(\mathbf{x},t)}\frac{\partial}{\partial z_\beta}\left(\rho(\mathbf{z},t)\frac{\delta G}{\delta M_\beta(\mathbf{z},t)}\right) - \frac{\delta G}{\delta\rho(\mathbf{x},t)}\frac{\partial}{\partial z_\beta}\left(\rho(\mathbf{z},t)\frac{\delta F}{\delta M_\beta(\mathbf{z},t)}\right)\right]\delta^3[\mathbf{z}-\mathbf{x}]\,d^3z\,d^3x\ ,\end{aligned} \tag{5.3-20}$$

but we know that $\mathbf{n}\cdot\delta F/\delta\mathbf{M}=0$ on $\partial\Omega'$, so that the surface term vanishes (besides, the variations are required to vanish on the boundary). Therefore, only the bulk term survives which, after integration over **z**, can be rewritten as

$$\{F,G\}_E^\rho = -\int_{\Omega'}\left[\frac{\delta F}{\delta\rho(\mathbf{x},t)}\frac{\partial}{\partial x_\beta}\left(\rho(\mathbf{x},t)\frac{\delta G}{\delta M_\beta(\mathbf{x},t)}\right) - \frac{\delta G}{\delta\rho(\mathbf{x},t)}\frac{\partial}{\partial x_\beta}\left(\rho(\mathbf{x},t)\frac{\delta F}{\delta M_\beta(\mathbf{x},t)}\right)\right] d^3x\ . \tag{5.3-21}$$

In an analogous fashion, we can write as well

$$\{F,G\}_E^s = -\int_{\Omega'}\left[\frac{\delta F}{\delta s(\mathbf{x},t)}\frac{\partial}{\partial x_\beta}\left(s(\mathbf{x},t)\frac{\delta G}{\delta M_\beta(\mathbf{x},t)}\right) - \frac{\delta G}{\delta s(\mathbf{x},t)}\frac{\partial}{\partial x_\beta}\left(s(\mathbf{x},t)\frac{\delta F}{\delta M_\beta(\mathbf{x},t)}\right)\right] d^3x\ . \tag{5.3-22}$$

Finally, we can split the last subbracket into two parts:

$$\{F,G\}_E^M = \int_{\Omega'}\int_{\Omega'} \frac{\delta F}{\delta M_\gamma(\mathbf{x},t)} \frac{\delta G}{\delta M_\beta(\mathbf{z},t)} M_\gamma(\mathbf{z},t) \frac{\partial \delta^3[\mathbf{z}-\mathbf{x}]}{\partial z_\beta} \,d^3z\, d^3x$$

$$- \int_{\Omega'}\int_{\Omega'} \frac{\delta F}{\delta M_\gamma(\mathbf{x},t)} \frac{\delta G}{\delta M_\beta(\mathbf{z},t)} M_\beta(\mathbf{x},t) \frac{\partial \delta^3[\mathbf{x}-\mathbf{z}]}{\partial x_\gamma} \,d^3x\, d^3z \,, \tag{5.3-23}$$

and then applying the above procedure to each half yields

$$\{F,G\}_E^M = -\int_{\Omega'} \left[\frac{\delta F}{\delta M_\gamma(\mathbf{x},t)} \frac{\partial}{\partial x_\beta} \left(M_\gamma(\mathbf{x},t) \frac{\delta G}{\delta M_\beta(\mathbf{x},t)} \right) - \frac{\delta G}{\delta M_\gamma(\mathbf{x},t)} \frac{\partial}{\partial x_\beta} \left(M_\gamma(\mathbf{x},t) \frac{\delta F}{\delta M_\beta(\mathbf{x},t)} \right) \right] d^3x \,. \tag{5.3-24}$$

By summing the three subbrackets via (5.3-18), we thus arrive at the spatial counterpart of the Poisson bracket in the material description:

$$\begin{aligned} \{F,G\}_E = &-\int_{\Omega'} \left[\frac{\delta F}{\delta \rho} \nabla_\beta \left(\frac{\delta G}{\delta M_\beta} \rho \right) - \frac{\delta G}{\delta \rho} \nabla_\beta \left(\frac{\delta F}{\delta M_\beta} \rho \right) \right] d^3x \\ &-\int_{\Omega'} \left[\frac{\delta F}{\delta M_\gamma} \nabla_\beta \left(\frac{\delta G}{\delta M_\beta} M_\gamma \right) - \frac{\delta G}{\delta M_\gamma} \nabla_\beta \left(\frac{\delta F}{\delta M_\beta} M_\gamma \right) \right] d^3x \\ &-\int_{\Omega'} \left[\frac{\delta F}{\delta s} \nabla_\beta \left(\frac{\delta G}{\delta M_\beta} s \right) - \frac{\delta G}{\delta s} \nabla_\beta \left(\frac{\delta F}{\delta M_\beta} s \right) \right] d^3x \,, \end{aligned} \tag{5.3-25}$$

where we have replaced $\partial/\partial x_\beta$ with ∇_β. This bracket is bilinear, antisymmetric, and satisfies the Jacobi identity. When integrated by parts, it takes the form that was originally postulated for ideal fluid flow by Morrison and Greene [1980, p. 793].

Now that we have the proper form of the Poisson bracket in terms of spatial variables, we only need to specify the spatial counterpart of the Hamiltonian, (5.2-9a). If we consider the arbitrary spatial observation point, **x**, with volume element d^3x, which corresponds to the material coordinate, **Y**, with volume element d^3Y at time t, then we can transfer the Lagrangian of $(5.1\text{-}9)_1$ directly over to the spatial coordinate at this instant, and hence via the Legendre transformation of the integrand, we obtain

$$H[\rho,\mathbf{M},s] = \int_{\Omega'} \left[\frac{|M(\mathbf{x},t)|^2}{2\rho(\mathbf{x},t)} + \rho(\mathbf{x},t)\, e_p^v(\mathbf{x},t) + \rho(\mathbf{x},t)\, \hat{U}[\rho,s] \right] d^3x \,. \tag{5.3-26}$$

p. 103: in (5.3-26), $e_p^v(\mathbf{x},t)$ should be $e_p^v(\mathbf{x})$

Thus the Hamiltonian is written as the total system energy

$$H = \int_{\Omega'} [e_k + e_p + u] \, \mathrm{d}^3x \; , \tag{5.3-27}$$

where the integrand expresses the total energy at any spatial position and time t.

In order to derive the spatial form of the equations of motion, we only need to consider an arbitrary functional, F, of the dynamical variables ρ, $\mathbf{M}$, and s:

$$F[\rho,\mathbf{M},s] = \int_{\Omega'} f(\rho,\mathbf{M},s) \, \mathrm{d}^3x \; . \tag{5.3-28}$$

Hence we can write the dynamical equation for this functional through (5.2-7) as

$$\frac{\mathrm{d}F}{\mathrm{d}t} = \int_{\Omega'} \left[\frac{\delta F}{\delta \rho}\frac{\partial \rho}{\partial t} + \frac{\delta F}{\delta M_\alpha}\frac{\partial M_\alpha}{\partial t} + \frac{\delta F}{\delta s}\frac{\partial s}{\partial t} \right] \mathrm{d}^3x \; . \tag{5.3-29}$$

Analogously to (5.2-12), we can also write the dynamical equation for F in terms of the bracket (5.3-25) as

$$\begin{aligned} \frac{\mathrm{d}F}{\mathrm{d}t} = \{F,H\}_E = \\ & - \int_{\Omega'} \left[\frac{\delta F}{\delta \rho} \nabla_\beta \left(\frac{\delta H}{\delta M_\beta} \rho \right) - \frac{\delta H}{\delta \rho} \nabla_\beta \left(\frac{\delta F}{\delta M_\beta} \rho \right) \right] \mathrm{d}^3x \\ & - \int_{\Omega'} \left[\frac{\delta F}{\delta M_\alpha} \nabla_\beta \left(\frac{\delta H}{\delta M_\beta} M_\alpha \right) - \frac{\delta H}{\delta M_\alpha} \nabla_\beta \left(\frac{\delta F}{\delta M_\beta} M_\alpha \right) \right] \mathrm{d}^3x \\ & - \int_{\Omega'} \left[\frac{\delta F}{\delta s} \nabla_\beta \left(\frac{\delta H}{\delta M_\beta} s \right) - \frac{\delta H}{\delta s} \nabla_\beta \left(\frac{\delta F}{\delta M_\beta} s \right) \right] \mathrm{d}^3x \; . \end{aligned} \tag{5.3-30}$$

This relation can be rewritten by integrating the second term in each integral by parts (realizing that the surface contribution vanishes), then combining into one integral:

$$\begin{aligned} \frac{\mathrm{d}F}{\mathrm{d}t} = - \int_{\Omega'} \Bigg[& \frac{\delta F}{\delta \rho} \nabla_\beta \Big(\frac{\delta H}{\delta M_\beta} \rho \Big) + \frac{\delta F}{\delta s} \nabla_\beta \Big(\frac{\delta H}{\delta M_\beta} s \Big) + \frac{\delta F}{\delta M_\alpha} \Bigg(\rho \nabla_\alpha \frac{\delta H}{\delta \rho} \\ & + \nabla_\beta \Big(\frac{\delta H}{\delta M_\beta} M_\alpha \Big) + M_\beta \nabla_\alpha \frac{\delta H}{\delta M_\beta} + s \nabla_\alpha \frac{\delta H}{\delta s} \Bigg) \Bigg] \mathrm{d}^3x \; . \end{aligned} \tag{5.3-31}$$

The two versions of the dynamical equation for F, Eqs. (5.3-29,31), can now be compared directly to obtain the equations of motion for the spatial variables:

$$\frac{\partial \rho}{\partial t} = -\nabla_\beta\left(\frac{\delta H}{\delta M_\beta}\rho\right), \tag{5.3-32}$$

$$\frac{\partial s}{\partial t} = -\nabla_\beta\left(\frac{\delta H}{\delta M_\beta}s\right), \tag{5.3-33}$$

$$\frac{\partial M_\alpha}{\partial t} = -\left[\rho\nabla_\alpha\frac{\delta H}{\delta \rho} + \nabla_\beta\left(\frac{\delta H}{\delta M_\beta}M_\alpha\right) + M_\beta\nabla_\alpha\frac{\delta H}{\delta M_\beta} + s\nabla_\alpha\frac{\delta H}{\delta s}\right]. \tag{5.3-34}$$

From the Hamiltonian (5.3-26), the necessary functional derivatives can be evaluated as

$$\frac{\delta H}{\delta M_\gamma} = \frac{M_\gamma}{\rho} = v_\gamma, \tag{5.3-35}$$

$$\frac{\delta H}{\delta s} = \frac{\partial u}{\partial s} = T, \tag{5.3-36}$$

$$\frac{\delta H}{\delta \rho} = -\frac{M_\gamma M_\gamma}{2\rho^2} + e_p^v + \frac{\partial u}{\partial \rho}, \tag{5.3-37}$$

where we have used the thermodynamic definition of the temperature from chapter 4. These relations can be substituted into Eqs. (5.3-32,33,34) to give

$$\frac{\partial \rho}{\partial t} = -\nabla_\beta(v_\beta\rho), \tag{5.3-38}$$

$$\frac{\partial s}{\partial t} = -\nabla_\alpha(v_\alpha s), \tag{5.3-39}$$

$$\begin{aligned}\frac{\partial(\rho v_\alpha)}{\partial t} &= \frac{\rho}{2}\nabla_\alpha v^2 - \rho\nabla_\alpha e_p^v - \rho\nabla_\alpha\frac{\partial u}{\partial \rho} \\ &\quad - \nabla_\beta(v_\beta\rho v_\alpha) - \rho v_\beta\nabla_\alpha v_\beta - s\nabla_\alpha\frac{\partial u}{\partial s}.\end{aligned} \tag{5.3-40}$$

The first two relations represent the conservation equations for mass and entropy, respectively; however, the third relation will require a little manipulation before we recognize it in its usual form.

First we note that $-\rho v_\beta \nabla_\alpha v_\beta = -(\rho/2)\nabla_\alpha v^2$, so that the fifth term on the right-hand side of (5.3-40) cancels the first. Furthermore, we can rewrite the third and the last terms using the product rule of differentiation as

$$-\rho\nabla_\alpha\frac{\partial u}{\partial \rho} - s\nabla_\alpha\frac{\partial u}{\partial s} = -\nabla_\alpha\left(\rho\frac{\partial u}{\partial \rho} + s\frac{\partial u}{\partial s}\right) + \frac{\partial u}{\partial \rho}\nabla_\alpha\rho + \frac{\partial u}{\partial \rho}\nabla_\alpha s\ . \tag{5.3-41}$$

The last two terms in this expression are immediately recognized as the differentiation by parts of $\nabla_\alpha u$, so that the above relation can be expressed as

$$-\rho\nabla_\alpha\frac{\partial u}{\partial \rho} - s\nabla_\alpha\frac{\partial u}{\partial s} = -\nabla_\alpha\left(\rho\frac{\partial u}{\partial \rho} + s\frac{\partial u}{\partial s} - u\right)\ , \tag{5.3-42}$$

or, in terms of the pressure, via a variation of (4-3-24), we obtain

$$-\rho\nabla_\alpha\frac{\partial u}{\partial \rho} - s\nabla_\alpha\frac{\partial u}{\partial s} = -\nabla_\alpha p\ . \tag{5.3-43}$$

Furthermore, by making use of the mass continuity equation, (5.3-38), we can rewrite (5.3-40) as the usual momentum equation:

$$\rho\frac{\partial v_\alpha}{\partial t} = -\rho v_\beta\nabla_\beta v_\alpha - \rho\nabla_\alpha e_p^v - \nabla_\alpha p\ . \tag{5.3-44}$$

Since the instant t is completely arbitrary, Eqs. (5.3-38,39,44) represent the equations of motion in the spatial description at all times. Therefore, we have succeeded in deriving the dynamical equations for an ideal fluid in terms of spatial variables directly from the Poisson bracket of (5.3-25), and implicitly from the alternate material description.

We can also obtain the energy equations for the inviscid fluid quite easily. Since we know that

$$\frac{\partial u(\rho,s)}{\partial t} = \frac{\partial u}{\partial \rho}\frac{\partial \rho}{\partial t} + \frac{\partial u}{\partial s}\frac{\partial s}{\partial t}\ , \tag{5.3-45}$$

we can substitute into this expression Eqs. (5.3-38,39) to obtain

$$\begin{aligned}\frac{\partial u}{\partial t} &= -\frac{\partial u}{\partial \rho}\nabla_\alpha(\rho v_\alpha) - \frac{\partial u}{\partial s}\nabla_\alpha(s v_\alpha)\\
&= -\nabla_\alpha\left(\frac{\partial u}{\partial \rho}\rho v_\alpha + \frac{\partial u}{\partial s}s v_\alpha\right) + \rho v_\alpha\nabla_\alpha\frac{\partial u}{\partial \rho} + s v_\alpha\nabla_\alpha\frac{\partial u}{\partial s}\\
&= -\nabla_\alpha(v_\alpha u) - \nabla_\alpha(v_\alpha p) + v_\alpha\nabla_\alpha\left(\rho\frac{\partial u}{\partial \rho} + s\frac{\partial u}{\partial s}\right)\\
&\quad - v_\alpha\left(\frac{\partial u}{\partial \rho}\nabla_\alpha\rho + \frac{\partial u}{\partial s}\nabla_\alpha s\right)\end{aligned}$$

$$= -\nabla_\alpha(v_\alpha u) - \nabla_\alpha(v_\alpha p) + v_\alpha\nabla_\alpha(p+u) - v_\alpha\nabla_\alpha u$$
$$= -\nabla_\alpha(v_\alpha u) - p\nabla_\alpha v_\alpha \,, \tag{5.3-46}$$

which is the usual equation for the internal energy. Likewise, we find the kinetic energy equation by dotting **v** with (5.3-44):

$$\rho\frac{\partial(\tfrac{1}{2}v^2)}{\partial t} = -\rho v_\alpha\nabla_\alpha(\tfrac{1}{2}v^2) - v_\alpha\nabla_\alpha p - \rho v_\alpha\nabla_\alpha e_p^v \,, \tag{5.3-47}$$

and finally, the energy equation may be obtained by the addition of Eqs. (5.3-46,47):

$$\frac{\partial e}{\partial t} = -\nabla_\alpha(v_\alpha e) - \nabla_\alpha(p v_\alpha) - \rho v_\alpha\nabla_\alpha e_p^v \,, \tag{5.3-48}$$

where $e=e_k+u$.

It is also possible to arrive at the energy relations directly through the Poisson bracket. For example, the kinetic energy relation may be obtained using the functional for the total kinetic energy,

$$E_k = \int_{\Omega'} \frac{1}{2\rho} M_\alpha M_\alpha \, d^3x \,, \tag{5.3-49}$$

as the arbitrary functional F in the Hamiltonian equation (5.3-30),[3]

$$\frac{dE_k}{dt} = \{E_k, H\}_E \,. \tag{5.3-50}$$

Similarly, one may obtain equations for the vorticity and other quantities of interest, in particular, specific results such as Reynolds's transport theorem and Kelvin's circulation theorem may also be derived.

This completes the derivation of the spatial equations of motion for an ideal fluid, which represents an explicit derivation of the formulae from the "material" variational problem. As such, it does not require application of variational calculus to a modified "spatial" version of Hamilton's principle, which implies the incorporation of constraints through Lagrange multipliers. (See Herivel [1955], Lin [1963], and Seliger and Whitham [1965], for examples.) In fact, the constraints that one needs to incorporate into Hamilton's principle in spatial form are actually just the conservation equations one wishes to derive. Thus, the approach presented here is more pleasing aesthetically. In §5.6, we shall return to this problem in order to ascertain the relationship of the above theory with the thermodynamics of chapter 4.

[3]When one declares the arbitrary F to be a specific functional, say E_k, one is no longer working with a global equation, but rather a local one in which the mathematical nuances in the manipulations are slightly different. Proper care will illicit the proper responses, however.

EXAMPLE 5.1: In making the transition from the material description to the spatial description, it was necessary to consider the spatial variables as per-volume (density) quantities. This specification allows the simplest possible representation of the spatial bracket, (5.3-25), and thus indicates that the density variables are indeed the natural variables for the spatial description of fluid flows; however, all of the characteristic variables are directly transferable from one description to the other. In §5.1 we noted that the pressure was given by

$$p = -\left.\frac{\partial \hat{U}}{\partial(1/\rho)}\right|_{\hat{S}} , \tag{5.3-51}$$

while from Eqs. (5.3-42,43), as well as §4.3, we know that

$$p = \rho \left.\frac{\partial u}{\partial \rho}\right|_{s} + s \left.\frac{\partial u}{\partial s}\right|_{\rho} - u \; . \tag{5.3-52}$$

Show how to transfer between the two pressure relations by noting the proper functional dependence of the quantities u and $\hat{U}$.

Since $u = \rho\hat{U}$, we can rewrite (5.3-51) as

$$p = -\left.\frac{\partial(u/\rho)}{\partial(1/\rho)}\right|_{\hat{S}} = -u - \frac{1}{\rho}\left.\frac{\partial u}{\partial(1/\rho)}\right|_{\hat{S}} = -u + \rho \left.\frac{\partial u}{\partial \rho}\right|_{\hat{S}} , \tag{5.3-53}$$

but, remembering that $u = u(\rho, s) = u(\rho, \rho\hat{S})$, we can use the rule of implicit differentiation to obtain

$$\left.\frac{\partial u}{\partial \rho}\right|_{\hat{S}} = \left.\frac{\partial u}{\partial s}\right|_{s} + \left.\frac{\partial u}{\partial s}\right|_{\rho} \left.\frac{\partial s}{\partial \rho}\right|_{\hat{S}} = \left.\frac{\partial u}{\partial \rho}\right|_{s} + \left.\frac{\partial u}{\partial s}\right|_{\rho} \hat{S} \; . \tag{5.3-54}$$

Thus, substituting (5.3-54) into (5.3-53), we see that

$$p = \rho \left.\frac{\partial u}{\partial \rho}\right|_{s} + s \left.\frac{\partial u}{\partial s}\right|_{\rho} - u \; , \tag{5.3-55}$$

which is identical to (5.3-52). Hence we see that the equations of fluid flow amount to the consideration of a local thermodynamic equilibrium with each fluid particle viewed as a simple thermodynamic system. ■

5.4 Ideal Fluid Flow with Constraints: The Incompressible Fluid

In this section, the dynamical equations of an incompressible, ideal fluid are derived through the Hamiltonian/Poisson formalism. This is a very interesting case since the fluid maintains a constant density, thus inducing the divergence-free condition on the velocity. This condition arises from the mass continuity equation,

p. 108: in (5.3-54), $\left.\frac{\partial u}{\partial s}\right|_{s}$ should be $\left.\frac{\partial u}{\partial \rho}\right|_{s}$

$$\frac{\partial \rho}{\partial t} + \nabla_\alpha(\rho v_\alpha) = 0 \; , \qquad (5.4\text{-}1)$$

since ρ is constant in both space and time:

$$\nabla_\alpha v_\alpha = 0 \; . \qquad (5.4\text{-}2)$$

Consequently, for an incompressible fluid, the divergence-free velocity represents an internal constraint which must be imposed upon the system from the outset.

In dealing with an incompressible fluid, the assumption is made that the time scale of the density fluctuations is so large that the density is a frozen variable. Hence no thermodynamic equation of state exists which relates the volume to the temperature and pressure, since the volume is constant: V=constant$\neq V(T,p)$. As such, ρ, and likewise s, are no longer variables of interest in the incompressible problem formulation. Consequently, p is no longer a state variable, but rather it is an arbitrary scalar field which guarantees that the divergence-free condition is automatically satisfied throughout space and time. The only dynamical variable present in this case is **M**, which is, as before, $\mathbf{M}(\mathbf{x},t)=\rho\mathbf{v}(\mathbf{x},t)$, where ρ is now a system constant. Hence, the operating space for the incompressible, ideal fluid reduces to

$$P \equiv \{\mathbf{M}(\mathbf{x},t) = \rho\,\mathbf{v}(\mathbf{x},t) \in \mathbb{R}^3; \;\; \boldsymbol{\nabla}\cdot\mathbf{M} = 0 \text{ in } \Omega, \; \mathbf{n}\cdot\mathbf{M} = 0 \text{ on } \partial\Omega'\} \; . \qquad (5.4\text{-}3)$$

The *incompressible Euler equation* represents the equations of motion for the ideal fluid in the spatial description, where $\mathbf{M}=\rho\mathbf{v}$ is a member of the space P:

$$\rho\frac{\partial v_\alpha}{\partial t} + \rho v_\beta \nabla_\beta v_\alpha = -\nabla_\alpha p \quad \text{in } \Omega' \; . \qquad (5.4\text{-}4)$$

The pressure p in (5.4-4) is a scalar field which can be specified by solving a Poisson equation,

$$\nabla^2 p = -\rho\nabla_\alpha(v_\beta\nabla_\beta v_\alpha) \quad \text{in } \Omega' \; , \qquad (5.4\text{-}5)$$

obtained by taking the divergence of (5.4-4) and using the divergence-free property of **v**. The boundary condition for p, defined as $\partial p/\partial \nu$ by tradition, is evaluated as

$$\frac{\partial p}{\partial \nu} \equiv n_\alpha \nabla_\alpha p = -\rho n_\alpha v_\beta \nabla_\beta v_\alpha \quad \text{on } \partial\Omega' \; , \qquad (5.4\text{-}6)$$

which is obtained by dotting the unit-normal vector, **n**, with (5.4-4) and then using the no-penetration boundary condition for **v**.

The above system of equations provides a well-posed set which specifies the flow of an incompressible, inviscid fluid. Eq. (5.4-4) is obtained in standard hydrodynamics textbooks by applying Newton's second law of motion to a fluid element subjected to only a normal pressure. The same system can also be expressed in the Hamiltonian/ Poisson formalism with no loss of rigor. Before the Hamiltonian/Poisson formalism can be applied to the incompressible fluid, however, two crucial properties of the operating space P, (5.4-3), must be recognized, as revealed by Marsden and Weinstein [1983, pp. 310,11], Holm *et al.* [1985, pp. 17,8], and in the following form by Beris and Edwards [1990a, pp. 60,1].

1. For every continuous, arbitrary function $f=f(\mathbf{M})$ and for every $\mathbf{M}\in P$,

$$\begin{aligned}\int_{\Omega'} \boldsymbol{\nabla} f\cdot\mathbf{M}\,\mathrm{d}^3x &= \int_{\Omega'} \boldsymbol{\nabla}\cdot(f\mathbf{M})\,\mathrm{d}^3x - \int_{\Omega'} f\,\boldsymbol{\nabla}\cdot\mathbf{M}\,\mathrm{d}^3x \\ &= \int_{\partial\Omega'} f\mathbf{M}\cdot\mathbf{n}\,\mathrm{d}^2x - 0 = 0\ ,\end{aligned} \tag{5.4-7}$$

where d^2x represents a surface element on $\partial\Omega$.

2. Every arbitrary vector field, $\mathbf{a}\in\mathbb{R}^3$, can be decomposed into two parts according to the Weyl/Hodge theorem [Ebin and Marsden, 1970, p. 121]:

$$\mathbf{a} = \mathbf{I}(\mathbf{a}) + \mathbf{K}(\mathbf{a})\ , \tag{5.4-8}$$

where $\mathbf{I}(\mathbf{a})\in P$ is thus a projection operator to the P space and $\mathbf{K}(\mathbf{a})$ is orthogonal to P. This decomposition is unique, with the orthogonality being defined through the use of the inner product, defined as

$$<\mathbf{a},\mathbf{b}> \equiv \int_{\Omega'} \mathbf{a}\cdot\mathbf{b}\,\mathrm{d}^3x\ . \tag{5.4-9}$$

Now one can demonstrate that the orthogonal-to-P component $\mathbf{K}(\mathbf{a})$ can be represented by the gradient of some, as yet undetermined, scalar function p. This choice certainly satisfies the orthogonality requirement according to (5.4-7); however, to be a proper definition for $\mathbf{K}(\mathbf{a})$, the corresponding $\mathbf{I}(\mathbf{a})$,

$$\mathbf{I}(\mathbf{a}) \equiv \mathbf{a} - \boldsymbol{\nabla} p\ , \tag{5.4-10}$$

has to belong to the P space, i.e., it has to be divergence free in Ω, and it has to have a zero normal component on $\partial\Omega$:

$$\boldsymbol{\nabla}\cdot\mathbf{I}(\mathbf{a}) = \boldsymbol{\nabla}\cdot\mathbf{a} - \boldsymbol{\nabla}\cdot\boldsymbol{\nabla} p = 0 \ \Rightarrow\ \nabla^2 p = \boldsymbol{\nabla}\cdot\mathbf{a} \quad \text{in } \Omega'\ , \tag{5.4-11a}$$

$$\mathbf{n}\cdot\mathbf{I}(\mathbf{a}) = \mathbf{n}\cdot\mathbf{a} - \frac{\partial p}{\partial n} = 0 \ \Rightarrow\ \frac{\partial p}{\partial n} = \mathbf{n}\cdot\mathbf{a} \quad \text{on } \partial\Omega'\ . \tag{5.4-11b}$$

Eqs. (5.4-11) constitute an elliptic (Poisson) equation, subject to flux (von Neumann) boundary conditions, which can be used to specify the scalar field p for a given vector field $\mathbf{a}$. Thus, the definition of the projection operator through (5.4-10) is completed.

The decomposition of Eq. (5.4-8) can be used together with the orthogonality property of $\mathbf{K}(\mathbf{a})= \nabla p$ to the P space in order to rewrite the inner product between the arbitrary vector field $\mathbf{a}\in\mathbb{R}^3$ and the vector field $\mathbf{M}\in P$ as

$$<\mathbf{a},\mathbf{M}> = <\mathbf{I}(\mathbf{a}),\mathbf{M}> , \quad \mathbf{M}\in P , \tag{5.4-12}$$

since $<\nabla p,\mathbf{M}>=0$ from (5.4-7). The significance of the substitution implied by (5.4-12) is that now an orthogonality condition expressed as

$$<\mathbf{a},\mathbf{M}> = 0 , \quad \text{for every } \mathbf{M}\in P , \tag{5.4-13}$$

implies that

$$\mathbf{I}(\mathbf{a}) = 0 , \tag{5.4-14}$$

since by (5.4-13) $\mathbf{a}$ is orthogonal to P, i.e., if $\mathbf{a}$ is orthogonal to P, it can have no component which is an element of P. Accordingly, if Eq. (5.4-13) is correct for every $\mathbf{M}\in P$, then (5.4-12) implies that

$$<\mathbf{I}(\mathbf{a}),\mathbf{M}> = 0 , \quad \text{for every } \mathbf{M}\in P , \tag{5.4-15}$$

and therefore

$$<\mathbf{I}(\mathbf{a}),\mathbf{I}(\mathbf{a})> = \int_{\Omega'} \mathbf{I}(\mathbf{a})\cdot\mathbf{I}(\mathbf{a})\, d^3x = 0 , \quad \text{Q.E.D.} \tag{5.4-16}$$

Given the two above-mentioned properties of the P space, the functional derivative can now be defined for an arbitrary functional,

$$F[\mathbf{M}] \equiv \int_{\Omega'} f(\mathbf{M})\, d^3x , \tag{5.4-17}$$

as $\delta F/\delta\mathbf{M}\in P$, such that

$$\left|F(\mathbf{M}+\delta\mathbf{M}) - F(\mathbf{M})\right|_{\delta\mathbf{M}\to 0} = <\frac{\delta F}{\delta\mathbf{M}},\delta\mathbf{M}> , \quad \delta\mathbf{M}\in P ; \tag{5.4-18a}$$

however,

$$\left|F(\mathbf{M}+\delta\mathbf{M}) - F(\mathbf{M})\right|_{\delta\mathbf{M}\to 0} = <\frac{\partial f}{\partial\mathbf{M}},\delta\mathbf{M}> , \tag{5.4-18b}$$

which, according to (5.4-12), is equivalent to

$$\left|F(\mathbf{M}+\delta\mathbf{M}) - F(\mathbf{M})\right|_{\delta\mathbf{M}\to 0} = <\mathbf{I}(\frac{\partial f}{\partial\mathbf{M}}),\delta\mathbf{M}> . \tag{5.4-18c}$$

To be a proper inner product, both variables must belong to the same operating space, in this case, P. In (5.4-18a), both $\delta F/\delta \mathbf{M}$ and $\delta \mathbf{M} \in P$, but not in (5.4-18b); however, this problem in the latter equation has been remedied in (5.4-18c). Thus, by comparing Eqs. (5.4-18a,c), we see immediately that

$$\frac{\delta F}{\delta \mathbf{M}} = \mathbf{I}(\frac{\partial f}{\partial \mathbf{M}}) \in P \ . \tag{5.4-19}$$

This expression holds in cases where other constraints in addition to, or instead of, the incompressibility constraint hold, provided that the expression for the projection operator, (5.4-10,11), is suitably altered.

The spatial Poisson bracket is as given by (5.3-25), where ρ and $s=\rho\hat{S}$ are constants, which implies that $H=H_m[\mathbf{M}]$ is only a functional of the momentum density (and therefore, so are F and G). Hence only $\delta F/\delta \mathbf{M}$ and $\delta G/\delta \mathbf{M}$ are non-vanishing, reducing the Poisson bracket of (5.3-25) to

$$\{F,G\}_E = -\int_{\Omega'} \left[\frac{\delta F}{\delta M_\alpha}\nabla_\beta\left(\frac{\delta G}{\delta M_\beta}M_\alpha\right) - \frac{\delta G}{\delta M_\alpha}\nabla_\beta\left(\frac{\delta F}{\delta M_\beta}M_\alpha\right)\right] \mathrm{d}^3x \ . \tag{5.4-20}$$

This bracket seems to have originated in another form from Arnold [1965, p. 1003; and 1966, p. 31], who was the first to express the incompressible Euler equation in Hamiltonian form (see also Arnold [1978, App. 2]). The Hamiltonian for this system, consisting solely of the mechanical (in this case, kinetic) energy, is given as

$$H_m[\mathbf{M}] = \int_{\Omega'} \frac{1}{2\rho} M_\gamma M_\gamma \ \mathrm{d}^3x \ , \tag{5.4-21}$$

so that the Hamiltonian/Poisson formalism is provided by an equivalent expression to (5.3-29,30):

$$\frac{\mathrm{d}F}{\mathrm{d}t} = \{F,H_m\}_E = <\frac{\delta F}{\delta \mathbf{M}}, \frac{\partial \mathbf{M}}{\partial t}> \ . \tag{5.4-22}$$

Substituting $\delta H_m/\delta \mathbf{M}$ into the Poisson bracket, (5.4-20), and integration by parts of the second term in the resulting integral gives

$$\begin{aligned} \frac{\mathrm{d}F}{\mathrm{d}t} &= -\int_{\Omega'} \frac{\delta F}{\delta M_\alpha}\left[I^{(\alpha)}\{\nabla_\beta(\rho v_\beta v_\alpha) + \rho v_\beta \nabla_\alpha v_\beta\}\right] \mathrm{d}^3x \\ &= -\int_{\Omega'} \frac{\delta F}{\delta M_\alpha}\left[I^{(\alpha)}\{\nabla_\beta(\rho v_\beta v_\alpha) + \tfrac{1}{2}\nabla_\alpha(\rho v_\beta v_\beta)\}\right] \mathrm{d}^3x \ . \end{aligned} \tag{5.4-23}$$

The second term in the integrand of (5.4-23) obviously vanishes since $\delta F/\delta \mathbf{M}$ is divergence free, i.e., integrating this term by parts gives

$\nabla_\alpha(\delta F/\delta M_\alpha)$, which is identically zero since $\delta F/\delta \mathbf{M} \in P$. Consequently, by comparing Eqs. (5.4-22,23), we have

$$\frac{\partial M_\alpha}{\partial t} = I^{(\alpha)}\{-\nabla_\beta(\rho v_\beta v_\alpha)\} = -\nabla_\beta(\rho v_\beta v_\alpha) - \nabla_\alpha p \ , \tag{5.4-24}$$

which is the incompressible Euler equation. Note that the pressure appears naturally in the equation as a consequence of the incompressibility condition.

5.5 Nonlinear Elasticity

In complex media such as viscoelastic fluids, polymeric liquids, liquid crystals, etc., one needs to describe not only the simple hydrodynamics of the fluid, but the internal microstructure as well. This structure evolves according to the kinematic history, and influences the internal stress of the material in addition to the overall hydrodynamic behavior. Here we use a second-rank deformation tensor to describe the internal structure of the elastic medium, which is a properly invariant tensor field, $\mathbf{c}(\mathbf{F},t)$. In general, the tensor field $\mathbf{c}(\mathbf{F},t)$ can be any arbitrary measure of the medium's structure; for instance, in chapter 8 we associate $\mathbf{c}$ with a weighted average over the usual orientational distribution function of kinetic theory. In this section, we define $c_{\alpha\beta} \equiv F_{\alpha\gamma}F_{\beta\gamma}$ and wish to obtain the equations of motion of the ideal elastic medium in terms of this quantity.

5.5.1 *The material description of nonlinear elasticity*

The Hamiltonian of the elastic system is written as

$$H[\mathbf{Y},\Pi] \equiv \int_\Omega \left[\frac{1}{2\rho_o}\Pi\cdot\Pi + \rho_o e_p^v(\mathbf{Y}) + \rho_o \hat{U}[\rho(\mathbf{F},t),s_o(\mathbf{r}),\mathbf{c}(\mathbf{F},t)]\right] d^3r \ , \tag{5.5-1}$$

where the internal energy now depends upon the structure of the medium as well as the density and entropy. Hence we can use the Poisson bracket of (5.2-13) in the dynamical relation (5.2-12) to derive the equations of motion in the material description. Thus we have

$$\dot{Y}_\alpha(\mathbf{r},t) = \frac{\delta H[\mathbf{Y},\Pi]}{\delta \Pi_\alpha(\mathbf{r},t)} = \frac{\Pi_\alpha(\mathbf{r},t)}{\rho_o(\mathbf{r})} \ , \tag{5.5-2a}$$

$$\dot{\Pi}_\alpha(\mathbf{r},t) = -\frac{\delta H[\mathbf{Y},\Pi]}{\delta Y_\alpha(\mathbf{r},t)} = -\rho_o\frac{\partial e_p^v}{\partial Y_\alpha} + \frac{\partial}{\partial r_\beta}\left(\rho_o\frac{\partial \hat{U}}{\partial F_{\alpha\beta}}\right) , \tag{5.5-2b}$$

where the second of these relations can easily be rewritten using similar arguments to those of §5.1 as

$$\rho\frac{\partial^2 Y_\alpha}{\partial t^2} = -\rho\frac{\partial e_p^v}{\partial Y_\alpha} - \frac{\partial p}{\partial Y_\alpha} + \frac{\partial}{\partial Y_\gamma}\left(2\rho c_{\gamma\varepsilon}\frac{\partial \hat{U}}{\partial c_{\alpha\varepsilon}}\right) . \tag{5.5-3}$$

This is the usual equation of motion for an ideal elastic fluid, with the third term on the right-hand side containing the elastic stress tensor.

5.5.2 *Spatial bracket derivation*

We may also now derive the spatial version of the Poisson bracket directly from the material version, (5.2-13), as was accomplished in §5.3 for the simple hydrodynamic fluid. In this case, little changes except that we have an additional spatial variable, $\mathbf{C}(\mathbf{x},t)\equiv\rho(\mathbf{x},t)\mathbf{c}(\mathbf{x},t)$, and therefore an additional transformation relation to (5.3-4,5,6),

$$C_{\alpha\beta}(\mathbf{x},t) = \int_\Omega \rho_o(\mathbf{r})\, c_{\alpha\beta}(\mathbf{F}(\mathbf{r},t),t)\; \delta^3[\mathbf{Y}-\mathbf{x}]\; d^3r . \tag{5.5-4}$$

Hence we need to redefine the operating space to include the new variable, $\mathbf{C}$, as

$$P \equiv \begin{cases} \rho(\mathbf{x},t) \in \mathbb{R}^+ \\ \mathbf{M}(\mathbf{x},t) \in \mathbb{R}^3, \quad \mathbf{n}\cdot\mathbf{M} = 0 \;\text{ on } \partial\Omega' \\ s(\mathbf{x},t) \in \mathbb{R} \\ \mathbf{C}(\mathbf{x},t) \in \mathbb{R}^3\times\mathbb{R}^{3T} \end{cases} , \tag{5.5-5}$$

which also includes the appropriate initial and boundary conditions on ρ, $\mathbf{M}$, s, and $\mathbf{C}$.

The derivatives of the conformation tensor,

$$\frac{\delta C_{\alpha\beta}(\mathbf{x},t)}{\delta Y_\gamma(\mathbf{r},t)} = \rho_o c_{\alpha\beta}\frac{\partial\delta^3[\mathbf{Y}-\mathbf{x}]}{\partial Y_\gamma(\mathbf{r},t)} - \frac{\partial}{\partial r_\eta}\left(\rho_o\delta^3[\mathbf{Y}-\mathbf{x}]\frac{\partial c_{\alpha\beta}}{\partial F_{\gamma\eta}}\right) , \tag{5.5-6a}$$

$$\frac{\delta C_{\alpha\beta}(\mathbf{x},t)}{\delta \Pi_\gamma(\mathbf{r},t)} = 0 , \tag{5.5-6b}$$

the former being calculated via (5.2-6), allow one to consider the arbitrary functional as $F=F[\rho(\mathbf{x},t),\mathbf{M}(\mathbf{x},t),s(\mathbf{x},t),\mathbf{C}(\mathbf{x},t)]$, such that

$$\frac{\delta F}{\delta Y_\alpha} = \int_{\Omega'} \left[\frac{\delta F}{\delta \rho(\mathbf{x},t)} \frac{\delta \rho(\mathbf{x},t)}{\delta Y_\alpha(\mathbf{r},t)} + \frac{\delta F}{\delta M_\beta(\mathbf{x},t)} \frac{\delta M_\beta(\mathbf{x},t)}{\delta Y_\alpha(\mathbf{r},t)} \right. \tag{5.5-7}$$

$$\left. + \frac{\delta F}{\delta s(\mathbf{x},t)} \frac{\delta s(\mathbf{x},t)}{\delta Y_\alpha(\mathbf{r},t)} + \frac{\delta F}{\delta C_{\gamma\beta}(\mathbf{x},t)} \frac{\delta C_{\gamma\beta}(\mathbf{x},t)}{\delta Y_\alpha(\mathbf{r},t)} \right] d^3x \, ,$$

$$\frac{\delta F}{\delta \Pi_\alpha} = \int_{\Omega'} \left[\frac{\delta F}{\delta \rho(\mathbf{x},t)} \frac{\delta \rho(\mathbf{x},t)}{\delta \Pi_\alpha(\mathbf{r},t)} + \frac{\delta F}{\delta M_\beta(\mathbf{x},t)} \frac{\delta M_\beta(\mathbf{x},t)}{\delta \Pi_\alpha(\mathbf{r},t)} \right. \tag{5.5-8}$$

$$\left. + \frac{\delta F}{\delta s(\mathbf{x},t)} \frac{\delta s(\mathbf{x},t)}{\delta \Pi_\alpha(\mathbf{r},t)} + \frac{\delta F}{\delta C_{\gamma\beta}(\mathbf{x},t)} \frac{\delta C_{\gamma\beta}(\mathbf{x},t)}{\delta \Pi_\alpha(\mathbf{r},t)} \right] d^3x \, .$$

Hence we obtain the additional non-zero subbracket over (5.3-10)

$$+ \int_{\Omega'} \int_{\Omega'} \left[\frac{\delta F}{\delta C_{\alpha\beta}(\mathbf{x},t)} \frac{\delta G}{\delta M_\gamma(\mathbf{z},t)} - \frac{\delta G}{\delta C_{\alpha\beta}(\mathbf{x},t)} \frac{\delta F}{\delta M_\gamma(\mathbf{z},t)} \right] \tag{5.5-9}$$

$$\times \, \{C_{\alpha\beta}(\mathbf{x},t), M_\gamma(\mathbf{z},t)\}_L \; d^3z \, d^3x \, ,$$

from which we calculate

$$\{C_{\alpha\beta}(\mathbf{x},t), M_\gamma(\mathbf{z},t)\}_L =$$

$$\int_\Omega \left[\rho_o \, c_{\alpha\beta} \frac{\partial \delta^3[\mathbf{Y}-\mathbf{x}]}{\partial Y_\gamma(\mathbf{r},t)} - \frac{\partial}{\partial r_\eta} \left(\rho_o \, \delta^3[\mathbf{Y}-\mathbf{x}] \frac{\partial c_{\alpha\beta}}{\partial F_{\gamma\eta}} \right) \right] \delta^3[\mathbf{Y}-\mathbf{z}] \; d^3r$$

$$= \int_{\Omega'} [\rho c_{\alpha\beta} \frac{\partial \delta^3[\mathbf{Y}-\mathbf{x}]}{\partial Y_\gamma(\mathbf{r},t)} - \frac{1}{J} \frac{\partial}{\partial r_\eta} (\rho J \delta^3[\mathbf{Y}-\mathbf{x}]$$

$$\times \, \{F_{\alpha\eta}\delta_{\beta\gamma} + F_{\beta\eta}\delta_{\alpha\gamma}\})] \; \delta^3[\mathbf{Y}-\mathbf{z}] \; d^3Y$$

$$= \int_{\Omega'} [\rho c_{\alpha\beta} \frac{\partial \delta^3[\mathbf{Y}-\mathbf{x}]}{\partial Y_\gamma} - \frac{1}{J} \frac{\partial Y_\varepsilon}{\partial r_\eta} \frac{\partial}{\partial Y_\varepsilon} (\rho J \delta^3[\mathbf{Y}-\mathbf{x}]$$

$$\times \, \{F_{\alpha\eta}\delta_{\beta\gamma} + F_{\beta\eta}\delta_{\alpha\gamma}\})] \; \delta^3[\mathbf{Y}-\mathbf{z}] \; d^3Y$$

$$= \int_{\Omega'} [\rho c_{\alpha\beta} \frac{\partial \delta^3[\mathbf{Y}-\mathbf{x}]}{\partial Y_\gamma} - \frac{\partial}{\partial Y_\varepsilon}(\rho F_{\varepsilon\eta}\, \delta^3[\mathbf{Y}-\mathbf{x}]$$

$$\times \{F_{\alpha\eta}\delta_{\beta\gamma} + F_{\beta\eta}\delta_{\alpha\gamma}\})]\, \delta^3[\mathbf{Y}-\mathbf{z}]\, \mathrm{d}^3Y$$

$$= C_{\alpha\beta}(\mathbf{z},t) \frac{\partial \delta^3[\mathbf{z}-\mathbf{x}]}{\partial z_\gamma} \tag{5.5-10}$$

$$- \frac{\partial}{\partial z_\nu}(\delta^3[\mathbf{z}-\mathbf{x}]\, \{C_{\nu\alpha}(\mathbf{z},t)\delta_{\beta\gamma} + C_{\nu\beta}(\mathbf{z},t)\delta_{\alpha\gamma}\}) \, ,$$

where the identity $(5.1\text{-}4)_2$ was used. Substitution of (5.5-10) into (5.5-9) then yields the spatial Poisson bracket as

$$\{F,G\}_E = - \int_{\Omega'} \left[\frac{\delta F}{\delta \rho} \nabla_\beta \left(\frac{\delta G}{\delta M_\beta} \rho \right) - \frac{\delta G}{\delta \rho} \nabla_\beta \left(\frac{\delta F}{\delta M_\beta} \rho \right) \right] \mathrm{d}^3x$$

$$- \int_{\Omega'} \left[\frac{\delta F}{\delta M_\gamma} \nabla_\beta \left(\frac{\delta G}{\delta M_\beta} M_\gamma \right) - \frac{\delta G}{\delta M_\gamma} \nabla_\beta \left(\frac{\delta F}{\delta M_\beta} M_\gamma \right) \right] \mathrm{d}^3x$$

$$- \int_{\Omega'} \left[\frac{\delta F}{\delta s} \nabla_\beta \left(\frac{\delta G}{\delta M_\beta} s \right) - \frac{\delta G}{\delta s} \nabla_\beta \left(\frac{\delta F}{\delta M_\beta} s \right) \right] \mathrm{d}^3x$$

$$- \int_{\Omega'} \left[\frac{\delta F}{\delta C_{\alpha\beta}} \nabla_\gamma \left(\frac{\delta G}{\delta M_\gamma} C_{\alpha\beta} \right) - \frac{\delta G}{\delta C_{\alpha\beta}} \nabla_\gamma \left(\frac{\delta F}{\delta M_\gamma} C_{\alpha\beta} \right) \right] \mathrm{d}^3x$$

$$- \int_{\Omega'} C_{\gamma\alpha} \left[\frac{\delta G}{\delta C_{\alpha\beta}} \nabla_\gamma \left(\frac{\delta F}{\delta M_\beta} \right) - \frac{\delta F}{\delta C_{\alpha\beta}} \nabla_\gamma \left(\frac{\delta G}{\delta M_\beta} \right) \right] \mathrm{d}^3x$$

$$- \int_{\Omega'} C_{\gamma\beta} \left[\frac{\delta G}{\delta C_{\alpha\beta}} \nabla_\gamma \left(\frac{\delta F}{\delta M_\alpha} \right) - \frac{\delta F}{\delta C_{\alpha\beta}} \nabla_\gamma \left(\frac{\delta G}{\delta M_\alpha} \right) \right] \mathrm{d}^3x \, . \tag{5.5-11}$$

The final three integrals in this expression represent the effects of elasticity in the medium, and originate in general form from Marsden *et al.* [1984, p. 171]. The above simplified form of the derivation was put forth by Edwards and Beris [1991b, §2], who obtained directly the bracket of (5.5-11).

With the Hamiltonian

$$H[\rho,\mathbf{M},s,\mathbf{C}] = \int_{\Omega'} \left[\frac{|M(\mathbf{x},t)|^2}{2\rho(\mathbf{x},t)} + \rho(\mathbf{x},t)\, e_p^v(\mathbf{x},t) + \rho(\mathbf{x},t)\,\hat{U}[\rho,s,\mathbf{C}] \right] d^3x \ , \tag{5.5-12}$$

it is now possible to derive the spatial equivalent of the equations of motion. Again following §5.3, it is an easy matter to show that

$$\frac{\partial \rho}{\partial t} = - \nabla_\beta(v_\beta \rho) \ , \tag{5.5-13a}$$

$$\frac{\partial s}{\partial t} = - \nabla_\beta (v_\beta s) \ , \tag{5.5-13b}$$

$$\rho \frac{\partial v_\alpha}{\partial t} = - \rho v_\beta \nabla_\beta v_\alpha - \rho \nabla_\alpha e_p^v - \nabla_\alpha p + 2\nabla_\gamma \left(C_{\gamma\beta} \frac{\partial u}{\partial C_{\alpha\beta}} \right) , \tag{5.5-13c}$$

$$\frac{\partial C_{\alpha\beta}}{\partial t} = - \nabla_\gamma(v_\gamma C_{\alpha\beta}) + C_{\gamma\alpha}\nabla_\gamma v_\beta + C_{\gamma\beta}\nabla_\gamma v_\alpha \ , \tag{5.5-13d}$$

where p is defined analogously to (4.4-6) as

$$p \equiv \rho \frac{\partial u}{\partial \rho} + s \frac{\partial u}{\partial s} + C_{\alpha\beta} \frac{\partial u}{\partial C_{\alpha\beta}} - u \ . \tag{5.5-13e}$$

The elasticity of the medium defines an extra stress in the momentum equation, expressed by the last term in (5.5-13c). The structure evolution equation, (5.5-13d), is a materially objective time derivative, as derived by Oldroyd [1950, p. 538]—Oldroyd's B (contravariant) derivative—and is commonly called the upper-convected derivative. We obtain the Oldroyd B derivative because **c** transforms as an absolute, second-rank, contravariant tensor; i.e., by considering a new coordinate $\bar{\mathbf{Y}}$, which may be related to the old coordinate system by a specific functional relationship $\bar{\mathbf{Y}}=\bar{\mathbf{Y}}(\mathbf{Y})$, we see that

$$\bar{c}_{\alpha\beta} \equiv \frac{\partial \bar{Y}_\alpha}{\partial r_\gamma}\frac{\partial \bar{Y}_\beta}{\partial r_\gamma} = \frac{\partial \bar{Y}_\alpha}{\partial Y_\varepsilon}\frac{\partial Y_\varepsilon}{\partial r_\gamma}\frac{\partial \bar{Y}_\beta}{\partial Y_\eta}\frac{\partial Y_\eta}{\partial r_\gamma} = \frac{\partial \bar{Y}_\alpha}{\partial Y_\varepsilon}\frac{\partial \bar{Y}_\beta}{\partial Y_\eta} c_{\varepsilon\eta} \ , \tag{5.5-14}$$

which is indeed a contravariant transformation. By starting with a second-rank, covariant or mixed tensor, we could just as easily have obtained the other Oldroyd derivatives as well (see Appendix C).

As before, the energy equations may be written as well. These are

$$\frac{\partial u}{\partial t} = -\nabla_\alpha(v_\alpha u) - p\nabla_\alpha v_\alpha + 2C_{\gamma\alpha}\frac{\partial u}{\partial C_{\alpha\beta}}\nabla_\gamma v_\beta \,, \tag{5.5-15a}$$

$$\rho\frac{\partial(\frac{1}{2}v^2)}{\partial t} = -\rho v_\alpha\nabla_\alpha(\tfrac{1}{2}v^2) + v_\alpha\nabla_\alpha p - \rho v_\alpha\nabla_\alpha e_p^v + 2v_\alpha\nabla_\gamma\left(C_{\gamma\beta}\frac{\partial u}{\partial C_{\alpha\beta}}\right), \tag{5.5-15b}$$

$$\frac{\partial e}{\partial t} = -\nabla_\alpha(v_\alpha e) - \nabla_\alpha(pv_\alpha) - \rho v_\alpha\nabla_\alpha e_p^v + 2\nabla_\gamma\left(v_\alpha C_{\gamma\beta}\frac{\partial u}{\partial C_{\alpha\beta}}\right). \tag{5.5-15c}$$

5.5.3 *Spatial bracket in terms of a tensor with unit trace*

In the previous subsection, the spatial version of the Poisson bracket for nonlinear elasticity was derived in terms of the second-rank deformation tensor **C**. This was a quite general description for an elastic medium since it specifies not only the orientation of the microstructure but its extension as well. Still, when one deals with anisotropic media, as in Part II, which are envisioned to consist of rigid, inextensible molecules, only the orientation is important and the second-rank deformation tensor is constrained to be a conformation tensor with unit trace. This introduces a mathematical complexity into the derivation of the equations of motion; particularly into the spatial version of the Poisson bracket, (5.5-11). Specifically, the problem is that the bracket of (5.5-11) is only applicable for a system free of constraints, for which it was derived, and one must be more resourceful when dealing with constrained systems, as we saw in §5.4. Here we wish to transform the bracket of (5.5-11) in terms of **C** into its equivalent expression in terms of a second-rank tensor of unit trace, **m**.

In order to proceed with this objective, a projection mapping, P_m, is defined which projects a tensor of arbitrary trace to one with unit trace. This mapping is given as [Edwards *et al.*, 1990a, p. 57]

$$P_m[C_{\alpha\beta}] = \frac{C_{\alpha\beta}}{\mathrm{tr}\,\mathbf{C}} \equiv m_{\alpha\beta} \,, \tag{5.5-16a}$$

which implies that

$$\nabla_\eta\left(\frac{C_{\alpha\beta}}{\mathrm{tr}\,\mathbf{C}}\right) = \nabla_\eta m_{\alpha\beta} \,, \tag{5.5-16b}$$

and we subsequently restrict our arbitrary functionals F and G to those which depend on **m** (and $\nabla\mathbf{m}$) only through their dependence on **C** (and $\nabla\mathbf{C}$): $F=F[\rho,\mathbf{M},s,\mathbf{m}]$ and $G=G[\rho,\mathbf{M},s,\mathbf{m}]$. Hence we can define the total time derivative of F through the sum of four inner products once again as

$$\frac{dF}{dt} = <\frac{\delta F}{\delta\rho},\frac{\partial\rho}{\partial t}> + <\frac{\delta F}{\delta\mathbf{M}},\frac{\partial\mathbf{M}}{\partial t}> + <\frac{\delta F}{\delta s},\frac{\partial s}{\partial t}> + <\frac{\delta F}{\delta\mathbf{m}},\frac{\partial\mathbf{m}}{\partial t}> . \tag{5.5-17}$$

Since both subjects of the inner product must belong to the same operating space, and therefore since $\partial(\mathrm{tr}\,\mathbf{m})/\partial t=0$, we require that $\mathrm{tr}\,(\delta F/\delta\mathbf{m})=0$ as well. (Actually, from physical considerations, **m** must also be symmetric, which requires that $\delta F/\delta\mathbf{m}$ be symmetric as well; but this amounts to only a trivial modification of the following analysis which we neglect for simplicity.) Hence we define $\delta F/\delta\mathbf{m}$ analogously to (5.2-6) as

$$\frac{\delta F}{\delta m_{\alpha\beta}} \equiv \frac{\partial f}{\partial m_{\alpha\beta}} - \frac{1}{3}\frac{\partial f}{\partial m_{\zeta\zeta}}\delta_{\alpha\beta} - \nabla_\eta\left(\frac{\partial f}{\partial(\nabla_\eta m_{\alpha\beta})}\right) + \frac{1}{3}\nabla_\eta\left(\frac{\partial f}{\partial(\nabla_\eta m_{\zeta\zeta})}\right)\delta_{\alpha\beta} , \tag{5.5-18}$$

which duly ensures that $\delta F/\delta\mathbf{m}$ is traceless. Of course, $\delta F/\delta\mathbf{C}$ has no restrictions placed upon it, so that it satisfies (5.2-6) directly:

$$\frac{\delta F}{\delta C_{\alpha\beta}} \equiv \frac{\partial f}{\partial C_{\alpha\beta}} - \nabla_\eta\left(\frac{\partial f}{\partial(\nabla_\eta C_{\alpha\beta})}\right) . \tag{5.5-19}$$

Now considering F to depend on **m** (and $\nabla\mathbf{m}$) only through its dependence on **C** (and $\nabla\mathbf{C}$), we can use the chain rule of differentiation to find that

$$\begin{aligned}\frac{\delta F}{\delta C_{\alpha\beta}} \equiv{}& \frac{\partial f}{\partial m_{\gamma\varepsilon}}\frac{\partial m_{\gamma\varepsilon}}{\partial C_{\alpha\beta}} + \frac{\partial f}{\partial(\nabla_\rho m_{\gamma\varepsilon})}\frac{\partial(\nabla_\rho m_{\gamma\varepsilon})}{\partial C_{\alpha\beta}} \\ &+ \frac{\partial f}{\partial m_{\gamma\varepsilon}}\frac{\partial m_{\gamma\varepsilon}}{\partial(\nabla_\eta C_{\alpha\beta})} + \frac{\partial f}{\partial(\nabla_\rho m_{\gamma\varepsilon})}\frac{\partial(\nabla_\rho m_{\gamma\varepsilon})}{\partial(\nabla_\eta C_{\alpha\beta})} .\end{aligned} \tag{5.5-20}$$

According to (5.5-16), it is straightforward, yet tedious, to show that

$$\frac{\partial m_{\gamma\varepsilon}}{\partial C_{\alpha\beta}} = \frac{1}{\mathrm{tr}\,\mathbf{C}}[\delta_{\alpha\gamma}\delta_{\beta\varepsilon} - m_{\gamma\varepsilon}\delta_{\alpha\beta}] , \tag{5.5-21a}$$

$$\frac{\partial m_{\gamma\varepsilon}}{\partial(\nabla_\eta C_{\alpha\beta})} = 0 , \tag{5.5-21b}$$

$$\frac{\partial(\nabla_\rho m_{\gamma\varepsilon})}{\partial C_{\alpha\beta}} = \nabla_\rho\left(\frac{\delta_{\alpha\gamma}\delta_{\beta\varepsilon} - m_{\gamma\varepsilon}\delta_{\alpha\beta}}{\mathrm{tr}\,\mathbf{C}}\right) , \tag{5.5-21c}$$

$$\frac{\partial(\nabla_\rho m_{\gamma\varepsilon})}{\partial(\nabla_\eta C_{\alpha\beta})} = \frac{1}{\mathrm{tr}\,\mathbf{C}}[\delta_{\eta\rho}\delta_{\alpha\gamma}\delta_{\beta\varepsilon} - m_{\gamma\varepsilon}\delta_{\alpha\beta}\delta_{\eta\rho}] , \tag{5.5-21d}$$

whereby we obtain

$$\frac{\delta F}{\delta C_{\alpha\beta}} = \frac{\delta F}{\delta m_{\gamma\varepsilon}}\frac{1}{\mathrm{tr}\,\mathbf{C}}[\delta_{\alpha\gamma}\delta_{\beta\varepsilon} - m_{\gamma\varepsilon}\delta_{\alpha\beta}] , \tag{5.5-22}$$

since

$$\delta_{\gamma\varepsilon}[\delta_{\alpha\gamma}\delta_{\beta\varepsilon} - m_{\gamma\varepsilon}\delta_{\alpha\beta}] = 0 , \tag{5.5-23}$$

which allows scalar multiples of (5.5-23) to be added to the right-hand side of (5.5-22) without altering its content.

Now we may substitute the expression (5.5-22) into the Poisson subbracket

$$\begin{aligned}\{F,G\}_E^C =& \\ &- \int_{\Omega'}\left[\frac{\delta F}{\delta C_{\alpha\beta}}\nabla_\gamma\left(\frac{\delta G}{\delta M_\gamma}C_{\alpha\beta}\right) - \frac{\delta G}{\delta C_{\alpha\beta}}\nabla_\gamma\left(\frac{\delta F}{\delta M_\gamma}C_{\alpha\beta}\right)\right]d^3x \\ &- \int_{\Omega'} C_{\gamma\alpha}\left[\frac{\delta G}{\delta C_{\alpha\beta}}\nabla_\gamma\left(\frac{\delta F}{\delta M_\beta}\right) - \frac{\delta F}{\delta C_{\alpha\beta}}\nabla_\gamma\left(\frac{\delta G}{\delta M_\beta}\right)\right]d^3x \\ &- \int_{\Omega'} C_{\gamma\beta}\left[\frac{\delta G}{\delta C_{\alpha\beta}}\nabla_\gamma\left(\frac{\delta F}{\delta M_\alpha}\right) - \frac{\delta F}{\delta C_{\alpha\beta}}\nabla_\gamma\left(\frac{\delta G}{\delta M_\alpha}\right)\right]d^3x ,\end{aligned} \tag{5.5-24}$$

in order to derive a new bracket in terms of **m**. For the last two integrals in the above expression, the transformation is obvious; however, for the first integral, we must be more resourceful. Due to (anti)symmetry, this integral may be split into two halves, of which only the first half needs to be considered, as the second half will follow suit. Upon substitution of (5.5-22) into this half, there results

$$- \int_{\Omega'}\frac{\delta F}{\delta m_{\gamma\varepsilon}}\frac{1}{\mathrm{tr}\,\mathbf{C}}[\delta_{\alpha\gamma}\delta_{\beta\varepsilon} - m_{\gamma\varepsilon}\delta_{\alpha\beta}]\nabla_\gamma\left(\frac{\delta G}{\delta M_\gamma}C_{\alpha\beta}\right)d^3x$$

$$= - \int_{\Omega'}\left[\frac{\delta F}{\delta m_{\alpha\beta}}\frac{1}{\mathrm{tr}\,\mathbf{C}}\nabla_\eta\left(\frac{\delta G}{\delta M_\eta}C_{\alpha\beta}\right) - \frac{\delta F}{\delta m_{\alpha\beta}}m_{\alpha\beta}\frac{1}{\mathrm{tr}\,\mathbf{C}}\nabla_\eta\left(\frac{\delta G}{\delta M_\eta}\mathrm{tr}\,\mathbf{C}\right)\right]d^3x$$

$$= - \int_{\Omega'} \left[\frac{\delta F}{\delta m_{\alpha\beta}} \nabla_\eta \left(\frac{\delta G}{\delta M_\eta} m_{\alpha\beta} \right) - \frac{\delta F}{\delta m_{\alpha\beta}} \frac{\delta G}{\delta M_\eta} C_{\alpha\beta} \nabla_\eta \left(\frac{1}{\operatorname{tr} \mathbf{C}} \right) \right. \mathrm{d}^3x$$

$$\left. - \frac{\delta F}{\delta m_{\alpha\beta}} m_{\alpha\beta} \nabla_\eta \frac{\delta G}{\delta M_\eta} - \frac{\delta F}{\delta m_{\alpha\beta}} m_{\alpha\beta} \frac{1}{\operatorname{tr} \mathbf{C}} \frac{\delta G}{\delta M_\eta} \nabla_\eta (\operatorname{tr} \mathbf{C}) \right] \mathrm{d}^3x \; ; \tag{5.5-25}$$

but since

$$\nabla_\eta \left(\frac{1}{\operatorname{tr} \mathbf{C}} \right) = \frac{\partial(\frac{1}{\operatorname{tr} \mathbf{C}})}{\partial(\operatorname{tr} \mathbf{C})} \nabla_\eta (\operatorname{tr} \mathbf{C}) = - \frac{1}{(\operatorname{tr} \mathbf{C})^2} \nabla_\eta (\operatorname{tr} \mathbf{C}) \; , \tag{5.5-26a}$$

$$\frac{\delta F}{\delta m_{\alpha\beta}} \nabla_\eta \left(\frac{\delta G}{\delta M_\eta} m_{\alpha\beta} \right) = \frac{\delta F}{\delta m_{\alpha\beta}} \left[\frac{\delta G}{\delta M_\eta} \nabla_\eta m_{\alpha\beta} + m_{\alpha\beta} \nabla_\eta \frac{\delta G}{\delta M_\eta} \right] , \tag{5.5-26b}$$

Eq. (5.5-25) becomes

$$- \int_{\Omega'} \frac{\delta F}{\delta m_{\alpha\beta}} \frac{\delta G}{\delta M_\eta} \nabla_\eta m_{\alpha\beta} \, \mathrm{d}^3x \; , \tag{5.5-27}$$

which, in combination with the second half of the integral, completes the Poisson subbracket transformation, with the final result

$$\{F,G\}_E^m = - \int_{\Omega'} \left[\frac{\delta F}{\delta m_{\alpha\beta}} \nabla_\gamma (m_{\alpha\beta}) \frac{\delta G}{\delta M_\gamma} - \frac{\delta G}{\delta m_{\alpha\beta}} \nabla_\gamma (m_{\alpha\beta}) \frac{\delta F}{\delta M_\gamma} \right] \mathrm{d}^3x$$

$$- \int_{\Omega'} m_{\gamma\alpha} \left[\frac{\delta G}{\delta m_{\alpha\beta}} \nabla_\gamma \left(\frac{\delta F}{\delta M_\beta} \right) - \frac{\delta F}{\delta m_{\alpha\beta}} \nabla_\gamma \left(\frac{\delta G}{\delta M_\beta} \right) \right] \mathrm{d}^3x$$

$$- \int_{\Omega'} m_{\gamma\beta} \left[\frac{\delta G}{\delta m_{\alpha\beta}} \nabla_\gamma \left(\frac{\delta F}{\delta M_\alpha} \right) - \frac{\delta F}{\delta m_{\alpha\beta}} \nabla_\gamma \left(\frac{\delta G}{\delta M_\alpha} \right) \right] \mathrm{d}^3x$$

$$+ 2 \int_{\Omega'} m_{\alpha\beta} m_{\gamma\varepsilon} \left[\frac{\delta G}{\delta m_{\gamma\varepsilon}} \nabla_\alpha \left(\frac{\delta F}{\delta M_\beta} \right) - \frac{\delta F}{\delta m_{\gamma\varepsilon}} \nabla_\alpha \left(\frac{\delta G}{\delta M_\beta} \right) \right] \mathrm{d}^3x \; . \tag{5.5-28}$$

Using the above bracket in combination with the one of (5.3-25), we obtain the evolution equations of (5.5-13) for ρ, *s*, and **M**, provided that we set the extra stress $\boldsymbol{\sigma}$ equal to

p. 121: in last line, σ should be $\boldsymbol{\sigma}$

$$\sigma_{\alpha\beta} = 2m_{\beta\gamma}\frac{\delta H}{\delta m_{\gamma\alpha}} - 2m_{\alpha\beta}m_{\gamma\varepsilon}\frac{\delta H}{\delta m_{\gamma\varepsilon}} , \tag{5.5-29}$$

and the pressure

$$p = -u + \rho\frac{\partial u}{\partial \rho} + s\frac{\partial u}{\partial s} . \tag{5.5-30}$$

The evolution equations for **m** now becomes

$$\begin{aligned}\frac{\partial m_{\alpha\beta}}{\partial t} = &- v_\gamma \nabla_\gamma m_{\alpha\beta} + m_{\gamma\alpha}\nabla_\gamma v_\beta + m_{\gamma\beta}\nabla_\gamma v_\alpha \\ &- 2m_{\alpha\beta}m_{\gamma\varepsilon}\nabla_\gamma v_\varepsilon ,\end{aligned} \tag{5.5-31}$$

which is indeed traceless.

5.6 The Relationship between Thermodynamics and Hydrodynamics

One concept from equilibrium thermodynamics which was unnecessary in chapter 4 is paramount for our discussion in this section: the notion of *adiabatic* and *diathermic walls* which enclose a given system Ω. Consider Ω to be separated from its surroundings by an encircling wall which, although flexible to allow changes in volume of the region it encloses, allows no exchange of heat ($dQ=0$). This is called an *adiabatic wall.* Conversely, a wall which permits the flow of heat is called a *diathermic wall.* Again, whether a wall is adiabatic or diathermic is largely dependent upon the time scale of the experimenter, τ, in comparison to the characteristic time for the system, dt, in response to changes in its environment. If $\tau \ll dt$, then the wall may be envisioned as adiabatic; and conversely, if $\tau \gg dt$, then the wall may be considered as diathermic.

5.6.1 *The global system*

Now if we consider a closed, adiabatic system Ω with fixed walls, then we realize that we have in effect isolated our system from its environment. Because of these requirements, we have at all times that $dQ=0$ and $dW=0$, so that the first law of thermodynamics requires

$$dU = 0 \Rightarrow U = \text{constant} , \tag{5.6-1}$$

since $dS=0 \Rightarrow S=\text{constant}$ and $dV=0 \Rightarrow V=\text{constant}$. Suppose, however, that we had chosen not to keep the wall fixed but we had the system always in contact with an environment (reservoir) of constant pressure, p; i.e., $dQ=0$ but $dW\neq 0$ so that

$$dU = -p dV \Rightarrow dU + p dV = 0 . \tag{5.6-2}$$

In this situation, we can no longer proclaim that the energy of the system is constant.[4] In fact, we have to be more ingenious to come up with a constant quantity at all. Let us define a differential quantity dZ as

$$dZ \equiv dU|_S + p dV \Rightarrow dZ = 0 . \tag{5.6-3}$$

Therefore, this quantity Z, along with S, is constant for all times. But what is Z? If we use the Euler relation, then we have

$$Z = U + pV , \tag{5.6-4}$$

which indicates that Z is the enthalpy. Taking the differential of this expression,

$$dZ = dU + V dp + p dV , \tag{5.6-5}$$

and subtracting (5.6-3) gives $V dp=0$. Hence from (4.1-5), we have that $dE=0$, which implies that the enthalpy of the system is constant in time. Using similar arguments, if we consider a system with a diathermic, fixed wall, and a diathermic, flexible wall, in contact with the appropriate reservoirs, then we find that A and B, respectively, must be constant in time.

In this chapter, we considered the flow of a fluid inside of a fixed, adiabatic wall. Although we had to generalize the internal energy to include the potential and kinetic energies (i.e., the Hamiltonian), we still required $dQ=dW=0$, which implied that $dH=0$. This fact was accounted for through the antisymmetry property of the Poisson bracket, $dH/dt=\{H,H\}=0$. Similarly, when $F=S$ or $F=V$, we must also have that $dS/dt=\{S,H\}=0$ and $dV/dt=\{V,H\}=0$. If, however, we had chosen to enclose the system in an adiabatic, flexible wall, then, in order to retain the antisymmetry property of the bracket, we would have had to require that the generalized enthalpy, E' (E' being the Legendre transformation of H), be the generating functional, so that $dE'/dt=\{E',E'\}=0$ and $dS/dt=\{S,E'\}=0$. However, in this case the limits of the integrals are dependent on the state of the system, subject to the requirement that $d\mathrm{P}/dt=0$, where P is the total mass of the system, i.e., there is no flux through the boundaries. This description is used in the formulation of free boundary value problems, which, due to size constraints, will not be discussed—see, for example, [Lewis *et al.*, 1986] for a description of the Hamiltonian structure of free boundary value problems. Similar arguments apply for the other

[4]In effect, we can reconstitute the system by considering not only Ω, but its immediate surroundings as well, thus considering a bigger system enclosed by an adiabatic, fixed wall in which case we have again that U(overall)=constant.

cases mentioned above as well. The point is that how one views the system thermodynamically greatly influences the form of the theory which one needs to arrive at the proper hydrodynamic equations. It would be interesting to pursue this line of reasoning further at this time, but it is beyond the scope of this introductory level book.

5.6.2 The compound system

Now that we have looked at the global hydrodynamic system from a thermodynamic perspective, let us do the same thing locally; i.e., let us consider the global system as being a compound system of phase type. Here we consider each fluid particle as a closed (constant mass) subsystem with an adiabatic, flexible wall. Consequently, for each particle, $dQ_r=0$ and $dW_r=-p_r d^3Y$. The constraint that $dQ_r=0$ implies that $S_r=$ constant, and since the system is closed, $\hat{S}_r$=constant as well.

This entire compound system is enclosed by an adiabatic, fixed wall, which requires that $U(H)=S=V$=constant. According to the additivity conditions of §4.5, we have that

$$U = \int_{\Omega'} U_r \, d^3Y \ , \quad S = \int_{\Omega'} S_r \, d^3Y \ , \quad \text{and} \quad V = \int_{\Omega'} d^3Y \ ; \tag{5.6-6}$$

however, in general both U_r and d^3Y may be varying with time as long as their integrals are constant for all time.

Now if we consider the global system enclosed by an adiabatic, flexible wall, then we have that U is still given by $(5.6\text{-}6)_1$ at any time t, but it is no longer required to be constant. Note that E_r is not (in general) constant either, despite the fact that the total E is, since locally the pressure p_r might vary from point to point. It would be interesting at this point to look at the situations when the subsystem walls are diathermic as well; however, this topic can be treated best after we introduce the concept of irreversibility.

6
Non-Equilibrium Thermodynamics

Laws of Thermodynamics:
1. *You cannot win.*
2. *You cannot break even.*
3. *You cannot get out of the game.*

—*Anonymous [Mackay, 1991, p. 7]*

After having devoted five chapters of this book to the discussion of equilibrium thermodynamics and conservative dynamic phenomena, it is now high time that we entered into the realm of irreversible transport processes. As mentioned in chapter 1, most of the physical systems which engineers wish to model exhibit dissipative phenomena. Therefore, although the techniques touched upon in the previous chapters are mathematically profound and well-suited for diverse analyses for conservative systems, it is in this chapter and the next that the major engineering applications will find their foundation. Granted, in describing irreversible phenomena on the continuum level a certain amount of phenomenology is necessarily introduced; yet we hope to illustrate here how the application of thermodynamic knowledge to the irreversible system can reduce this phenomenology to the bare minimum.

The objective of this chapter is similar to that of chapter 4; we wish to present a brief, yet sufficiently thorough, discussion concerning the theory of non-equilibrium thermodynamics applied to irreversible processes. There already exist several outstanding references on the subject [De Groot and Mazur, 1962; Yourgrau *et al.*, 1966; Prigogine, 1967; Gyarmati, 1970; Woods, 1975; Lavenda, 1978; Truesdell, 1984]. Thus, the objective of our discussion here is mainly to introduce the principles that are subsequently used to formulate the dissipative bracket, as outlined in the next chapter. Moreover, the presentation of the subject is biased towards the presentation of the concepts that we consider as most helpful to continuum modeling. For example, the notion of internal variables is introduced early on, in §6.2. As we shall see, the inclusion of internal variables in the non-equilibrium description of the system has profound implications concerning the roles of the various thermodynamic variables and the definitions of the various state functions, in particular, the entropy. Indeed, the definitions of these functions hinge upon the notion of time scales which become of chief importance in the discussion of irreversible thermodynamics.

In the philosophy of equilibrium thermodynamics, it is assumed that the time scale for changes in the system is sufficiently large as compared to the intrinsic time scales of any internal variables within the system. In this case, we can safely assume that the internal variables remain (at any time) at their equilibrium values, which are dictated by the dynamic variables of the system. However, if the system depends on variables with intrinsic relaxation times on the order of the time scales for changes within the system, then it may still be possible to describe the system accurately enough through an extended equilibrium relation, if we extend the state space to include internal variables which account for intermediate structural states. In addition, irreversibility is introduced into the system as these internal variables progress towards equilibrium within a shorter time increment than the one characterizing the overall system response time. Thus, we now have to cope with the uncertainties of an irreversible process and try to fathom what implications arise in the fundamental description of the system processes.

6.1 Irreversibility and Stability

Let us consider a simple system Ω surrounded by a flexible, adiabatic wall. If we impose an infinitesimal, reversible deformation, $\mathrm{d}V$, upon the system, then the work done is given by $\mathrm{d}W = -p\mathrm{d}V$. Yet if the deformation is applied at a rate exceeding the capability of the system to react to it, then the process becomes a *non-equilibrium process*, and the work done on the system must be given by

$$\mathrm{d}W = -p\mathrm{d}V + \mathrm{d}W_i , \tag{6.1-1}$$

since we would be unable to regain all of the work input into the system by reversing the process. Hence $\mathrm{d}W_i$ represents the amount of work lost in recovering the initial deformation state of the system, passing through an unspecified intermediate deformation state, via some definite process. We may therefore write the *second law of thermodynamics* as

$$\mathrm{d}W_i \geq 0 , \tag{6.1-2}$$

which is a sufficient definition of this law for the purposes of this book.[1] Note that here, and in the following paragraphs, the equality is valid only for a reversible process.

[1]It may well be that the reader does not endorse our definition of the second law; however, quoting K. Hutter [Kestin, 1990, p. 193], "There are as many second laws of thermodynamics as there are thermodynamicists."

If we apply a pressure, say p_a, to the outside of the wall of the system, which is appreciably different from the internal pressure of the system, p, then a certain amount of work will be transferred into the system in a small time period—a non-equilibrium process. It is obvious that if $p_a < p$, then Ω will expand ($\mathrm{d}V > 0$), and vice versa. In this process, we thus apply $\mathrm{d}W = -p\mathrm{d}V$ amount of work to Ω, which is transferred inside the system as (6.1-1). Therefore, we have that

$$\mathrm{d}W_i = (p - p_a)\,\mathrm{d}V \,, \tag{6.1-3}$$

which, via the second law, is always positive. When $p_a < p$ and Ω expands ($\mathrm{d}V > 0$), then in order to ensure stability (i.e., that the system moves such as to approach and restore equilibrium) the internal pressure must be reduced until a mechanical equilibrium is attained at $p = p_a$. Consequently, $\mathrm{d}p < 0$, and for the opposite situation ($p_a > p$, $\mathrm{d}V < 0$) $\mathrm{d}p > 0$. Since $\mathrm{d}V$ and $\mathrm{d}p$ are always of opposite sign, we can express the condition for mechanical stability as

$$\left.\frac{\partial V}{\partial p}\right|_S < 0 \;\Rightarrow\; \kappa_S > 0 \,. \tag{6.1-4}$$

Now let us consider that the system Ω is transferred from one state to another ($A \to B$) via an unspecified adiabatic irreversible process, but we require that both states have the same deformation coordinates. (This is true only of the states A and B, and not of any intermediate state.) In this process, in accordance with the second law, $\mathrm{d}W_i > 0$, the energy of state B can only be higher than the energy of state A, $U_B > U_A$. If we now allow for a heat exchange with the environment kept at the initial system temperature T_A, then, for the system to return to its initial state (and therefore to lose energy) it will have to be at a higher temperature $T_B > T_A$. Hence we conclude that the condition for *thermal stability* may be written as

$$\left.\frac{\partial U}{\partial T}\right|_V > 0 \;\Rightarrow\; C_v > 0 \,. \tag{6.1-5}$$

It is therefore allowable to extend the second law to general non-adiabatic processes as

$$\mathrm{d}Q \le T\mathrm{d}S \quad \text{and} \quad \mathrm{d}W_i \ge 0 \,, \tag{6.1-6}$$

which allows us to extend the fundamental equation for an irreversible process:

$$\mathrm{d}U \le T\mathrm{d}S - p\mathrm{d}V \,. \tag{6.1-7}$$

Via Eqs. (4.2-25,26), it is possible to extend the conditions for thermal and mechanical stability to

$$C_p > C_v > 0 \quad \text{and} \quad \kappa_T > \kappa_S > 0 \,. \tag{6.1-8}$$

If a system is in a stable equilibrium state, then after any arbitrary, infinitesimal perturbation away from equilibrium, the system will move so as to restore its initial state. Thus any perturbation of an equilibrium state into non-equilibrium states is not thermodynamically allowed, which implies, according to (6.1-7), that for constant entropy and volume, the energy is a minimum, or, for constant energy and volume, that the entropy is a maximum. In turn this implies that the first-order differentials of the extensive quantities in the fundamental equation vanish, i.e., $dU=dS=dV=0$, and, in accordance with the stability conditions given above,

$$d^2U\big|_{S,V} > 0 \quad \text{and} \quad d^2S\big|_{U,V} < 0 \; . \tag{6.1-9}$$

The former represents the *energy minimum principle* and the latter the *entropy maximum principle*. Regardless of what process a system undergoes, the final equilibrium state must satisfy both of these conditions; i.e., any equilibrium value of any unconstrained internal variable of the system is such that it minimizes the energy *and* maximizes the entropy [Callen, 1960, p. 88].

6.2 Systems with Internal Variables

In this section, we wish to illustrate how some of the fundamental relations given above generalize to a situation where the thermodynamic state space is extended to include various internal variables. We consider internal variables in the sense as outlined by Woods [1975, p. 44]: unlike U and V, internal variables can alter without any corresponding changes occurring outside of Ω. Hence these additional variables extend the state space from (U,V) to $(U,V,\mathbf{X})$, where the ν variables X_i are written as the components of the vector $\mathbf{X}$. These variables have intrinsic time scales of their own, and the experimenter usually has no control over their behavior, especially when these time scales are short relative to the overall system response time.

Internal variables are introduced in non-equilibrium thermodynamics in an effort to define the non-equilibrium state of a system at a specific time, t_0, in terms of a set of variables all of which are taken at the same time, t_0. There are of course other alternatives, as for example the use of the previous history of the system specified in terms of the external action which the system has endured from an equilibrium state (at time $t=-\infty$) up to the observation time, t_0 [Meixner, 1974, p. 134]. However, we shall agree at this point with Kestin, who was one of the first advocates of the internal variable approach, in stating [Kestin, 1974, p. 145]

> As far as I am concerned, only statement (4)—i.e. the description of the state of a system through internal variables—is acceptable

as a definition of state. In other words, the concept of state refers to a set of variables which one imagines are measured over the system *instantaneously*. The words *history, set of data referring to times* $t \leq t_0$; *time-dependent properties, as well as functionals of functions of time* conjure up in my mind the concept of a *process*. I believe it is essential to keep the two apart.

Of course, the use of the internal variables approach presupposes more detailed "inside the black box" information about the system. This involves consideration of the internal variables, $\mathbf{X}$, as well as the specification of a fundamental equation of state, say of the form $U=U(s,\boldsymbol{\alpha},\mathbf{X})$ in terms of the entropy density, s, the equilibrium thermodynamic variables, $\boldsymbol{\alpha}$, and the internal variables, $\mathbf{X}$ [Kestin, 1974, p. 149]. Although we agree, in principle, with Domingos [1974, p. 4] in stating that "neither a principle nor even a rule exists to guide in the choice or proposal of such hidden variables or coordinates," we, however, can see several examples (some of them discussed in Part II of this book) where the selection of the internal variables is naturally offered from an analysis of the microstructure. Thus, we consider the investigation of the microstructure an integral part of the macroscopic (continuum) modeling which blends naturally with the internal variables approach followed here.

In §4.4, we talked a little about the extension of the fundamental equation to systems with internal microstructure. At that time, we used the tensorial parameters $\mathbf{Z}_i$ in a sense as internal variables which described the inherent microstructure of the system. Here we consider the components of $\mathbf{X}$ to be scalars, for simplicity, which represent some general characteristic responses of the system to imposed processes. In an approach similar to the one used in describing the structural potential, we may also define a ν-component vector of conjugate variables,[2] also scalars, as $\mathbf{x}=(x_1,x_2,\ldots,x_\nu)^{\mathrm{T}}$. Consequently, we can rewrite the fundamental equation in terms of $S=S(U,V,\mathbf{X})$ as

$$T\mathrm{d}S(U,V,\mathbf{X}) = \mathrm{d}U + p\mathrm{d}V - \mathbf{x}^{\mathrm{T}}\cdot\mathrm{d}\mathbf{X} \,, \tag{6.2-1}$$

with

$$\mathbf{x}^{\mathrm{T}}\cdot\mathrm{d}\mathbf{X} \leq 0 \,, \tag{6.2-2}$$

the equality holding only for a system in *internal equilibrium*. If the X_i are independent, then at equilibrium $\mathbf{x}^{\mathrm{T}}\cdot\mathrm{d}\mathbf{X}=0$ requires that

[2]In extreme situations (common when the internal variable represents the statistical distribution of a continuous internal quantity like the end-to-end distance of a polymer chain) $\nu\rightarrow\infty$, i.e., the internal variable space becomes a continuum [Prigogine, 1955, p. 36]—see also §3.5 and Example 6.2.

$$\mathbf{x} = 0 \Rightarrow \mathbf{X}^{eq} = \mathbf{X}^{eq}(S,V) \ , \tag{6.2-3}$$

so that $S(U,V,\mathbf{X})$ becomes simply $S(U,V)$ [Woods, 1975, p. 49]. Indeed, it is because of the inequality (6.2-2) that the process described by (6.2-1) is made irreversible. It is now apparent that

$$T\mathrm{d}S(U,V,\mathbf{X}) \geq T\mathrm{d}S(U,V) = \mathrm{d}U + p\mathrm{d}V \ . \tag{6.2-4}$$

However, we have the opposite inequalities in the corresponding hierarchy of entropy functions,

$$\ldots \leq S(U,V,\mathbf{X},\mathbf{Y}) \leq S(U,V,\mathbf{X}) \leq S(U,V) \ , \tag{6.2-5}$$

which indicates that as more internal variables reach equilibrium (and therefore are not necessary for the description of the state of the system), the entropy monotonically increases. Hence a more detailed description of state implies an entropy function of smaller magnitude. Eqs. (6.2-4) and (6.2-5) are schematically represented in Figure 6.1.

Another interesting type of process, intermediate between reversible and irreversible ones, that could be described is the *natural process*, which corresponds to a reversible process, but only in an extended space with the internal variables left unspecified. Thus, it appears as an irreversible process in the (U,V) subspace (see the second inequality in Eq. (6.2-4)). The implications of processes of this type are not pursued here; however, the reader may refer to the intriguing account of Woods [1975, pp. 49,50] for a detailed description.

EXAMPLE 6.1: In following up Example 4.2, show that if the original composite system described therein, with subsystems initially at T_1 and T_2, is allowed to reach thermal equilibrium (assuming that the barrier separating the two subsystems is diathermal and rigid), then the total entropy of the composite system is increased.

Using expression (4.5-4), we can readily calculate the entropy difference between the final equilibrium (where $T = (T_1 + T_2)/2$) and the initial states as

$$\Delta S = \alpha RN \ln[(T_1 + T_2)^2 / 4T_1T_2] \ , \tag{6.2-6}$$

which is indeed always non-negative since $(T_1 - T_2)^2 = T_1^2 + T_2^2 - 2T_1T_2 \Rightarrow T_1^2 + T_2^2 > 2T_1T_2$ and $(T_1 + T_2)^2 = T_1^2 + T_2^2 + 2T_1T_2 \Rightarrow (T_1 + T_2)^2 > 4T_1T_2$. ■

EXAMPLE 6.2: When the number of internal variables tends towards infinity, it is possible to replace them by a continuous descriptive function, as in §3.5. The best example of an application of this concept is in the kinetic theory description of the energy distribution within a system of particles through the Maxwell/Boltzmann approach.

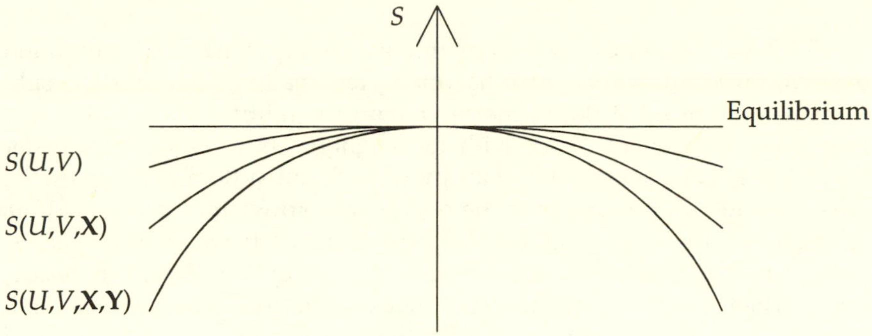

Figure 6.1: The entropy functions for various thermodynamic state spaces involving the internal variables . Extending the state space decreases the entropy function, which is always maximum at equilibrium, but results in a greater rate of change in S.

The starting point of this analysis is the definition of the corresponding entropy. Let us consider a system of particles consisting of multiple species, i=1,2,...,ν, each characterized by a velocity represented by the vector $\mathbf{u}$. The state of this system is described by a distribution function, $f_i = f_i(\mathbf{u},t)$, in the sense that the number of particles of species i with velocities between $\mathbf{u}$ and $\mathbf{u}+\mathrm{d}\mathbf{u}$ is proportional to $\int f_i\, \mathrm{d}^3u$. Consequently, it can be shown that (6.2-1) is given by [Woods, 1975, p. 77]

$$T\mathrm{d}\hat{S} = \mathrm{d}\hat{U} + p\mathrm{d}\hat{V} - \sum_{i=1}^{\nu} \mu_i^{+}\mathrm{d}\hat{n}_i - k_B T \hat{V} \sum_{i=1}^{\nu} \int \left[\mathrm{d}f_i \ln\left(\frac{f_i}{f_i^{eq}}\right) \right] \mathrm{d}^3u \;, \tag{6.2-7}$$

where f_i^{eq} is the Maxwellian distribution function provided through kinetic theory as

$$f_i^{eq} = N_i \left(\frac{m_i}{2\pi k_B T} \right)^{3/2} \exp(-\; m_i \mathbf{u}\cdot\mathbf{u}/2k_B T) \;. \tag{6.2-8}$$

Note that the last factor on the right-hand side of (6.2-7) must be positive via (6.2-2). This leads to the following definition for the entropy:

$$S = -\, k_B \sum_{i=1}^{\nu} \int f_i (\ln f_i - 1)\, \mathrm{d}^3u \;. \tag{6.2-9}$$

If the entropy is maximized for constant energy, volume, and moles of species i, then it can be shown [Woods, 1975, pp. 85-7] that from (6.2-9) the corresponding equilibrium probability distribution function is the Maxwellian distribution, (6.2-8). ■

6.3 The Clausius Inequality

In §5.6.2, we viewed the hydrodynamic system of interest as a compound system, enclosed within an adiabatic, fixed wall, in which each subsystem, considered at (local) thermodynamic equilibrium, could transfer expansion/contraction work with its neighbors but could not transfer heat. As a consequence, the entropy of each subsystem (fluid particle) was constant, and therefore in the overall system we had $dU=dS=dV=0$. If we relax the constraint on the heat transfer between subsystems, however, then it is necessary that the entropy of the system increases, even when internally within each subsystem the processes are still reversible (i.e, the assumption of local thermodynamic equilibrium is still valid).

In order to see that the above statement is true, let us consider a simple compound system composed of just two subsystems. Again, the overall system Ω is enclosed within a fixed, adiabatic wall which allows no energy transfer with the environment; however, we consider each subsystem, Ω_A and Ω_B, to be surrounded by an inflexible, diathermic wall. If during some process an infinitesimal amount of heat $-dQ_A=dQ_B>0$ transfers from Ω_A to Ω_B, then via the second law,

$$T_A dS_A \geq dQ_A \ , \quad T_B dS_B \geq dQ_B \ . \tag{6.3-1}$$

Furthermore, the thermal stability condition ($C_v > 0$) requires that $T_A \geq T_B$, so that

$$dS = dS_A + dS_B \geq dQ_B \left(\frac{1}{T_B} - \frac{1}{T_A}\right) \geq 0 \ . \tag{6.3-2}$$

In view of the representation of the compound system by an assembly of interacting subsystems, each one locally equilibrated, it is important to realize that the total entropy of the system will increase even when the processes within A and B are reversible, as long as there is a temperature difference between the subsystems. Consequently, the entropy of the system is always increased by internal heat transfers due to the existence of non-constant temperature profiles.[3]

We may generalize the above arguments by considering a compound system, Ω, composed of $\nu+1$ subsystems, Ω_i, $i=0,1,2,\ldots,\nu$, each at a local equilibrium at its own temperature T_i. Let us focus on subsystem Ω_0, assuming that it thermally interacts with the remaining subsystems, Ω_i,

[3]Similar arguments hold when we allow other forms of dissipative transport processes between the subsystems, such as mass or viscous momentum, as well as internal relaxation processes for a structured system characterized by internal configuration parameters. These entropy increases are a manifestation of internal dissipative processes taking place well beyond the dynamic evolution phenomena investigated in chapter 5; they are further investigated in chapter 7.

i=1,2,...,ν (which are not allowed to interact with each other), exchanging heat dQ between two (infinitesimally close) equilibrium states. Then, the entropy change of Ω_0, dS, is

$$\mathrm{d}S = \frac{\mathrm{d}Q}{T_0} = \sum_{i=1}^{\nu} \frac{\mathrm{d}Q_i}{T_i} + \sum_{i=1}^{\nu} \mathrm{d}Q_i \left(\frac{1}{T_0} - \frac{1}{T_i}\right) , \tag{6.3-3}$$

where $\mathrm{d}Q=\Sigma \mathrm{d}Q_i$ and $\mathrm{d}Q_i$ is the infinitesimal amount of heat transferred between subsystem Ω_0 and Ω_i. Since when $T_i \geq T_0$, $\mathrm{d}Q_i \geq 0$, and when $T_i \leq T_0$, $\mathrm{d}Q_i \leq 0$, the second term on the right-hand side of (6.3-3) is always non-negative, so that

$$\mathrm{d}S \geq \sum_{i=1}^{\nu} \frac{\mathrm{d}Q_i}{T_i} . \tag{6.3-4}$$

Expression (6.3-4) is called the *Clausius inequality* [Woods, 1975, p. 46], and the equality holds only for a reversible process when $T_0=T_i$, for all i, corresponding to the absence of any irreversible heat transfer. Explicitly, this requirement is only met when the temperature is the same within all the subsystems.

When the internal processes within the subsystems are reversible, (6.3-4), used as an equality, allows us to express entropy changes within each subsystem that are consistent with the local equilibrium hypothesis. In contrast to §5.6.2, however, we can now allow thermal interactions between the subsystems by rewriting the entropy change in the most general form,

$$\mathrm{d}S = \mathrm{d}S_r + \mathrm{d}S_i , \tag{6.3-5}$$

where we have defined $\mathrm{d}S_r$, $\mathrm{d}S_i$ as the reversible, irreversible entropy changes occurring within Ω_0, respectively:

$$\mathrm{d}S_r \equiv \sum_{i=1}^{\nu} \frac{\mathrm{d}Q_i}{T_i} , \quad \mathrm{d}S_i \equiv \mathrm{d}S - \mathrm{d}S_r = \sum_{i=1}^{\nu} \mathrm{d}Q_i \left(\frac{1}{T_0} - \frac{1}{T_i}\right) . \tag{6.3-6}$$

Then, the inequality (6.3-4) can be expressed simply as

$$\mathrm{d}S_i \geq 0 . \tag{6.3-7}$$

Hence, heat transfer caused by temperature differences in any system always corresponds to an increase in entropy beyond the changes dictated from equilibrium (reversible) thermodynamic considerations and induced from the interactions with the environment.[4]

In view of the nature of the system Ω considered when developing the bracket formalism (i.e., Ω is isolated), Eq. (6.3-1) forces us to con-

[4]In fact, we can even extend the validity of inequality (6.3-7) for open systems as well, defining $\mathrm{d}S_i$ from (6.3-5) as $\mathrm{d}S_i \equiv \mathrm{d}S - \mathrm{d}S_r$, although within $\mathrm{d}S_r$ we need to take into account the reversible entropy changes connected with the interchange of mass between the subsystem under consideration and its neighbors.

template the effects of internal heat transfers upon our overall system. Let us consider initially that the compound system Ω is composed of ν subsystems, each surrounded by a flexible, adiabatic wall. These subsystems may all have their own temperatures and pressures. If, at some instant, we replace all of the adiabatic walls with diathermic walls, then we shall induce a heat transfer process which will result, via (6.3-1), in an increase in the entropy of the overall system. Yet, Ω is still isolated, and so the overall energy is forced to remain constant. Hence any increase in energy (TdS) which occurs due to heat transfer within the system must be made up for by a rearrangement of the pressures of the substituent subsystems in such a way that the overall volume remains constant. Still, what if the fluid is incompressible? In this case, the volumes V_i cannot change and the increase in energy must impart a kinetic energy (thermal convection) to the subsystems to keep the total energy constant.

6.4 Non-Equilibrium Thermodynamics of Flowing Systems

In this section, we begin discussion of the non-equilibrium thermodynamics description of rate processes within a compound system whose component subsystems are allowed to convect in time, i.e., flowing systems. Although the overall system is still surrounded by a fixed, adiabatic wall, the subsystems are allowed to undergo both heat transfer and expansion/compression work. Consequently, we now wish to generalize the results of preceding sections to include the more exotic systems encountered in engineering applications. In order to accomplish this, we shall examine a continuum fluid system in the spatial description, where the major engineering interests lie. It is therefore necessary to assume that characteristic dynamic variables, in particular the entropy, vary continuously in a small neighborhood of each thermodynamic subsystem (fluid particle); adjacent subsystems have values for the dynamic variables which are differentially close to each other. In addition, we make use of the so-called *local equilibrium hypothesis*, which declares that, although the overall system is not in equilibrium, the time scales of the subsystems are small enough so that each fluid particle may be considered as a thermodynamic system at (a possibly metastable) equilibrium.

6.4.1 *The entropy balance equation*

In view of the local equilibrium hypothesis, we may consider that the fundamental equation, (4.1-2), applies to every fluid particle, which in the spatial description is located at position $\mathbf{x}$ at time t. For the fluid particle subsystem it is necessary to use specific (per mass) variables, so that dividing the Gibbs relation by m_r gives

$$T_r \mathrm{d}\hat{S}_r = \mathrm{d}\hat{U}_r + p_r \mathrm{d}(1/\rho) \;, \quad r=1,2,3,... \; ; \tag{6.4-1}$$

however, since the thermodynamic variables are assumed to vary continuously, it is sufficient to write (6.4-1) for the overall system as

$$T\mathrm{d}\hat{S}(\mathbf{Y},t) = \mathrm{d}\hat{U}(\mathbf{Y},t) + p\mathrm{d}(1/\rho(\mathbf{Y},t)) \;, \tag{6.4-2}$$

where $\mathbf{Y}$ is the material (Lagrangian) coordinate. Now we may divide this expression by a time scale on the order of the experimenter's, $\mathrm{d}t$, so as to obtain a rate expression for the entropy:

$$T\frac{\mathrm{d}\hat{S}(\mathbf{Y},t)}{\mathrm{d}t} = \frac{\mathrm{d}\hat{U}(\mathbf{Y},t)}{\mathrm{d}t} + p\frac{\mathrm{d}(1/\rho(\mathbf{Y},t))}{\mathrm{d}t} \;. \tag{6.4-3}$$

Hence the local equilibrium hypothesis is valid provided that $\mathrm{d}t \geq \tau_i$, τ_i being the longest transit time for a signal containing thermodynamic information to propagate throughout the subsystem. Note that in the spatial description, we must map the coordinate function $a(\mathbf{Y},t) \rightarrow a(\mathbf{x}(\mathbf{Y},t),t)$ in that

$$\frac{\mathrm{d}a}{\mathrm{d}t} = \left.\frac{\partial a(\mathbf{Y},t)}{\partial t}\right|_{\mathbf{Y}} = \left.\frac{\partial a}{\partial t}\right|_{\mathbf{x}} + \left.\frac{\partial a}{\partial x_\alpha}\frac{\partial x_\alpha}{\partial t}\right|_{\mathbf{Y}} = \frac{\partial a}{\partial t} + v_\alpha \nabla_\alpha a \;. \tag{6.4-4}$$

Although we now have a rate equation for the entropy (per mass), we still need to know the energy and density rate expressions in order to apply it.[5] Therefore, although we may use (6.4-3) as a consistency check once we obtain the desired result, it becomes necessary in order to continue that we make some assumptions about the form of the rate equation that the entropy should possess. It is at this point that phenomenology is introduced into the program.

Since we have the additivity property of the entropy function, we can write the differential of the entropy at any location $\mathbf{x}$ and time t as the sum of two contributions:

$$\mathrm{d}S(\mathbf{x},t) = \mathrm{d}S_r(\mathbf{x},t) + \mathrm{d}S_i(\mathbf{x},t) \;. \tag{6.4-5}$$

In this relation, $\mathrm{d}S_r$ is the entropy change which is due to convection of the fluid particle at location $\mathbf{x}-\mathrm{d}\mathbf{x}$ at time $t-\mathrm{d}t$ into $\mathbf{x}$ at time t and convection of the fluid particle at $\mathbf{x}$ at $t-\mathrm{d}t$ into $\mathbf{x}+\mathrm{d}\mathbf{x}$ at t. As such, it is completely reversible, and may be a positive, zero, or negative quantity. Furthermore, $\mathrm{d}S_r$ also contains any other reversible entropy changes that may occur in the fluid particle at $\mathbf{x},t$. The other term, $\mathrm{d}S_i$, represents the irreversible heat transfer between the subsystems (fluid particles)

[5]Actually, for systems of constant mass, which certainly applies for the vast majority of engineering applications, the density must satisfy the continuity equation of (5.3-38), which at this stage of the development may be taken as a given. However, the rate expression for the internal energy, (5.3-46), is no longer consistent with the ideas of irreversible thermodynamics, forcing an indeterminacy in (6.4-3).

evaluated at time t between the fluid particle at $\mathbf{x}$ and its immediate neighbors. As we have seen, $dS_i \geq 0$, with the equality valid only for a reversible heat transfer. In general, since we do not know the sign of dS_r, we can make no claims about the sign of dS; however, since the overall system Ω is enclosed within a fixed, adiabatic wall, there are no externally induced reversible entropy changes and the integral of the convected entropy over the entire volume of Ω must vanish (ergo, $dS_r^{Tot}/dt = 0$). Consequently, because of internal heat transfers, we must have that $dS^{Tot}/dt \geq 0$.

At this point, we introduce two new quantities of the system: the total entropy flux (per unit area), $\mathbf{J}_s^{Tot}$, and the entropy production rate (per unit volume), σ. These two additional variables allow us to write rate expressions for the components of (6.4-5) as

$$\frac{dS_r}{dt} = -\int_{\partial\Omega} \mathbf{J}_s^{Tot} \cdot \mathbf{n}\, d^2x \;, \tag{6.4-6a}$$

$$\frac{dS_i}{dt} = \int_{\Omega} \sigma\, d^3x \;; \tag{6.4-6b}$$

however, in order to further use these expressions, it is necessary to introduce phenomenology to provide relations for $\mathbf{J}_s^{Tot}$ and σ. This we shall put off until the next subsection; for the time being, we shall only consider general characteristics of the fluid system.

If we express the total system entropy as

$$S = \int_{\Omega} s(\mathbf{x},t)\, d^3x \;, \tag{6.4-7}$$

and use the divergence theorem, it is possible to rewrite (6.4-6) through (6.4-5) as

$$\int_{\Omega} \left[\frac{\partial s}{\partial t} + \nabla\cdot \mathbf{J}_s^{Tot} - \sigma \right] d^3x = 0 \;. \tag{6.4-8}$$

Consequently, since (6.4-8) must hold for any d^3x, we have

$$\frac{\partial s}{\partial t} = -\nabla\cdot \mathbf{J}_s^{Tot} + \sigma \;, \quad \sigma \geq 0 \;, \tag{6.4-9}$$

or, alternatively,

$$\frac{\partial s}{\partial t} = -\nabla\cdot (\mathbf{v}s) - \nabla\cdot \mathbf{J}_s + \sigma \;, \tag{6.4-10}$$

with

$$\mathbf{J}_s \equiv \mathbf{J}_s^{Tot} - s\mathbf{v} \;. \tag{6.4-11}$$

Eq. (6.4-10) represents the extension of the entropy equation, (5.3-39), to a non-conservative system.

6.4.2 *The linear phenomenological relations*

If we accept that the relations of (6.4-6) represent physical laws, then we still have not introduced any phenomenology into the system description.[6] Yet, in order to continue any further, we cannot avoid doing so; we must now specify the phenomenological relations for the fluxes and the entropy production rate. We may write down expressions[7] for the J_i (i=1,2,...), and σ which are, in essence, merely postulates which we shall try to justify to some extent in §6.4.5 for a couple of simple cases. Yet, we must never forget that all of our ignorance of non-equilibrium systems is contained in these relations. Indeed, the extent of our ignorance is such that in order to preserve the generality of our presentation we presently limit our discussion to systems close enough to equilibrium where these relations can be safely assumed to be linear [Onsager, 1931a,b]. Further away from equilibrium this is not necessarily the case, and we shall consider nonlinear relationships later on once we have introduced the concepts behind the dissipation bracket—see §7.1.1. An additional complication ensues from the fact that various J_i appear in different evolution equations for the dynamic vari-ables (e.g., the momentum and energy equations), and the fluxes are therefore highly coupled with each other, as is evident from (6.4-3). Note that the $\mathbf{J}_s$ of (6.4-11) may include one or more of the J_i.

The relations which we postulate are thus:

$$J_i = \sum_k \alpha_{ik} \Lambda_k \ , \quad i = 1,2,\dots \ , \tag{6.4-12}$$

$$\sigma = \sum_i J_i \Lambda_i \ , \tag{6.4-13}$$

with $\boldsymbol{\alpha}$ being a matrix of phenomenological coefficients and $\boldsymbol{\Lambda}$ being the affinity or thermodynamic driving force. (We shall leave the nature of these affinities unspecified until §6.4.5.) Substitution of (6.4-12) into (6.4-13) yields

$$\sigma = \sum_i \sum_k \Lambda_k \alpha_{ik} \Lambda_i = \frac{1}{2} \sum_i \sum_k \Lambda_k (\alpha_{ik} + \alpha_{ki}) \Lambda_i \ , \tag{6.4-14}$$

with the latter equality being necessary in the following since we do not know (yet) that $\boldsymbol{\alpha}$ is a symmetric matrix. (Remember, any matrix may be written as a sum of a symmetric matrix and an antisymmetric matrix, $\mathbf{A}$, but for any $\mathbf{x}$, $\mathbf{x}^T \cdot \mathbf{A} \cdot \mathbf{x} = 0$.) Since $\sigma \geq 0$, it follows from (6.4-14) that $\det(\boldsymbol{\alpha} + \boldsymbol{\alpha}^T)$ and all its principal minors must be non-negative (more details of this calculation will be given in later chapters):

[6]Actually, as we shall see in chapter 8, we can envision another contribution to (6.4-6) for a material with internal microstructure. This additional contribution is due to entropic differences between various configurations, and is also reversible in nature.

[7]Although the flux of (6.4-11) is a vector, we choose to work here in terms of scalar fluxes for simplicity; the more complicated tensorial expressions following analogously.

$$\alpha_{ii} \geq 0 \ , \ \ \alpha_{ii}\alpha_{kk} - (\alpha_{ik} + \alpha_{ki})^2/4 \geq 0 \ , \ \ \ldots \ , \ i,k = 1,2,\ldots \ . \tag{6.4-15}$$

In reality, the α_{ik} are phenomenological coefficients, such as the viscosity, which need to be specified, in general, through the techniques of physical chemistry and statistical mechanics. All that we know about them is, in most cases, nothing more than (6.4-15) and a few other items to be discussed in the remainder of this chapter.

The bulk of non-equilibrium thermodynamics is concerned with the above type of linear phenomenological relations, and to this date has not progressed considerably beyond this point—see §7.1.1. Still, in a large number of cases, it is accepted that the linear relations do indeed yield correct results; i.e., the equations so written seem to adequately approximate the experimentally observed behavior.

6.4.3 *Frame indifference and material invariance*

Two more points need to be considered in this subsection. First, in expressing the above relations, we have assumed for simplicity that the fluxes and affinities were scalars. In so doing, we have guaranteed that the relations are invariant of the particular coordinate system with respect to which we are working. In general, however, we shall have to deal with tensorial quantities, and, in such cases, it is essential that the above relations be defined with the proper transformation properties in mind. The consistency of the equations with respect to changes in the coordinate frame is called the *principle of frame indifference* [Truesdell, 1984, p. 230], and its application follows the usual rules of tensor calculus and defines the proper mathematical character of each phenomenological coefficient α_{ik}, depending on the tensorial character of the corresponding flux and affinity, J_i and Λ_k, respectively [Gyarmati, 1970, §III.4a]. (We shall have more to say about this principle in chapter 7.) It is also advisable to consider whether or not the properties of the system will change upon certain coordinate transformations, such as rotation and central inversion. If not, then the number of independent coefficients can be greatly reduced. For example, many times, particularly for fluids, it is safe to assume that particular characteristic properties of the system are the same in all directions. The invariance of the equations with respect to specific symmetry transformations, in accordance to the assumed symmetry of the material, is called the *principle of material invariance*, and it is, when applicable, the single greatest factor in limiting the number of independent phenomenological coefficients appearing in the matrix $\boldsymbol{\alpha}$. Both principles are, of course, applicable to general nonlinear constitutive equations as well—see Rivlin [1980] for a lucid description of the restrictions imposed by the above principles to general constitutive equations.

There are several types of symmetries which we shall use as material invariance conditions throughout the remainder of the book. The first of these is *central symmetry*, which applies for achiral systems and implies that the material properties are invariant during an inversion of the coordinate axes; i.e., as the Cartesian base vectors **i**, **j**, and **k** become $-\mathbf{i}$, $-\mathbf{j}$, and $-\mathbf{k}$. An *isotropic* material is one in which the characteristic properties are invariant to arbitrary rotations of the coordinate axes in space;[8] and a *transversely isotropic* material is one in which the properties are invariant under rotations about a particular direction in space. Most simple fluids are isotropic, yet we shall see in Part II how transversely isotropic and anisotropic fluid systems result from the same thermodynamic formalism.

As a consequence of the material invariance principle, for isotropic materials we may derive a result which is commonly referred to as the *Curie principle* [de Groot and Mazur, 1962, ch. VI, §2; Yourgrau *et al.*, 1966, p. 21; Prigogine, 1967, p. 54; Gyarmati, 1970, §III.4; Woods, 1975, ch. 8; Lavenda, 1978, pp.38,9].[9] Let us consider a fluid particle system where

[8]Many authors [e.g., Woods, 1975, §40.5; Truesdell, 1984, p. 390] consider isotropy equivalent to property invariance under general orthogonal transformations, $\mathbf{x}'=\mathbf{Q}\cdot\mathbf{x}$, $\mathbf{Q}^T=\mathbf{Q}^{-1}$. Others [Rivlin, 1980, p. 333], however, make the distinction between invariance with respect to *proper* orthogonal transformations ($\mathbf{Q}^T=\mathbf{Q}^{-1}$ and $\det(\mathbf{Q})=1$, i.e., rotations) as opposed to invariance with respect to the *full* orthogonal group (i.e., both rotations and central inversion). Since the applications addressed here involve achiral systems, (i.e., systems possessing central symmetry) the first, more restricted, definition for isotropy is the one used here.

[9]Actually, Woods [1975, p. 189] notes that there is some question as to the origin of this principle. Indeed, Truesdell [1984, p. 389] presents the conclusions from the quoted paper of Curie [1894, pp. 413,4], noting that there is little in common with what is usually attributed in this excerpt, and therefore dubs it the "non-existent theorem of algebra"—see [Truesdell, 1984, p. 391]. However, a more cautious search reveals the assertion "*Il n'est pas d'effet sans causes.* Les effets, ce sont des phénomènes qui nécessitent toujours, pour se produire, une certaine dissymétrie. Si cette dissymétrie n'existe pas, le phénomène est impossible" [Curie, 1894, p. 414], which more or less corresponds to that usually attributed to Curie: "macroscopic causes can not have more elements of symmetry than the effects they produce" [Yourgrau *et al.*, 1966, p. 21]. There are, however, some problems associated in placing excessive weight on the value of this statement. First, it is not, in general, true, as the breaking of symmetries during bifurcations indicates. Second, to be useful, it has been reinterpreted over the years to finally be represented by more mathematical statements of which a more cautious example is: "if a system is isotropic, it may be shown that those terms which correspond to a coupling of tensors whose orders differ by an odd number do not occur" [Hirschfelder *et al.*, 1954]. As Truesdell has shown, this is but an oversimplification of the rules that can be deduced for the behavior of an isotropic material [Truesdell, 1984, p. 390]. Consequently, we agree with Truesdell in stating that this "theorem" is best expressed in its full generality using the tools of continuum mechanics, instead. However, we need to recognize in Curie the insight that eventually led to the mathematical representations now employed by continuum mechanics—see Altmann [1992, ch. 1] for an insightful discussion on the origin of the symmetry principle. Thus, we shall continue to designate this result as the "Curie principle."

the local entropy production rate, i.e., σ at any $\mathbf{x}$ and t corresponding to a single fluid particle subsystem, may be expressed by the fully-general relation:

$$\sigma = \sum_{i=1}^{\nu_o} J_i^o \Lambda_i^o + \sum_{i=1}^{\nu_v} \mathbf{J}_i^v \cdot \boldsymbol{\Lambda}_i^v + \sum_{i=1}^{\nu_t} \mathbf{J}_i^t : \boldsymbol{\Lambda}_i^t \; . \qquad (6.4\text{-}16)$$

In this expression, we generalize the relation of (6.4-13) to include any given number of fluxes and affinities of higher-rank tensorial character than the simple scalar fluxes and affinities of the previous example. Here $\mathbf{J}_i^v$ and $\boldsymbol{\Lambda}_i^v$ are the usual (polar) vectors evaluated in a Cartesian space and $\mathbf{J}_i^t$ and $\boldsymbol{\Lambda}_i^t$ are Cartesian, second-rank (general) tensors. There is no reason why we cannot include even higher-rank tensors into this expression as well, yet this is unnecessary for our discussion at the moment.

Although (6.4-16) is completely general, it is not in the most useful form for our analysis. As seen below, it is helpful to decompose the general second-rank tensor in terms of its polar and axial components. A tensor is polar (respectively, axial) if its components do not change sign (respectively, change sign) upon a central inversion. Any second-order Cartesian tensor, $\mathbf{a}$, may be trivially decomposed into symmetric (polar) and antisymmetric (axial) tensors according to

$$a_{\alpha\beta} = \tfrac{1}{2}(a_{\alpha\beta} + a_{\beta\alpha}) + \tfrac{1}{2}(a_{\alpha\beta} - a_{\beta\alpha}) \equiv a^s_{\alpha\beta} + a^a_{\alpha\beta} \; . \qquad (6.4\text{-}17a)$$

Furthermore, it is immediately obvious that $\mathrm{tr}\,\mathbf{a}^a = 0$. The symmetric component can also be made traceless by subtracting $\tfrac{1}{3}(\mathrm{tr}\,\mathbf{a})\boldsymbol{\delta}$ from its expression,

$$a^z_{\alpha\beta} \equiv a^s_{\alpha\beta} - \tfrac{1}{3}(\mathrm{tr}\,\mathbf{a})\delta_{\alpha\beta} \; , \qquad (6.4\text{-}17b)$$

so that $\mathbf{a}$ can be uniquely decomposed as the linear superposition of three contributions

$$a_{\alpha\beta} = \tfrac{1}{3}(\mathrm{tr}\,\mathbf{a})\delta_{\alpha\beta} + a^z_{\alpha\beta} + a^a_{\alpha\beta} \; , \qquad (6.4\text{-}18)$$

each one of which has unique symmetry properties: the first term, $\tfrac{1}{3}(\mathrm{tr}\,\mathbf{a})\boldsymbol{\delta}$, is isotropic, the second is a traceless and symmetric (polar) second-rank tensor, and the third is an antisymmetric (axial) second-rank tensor. What we have accomplished is to decompose $\mathbf{a}$ into three orthogonal components. This is evident from the orthogonality relations

$$\boldsymbol{\delta} : \mathbf{a}^a = 0 \; , \quad \boldsymbol{\delta} : \mathbf{a}^z = 0 \; , \quad \mathbf{a}^z : \mathbf{a}^a = 0 \; , \qquad (6.4\text{-}19a)$$

which lead to the following decomposition of the inner product between two arbitrary tensors $\mathbf{a}$, $\mathbf{b}$:

$$\mathbf{a} : \mathbf{b} = \tfrac{1}{3}(\mathrm{tr}\,\mathbf{a})(\mathrm{tr}\,\mathbf{b}) + \mathbf{a}^z : \mathbf{b}^z + \mathbf{a}^a : \mathbf{b}^a \; , \qquad (6.4\text{-}19b)$$

in terms of their corresponding representations. In addition, we can take advantage of the three-dimensionality of space by representing the antisymmetric tensor $\mathbf{a}^a$ in terms of an axial vector, $\mathbf{z}^a$, through

$$z_\alpha^a = \tfrac{1}{2}\varepsilon_{\alpha\beta\gamma}a_{\beta\gamma}^a \ . \tag{6.4-20}$$

(As a physical example, consider the vorticity arising similarly from the antisymmetric contribution to the velocity gradient tensor.) Note that (6.4-20) implies that the cross product of two polar vectors, manifesting as the dyad giving $\mathbf{a}^a$, is an axial vector. Hence, under a coordinate inversion when two polar vectors (written in the form $\mathbf{a}^v = a_1\mathbf{i}+a_2\mathbf{j}+a_3\mathbf{k}$) change sign, an axial vector will not change its sign. With the above properties now defined, we may rewrite (6.4-16) in a more suitable (irreducible) form for our analysis:

$$\sigma = \sum_{i=1}^{\nu_o} J_i^o\Lambda_i^o + \sum_{i=1}^{\nu_v} \mathbf{J}_i^v\cdot\boldsymbol{\Lambda}_i^v + \sum_{i=1}^{\nu_a} \mathbf{J}_i^a\cdot\boldsymbol{\Lambda}_i^a + \sum_{i=1}^{\nu_s} \mathbf{J}_i^s:\boldsymbol{\Lambda}_i^s\,, \tag{6.4-21}$$

in terms of scalar, polar, and axial vectors, and traceless polar tensor contributions. In the following analysis, for simplicity, we shall consider only a single term in each summation of (6.4-21):

$$\sigma = J^o\Lambda^o + \mathbf{J}^v\cdot\boldsymbol{\Lambda}^v + \mathbf{J}^a\cdot\boldsymbol{\Lambda}^a + \mathbf{J}^s:\boldsymbol{\Lambda}^s\ . \tag{6.4-22}$$

In order to represent the most general constitutive relationships for the fluxes which depend in a linear fashion in terms of the affinities, let us consider all possible linear couplings between flux/affinity pairs. Hence,

$$\begin{aligned}
J^o &= \alpha^{oo}\Lambda^o + \alpha_\alpha^{ov}\Lambda_\alpha^v + \alpha_\alpha^{oa}\Lambda_\alpha^a + \alpha_{\alpha\beta}^{os}\Lambda_{\alpha\beta}^s\,,\\
J_\alpha^v &= \alpha_\alpha^{vo}\Lambda^o + \alpha_{\alpha\beta}^{vv}\Lambda_\beta^v + \alpha_{\alpha\beta}^{va}\Lambda_\beta^a + \alpha_{\alpha\beta\gamma}^{vs}\Lambda_{\beta\gamma}^s\,,\\
J_\alpha^a &= \alpha_\alpha^{ao}\Lambda^o + \alpha_{\alpha\beta}^{av}\Lambda_\beta^v + \alpha_{\alpha\beta}^{aa}\Lambda_\beta^a + \alpha_{\alpha\beta\gamma}^{as}\Lambda_{\beta\gamma}^s\,,\\
J_{\alpha\beta}^s &= \alpha_{\alpha\beta}^{so}\Lambda^o + \alpha_{\alpha\beta\gamma}^{sv}\Lambda_\gamma^v + \alpha_{\alpha\beta\gamma}^{sa}\Lambda_\gamma^a + \alpha_{\alpha\beta\gamma\delta}^{ss}\Lambda_{\gamma\delta}^s\ .
\end{aligned} \tag{6.4-23}$$

Note that the Greek subscripts refer to the three spatial Cartesian coordinates over which repeated indices imply a summation. These expressions obviously contain phenomenological coefficients whose tensorial rank, n, ranges from zero to four and being either polar or axial, as determined from the known tensorial ranks and characters of the fluxes and the affinities, according to the principle of frame indifference [Gyarmati, 1970, p. 80]. Moreover, occasionally, their properties upon an interchange and/or contraction of selective indices are also known.

Now that we have established the most general constitutive relations for the fluxes, we may examine the implications of material invariance upon the system (6.4-23). In this subsection, we shall restrict our attention to an isotropic system, relegating the discussion of other types of materials to the appropriate section of the book. For an arbitrary orthogonal transformation represented by the tensor **A**,[10] a general (relative) n-th rank tensor, **T**, transforms to **T**′ via the relation[11] [Lodge, 1974, §5.1]

$$T'_{ab...n} = (\det \mathbf{A})^W A_{a\alpha} A_{b\beta} \times \ldots \times A_{n\nu} T_{\alpha\beta...\nu} , \tag{6.4-24}$$

where W is its weight. For polar tensors, W is even, and for axial tensors, W is odd. Consequently, if we consider an inversion of the coordinates in Cartesian 3-space, where $A_{\alpha\alpha}=-1$ and $A_{\alpha\beta}=0$ for all $\alpha\neq\beta$, then (6.4-24) becomes

$$T'_{ab...n} = (-1)^W(-1)^n \left|A_{a\alpha}\right| \left|A_{b\beta}\right| \times \ldots \times \left|A_{n\nu}\right| T_{\alpha\beta...\nu} . \tag{6.4-25}$$

For an isotropic material, **T**′ must be the same as **T**, and therefore this equation implies that all isotropic polar tensors of odd rank and all isotropic axial tensors of even rank (i.e., when $W+n$ is odd) vanish since they change sign under the inversion. Thus, the phenomenological coefficients $\boldsymbol{\alpha}^{ov}$, $\boldsymbol{\alpha}^{vo}$, $\boldsymbol{\alpha}^{va}$, $\boldsymbol{\alpha}^{av}$, $\boldsymbol{\alpha}^{vs}$, and $\boldsymbol{\alpha}^{sv}$ are all identically equal to zero.

Now let us consider an arbitrary rotation of the coordinate axes and deduce what other consequences will result. Following the well-known rules of tensor calculus, we know that all scalars are invariants, while no vectors exist (except the trivial one, **0**) which is invariant under rotation. Similarly, the restrictions imposed by the rotational invariance can be used to define higher-order invariant tensors. Among second-rank tensors, only the Kronecker $\boldsymbol{\delta}$ tensor is rotationally invariant, among third-rank tensors only the permutation symbol, $\boldsymbol{\epsilon}$, and among fourth-rank tensors, the three dyadic $\boldsymbol{\delta\delta}$ tensors, $\delta_{\alpha\beta}\delta_{\gamma\varepsilon}$, $\delta_{\alpha\gamma}\delta_{\beta\varepsilon}+\delta_{\alpha\varepsilon}\delta_{\beta\gamma}$, and $\delta_{\alpha\gamma}\delta_{\beta\varepsilon}-\delta_{\alpha\varepsilon}\delta_{\beta\gamma}$ [Aris, 1962, §2.7].

This information can be used to specify the form of all the remaining phenomenological coefficients. In particular, since all isotropic vectors vanish, $\boldsymbol{\alpha}^{oa}$ and $\boldsymbol{\alpha}^{ao}$ vanish. Moreover, since the Kronecker $\boldsymbol{\delta}$ tensor is orthogonal to traceless symmetric tensors we can consider without loss of generality $\boldsymbol{\alpha}^{os}=\boldsymbol{\alpha}^{so}=0$. Similarly, since the contraction between an antisymmetric and a symmetric tensor vanish, we can consider, also without loss of generality, $\boldsymbol{\alpha}^{as}=\boldsymbol{\alpha}^{sa}=\mathbf{0}$.

[10]Remember that an orthogonal transformation is one where the transformation tensor satisfies the relation $\mathbf{A}^T=\mathbf{A}^{-1}$. In consequence, detA=1 or –1.

[11]For Cartesian tensors, it is not necessary to specify contravariant and covariant indices since the two tensorial characters coincide.

The above arguments resulting from the application of the principles of frame indifference and material invariance for an *isotropic* system, result in the statement that coupling between affinities and fluxes of differing tensorial character does not occur (strong formulation of the *Curie principle* [Lavenda, 1978, p. 38]:

$$J^o = \alpha^{oo}\Lambda^o \ , \ \mathbf{J}^v = \boldsymbol{\alpha}^{vv}\cdot\boldsymbol{\Lambda}^v \ , \ \mathbf{J}^a = \boldsymbol{\alpha}^{aa}\cdot\boldsymbol{\Lambda}^a \ , \ \mathbf{J}^s = \boldsymbol{\alpha}^{ss}:\boldsymbol{\Lambda}^s \ . \tag{6.4-26}$$

However, it is important to realize that the above result does not necessarily apply to arbitrary (non-isotropic) systems, neither when the constitutive relations are allowed to be nonlinear, in which case Curie's statement is applicable in its usual "weak form" only (see footnote 9).

Actually, the above results do not represent the entire catalog of information to be gleaned from a consideration of the material invariance. Indeed, using the information on isotropic tensors provided above we can arrive at more specific expressions for the remaining phenomenological coefficients:

$$\boldsymbol{\alpha}^{vv} = \kappa_1\,\boldsymbol{\delta} \ , \quad \boldsymbol{\alpha}^{aa} = \kappa_2\,\boldsymbol{\delta} \ , \tag{6.4-27}$$

$$\alpha^{ss}_{\alpha\beta\gamma\varepsilon} = \tfrac{1}{2}\mu(\delta_{\alpha\gamma}\delta_{\beta\varepsilon} + \delta_{\alpha\varepsilon}\delta_{\beta\gamma} - \tfrac{2}{3}\delta_{\alpha\beta}\delta_{\gamma\varepsilon}) \ , \tag{6.4-28}$$

which reduces the dependencies of the phenomenological coefficients to only four scalar constants (including α^{oo}). Note that the symmetry properties with respect to an interchange of the indices and the traceless character of $\boldsymbol{\Lambda}^s$ have been used to reduce the dependence of $\boldsymbol{\alpha}^{ss}$ to one single constant, μ. Finally, given the independence of the thermodynamic forces (affinities), when the above-presented linear constitutive relations are used in (6.4-22), inequality (6.4-15) introduces the additional requirement that all four material constants be non-negative, which we term "thermodynamic admissibility conditions."

6.4.4 *The principle of microscopic reversibility*

Now suppose that we have a dynamic system with various affinities, Λ_i, i=1,2,..., and that at some instant in time we reverse the motion of all fluid particles in the system. In this situation, for an isolated system, any dynamic variable a_i and any affinity Λ_i which are fundamental properties of the system will either remain constant or change their signs (but not their magnitudes) in some time increment, $d\tau$, which is small enough that no information from the surroundings will upset the reversal. This is called the *principle of microscopic reversibility*, and is a direct result of the axiomatic principle that the laws of physics remain unchanged whether or not the system is moving forward or backward in time (i.e., if t is replaced everywhere with $-t$).[12] Following this principle, we can write a parity relation as

$$\Lambda_i(t) = \eta_i \Lambda_i(-t) , \quad i = 1,2,\dots , \tag{6.4-29}$$

where the parity η_i is either 1 or -1, depending on the nature of Λ_i. All of the thermodynamic state variables which we have discussed so far have even parities ($\eta_i = 1$), and it is therefore obvious that their time derivatives have odd parities ($\eta_i = -1$). Accordingly,

$$\sigma(t) = -\sigma(-t) , \tag{6.4-30}$$

and provided that $\boldsymbol{\alpha}$ is non-negative definite, we see from (6.4-12,13) that

$$J_i(t) = -\eta_i J_i(-t) , \quad i = 1,2,\dots . \tag{6.4-31}$$

Note, however, that if the time interval is expanded such that $\mathrm{d}t \gg \mathrm{d}\tau$, then σ can only be non-negative on such a time scale, $\sigma(\mathrm{d}t) = \sigma(-\mathrm{d}t)$, since the irreversible effects (such as particle collisions, etc.) will rapidly suppress what then amounts to a "reversal fluctuation" [Woods, 1975, p. 158].

6.4.5 *The nature of affinities and fluxes*

Let us consider the isolated system Ω as being composed of two subsystems, Ω_A and Ω_B, separated by a flexible, diathermic wall. The entropy of the overall system is considered to be a function of the extensive parameters of the system, such as U and V, which are here denoted with the generalized coordinates X_i: $S=S(X_1,X_2,\dots)$. The isolation of Ω requires that $X_i^A+X_i^B=X_i$=constant, for all i. If we calculate the rate of entropy production in Ω_A by differentiating S^A with respect to time,

$$\sigma^A = \frac{\mathrm{d}S^A}{\mathrm{d}t} = \sum_i \frac{\partial S^A}{\partial X_i^A}\frac{\mathrm{d}X_i^A}{\mathrm{d}t} , \tag{6.4-32}$$

then we can identify the flux as the time rate of change of the extensive parameters,

$$J_i^A \equiv \frac{\mathrm{d}X_i^A}{\mathrm{d}t} , \quad i = 1,2,\dots , \tag{6.4-33}$$

which, via (6.4-13), implies that

[12]Strictly speaking, the time invariance is violated, in general, and only the combined charge/parity/time (CPT) invariance is valid. However, the deviations from the time invariance are rare and small enough to be safely neglected for all practical purposes [de Groot, 1974, p. 160].

$$\Lambda_i^A = \frac{\partial S^A}{\partial X_i^A} , \quad i = 1,2,\ldots . \tag{6.4-34}$$

Hence, in this case, we find that the intrinsic affinities of each subsystem are the derivatives of the entropy function, for that subsystem, with respect to the extensive parameters of that subsystem. For example, if we consider the internal energy as the extensive parameter of interest, then we can express the overall system affinity as

$$\Lambda_u \equiv \Lambda_u^B - \Lambda_u^A = \frac{1}{T_B} - \frac{1}{T_B} \equiv \Delta(\frac{1}{T}) , \tag{6.4-35}$$

since any change in U^B corresponds to an equal, but opposite, change in U^A. Therefore, the affinity can be viewed as a thermodynamic force which wishes to restore the overall system to equilibrium. Other affinities may be defined analogously for this system by consideration of the other extensive variables. (In §7.1 we shall see how the affinities are interpreted in terms of the generalized bracket.) Note that for a continuous system, we must replace the $\Delta(\cdot)$ in (6.4-35) by the gradient operator $\nabla(\cdot)$ in order to generalize this expression to fluid particle subsystems. It also follows from (6.4-34) that the differential change in the local entropy, $S^A(X_1^A, X_2^A, \ldots)$, is

$$dS^A = \sum_i \Lambda_i^A \, dX_i^A , \tag{6.4-36}$$

since we have assumed that each subsystem is at internal equilibrium.

In this type of analysis, we can distinguish between two types of affinities which occur in compound systems: one resulting from changes in the internal variables within a system, called *relaxational affinities*, and the other resulting from transfers *between* subsystems, called *transport affinities*. Some extensive variables, such as U and V (and, later on, $\mathbf{M}$ as well in extended state space), give rise only to transport affinities, and always appear with a $\Delta(\cdot)$ operator (or a gradient operator for continua) and thus are always affinities of, at least, vectorial nature. These variables are restricted in the sense that an increase in, say, U at time t in one subsystem corresponds to a decrease in U of the surrounding subsystems by an equal amount at time t. Still other variables may give rise to both types of affinities; for example, N_i may be associated with a relaxational affinity in a reacting system, yet still have a transport affinity due to diffusion. (A similar statement applies to a material with internal microstructure.) Hence these latter variables may be associated with affinities which display *any* tensorial character, and the coupling therein must be handled very carefully.

In the above scenario, the affinities popped out fairly easily from the consideration of the thermodynamics of the system, yet we shall not always be so lucky in that it may require more ingenuity to find the proper affinity for a given system process. This is particularly important since the affinities are coupled together tenuously through the Onsager/Casimir reciprocity theorem, which we wish to use to restrict the number of independent coefficients of the matrix $\boldsymbol{\alpha}$ even further than discussed in §6.4.2. As discussed later in §6.5, in order to apply the symmetry relations arising from this theorem, the fluxes and affinities must represent an independent set.

In addition, the selection among all possible sets of independent affinities for describing the non-equilibrium state of the system may be provided by experimental considerations. An illustration of this point is given below, where we show that the affinities obtained using the simple ideas presented above, though thermodynamically correct, are inadequate for an experimental description of simultaneous heat and mass transfer between two subsystems because of the "unmeasurable (non-sensible) heat" associated with the enthalpy of the transferred mass.[13]

Let us consider an adiabatic, closed system, Ω, composed again of two subsystems, Ω_A and Ω_B, separated by a flexible, permeable, diathermic partition. This time, however, we allow the subsystems to exchange both heat and matter in an unspecified process, so that Ω_A and Ω_B are now considered as containing N_i^A, N_i^B, i=1,2,...,ν, moles of component i, respectively. Furthermore, in order to simplify the following analysis, we assume that each subsystem is in contact with a pressure reservoir, which maintains a constant pressure, p_A and p_B, in each subsystem. In general, there will be a flow of heat between the two subsystems, $d\Xi$, and a flow of mass, $dN_i^B = -dN_i^A \equiv dN_i$.

If we allow an infinitesimal amount of compression/expansion work to be performed on the system, then, according to the first law, the internal energy will change by an amount

$$dU_A + dU_B = -p_a dV_A - p_B dV_B \,, \tag{6.4-37a}$$

or

$$dU_A + p_A dV_A = -(dU_B + p_B dV_B) \,. \tag{6.4-37b}$$

Such a process will, in general, require a transfer of both heat and mass within the system to re-establish the appropriate equilibrium. If we assume that the subsystems are in local equilibrium, then we can exploit the Gibbs relation for Ω_A and Ω_B,

$$dS_A = \frac{1}{T_A}(dU_A + p_A dV_A) - \frac{1}{T_A}\sum_{i=1}^{\nu} \mu_i^A \, dN_i^A \,, \tag{6.4-38a}$$

[13]This example is taken in modified form from Førland *et al.* [1988, §2.1.2].

$$dS_B = \frac{1}{T_B}(dU_B + p_B dV_B) - \frac{1}{T_B}\sum_{i=1}^{\nu} \mu_i^B \, dN_i^B \,, \qquad (6.4\text{-}38b)$$

in order to express the total entropy change as

$$dS = dS_A + dS_B = (dU_B + p_B dV_B)\,\Delta\left(\frac{1}{T}\right) - \sum_{i=1}^{\nu} \Delta\left(\frac{\mu_i^+}{T}\right) dN_i \,. \qquad (6.4\text{-}39)$$

The quantity $dU_B + p_B \, dV_B$ is obviously just the total heat change in Ω_B, which we define as $d\Xi$. Also obviously, this is equal but opposite to the total heat change in Ω_A.

It might now seem that (6.4-39) has defined the affinities for us; however, we know that the affinities which are the most useful to us (i.e., those which apply to the experimental considerations) have to be associated with measurable quantities. Yet it is apparent that the total heat change embodied by $d\Xi$ is composed not only of the "measurable heat" which we commonly recall, but also of heat associated with the enthalpy of the transferred mass. This enthalpy change during the process between the two subsystems may be unequal, and hence so will be the change in the measurable heat. Consequently, the affinities of (6.4-39), although independent, are coupled (i.e., they are not orthogonal) as shown below, and we therefore wish to decouple them into a new set of independent affinities.

In order to accomplish this, we start by rewriting (6.4-38b) as

$$\frac{1}{T_B} d\Xi = dS_B + \frac{1}{T_B}\sum_{i=1}^{\nu} \mu_i^B \, dN_i^B \,. \qquad (6.4\text{-}40)$$

Now if we treat S_B as a state variable of p_B, T_B, and the N_i^B, then we can calculate the differential as

$$dS_B = \left.\frac{\partial S_B}{\partial T_B}\right|_{p_B, N_i^B} dT_B + \left.\frac{\partial S_B}{\partial p_B}\right|_{T_B, N_i^B} dp_B + \sum_{i=1}^{\nu} \left.\frac{\partial S_B}{\partial N_i^B}\right|_{p_B, T_B, N_j^B} dN_i^B \,. \qquad (6.4\text{-}41)$$

Yet the second term on the right-hand side vanishes due to the pressure reservoir, and if we use some relations from chapter 4, then we can rewrite this expression as

$$dS_B = \frac{C_p^B}{T_B} dT_B - \sum_{i=1}^{\nu} \left.\frac{\partial \mu_i^B}{\partial T_B}\right|_{p_B, N_i^B} dN_i^B \,. \qquad (6.4\text{-}42)$$

Substituting (6.4-42) into (6.4-40) then yields

$$\frac{1}{T_B} d\Xi = \frac{dq_B}{T_B} - \sum_{i=1}^{\nu} \left.\frac{\partial \mu_i^B}{\partial T_B}\right|_{p_B, N_i^B} dN_i^B + \frac{1}{T_B}\sum_{i=1}^{\nu} \mu_i^B \, dN_i^B \,, \qquad (6.4\text{-}43)$$

with dq_B now being the measurable heat change in Ω_B during the process.

Let us now assume that for an infinitesimal transfer we have $T_B = T_A + \Delta T$ such that $\Delta T \ll T_A, T_B$, as well as $\mu_i^B = \mu_i^A + \Delta\mu_i^+$ such that $|\Delta\mu_i^+| \ll |\mu_i^A|, |\mu_i^B|$. This allows us to write

$$\Delta(\frac{\mu_i^+}{T}) \approx \mu_i^+ \Delta(\frac{1}{T}) + \frac{1}{T}\Delta(\mu_i^+) , \tag{6.4-44a}$$

$$\Delta(\frac{1}{T}) \approx - \frac{\Delta T}{T^2} . \tag{6.4-44b}$$

Substituting Eqs. (6.4-43,44) back into (6.4-39) then gives

$$dS = dq_B \Delta(\frac{1}{T}) - \frac{1}{T}\sum_{i=1}^{\nu} \left[\Delta(\mu_i^+) - \left.\frac{\partial \mu_i^+}{\partial T}\right|_{p,N_i} \Delta T \right] dN_i , \tag{6.4-45}$$

so that we can express the dissipation in the form

$$dS = -\frac{1}{T}\left[\frac{dq_B}{T}\Delta T + \sum_{i=1}^{\nu} \Delta(\Lambda_i^+)\, dN_i \right] , \tag{6.4-46}$$

where $\Delta\Lambda_i^+$ is defined as the quantity within the square brackets in (6.4-45); i.e., $\Delta\Lambda_i^+ \equiv \Delta(\mu_i^+) + M_w^i \overline{S}_i \Delta T$. Hence these new affinities are independent and each is associated with a measurable quantity, dq_B or the dN_i. It is now obvious from (6.4-46) that the proper driving force for the mass transfer is the *difference in the chemical potential at constant temperature*. In order to see this more directly, consider the chemical potential as the function $\mu_i^+ = \mu_i^+(T,p,N_1,\ldots,N_\nu)$. In this case,

$$\Delta(\mu_i^+) = \left.\frac{\partial \mu_i^+}{\partial T}\right|_{p,N_i} \Delta T + \left.\frac{\partial \mu_i^+}{\partial p}\right|_{T,N_i} \Delta p + \sum_{i=1}^{\nu} \left.\frac{\partial \mu_i^+}{\partial N_i}\right|_{T,p,N_j} \Delta N_i , \tag{6.4-47}$$

but

$$\Delta(\Lambda_i^+) = \Delta(\mu_i^+) + M_w^i \overline{S}_i \Delta T = \left.\frac{\partial \mu_i^+}{\partial p}\right|_{T,N_i} \Delta p + \sum_{i=1}^{\nu} \left.\frac{\partial \mu_i^+}{\partial N_i}\right|_{T,p,N_j} \Delta N_i . \tag{6.4-48}$$

Note that $\partial\mu_i^+/\partial p$ is the partial molar volume. The above arguments transfer over directly to a continuum system by replacing the $\Delta(\cdot)$ operator with $\nabla(\cdot)$. In such a case, the similarity of this expression with (6.4-4) is interesting.

For one particular subsystem, say Ω_B, we can inspect in more detail the exact nature of the total heat transfer during the process. In Ω_B, the total change in entropy is given by

$$dS_B = \frac{d\Xi}{T_B} - \frac{1}{T_B}\sum_{i=1}^{\nu} \mu_i^B \, dN_i^B = \frac{dq_B}{T_B} + \sum_{i=1}^{\nu} M_w^i \overline{S}_i^B \, dN_i^B \,, \qquad (6.4\text{-}49)$$

so that

$$d\Xi = dq_B + \sum_{i=1}^{\nu} (\mu_i^B + T_B M_w^i \overline{S}_i^B)\, dN_i^B = dq_B + \sum_{i=1}^{\nu} M_w^i \overline{E}_i^B \, dN_i^B \,, \qquad (6.4\text{-}50)$$

($M_w^i \overline{E}_i^B$ being the partial molar enthalpy of Ω_B) which verifies explicitly that $d\Xi$ is composed of two very different terms, one being associated with the enthalpy of the transferred mass.

Let us again emphasize that the above arguments have an effect on the experimental determination of the phenomenological coefficients only. As we shall formally prove in §6.5.3, any independent set of fluxes and forces can be used to describe dissipative phenomena without affecting the validity of the thermodynamic implications, namely, the non-negative character of the entropy production rate and the Onsager reciprocity relations. However, from an experimental viewpoint, the determination of the components of $\boldsymbol{\alpha}$ necessitates the establishment of experimentally realizable individual affinities such as those defined above. Furthermore, in the above analysis, it was assumed that in addition to mass transfer, the only other quantity to change between the two systems was the temperature corresponding to an internal heat transfer. If additional quantities are allowed to vary in the state space of the problem, as, for example, the momentum density $\mathbf{M}=\rho\mathbf{v}$ due to a momentum transfer, the corresponding affinity associated with the mass transfer may be suitably modified as

$$\Delta\Lambda_i^* \equiv \left|\Delta\mu_i^*\right|_{T,\mathbf{v}} = \Delta\mu_i^* - \frac{\partial\mu_i^*}{\partial T}\Delta T - \frac{\partial\mu_i^*}{\partial \mathbf{v}}\cdot\Delta\mathbf{v} \,, \qquad (6.4\text{-}51)$$

where $\mu_i^* = \mu_i^*(T,p,N_1,\ldots,N_\nu,\mathbf{v})$ is now defined as the generalized chemical potential (on a unit mole basis). This procedure effectively decouples the enthalpy associated with the mass transfer from the measurable heat transfer and the momentum transfer.

6.5 The Onsager/Casimir Reciprocal Relations

In order to reduce the number of independent phenomenological coefficients appearing in the matrix $\boldsymbol{\alpha}$ even further than discussed in §6.4, we discuss in this section a result known as the *Onsager/Casimir reciprocity theorem* [de Groot and Mazur, 1962, p. 2]. This theorem in essence declares that $\boldsymbol{\alpha}$ must be a symmetric (in some cases, antisymmetric) matrix, although the proof of this statement is not necessarily an obvious consequence of the phenomenological relations, but a subtle one based upon the principle of microscopic reversibility. Even so, to designate these relationships with the word "theorem" is probably tenuous, at best,[14] since they have only been generally proven for relaxational affinities, and, even in this simpler case, there is still some semblance of hypothesis in the necessary arguments.[15] Consequently, although we present in the following section the standard derivation of these reciprocal relations, we still must realize that assuming their validity, especially for transport affinities, is an act of faith.[16] The saving grace behind such an assumption is that from an experimental standpoint there seems to be no doubt as to their validity [Miller, 1974]. Regardless, we want to explore the Onsager /Casimir reciprocal relations in their most complete generality, and it is therefore necessary to show as well the validity of the reciprocity theorem under arbitrary redefinitions of the independent affinities; i.e., we must show that $\boldsymbol{\alpha}$ remains symmetric under certain transformations of the system affinities. (This point will become of importance in §7.1.)

6.5.1 Einstein's fluctuation theory

Before presenting the derivation of the reciprocal relations, it is necessary to discuss in general the principles upon which it is based. We have already discussed in §6.4.4 the principle of microscopic reversibility. Yet this by itself is not enough, and we shall take recourse to the fluctuation theory, as developed by Einstein [1905], as a requisite mathematical tool for the derivation. The following analysis is standard, and the interested reader may turn to any modern textbook on statistical mechanics for a deeper discussion.

[14]Indeed, as shown by Geigenmüller *et al.* [1983b], the Onsager/Casimir reciprocal relations do not hold but in an approximate fashion due to the approximate nature of the macroscopic descriptions. Those descriptions typically arise by eliminating rapidly evolving variables in favor of slowly evolving ones. The Onsager symmetry violation, observed during the elimination process of fast variables in a linear system [Geigenmüller *et al.*, 1983a], has been attributed by Geigenmüller *et al.* [1983b] to the initial slip of the correlation functions on the reduced level of description. However, the observed effect is small, which explains the success of the Onsager approximation.

[15]Proofs involving transport processes are available through microscopic theories when additional assumptions are made, as, for example, through the kinetic theory of dilute gases [de Groot, 1974, §9.3].

[16]The most extroverted critic of the Onsager/Casimir relations is undoubtedly Truesdell [1984, Lecture 7], who devotes forty pages of his treatise to debunking "Onsagerism." Note, however, that by mentioning this reference we do not wish to imply that we condone this terminology. Indeed, we are not even certain whether or not the author of the above-cited treatise would consider us as "Onsagerists." The reader is left to form his own opinion.

Einstein's fluctuation theory is based on the concept from statistical mechanics that the entropy of a system, S, at a specific energy, U (more precisely, between U and $U+dU$), is correlated with the number of microscopic states $\Gamma(U)$ through the Boltzmann relationship [Reichl, 1980, p. 309],

$$S = S(U) = k_B \ln \Gamma(U) . \tag{6.5-1}$$

This fundamental relation of statistical mechanics can also be used in order to define the entropy associated with a non-equilibrium macroscopic state identified through a set of parameters $\{A_1,A_2,\ldots,A_n\}$ as

$$S(U,A_1,A_2,\ldots,A_n) = k_B \ln \Gamma(U,A_1,A_2,\ldots,A_n) . \tag{6.5-2}$$

Moreover, $\Gamma(U,A_1,A_2,\ldots,A_n)$ is proportional to the probability of simultaneously observing the parameters $A_1,A_2,\ldots,A_n$, which is proportional to the number of microstates which are compatible with their macroscopic observation. Therefore, we can obtain an expression for the probability distribution function $f(U,A_1,A_2,\ldots,A_n) \equiv \Gamma(U,A_1,A_2,\ldots,A_n)/\Gamma(U)$ for the observation of $U,A_1,A_2,\ldots,A_n$ from the corresponding expression for the associated entropy S by inverting (6.5-2):

$$f(U,A_1,A_2,\ldots,A_n) = \frac{1}{\Gamma(U)} \exp\left[\frac{1}{k_B} S(U,A_1,A_2,\ldots,A_n)\right] . \tag{6.5-3}$$

Furthermore, around a stable equilibrium state characterized by A_i^o, $i=1,\ldots,n$, the entropy can be further approximated by its truncated Taylor expansion with respect to fluctuations of the defining variables

$$a_i \equiv A_i - A_i^o , \quad i = 1,2,\ldots,n , \tag{6.5-4}$$

as

$$S(U,A_1,A_2,\ldots,A_n) \approx S(U,A_1^o,A_2^o,\ldots,A_n^o) - \sum_{i=1}^{n} \sum_{j=1}^{n} \tfrac{1}{2} g_{ij} a_i a_j , \tag{6.5-5}$$

where

$$g_{ij} \equiv - \frac{\partial^2 S}{\partial a_i \partial a_j} , \tag{6.5-6}$$

corresponds to a symmetric, non-negative-definite matrix from the requirements of stable equilibrium; i.e., the entropy must be a maximum at equilibrium. Therefore, a simplified expression of the probability distribution function $f(U,a_1,a_2,\ldots,a_n)$ for the observation of the fluctuations $\mathbf{a}=(a_1,a_2,\ldots,a_n)^T$ can be obtained from (6.5-3) as

$$f(U,a_1,a_2,\ldots,a_n) = C \exp\left[- \frac{1}{2k_B} \sum_{i=1}^{n} \sum_{j=1}^{n} g_{ij} a_i a_j \right] , \tag{6.5-7}$$

where C is a normalization constant

$$C \equiv \left[\frac{\det \mathbf{g}}{(2\pi k_B)^n} \right]^{1/2} . \tag{6.5-8}$$

This is a generalized Gaussian distribution function. A critical property of this type of probability function is that moment averages are very easily calculated in explicit form [Reichl, 1980 p. 310]. For example,

$$<a_i,a_j> \equiv \int_{-\infty}^{\infty} a_i a_j\, f(U,a_1,a_2,\ldots,a_n)\, \mathrm{d}^n a = k_B\, g_{ij}^{-1} , \tag{6.5-9}$$

and therefore

$$<a_i , \frac{\partial S}{\partial a_j}> = k_B\, \delta_{ij} . \tag{6.5-10}$$

The matrix $\mathbf{g}$ can be determined based on the microstructure of the system under investigation and the relevant assumptions. Several examples are provided in standard textbooks on irreversible thermodynamics, such as de Groot and Mazur [1962].

EXAMPLE 6.3: Here we discuss the simple example of the above principles presented by Reichl [1980, pp. 311-3]. Consider m subsystems of a compound system, each containing the same amount of a single-component fluid. In this case, the associated total entropy change, ΔS_T, in response to a random set of independent fluctuations in temperature and volume in the various subsystems is given as

$$\Delta S_T = \frac{1}{2T} \sum_{i=1}^{m} [-\Delta T_i \Delta S_i + \Delta p_i \Delta V_i] , \tag{6.5-11}$$

where ΔT_i, etc. are the fluctuations monitored in the i-th subsystem and T is the (average) system temperature. If now for every subsystem we consider the local temperature and volume as the independent variables, we can express the changes in entropy and pressure in that subsystem as

$$\Delta S_i = \frac{\partial S}{\partial T} \Delta T_i + \frac{\partial S}{\partial V} \Delta V_i = \frac{C_v}{T} \Delta T_i + \frac{\partial p}{\partial T} \Delta V_i , \tag{6.5-12a}$$

$$\Delta p_i = \frac{\partial p}{\partial T} \Delta T_i + \frac{\partial p}{\partial V} \Delta V_i = \frac{\partial p}{\partial T} \Delta T_i - \frac{1}{\kappa_T V} \Delta V_i . \tag{6.5-12b}$$

Therefore

$$\Delta S_T = -\frac{1}{2T} \sum_{i=1}^{m} \left[\frac{C_v}{T} (\Delta T_i)^2 + \frac{1}{\kappa_T V} (\Delta V_i)^2 \right] , \tag{6.5-13a}$$

and consequently,

$$g_{ij} = \begin{cases} C_v/T^2 & \text{for } i=j,\ 1 \le i \le m \\ 1/(\kappa_T T V) & \text{for } i=j,\ m+1 \le i \le 2m\ , \\ 0 & \text{otherwise} \end{cases} \tag{6.5-13b}$$

where $\mathbf{a}=(\Delta T_1,\dots,\Delta T_m,\Delta V_1,\dots,\Delta V_m)^{\mathrm{T}}$. This implies that in this particular example the entropy matrix $\mathbf{g}$ is diagonal; i.e., there is no correlation between the various thermodynamic subsystems. This is the result of our assumptions in developing Eqs. (6.5-11,12), where we considered the fluctuations ΔT_i and ΔV_i, $i=1,2,\dots,m$, to be independent from each other. This is a valid approximation when the overall system is at thermal and mechanical equilibrium, which implies that all the subsystem temperatures and volumes are equal to each other. A different matrix $\mathbf{g}$ will result if the fluctuations in one subsystem are correlated with the fluctuations in the neighboring subsystems, corresponding to an interchange of heat and/or volume; however, the more general expression necessitates information about the coupling between different subsystems. It is this more general development, applicable when temperature and density gradients are present, which is of practical importance in the hydrodynamics of real systems. ■

6.5.2 *The Onsager/Casimir reciprocal relations*

The Onsager/Casimir reciprocal relations [Onsager, 1931a,b; Casimir, 1945] are based on the principle of microscopic reversibility and the ideas of fluctuation theory presented above. Consider the joint probability distri-bution function $f(\mathbf{a},\mathbf{a}';t,t+\tau)$ defined as the probability density of observing the set of fluctuations $\mathbf{a}$ and $\mathbf{a}'$ at times t and $t+\tau$, respectively. Then we can define the conditional probability distribution function $f(\mathbf{a},t\,|\,\mathbf{a}',t+\tau)$ of observing the fluctuation vector $\mathbf{a}'$ at time $t+\tau$ given that the fluctuation vector $\mathbf{a}$ has been observed at time t as

$$f(\mathbf{a},t\,|\,\mathbf{a}',t+\tau)\ \mathrm{d}^n a' \equiv \frac{f(\mathbf{a},\mathbf{a}';t,t+\tau)\ \mathrm{d}^n a'}{f(\mathbf{a},t)}\ , \tag{6.5-14}$$

where the overall probability density function $f(\mathbf{a},t)$ is defined as

$$f(\mathbf{a},t) \equiv \int f(\mathbf{a},\mathbf{a}';t,t+\tau)\ \mathrm{d}^n a'\ . \tag{6.5-15}$$

These definitions follow analogously to those of §3.5. The principle of microscopic reversibility (provided that all the quantities A_i, $i=1,\dots,n$, are even functions of the velocities) can be written as

$$f(\mathbf{a},\mathbf{a}';t,t+\tau) = f(\mathbf{a}',\mathbf{a};t,t+\tau) , \tag{6.5-16}$$

i.e., expressing that the joint probability of observing **a** at time t and **a**′ at time $t+\tau$ is the same as the joint probability of observing **a**′ at time t and **a** at time $t+\tau$. With the help of the definitions of the conditional probabilities in (6.5-14,15), Eq. (6.5-16) may be rewritten as

$$f(\mathbf{a},t)\, f(\mathbf{a},t\,|\mathbf{a}',t+\tau) = f(\mathbf{a}',t)\, f(\mathbf{a}',t\,|\mathbf{a},t+\tau) . \tag{6.5-17}$$

This relationship can be used to derive the Onsager reciprocal relations as follows.

Let us assume that a substantial variation in the fluctuations has taken place (in the sense that the dyad $\mathbf{aa}' \gg k_B \mathbf{g}^{-1}$). In this case, a phenomenological law usually covers the dynamics of the change of **a**, which close to equilibrium is provided by the form

$$\frac{\mathrm{d}\langle\mathbf{a}\rangle_{(\mathbf{a}_\circ)}}{\mathrm{d}t} = \boldsymbol{\alpha}\cdot\langle\boldsymbol{\Lambda}\rangle_{(\mathbf{a}_\circ)} , \tag{6.5-18}$$

where the conditional averages $\langle\cdot\rangle_{(\mathbf{a}_\circ)}$ are defined by

$$\langle\mathbf{a}\rangle_{(\mathbf{a}_\circ)} \equiv \int \mathbf{a}\, f(\mathbf{a}_\circ,t_\circ\,|\mathbf{a},t)\,\mathrm{d}^n a , \tag{6.5-19}$$

and

$$\boldsymbol{\Lambda} \equiv \frac{\partial(\Delta S)}{\partial \mathbf{a}} = -\,\mathbf{g}\cdot\mathbf{a} , \tag{6.5-20}$$

with $\Delta S \equiv S - S^{\circ}$. Thus

$$\frac{\mathrm{d}\langle\mathbf{a}\rangle_{(\mathbf{a}_\circ)}}{\mathrm{d}t} = -\,\boldsymbol{\alpha}\cdot\mathbf{g}\cdot\langle\mathbf{a}\rangle_{(\mathbf{a}_\circ)} \equiv -\,\mathbf{M}\cdot\langle\mathbf{a}\rangle_{(\mathbf{a}_\circ)} , \tag{6.5-21}$$

where

$$\mathbf{M} \equiv \boldsymbol{\alpha}\cdot\mathbf{g} . \tag{6.5-22}$$

This leads to a formal solution to the differential equation of (6.5-21) of the form

$$\langle\mathbf{a}\rangle_{(\mathbf{a}_\circ)}(t) = \exp[-\mathbf{M}(t-t_\circ)]\cdot\mathbf{a}_\circ . \tag{6.5-23}$$

If (6.5-17) is multiplied by $\mathbf{aa}'$ and integrated over $\mathrm{d}^n a$ and $\mathrm{d}^n a'$, then, using Eqs. (6.5-19,23), we find the equality

$$\int \mathbf{a}\,(\mathrm{e}^{-\mathbf{M}\tau}:\mathbf{a})\, f(\mathbf{a},t)\,\mathrm{d}^n a = \int \mathbf{a}'\,(\mathrm{e}^{-\mathbf{M}\tau}:\mathbf{a}')\, f(\mathbf{a}',t)\,\mathrm{d}^n a' , \tag{6.5-24}$$

which upon evaluation of the integrals leads to

$$\mathbf{g}^{-1} \cdot (e^{-\mathbf{M}\tau})^{\mathrm{T}} = e^{-\mathbf{M}\tau} \cdot \mathbf{g}^{-1} , \tag{6.5-25}$$

or

$$\boldsymbol{\alpha} = \boldsymbol{\alpha}^{\mathrm{T}} . \tag{6.5-26}$$

These relations can be generalized for variables which are odd functions of velocities, as well as for systems with magnetic fields or angular velocities, $\mathbf{H}$ and $\boldsymbol{\omega}$, respectively, as [Woods, 1975, p. 168]

$$\alpha_{ij}(\mathbf{H},\boldsymbol{\omega}) = \eta_i \eta_j \alpha_{ji}(-\mathbf{H},-\boldsymbol{\omega}) , \quad i,j = 1,2,\ldots , \tag{6.5-27}$$

where

$$\eta_i = \begin{cases} 1 \text{ if } A_i \text{ is an even function of the velocities} \\ -1 \text{ if } A_i \text{ is an odd function of the velocities} . \end{cases} \tag{6.5-28}$$

In addition, it is implicit that $\boldsymbol{\alpha}$ is a non-negative definite matrix, since, via (6.5-20), $d\Delta S/dt = -\mathbf{g}\cdot\mathbf{a}:(d\mathbf{a}/dt) = \boldsymbol{\Lambda}:(d\mathbf{a}/dt) = \boldsymbol{\Lambda}:\boldsymbol{\alpha}\cdot\boldsymbol{\Lambda} \geq 0$, and the same follows for $\mathbf{M}$. It is interesting to note that the symmetries of $\boldsymbol{\alpha}$ ($\mathbf{M}$) are also preserved under a general transformation of the fluctuation vector $\mathbf{a}$,

$$\mathbf{a}^* \equiv \mathbf{A} \cdot \mathbf{a} , \tag{6.5-29}$$

where the matrix $\mathbf{A}$ is non-singular,[17] provided, of course, that the corresponding fluxes, affinities, and ΔS are expressed in terms of the new vector accordingly [de Groot and Mazur, 1962, pp. 107,8].

The Onsager relations can be generalized [de Groot and Mazur, 1962, p. 76; Kremer, 1987, p. 108] when the affinities are represented as gradients of the partial derivatives of S with respect to the extensive variables of the system, corresponding to transport fluxes:

$$J^i_\alpha = \sum_j \alpha^{ij}_{\alpha\beta} \nabla_\beta \frac{\partial S}{\partial X_j} , \quad i = 1,2,\ldots , \tag{6.5-30}$$

which implies that we are again dealing with the continuum compound system discussed previously. Although this form for the affinities has been used by many authors in the literature [de Groot and Mazur, 1962], when the local equilibrium assumption is consistently used in conjunction with a spatial partitioning of the continuum system, the affinities should be slightly modified, as explained in more detail in §6.6. We believe that

[17]Another requirement is that the transformation provided by (6.5-29) does not mix even and odd functions of the velocities.

these latter expressions for the affinities, also used by some authors [Hirschfelder *et al.*, 1954, ch. 11; Førland *et al.*, 1988, ch. 2], represent a more natural selection, as demonstrated below. However, irrespective of which type of expressions are used for the affinities, there seems to be agreement on the symmetry relations that are proposed for the phenomenological transport coefficients, $\alpha^{ij}_{\alpha\beta}$. We would like to alert the reader, though, to the fact that although these coefficients are symmetric for one selection of affinities, they might not necessarily be symmetric for the other, an issue elaborated upon in the following paragraphs. However, the differences are probably negligible for small departures from equilibrium, which could lead to confusion concerning experimental results.

The directional nature of both the affinities and the fluxes for transport processes led to the establishment of two symmetry conditions for the phenomenological coefficients associated with either the spatial components, α and β, or the coupling indices, i and j. It is interesting to note that, although these relationships have appeared in many references in the literature over the years, a formal proof is available only for the case of the spatial components [Casimir, 1945, p. 348]. For the symmetry requirements on the coupling indices, we are limited at the present time to a "by faith" extension of the microscopic reversibility arguments based on the entropy associated with the fluctuations in the system, coupled with a calculation of the total entropy production using a spatial partitioning of the continuum system—an approach followed in §6.6.

Based on the original analysis by Casimir [1945], the most general expression for the symmetry involving the spatial indices is [de Groot and Mazur, 1962, p. 73]

$$\frac{\partial \alpha^{ij}_{\alpha\beta}(\mathbf{H})}{\partial x_\alpha} - \frac{\partial \alpha^{ij}_{\beta\alpha}(\mathbf{H})}{\partial x_\alpha} = -\left[\frac{\partial \alpha^{ij}_{\alpha\beta}(-\mathbf{H})}{\partial x_\alpha} - \frac{\partial \alpha^{ij}_{\beta\alpha}(-\mathbf{H})}{\partial x_\alpha}\right], \quad i,j=1,2,\dots \,. \tag{6.5-31}$$

As shown by Casimir, for coefficients which are independent of an applied magnetic field, together with the fact that each transport flux remains unchanged upon a divergence-free field, the above relationship is practically equivalent to the symmetry condition

$$\alpha^{ij}_{\alpha\beta} = \alpha^{ij}_{\beta\alpha}\,, \quad i,j=1,2,\dots\,, \tag{6.5-32}$$

in the sense that the addition of an antisymmetric component would not lead to any measurable effects [Casimir, 1945, p. 348]. The symmetries on the coupling indices i and j are exactly the same as for the coefficients governing relaxational phenomena, (6.5-27). As a consequence of (6.5-27) and (6.4-14), it is apparent that when $\eta_i\eta_j=1$ and $\alpha^{ij}_{\alpha\beta}=-\alpha^{ji}_{\alpha\beta}$ or $\eta_i\eta_j=-1$ and $\alpha^{ij}_{\alpha\beta}=\alpha^{ji}_{\alpha\beta}$, the corresponding process makes no contribution to the entropy production within the system.

6.5.3 *Transformation properties of the reciprocal relations*

Due to the linearity of the phenomenological relations and the independence of the affinities, the reciprocity theorem has a couple of interesting transformation properties which we shall find fairly important later on. The first of these is that the symmetry of the phenomenological coefficient matrix is unaffected by arbitrary rescaling of the independent affinities, provided that this rescaling is applied consistently throughout (including the definition of the appropriate fluxes). In particular, the rate of entropy production, being an intrinsic physical property of the system, is required to remain invariant with respect to the scaling.

Consider a system which is described by the ν independent affinities $\Lambda_1,\Lambda_2,\ldots,\Lambda_\nu$, with the entropy production rate given by (6.4-13) and the ν fluxes by (6.4-12). A phenomenological coefficient matrix is thus specified which, according to the arguments of the preceding subsection, is assumed to be symmetric. If we now transform the affinities arbitrarily to a new set of independent affinities, $\tilde{\Lambda}_i$, as

$$\begin{aligned} \Lambda_1 &\rightarrow \tilde{\Lambda}_1 = \Lambda_1/b_1 \\ \Lambda_2 &\rightarrow \tilde{\Lambda}_2 = \Lambda_2/b_2 \\ &\;\;\vdots \\ \Lambda_\nu &\rightarrow \tilde{\Lambda}_\nu = \Lambda_\nu/b_\nu \,, \end{aligned} \tag{6.5-33}$$

where the b_i are arbitrary multiplicative factors, then in order for σ to be invariant, we must multiply each new flux, $\tilde{J}_i$, by b_i in order to preserve the quadratic form

$$\sigma = \sum_{i=1}^{\nu} \tilde{J}_i \tilde{\Lambda}_i \,. \tag{6.5-34}$$

Doing so yields the flux relations

$$\begin{aligned} \tilde{J}_1 &= \alpha_{11} b_1 b_1 \tilde{\Lambda}_1 + \alpha_{12} b_1 b_2 \tilde{\Lambda}_2 + \ldots + \alpha_{1\nu} b_1 b_\nu \tilde{\Lambda}_\nu \\ \tilde{J}_2 &= \alpha_{21} b_2 b_1 \tilde{\Lambda}_1 + \alpha_{22} b_2 b_2 \tilde{\Lambda}_2 + \ldots + \alpha_{2\nu} b_2 b_\nu \tilde{\Lambda}_\nu \\ &\;\;\vdots \\ \tilde{J}_\nu &= \alpha_{\nu 1} b_\nu b_1 \tilde{\Lambda}_1 + \alpha_{\nu 2} b_\nu b_2 \tilde{\Lambda}_2 + \ldots + \alpha_{\nu\nu} b_\nu b_\nu \tilde{\Lambda}_\nu \,, \end{aligned} \tag{6.5-35}$$

which reveals that $\tilde{\boldsymbol{\alpha}}$ is a symmetric matrix. Hence the reciprocity theorem remains valid under arbitrary re-scaling of the system affinities. Yet we must realize, as pointed out by Coleman and Truesdell [1960], that the requirements for such reasoning to be valid are not only (6.4-12), but (6.4-32) through (6.4-34) as well.

Another interesting transformation property is that the affinities do not have to be orthogonal for the reciprocity theorem to be valid. For example, if we transform the Λ_i according to

$$\begin{aligned}
\Lambda_1 &\to \tilde{\Lambda}_1 = \Lambda_1 + \sum_{i=2}^{\nu} b_i \Lambda_i \\
\Lambda_2 &\to \tilde{\Lambda}_2 = \Lambda_2 \\
&\vdots \\
\Lambda_\nu &\to \tilde{\Lambda}_\nu = \Lambda_\nu ,
\end{aligned} \tag{6.5-36}$$

then the original fluxes become

$$\begin{aligned}
J_1 &= \alpha_{11}(\tilde{\Lambda}_1 - \sum_{i=2}^{\nu} b_i \tilde{\Lambda}_i) + \alpha_{12}\tilde{\Lambda}_2 + \ldots + \alpha_{1\nu}\tilde{\Lambda}_\nu \\
J_2 &= \alpha_{21}(\tilde{\Lambda}_1 - \sum_{i=2}^{\nu} b_i \tilde{\Lambda}_i) + \alpha_{22}\tilde{\Lambda}_2 + \ldots + \alpha_{2\nu}\tilde{\Lambda}_\nu \\
&\vdots \\
J_\nu &= \alpha_{\nu 1}(\tilde{\Lambda}_1 - \sum_{i=2}^{\nu} b_i \tilde{\Lambda}_i) + \alpha_{\nu 2}\tilde{\Lambda}_2 + \ldots + \alpha_{\nu\nu}\tilde{\Lambda}_\nu ,
\end{aligned} \tag{6.5-37}$$

which in turn requires, for σ to be invariant, that

$$\begin{aligned}
\tilde{J}_1 &= J_1 \\
\tilde{J}_2 &= J_2 - b_2 J_1 \\
&\vdots \\
\tilde{J}_\nu &= J_\nu - b_\nu J_1 .
\end{aligned} \tag{6.5-38}$$

Under such a transformation, it is easy to see that the new phenomenological coefficient matrix will again be symmetric. Note, however, that in both of the above cases, the fluxes are different in the old and transformed systems.

Although in the above calculations we have considered only scalar affinities and fluxes for simplicity, corresponding to relaxational affinities, all of the results carry over directly to higher-rank tensorial quantities. Consequently, the above results apply to transport affinities as well as relaxational affinities; however, the requirement on the flux in this case is that

$$\frac{\mathrm{d}X_i}{\mathrm{d}t} = -\frac{\partial J_i^\alpha}{\partial x_\alpha}, \quad i = 1,2,\ldots \,. \tag{6.5-39}$$

A deeper explanation as to why we must have (6.5-39) follows in the next section.

EXAMPLE 6.4: In combined heat and mass transfer, different investigators often use different sets of affinities. Show that the symmetry relations between the two systems are compatible with each other only if the fluxes defined in the two systems are not the same.

Let $\mathbf{J}_1$ and $\mathbf{J}_2$ be the heat and mass flux vectors for the single-component system, respectively, associated with the affinities $\boldsymbol{\Lambda}_1 \equiv \boldsymbol{\nabla} T$ and $\boldsymbol{\Lambda}_2 \equiv \boldsymbol{\nabla}\mu^+$. Furthermore, assume that the medium is isotropic, in which case the linear relationship between the fluxes and the affinities can be simply expressed in terms of scalar phenomenological coefficients (see §6.4.3) as

$$\begin{aligned} \mathbf{J}_1 &= \alpha_{11}\,\boldsymbol{\nabla} T + \alpha_{12}\,\boldsymbol{\nabla}\mu^+ , \\ \mathbf{J}_2 &= \alpha_{21}\,\boldsymbol{\nabla} T + \alpha_{22}\,\boldsymbol{\nabla}\mu^+ , \end{aligned} \tag{6.5-40}$$

with the corresponding rate expression for the entropy production arising from (6.4-13) as

$$\sigma = \mathbf{J}_1 \cdot \boldsymbol{\nabla} T + \mathbf{J}_2 \cdot \boldsymbol{\nabla}\mu^+ \,. \tag{6.5-41}$$

Now, in terms of a new set of affinities, $\tilde{\boldsymbol{\Lambda}}_1$ and $\tilde{\boldsymbol{\Lambda}}_2$, defined as

$$\tilde{\boldsymbol{\Lambda}}_1 \equiv \boldsymbol{\nabla}\left(\frac{1}{T}\right) = -\left(\frac{1}{T^2}\right)\boldsymbol{\nabla} T , \tag{6.5-42a}$$

$$\tilde{\boldsymbol{\Lambda}}_2 \equiv \boldsymbol{\nabla}\left(\frac{\mu^+}{T}\right) = -\left(\frac{\mu^+}{T^2}\right)\boldsymbol{\nabla} T + \left(\frac{1}{T}\right)\boldsymbol{\nabla}\mu^+ , \tag{6.5-42b}$$

corresponding to the fluxes $\tilde{\mathbf{J}}_1$ and $\tilde{\mathbf{J}}_2$, respectively, the corresponding expression for the entropy production rate is

$$\sigma = \tilde{\mathbf{J}}_1 \cdot \tilde{\boldsymbol{\Lambda}}_1 + \tilde{\mathbf{J}}_2 \cdot \tilde{\boldsymbol{\Lambda}}_2 = \left[-\left(\frac{1}{T^2}\right)\tilde{\mathbf{J}}_1 - \left(\frac{\mu^+}{T^2}\right)\tilde{\mathbf{J}}_2\right] \cdot \boldsymbol{\nabla} T + \left(\frac{1}{T}\right)\tilde{\mathbf{J}}_2 \cdot \boldsymbol{\nabla}\mu^+ \,. \tag{6.5-43}$$

This expression implies, upon a direct comparison with the corresponding expression given by (6.5-41), that, in order for the entropy production to be invariant, the fluxes need to be correlated as

$$\tilde{\mathbf{J}}_1 = -T^2\mathbf{J}_1 - \mu^+ T\mathbf{J}_2$$
$$\tilde{\mathbf{J}}_2 = T\mathbf{J}_2 \,, \tag{6.5-44}$$

which implies that the corresponding fluxes cannot be the same! Notice, however, that if the fluxes are changed according to (6.5-44) and the new fluxes are expressed in terms of the new affinities using linear relationships analogous to (6.5-40) through a new matrix $\tilde{\alpha}$, then the resulting relationships are still linear and symmetric. The symmetry is, of course, destroyed if the fluxes are not expressed as (6.5-44). This simple example gives an illustration of the care that has to be exercised in the selection of the proper affinities to apply toward the reciprocal relations, a caveat which was expressed explicitly by Coleman and Truesdell [1960]. ■

6.6 Affinities and Fluxes for Continua

Now that we have examined superficially the basic structure of non-equilibrium thermodynamics, we must begin the specialization of this knowledge toward the systems for which we are interested in this book. Specifically, we need to look into the details of applying our knowledge of irreversible processes to hydrodynamic continuum systems. As we shall see, in order to do just this, it is first necessary to re-express some of the above ideas in a form suitable for a continuum analysis, as well as to extend these concepts to more complex situations.

6.6.1 *Density description of the entropy production rate*

In the previous sections of this chapter, the analysis was based on a thermodynamic description of the system of interest in terms of extensive quantities. While this policy is adequate for homogeneous systems, it is cumbersome for generally non-homogeneous continua because of the variations from point to point between the subsystems. The description which appears to be the natural one in this case is the one in terms of densities (see also §4.3.3). This description can be visualized to arise from a spatial partitioning of the system into a large number of subsystems, in the limit where the number of the subsystems goes to infinity. It has also been found to be the proper one for the description of the Poisson bracket in conservative continua (see §5.3). Indeed, as we shall see in the next chapter, the dissipation bracket must be written in terms of the density variables as well, as the entire problem must be formulated in terms of these quantities. For consistency, as we define the affinities through the entropy production rate, (6.4-32), which is expressed in terms of the absolute (extensive) variables, and as this quantity must be an invariant, it is necessary that we arrive at the same entropy production rate in the density variable formulation. This equality is not readily apparent, however, and so we must set about proving it.

Let us consider the absolute entropy of a homogeneous (sub)system to be a function of the extensive system variables: $S=S(U,V,N_1,\ldots N_\nu)$.[18] The entropy production rate for this system may then be written as

$$\frac{dS}{dt} = \left.\frac{\partial S}{\partial U}\right|_{V,N_i} \frac{dU}{dt} + \left.\frac{\partial S}{\partial V}\right|_{U,N_i} \frac{dV}{dt} + \sum_{i=1}^{\nu} \left.\frac{\partial S}{\partial N_i}\right|_{U,V,N_j} \frac{dN_i}{dt} . \tag{6.6-1}$$

In the above equation, if we replace S with sV, etc. and then rearrange yields

$$\begin{aligned} \frac{ds}{dt} &= \left.\frac{\partial s}{\partial u}\right|_{n_i} \frac{du}{dt} + \sum_{i=1}^{\nu} \left.\frac{\partial s}{\partial n_i}\right|_{u,n_j} \frac{dn_i}{dt} \\ &+ \frac{dV}{dt}\left[- s + \frac{u}{T} + \sum_{i=1}^{\nu} \mu_i^+ n_i + \left.\frac{\partial S}{\partial V}\right|_{U,N_i} \right] . \end{aligned} \tag{6.6-2}$$

However, from (4.3-4a), (4.3-23) and (4.3-24), it is clear that the last term in (6.6-2) vanishes, so that the entropy production rate in terms of the entropy density is exactly what we would expect from the functional dependence of s: $s(u,n_1,\ldots,n_\nu)$. Thus the affinities, and therefore the Onsager relations, should carry over directly from the extensive variable formulation. In fact, working in terms of the density variables has deleted the redundant affinity associated with the pressure which arises in the absolute variable framework as a result of the dependency relation represented by the Gibbs/Duhem equation.

For a homogeneous medium, when only relaxational effects are taken into account, the Onsager reciprocal relations, (6.5-27), and the quantities which they involve, can be re-expressed in terms of density fluctuations as follows. The first assumption is that the entropy density, s, can be expressed in terms of the fluctuations of the thermodynamic densities ω_i, close to equilibrium, as

$$s \approx - \sum_{i,j} \tfrac{1}{2} g_{ij} \omega_i \omega_j \, , \tag{6.6-3}$$

where $\mathbf{g}$ is a non-negative-definite, symmetric tensor,

[18] Here we consider the multicomponent system of §4.3 for the purposes of illustration. In general, however, the analysis applies to any system by incorporation of additional dependencies into the function S.

$$\mathbf{g} \equiv -\frac{\partial^2 s}{\partial \boldsymbol{\omega} \partial \boldsymbol{\omega}} \,. \tag{6.6-4}$$

The corresponding affinity is then defined as

$$\Lambda_i \equiv \frac{\partial s}{\partial \omega_i} \,, \quad i = 1,2,\ldots \,. \tag{6.6-5}$$

Note that the affinities described in terms of the densities are exactly the same quantities as the ones described in terms of their extensive counterparts because of the properties of partial differentiation[19] (see §4.3.3). The corresponding dynamic equations for the density variables are then expressed in terms of volumetric fluxes, J_i^r, as

$$\frac{\mathrm{d}\omega_i}{\mathrm{d}t} = J_i^r \,, \quad i = 1,2,\ldots \,, \tag{6.6-6}$$

for which the linear phenomenological relations can be postulated, close to equilibrium, as

$$J_i^r = \sum_j \alpha_{ij} \Lambda_i \,, \quad j = 1,2,\ldots \,. \tag{6.6-7}$$

The corresponding total rate for the entropy production in the system, $\mathrm{d}S/\mathrm{d}t$, is then given according to (6.6-2) as

$$\frac{\mathrm{d}S}{\mathrm{d}t} = \int_\Omega \frac{\mathrm{d}s}{\mathrm{d}t} \mathrm{d}^3x = \int_\Omega \sum_i \frac{\partial s}{\partial \omega_i} \frac{\mathrm{d}\omega_i}{\mathrm{d}t} \mathrm{d}^3x = \int_\Omega \sum_i \Lambda_i J_i^r \, \mathrm{d}^3x \,. \tag{6.6-8}$$

6.6.2 *Description of the entropy production rate in terms of fluxes*

Now that we have expressed the entropy production rate in terms of the density variables for relaxational processes, it is necessary to examine the fluxes $\mathrm{d}\omega_i/\mathrm{d}t$ (and the affinities as well) for transport processes. Indeed, such an analysis is necessary in order to find the appropriate form for the application of the reciprocal relations. When we include transport affinities in the system description, the simple relations of (6.6-6) are no longer valid and must be generalized to

[19]Except, of course, for the affinity associated with the partial derivative of the absolute entropy with respect to the absolute volume.

$$\frac{d\omega_i}{dt} = J_i^r - \boldsymbol{\nabla}\cdot\mathbf{J}_i^t \ , \quad i = 1,2,\ldots \ , \tag{6.6-9}$$

where J_i^r indicates a relaxational flux and $\mathbf{J}_i^t$ a transport flux. The transport flux is therefore a surface flux, analogous to the one of (6.4-6a). If we now substitute (6.6-9) into the second integral in (6.6-8), then we get the following expression for the total production rate of entropy:

$$\frac{dS}{dt} = \int_\Omega \sum_i \frac{\partial s}{\partial \omega_i}(J_i^r - \boldsymbol{\nabla}\cdot\mathbf{J}_i^t)\, d^3x \ . \tag{6.6-10}$$

Upon integration by parts of the second term on the right-hand side of (6.6-10), this equation becomes

$$\frac{dS}{dt} = \int_\Omega \sum_i \left[\frac{\partial s}{\partial \omega_i} J_i^r + \boldsymbol{\nabla}\left(\frac{\partial s}{\partial \omega_i}\right)\cdot\mathbf{J}_i^t\right] d^3x \ , \tag{6.6-11}$$

which is very similar in form to the corresponding expression appearing as the last term on the right-hand side of (6.6-8), provided that the affinities corresponding to the fluxes J_i^r and $\mathbf{J}_i^t$ are appropriately defined. Consequently, it is apparent that the proper framework for postulating the extension of the reciprocal relations to transport fluxes is that in which

$$\begin{aligned} \Lambda_i &\equiv \frac{\partial s}{\partial \omega_i} \quad \text{for relaxation} \\ \Lambda_i^\alpha &\equiv \nabla_\alpha\left(\frac{\partial s}{\partial \omega_i}\right) \quad \text{for transport} \ . \end{aligned} \tag{6.6-12}$$

If now a direct linear relationship between an affinity and the corresponding flux, similar to (6.6-7), is postulated, then—and only for this set of affinities—we can assume the corresponding reciprocal relations expressed by Eqs. (6.5-27,31,32) to be valid. In the next chapter, we shall see how the proper affinities and fluxes are derived in another description, one using the energy rather than the entropy as the fundamental equation of state, so that the assumed validity of the reciprocal relations is preserved.

7
The Dissipation Bracket

The first discovery of a new law is the discovery of a similarity which has hitherto been concealed in the course of natural processes. It is a manifestation of that which our forefathers in a serious sense described as "wit," it is of the same quality as the highest performances of artistic perception in the discovery of new types of expression.

—*Hermann Ludwig Ferdinand von Helmholtz*

We are all agreed that your theory is crazy. The question that divides us is whether or not it is crazy enough to stand a chance of being correct.

—*Niels Bohr*

In the opening chapters of the book, we saw how a variety of conservative phenomena could be described through the Poisson bracket of classical physics. As already mentioned, the majority of systems with which engineers and physicists must deal reveal dissipative phenomena inherent to their nature. In the last chapter we offered a concise overview of non-equilibrium thermodynamics as traditionally applied to describe, close to equilibrium, dissipative dynamic phenomena. In this chapter, we lay the groundwork for the incorporation of non-conservative effects into an equation of Hamiltonian form, and show several well-known examples in way of proof that such a thesis is, in fact, tenable.

The unified formulation of conservative and dissipative processes based on the fundamental equations of motion is not a new idea. Among the most complete treatments is the one offered by the Brussels school of thermodynamics [Prigogine *et al.*, 1973; Prigogine, 1973; Henin, 1974] based on a modified Liouville/von Neumann equation. This is a seminal work where, using the mathematical approach of projection operators and relying only on first principles, it is demonstrated how large isolated dynamic systems may present dissipative properties in some asymptotic limit. A key characteristic of their theory is that dissipative phenomena arise spontaneously without the need of any macroscopic assumptions, including that of local equilibrium [Prigogine *et al.*, 1973, p. 6]. The main value, however, is mostly theoretical, demonstrating the compatibility of dissipative, irreversible, processes with the reversible dynamics of elemental processes through a "symmetry-breaking process." The value of the theory in applications is limited since it relies on a quantum-mechanical equation for the density matrix ρ which is, in general, very difficult to solve except for highly simplified problems [Henin, 1974]. In a nutshell, we hope to offer with the present work a macroscopic equivalent of the Brussels school theory which, at the expense of the introduction of the local equilibrium assumption, attempts to unify the description of dynamic and dissipative phenomena from a continuum, macroscopic viewpoint. The main tool to achieve that goal is the extension of the Poisson bracket formalism, analyzed in chapter 5, to dissipative continua.

Kaufman [1984], Morrison [1984], and Grmela [1984] almost simultaneously became the first researchers to study the inclusion of dissipative phenomena in Hamiltonian equations through a dissipation bracket, analogous to the Poisson bracket. Kaufman required his bracket to be bilinear, symmetric, and positive semi-definite, and to use an "entropy functional" as the generating functional in a Hamiltonian equation; this functional is similar to the Hamiltonian functional, H, which is the generating functional of the Poisson bracket. Morrison [1984] offered a more general formulation in terms of either the energy, entropy, or Helmholtz free energy. In a later work, Morrison [1986] analyzed further the structure of the (bilinear) generalized bracket into an antisymmetric (Poisson bracket) and a symmetric (dissipative) component. The evolution equations are then obtained utilizing a generalized free energy defined as the difference of the Hamiltonian minus a generalized entropy, which is considered to be a Casimir function of the dissipationless (Poisson) bracket. Grmela [1984, p. 357] uses the Helmholtz free energy as his generating functional in his initial papers (see also Grmela [1985 and 1986]), then later uses a nonlinear dissipative potential in subsequent works [Grmela, 1989a].

These brackets come very close to the form which is proposed in the present work, below. They already have found uses in describing various diverse physical phenomena, such as a plasma collision operator [Morrison, 1986], diffusion/convection equations [Grmela, 1986], the Enskog kinetic equation [Grmela, 1985], wave/particle resonances [Kaufman, 1984], etc. The bracket utilized in the present work is a nonlinear generalization of the one introduced by Edwards and Beris [1991a,b]. It differs from the ones mentioned before mainly in its generality (the previous forms being recovered as particular cases). Thus, following the example of Edwards and Beris [1991a,b], a more general description has been chosen to allow for the simplest continuum description of general hydrodynamic phenomena. The resulting bracket has the advantage of using the Hamiltonian as its generating functional, just like the Poisson bracket. In the case of linear, close-to-equilibrium thermodynamics, this bracket has an especially simple form, which may be directly correlated with the ideas presented in chapter 6.

In the presentation of chapter 5, we discussed the hydrodynamics of simple fluids using both a material and a spatial description of the fluid flow. In the material description, since the mass and entropy of each fluid particle (thermodynamic system) was assumed to be constant, we only required evolution equations to be developed for the canonical coordinates, supplemented by a constitutive equation for the density, (5.1-6), and one for the entropy density, (5.3-7). These constitutive relationships correlated the above quantities to their initial values through the Jacobian of the time-dependent coordinate transformation. In order to describe the same flow in the spatial description, it was necessary to write evolution equations for all three dynamic variables, ρ, s, and $\mathbf{M}$, since the same fluid particle is not always occupying the same position $\mathbf{x}$. When we incorporate dissipation into the system, however, we lose the ability to describe the system in terms of the canonical coordinates alone, as we did in the material description, since the entropy density can no longer be given by a simple constitutive equation, but must satisfy the more elaborate rate expression of (6.4-3). Consequently, in the presence of dissipation, the two descriptions become more alike, in that coupled evolution equations appear for the dynamic variables in the material description, although the mass density still obeys the constitutive relation of (5.1-6). Therefore, no advantage is gained working in the material description.

In addition to the arguments in the preceding paragraph, the physical significance of the material description after the introduction of mass diffusion becomes questionable since the material particles follow the pathlines corresponding to the flow kinematics only on the average. Consequently, when discussing the dissipation bracket, we always work in terms of the spatial description where each of the dynamic variables of the system satisfies an evolution equation. This allows for a ready extension of the continuum ideas, developed for conservative systems in chapter 5, to incorporate a dissipative thermodynamic description into the coupled dynamic equations. Furthermore, in the past, dissipation in hydrodynamic systems has almost always been discussed in the spatial description, and, indeed, it is in this framework that the experiments are conducted.

7.1 The General Dissipation Bracket

Due to the phenomenological nature of the dissipation, it is as yet impossible to start with underlying first principles and proceed directly through mathematical manipulation to a final "working" dissipation bracket, as was done for the Poisson bracket, although substantial "indirect" evidence for its correctness exists as indicated below and as seen from the examination of its predictions. Hence the starting point for the following development is a mere postulate; however, it so happens that once the initial postulate is made, everything else follows automatically without arbitrariness. The postulate which we start with is as follows:

Postulate: The internal dynamics of an isolated system is completely described by the equation

$$\frac{dF}{dt} = \{[F,H]\} \ , \tag{7.1-1}$$

where F is an arbitrary function(al) of the proper dynamic variables, H is the Hamiltonian function(al) of the system, and $\{[F,H]\}$ is the generalized bracket, defined as

$$\{[F,H]\} \equiv \{F,H\} + [F,H] \ . \tag{7.1-2}$$

In the above expression, $\{\cdot,\cdot\}$ is the Poisson bracket describing the conservative (convective) effects and $[\cdot,\cdot]$ is the dissipation bracket accounting for non-conservative phenomena. Note, however, that the two brackets of (7.1-2) do not necessarily have any properties in common. The Poisson bracket of conservative media necessarily satisfies the properties discussed in chapters 2, 3, and 5, but the dissipation bracket, in accordance with the theory of irreversible thermodynamics, has properties all its own which are completely different from those of the Poisson bracket. Such being the case, the generalized bracket, $\{[\cdot,\cdot]\}$, has no specific properties which can be assigned to it, and therefore it represents a somewhat artificial construction. Yet we employ it simply for notational convenience where appropriate in the following chapters.

Partial justification for Eqs. (7.1-1) and (7.1-2) can be construed from their similarity to the equations (rather their asymptotic approximation) offered for the unified formulation of dynamics and thermodynamics by the Brussels school of thermodynamics [Prigogine, 1973, Eq. (25); Prigogine *et al.*, 1973, p. 16]:

$$i\frac{d\rho_o}{dt} = \Psi(0)\rho_o \ , \tag{7.1-3}$$

where ρ_o represents the diagonal elements of the density matrix and $\Psi(0)$ represents an operator which can be split into two types of "star-Hermitian" operators:

$$\Psi(0) \equiv \Psi^e(0) + \Psi^o(0) \ , \tag{7.1-4}$$

where the odd part is anti-Hermitian and the even part Hermitian and a negative operator,

$$-i\,\Psi^{e}(0) \le 0 \; . \tag{7.1-5}$$

All of the above properties have their equivalent in the bracket equations (7.1-1) and (7.1-2) as we shall see further by examining the properties of the dissipation bracket below. The similarity becomes even more striking when one considers that the "odd" and "even" character of the operators reflects their corresponding symmetry and antisymmetry upon time inversion [Prigogine, 1973, Eq. (30)]. Moreover, (7.1-3) originates from a projection of the Liouville/von Neumann equation for the density matrix ρ [Prigogine, 1973, Eq. (7)]:

$$i\,\frac{\mathrm{d}\rho}{\mathrm{d}t} = L(\rho) \; , \tag{7.1-6}$$

where $L(\cdot)$ is a Hermitian operator (better called super-operator because it acts on matrices in a quantum mechanical sense) identified with the Poisson bracket [Prigogine, 1973, Eq. (8)]:

$$L(\rho) \equiv -i\,\{H,\rho\} \; . \tag{7.1-7}$$

In direct analogy with (7.1-2), $\Psi^{o}(0)$ describes the conservative dynamic effects while $\Psi^{e}(0)$ accounts, to a first approximation, for the non-conservative dissipative phenomena.

Further justification for Eqs. (7.1-1) and (7.1-2) comes through a direct comparison with experiments as well as through comparisons with previous theories, namely Hamiltonian dynamics (chapter 5) and irreversible thermodynamics (chapter 6). This comparison is feasible in the cases where both the generalized bracket description and either one of these two limiting theories are applicable, i.e., when only conservative or only dissipative effects are present. By definition, the comparison at the conservative limit shows that the generalized bracket is identical with the Hamiltonian dynamics of previous chapters. Since the properties of the Poisson bracket have already been discussed extensively, we ignore it completely in the following analysis and devote the remainder of this section to providing the proper form for the dissipation bracket.

7.1.1 *Definition and properties of the dissipation bracket*

In order to come to grips with the form that the dissipation bracket must take, it is first necessary to write down as many principal requirements as possible. As a start, we have the first law of thermodynamics, which states that the total energy of the (scleronomic[1]) system must be conserved. Since the Hamiltonian represents the total energy of the system, we may replace F with H in (7.1-1) to find that

$$\frac{\mathrm{d}H}{\mathrm{d}t} = \{H,H\} + [H,H] = 0 \,, \tag{7.1-8}$$

or, due to the antisymmetry of the Poisson bracket,

$$[H,H] = 0 \,. \tag{7.1-9}$$

Hence we have the first restriction upon the form of the dissipation bracket, which arises from placing the system inside of a rigid, adiabatic wall.

Also, we have the second law of thermodynamics, which states that the total rate of entropy production within the system must be nonnegative. If we define the entropy functional as before, by (6.4-7),

$$S \equiv \int_{\Omega} s \, \mathrm{d}^3x \,, \tag{7.1-10}$$

then we know that

$$\frac{\mathrm{d}S}{\mathrm{d}t} = \{S,H\} + [S,H] \geq 0 \,, \tag{7.1-11}$$

or, since $\{S,H\}=0$, we have that

$$[S,H] \geq 0 \,. \tag{7.1-12}$$

By the same token, we can write the total mass of the system as

$$\wp \equiv \int_{\Omega} \rho \, \mathrm{d}^3x \,, \tag{7.1-13}$$

so that we may require that the total mass of the system be conserved as well:

$$\frac{\mathrm{d}\wp}{\mathrm{d}t} = \{\wp,H\} + [\wp,H] = 0 \;\Rightarrow\; [\wp,H] = 0 \,. \tag{7.1-14}$$

The only other requirement, other than frame indifference, which we place on $[\cdot,\cdot]$ from the outset, is that the bracket must be linear in F, as is the Poisson bracket. The left-hand side of (7.1-1) is by definition linear in F as the rate of change in time of a linear superposition of functionals is the same as the linear superposition of the individual rates of time

[1]A "scleronomic" system is one in which the energy is independent of time, as opposed to "rheonomic" system, in which the energy of the system does depend on time [Lanczos, 1970, p. 32].

change. In general, however, there is no restriction placed on H, and $[\cdot,\cdot]$ may well be nonlinear in H.

With these properties in mind, we can now write the most general expression possible for the dissipation bracket from a continuum perspective as

$$[F,H] = \int_\Omega \left[\Xi\left(L[\frac{\delta F}{\delta \boldsymbol{\omega}}, \boldsymbol{\nabla}\frac{\delta F}{\delta \mathbf{w}}]; \frac{\delta H}{\delta \mathbf{w}}, \boldsymbol{\nabla}\frac{\delta H}{\delta \mathbf{w}} \right) + \frac{\delta F}{\delta s} \Psi\left(\frac{\delta H}{\delta \mathbf{w}}, \boldsymbol{\nabla}\frac{\delta H}{\delta \mathbf{w}} \right) \right] d^3x \ , \tag{7.1-15}$$

where $L[\cdot]$ denotes that the dependence of Ξ is linear with respect to $\cdot$, $\mathbf{w}=(a,b,c,\ldots,\mathbf{M},s)^T$, and $\boldsymbol{\omega}=(a,b,c,\ldots,\mathbf{M})^T$; i.e., $\mathbf{w}=(\boldsymbol{\omega},s)^T$, s being the entropy density and $a,b,c,\ldots$ being the other dynamic system variables, e.g., ρ, $\mathbf{C}$, etc. Thus (7.1-15) is the most general expression possible where we have pulled out the linear dependence on $\delta F/\delta s$ and written it separately. However, it must be noted here that the linear dependence of Ξ on certain components $\delta F/\delta w_i$, $\nabla_\alpha(\delta F/\delta w_i)$ (or combinations thereof) might be prohibited due to physical considerations; i.e., some variables may or may not exhibit relaxational or transport behavior. For example, since the total mass is conserved for a single-component system, this implies (as we will see later on) that Ξ is independent of $\delta F/\delta\rho$ and $\boldsymbol{\nabla}(\delta F/\delta\rho)$. For simplicity in the following analysis, we shall still use the most general description provided by (7.1-15) in terms of all of the available variables, but the reader is cautioned as to the constraints imposed by the appropriate conservation laws before applying this expression to specific applications.

Now if we use the properties described in the opening paragraphs of this section, we can learn a little more about the above expression. For instance, using the property of (7.1-12) (which indicates why we broke the expression up the way we did), we immediately see that the second law of thermodynamics amounts to

$$\frac{dS}{dt} \geq 0 \ \Rightarrow \ \Psi\left(\frac{\delta H}{\delta \mathbf{w}}, \boldsymbol{\nabla}\frac{\delta H}{\delta \mathbf{w}} \right) \geq 0 \ . \tag{7.1-16}$$

Hence we see that Ψ is a convex functional, with an absolute minimum equal to zero at equilibrium where $\delta H/\delta\boldsymbol{\omega}=\mathbf{0}$ and $\boldsymbol{\nabla}(\delta H/\delta\mathbf{w})=\mathbf{0}$. Therefore, in order for the inequality (7.1-16) to be valid for arbitrary Hamiltonians, we have

$$\Psi\left(\frac{\delta H}{\delta \omega} = 0, \boldsymbol{\nabla}\frac{\delta H}{\delta \mathbf{w}} = 0 \right) = 0 \ , \tag{7.1-17a}$$

with

$$\left.\frac{\partial \Psi}{\partial(\delta H/\delta\boldsymbol{\omega})}\right|_{(0,0)} = 0 \ , \quad \left.\frac{\partial \Psi}{\partial(\boldsymbol{\nabla}[\delta H/\delta\mathbf{w}])}\right|_{(0,0)} = 0 \ , \tag{7.1-17b}$$

p. 171: in (7.1-17a), ω should be $\boldsymbol{\omega}$

and non-negative definite second derivatives. (Note that $|_{(0,0)}$ indicates that the derivatives are evaluated at $\delta H/\delta\boldsymbol{\omega}=\mathbf{0}$ and $\boldsymbol{\nabla}(\delta H/\delta\mathbf{w})=\mathbf{0}$.)

We may also use (7.1-9), [H,H]=0, to immediately see that

$$\Psi = -\frac{1}{\delta H/\delta s}\,\Xi\left(L[\frac{\delta H}{\delta\boldsymbol{\omega}},\boldsymbol{\nabla}\frac{\delta H}{\delta\mathbf{w}}]\,;\,\frac{\delta H}{\delta\mathbf{w}},\boldsymbol{\nabla}\frac{\delta H}{\delta\mathbf{w}}\right). \tag{7.1-18}$$

The reason that the two terms of the integrand on the right-hand side of (7.1-15) cancel each other on a point-by-point basis when $F=H$ arises from the assumed universality of Eq. (7.1-15), valid for arbitrary functionals F and H, including the limiting case when they are just the value of a functional at a point (i.e., corresponding to integrands of delta functions). Hence we may rewrite the dissipation bracket of (7.1-15) using the thermodynamic definition of the absolute temperature, $\delta H/\delta s\equiv T$, and (7.1-18), to give (in terms of arbitrary G as well as F)

$$\begin{aligned}[F,G] = \int_\Omega\Bigg[&\Xi\left(L[\frac{\delta F}{\delta\boldsymbol{\omega}},\boldsymbol{\nabla}\frac{\delta F}{\delta\mathbf{w}}]\,;\,\frac{\delta G}{\delta\mathbf{w}},\boldsymbol{\nabla}\frac{\delta G}{\delta\mathbf{w}}\right)\\ &-\frac{1}{T}\frac{\delta F}{\delta s}\,\Xi\left(L[\frac{\delta G}{\delta\boldsymbol{\omega}},\boldsymbol{\nabla}\frac{\delta G}{\delta\mathbf{w}}]\,;\,\frac{\delta G}{\delta\mathbf{w}},\boldsymbol{\nabla}\frac{\delta G}{\delta\mathbf{w}}\right)\Bigg]\,d^3x\,.\end{aligned} \tag{7.1-19}$$

This is the most general possible expression for the dissipation bracket which is consistent with the first and second laws of thermodynamics.

The physical significance of the dissipation bracket becomes immediately obvious when we consider the evolution equation, (7.1-1), in conjunction with the definition of the dissipation bracket, (7.1-19), for purely dissipative processes; i.e., under these conditions, the contribution of the Poisson bracket to the evolution equation for F is zero. If, in parallel to the analysis for conservative continua (see chapter 5), we express the evolution of an arbitrary functional F in terms of its Volterra derivatives through a straightforward integration by parts, then we obtain

$$\frac{dF}{dt} = \int_\Omega\sum_{i=0}^{m}\frac{\delta F}{\delta w_i}\frac{\partial w_i}{\partial t}\,d^3x\,, \tag{7.1-20}$$

where we consider a system with $m+1$ dynamic variables, $i=0$ corresponding to the entropy density. Now if we compare the final answer to the expression derived after substitution of (7.1-19) for the dissipation bracket into the constitutive equation (7.1-1), then the evolution equation for each one of the primary variables can be obtained as

$$\frac{\partial w_i}{\partial t} = \frac{\partial\Xi}{\partial\left(\dfrac{\delta F}{\delta w_i}\right)} - \nabla_\alpha\left[\frac{\partial\Xi}{\partial\left(\nabla_\alpha\dfrac{\delta F}{\delta w_i}\right)}\right],\quad\text{for } i\neq 0\,, \tag{7.1-21a}$$

p. 172: in last line of text, "primary" should be "dynamic"

$$\frac{\partial w_0}{\partial t} = -\frac{1}{T}\Xi\left(L[\frac{\delta H}{\delta \boldsymbol{\omega}}, \boldsymbol{\nabla}\frac{\delta H}{\delta \mathbf{w}}] ; \frac{\delta H}{\delta \mathbf{w}}, \boldsymbol{\nabla}\frac{\delta H}{\delta \mathbf{w}}\right) - \nabla_\alpha\left[\frac{\partial \Xi}{\partial\left(\nabla_\alpha \frac{\delta F}{\delta w_0}\right)}\right] , \tag{7.1-21b}$$

where, unless otherwise indicated, Ξ is considered as a linear function of $\delta F/\delta w_i$ and $\boldsymbol{\nabla}(\delta F/\delta w_i)$ and a nonlinear (in general) function of $\delta H/\delta w_i$ and $\boldsymbol{\nabla}(\delta H/\delta w_i)$.

Furthermore, for a generic variable w_i, (7.1-21) can always be interpreted in terms of the two (traditional) fluxes: a volumetric (relaxational) flux J_i^r and a surface (transport) flux $\mathbf{J}_i^t$ defined from rewriting (7.1-21a) as

$$\frac{\partial w_i}{\partial t} \equiv J_i^r - \boldsymbol{\nabla}\cdot \mathbf{J}_i^t . \tag{7.1-22}$$

Notice that in so doing the transport flux is only defined up to a divergence-free field $\mathbf{X}$:

$$\mathbf{J}_i^t = \frac{\partial \Xi}{\partial\left(\boldsymbol{\nabla}\frac{\delta F}{\delta w_i}\right)} + \mathbf{X} , \tag{7.1-23}$$

which, as having no physical consequences, may be taken as zero in the present work.

Even when, as later seen, we are occasionally employing the full nonlinear form of the dissipation bracket (see chapter 12 on applications to chemical reactions), it is important to examine its linearized form, which is expected to be valid for systems which are close to equilibrium. There, Ξ is approximated by a bilinear function with respect to F and G. Hence we can express Ξ in terms of the functional derivatives simply enough as

$$\Xi = \sum_{i,j}\left[A_{ij}\frac{\delta F}{\delta \omega_i}\frac{\delta G}{\delta w_j} + B_{ij}^{\alpha}\frac{\delta F}{\delta \omega_i}\nabla_\alpha\frac{\delta G}{\delta w_j} \right. \tag{7.1-24}$$
$$\left. + C_{ij}^{\alpha}\nabla_\alpha\frac{\delta F}{\delta w_i}\frac{\delta G}{\delta w_j} + D_{ij}^{\alpha\beta}\nabla_\alpha\frac{\delta F}{\delta w_i}\nabla_\beta\frac{\delta G}{\delta w_j}\right] ,$$

where **A**, **B**, **C**, and **D** are phenomenological coefficient matrices (analogous to $\boldsymbol{\alpha}$ defined in §6.4), which, in general, may depend on $\delta G/\delta s$ and the dynamic variables of the system, where the indices α and β run from one to three (spatial coordinates—Einstein summation convention applies), and where i and j run over the components of $\boldsymbol{\omega}$ and $\mathbf{w}$. Indeed, by comparison of Eqs. (7.1-21a) and (7.1-22), the fluxes are defined as linear functions of the affinities, $\delta H/\delta\boldsymbol{\omega}$ and $\boldsymbol{\nabla}(\delta H/\delta\mathbf{w})$, upon substitution of the bilinear expression (7.1-24) for Ξ. One may impose further constraints upon the phenomenological coefficients by considering the conservation of mass (as we shall see later on—see the remark following Eq. (7.1-15)) and material invariance. Indeed, in defining Ξ by (7.1-24), one must also consider the implications of the Curie principle for isotropic media, as discussed in §6.4.3.

The quantities $\delta H/\delta\mathbf{w}$ and $\boldsymbol{\nabla}(\delta H/\delta\mathbf{w})$ which appear in the definition of the fluxes as they automatically arise in the bracket description, (7.1-21a), represent the system affinities, the former being associated with relaxational phenomena and the latter with the transport fluxes of the dynamic variables. For example, $\boldsymbol{\nabla}(\delta H/\delta s)=\boldsymbol{\nabla}T$, is the temperature gradient field. Although in traditional irreversible thermodynamics the affinities appear in simple cases to be the derivatives of the entropy with respect to the extensive variables of the system, the present theory requires an energy representation (due to the incorporation of the momentum density into the thermodynamic state space), and the affinities appear naturally as the *functional derivatives of the Hamiltonian with respect to the dynamic (density) variables* (and the gradients thereof for transport affinities) listed as the components of the vector $\mathbf{w}$.

In order to see an informal mathematical justification of the above argument (a formal equivalence with the entropy-based description follows in the next subsection), let us consider the fictitious (unrealizable) process according to which we can establish a volumetric entropic production rate in the system of equal magnitude and of opposite sign from that prescribed in (7.1-21b). This implies that in this fictitious process the net volumetric rate of entropy production is zero, i.e., $\Psi=0$ in (7.1-15). Therefore, if the system is still assumed to be isolated at the surface (implying no surface fluxes), then the total entropy has to remain constant. Thus, according to the second law of thermodynamics, the total energy of the system can only decrease, becoming minimum at equilibrium.[2] The corresponding rate of energy production is then given by the integral of $\Xi(H,H)$, which, according to the time-derivative interpretation of the bracket, can be equated to

$$\frac{dH}{dt}=\int_\Omega \Xi\, d^3x=\int_\Omega \sum_i \frac{\delta H}{\delta\omega_i}\frac{d\omega_i}{dt}\, d^3x=\int_\Omega \sum_i \left[J_i^r\frac{\delta H}{\delta\omega_i}+\mathbf{J}_i^t\cdot\boldsymbol{\nabla}\left(\frac{\delta H}{\delta\omega_i}\right)\right] d^3x\ , \tag{7.1-25}$$

[2]In reality, the isolation of the system requires that the energy be constant. However, the mathematical argument given here requires us to "forget" about this apparent contradiction.

where the last equality has been obtained after substitution of the time derivative of the density ω_i with the sum of the volumetric rate of production, J_i^r, and the divergence of the surface flux, $\mathbf{J}_i^t$, from (7.1-22). The final expression is analogous to the expression for the entropy production in an entropy-based formalism, represented by (6.6-8). In parallel to the interpretation offered for this expression in terms of the product of the fluxes and their corresponding affinities, we can then identify $\delta H/\delta\omega_i$ and $\nabla(\delta H/\delta\omega_i)$ as the appropriate affinities corresponding to the fluxes, J_i^r and $\mathbf{J}_i^t$, respectively.

From the above arguments, we see that we can view the system in two ways: one physically realizable, which allows the entropy to vary but requires that the energy remain constant, and one "fictitious" process which requires the constancy of the entropy, implying a decreasing energy. The latter viewpoint makes obvious the nature of the affinities and fluxes in that we are now expressing them in terms of the energy of the system rather than the entropy. For dynamical systems, the former description is necessary because of the incorporation of the momentum density into the state space of the problem. Although the introduction of this quantity into the energy expression is straightforward, through the addition of the kinetic energy as evident from chapter 5, an equivalent extension of the entropy expression is not available (or, at least, not obvious).

Additional constraints on Ξ can be imposed through consideration of the Onsager/Casimir reciprocal relations and the non-negative character of the entropy production rate. For the remainder of this book, we shall assume that the reciprocal relations are valid *for systems close to equilibrium*, i.e., when the function Ξ in the dissipation bracket, (7.1-15), depends linearly on both F and G, as indicated by (7.1-24). This assumption imposes symmetry constraints on the phenomenological coefficient matrices **A**, **B**, **C**, and **D**, as seen in the next subsection and through the rest of the book. These restrictions imply that close to equilibrium the lowest-order representation of the dissipation bracket $[F,G]$ is provided by a symmetric bilinear functional with respect to F, G. Since, through (7.1-21b), Ξ is connected to the entropy production term, σ,

$$\sigma = -\frac{1}{T}\,\Xi\left(L[\frac{\delta H}{\delta\boldsymbol{\omega}},\nabla\frac{\delta H}{\delta\mathbf{w}}];\frac{\delta H}{\delta\mathbf{w}},\nabla\frac{\delta H}{\delta\mathbf{w}}\right), \tag{7.1-26}$$

close to equilibrium (when (7.1-24) is substituted for Ξ), (7.1-26) gives rise to a symmetric, biquadratic dissipation function with respect to $\delta H/\delta\boldsymbol{\omega}$ and $\nabla(\delta H/\delta\mathbf{w})$, in accordance with linear irreversible thermodynamics [Gyarmati, 1970, Eq. (4.4)] (see also §7.1-2). Thus, application of the inequality (6.3-7) imposes further constraints on the phenomenological coefficient matrices. These constraints constitute the equivalent of thermodynamic admissibility criteria in the bracket formalism and are widely used throughout the rest of the book.

In the limit close to equilibrium where the linear phenomenological constitutive relations and the Onsager/Casimir reciprocal relations apply, we saw that Eqs. (7.1-25) or (7.1-26) give rise to symmetric, non-negative, bilinear expressions for the energy or entropy dissipation, respectively. These are the same expressions as the *dissipation functions*, $\Phi(J_i,J_i)$, of Onsager [1931b, p. 2276] (in terms of the generic fluxes, J_i) or the *dissipation potentials*, $\Phi(J_i,J_i)$ and $\Psi(\Lambda_i,\Lambda_i)$, of Gyarmati [1970, p. 91] (in terms of either the generic fluxes, J_i, or the generic affinities, Λ_i):

$$\Psi = \frac{1}{2}\sum_{i,j} \alpha_{ij}\Lambda_i\Lambda_j \ , \tag{7.1-27a}$$

$$\Phi = \frac{1}{2}\sum_{i,j} \beta_{ij}J_iJ_j \ , \tag{7.1-27b}$$

where $\boldsymbol{\alpha}$ is the phenomenological coefficient matrix and $\boldsymbol{\beta}$ its inverse. Indeed, in the linear limit we could identify Ξ with twice the dissipation potential of Gyarmati. These dissipation functions, in addition to representing the energy and/or entropy dissipation rate, serve to express the fluxes in the concise form

$$J_i = \frac{\partial\Psi}{\partial\Lambda_i} \ , \quad \text{for all } i \ , \tag{7.1-28}$$

which automatically guarantees satisfaction of the Onsager reciprocal relations [Gyarmati, 1970, Eq. (4.11)].

Driven by the symmetry and economy in the description of the fluxes presented by the expressions given by Eqs. (7.1-27) and (7.1-28) for the linear phenomenological coefficients, several attempts to generalize them are reported in the literature for conditions far away from equilibrium where the dependence of the fluxes on the affinities is, in general, expected to be nonlinear [Ziegler, 1958, pp. 757,8; Van Rysselberghe, 1962, Eq. (11); Edelen, 1972, p. 488; Van Kampen, 1973, p. 8]. The first two generalizations, although leading to mathematically elegant and aesthetically pleasing expressions, have been postulated in an *ad hoc* fashion without microscopic justification [Wei and Zahner, 1965]. As a result, in general, they overconstrain the allowed form of the resulting equations. For example, they are not compatible, in general, with the expressions used to describe the rate equations of chemical kinetics [Wei and Pratter, 1962, pp. 354,5; Wei and Zahner, 1965; Bataille *et al.*, 1978; Lengyel, and Gyarmati, 1981, §7]. The Van Kampen theory [1973] is presently the only nonlinear theory which is compatible with chemical kinetics, basically because it has been motivated by the underlying microscopic phenomena.

According to the Van Kampen [1973] model, the fluxes appear as a combination of unilateral transfer flows. The underlying physical picture associated with this idea is that the macroscopic system is a combined system consisting of a variety of ν subsystems (not necessarily occupying different locations in space—they can be interpenetrating). A pair of unilateral fluxes, $F^{r,s}$ and $F^{s,r}$, can always be associated with each pair r,s of subsystems, which are, in general, non-zero even at equilibrium, and which describe the unilateral flux (of particles, energy, etc.) from s to r and from r to s, respectively. Thus, the overall flux, J^r, to the (receiving) system r from its environment (i.e., from the other subsystems, $s=1,\ldots,\nu$, $s\neq r$) is [Van Kampen, 1973, Eq. (6)]

$$J^r = \sum_{s=1}^{\nu} (F^{r,s} - F^{s,r}) \ . \tag{7.1-29}$$

The generalization of the Onsager/Casimir relations is then accomplished by considering constitutive relationships similar in form to (7.1-28) for each one of the unilateral fluxes [Van Kampen, 1973, Eq. (12)],

$$F^{r,s} = \frac{\partial K^{rs}}{\partial \Lambda^s} \ , \tag{7.1-30}$$

where $K^{rs}=K^{sr}$ is a nonlinear generalization of the dissipation function, Ψ, considered to be a function of the affinities of the source subsystem, i.e., $K^{rs}=K^{rs}(\Lambda^s)$. Of course, although for simplicity only one flux/affinity has been considered for each subsystem, Eqs. (7.1-27,28) can be trivially generalized for many. In addition to the symmetry condition $K^{rs}=K^{sr}$, the functionality of K^{rs} is further constrained from the requirements of a non-negative entropy production rate. The Van Kampen formulation has been shown to be compatible with various nonlinear thermodynamic models such as Knudsen flow through a membrane and the Marcelin/De Donder-type rate equations in chemical kinetics [Van Kampen, 1973, §6]. Recently, it has been rediscovered and further advocated as the natural nonlinear generalization of the Onsager/Casimir relations [Oláh, 1987; Lengyel, 1989b], primarily because of its success in describing the non-linear models of reaction rates from chemical kinetics.

Thus, chemical kinetics has been the premier "test-bed" for the evaluation of any nonlinear irreversible thermodynamic theory, primarily because of the confidence on the validity of the equations. However, whether the Van Kampen formulation (or any other formulation, as a matter of fact, which is compatible with the equations of chemical kinetics) is also compatible with more general nonlinear phenomena remains to be seen. One thing that is certain though, and we reiterate this with the danger of becoming repetitious, is that both the original Onsager/Casimir relations and the Van Kampen formulation are the result of a specific understanding of the underlying microscopic physics. Thus, in principle, we rather consider the restrictions that they place on the dissipation bracket as part of the case-by-case specific modeling process—which, anyhow, needs to be based on a careful examination of the underlying microscopic dynamics—instead of elevating them as "theorems" or "principles." Using this approach, we obtain the best of both worlds: the benefit from the use of the additional information they contain (when they are applicable) and the flexibility to accommodate a different microscopic physics, when they are invalid.

In practice, for most applications (certainly for those discussed in this book), the principle of microscopic reversibility is expected to be valid. As such, it is helpful to consider in all these cases the restrictions imposed on the dissipation function Ξ by the Onsager/Casimir relations *a priori*. However, since little is known for the more general nonlinear case (except for chemical reactions) we prefer, in general, that the only constraint which we *a priori* impose on the dissipation function Ξ be that it should always result in a non-negative rate of entropy production (no matter how close or how far away from equilibrium the system is). Of course, even in that case, Ξ should always reduce, close to equilibrium, to an expression compatible with the Onsager/Casimir relations.

7.1.2 *Equivalence of the energetic and entropic formalisms*

In order to justify thoroughly the use of the Volterra derivatives of the Hamiltonian with respect to the dynamic variables (and gradients thereof) as the proper expressions of the affinities in our system, and to establish the reciprocal relations and the non-negative nature of $\boldsymbol{\alpha}$ in terms of these quantities as well, it is necessary to formally show the equivalence of the energetic approach introduced above with that of §6.6. We do this in the present subsection, where we demonstrate that for purely dissipative processes (zero velocities) both formalisms are applicable. Hence the same entropy production is obtained, and the symmetry and non-negative character of the phenomenological coefficient matrix $\boldsymbol{\alpha}$ (in the entropic formalism) implies the symmetry and non-positive character of the phenomenological matrix $\tilde{\boldsymbol{\alpha}}$ (in the energetic formalism), and vice versa.

Let us suppose that a given system can be described by m+1 thermodynamic variables ω_i, i=0,1,...,m, with $\omega_0 \equiv u$, and that close to equilibrium the fluctuations of the corresponding densities satisfy a general convection/diffusion equation of the form indicated by (6.6-9). In addition, let us express the corresponding fluxes as a linear superposition of the corresponding affinities, $\delta S/\delta\omega_i$ and $\boldsymbol{\nabla}(\delta S/\delta\omega_i)$, using as proportionality coefficients the phenomenological coefficients α_{ij} and $\alpha_{\alpha\beta}^{ij}$, through Eqs. (6.4-12) and (6.5-30), respectively. According to the reciprocal relations, these coefficients have a special symmetry, which in its simplest form reduces to invariance symmetry relations obtained by interchanging i with j and/or α with β (the most general case has been discussed in detail in §6.5.2). Furthermore, from (6.6-11) the rate of the entropy production is

$$\frac{dS}{dt} = \int_\Omega \sum_i \left[\frac{\partial s}{\partial \omega_i} J_i^r + \boldsymbol{\nabla}\left(\frac{\partial s}{\partial \omega_i}\right) \cdot \mathbf{J}_i^t \right] d^3x \,, \tag{7.1-31}$$

which becomes, using the linear phenomenological relations for the fluxes,

$$\frac{dS}{dt} = \int_\Omega \sum_{i,j} \left[\frac{\partial s}{\partial \omega_i} \alpha_{ij} \frac{\partial s}{\partial \omega_j} + \nabla_\alpha \left(\frac{\partial s}{\partial \omega_i}\right) \alpha_{\alpha\beta}^{ij} \nabla_\beta \left(\frac{\partial s}{\partial \omega_j}\right) \right] d^3x \,. \tag{7.1-32}$$

Eq. (6.6-11) and the non-negative character of the entropy production for arbitrary values of the affinities also imply the non-negative character of the phenomenological matrix $\boldsymbol{\alpha}$.

Let us first express the entropy production rate in terms of the affinities $\delta U/\delta\omega_i$ and $\boldsymbol{\nabla}(\delta U/\delta\omega_i)$ as well as in terms of the corresponding fluxes, $\tilde{J}_i^r$ and $\tilde{\mathbf{J}}_i^t$, where now $\omega_0 \equiv s$. In order to achieve this goal, we first need to correlate the two expressions for the affinities with each other, and then do the same for the fluxes. The relationship between the affinities is provided through a straightforward change of variables (see §4.2) as

$$\frac{\partial s}{\partial \omega_i} = \begin{cases} -\dfrac{\partial s}{\partial u}\dfrac{\partial u}{\partial \omega_i} = -\dfrac{1}{T}\dfrac{\partial u}{\partial \omega_i} \,, & \text{for } \omega_i \neq u \ (i \neq 0) \\[2ex] \dfrac{\partial s}{\partial u} = \dfrac{1}{T} \,, & \text{for } \omega_i = u \ (i = 0) \end{cases} \tag{7.1-33}$$

The fluxes corresponding to the same density, i.e., for $\omega_i \neq u, s$, are the same in the two formulations, since they correspond to the same physical quantity. The transport flux corresponding to the internal energy density, u, in the entropic formalism, $\mathbf{J}_u^t$, is physically equal to the rate of transfer of energy per unit area, which, according to the Gibbs relation, can be expressed as

$$\mathbf{J}_u^t = T\mathbf{J}_s^t + \sum_{i=1}^m \frac{\partial u}{\partial \omega_i} \mathbf{J}_i^t \,. \tag{7.1-34}$$

If now we use the equality of the fluxes (except for i=0), as well as Eqs. (7.1-32) through (7.1-34), then the rate of the entropy production can be written as

$$\frac{dS}{dt} = -\int_\Omega \frac{1}{T} \left[\sum_{i=1}^m \frac{\partial u}{\partial \omega_i} \tilde{J}_i^r + \sum_{i=0}^m \boldsymbol{\nabla}\left(\frac{\partial u}{\partial \omega_i}\right) \cdot \tilde{\mathbf{J}}_i^t \right] d^3x \,, \tag{7.1-35}$$

which is identical with the expression obtained from the dissipation bracket $[S,H]$, given by (7.1-19) using $H=\int u d^3x$. Note that this equality of the entropy production rates obtained with the two expressions does not imply that the fluxes depend linearly on the affinities, but is valid even in the nonlinear case.

Let us now evaluate the relationship between the fluxes $\tilde{J}_i^r$ and $\tilde{\mathbf{J}}_i^t$ in the energetic framework with respect to their corresponding affinities, $\delta U/\delta\omega_i$ and $\nabla(\delta U/\delta\omega_i)$, respectively. If a linear phenomenological relationship is assumed between the corresponding fluxes and affinities in the entropic formalism, the use of the equality of the fluxes (except for i=0), as well as of Eqs. (7.1-32) through (7.1-34) provides a similar linear relationship between the corresponding fluxes and affinities in the energetic framework provided by new proportionality coefficients $\tilde{\alpha}_{ij}$ and $\tilde{\alpha}_{\alpha\beta}^{ij}$. It is straightforward then to show that these coefficients are related to the previous ones, α_{ij} and $\alpha_{\alpha\beta}^{ij}$, through the relationships

$$
\begin{aligned}
\tilde{\alpha}_{ij} &= -\frac{1}{T}\alpha_{ij}\ , \quad i,j=1,\ldots,m\ ,\\
\tilde{\alpha}_{\alpha\beta}^{00} &= -\frac{1}{T^3}\left[\alpha_{\alpha\beta}^{00} - 2\sum_{i=1}^{m}\alpha_{\alpha\beta}^{0i}\frac{\partial u}{\partial\omega_i} + \sum_{i,j=1}^{m}\alpha_{\alpha\beta}^{ij}\frac{\partial u}{\partial\omega_i}\frac{\partial u}{\partial\omega_j}\right],\\
\tilde{\alpha}_{\alpha\beta}^{0j} &= -\frac{1}{T^2}\left[\alpha_{\alpha\beta}^{0j} - \sum_{i=1}^{m}\alpha_{\alpha\beta}^{ij}\frac{\partial u}{\partial\omega_i}\right],\quad j=1,\ldots,m\ ,\\
\tilde{\alpha}_{\alpha\beta}^{i0} &= -\frac{1}{T^2}\left[\alpha_{\alpha\beta}^{i0} - \sum_{j=1}^{m}\alpha_{\alpha\beta}^{ij}\frac{\partial u}{\partial\omega_j}\right],\quad i=1,\ldots,m\ ,\\
\tilde{\alpha}_{\alpha\beta}^{ij} &= -\frac{1}{T}\alpha_{\alpha\beta}^{ij}\ , \quad i,j=1,\ldots,m\ .
\end{aligned}
\tag{7.1-36}
$$

Notice that the symmetry of the new coefficients is a direct consequence of the symmetry of the old (provided that they are symmetric) and vice versa. In addition, it can be shown that the new matrices have a non-positive character. The complete proof for the most general case is a rather lengthy one, and, consequently, is omitted here; however, the proof for a special case follows as an example.

EXAMPLE 7.1: Show that the coefficient matrix corresponding to the transport fluxes, $\tilde{\alpha}_{\alpha\beta}^{ij}$, is non-positive for the particular case when m=1.

When m=1 and for a particular pair of spatial components, α and β, the coefficient matrix reduces to a 2×2 matrix. Using the relationships provided by (7.1-36), we can easily evaluate the diagonal components of $\tilde{\boldsymbol{\alpha}}$ as $-\alpha_{11}/T$ and $-(\alpha_{00}/T^3)[\{1-(\partial u/\partial\omega_1)(\alpha_{01}/\alpha_{00})\}^2 + (\partial u/\partial\omega_1)^2(\det\boldsymbol{\alpha})/\alpha_{00}{}^2]$, whereas the determinant of the new matrix is $(\det\boldsymbol{\alpha})/T^4$. It therefore trivially follows from the non-negative-definite criterion of the matrix

$\boldsymbol{\alpha}$—see (6.4-15)—that the diagonal terms of $\tilde{\boldsymbol{\alpha}}$ are non-positive and its determinant is non-negative, i.e., $\tilde{\boldsymbol{\alpha}}$ is non-positive definite. ■

In §6.4.5, we saw that the appropriate choice of the flux/affinity pairs is often determined by experimental considerations. For instance, we have analyzed at length the case of coupled heat and mass transfer, whereupon it became obvious that the transferred mass had an enthalpy associated with it. This additional enthalpy was unmeasurable as heat, and we were therefore required to restructure the affinities so as to take this effect into consideration. Consequently, it was necessary to use the gradients of the chemical potentials at constant temperature as the proper affinities in this circumstance. It is straightforward to show—through the Jacobian of the transformation of the affinities—that the new affinities also constitute an independent set of affinities.

What remains to be shown at this point is that the phenomenological matrix retains its symmetry and non-negative character regardless of whether or not we use the original affinities or the affinities at constant temperature (or, for that matter, in terms of holding other variables constant as well), provided that we are consistent in the definitions of the corresponding fluxes. Indeed, the interrelationship of the fluxes is determined following the Gibbs relation through (7.1-34), which needs to be modified in order to accommodate the fact that part of the entropy change associated with the species fluxes is now accounted for explicitly. Thus for a multicomponent mixture which consists of ν species, $\omega_i=\rho_i$, $i=1,\ldots,\nu\leq m$, we have

$$\mathbf{J}_u^t = T\tilde{\mathbf{J}}_s^t + \sum_{i=1}^{\nu} (\overline{S}_i + \frac{\partial u}{\partial \omega_i})\mathbf{J}_i^t + \sum_{i=\nu+1}^{m} \frac{\partial u}{\partial \omega_i}\mathbf{J}_i^t \,, \tag{7.1-37}$$

where $\overline{S}_i$ is the partial specific entropy for species i. By comparing Eqs. (7.1-34) and (7.1-37), it is easy to show that the new entropy flux needs to be modified as

$$\tilde{\mathbf{J}}_s^t = \mathbf{J}_s^t - \sum_{j=1}^{\nu} \overline{S}_j \mathbf{J}_j^t \,. \tag{7.1-38}$$

Note that a similar modification can occur to any one of the remaining fluxes, $\mathbf{J}_i^t$, $i=\nu+1,\ldots,m$, subject to a further modification of the corresponding affinities for the mass fluxes. If we now use (7.1-38) in conjunction with the new definition for the mass flux affinities, one can show, in general, the preservation of the symmetry and non-positive characteristics of the phenomenological proportionality coefficients associated with the new affinities. Moreover, one can show this to be generally the case when a non-singular transformation of the affinities is performed together with a corresponding change of the fluxes; however, since the formal proof in the general case, although straightforward, is again rather long, we content ourselves with demonstrating the validity of this statement in a simplified subcase with the following example.

EXAMPLE 7.2: Show that for a one-component, isotropic, open system the phenomenological matrix $\tilde{\boldsymbol{\alpha}}$ retains its symmetry and non-positive character when we transform the affinities from ∇T and $\nabla\mu^{+}$, corresponding to $\boldsymbol{\alpha}$, to the new affinities ∇T and $|\nabla\mu^{+}|_{T} \equiv \nabla\mu^{+} + \overline{S}\nabla T$.

Let us use the tilde to denote the quantities pertaining to the second (constant T) formalism. As before, we may easily calculate the new fluxes from the rate of the entropy production as

$$\tilde{\mathbf{J}}_0 = \mathbf{J}_0 - \overline{S}\mathbf{J}_1 \,, \qquad \tilde{\mathbf{J}}_1 = \mathbf{J}_1 \,, \tag{7.1-39}$$

whereupon we can find that the new proportionality coefficients are

$$\tilde{\alpha}_{00} = \alpha_{00} - 2\overline{S}\alpha_{10} + \overline{S}^2\alpha_{11} \,, \qquad \tilde{\alpha}_{10} = \tilde{\alpha}_{01} = \alpha_{10} - \alpha_{11}\overline{S} \,, \quad \text{and} \quad \tilde{\alpha}_{11} = \alpha_{11} \,. \tag{7.1-40}$$

This matrix is obviously symmetric, whereas its non-positive definite character may be noted directly as in Example 7.1. ■

Based on the last two examples, it is interesting to note the existence of multiple descriptions where the phenomenological coefficients have similar symmetry properties and the corresponding matrices a non-negative or non-positive character. This phenomenon is analogous to the multiple formulations of the fundamental thermodynamic potentials and may be responsible for the impunity observed when loose and non-rigorous expressions have been used in the past, either theoretically or experimentally (in conjunction with the application of Onsager's reciprocal relations), which seemingly led to correct results. However, we can also understand the frustration of theoreticians when such an event occurs [Coleman and Truesdell, 1960]. Hopefully, the above detailed discussion has somehow clarified this issue.

7.2 The Hydrodynamic Equations for a Single-Component System

As a first example of the generality and applicability of the generalized bracket of (7.1-2), we shall now present a derivation of the spatial equations of motion for a single-component, dissipative fluid. This is done in its fullest generality in §7.2.1 for a compressible, non-isothermal, isotropic fluid, and in §7.2.2 for the incompressible, isothermal, Navier/Stokes equations.

7.2.1 *The general isotropic fluid*

In this subsection, we wish to describe the transport processes within a compressible, isotropic fluid with shear and bulk viscosities, $\mu=\mu(\rho,T)$ and $\kappa=\kappa(\rho,T)$, respectively, and (Fourier) thermal conductivity $k=k(\rho,T)$, which may be considered as functions of the local fluid density and temperature. In this case, the dynamic variables of the system are the same as for the non-dissipative system of §5.3: $F=F[\rho,\mathbf{M},s]$. These variables belong to the same operating space as before, (5.3-2), with the exception that the no-penetration condition on the boundary $\partial\Omega'$ is replaced with the no-slip condition, $\mathbf{M}=\mathbf{0}$ on $\partial\Omega'$. This substitution is necessary for the problem in terms of the following dissipation bracket to be well posed.

The Poisson bracket of (7.1-2) is exactly the same as for the non-dissipative case, (5.3-25); and the dissipation bracket is as expressed by (7.1-19), with Ξ given by the bilinear expansion (7.1-24). Written explicitly in terms of the dynamic variables ρ, $\mathbf{M}$, and s, this translates into the expression

$$
\begin{aligned}
[F,G] = &- \int_{\Omega'} Q_{\alpha\beta\gamma\varepsilon} \nabla_\alpha\left(\frac{\delta F}{\delta M_\beta}\right) \nabla_\gamma\left(\frac{\delta G}{\delta M_\varepsilon}\right) d^3x \\
&+ \int_{\Omega'} Q_{\alpha\beta\gamma\varepsilon} \frac{1}{T}\frac{\delta F}{\delta s} \nabla_\alpha\left(\frac{\delta G}{\delta M_\beta}\right) \nabla_\gamma\left(\frac{\delta G}{\delta M_\varepsilon}\right) d^3x \\
&- \int_{\Omega'} \sum_{i,k=0}^{1} \alpha_{ik}^{\gamma\beta} \nabla_\gamma\left(\frac{\delta F}{\delta \rho_i}\right) \nabla_\beta\left(\frac{\delta G}{\delta \rho_k}\right) d^3x \\
&+ \int_{\Omega'} \sum_{i,k=0}^{1} \alpha_{ik}^{\gamma\beta} \frac{1}{T}\frac{\delta F}{\delta s} \nabla_\gamma\left(\frac{\delta G}{\delta \rho_i}\right) \nabla_\beta\left(\frac{\delta G}{\delta \rho_k}\right) d^3x \ ,
\end{aligned}
\tag{7.2-1}
$$

which, obviously, requires a little explanation.[3] For notational convenience, we have defined the entropy density as the variable ρ_0, i.e., $s\equiv\rho_0$. This convention allows a combination of terms to be expressed as a

[3]Note that we have defined (7.2-1) as the negative of (7.1-19), which requires that the coefficient matrices be non-negative rather than non-positive.

simple summation over the variables ρ_0 and $\rho_1 \equiv \rho$. The significance of this simplification will become more pronounced when we deal with a multiple component fluid in §7.3. We have also used the conservation of mass, (see the remark after Eq. (7.1-15), to eliminate a number of terms required by (7.1-24). Also, it is obvious that several of the dynamic variables (actually, in this case, *all* dynamic variables) do not have relaxational affinities; i.e., they can only change by transfers between subsystems.[4] In addition, conservation of mass requires that $\alpha_{01}^{\gamma\beta}=\alpha_{10}^{\gamma\beta}=\alpha_{11}^{\gamma\beta}=0$ as well, yet we include these terms in (7.2-1) so that the generalization to the multicomponent fluid will be as straightforward as possible. Furthermore, due to the Onsager relations we require that

$$\alpha_{00}^{\gamma\beta} = \alpha_{00}^{\beta\gamma} \, , \tag{7.2-2a}$$

$$Q_{\alpha\beta\gamma\varepsilon} = Q_{\gamma\varepsilon\alpha\beta} \, . \tag{7.2-2b}$$

For an isotropic fluid, i.e., a material whose properties are not dependent on any direction in space, we have

$$\alpha_{00}^{\gamma\beta} \equiv \alpha_{00} \delta_{\gamma\beta} \, , \tag{7.2-3a}$$

$$Q_{\alpha\beta\gamma\varepsilon} \equiv \mu\,(\delta_{\alpha\gamma}\delta_{\beta\varepsilon} + \delta_{\alpha\varepsilon}\delta_{\beta\gamma}) + \eta\,(\delta_{\alpha\gamma}\delta_{\beta\varepsilon} - \delta_{\alpha\varepsilon}\delta_{\beta\gamma}) + \kappa'\delta_{\alpha\beta}\delta_{\gamma\varepsilon} \, , \tag{7.2-3b}$$

the latter of which accounts for all possible second-rank, isotropic contributions to **Q** [Aris, 1962, p. 34]. Alternatively, in accordance with §6.4.3, we could have associated part of κ' with a scalar dependency (through ⅓ tr **Q**), but this is unnecessary. The viscosities μ, η, and κ' may, in general, depend upon the density and the temperature, yet are required explicitly to have no velocity dependence by postulate. Note that the Onsager reciprocal relations, (7.2-2b), are automatically satisfied. In order to proceed further, we shall make use of another concept of key usefulness in continuum mechanics, that of *frame indifference*.

[4]That the term proportional to $(\delta F/\delta \mathbf{M})(\delta G/\delta \mathbf{M})$ does not appear in this expression is due to the assumption of conservation of momentum, which is implied directly by the hypothesis of local equilibrium. Consider that each thermodynamic subsystem is composed of many "molecules," each with its own velocity, which, macroscopically, behave as a single entity, the fluid particle. Due to the local equilibrium hypothesis, the distribution of the molecular velocities is Maxwellian, of which the *H*-theorem implies the stability. Disregarding fluctuations from this state, which are basically irrelevant on a macroscopic time scale, each fluid particle will have the velocities of its constituent molecules governed by this distribution, so that no further internal relaxation, caused by collisions, can change the overall subsystem momentum. Hence (in the absence of external forces) the only way to change the momentum of a fluid particle is by a transferral of momentum from one particle to another. Note that the average velocity of the constituent molecules of any fluid particle is the local fluid velocity, $\mathbf{v}(\mathbf{x},t)$.

The application of the frame-indifference constraint in the modeling of continuum systems has stirred considerable controversy, as can be easily documented from the volume of the literature that has been produced to deal with this subject [Müller, 1972; Edelen and McLennan, 1973; Wang, 1975; Truesdell, 1976; Murdoch, 1983; Ryskin, 1985 and 1987; Speziale, 1987 and 1988].[5] As it turns out, there are two ways of expressing it: a broad one, the validity of which nobody contests but which is of little use, and a much stricter one of approximative (at best) character which, nevertheless, can considerably simplify otherwise complex general phenomenological material expressions. The broad formulation of the frame indifference states [Ryskin, 1985, p. 1239]: "Any physical phenomenon (including material behavior) is independent of the motion of an observer (a frame of reference)." The much stricter one states [Ryskin, 1985, p. 1239]: "Any physical phenomenon is described by a mathematical relation, which, written in terms of quantities measured by each observer according to the same rules (i.e., *frame-indifferently defined* quantities), has the same *functional form* for all observers." Note that the allowed motions for the observers include not only arbitrary uniform translation, but also uniform rotation and inversion, i.e., any rigid-body motion.

Given that an inertial framework is the only appropriate one to describe the dynamics of individual particles, the second (and more useful) formulation of the frame indifference, which accepts no influence on the form and mathematical expression of rotation as expressed by Truesdell [1976] and others, can be but merely approximate.[6] However, it is a very useful one based on the fact that the neglected terms are indeed, in many situations, small enough to be negligible. For example, for flow of particles, the neglected microscopic inertia terms are of the order $\omega d^2/\nu$, which for small particle angular velocities, ω, small particle diameters, d, and large fluid kinematic viscosity, ν, can be as low as 10^{-10}.

With this caveat put forth, let us see how the principle of frame indifference is applied to the modeling of the viscous stress within an isotropic fluid.[7] In order to see its implications on the modeling of the viscosity, it is instructive to apply the principle to a more general stress expression [Woods, 1975, pp. 160,1]. For example, assume that the

[5]Indeed, there is even a controversy among the authors who have attempted to clear up the original controversy—see Speziale [1987] and Ryskin [1987].

[6]As an afterthought, we include a quote by Speziale [1988, p. 219] since it applies to several concepts discussed in this and the previous chapter: "The history of science has taught us that almost all physical postulates which were once thought to be 'laws of nature' are only good approximations for a large subset of the natural phenomena that they were formulated to describe."

[7]In this case, it amounts to the postulate of rotational invariance for both the viscous stress tensor and the entropy production rate.

general viscous stress in an isotropic fluid is $\boldsymbol{\pi}$. Then, upon a change to a rotating coordinate frame (which is used for the *description* of the phenomena), based on the transformation

$$\mathbf{v}' \equiv \mathbf{v} + \mathbf{c}^a \times \mathbf{x} \,, \tag{7.2-4}$$

where $\mathbf{c}^a$ is a constant axial vector and $\mathbf{x}$ is the position vector, neither the viscous stress, $\boldsymbol{\pi}$, nor the entropy production rate, Φ, (as representing diffusive or irreversible processes) should change. Furthermore (and this is where the arbitrary nature of the postulate enters), the *form* expressing the relationship between the rate of the entropy production and the velocity gradient

$$\Phi = \boldsymbol{\pi}^{\mathrm{T}} : \boldsymbol{\nabla}\mathbf{v} \,, \tag{7.2-5}$$

should also stay the same after the transformation. However,

$$\Phi' = (\boldsymbol{\pi}')^{\mathrm{T}} : \boldsymbol{\nabla}\mathbf{v}' = \Phi - \boldsymbol{\pi}^{\mathrm{T}} : \boldsymbol{\delta} \times \mathbf{c}^a = \Phi + \varepsilon_{\alpha\beta\gamma}\pi_{\alpha\beta}c^a_\gamma \,, \tag{7.2-6}$$

which, in order to be valid for arbitrary $\mathbf{c}^a$, implies that the viscous stress must be symmetric. Note that in order to derive (7.2-6) the identity

$$\boldsymbol{\nabla}(\mathbf{c}^a \times \mathbf{x}) = -\,\boldsymbol{\nabla}\mathbf{x} \times \mathbf{c}^a = -\,\boldsymbol{\delta} \times \mathbf{c}^a \,, \tag{7.2-7}$$

was used. An immediate consequence of (7.2-7), in conjunction with the general expression provided by (7.2-3b), is that

$$\eta = 0 \,, \tag{7.2-8}$$

for an isotropic, structureless, viscous medium.

Note that this constraint is not applicable if there is an *axial* vector $\boldsymbol{\omega}$ (like an intrinsic angular velocity—see Example 7.3 and §7.2.3) embedded in the fluid [Woods, 1975, pp. 160,1]. Then, the frame indifference can be re-established by using for the construction of the proper affinity for the stress, the gradient of the relative velocity with respect to the internal rotation rate $\boldsymbol{\omega}$, i.e., $\mathbf{v} - \boldsymbol{\omega} \times \mathbf{x}$. Upon a change to a rotating coordinate frame,

$$\boldsymbol{\omega}' = \boldsymbol{\omega} + \mathbf{c}^a \,, \tag{7.2-9}$$

$$\begin{aligned} \Phi' &= (\boldsymbol{\pi}')^{\mathrm{T}} : \boldsymbol{\nabla}(\mathbf{v} - \boldsymbol{\omega}' \times \mathbf{x}) \\ &= \boldsymbol{\pi}^{\mathrm{T}} : \boldsymbol{\nabla}(\mathbf{v} + \mathbf{c}^a \times \mathbf{x} - (\boldsymbol{\omega} + \mathbf{c}^a) \times \mathbf{x}) \\ &= \boldsymbol{\pi}^{\mathrm{T}} : \boldsymbol{\nabla}(\mathbf{v} - \boldsymbol{\omega} \times \mathbf{x}) = \Phi \,. \end{aligned} \tag{7.2-10}$$

Using the Hamiltonian of (5.3-26), the Poisson bracket of (5.3-25), and the above dissipation bracket, it is possible to derive the dissipative hydrodynamic equations through (7.1-1,2) and the procedure of §5.3 as

p. 186: all Φ's on this page should be σ's

$$\frac{\partial \rho}{\partial t} = - \nabla_\alpha (\rho v_\alpha) \,, \tag{7.2-11a}$$

$$\rho \frac{\partial v_\alpha}{\partial t} = - \rho v_\beta \nabla_\alpha v_\alpha - \nabla_\alpha p - \rho \nabla_\alpha e_p^v + \nabla_\gamma [\mu (\nabla_\gamma v_\alpha + \nabla_\alpha v_\gamma) + \kappa' \delta_{\alpha\beta} \nabla_\varepsilon v_\varepsilon] \,, \tag{7.2-11b}$$

$$\frac{\partial s}{\partial t} = - \nabla_\alpha (s v_\alpha) + \frac{1}{T} \nabla_\gamma (k \nabla_\gamma T) + \frac{1}{T} [\mu (\nabla_\gamma v_\alpha + \nabla_\alpha v_\gamma) + \kappa' \delta_{\alpha\gamma} \nabla_\varepsilon v_\varepsilon] \nabla_\alpha v_\gamma \,, \tag{7.2-11c}$$

where the pressure is given by (5.3-52) and the thermal conductivity, k, is defined as $k \equiv \alpha_{00} T$. As a consequence of (7.2-11b), we may denote the extra stress of the system as

$$\sigma_{\alpha\gamma} \equiv \mu (\nabla_\gamma v_\alpha + \nabla_\alpha v_\gamma) + \kappa' \delta_{\alpha\gamma} \nabla_\varepsilon v_\varepsilon \,, \tag{7.2-11d}$$

and through arguments based upon the mean of the principal (total) stress, it results that [Aris, 1962, pp. 111,2] $\kappa' = \kappa - \tfrac{2}{3}\mu$, where κ is the bulk viscosity. (We shall soon see how the bulk viscosity coefficient arises through the dissipation inequality.) Hence we can visualize the fluid motion as consisting of two types of deformation [Hirschfelder *et al.*, 1954, p. 711]: (1) a deformation in which adjacent planes of the fluid move with different velocities, characterized by the shear viscosity, μ; and (2) a pure expansion/compression of the fluid characterized by the bulk viscosity, κ. Using an approach similar to the non-dissipative procedure of §5.3, it is also easy to obtain the dissipative form of the energy equation:

$$\frac{\partial e}{\partial t} = - \nabla_\alpha (v_\alpha e) - \nabla_\alpha (p v_\alpha) - \rho v_\alpha \nabla_\alpha e_p^v + \nabla_\alpha (k \nabla_\alpha T) + \nabla_\gamma (\sigma_{\alpha\gamma} v_\alpha) \,. \tag{7.2-12}$$

Other forms of the energy equation could also be obtained.

Lastly, we may use the second law of thermodynamics, as embodied in (6.4-9), to arrive at consistency requirements for the phenomenological coefficients. In this case,

$$\sigma = \frac{1}{T} Q_{\alpha\beta\gamma\varepsilon} (\nabla_\alpha v_\beta)(\nabla_\gamma v_\varepsilon) + \frac{1}{T} \alpha_{00}^{\gamma\beta} (\nabla_\gamma T)(\nabla_\beta T) \geq 0 \,. \tag{7.2-13}$$

': 2nd line from bottom: (6.4-9) should be (7.1-12)

p. 187

Yet if we rewrite the first term, using the fact that $Q_{\alpha\beta\gamma\varepsilon}$, as defined from Eqs. (7.2-2b) and (7.2-3b), is symmetric with respect to its first two and its last two indices, we have (sans $1/T$)

$$\begin{aligned} Q_{\alpha\beta\gamma\varepsilon}(\nabla_\alpha v_\beta)(\nabla_\gamma v_\varepsilon) &= \tfrac{1}{2} Q_{\alpha\beta\gamma\varepsilon}(\nabla_\alpha v_\beta)(\nabla_\gamma v_\varepsilon) + \tfrac{1}{2} Q_{\alpha\beta\varepsilon\gamma}(\nabla_\alpha v_\beta)(\nabla_\varepsilon v_\gamma) \\ &= \tfrac{1}{2} Q_{\alpha\beta\gamma\varepsilon}(\nabla_\alpha v_\beta)(\nabla_\gamma v_\varepsilon + \nabla_\varepsilon v_\gamma) \\ &= \tfrac{1}{4} Q_{\alpha\beta\gamma\varepsilon}(\nabla_\alpha v_\beta + \nabla_\beta v_\alpha)(\nabla_\gamma v_\varepsilon + \nabla_\varepsilon v_\gamma) \\ &= Q_{\alpha\beta\gamma\varepsilon} A_{\alpha\beta} A_{\gamma\varepsilon} \, , \end{aligned} \tag{7.2-14a}$$

where **A** is the symmetric contribution to the velocity gradient tensor, defined by

$$A_{\alpha\beta} \equiv \tfrac{1}{2}(\nabla_\alpha v_\beta + \nabla_\beta v_\alpha) \, . \tag{7.2-14b}$$

Then we can express (7.2-13) as

$$Q_{\alpha\beta\gamma\varepsilon} A_{\alpha\beta} A_{\gamma\varepsilon} + \alpha_{00}^{\gamma\beta}(\nabla_\gamma T)(\nabla_\beta T) \geq 0 \, , \tag{7.2-15}$$

since $T > 0$. Eq. (7.2-15) is, in effect, the Clausius inequality which states the thermodynamic requirements for the non-negative rate of entropy production within the system. For the isotropic fluid defined by Eqs. (7.2-3) and (7.2-11), this inequality becomes

$$2\mu A_{\alpha\beta} A_{\alpha\beta} + \kappa' A_{\alpha\alpha} A_{\beta\beta} + \alpha_{00}(\nabla_\gamma T)(\nabla_\gamma T) \geq 0 \, , \tag{7.2-16}$$

so that if we define the vector **V** as

$$\mathbf{V}^{\mathrm{T}} \equiv (A_{11}, A_{22}, A_{33}, A_{12}, A_{13}, A_{23}, \nabla_1 T, \nabla_2 T, \nabla_3 T) \, , \tag{7.2-17}$$

then we can express the inequality (7.2-16) in matrix form (remembering that **A** is symmetric) as[8]

$$\mathbf{V}^{\mathrm{T}} \cdot \mathbf{B} \cdot \mathbf{V} \geq 0 \, . \tag{7.2-18}$$

In the case of (7.2-16), however, the dissipation inequality has a particularly simple form in that by defining $E_{\alpha\beta} \equiv A_{\alpha\beta} - \tfrac{1}{3}\delta_{\alpha\beta}A_{\gamma\gamma}$, we find that

[8]When applying this technique, one must be careful to decouple all of the different tensorial quantities and to take any system constraints into account.

$$2\mu E_{\alpha\beta} E_{\alpha\beta} + (\kappa' + ⅔\mu) A_{\alpha\alpha} A_{\beta\beta} + \alpha_{00}(\nabla_\gamma T)(\nabla_\gamma T) \geq 0 \,. \tag{7.2-19}$$

Since these terms are independent (i.e., each may be set to zero independently of the others[9]), we have that

$$\kappa' + ⅔\mu \geq 0 \,, \quad \mu \geq 0 \,, \tag{7.2-20}$$

$$\alpha_{00} \geq 0 \,. \tag{7.2-21}$$

These inequalities are often called the *Stokes relations*. Note that since $\kappa' + ⅔\mu \geq 0$, we naturally define as a coefficient of viscosity the bulk viscosity, for which $\kappa \geq 0$.

7.2.2 The incompressible Navier/Stokes equations

The derivation of the incompressible Navier/Stokes equations represents a trivial generalization of the procedure used in §5.4 to derive the incompressible Euler equations. In this case, as one restricts the operating space to include only the momentum density (and the replacement of the no-penetration boundary condition with the no-slip condition), the dissipation bracket of (7.2-1) nicely resolves itself into the simple expression

$$[F,G] = -\int_{\Omega'} Q_{\alpha\beta\gamma\varepsilon} \nabla_\alpha\left(\frac{\delta F}{\delta M_\beta}\right) \nabla_\gamma\left(\frac{\delta F}{\delta M_\varepsilon}\right) d^3x \,. \tag{7.2-22}$$

The expression for the Hamiltonian which we shall use here is

$$H_m[\mathbf{M}] = \int_{\Omega'} \left[\frac{1}{2\rho}\mathbf{M}\cdot\mathbf{M} - \rho\mathbf{g}\cdot\mathbf{x}\right] d^3x \,, \tag{7.2-23}$$

which includes the external gravitational field associated with the Navier/Stokes equations, as represented by the acceleration vector, **g**. In the obvious manner via (5.4-22), the procedure of §5.4, generalized to include (7.2-22), yields the Navier/Stokes equations directly as

$$\rho\frac{\partial v_\alpha}{\partial t} = -\rho v_\beta \nabla_\alpha v_\alpha - \nabla_\alpha p + \rho g_\alpha + \nabla_\gamma[\mu(\nabla_\gamma v_\alpha + \nabla_\alpha v_\gamma)] \,, \tag{7.2-24}$$

where the divergence-free condition on the velocity field, (5.4-2), was used.

[9]It is obvious that ∇T may be set independently to the zero vector, but the other two terms are not so obvious. Let the diagonal components of $\nabla\mathbf{v}$ vanish, implying that $\mu \geq 0$, then let the off-diagonal components vanish and let $A_{11}=A_{22}=A_{33}$ to see that $\kappa'+⅔\mu \geq 0$.

In this case, we must have a degradation of the mechanical energy of the system, H_m, which implies that $[H_m,H_m] \leq 0$, as in §1.4, so that

$$\frac{dH_m}{dt} = [H_m,H_m] = -\int_{\Omega'} Q_{\alpha\beta\gamma\varepsilon} (\nabla_\alpha v_\beta)(\nabla_\gamma v_\varepsilon)\, d^3x \leq 0 \,. \qquad (7.2\text{-}25)$$

Consequently, via (7.2-14a),

$$Q_{\alpha\beta\gamma\varepsilon} A_{\alpha\beta} A_{\gamma\varepsilon} \geq 0 \qquad (7.2\text{-}26a)$$

implies from (7.2-3b) that

$$2\mu A_{\alpha\beta} A_{\alpha\beta} \geq 0 \;\Rightarrow\; \mu \geq 0 \,, \qquad (7.2\text{-}26b)$$

A being traceless due to the incompressibility constraint. Hence we again recover the requirement that the shear viscosity must be non-negative.

EXAMPLE 7.3: Show that the anisotropic part of the viscous stress tensor naturally arises in systems consisting of particles with internal rotation (spin) in non-equilibrium cases before the internal rotational rate $\boldsymbol{\omega}$ relaxes to its (dynamic) equilibrium value

$$\boldsymbol{\omega}_0 = \tfrac{1}{2}\mathbf{w} \,, \qquad (7.2\text{-}27)$$

where $\mathbf{w}$ is the vorticity of the flow,

$$w_\alpha = \varepsilon_{\alpha\beta\gamma} (\nabla_\beta v_\gamma - \nabla_\gamma v_\beta) \,. \qquad (7.2\text{-}28)$$

For simplicity, let us model an isothermal and incompressible system, since the more general case may be treated in a similar fashion. In order to describe a system of microscopic particles with internal rotation, we first need to incorporate in the analysis described above for the Navier/Stokes equations the internal rotational energy of the particles. Thus, the expression for the Hamiltonian which we shall use here is

$$H_m[\mathbf{M}] = \int_{\Omega'} \left[\frac{1}{2\rho} \mathbf{M}\cdot\mathbf{M} + \frac{1}{2\rho_s} \boldsymbol{\Omega}\cdot\boldsymbol{\Omega} - \rho\mathbf{g}\cdot\mathbf{x} \right] d^3x \,, \qquad (7.2\text{-}29)$$

which, as compared to the expression (7.2-23) used for the Navier/Stokes equations, includes an additional term describing the internal rotational energy. This term involves the particle internal moment density, which for spherical particles of radius a is [Lighthill, 1986, p. 46]

$$\rho_s \equiv 2\rho a^2/5 \,, \qquad (7.2\text{-}30)$$

and the angular momentum density

p. 190: in (7.2-29), $H_m[\mathbf{M}]$ should be $H_m[\mathbf{M},\boldsymbol{\Omega},\rho]$

$$\boldsymbol{\Omega} \equiv \rho_s \boldsymbol{\omega} \; . \tag{7.2-31}$$

The Poisson bracket also needs to be changed in order to account for the presence of an additional (axial) vector variable, $\boldsymbol{\Omega}$. The effect of this new variable is to introduce an additional term in the bracket (5.4-20) as

$$\begin{aligned}\{F,G\} = &-\int_{\Omega'}\left[\frac{\delta F}{\delta M_\gamma}\nabla_\beta\left(\frac{\delta G}{\delta M_\beta}M_\gamma\right) - \frac{\delta G}{\delta M_\gamma}\nabla_\beta\left(\frac{\delta F}{\delta M_\beta}M_\gamma\right)\right]d^3x \\ &-\int_{\Omega'}\left[\frac{\delta F}{\delta \Omega_\gamma}\nabla_\beta\left(\frac{\delta G}{\delta M_\beta}M_\gamma\right) - \frac{\delta G}{\delta \Omega_\gamma}\nabla_\beta\left(\frac{\delta F}{\delta M_\beta}M_\gamma\right)\right]d^3x \; ,\end{aligned} \tag{7.2-32}$$

where we have replaced $\partial/\partial x_\beta$ with ∇_β. This Poisson bracket leads to the conservation of vorticity in the absence of any dissipation, i.e.,

$$\dot{\boldsymbol{\Omega}} \equiv \frac{\partial \boldsymbol{\Omega}}{\partial t} + \mathbf{v}\,\boldsymbol{\nabla}\cdot\boldsymbol{\Omega} = 0 \; , \tag{7.2-33}$$

for an incompressible, isothermal, and inviscid fluid.

The final change that needs to be implemented is in the dissipation bracket. Following the guidelines of the analysis described in the above section, Eqs. (7.2-1) and (7.2-3), the most general viscous dissipation for an isotropic (but rotational) incompressible, isothermal continuum, constrained by the equilibrium condition (7.2-27), can be written as

$$\begin{aligned}[F,G] = &-\int_{\Omega'}\frac{\mu}{2}\left[\nabla_\alpha\left(\frac{\delta F}{\delta M_\beta}\right) + \nabla_\beta\left(\frac{\delta F}{\delta M_\alpha}\right)\right]\left[\nabla_\alpha\left(\frac{\delta G}{\delta M_\beta}\right) + \nabla_\beta\left(\frac{\delta G}{\delta M_\alpha}\right)\right]d^3x \\ &-\int_{\Omega'}\frac{\eta}{2}\left[\nabla_\alpha\left(\frac{\delta F}{\delta M_\beta}\right) - \nabla_\beta\left(\frac{\delta F}{\delta M_\alpha}\right) - 2\varepsilon_{\alpha\beta\gamma}\frac{\delta F}{\delta\Omega_\gamma}\right] \\ &\times\left[\nabla_\alpha\left(\frac{\delta G}{\delta M_\beta}\right) - \nabla_\beta\left(\frac{\delta G}{\delta M_\alpha}\right) - 2\varepsilon_{\alpha\beta\delta}\frac{\delta G}{\delta\Omega_\delta}\right]d^3x \; .\end{aligned} \tag{7.2-34}$$

Note that the non-negative definite nature of the entropy production implies that $\mu, \eta \geq 0$.

The implications of the additional term over the expression used for the Navier/Stokes equations are dual. First, the stress tensor, (7.2-11d), acquires an anisotropic term

$$\sigma_{\alpha\gamma} \equiv \mu\left(\nabla_\alpha v_\gamma + \nabla_\gamma v_\alpha\right) + \eta\left(\nabla_\alpha v_\gamma - \nabla_\gamma v_\alpha - 2\varepsilon_{\alpha\beta\gamma}\omega_\beta\right) \; . \tag{7.2-35}$$

Second, the evolution equation for the density of the internal angular momentum, (7.2-33), involves a relaxation term

$$\frac{\partial \boldsymbol{\Omega}}{\partial t} + \mathbf{v} \cdot \nabla \boldsymbol{\Omega} = 2\eta(\mathbf{w} - 2\boldsymbol{\omega}) \ . \tag{7.2-36}$$

It is interesting to note that there is only one additional material parameter, η, which enters into both the stress expression and the evolution equation for the internal angular momentum density. Furthermore, when rotational equilibrium is established, i.e., when $\boldsymbol{\omega}=\boldsymbol{\omega}_0$, both additional terms become zero and the equations reduce (as they should) to the Navier/Stokes equations. For a more complete description, which also introduces additional dissipation in the evolution equation for $\boldsymbol{\Omega}$, the reader is referred to the next section on polar fluids, §7.2.3. ■

EXAMPLE 7.4: Up to now, the discussion of the viscous dissipation bracket, $[F,G]$, has been limited to expressions which are bilinear in F, G—excluding the entropy correction (see, for example, Eqs. (7.2-1) and (7.2-22))—which led to viscous stresses represented by linear functions of the velocity gradient, $\nabla\mathbf{v}$—see Eq. (7.2-11d). This linear dependence of the stresses on the velocity gradient considerably simplifies the modeling. For example, for an incompressible, isotropic fluid, and with the effects of internal rotation neglected, the viscous stresses are modeled by a single positive constant, the (Newtonian) viscosity, μ, which represents the coefficient of proportionality between the stress, $\boldsymbol{\sigma}$, and the rate of deformation, $\mathbf{A} \equiv \nabla\mathbf{v} + \nabla\mathbf{v}^T$, tensors:

$$\boldsymbol{\sigma} = \mu \mathbf{A} \ . \tag{7.2-37}$$

Although this representation is, in general, expected to remain valid only for conditions close to static equilibrium, for low molecular weight (structureless) homogeneous fluids, it has a surprisingly large (in terms of the magnitude of the rate of deformation tensor, $\mathbf{A}$) domain of validity. Thus, the development of *Fluid Mechanics* as a separate discipline for the study of the flow dictated by the Navier/Stokes equations (also called *Newtonian Fluid Mechanics*)—corresponding to Newton's law for the viscous stress behavior, (7.2-3b), and including the limiting case of ideal (dissipationless) fluids and free-surface flows—is justified [Lamb, 1932; Landau and Lifshitz, 1959; Batchelor, 1967]. Furthermore, aided by the advance of powerful computers, the extended validity of the Navier/Stokes equations to more complex flow geometries has led to the explosive development of the field of *Computational Fluid Dynamics* (CFD), with a very wide range of applications.

However, with fluids endowed with an internal microstructure, such as particle suspensions or polymer solutions and melts, the linear relationship implied by (7.2-37) ceases, in general, to be valid. Although the most appropriate way to describe the flow behavior of these systems, especially in transient cases, is through the introduction of an appropriate internal variable(s) characterizing the time evolution of the system's internal microstructure—see chapters 8 through 11 for more details—a simpler, approximate model (the "incompressible, generalized Newtonian fluid" [Bird *et al.*, 1987a, p. 170]) can be built based on a phenomenological nonlinear relationship between the stress and the rate of deformation:

$$\boldsymbol{\sigma} = \eta(\mathbf{A})\,\mathbf{A} \;, \tag{7.2-38}$$

where the non-Newtonian viscosity, η, is now assumed to depend on the invariants of the rate of deformation tensor. The purpose of the present example is to demonstrate the utility of the general nonlinear expression for the dissipation bracket, (7.1-19), in the modeling of generalized non-Newtonian flow behavior.

For simplicity, let us again focus on the momentum transport, considering an isothermal and incompressible system. In this case, the modeling of a generalized non-Newtonian flow behavior can be achieved through a generalization of the dissipation bracket proposed for the derivation of Navier/Stokes equations, (7.2-22), along the lines of the more general description of (7.1-19). Thus, ignoring the entropy correction term, we have

$$[F,G] = -\int_{\Omega'} Q_{\alpha\beta}\left[\nabla_\alpha\Big(\frac{\delta F}{\delta M_\beta}\Big) + \nabla_\beta\Big(\frac{\delta F}{\delta M_\alpha}\Big)\right] d^3x \;, \tag{7.2-39}$$

where $\mathbf{Q}$ is, in general, a nonlinear (odd) tensor function of $\boldsymbol{\nabla}(\delta G/\delta\mathbf{M}) + \boldsymbol{\nabla}(\delta G/\delta\mathbf{M})^{\mathrm{T}}$ which, close to static equilibrium, has the value

$$\left[Q_{\alpha\beta} = \eta_0\Big(\nabla_\alpha\frac{\delta G}{\delta M_\beta} + \nabla_\beta\frac{\delta G}{\delta M_\alpha}\Big)\right]_{\left(\nabla_\alpha\frac{\delta G}{\delta M_\beta} + \nabla_\beta\frac{\delta G}{\delta M_\alpha} \to 0\right)} \;, \tag{7.2-40}$$

where η_0 is a non-negative constant representing the zero-shear-rate viscosity. Note that the form of the generalized dissipation bracket provided by Eqs. (7.2-39) and (7.2-40) is dictated from the principle of material invariance and the requirements to satisfy the non-negative entropy production and the Onsager reciprocal relations for an isotropic material. In particular, to obtain the generalized Newtonian fluid behavior indicated by (7.2-38), the tensorial parameter $\mathbf{Q}$ must take the form

$$\left[Q_{\alpha\beta} = \eta\Big(\nabla_\alpha\frac{\delta G}{\delta M_\beta} + \nabla_\beta\frac{\delta G}{\delta M_\alpha}\Big)\right]_{\left(\nabla_\alpha\frac{\delta G}{\delta M_\beta} + \nabla_\beta\frac{\delta G}{\delta M_\alpha} \to 0\right)} \;, \tag{7.2-41}$$

where η is, in general, a nonlinear function of the invariants I_2 and I_3 of $\boldsymbol{\nabla}(\delta G/\delta\mathbf{M}) + \boldsymbol{\nabla}(\delta G/\delta\mathbf{M})^{\mathrm{T}}$:

$$I_2^2 \equiv \frac{1}{2}\left(\nabla_\alpha \frac{\delta G}{\delta M_\beta} + \nabla_\beta \frac{\delta G}{\delta M_\alpha}\right)\left(\nabla_\alpha \frac{\delta G}{\delta M_\beta} + \nabla_\beta \frac{\delta G}{\delta M_\alpha}\right), \tag{7.2-42a}$$

$$I_3 \equiv \det\left[\boldsymbol{\nabla}\frac{\delta G}{\delta \mathbf{M}} + \left(\boldsymbol{\nabla}\frac{\delta G}{\delta \mathbf{M}}\right)^{\mathrm{T}}\right]. \tag{7.2-42b}$$

Note that the behavior close to static equilibrium imposed by (7.2-40) is inconsistent with the use of the power-law functional as the expression of the generalized Newtonian behavior

$$\eta = K(I_2)^{n-1}, \tag{7.2-43}$$

where K is a positive constant and n a dimensionless exponential factor, except for the (trivial) Newtonian case, n=1. This is a well-known defect of the power-law model [Bird *et al.*, 1987a, p. 175]. It can be rectified by resorting to a bounded expression for the viscosity as, for example, obtained from the five-parameter Bird/Carreau/Yasuda model [Bird *et al.*, 1987a, Eq. (4.1-9)]

$$\eta = \eta_\infty + (\eta_0 - \eta_\infty)(1 + (\lambda I_2)^a)^{(n-1)/a}, \tag{7.2-44}$$

where η_∞, λ, and a are three additional scalar parameters, representing the (positive) infinite-shear-rate viscosity, a characteristic (positive) time constant, and a dimensionless parameter characterizing the transition from a constant to a power-law-type dependence in the viscosity versus I_2 profile, respectively. This five-parameter model has sufficient flexibility to fit a wide variety of experimental $\eta(I_2)$ curves [Bird *et al.*, 1987a, §4.1a]. However, we should not forget that Eqs. (7.2-43) and (7.2-44) represent, at best, empirical fits to experimentally measured stress-versus-deformation rate data, typically obtained for steady-state (simple shear) viscometric flows. As such, they can lead to predictions which contain considerable error when applied to more general transient and/or non-viscometric flows.

In conclusion, although the nonlinear generalization of the bracket allows generic expressions such as the one given by (7.2-38) to be obtained for the stress constitutive equation, they should be used with caution given the existing limitations (steady state, simple shear, etc.) typically constraining the validity of the underlying assumptions. As we shall see in the next chapter, the same phenomenological behavior as that represented by (7.2-38) can also be obtained through the use of the bilinear form for the dissipation bracket when one or more internal structural parameters are used in conjunction with a suitable model for the Hamiltonian and the system dissipation. This approach provides, in general, a more robust modeling of the system which remains, in principle, valid over a wider range of conditions. ■

7.2.3 The hydrodynamic equations for a polar fluid

The analysis in Example 7.3 is intimately related to the notion of a "polar fluid" [Grad, 1952; Aero *et al.*, 1965; Eringen, 1966; Cowin, 1968; Pennington and Cowin, 1969; Cowin, 1974]. A polar fluid is formally defined as a fluid for which the kinematics are described through the use of two independent kinematical vector fields: the (polar) vector field representing the velocity of the fluid particle, **v**, and an axial vector field representing the angular velocity of the fluid particle, $\boldsymbol{\omega}$. Thus, in a polar fluid the rotational motion of the fluid particles is not completely determined by the vorticity field. The application of the polar fluid theory has been to suspensions, to blood flow, to mean turbulent flow, and, in general, to problems such as lubrication or thin films, with a characteristic geometric dimension of the same order of magnitude as the size of the microstructure. One of the principal predictions of the theory is the increased viscosity close to solid surfaces and in narrow geometries such as bearings or capillary tubes [Cowin, 1974, pp. 279,80]. In this subsection, we wish to show how the most general description for an isotropic, polar fluid arises naturally from the bracket formalism through a straightforward generalization of the analysis used in Example 7.3.

Crucial for the understanding of the development of the equations is the use of the principle of frame indifference which, in this particular application, relies on the distinction between "absolute" and "relative" tensors and between "polar" and "axial" vectors [Lodge, 1974, §5.1]. As explained in more detail in §6.4.3, a tensor is characterized as relative or absolute depending on whether the transformation of its components from one coordinate system to another involves or does not involve a power of the determinant of the Jacobian of the transformation. For relative tensors in an orthogonal (Cartesian) coordinate system (where, in consequence, the determinant of the transformation is either 1 or -1), the odd or even character of the power exponent W of the transformation is of significance. All odd powers will give the same transformation law, as will all even powers (including the absolute tensors which can also be considered relative with weight W=0). Practically speaking, the Cartesian components of an absolute or an even-power relative tensor are independent of the order of the axes whereas the Cartesian components of a relative tensor of odd power are not. An absolute vector is usually called *polar*, whereas a relative vector of power one is called *axial*. The velocity, **v**, and the Kronecker delta tensor, $\boldsymbol{\delta}$, are examples of even-power relative tensors. The angular velocity, $\boldsymbol{\omega}$, and the permutation third-rank tensor, $\boldsymbol{\epsilon}$, are examples of odd power relative tensors.

Although most tensor operations preserve the even or odd character of a tensor, some do not. Examples of the latter are the multiplication of a tensor with the permutation symbol and the *curl* operation. The distinction between even and odd relative tensors is both mathematically and physically important, as discussed in §6.4.3. Mathematically, the different transformation characteristics make invalid expressions where tensors of different kind are equated—principle of frame indifference. Physically, a solid-body rotation is quantified with an axial vector, the angular velocity, which is locally correlated with the antisymmetric part of the velocity gradient. Thus, the physical aspect of frame indifference becomes equivalent to the mathematical requirement that the stress tensor (an even power relative tensor) depends only on the symmetric component of the velocity gradient (also an even tensor). Indeed, as we saw before in §7.2.1, the requirement of frame indifference excludes the antisymmetric component from the phenomenological transport parameter **Q** for a simple, isotropic fluid. As a result, the stress tensor is proportional only to the symmetric part of the velocity gradient, **A** (see Eq. (7.2-11d)).

In this subsection, we wish to describe more rigorously the flow transport processes within an incompressible, isotropic, polar fluid, thereby generalizing the analysis offered in Example 7.3. As in that case, for simplicity, we shall restrict the analysis to the isothermal case. Then, the dynamic variables of the system are the momentum density **M**, corresponding to the velocity, **v** (a polar vector), and the total angular momentum density $\boldsymbol{\Omega}$, corresponding to the total angular velocity, $\boldsymbol{\omega}$ (an axial vector), whereas the density, ρ, is assumed constant.

The Hamiltonian and the Poisson bracket are exactly the same as given before by Eqs. (7.2-29) and (7.2-32), respectively. However, the dissipation bracket acquires three extra terms resulting, for an isotropic medium, in the following expression

$$[F,G] =$$

$$-\int_{\Omega'} \frac{\mu}{2}\left[\nabla_\alpha\left(\frac{\delta F}{\delta M_\beta}\right)+\nabla_\beta\left(\frac{\delta F}{\delta M_\alpha}\right)\right]\left[\nabla_\alpha\left(\frac{\delta G}{\delta M_\beta}\right)+\nabla_\beta\left(\frac{\delta G}{\delta M_\alpha}\right)\right] d^3x$$

$$-\int_{\Omega'} \frac{\eta}{2}\left[\nabla_\alpha\left(\frac{\delta F}{\delta M_\beta}\right)-\nabla_\beta\left(\frac{\delta F}{\delta M_\alpha}\right)-2\varepsilon_{\alpha\beta\gamma}\frac{\delta F}{\delta \Omega_\gamma}\right]$$

$$\times\left[\nabla_\alpha\left(\frac{\delta G}{\delta M_\beta}\right)-\nabla_\beta\left(\frac{\delta G}{\delta M_\alpha}\right)-2\varepsilon_{\alpha\beta\delta}\frac{\delta G}{\delta \Omega_\delta}\right] d^3x$$

$$-\int_{\Omega'} \frac{\alpha'}{2}\left[\nabla_\alpha\left(\frac{\delta F}{\delta \Omega_\alpha}\right)\right]\left[\nabla_\beta\left(\frac{\delta G}{\delta \Omega_\beta}\right)\right] d^3x$$

$$
\begin{aligned}
&-\int_{\Omega'} \frac{\beta'}{2}\left[\nabla_\alpha(\frac{\delta F}{\delta \Omega_\beta}) + \nabla_\beta(\frac{\delta F}{\delta \Omega_\alpha})\right]\left[\nabla_\alpha(\frac{\delta G}{\delta \Omega_\beta}) + \nabla_\beta(\frac{\delta G}{\delta \Omega_\alpha})\right] d^3x \\
&-\int_{\Omega'} \frac{\gamma'}{2}\left[\nabla_\alpha(\frac{\delta F}{\delta \Omega_\beta}) - \nabla_\beta(\frac{\delta F}{\delta \Omega_\alpha})\right]\left[\nabla_\alpha(\frac{\delta G}{\delta \Omega_\beta}) - \nabla_\beta(\frac{\delta G}{\delta \Omega_\alpha})\right] d^3x \;,
\end{aligned}
\tag{7.2-45}
$$

which involves three additional scalar parameters, α', β', and γ'. Expres-sion (7.2-45) is the most general bilinear form with respect to F, G, corresponding to an incompressible and isotropic fluid characterized by the previously defined dynamic variables **M** and $\boldsymbol{\Omega}$. The parameters α', β' and γ' arise naturally from the decomposition of a fourth-rank ten-sorial parameter in parallel to the derivation of the viscosity parameters μ, η, and κ' in (7.2-3b).

The even parity (under time inversion), the frame indifference, and the even character of the dissipation are duly preserved in the form of the expression (7.2-45) and the previously defined properties of **M** and $\boldsymbol{\Omega}$. In addition, the Onsager reciprocal relations, given equivalently to (7.2-2b), are automatically satisfied. Moreover, the non-negative definite nature of the entropy production implies that $(\alpha' + ⅔\beta')$, β', and γ' must all be greater than or equal to zero. This is obtained, in parallel to the develop-ment in (7.2-19), by decomposing the (in general) arbitrary second-rank tensor $\boldsymbol{\nabla\omega}$ into three independent modes, the first one isotropic, the second symmetric and traceless, and the third antisym-metric. Then, the above-mentioned inequalities are obtained by consider-ing the dissipation $[H,H]$ for each one of the modes separately. Note that the same inequalities, together with the previously obtained $\mu, \eta \geq 0$, have also been obtained by Pennington and Cowin [1969, p. 390] where the parameters $(\mu,\eta,\alpha',\beta',\gamma')$ correspond to their $(\mu,\tau,\alpha,\beta,\gamma)$. As also mentioned in the same article, these inequalities are slightly different from the ones obtained earlier by Eringen [1966, Eq. (5.8)], which therefore must involve an error [Pennington and Cowin, 1969, pp. 398-400].

The implication from the additional terms in (7.2-45) over the expression used in Example 7.3, is that although the equation for the stress tensor, (7.2-35), remains unchanged, the evolution equation for the density of the internal angular momentum, (7.2-36), involves additional dissipative terms:

$$
\begin{aligned}
\frac{\partial \boldsymbol{\Omega}}{\partial t} + \mathbf{v}\cdot\boldsymbol{\nabla\Omega} &= 2\eta\,(\mathbf{w} - 2\boldsymbol{\omega}) + \boldsymbol{\nabla}(\alpha'\,\boldsymbol{\nabla}\cdot\boldsymbol{\omega}) \\
&\quad - \boldsymbol{\nabla}\cdot[\gamma'(\boldsymbol{\nabla\omega} - \boldsymbol{\nabla\omega}^{\mathrm{T}}) - \beta'(\boldsymbol{\nabla\omega} + \boldsymbol{\nabla\omega}^{\mathrm{T}})] \;.
\end{aligned}
\tag{7.2-46}
$$

p. 197: in middle of the page, $[H,H]$ should be $[H_m,H_m]$

This expression is exactly the same as Eq. (11) in Pennington and Cowin [1969, p. 390] if translated into their notation. It is interesting to note that in this case, as opposed to the case examined in Example 7.3, the right-hand side of (7.2-46) is, for a general non-homogeneous flow, non-zero, even when rotational equilibrium is established, i.e., when $\boldsymbol{\omega}=\boldsymbol{\omega}_0$. Thus, the condition $\eta=0$ is necessary in order for the Navier/Stokes equations to be recovered.

7.3 The Hydrodynamic Equations for a Multicomponent Fluid

In this section,[10] we shall demonstrate further the generality of the bracket formulation by considering a ν-component, non-reacting, fluid system for which the dynamic variables are **M**, s, and ρ_i, $i=1,2,\ldots,\nu$, where ρ_i is the mass density of the i-th component. (We shall consider the reacting fluid system in chapter 12.) Although we have the relation

$$\rho = \sum_{i=1}^{\nu} \rho_i \,, \tag{7.3-1}$$

ρ is no longer a variable of the problem. Still, the total mass density of the system must be given by (7.3-1) since

$$\rho \equiv \frac{m_t}{V} = \frac{m_1}{V} + \frac{m_2}{V} + \ldots + \frac{m_\nu}{V} \,. \tag{7.3-2}$$

Consequently, the operating space for this problem is

$$P \equiv \begin{cases} \rho_i(\mathbf{x},t) \in \mathbb{R}^+; & i=1,2,\ldots,\nu \\ \mathbf{M}(\mathbf{x},t) \in \mathbb{R}^3; & \mathbf{M}=\mathbf{0} \text{ on } \partial\Omega' \\ s(\mathbf{x},t) \in \mathbb{R} \end{cases} \tag{7.3-3}$$

(with initial and boundary conditions on **M**, s and the ρ_i as well), so that $F=F[\mathbf{M},s,\rho_1,\rho_2,\ldots,\rho_\nu]$ and the Hamiltonian may be generalized from (5.3-26) to

$$H[\mathbf{M},s,\rho_1,\rho_2,\ldots,\rho_\nu] = \int_{\Omega'} \frac{1}{2\rho}\mathbf{M}\cdot\mathbf{M} + \rho e_p^v + u]\, d^3x \,, \tag{7.3-4}$$

with ρ denoting the sum of (7.3-1) as a shorthand notation. The functional derivatives of the Hamiltonian may now be evaluated via (5.2-2) as

[10]The material presented in the core of this section follows closely the analysis presented in §4 of Edwards and Beris [1991a].

$$\frac{\delta H}{\delta M_\alpha} = v_\alpha \ , \tag{7.3-5a}$$

$$\frac{\delta H}{\delta s} = \frac{\partial u}{\partial s} = T \ , \tag{7.3-5b}$$

$$\frac{\delta H}{\delta \rho_i} \equiv \mu_i^* = \frac{\partial u}{\partial \rho_i} - \tfrac{1}{2} v^2 + e_p^v \ , \tag{7.3-5c}$$

where we define μ_i^* as the generalized chemical potential (on a per mass basis):

$$\mu_i^* \equiv \mu_i - \tfrac{1}{2} v^2 + e_p^v \ . \tag{7.3-5d}$$

Following the procedure of §5.3 with minor modifications resulting from the substitution of the variables ρ_i for ρ, it is easy to arrive at the spatial Poisson bracket for the multicomponent fluid directly from a corresponding material bracket, (5.2-13), as

$$\begin{aligned}\{F,G\} = &- \sum_{i=1}^{\nu} \int_{\Omega'} \left[\frac{\delta F}{\delta \rho_i} \nabla_\beta \left(\frac{\delta G}{\delta M_\beta} \rho_i \right) - \frac{\delta G}{\delta \rho_i} \nabla_\beta \left(\frac{\delta F}{\delta M_\beta} \rho_i \right) \right] d^3x \\ &- \int_{\Omega'} \left[\frac{\delta F}{\delta M_\alpha} \nabla_\beta \left(\frac{\delta G}{\delta M_\beta} M_\alpha \right) - \frac{\delta G}{\delta M_\alpha} \nabla_\beta \left(\frac{\delta F}{\delta M_\beta} M_\alpha \right) \right] d^3x \\ &- \int_{\Omega'} \left[\frac{\delta F}{\delta s} \nabla_\beta \left(\frac{\delta G}{\delta M_\beta} s \right) - \frac{\delta G}{\delta s} \nabla_\beta \left(\frac{\delta F}{\delta M_\beta} s \right) \right] d^3x \ .\end{aligned} \tag{7.3-6}$$

The non-conservative contributions to the system behavior can be picked up through a dissipation bracket similar to the one defined by (7.2-1):

$$\begin{aligned}[F,G] = &- \int_{\Omega'} Q_{\alpha\beta\gamma\varepsilon} \nabla_\alpha \left(\frac{\delta F}{\delta M_\beta} \right) \nabla_\gamma \left(\frac{\delta G}{\delta M_\varepsilon} \right) d^3x \\ &+ \int_{\Omega'} Q_{\alpha\beta\gamma\varepsilon} \frac{1}{T} \frac{\delta F}{\delta s} \nabla_\alpha \left(\frac{\delta G}{\delta M_\beta} \right) \nabla_\gamma \left(\frac{\delta G}{\delta M_\varepsilon} \right) d^3x \\ &- \int_{\Omega'} \sum_{i,k=0}^{\nu} \alpha_{ik}^{\gamma\beta} \left| \nabla_\gamma \left(\frac{\delta F}{\delta \rho_i} \right) \right|^* \left| \nabla_\beta \left(\frac{\delta G}{\delta \rho_k} \right) \right|^* d^3x \\ &+ \int_{\Omega'} \sum_{i,k=0}^{\nu} \alpha_{ik}^{\gamma\beta} \frac{1}{T} \frac{\delta F}{\delta s} \left| \nabla_\gamma \left(\frac{\delta G}{\delta \rho_i} \right) \right|^* \left| \nabla_\beta \left(\frac{\delta G}{\delta \rho_k} \right) \right|^* d^3x \ ,\end{aligned} \tag{7.3-7}$$

where the development of the dissipation is in terms of the constant temperature affinities described in §6.4.5:

$$\left.\nabla_\gamma\left(\frac{\delta F}{\delta\rho_0}\right)\right|^* \equiv \nabla_\gamma\left(\frac{\delta F}{\delta s}\right) , \tag{7.3-8a}$$

$$\left.\nabla_\gamma\left(\frac{\delta F}{\delta\rho_i}\right)\right|^* \equiv \nabla_\gamma\left(\frac{\delta F}{\delta\rho_i}\right) + \left.\frac{\partial s}{\partial\rho_i}\right|_T \nabla_\gamma\left(\frac{\delta F}{\delta s}\right) , \quad i=1,\ldots,\nu , \tag{7.3-8b}$$

where $(\partial s/\partial\rho_i)|_T \equiv \overline{S}_i$ is the partial specific entropy from (4.3-30). The above definitions are used in order to facilitate the comparison with previous works [Hirschfelder *et al.* 1954, p. 715]. As shown in Example 7.2, the description of the dissipation in terms of these affinities is not really necessary in order to preserve their symmetry and non-negative definite characteristics.

In (7.3-7), $\boldsymbol{\alpha}$ is a $(\nu+1)\times(\nu+1)$ symmetric matrix of the phenomenological transport coefficients; however, these coefficients are not all independent. For a general system, the symmetry requirements are the same as discussed in §6.5.2:

$$\alpha_{ik}^{\gamma\beta} = \alpha_{ki}^{\beta\gamma} ; \tag{7.3-9a}$$

however, for an isotropic system with which we are here concerned,

$$\alpha_{ik}^{\gamma\beta} = \alpha_{ik}\,\delta_{\gamma\beta} , \tag{7.3-9b}$$

so (7.3-9a) implies that

$$\alpha_{ik} = \alpha_{ki} . \tag{7.3-9c}$$

Furthermore, conservation of total mass imposes the constraint that the α_{ik} must satisfy the linear relations

$$\sum_{i=1}^{\nu} \alpha_{ik} = \sum_{i=1}^{\nu} \alpha_{ki} = 0 , \quad k=0,1,\ldots,\nu , \tag{7.3-10}$$

thus indicating that there are $\nu^2-\nu-\frac{1}{2}\nu(\nu-1)=\frac{1}{2}\nu(\nu-1)$ independent coefficients α_{ik}, $i\neq 0$ and/or $k\neq 0$.[11] In effect, all of the α_{ik} (except α_{00}) are diffusion coefficients which govern the rate of mass flux within the system. The exact functional form in terms of the parameters of the problem, T, m_i, etc., which these coefficients take, is entirely system

[11]Note that in effect the imposition of this constraint can be avoided if the dissipation bracket is expressed as a quadratic form involving gradients of the differences $\mu_i^*-\mu_\nu^*$.

dependent. For instance, Hirschfelder *et al.* [1954, p. 715] give exact expressions for the α_{ik} based upon the kinetic theory of dilute gases. Yet for each individual system, the proper functional relationship must be determined through appropriate arguments from physical chemistry or statistical thermodynamics. This is not necessarily an easy thing to do, but one may again return to the thermodynamic and conservation requirements of §7.1 for guidance on this matter.

We may now insert the Hamiltonian derivatives of (7.3-5) into the dynamical equation for F, in terms of the generalized bracket of (7.3-6,7), in order to obtain the evolution equations for the dynamic variables:

$$\frac{\partial \rho_i}{\partial t} = -\nabla_\alpha (v_\alpha \rho_i) + \nabla_\gamma (\alpha_{i0} \nabla_\gamma T) + \sum_{k=1}^{\nu} \nabla_\gamma (\alpha_{ik} \Lambda_\gamma^k) , \quad i=1,2,\ldots,\nu , \tag{7.3-11a}$$

$$\rho \frac{\partial v_\alpha}{\partial t} = -\rho v_\beta \nabla_\beta v_\alpha - \nabla_\alpha p - \rho \nabla_\alpha e_p^v + \nabla_\beta \sigma_{\alpha\beta} , \tag{7.3-11b}$$

$$\begin{aligned} \frac{\partial s}{\partial t} = &- \nabla_\alpha (v_\alpha s) + \frac{1}{T} \nabla_\gamma (k \nabla_\gamma T) + \frac{1}{T} \sigma_{\alpha\beta} \nabla_\alpha v_\beta \\ &+ \frac{1}{T} \sum_{k=1}^{\nu} \nabla_\gamma (\alpha_{0k} T \Lambda_\gamma^k) + \sum_{i=1}^{\nu} \nabla_\gamma (\alpha_{i0} \overline{S}_i \nabla_\gamma T) + \sum_{i,k=1}^{\nu} \nabla_\gamma (\alpha_{ik} \overline{S}_i \Lambda_\gamma^k) \\ &+ \frac{1}{T} \sum_{i=1}^{\nu} \alpha_{i0} \Lambda_\gamma^i \nabla_\gamma T + \frac{1}{T} \sum_{i,k=1}^{\nu} \alpha_{ik} \Lambda_\gamma^i \Lambda_\gamma^k . \end{aligned} \tag{7.3-11c}$$

In this expression, $\boldsymbol{\sigma}$ is as given by (7.2-11d), the affinity, Λ_γ^k, is defined as

$$\Lambda_\gamma^k \equiv \nabla_\gamma (\mu_k^*) + \overline{S}_k \nabla_\gamma T , \tag{7.3-12a}$$

and the pressure is given by the close analog of (4.3-24) as

$$p = -u + \frac{\partial u}{\partial s} s + \frac{\partial u}{\partial \rho_1} \rho_1 + \ldots + \frac{\partial u}{\partial \rho_\nu} \rho_\nu . \tag{7.3-12b}$$

Because of the relations (7.3-10), summing together all ν evolution equations for the ρ_i results in the usual continuity equation for the overall mass density, as given by (7.2-11a). The energy equations may also be obtained; for instance, expressing $\partial u/\partial t$ as

$$\frac{\partial u}{\partial t}(\rho_1,\ldots,\rho_\nu,s) = \sum_{i=1}^{\nu} \frac{\partial u}{\partial \rho_i}\frac{\partial \rho_i}{\partial t} + \frac{\partial u}{\partial s}\frac{\partial s}{\partial t} , \tag{7.3-13a}$$

we find that the evolution equation for the internal energy is

$$\frac{\partial u}{\partial t} = -\nabla_\alpha(v_\alpha u) - p\nabla_\alpha v_\alpha + \nabla_\alpha(k\nabla_\alpha T) + \sigma_{\alpha\beta}\nabla_\alpha v_\gamma$$

$$+ \sum_{k=1}^{\nu} \nabla_\gamma(\alpha_{0k} T\Lambda_\gamma^k) + \sum_{i=1}^{\nu} [\mu_i^+\nabla_\gamma(\alpha_{i0}\nabla_\gamma T) + T\nabla_\gamma(\alpha_{i0}\overline{S}_i\nabla_\gamma T)$$

$$+ \alpha_{i0}\Lambda_\gamma^i\nabla_\gamma T] + \sum_{i,k=1}^{\nu} [\mu_i^+\nabla_\gamma(\alpha_{ik}\Lambda_\gamma^i) + T\nabla_\gamma(\alpha_{ik}\overline{S}_i\Lambda_\gamma^k) + \alpha_{ik}\Lambda_\gamma^i\Lambda_\gamma^k] . \tag{7.3-13b}$$

As in §7.2.1, we may use the second law of thermodynamics to arrive at consistency requirements for the phenomenological coefficients. In this case, the affinities involving the momentum density are completely decoupled from the others, and consequently we arrive at the same conditions upon μ and κ as in §7.2.1. The remaining affinities are coupled, however, yet we can use the requirement that $[S,H]\geq 0$ and the dissipation bracket of (7.3-7) to obtain the inequality

$$\sum_{i,k=1}^{\nu} \alpha_{ik}\Lambda_\gamma^i\Lambda_\gamma^k + 2\sum_{i=1}^{\nu} \alpha_{i0}\Lambda_\gamma^i\nabla_\gamma T + \alpha_{00}(\nabla_\gamma T)(\nabla_\gamma T) \geq 0 . \tag{7.3-14}$$

By defining the vector **V** as

$$\mathbf{V}^{\mathrm{T}} \equiv (\Lambda_1^1,\Lambda_2^1,\Lambda_3^1,\Lambda_1^2,\ldots,\Lambda_3^\nu,\nabla_1 T,\nabla_2 T,\nabla_3 T) , \tag{7.3-15}$$

and using the procedure of §7.2.1, we can arrive at the inequalities

$$\alpha_{ii} \geq 0 , \quad i=0,1,\ldots,\nu ,$$

$$(\alpha_{ii}\,\alpha_{kk} - 4\,\alpha_{ik}^2) \geq 0 , \quad i,k=0,1,\ldots,\nu , \quad i\neq k , \quad \text{etc.} , \tag{7.3-16}$$

which express the requirements for the α_{ik} to be compatible with thermodynamics.

EXAMPLE 7.5: The rheological behavior of dilute suspensions of spherical particles has excited scientific interest since the time Einstein first described the apparent viscosity of a dilute suspension of spheres in terms of the viscosity of the solvent, μ, and the particle volume fraction, ϕ [Einstein, 1906; 1911]. Since then, interest has been primarily focused on the description of the apparent viscosity as a function of the particle concentration and the shear rate [Jeffrey and Acrivos, 1976; Metzner, 1985]. However, about the same time, it became apparent that the rheological properties of suspensions are influenced by a large number of factors which we are only starting to understand [Jeffrey and Acrivos, 1976]. In particular, the interest in the fundamental understanding of the dynamics of particles in suspensions was revived with the discovery of new and exciting phenomena such as shear-induced migration and diffusion [Ho and Leal, 1974; Leighton and Acrivos, 1987]. The purpose of the present example is to demonstrate one of the possible mechanisms for particle migration in dilute particle suspensions as it arises naturally from the bracket formalism applied for a two-component, rotationally relaxing, viscous medium. This is achieved by suitably extending Example 7.3 to accommodate the presence of rigid particles which can be safely assumed to carry all of the internal rotational energy of the medium.

Let us extend Example 7.3 by considering as a second phase a suspension of rigid, non-interacting particles of mass density ρ_p and radius a_p, with a rotational angular velocity, $\boldsymbol{\omega}$, within a low molecular weight medium of density ρ_m. Then, the expression for the Hamiltonian becomes

$$H_m[\mathbf{M}] = \int_{\Omega'} \left[\frac{1}{2\rho} \mathbf{M} \cdot \mathbf{M} + \frac{1}{2\rho_s} \boldsymbol{\Omega} \cdot \boldsymbol{\Omega} - \rho \mathbf{g} \cdot \mathbf{x} \right] d^3x \ , \tag{7.3-17}$$

which is a slight modification of the expression (7.2-29) used for a single-component rotational medium manifested in the definitions of the total mass density, ρ, and moment density, ρ_s, as

$$\rho \equiv \rho_m + \rho_p \ , \tag{7.3-18}$$

$$\rho_s \equiv 2\rho_p a_p^2/5 \equiv \rho k^2 \ . \tag{7.3-19}$$

The angular momentum density and the Poisson bracket are defined exactly as before by Eqs. (7.2-31) and (7.2-32), respectively. Then, a diffusive term similar to the ones introduced into the dissipation bracket for a general multicomponent fluid, (7.3-7), is added to the dissipation bracket defined by (7.2-34) to give

$$[F,G] =$$

$$-\int_{\Omega'} \frac{\mu}{2} \left[\nabla_\alpha \left(\frac{\delta F}{\delta M_\beta} \right) + \nabla_\beta \left(\frac{\delta F}{\delta M_\alpha} \right) \right] \left[\nabla_\alpha \left(\frac{\delta G}{\delta M_\beta} \right) + \nabla_\beta \left(\frac{\delta G}{\delta M_\alpha} \right) \right] d^3x$$

p. 203: in (7.3-17), $H_m[\mathbf{M}]$ should be $H_m[\mathbf{M},\boldsymbol{\Omega},\rho_m,\rho_p]$

$$-\int_{\Omega'} \frac{\eta}{2} \left[\nabla_\alpha (\frac{\delta F}{\delta M_\beta}) - \nabla_\beta (\frac{\delta F}{\delta M_\alpha}) - 2\varepsilon_{\alpha\beta\gamma} \frac{\delta F}{\delta \Omega_\gamma} \right]$$

$$\times \left[\nabla_\alpha (\frac{\delta G}{\delta M_\beta}) - \nabla_\beta (\frac{\delta G}{\delta M_\alpha}) - 2\varepsilon_{\alpha\beta\delta} \frac{\delta G}{\delta \Omega_\delta} \right] d^3x \tag{7.3-20}$$

$$-\int_{\Omega'} D \left[\nabla_\alpha (\frac{\delta F}{\delta \rho_m}) - \nabla_\alpha (\frac{\delta F}{\delta \rho_p}) \right] \left[\nabla_\alpha (\frac{\delta G}{\delta \rho_m}) - \nabla_\alpha (\frac{\delta G}{\delta \rho_p}) \right] d^3x \ .$$

The additional term in the dissipation bracket is the most general one (for isotropic media) which is compatible with the conservation of total mass constraint and satisfies the Onsager reciprocal relations. Furthermore, note that due to the non-negative nature of the rate of entropy production, $D \geq 0$. The new term affects neither the equation for the stress tensor, (7.2-35), nor the evolution equation for the density of the angular momentum, (7.2-36), which remain the same as the ones described in Example 7.3. However, there is an additional evolution equation for the particles density, ρ_p,

$$\frac{\partial \rho_p}{\partial t} + \mathbf{v} \cdot \boldsymbol{\nabla} \rho_p = - \boldsymbol{\nabla} \cdot \mathbf{J}_p \ , \tag{7.3-21}$$

where $\mathbf{J}_p$ is the particle flux, defined as

$$\mathbf{J}_p = - D \, \boldsymbol{\nabla} \left[\frac{\delta H}{\delta \rho_p} - \frac{\delta H}{\delta \rho_m} \right] \tag{7.3-22}$$

$$= (D a_p^2 / 5) \, \boldsymbol{\nabla} (\boldsymbol{\omega} \cdot \boldsymbol{\omega}) \ ,$$

where (7.3-17) has been used for the Hamiltonian. Of course, in reality we also have a traditional diffusive flux proportional to the gradient of the particle density in (7.3-21) arising from an (usual) entropic term in the Hamiltonian which will eventually, at steady state, balance the flux induced by (7.3-21).

The important issue here is that we automatically get a "particle migration" term which will drive the particles towards areas in the flow corresponding to a high magnitude of the rotation $\boldsymbol{\omega}$. Considering that the equilibrium value for the rotation rate is half the vorticity of the flow, which, for a rectilinear flow, is proportional to the shear rate, we can now explain the migration of particles away from the centerline in a planar Poiseuille flow. Indeed, detailed analysis of the forces on isolated spherical particles in Poiseuille flow does indicate the presence of two forces on the particles. One tends to push the spheres to the wall and is proportional to the gradient of the shear rate squared, which is exactly of the form indicated by (7.3-21) [Ho and Leal, 1974, p. 386]. The other one tends to push the spheres to the center and arises due to wall interactions which are not captured in the present, simplified model. Finally, a word of caution is warranted here in that there exist several possible mechanisms which can explain shear-enhanced diffusion and shear-induced particle migration phenomena, each one operating under different conditions. In particular, the inertial mechanism explained above should not be confused with the collision-frequency and viscosity-gradient based ones which are the primary operating mechanisms in concentrated suspensions [Leighton and Acrivos, 1987; Phillips *et al.*, 1992]. ■

At this point, as Part I concludes, we feel that an additional comment needs to be made. So far, we have emphasized most the development of model equations which govern the material behavior *in the bulk*. Obviously, these equations alone do not lead to a well-defined mathematical problem: boundary conditions are also needed. This step, which has only briefly been mentioned for the simple cases studied so far—mainly in setting up the functional space for the variables involved in the bracket formalism—should not be neglected. It is important to realize that boundary conditions are an integral part of the modeling effort and do not belong to an *a posteriori* process, as has been the common practice.

The traditional approach has been to follow the development of the governing equations during the mathematical modeling of the underlying physics by posing *a posteriori* the *mathematical* question:

> Here are the equations that I wish to solve. What boundary and initial conditions are required to solve them?

This approach works adequately for most of the simple cases studied to date.

For the future of complex applications, however, we must realize that this approach is inadequate. Instead, another paradigm emerges through the implementation of the bracket formalism where the boundary conditions are imposed *a priori* before the development of the model equations. In this approach, we are rather asked, before the modeling, the following *physical* question:

> Here are the boundary and initial conditions which I arrived at based upon realistic, physical assumptions. What are the corresponding governing equations?

As we shall see in Part II, this approach facilitates the development of complex model equations. It is even more crucial when discussing free-surface problems. However, although the bracket formalism can, in principle, be extended to handle free surface problems—see Lewis *et al.* [1986] and Abarbanel *et al.* [1988] where the Poisson bracket is developed for free surface problems involving ideal inviscid fluids—free surface applications of the bracket formalism are not addressed here. The interested reader is referred to a number of recent monographs and textbooks on this subject [Rowlinson and Widom, 1989; Slattery, 1990; Edwards (D.A.) *et al.*, 1991].

Part II

Applications

He who seeks for methods without having a definite problem in mind seeks for the most part in vain.

—*David Hilbert*

8
Incompressible Viscoelastic Fluids

If you want to understand function, study structure.
—Francis Harry Compton Crick

In Part I, we discussed in detail the foundations of the bracket description of dynamical behavior, demonstrating how the generalized bracket is linked to the theories of both Hamiltonian mechanics and irreversible thermodynamics. Now it is time to discuss the various applications towards seemingly complex systems which are the main focus of this book. Specifically, we want to look at a variety of microstructured media of immediate concern in science and industry, and to illustrate the advantages of using the generalized bracket formalism over traditional techniques when developing system-particular models. As we shall also see, there are certain advantages to be gained even when we are simply expressing existing models in Hamiltonian form.

The first subject that we wish to address is that of viscoelastic fluid dynamics. As the name implies, viscoelasticity characterizes the materials that possess properties intermediate to those of an elastic solid and a viscous fluid. The most characteristic property is that of limited ("fading") memory: viscoelastic materials partially resume their previous deformation state upon removal of the externally applied forces; the smaller the duration of the application of the forces, the better the recovery. Materials of this type contain a certain degree of internal microstructure (e.g., polymeric solutions and melts, advanced composites, liquid crystals, etc.), and are very important in the processing industry where one wishes to combine the "processability" of the medium's fluidity with the "structural quality" of the internal architecture to obtain high-strength/low-weight final products. We can distinguish two types of viscoelasticity: viscoelastic solids and viscoelastic fluids characterized by the ability or lack of ability respectively, to support shear stresses at finite deformations. In the following we shall focus on the analysis of viscoelastic fluids although the approach followed applies and/or can be extended in a straightforward fashion to viscoelastic solids as well. For a description of solid viscoelasticity, the interested reader may consult one of the many excellent monographs in the area [Eringen, 1962, chs. 8, 10; Ferry, 1980; Sobotka, 1984; see also Tschoegl, 1989].

The bulk of viscoelastic fluids are polymeric solutions and melts, which, due to the macromolecular dimensions of the constituent molecules (composed of long chains or branched structures), show a high degree of anisotropy and structural changes under flow conditions. These are just the structural characteristics that one wishes to describe when modeling viscoelastic media, as they are responsible for not only the mechanical properties of the final product, but also govern the rheological response of the material while processing.

Therefore, the main issue in working with polymeric materials is to know how to process these fluids so as to obtain the final desired condition. This is not an easy task, and in the absence of any realistic models to predict the structural characteristics of such materials under flowing, processing conditions, many man-hours of labor must be spent in trial-and-error design procedures. When one considers that the kinematics of the processing geometry greatly affect the structural characteristics of the finished product, it becomes crucial to develop reliable models for describing not only the structural state of the medium during flow, but also its rheological behavior. Indeed, the two are intimately coupled, and in designing products using materials such as these, one must take into account the proper coupling between the rheological properties of the medium, the flow geometry, and the properties of the finished product.

Thus, it comes as no surprise that considerable work has already been devoted to the analysis, modeling, and simulations of viscoelastic flow behavior—see, for example, Lodge [1964], Huilgol [1975], Schowalter [1978], Ferry [1980], Tanner [1985], Doi and Edwards [1986], Bird *et al.* [1987a,b], Larson [1988], White [1990], Bird and Öttinger [1992], and, for a general introduction to rheology, Barnes *et al.* [1989]. Very briefly, among these and other works we can distinguish between several modeling approaches. First, there are two diametrically opposing views. In one, the model of the most general viscoelastic behavior is sought, but for limited types of deformations—such as very small displacement gradients in linear viscoelasticity [Gurtin and Sternberg, 1962; Pipkin, 1986, chs. 1-5; Bird *et al.*, 1987a, ch. 5; Tschoegl, 1989] or small magnitude for the rate-of-strain tensor and its derivatives in the slow viscoelastic flow "ordered fluids" models [Rivlin and Ericksen, 1955; Pipkin, 1986, ch. 8; Bird *et al.*, 1987a, ch. 6]—and/or limited types of flow, such as the theory for viscometric flows [Coleman *et al.*, 1966; Pipkin, 1986, ch. 9; Bird *et al.*, 1987a, p. 503; Larson, 1988, ch. 9]. In the other view, the rheological behavior of general flow deformations is described, but for a limited class of fluids.

It is this second view that it is of interest here. Within this view, the phenomenological (continuum mechanics) description was historically the first one to appear [Maxwell, 1867; for an excellent historical perspective of the origins of modern rheology see White, 1990, pp. 210-14]. However, it was not until Oldroyd's seminal work on the proper tensorial representation of viscoelastic fluid models [Oldroyd, 1950] that the continuum models could be applied to general deformations. Another approach started from the development of microscopic models using ideas from kinetic theory [Bird *et al.*, 1987b; Bird and Öttinger, 1992]. Yet another microscopic approach, utilizing non-equilibrium statistical mechanics ideas (random walkers) and scaling theory provided both microscopic as well as macroscopic models—reptation theories [Doi and Edwards, 1986; Larson, 1988, ch. 4]. Finally, several semi-empirical, semi-phenomenological approaches based on the idea of a network, starting with the work of Treloar [1975] on rubber elasticity and that of Green and Tobolsky [1946] which for the first time applied the Maxwell model to viscoelasticity, finally led to a series of continuum models for polymer melts [Phan-Thien and Tanner, 1977; Giesekus, 1982; see also Larson, 1988, ch. 6].

As mentioned in chapter 1, no matter which modeling approach is taken, there are two completely equivalent descriptions of the dynamic behavior of a complex system based on an integral or internal variables formulation. For the reasons also mentioned in that chapter, we shall focus on the second formulation here. The interested reader can consult one of the many available references on integral viscoelastic formulations [Lodge, 1956; Bernstein *et al.*, 1963; Wagner, 1979; Doi and Edwards, 1978 and 1979; Curtiss and Bird, 1981a,b; see also Bird *et al.*, 1987a, ch. 8; Larson, 1988, pp. 75-125]. The main remaining question is then how to choose the internal variables that best describe the structural state of the viscoelastic medium. In general, there are many levels on which we could do this, which are illustrated in Figure 8.1.

The most fundamental and exact method of describing materials with internal microstructure is through the distribution (density) function, Ψ. This is a density function in the phase space generated by considering all possible locations and velocities for each degree of freedom of the system under consideration. For example, the phase space variables for n independent degrees of freedom in the system are the position vector for each point of mass or resistance of the molecules, $\mathbf{r}_\nu$, $\nu=1,2,\ldots,n$, the corresponding velocity vectors, $\dot{\mathbf{r}}_\nu$, $\nu=1,2,\ldots,n$, and the time, t. A kinetic equation for the distribution function may then be formulated to describe the dynamics of the system in arbitrary kinematics. The accuracy of this method depends upon a judicious choice for the number of points and upon the physics incorporated into the kinetic equation for their motion. As the system becomes increasingly complex, more and more points are required to incorporate the appropriate physics into the kinetic equation. Therefore, although this description is the most fundamental, with molecular dynamics as an extreme, one can easily see that the number of variables quickly becomes too large to handle in any practical manner in its full rigor.

Molecular Dynamics

$$\Psi = \Psi\,(\,\mathbf{r}_\nu, \dot{\mathbf{r}}_\nu; t\,) \qquad \nu = 1,2,\ldots,n$$

• Most fundemental description of materials with microstructure
• Multidimensional problem

Mechanical Models

$$\Psi = \Psi\,(\,\mathbf{R}\,,\mathbf{r},t\,)$$

• 7-dimensional problem

Higher-Order Approximative Theories

R moments of Ψ, higher than second

• Many variables problem

Conformation Tensor Theory

$$c = \int \mathbf{R}\mathbf{R}\,\Psi\,d^3\mathbf{R}$$

• Retains validity for general flows

"Order Fluid" Models

$$S = f(\,Q, Q_{(1)},\ldots,Q_{(n)})$$

• Least fundemental description
• Applicable only for flows of special type

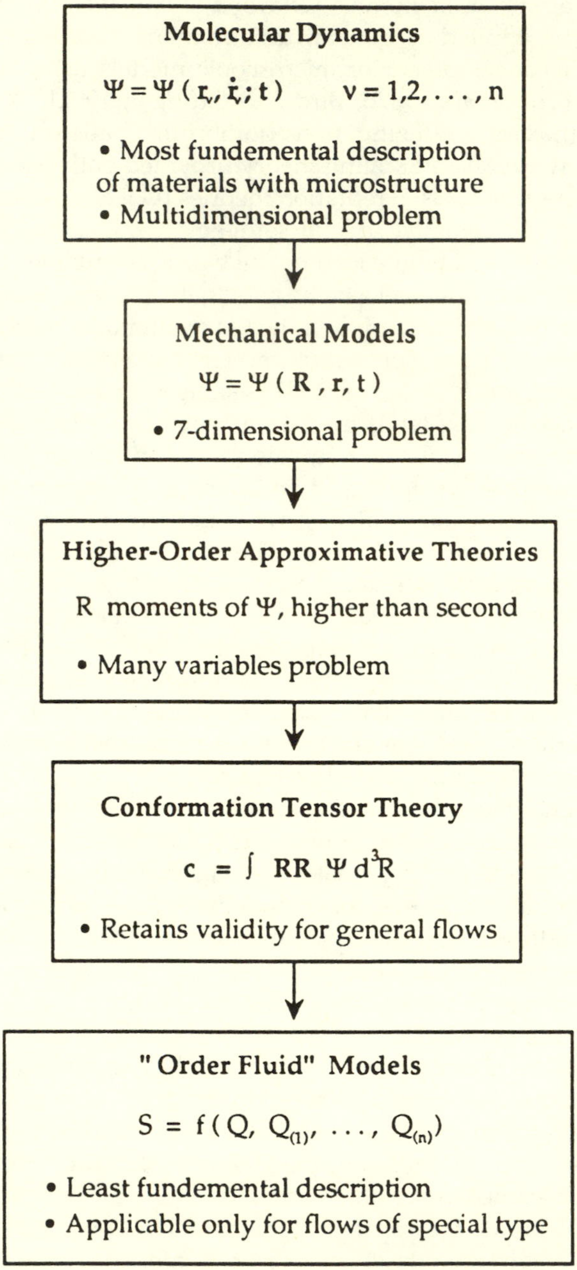

Figure 8.1: Levels of description of materials with internal microstructure (adapted from Edwards, 1991, p. 9.)

Thus, the need for approximations to the above description arises, resulting in the construction of mechanical analogues (e.g., rigid rods, dumbbells, etc.) aimed at integrating the many degrees of freedom involved in molecular physics to the much fewer degrees of freedom at the next level of description. With the reduction of the degrees of freedom, the fundamental laws describing the dynamics at the molecular level also need to be complemented by new relationships and laws, expressing the effect of the neglected degrees of freedom on the remaining ones at the next level of abstraction. This necessary complication has given rise to notions aimed at specific accounting of various physical phenomena (such as excluded volume, hydrodynamic interaction, and viscous solvent effects). Used judiciously, this approach can significantly reduce the dimensionality of the problem; however, this dimensionality reduction comes along with a more complex set of evolution equations to solve. Even when the simplest two-point (dumbbell) mechanical analog is used, the mechanical-energy-equilibration assumption is used and all correlations neglected (assuming an infinitely dilute polymer solution), the resulting system of equations still involves a seven-dimensional distribution function, $\Psi(\mathbf{r},\mathbf{R},t)$, where $\mathbf{r}$ is the position vector and $\mathbf{R}$ is the end-to-end macromolecular distance vector.

Consequently, further approximations are needed in order to restrict the information contained in the distribution function, so that the model may be practically applied to arbitrary situations. Typically, this results in a decrease in the dimensionality of the problem. Yet, it is essential that the model retain its ability to predict the desired physical phenomena, and so a compromise is in order to determine the simplest level upon which one can operate without too much loss in descriptive capability. As foreshadowed in §4.4, the best compromise appears to be the use of internal structural parameters which represent the various moments of the distribution function. In this way, the spatial dimensionality of the problem is reduced from six to three at the (relatively minor) expense of introducing a few additional (internal) variables.

A schematic account of the possible levels of description for viscoelastic fluids is shown in Figure 8.1. The conformation tensor theories, in terms of a second-rank tensor $\mathbf{c}$,[1] emerge as the simplest ones

[1]It is common in the literature—see, for example, [Flory, 1969, p. 15]—for $\mathbf{c}$ to be termed the "configuration tensor" as well as the "conformation tensor," and indeed the two expressions are often used interchangeably. In classical mechanics, however, the term "configuration" has a very special meaning: it is used in Lagrangian mechanics to represent the state of the system at any point in time through specification of the variables $\mathbf{x}_i$, $i=1,2,\ldots,N$. Hence, "configuration space" denotes the $3N$-dimensional coordinate system $(\mathbf{x}_1, \mathbf{x}_2, \ldots, \mathbf{x}_N)$, which is a great aid to the visualization of the system's trajectory in time. Similarly, Gibbs introduced "phase space" as the $6N$-dimensional space consisting of $(\mathbf{x}_1, \ldots,$

able to provide models representing the polymer rheology under a broad variety of flow conditions. Conformation tensor models can be formally obtained from kinetic theory simply by taking the second moment of the distribution function, where **R** is the orientation/extension vector (end-to-end distance between neighboring points). Usually, some information that is contained in the distribution function is lost in the "decoupling approximation" which is used in order to close the resulting system of equations from the averaged one; however, hopefully not too much of what is needed to describe the essential physics of viscoelastic fluids is neglected. The advantage gained from this procedure is the ability to actually solve the model equations in a wider variety of complex flow situations. This is the highest-order approximation to kinetic theory which is practical for use in numerical simulations, at least at this point in time. Moreover, the interpretation of **c** as the second moment of the end-to-end distance vector, **R**, is neither unique nor necessary. For example, as seen in §5.5, **c** can also be identified with the elastic Cauchy strain tensor, thus allowing the conformation tensor theories to describe a much broader material flow behavior than allowed by the kinetic theory.

The most difficult part of deriving a model based on the level of a second-rank tensor from a kinetic theory standpoint is deciding what form of approximation should be used to abstract the most essential information from the more detailed kinetic theory description. This usually involves deciding arbitrarily on complex "decoupling approximations" which allow one to restrict the excess information. The problem with these is that there is a large number of possibilities for decoupling approximations, and no one has yet discovered any objective criteria for selecting the best. A given selection is critical with respect to what physics is kept within the resulting approximation, but even more

x_N, $\mathbf{p}_1$, ..., $\mathbf{p}_N$) to represent the system evolution in time in Hamiltonian mechanics. Since the term "configuration" has been associated historically with Lagrangian mechanics, we feel that we have no justification to preempt it for our purposes here. Hence we prefer the terms "conformation tensor" and "conformation space" to denote the ideas in this and the following chapters. This is especially appropriate for the simple dumbbell-type models discussed below which are characterized solely by the orientation end-to-end distance, **R**. "Conformation space" then naturally refers to the space consisting of the components of the conformation tensor. When we deal with bead/spring-chain models, however, the two terms become essentially synonymous since the orientation vectors between adjacent beads can be mapped onto position coordinates for the beads in configuration space (up to within an arbitrary translation of the center of mass) via one-to-one mappings. Therefore, it is not surprising that physical chemists, dealing primarily with the later concept, have primarily selected the term "configuration" as opposed to "conformation," however, not until after a lot of ambivalence—see arguments in Flory, 1969, p. 15.

important is the issue of unwittingly introducing a pathological behavior into the approximate equations, resulting in aphysical predictions and/or a mathematically ill-posed problem. In complex flows, even when these unwanted features do not result in large errors, they typically lead to numerical instabilities and loss of convergence of the numerical solution of the problem. This issue is closely related to the question of thermodynamic admissibility of a model.

It is often the case in the literature that models for viscoelastic fluids are developed using only limited thermodynamic information, or none at all, in which case they both result in a number of redundant parameters, and, what is worse, the possibility of predicting an aphysical behavior expressed mathematically by either an ill-posed or a totally unstable mathematical problem. As an example, the oft-cited classical article by Hand [1962, §2] contains exactly the above-mentioned shortcomings, with severe implications as to the realism of the model. Indeed, we are not the only ones to notice this, as Maugin and Drouot [1983, pp. 712-13] write:

> But his (Hand's) theory, although mathematically sound and rigorous (it respects invariance requirements), has no thermodynamical basis so that this author does not recognize the necessary relationships which exist between the coefficients involved in the stress constitutive equation and the evolution equation of the microstructure.

The above being the case, there is much to be gained in developing model equations through generalized brackets, even when this simply amounts to redeveloping the same equations obtained before through a different method. The generalized bracket procedure allows one to obtain the governing equations at any level of description, including the conformation tensor level, straight from a knowledge of the symmetries and structure of the "Hamiltonian form" of the dynamic equations. This method, through the definition of the dissipation bracket, leads itself to an upfront, direct verification of resulting model physical consistencies through the arguments of Hamiltonian mechanics and non-equilibrium thermodynamics, as discussed in Part I. Thus, even if the necessity of using a distribution function and the associated decoupling approximations in not entirely eliminated, at least there exists a systematic approach for selecting physically meaningful approximations. However, it is equally valid to use another, consistent with a conformation tensor, continuum approach, starting with Eqs. (7.1-1,2) as a completely independent description of the dynamical phenomena of the system. Thus, the issue of decoupling approximations is entirely avoided.

In addition, the inherent thermodynamic information necessary to the modeling procedure, such as the satisfaction of the Onsager/Casimir symmetry constraints, is already imbedded in the structure of the bracket. This allows one to obtain a model where all the proper interrelationships assumed to hold between its parameters are *a priori* enforced. This is particularly important in viscoelastic flow equations where there is a considerable duplication of parameters between the constitutive equation for the stresses and the evolution equation for the conformation tensor. Thus, models with the minimum number of allowable parameters are obtained. Finally, this uniformly valid and systematic approach allows a more critical look at the models from yet another (compared to, for example, kinetic theory [Bird *et al.*, 1987b], or mathematical characteristics [Joseph, 1990]) point of view. From the point of view of extended thermodynamics, models are classified with respect to the dissipation and relaxation mechanisms that they incorporate. This offers a totally new understanding of the incorporated physics and underlying assumptions. Models which have been derived based on totally different assumptions can surprisingly appear very similar when examined from this perspective. In addition, new opportunities arise as, for example, the incorporation of detailed information on polymer thermodynamics (see the comprehensive treatise of des Cloizeaux and Jannink [1989] for a discussion of thermodynamic equilibrium properties of polymers in solution) or the extension of previous modeling approaches (some are discussed already in this chapter; more follow in chs. 9 and 10).

We devote this chapter to demonstrating the validity and usefulness of the bracket approach in describing incompressible and isothermal viscoelastic fluid behavior. We do this by deriving through the bracket formalism a set of general dynamical evolution equations, both for single- and multi-variable systems, and thereafter showing how these general equations correspond to common rheological models under special circumstances. Extensions of these models to more complex systems are also discussed, as well as their relation to linear viscoelasticity and the thermodynamic restrictions which are placed upon the various parameters appearing in the resulting model equations.

8.1 Incompressible and Isothermal Viscoelastic Fluid Models in Terms of a Single Conformation Tensor

Over the years, many differential-type viscoelastic fluid models have appeared in the literature.[2] These models are often derived in widely

[2]The work presented in this section is largely derived from the theory of conformation tensor models, as initially developed by Maugin and Drouot [1983], Grmela and Carreau

different circumstances, and at first sight it appears that various fundamental differences separate the various formalisms.[3] However, if we make a close inspection of the underlying mathematical structure of these models, then it is possible to illustrate that the same principles do actually support the vastly different constructions. It is our purpose in this section to demonstrate the above assertion by expressing many of the most common single-variable viscoelastic fluid models in Hamiltonian form. In fact, as we shall see, all of these models involve only slight variations of the same general Hamiltonian and phenomenological coefficient matrices. Furthermore, once a model is expressed in Hamiltonian form, it becomes readily apparent, through the ideas of chapters 6 and 7, when the model is consistent with the concepts developed through irreversible thermodynamics. Indeed, one can establish thermodynamic admissibility criteria for a given model in such a fashion, which, in complex flow situations, may help to shed light upon the range of validity and/or the faithfulness of numerical simulations involving the model. Thus, we wish to narrow the gap between the detailed microscopic descriptions of the motions of polymer chains and the macroscopic, continuum descriptions which the engineer is forced to use in actual practice.

The first step in describing an incompressible viscoelastic fluid through a generalized bracket is to establish which variables are to be used to describe completely the state of the system. As in §5.4, we may in this case neglect the variables ρ and s, and only consider the contribution of the mechanical energy to the Hamiltonian. This gives us the variables (from §5.5) $\mathbf{M}(\mathbf{x},t)=\rho\,\boldsymbol{v}(\mathbf{x},t)$ and $\mathbf{C}(\mathbf{x},t)=\rho\mathbf{c}(\mathbf{x},t)$; however, if we inspect the Poisson bracket according to (5.5-11),[4]

$$\{F,G\} = -\int_\Omega \left(\frac{\partial F}{\partial M_\gamma}\nabla_\beta\Big(\frac{\partial G}{\partial M_\beta}M_\gamma\Big) - \frac{\partial G}{\partial M_\gamma}\nabla_\beta\Big(\frac{\partial F}{\partial M_\beta}M_\gamma\Big)\right) d^3x$$

$$-\int_\Omega \left(\frac{\partial F}{\partial C_{\alpha\beta}}\nabla_\gamma\Big(C_{\alpha\beta}\frac{\partial G}{\partial M_\gamma}\Big) - \frac{\partial G}{\partial C_{\alpha\beta}}\nabla_\gamma\Big(C_{\alpha\beta}\frac{\partial F}{\partial M_\gamma}\Big)\right) d^3x$$

[1987], and Grmela [1988; 1989b], and extended into the form presented below by Beris and Edwards [1990a,b].

[3]Extensive reviews of the most popular incompressible viscoelastic fluid models may be found in Bird *et al.* [1987a,b] and Larson [1988].

[4]For the remainder of the book, we shall only be considering the spatial description, and hence we drop the subscript "E" on the Poisson bracket.

$$-\int_{\Omega} C_{\gamma\alpha}\left(\frac{\partial G}{\partial C_{\alpha\beta}}\nabla_{\gamma}(\frac{\partial F}{\partial M_{\beta}}) - \frac{\partial F}{\partial C_{\alpha\beta}}\nabla_{\gamma}(\frac{\partial G}{\partial M_{\beta}})\right)d^3x$$

$$-\int_{\Omega} C_{\gamma\beta}\left(\frac{\partial G}{\partial C_{\alpha\beta}}\nabla_{\gamma}(\frac{\partial F}{\partial M_{\alpha}}) - \frac{\partial F}{\partial C_{\alpha\beta}}\nabla_{\gamma}(\frac{\partial G}{\partial M_{\alpha}})\right)d^3x \ , \tag{8.1-1}$$

then we find that **C** always appears both in the numerator and the denominator so that we may cancel the constant ρ, thereby reducing the variables of interest to **M** and **c**. Moreover, one may also find it convenient to work in terms of a dimensionless structural tensor, whereas **c**, with the kinetic theory interpretation, **c**≡<**RR**>, has units of (length)^2. Consequently, whenever we associate **c** with <**RR**>, it is more convenient to work in terms of a dimensionless **c**, $\tilde{\mathbf{c}}\equiv\mathbf{c}K/k_BT$, where K is a characteristic elastic constant with units of (energy)/(length)^2. For polymeric systems, the parameter K arises naturally as a characterization of the (internal) entropic contribution of the equilibrium free energy of a polymer chain in the limit of small deformations [Marrucci, 1972, p. 322]:

> According to Kuhn and Grün [1942],[5] a flexible macromolecule in solution can be thought as equivalent to a freely-jointed chain of N statistical segments of equal length l. Such a chain has an entirely entropic free energy which, if the end-to-end distance R is small with respect to the extended length Nl, is given by

[5]The development of the theory for the entropic origin of the internal free energy seems to have emerged gradually in a series of seminal papers by Kuhn and his coworkers. The statistical analysis of conformations of individual chains was first investigated by Kuhn [1934]. It led to calculations for the entropy correction, which in a form similar to the one which appears in Eq. (8.1-3) has been applied to calculate the tension of stretched rubbers [Kuhn, 1936]. Kuhn and Grün [1942] extended the previous analysis with the addition into the previous entropy expression (which is only taking into account the contribution from the conformation of single chains) terms arising due to the statistical distribution of orientation and the mixing in the solution of many single chains. Independently working, James and Guth [1941], Wall [1942a,b], and Treloar [1942; 1943] have developed similar extensions to Kuhn's original entropic expression; see also [Kuhn and Grün, 1946] for a review in English of this early work. In the latter works [Kuhn and Grün, 1942; Kuhn and Grün, 1946, p. 185], a more accurate for high deformation expression for the entropy contribution of the single-chain conformation also appears, which gave rise later on to the various finite-extensibility, nonlinear elastic (FENE) extensions of the original linearly elastic (Hookean) model [Peterlin, 1961a,b; Peterlin, 1966; Warner, 1972; Bird *et al.*, 1987b, pp. 76-91; see also Wedgewood *et al.*, 1991 and references therein]—see also §8.1.2B—and to the development of the theory of rubber elasticity [Treloar, 1975]. The additional entropic terms arising from the statistical consideration of the assembly of many polymer chains are taken into account, in their proper form for dilute polymer solutions, in the evaluation of the Helmholtz free energy later in the present work—see Eqs. (8.1-13) and (9.2.5).

$$A = -ST = k_B T \mu R^2 + constant \;, \tag{8.1-2}$$

where μ is related to the mean square distance in a stagnant solution $\langle R^2 \rangle_0$ by

$$\mu = \frac{3}{2\langle R^2 \rangle_0} = \frac{3}{2Nl^2} \;. \tag{8.1-3}$$

Note that this expression for the free energy is equivalent to the free energy of a Hookean spring, KR^2, corresponding to a chain force which increases linearly with the chain length, R:

$$F = \frac{\mathrm{d}A}{\mathrm{d}R} = 2k_B T \mu R = 2KR \;. \tag{8.1-4}$$

Thus, $K/k_BT = \mu = 3/(2Nl^2)$. Again, the bracket (8.1-1) is easily expressed in terms of $\tilde{\mathbf{c}}$, so that if the Hamiltonian is written as $H_m = H_m[\mathbf{M}, \tilde{\mathbf{c}}]$, then everything will be consistent. The operating space for these variables is simply that of (5.4-3), expanded to include $\mathbf{c}$ as well as $\mathbf{M}$.[6] For the time being, we shall work solely in terms of $\mathbf{c}$; however, we shall later take recourse to working with $\tilde{\mathbf{c}}$ when appropriate.

Since we have already expressed the Poisson bracket as (8.1-1), the next step is to specify the general form of the dissipation bracket for an incompressible/isothermal viscoelastic medium. In accordance with the most general form for this bracket as presented in chapter 7, we have

$$\begin{aligned}
[F,G] = &- \int_\Omega \Lambda_{\alpha\beta\gamma\varepsilon} \frac{\delta F}{\delta c_{\alpha\beta}} \frac{\delta G}{\delta c_{\gamma\varepsilon}} \mathrm{d}^3x \\
&- \int_\Omega \mathrm{B}_{\alpha\beta\gamma\varepsilon\eta\nu} \nabla_\gamma \frac{\delta F}{\delta c_{\alpha\beta}} \nabla_\nu \frac{\delta G}{\delta c_{\varepsilon\eta}} \mathrm{d}^3x \\
&- \int_\Omega Q_{\alpha\beta\gamma\varepsilon} \nabla_\alpha \frac{\delta F}{\delta M_\beta} \nabla_\gamma \frac{\delta G}{\delta M_\varepsilon} \mathrm{d}^3x \\
&- \int_\Omega L_{\alpha\beta\gamma\varepsilon} \left(\nabla_\alpha \frac{\delta F}{\delta M_\beta} \frac{\delta G}{\delta c_{\gamma\varepsilon}} - \nabla_\alpha \frac{\delta G}{\delta M_\beta} \frac{\delta F}{\delta c_{\gamma\varepsilon}} \right) \mathrm{d}^3x \;,
\end{aligned} \tag{8.1-5}$$

[6]Note that the continuum parameter $\mathbf{c}$ defined as the Cauchy strain tensor, $\mathbf{c} \equiv \mathbf{F}^{\mathrm{T}} \cdot \mathbf{F}$, where $\mathbf{F}$ is the displacement gradient tensor [Bird *et al.*, 1987, §8.1], is already dimensionless; however, $\mathbf{c}$ derived from the kinetic theory description has the units of (length)^2 since we take the average over the distribution function of the dyad $\mathbf{RR}$: $\mathbf{c} \equiv \langle \mathbf{RR} \rangle$.

which duly neglects the contributions to the entropy production. Eq. (8.1-5) represents the most general expression for the dissipation bracket $[F,G]$ which remains bilinear with respect to F,G; i.e., nonlinear terms are presently neglected. The first integral in this expression represents the relaxational effects upon the viscoelastic system. As such, $\boldsymbol{\Lambda}$ is essentially an inverse relaxation time (or a rotational diffusivity) intrinsic to the material. The transport affinity/flux associated with the structural parameter is represented by the second integral, which has the effect of an anisotropic translational diffusivity; however, in common viscoelastic fluid models this effect is never considered, so that for the present section we shall set $\mathbf{B}=\mathbf{0}$ henceforth.[7] The third integral is the usual viscous dissipation, correlated with an (in general) anisotropic viscosity matrix, $\mathbf{Q}$. Again, following the usual practice in the viscoelastic fluid mechanics literature, we shall limit ourselves in this chapter to the consideration of an isotropic viscous dissipation: $\mathbf{Q} \propto \boldsymbol{\delta\delta}$.[8]

The fourth integral accounts for additional coupling between the velocity gradient field and the structural tensor field, over and above the one specified through the Poisson bracket, Eq. (8.1-1). As such, it represents in effect the degree of non-affine motion in the system; i.e., the degree to which the structure of the fluid is allowed to move ("slip") relative to the flow field. The minus sign in this integral appears because the affinities $\boldsymbol{\nabla}(\delta H_m/\delta \mathbf{M})$ and $\delta H_m/\delta \mathbf{c}$ have opposite parities—see Eq. (6.5-27). Thus, the fourth integral is distinct in its nature (antisymmetric versus symmetric) from the other integrals participating in the dissipation bracket. In a way, its effects on the time evolution of the system variables are intermediate to those arising from the Poisson bracket and the symmetric (purely dissipative) terms in the dissipation bracket. For one thing, this integral makes no contribution to the entropy production within the system (see the discussion following Eq. (6.5-32)), indicating that these effects could have been characterized, in principle, as reversible. However, these effects can not be incorporated into the Poisson bracket since this addition can be shown to result in a bracket that violates the Jacobi identity [Beris and Edwards, 1990a, App. B]. This is in agreement with Larson [1988, ch. 5], who argues that non-affine motion is irreversible. This latter assertion, however, appears to be

[7]Recently, however, researchers have begun to recognize the importance of these inhomogeneous effects in polymer systems [El-Kareh and Leal, 1989, pp. 264,65; Edwards *et al.*, 1990b, §2; Edwards, 1991, §4.3.4; Bhave *et al.*, 1991]. We shall discuss effects such as these in chapters 9 (for flexible polymers) and 11 (for liquid crystals.)

[8]Anisotropic viscous dissipation may be very important in some systems, particularly those materials which are very viscous by nature and possess a high degree of internal microstructure. These effects are very important in the study of liquid-crystalline dynamics—see chapter 11.

inconsistent with the Onsager/Casimir reciprocity theorem, (6.5-27), although what the exact form of this theorem should be for a coupling between a transport affinity and a relaxational affinity is wide open to debate. We assume that (6.5-27) is valid based upon the fact that, as we shall see later, (8.1-5) does indeed yield well-known and well-accepted results. However, the exact form of the coefficients and the range of the applicability of the resulting model equations should be carefully monitored as they can easily lead to an excessively oscillatory behavior even in simple flows, such as the application of sudden shear [Larson, 1983, p. 292].

Now that we have specified the general form of the generalized bracket, let us do the same for the mechanical part of the Hamiltonian, H_m. This we express as

$$H_m[M,c] = K[M] + E_\phi[\mathbf{c}] = K[M] + A[c] \ , \tag{8.1-6a}$$

where $K[\mathbf{M}]$ is the kinetic energy, given as

$$K[\mathbf{M}] = \int_\Omega \frac{1}{2\rho}\mathbf{M}\cdot\mathbf{M}\,d^3x \ , \tag{8.1-6b}$$

$E_\phi[\mathbf{c}]$ is the *potential energy* associated with the elasticity of the medium (accounting for the contribution of the internal—i.e., within the polymer chain—degrees of freedom to the system's entropy, see Eq. (8.1-2)), and $S[\mathbf{c}]$ is the *interchain entropy* associated with the external degrees of freedom, i.e., with the distribution of the end-to-end distances among the chains within the solution.[9] Both E_ϕ and S are assumed to be functions of $\mathbf{c}$, which in the present approximation is the sole device used to characterize the internal microstructure of the system. Consequently, through (8.1-6), we can express $E_\phi - TS$ as the Helmholtz free energy, $A[\mathbf{c}]$. Although $K[\mathbf{M}]$ is always given by (8.1-6b), $A[\mathbf{c}]$ may vary between the different models.

Now that we have the appropriate generalized bracket and Hamiltonian, we may derive the equations of motion for the viscoelastic medium. Following the procedure of Part I, we find that

$$\rho\frac{\partial v_\alpha}{\partial t} = -\rho v_\beta \nabla_\beta v_\alpha - \nabla_\alpha p + \nabla_\beta \sigma_{\alpha\beta} \ , \tag{8.1-7a}$$

and

[9]Kuhn and Grün [1942; 1946] were the first ones to distinguish between the internal and external degrees of freedom of entropic contributions to the free energy—see also footnote 5. To distinguish between the two, as is customarily done, we count the internal mode contribution in the "potential energy," E_ϕ, and the external mode (orientational) in the entropic one, TS.

$$\frac{\partial c_{\alpha\beta}}{\partial t} = -\rho\upsilon_\gamma\nabla_\gamma c_{\alpha\beta} + c_{\alpha\gamma}\nabla_\gamma\upsilon_\beta + c_{\gamma\beta}\nabla_\gamma\upsilon_\alpha$$
$$- \Lambda_{\alpha\beta\gamma\varepsilon}\frac{\delta A}{\delta c_{\gamma\varepsilon}} + L_{\alpha\beta\gamma\varepsilon}\nabla_\gamma\upsilon_\varepsilon \,, \tag{8.1-7b}$$

where

$$\sigma_{\alpha\beta} = Q_{\alpha\beta\gamma\varepsilon}\nabla_\gamma\upsilon_\varepsilon + 2c_{\beta\gamma}\frac{\delta A}{\delta c_{\alpha\gamma}} + 2L_{\alpha\beta\gamma\varepsilon}\frac{\delta A}{\delta c_{\gamma\varepsilon}} \,. \tag{8.1-7c}$$

Note that in order to arrive at these final expressions, we have used implicitly the symmetry properties of the phenomenological matrices (e.g., $Q_{\alpha\beta\gamma\varepsilon}=Q_{\beta\alpha\gamma\varepsilon}$) and the fact that, provided

$$A[\mathbf{c}] = \int_\Omega a(\mathbf{c})\,\mathrm{d}^3x \,, \tag{8.1-8}$$

with the Helmholtz free energy density, a, depending on **c** only (i.e., a does not depend on the gradient of **c** as well),

$$\int_\Omega \frac{\delta A}{\delta c_{\alpha\beta}}\nabla_\gamma(c_{\alpha\beta}\frac{\delta F}{\delta M_\gamma})\,\mathrm{d}^3x = \int_\Omega \frac{\partial a}{\partial c_{\alpha\beta}}\left(\nabla_\gamma c_{\alpha\beta}\right)\frac{\delta F}{\delta M_\gamma}\,\mathrm{d}^3x$$
$$= \int_\Omega \left(\nabla_\gamma a\right)\frac{\delta F}{\delta M_\gamma}\,\mathrm{d}^3x \tag{8.1-9}$$
$$= - \int_\Omega a\nabla_\gamma(\frac{\delta F}{\delta M_\gamma})\,\mathrm{d}^3x = 0 \,.$$

The above evolution equations embodied in (8.1-7) form a closed set which, together with the appropriate boundary and initial conditions on both $\boldsymbol{\upsilon}$ *and* **c**, may be solved in arbitrary flow situations. As we shall see, these equations are fully capable of encompassing most viscoelastic fluid models on this (conformation tensor) level of description simply by varying the Hamiltonian or the phenomenological matrices $\boldsymbol{\Lambda}$, **L**, and **Q**.

8.1.1 Simple phenomenological spring/dashpot-type models

A. The Upper-Convected Maxwell Model

Several viscoelastic fluid models have been derived by considering mechanical analogues for the elastic and viscous effects of the material

p. 222: in (8.1-7b), delete "ρ" from the first term of the R.H.S.

interconnected in various manners. The Maxwell model arises by considering a Hookean spring, associated with the elasticity of the medium, connected in series to a dashpot, representing the viscous damping within the fluid.[10] With this physical system in mind, one may easily derive the constitutive equation for the (objective) extra stress of the medium as

$$\sigma_{\alpha\beta} + \lambda\check{\sigma}_{\alpha\beta} = 2\eta A_{\alpha\beta} , \tag{8.1-10}$$

where we have used the inverse hat to denote the upper-convected time derivative of the extra stress, corresponding to the convected differentiation with respect to time of a contravariant second-order tensor first proposed by Oldroyd [1950, p. 530]:

$$\check{\sigma}_{\alpha\beta} \equiv \frac{\partial\sigma_{\alpha\beta}}{\partial t} + \upsilon_\gamma\nabla_\gamma\sigma_{\alpha\beta} - \sigma_{\alpha\gamma}\nabla_\gamma\upsilon_\beta - \sigma_{\gamma\beta}\nabla_\gamma\upsilon_\alpha . \tag{8.1-11}$$

Alternatively, from a continuum mechanics point of view, other objective time derivatives could have been used (see the discussion about the Johnson/Segalman model, §8.1.1C); however, the upper-convected model is the one which arises naturally from microscopic models (see the discussion on Hookean dumbbell models, §8.1.2A). In the above expression, **A** is the rate of deformation tensor which is defined as the symmetric part of the velocity gradient, $\mathbf{A}\equiv\frac{1}{2}(\nabla\boldsymbol{v}+\nabla\boldsymbol{v}^T)$, λ is a characteristic relaxation time for the material, and η is the viscosity associated with the dashpot. The modulus of elasticity may then be defined in the usual manner as $G_0\equiv\eta/\lambda$ $(=nk_BT)$ [Maxwell, 1867, p. 83; Larson, 1988, p. 25]. Eq. (8.1-10), coupled with the momentum equation, (8.1-7a), may be solved for arbitrary flow situations to yield the rheological properties of this simple viscoelastic medium.

Now let us see how this model may be expressed in terms of a Hamiltonian and a generalized bracket. Since the elasticity of the material is represented by a Hookean spring, the elastic potential energy is written analogously to (8.1-2) as

$$E_\phi[\mathbf{c}] = \int_\Omega \frac{1}{2}nK\,\mathrm{tr}\,\mathbf{c}\,d^3x , \tag{8.1-12}$$

where n characterizes the "elasticity density" of the medium with units of $(\text{volume})^{-1}$. The entropy functional is given by the Boltzmann entropy (expressed in terms of a conformation tensor—see Grmela and Carreau [1987, p. 274] and Grmela [1989b, p. 219]) as

[10]Actually, the Maxwell model was derived by considering the viscoelastic nature of systems of gases [Maxwell, 1867, pp. 80-83], without recourse to mechanical analogies, but for our discussion it is natural to present it in the form stated.

$$S[\mathbf{c}] = \int_\Omega \tfrac{1}{2} n k_B \ln \det\left(\frac{K}{k_B T}\mathbf{c}\right) d^3x \,. \tag{8.1-13}$$

Note that in the above expression, det($\tilde{\mathbf{c}}$), being the multiplicative factor of the three eigenvalues of $\tilde{\mathbf{c}}$, is correlated to the "sphericity" or "conformational probability" of the molecular structure, and hence the entropy has the same form as expected from statistical mechanics—see Eq. (6.5-1). Hence, we arrive at the same free energy expression in terms of the conformation tensor as derived by Booij [1984, p. 4572]. A similar expression can also be derived from the statistical theory of rubber elasticity [Flory, 1953, §9-3]. It is a characteristic of the upper-convected Maxwell model to arise as the lowest-order approximation of both a dilute polymer solution and a polymer (or rubber) melt. With this free energy, we may easily calculate the functional derivative of the Hamiltonian with respect to **c** as

$$\begin{aligned}\frac{\delta H_m}{\delta c_{\alpha\beta}} = \frac{\delta A}{\delta c_{\alpha\beta}} &= \tfrac{1}{2} nK\delta_{\alpha\beta} - \tfrac{1}{2} n k_B T \frac{1}{\det \mathbf{c}} \frac{\partial(\det \mathbf{c})}{\partial c_{\alpha\beta}} \\ &= \tfrac{1}{2} nK\delta_{\alpha\beta} - \tfrac{1}{2} n k_B T c^{-1}_{\alpha\beta} \,,\end{aligned} \tag{8.1-14}$$

where we have used the algebraic identity $c^{-1}_{\alpha\beta}$=(cofactor $c_{\alpha\beta}$)/detc (e.g., [Bedford, 1985, p. 17]) and the fact that ln(det$\tilde{\mathbf{c}}$)=3ln(K/k_BT)+ln(detc) while maintaining the proper units. At equilibrium where the distribution is spherical, $\tilde{\mathbf{c}}=\boldsymbol{\delta}$ and $\delta H_m/\delta\mathbf{c}=\mathbf{0}$.

If we set **L**=**Q**=**0** and pick up the relaxation matrix $\boldsymbol{\Lambda}$ in its simplest anisotropic form consistent with the Onsager symmetry relations,

$$\Lambda_{\alpha\beta\gamma\varepsilon} = \frac{1}{2\lambda nK}\left(c_{\alpha\gamma}\delta_{\beta\varepsilon} + c_{\alpha\varepsilon}\delta_{\beta\gamma} + c_{\beta\gamma}\delta_{\alpha\varepsilon} + c_{\beta\varepsilon}\delta_{\alpha\gamma}\right) , \tag{8.1-15}$$

then we obtain the expressions from (8.1-7b,c), respectively,

$$\overset{\nabla}{c}_{\alpha\beta} = -\frac{1}{\lambda} c_{\alpha\beta} + \frac{k_B T}{\lambda K}\delta_{\alpha\beta} \,, \tag{8.1-16}$$

and

$$\sigma_{\alpha\beta} = nKc_{\alpha\beta} - G_0\delta_{\alpha\beta} \,. \tag{8.1-17}$$

Now if we solve (8.1-17) for **c**,

$$c_{\alpha\beta} = \frac{1}{nK}\sigma_{\alpha\beta} + \frac{k_B T}{K}\delta_{\alpha\beta} \,, \tag{8.1-18}$$

and substitute into (8.1-16), we do indeed arrive back at the constitutive relationship for the extra stress, (8.1-10).

B. The Voigt and Oldroyd-B Models

The Voigt model arises similarly to the Maxwell model, but through considering the spring and dashpot in parallel rather than in series. Consequently, the Hamiltonian remains the same as in the preceding case, but this time we set $\boldsymbol{\Lambda}=\mathbf{L}=\mathbf{0}$ and

$$Q_{\alpha\beta\gamma\varepsilon} = \eta_s(\delta_{\alpha\gamma}\delta_{\beta\varepsilon} + \delta_{\alpha\varepsilon}\delta_{\beta\gamma}) \, , \tag{8.1-19}$$

where η_s is the viscosity associated with the solvent of a polymer solution. Hence the evolution equations for the conformation tensor and the extra stress become

$$\begin{aligned} \check{c}_{\alpha\beta} &= 0 \, , \\ \check{\sigma}_{\alpha\beta} &= nKc_{\alpha\beta} - G_0\delta_{\alpha\beta} + 2\eta_s A_{\alpha\beta} \, , \end{aligned} \tag{8.1-20}$$

so that a similar substitution as before gives the constitutive relationship for the extra stress tensor as

$$\check{\sigma}_{\alpha\beta} = 2\eta_s \check{A}_{\alpha\beta} + G_0 A_{\alpha\beta} \, . \tag{8.1-21}$$

The use of the upper-convected time derivative here is imposed naturally from the physical interpretation of the conformation tensor **c** in elasticity theory as the Finger strain tensor (see the discussion on the formulation of large-deformation elasticity in §5.5).

Now the Oldroyd-B model may be obtained by combining the above two models, using both **Q** from (8.1-19) and $\boldsymbol{\Lambda}$ from (8.1-15) to obtain $(8.1\text{-}20)_2$ and (8.1-16). Substitution of (8.1-20b) into (8.1-16) then gives the Oldroyd-B constitutive relation [Oldroyd, 1950, p. 537]:

$$\sigma_{\alpha\beta} + \lambda\check{\sigma}_{\alpha\beta} = 2\eta A_{\alpha\beta} + 2\eta_s\lambda\check{A}_{\alpha\beta} \, . \tag{8.1-22}$$

C. The Johnson/Segalman and Oldroyd-A Models

Johnson and Segalman [1977, pp. 259,60] introduced an extended Maxwell model by replacing the upper-convected derivative in (8.1-10) with the Gordon/Schowalter derivative [Gordon and Schowalter, 1972, p. 86], defined as

$$\begin{aligned} \frac{\mathrm{D}^{(a)}}{\mathrm{D}t}\sigma_{\alpha\beta} \equiv \frac{\partial\sigma_{\alpha\beta}}{\partial t} &+ \upsilon_\gamma\nabla_\gamma\sigma_{\alpha\beta} - \frac{a+1}{2}\left(\sigma_{\alpha\gamma}\nabla_\gamma\upsilon_\beta + \sigma_{\beta\gamma}\nabla_\gamma\upsilon_\alpha\right) \\ &- \frac{a-1}{2}\left(\sigma_{\alpha\gamma}\nabla_\beta\upsilon_\gamma + \sigma_{\beta\gamma}\nabla_\alpha\upsilon_\gamma\right) \, . \end{aligned} \tag{8.1-23}$$

Alternatively, this derivative may be expressed in terms of the parameter ξ,

$$\xi \equiv 1 - a \, , \tag{8.1-24}$$

as

$$\frac{\mathrm{D}^{(1-\xi)}}{\mathrm{D}t}\sigma_{\alpha\beta} \equiv \frac{\partial \sigma_{\alpha\beta}}{\partial t} + \upsilon_\gamma \nabla_\gamma \sigma_{\alpha\beta} - \sigma_{\beta\gamma}\left(\nabla_\gamma \upsilon_\alpha - \xi A_{\gamma\alpha}\right) - \sigma_{\alpha\gamma}\left(\nabla_\gamma \upsilon_\beta - \xi A_{\gamma\beta}\right) . \tag{8.1-25}$$

This derivative allows non-affine motion into the system description by representing the "slippage" of the polymer chains with respect to the surrounding solvent. Since the polymer chains slip through the solvent which is normally observed with rigid particles as a result of the torque they are experiencing in the flow, slip can be considered as a measure of molecular rigidity [Larson, 1988, p. 266]. However, it is also a strong function of the particles' aspect ratio—see discussion in §11.3A.2 and Eq. (11.3-54). Consequently, we would expect when using this derivative that we would not see as great elastic forces induced by the flow as with the upper-convected derivative. Note that a is generally considered as lying in the range $-1 \leq a \leq 1$, where $a=1$ corresponds to the upper-convected derivative, $a=0$ to the corotational derivative, and $a=-1$ to the lower-convected derivative. Correspondingly, $a=1$ describes the affine motion of either a flexible or infinite aspect ratio solid ellipsoid, whereas as the aspect ratio decreases, $a=0$ in the extreme limit of spherical solid particles, with the elastic contribution to the stress decreasing to zero.

The Johnson/Segalman model may be obtained through the general equations (8.1-7) simply by using **Q**=0, $\boldsymbol{\Lambda}$ its value for the Maxwell model, (8.1-15), and a similar expression for **L**:

$$L_{\alpha\beta\gamma\varepsilon} = \frac{a-1}{2}\left(c_{\alpha\gamma}\delta_{\beta\varepsilon} + c_{\alpha\varepsilon}\delta_{\beta\gamma} + c_{\beta\gamma}\delta_{\alpha\varepsilon} + c_{\beta\varepsilon}\delta_{\alpha\gamma}\right) . \tag{8.1-26}$$

Hence we obtain

$$\sigma_{\alpha\beta} = (a+1)c_{\beta\gamma}\frac{\delta A}{\delta c_{\gamma\alpha}} + (a-1)c_{\alpha\gamma}\frac{\delta A}{\delta c_{\gamma\beta}} \, , \tag{8.1-27a}$$

and

$$\frac{\partial c_{\alpha\beta}}{\partial t} = -\rho \upsilon_\gamma \nabla_\gamma c_{\alpha\beta} + \frac{a+1}{2}\left(c_{\alpha\gamma}\nabla_\gamma \upsilon_\beta + c_{\beta\gamma}\nabla_\gamma \upsilon_\alpha\right) + \frac{a-1}{2}\left(c_{\alpha\gamma}\nabla_\beta \upsilon_\gamma + c_{\beta\gamma}\nabla_\alpha \upsilon_\gamma\right)$$

$$- \frac{1}{\lambda n K}\left(c_{\beta\gamma}\frac{\delta A}{\delta c_{\gamma\alpha}} + c_{\alpha\gamma}\frac{\delta A}{\delta c_{\gamma\beta}} \right) , \tag{8.1-27b}$$

which, after substituting for $\delta A/\delta \mathbf{c}$ with the same expression as that for the Maxwell model, Eq. (8.1-14), reduce to

$$\sigma_{\alpha\beta} = an(Kc_{\alpha\beta} - k_B T\delta_{\alpha\beta}) , \tag{8.1-28a}$$

and

$$\frac{\partial c_{\alpha\beta}}{\partial t} = - \rho \upsilon_\gamma \nabla_\gamma c_{\alpha\beta} + \frac{a+1}{2}\left(c_{\alpha\gamma}\nabla_\gamma \upsilon_\beta + c_{\beta\gamma}\nabla_\gamma \upsilon_\alpha \right)$$

$$+ \frac{a-1}{2}\left(c_{\alpha\gamma}\nabla_\beta \upsilon_\gamma + c_{\beta\gamma}\nabla_\alpha \upsilon_\gamma \right) \tag{8.1-28b}$$

$$- \frac{1}{\lambda}\left(c_{\alpha\beta} - \frac{k_B T}{K}\delta_{\alpha\beta} \right) ,$$

which are equivalent to the Johnson/Segalman equations[11] with the viscosity, η, modulus of elasticity, G, and the (dimensionless) conformation tensor, $\bar{\mathbf{c}}$ defined as [Hulsen, 1990, Eq. (9)]

$$\eta \equiv G\lambda = a^2 nk_B T\lambda \; ; \quad \bar{\mathbf{c}} = \frac{a}{G}\left(\boldsymbol{\sigma} + \frac{G}{a}\boldsymbol{\delta} \right) ; \quad \boldsymbol{\sigma} = \frac{G}{a}(\bar{\mathbf{c}} - \boldsymbol{\delta}) \; . \tag{8.1-29}$$

In the Johnson/Segalman model, the parameter a characterizing the polymer slip has been assumed to be constant. However, the polymer molecules rigidity/aspect ratio varies strongly with deformation, which implies that a variable parameter a might be a more appropriate choice. In fact, Hinch [1977] and Rallison and Hinch [1988] have proposed models based on a modification of the finitely-extensible, nonlinear elastic dumbbell (FENE dumbbell, see §8.1.2) which utilizes a mixed-convected derivative, such as the one in Eq. (8.1-28), with the parameter a given as [Rallison and Hinch, 1988, p. 43]

[11]However, note that the so-called *corotational* Maxwell model which corresponds to the corotational or Jaumann derivative [Bird *et al.*, 1987a, p. 507] obtained when a=0 in Eq. (8.1-23), cannot be obtained within the context of the generalized bracket formalism. This happens because the model defined by Eqs. (8.1-24) through (8.1-28) has extra stress which is identically equal to zero (obtained from Eq. (8.1-28a) after substituting a=0 and taking into account that $\mathbf{c}\cdot\delta H/\delta \mathbf{c}$ is symmetric for the Maxwell Hamiltonian). The implications of this observation are further discussed in §8.1.6.

$$a = \frac{\mathrm{tr}\mathbf{c}}{\beta R_G^2 + \mathrm{tr}\mathbf{c}} , \qquad (8.1\text{-}30)$$

where β is a numerical factor of order of unity and R_G is the undistorted radius of gyration of the chain, $R_G = l\,(N/6)^{1/2}$, with l representing the size of a monomer and N the number of monomers per chain.

Note that models like the Johnson/Segalman model which make use of a mixed-convected derivative, $-1<a<1$, are characterized by a non-monotonic shear-stress versus shear-rate simple shear behavior (for small enough values of solvent viscosity, if any). This has been attributed as the cause for results which are considered aphysical, such as large oscillations in the stress tensor in sudden shear flows (with the shear stress as a function of the shear strain, oscillating between large positive and negative values during start-up of steady shearing [Larson, 1983, pp. 292-93]), and the violation of the experimentally supported Lodge-Meissner relationship [Larson, 1983, p. 305]. Moreover, depending on the magnitude of the components of the conformation tensor, the linearized governing equations might not be evolutionary in time [Dupret and Marchal, 1986b, §7; Joseph and Saut, 1986, p. 135; Joseph, 1990, p. 78; see also §8.1.6]. The loss of the evolutionarity implies that the solution to the equations can exhibit catastrophic, high-frequency, Hadamard instabilities [Joseph, 1990, p. 70]—see also §8.1.6. For the Johnson/Segalman model with a mixed-convected derivative ($|a| \neq 1$), it has been shown that loss of evolution occurs for sink flow and can take place despite the fact that the conformation tensor is positive definite and the corresponding thermodynamic dissipation inequalities are satisfied [Dupret and Marchal, 1986b, p. 170]. On the other hand, Johnson/Segalman's rich dynamic behavior can be used to advantage in order to explain certain anomalous phenomena like "latency" and "spurt" in shear flows [Kolkka *et al.*, 1988; Malkus *et al.*, 1990]—see also footnote 13 and the remark at the end of §8.1.6. Of course, the solution obtained during dynamic simulations remains always evolutionary (otherwise, it could not have been calculated) [Kolkka *et al.*, 1988, p. 334].

Other models can similarly be obtained by different combinations of the above phenomenological matrices. For instance, the Oldroyd-A model [1950, p. 537] is obtained by setting $a=-1$, $\boldsymbol{\Lambda}$ as (8.1-15), and $\mathbf{Q}$ as in (8.1-19).

D. Models with a Variable Relaxation Time

D.1 The White/Metzner model

In an attempt to at least partially account for the relaxation spectra observed with real polymer solutions and melts, phenomenological viscoelastic fluid models have been introduced in the literature which explored the consequences of a non-constant relaxation time and viscosity. The first and most popular among these models is the nonlinear generalization of the upper-convected Maxwell model proposed by White and Metzner [1963; White, 1990, pp. 227-29]

$$\check{\sigma}_{\alpha\beta} + \frac{1}{\lambda(\mathrm{II_A})}\sigma_{\alpha\beta} = 2G_0 A_{\alpha\beta} \ , \tag{8.1-31}$$

allowing for the relaxation to be a function of the second invariant[12] of the deformation rate, **A**,

$$\mathrm{II_A} \equiv \sqrt{2\mathbf{A}:\mathbf{A}} \ . \tag{8.1-32}$$

In later applications, the modulus of elasticity was also allowed to be a function of $\mathrm{II_A}$, but for the purposes of our discussion we shall limit ourselves to the form provided by Eq. (8.1-31). The White/Metzner modification arose along the lines of development of the various generalized Newtonian constitutive equations,

$$\boldsymbol{\sigma} = 2\eta(\mathrm{II_A})\mathbf{A} \ , \tag{8.2-33}$$

such as the power-law

$$\eta = K\, \mathrm{II_A}^{\frac{n-1}{2}} \ , \tag{8.1-34}$$

and its generalization, the Bird/Carreau or Carreau/Yasuda models [Bird *et al.*, 1987a, ch. 4],

$$\frac{\eta - \eta_\infty}{\eta_0 - \eta_\infty} = \left(1 + (\lambda \mathrm{II_A})^a\right)^{\frac{n-1}{a}} \ , \tag{8.2-35}$$

where η_0, η_∞, λ, n, and a are characteristic model parameters. Thus, similar forms to Eqs. (8.1-34) and (8.1-35) have been proposed for the White/Metzner model in order to describe the dependence of the relaxation time on the deformation rate [White, 1990, p. 229].

As compared to the upper-convected Maxwell model, the White/Metzner model has additional flexibility through the dependence on the deformation rate of the relaxation time which allows a quantitative fit to the shear-thinning viscosity and first normal-stress difference coefficient of polymer melts and concentrated polymer solutions. However, like the upper-convected Maxwell model, it predicts an infinite extensional viscosity for a finite extension rate (unless a specific functional dependence for $\lambda(\mathrm{II_A})$ is used in a form analogous to Eq. (8.1-35) with

[12]The other invariants may, in general, be utilized as well but, in practice, rarely are, since the first invariant of **A** is zero for incompressible flows and the third is zero for planar flows.

λ_∞=0, a=1, n=0 [Ide and White, 1977; Larson, 1988, p. 152]). Moreover, the direct dependence on the shear rate of its parameters allows for the possibility of the loss of evolutionarity in the governing equations [Dupret and Marchal, 1986b; Verdier and Joseph, 1989]—see also the discussion in §8.1.6. It is exactly because of this direct dependence that this model cannot be represented through the bracket formulation using a bilinear form for the dissipation function, Ξ. A model originating from the White/Metzner idea but involving instead conformation tensor dependent parameters (the extended White/Metzner [Souvaliotis and Beris, 1992]) is discussed further below.

It is of interest to point out here that in the original paper, White and Metzner explicitly state "the relaxation time is a function of the invariants of the stress matrix" [White and Metzner, 1963, p. 1871], and it was only for convenience and due to the fact that they have mostly studied steady simple shear flows (where there is a monotonic—for power law index n, $0<n<1$—relationship between stress and shear rate) that they used the direct dependence on the rate of deformation. Shortly after White and Metzner, Kaye [1966; Eq. (15)] came up with a similar generalization of Lodge's rubber-like model [Lodge, 1956] which corresponds to the integral equivalent of the Maxwell model [Larson, 1988, pp. 22-25]. Since for a Hookean free energy a linear relationship exists between the stress and the conformation parameter, as Eq. (8.1-17) shows, the dependence on the stress equivalently expresses a dependence on the conformation tensor and in the following no distinction between the two will be made. After Kaye, other researchers followed a similar approach in modifying the differential form of the Maxwell equations.

The first of these models is the Phan-Thien/Tanner model [1977, §3] which, although originally developed based on network theory, falls naturally into the spring/dashpot category once λ is considered as a function of $\mathbf{c}$. More recently, enhanced interest has been shown toward improving the original Hookean elastic dumbbell (i.e., upper-convected Maxwell) model through the introduction of stress- or conformation-dependent relaxation and viscosity properties. Apelian *et al.* [1988, §2.1] introduced a modified Maxwell model, and also based their model on network theory. Chilcott and Rallison [1988, §2] based their development on the theory for the finitely-extensible, nonlinear elastic dumbbell—see also §8.1.2. Finally, Souvaliotis and Beris [1992, §3,4] introduced an extended White/Metzner model founded upon the same type of phenomenology as the original White/Metzner model. All of these models exhibit striking similarities to the Phan-Thien/Tanner model, although the predictions of the three different models in various flow fields are not necessarily equivalent because of the particular form chosen for the functional dependence of λ on $\mathbf{c}$.

D.2 The Phan-Thien and Tanner model

Phan-Thien and Tanner [1977, p. 358] and Phan-Thien [1978, p. 266] actually proposed two similar models, written in terms of the extra stress tensor as

$$\frac{\mathrm{D}^{(a)}}{\mathrm{D}t}\sigma_{\alpha\beta} + \frac{1}{\lambda_0}\left(1 + \frac{\varepsilon}{G_0}\mathrm{tr}\,\boldsymbol{\sigma}\right)\sigma_{\alpha\beta} = 2G_0 A_{\alpha\beta}\ , \qquad (8.1\text{-}36a)$$

and

$$\frac{\mathrm{D}^{(a)}}{\mathrm{D}t}\sigma_{\alpha\beta} + \frac{1}{\lambda_0}\exp\left(\frac{\varepsilon}{G_0}\mathrm{tr}\,\boldsymbol{\sigma}\right)\sigma_{\alpha\beta} = 2G_0 A_{\alpha\beta}\ . \qquad (8.1\text{-}36b)$$

In these expressions, ε is a parameter which governs the strength of the material's dependence upon the extra stress tensor and λ_0 is a constant relaxation time. Moreover, model (8.1-36a) can be recovered from (8.1-36b) in the limit of small $(\varepsilon/G_0)\mathrm{tr}\,\boldsymbol{\sigma}$. Since these models arise as nonlinear generalizations of the Johnson/Segalman model (which is recovered from either (8.1-36a) or (8.1-36b) in the limit ε=0, the corresponding conformation parameter is related to the stress through Eq. (8.1-29), for G=G_0 [Hulsen, 1990, Eq. (9)]. Then, by extending the analysis reported in the Johnson/Segalman case, §8.1.1C, it can be shown that either of these models can be obtained through the general equations (8.1-7) simply by using **Q=0**, **L** given by (8.1-26), and $\boldsymbol{\Lambda}$ given by

$$\Lambda_{\alpha\beta\gamma\varepsilon} = \frac{1}{2\lambda_0 nK}\left(1 + \frac{\varepsilon}{aG_0}\mathrm{tr}(nK\mathbf{c} - G_0\boldsymbol{\delta})\right) \qquad (8.1\text{-}37a)$$
$$\times\left(c_{\alpha\gamma}\delta_{\beta\varepsilon} + c_{\alpha\varepsilon}\delta_{\beta\gamma} + c_{\beta\gamma}\delta_{\alpha\varepsilon} + c_{\beta\varepsilon}\delta_{\alpha\gamma}\right)\ ,$$

or

$$\Lambda_{\alpha\beta\gamma\varepsilon} = \frac{1}{2\lambda_0 nK}\exp\left(\frac{\varepsilon}{aG_0}\mathrm{tr}(nK\mathbf{c} - G_0\boldsymbol{\delta})\right) \qquad (8.1\text{-}37b)$$
$$\times\left(c_{\alpha\gamma}\delta_{\beta\varepsilon} + c_{\alpha\varepsilon}\delta_{\beta\gamma} + c_{\beta\gamma}\delta_{\alpha\varepsilon} + c_{\beta\varepsilon}\delta_{\alpha\gamma}\right)\ .$$

Notice that either formula (8.1-37a) or (8.1-37b) breaks down in the singular limit of the corotational derivative (i.e., when a=0).

D.3 The modified upper-convected Maxwell model

The constitutive equation for the modified Maxwell model [Apelian *et al.*, 1988, p. 305] is given by the expression

$$\breve{\sigma}_{\alpha\beta} + \frac{1}{\lambda(\mathrm{tr}\boldsymbol{\sigma})}\sigma_{\alpha\beta} = \frac{2\eta}{\lambda(\mathrm{tr}\boldsymbol{\sigma})}A_{\alpha\beta} , \tag{8.1-38}$$

where $\lambda(\mathrm{tr}\boldsymbol{\sigma})$ indicates that the relaxation time is again dependent upon the first invariant of the extra stress tensor, and, consequently, the first invariant of the conformation tensor. This relaxation time is given in the functional form [Apelian *et al.*, 1988, p. 305]

$$\lambda(\mathrm{tr}\boldsymbol{\sigma}) = \frac{\lambda_0}{1 + (F(\lambda_0)\mathrm{tr}\boldsymbol{\sigma})^{\alpha-1}} , \tag{8.1-39}$$

with $F(\lambda_0)$ being inversely proportional to λ_0. Hence, as λ_0 increases, F decreases. The overall relaxation time, λ, is equivalent to λ_0 for small stress values, and approaches zero as the extra stress tensor becomes very large, and hence a Newtonian constitutive equation arises in this limit. α represents an adjustable parameter of the problem which allows fitting of experimental data. Physically, it represents the sensitivity of the fluid relaxation time to increasing stress beyond a critical stress level. This model requires the coefficient matrix

$$\Lambda_{\alpha\beta\gamma\varepsilon} = \frac{1}{2\lambda_0 nK}\left(1 + [F(\lambda_0)\mathrm{tr}(nK\mathbf{c} - G_0\boldsymbol{\delta})]^{\alpha-1}\right)$$
$$\times \left(c_{\alpha\gamma}\delta_{\beta\varepsilon} + c_{\alpha\varepsilon}\delta_{\beta\gamma} + c_{\beta\gamma}\delta_{\alpha\varepsilon} + c_{\beta\varepsilon}\delta_{\alpha\gamma}\right) . \tag{8.1-40}$$

D.4 The extended White/Metzner model

The extended White/Metzner (EWM) constitutive relation of Souvaliotis and Beris [1992, §3,4] is given by

$$\breve{\sigma}_{\alpha\beta} + \frac{1}{\lambda(c^*)}\sigma_{\alpha\beta} = \frac{2\eta}{\lambda(c^*)}A_{\alpha\beta} , \tag{8.1-41}$$

where $\lambda(c^*)$ denotes that the relaxation time is allowed to depend on any one of the invariants of $\mathbf{c}$ (or a combination thereof). In particular, $\lambda(c^*)$ was chosen to be a simple power law of the first invariant, which is written as

$$\lambda(c^*) = \lambda_0(\tfrac{1}{3}\mathrm{tr}\tilde{\mathbf{c}})^k . \tag{8.1-42}$$

Therefore the coefficient matrix takes the form

$$\Lambda_{\alpha\beta\gamma\varepsilon} = \frac{1}{2\lambda_0 nK}\left(\tfrac{1}{3}\mathrm{tr}\tilde{\mathbf{c}}\right)^{-k}\left(c_{\alpha\gamma}\delta_{\beta\varepsilon} + c_{\alpha\varepsilon}\delta_{\beta\gamma} + c_{\beta\gamma}\delta_{\alpha\varepsilon} + c_{\beta\varepsilon}\delta_{\alpha\gamma}\right) . \tag{8.1-43}$$

As Figures 8.2, 8.3, and 8.4 show, the capability of varying the power-law index k gives (at moderate and high shear rates) the flexibility of controlling the slope, in a log-log plot, of both the shear and extensional viscosity with respect to the shear and extension rate, respectively.

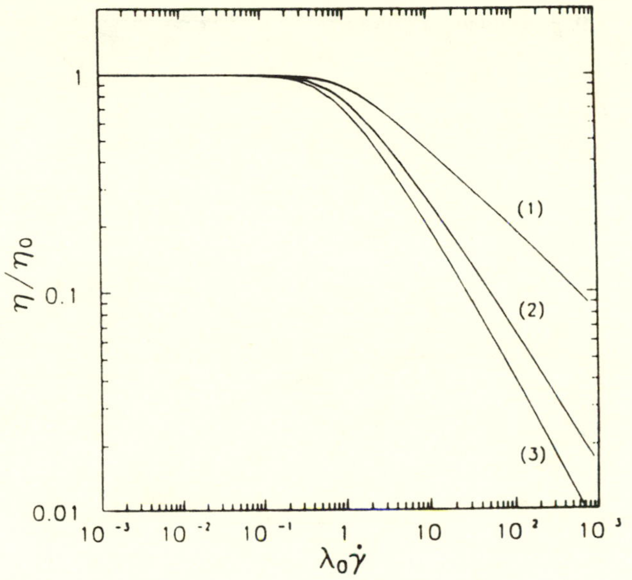

Figure 8.2: Dimensionless viscosity, η/η_0, as a function of the dimensionless shear rate for the power-law EWM model: (1) $k= -0.3$, (2) $k= -0.8$, and (3) $k= -1.3$. (Reprinted with permission from Souvaliotis and Beris [1992, p. 252].)

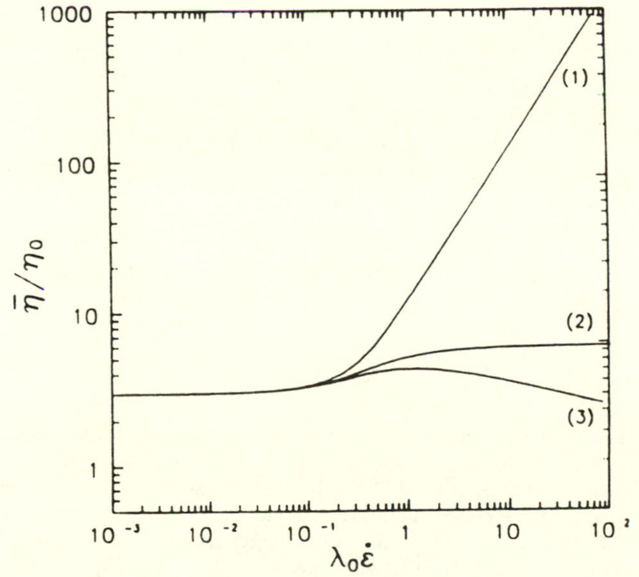

Figure 8.3: Dimensionless extensional viscosity, $\bar{\eta}/\eta_0$, in elongational flow ($\dot{\varepsilon}>0$) for the power-law EWM model: (1) $k= -0.3$, (2) $k= -0.8$, and (3) $k= -1.3$. (Reprinted with permission from Souvaliotis and Beris [1992, p. 254].)

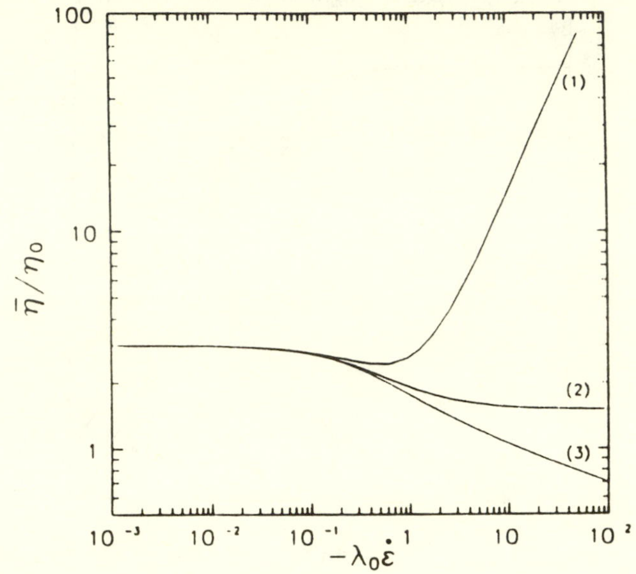

Figure 8.4: Dimensionless extensional viscosity, $\bar{\eta}/\eta_0$, in biaxial stretching ($\dot{\varepsilon}<0$) for the power-law EWM model: (1) $k=-0.3$, (2) $k=-0.8$, and (3) $k=-1.3$. (Reprinted with permission from Souvaliotis and Beris [1992, p. 254].)

Thus, the major advantage of the original White/Metzner equation is recovered. In addition, the predictions are well-behaved (no maximum in the shear stress) for all negative values of the exponent k, in contrast to the behavior of the original White/Metzner model, where the constraint $k\geq-1$ needs to be satisfied for a monotonically increasing value for the shear stress with respect to the shear rate.[13] Asymptotically, for large shear and extensional rates, the model predictions for the shear viscosity, normal stress, and extensional viscosity are [Souvaliotis and Beris, 1992, §IV]

[13]Whether the non-monotonic shear stress behavior with respect to the shear rate is physical or aphysical is a highly debated and currently unresolved issue among researchers on polymer rheology [Malkus *et al.*, 1991]. The fact of the matter is that there are only limited data to support the observation of a non-monotonic shear stress behavior [Vinogradov *et al.*, 1972]—see the pertinent discussion in §8.1.6A. It might very well be that the shear stress loses its monotonicity only when a catastrophic event happens in the system, appearing possibly in the form of a (phase) transition. Thus, it seems prudent to require from simple equations of the form discussed in this chapter, to preserve the monotonicity of the shear stress, relegating the discussion of non-monotonic shear stress behavior to more complex models of the fluid microstructure—see chapters 9 and 11.

$$\eta \propto \dot{\gamma}^{\frac{2k}{1-2k}}, \quad \Psi_1 \propto \eta^2, \quad \bar{\eta} \propto \dot{\varepsilon}^{-\frac{k+1}{k}}. \tag{8.1-44}$$

Note that the rather restricted behavior for the first normal-stress coefficient Ψ_1 can be improved by using a different power-law dependence for the relaxation time and the viscosity [Souvaliotis and Beris, 1992, p. 252], as is currently done in practice with the White/Metzner model.

In further contrast to the White/Metzner model, the EWM model, because of its explicit dependence on the conformation tensor rather than on the deformation rate, is devoid of notorious problems that have hindered extensive applications of the first model, such as a singularity in the extensional viscosity for finite extensional rates and a possibility for a loss of the evolutionary character of the governing equations [Verdier and Joseph, 1989; see also discussion in §8.1.6B]. Of course, the singularity in the extensional viscosity is also a feature of the upper-convected Maxwell model as well. Yet, as seen from Figures 8.3 and 8.4, it disappears in the power-law EWM fluid when a power-law exponent different than zero is used. The same simple law of dependence of the relaxation time is also seen from Figure 8.2 to be adequate for the representation of another characteristic of viscoelastic fluids behavior, i.e., the approach to a constant value for the shear viscosity as the shear rate goes to zero. It is possible to obtain both these effects from the original White/Metzner model at the expense of using more involved (thus, also containing more adjustable parameters) expressions to represent the shear-rate dependence of the viscosity and the relaxation time [Ide and White, 1977; Larson, 1988, p. 152]—see also the discussion in §8.1.1D. Finally from Figure 8.5, note that in contrast to the power-law EWM fluid for which the power-law adjustable parameter k controls the increasing ($k>-1$) or the decreasing ($k<-1$) part of the $\bar{\eta}$ versus $\dot{\varepsilon}$ curve (see Figures 8.3 and 8.4), the adjustable parameter of the PTT model basically controls the maximum of the curve. Although this feature might be closer to the actual behavior of certain melts, the power-law EWM is more useful in the modeling of polymer solutions. Also note that an uncoupled, multiple-mode version of the extended White/Metzner model (following §8.2.1) has been used recently in fitting experimental data obtained for a variety of polymer melts, and the predictions compared well with those of the Phan-Thien and Tanner model [Souvaliotis and Beris, 1992].

Models of this type are the first step towards the development of better constitutive models for viscoelastic fluids. Although these models basically propose just a phenomenological expression for the relaxation time in terms of the conformation of the medium, they still result in predictions which compare favorably with experimental trends. In addition, by insisting that there is not a direct dependence of the material parameters (such as viscosity, relaxation time, etc.) with the (invariants

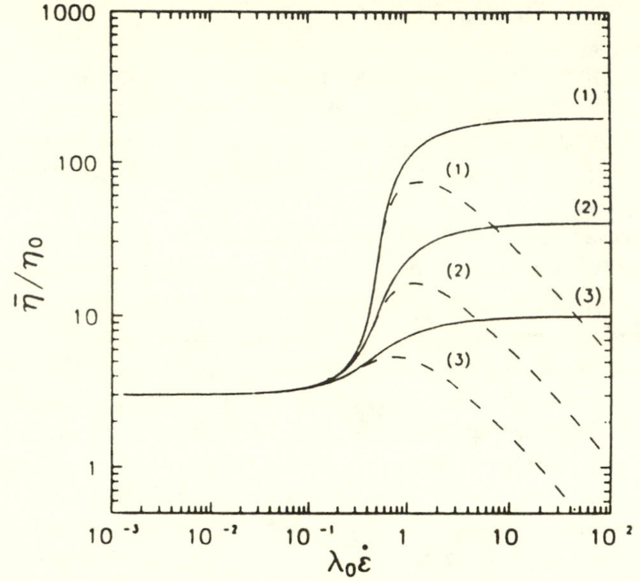

Figure 8.5: Dimensionless extensional viscosity, $\bar{\eta}/\eta_0$, in elongational flow ($\dot{\varepsilon}$>0) for the PTT (ξ=0) model: (——) Linear function, (1) ε=0.01, (2) ε=0.05, and (3) ε=0.2; (- - -) Exponential function, (1) ε=0.01, (2) ε=0.05, and (3) ε=0.2. (Reprinted with permission from Souvaliotis and Beris [1992, p. 256].)

of the) rate-of-strain tensor but rather an indirect one through the conformation tensor **c**, we obtain models which usually satisfy both thermodynamic (i.e., corresponding to a positive entropy production) and mathematical (i.e., giving rise to evolutionary governing equations) admissibility criteria—see the discussion in §8.1.4A. The most important outstanding issue, especially as more complicated nonlinear models are developed, is to guarantee the positive-definiteness of the internal variable, **c**, so as to retain the possibility of extracting information from an underlying internal microstructure—see also the pertinent discussion in §8.1.4B. This bodes well for future research in this area, focusing on the more detailed theoretical aspects of the relaxation-time/microstructure interdependence.

8.1.2 Elastic dumbbell models of dilute polymer solutions arising through kinetic theory

Up until this point, we have only considered simple continuum models based upon spring/dashpot mechanical analogues. Now we wish to consider those simple models which are directly derived from kinetic theory by taking the second moment of the diffusion equation, thus reducing the variable of interest from the distribution function to the conformation tensor through the average

$$c_{\alpha\beta} = \int R_\alpha R_\beta \Psi \mathrm{d}^3 R \equiv \langle R_\alpha R_\beta \rangle \ . \tag{8.1-45}$$

As we shall see, many kinetic theory models so written have counterparts discussed in the preceding subsection. Only the simplest 2-bead kinetic models (dumbbells) are discussed here, the presentation of some of the more general N-bead chain models being offered in §8.2.2. The interested reader is advised to consult the treatise by Bird *et al.* [1987b] and the recent progress review by Bird and Öttinger [1992] on the subject.

A. Hookean (linearly elastic) Dumbbell Models

The simplest example of dumbbell models is the Hookean dumbbell, consisting of two beads attached to opposite ends of an elastic connector of constant spring coefficient. This mechanical analogue provides the simplest microscopic model for a polymer chain with its internal degrees of freedom locked within the law used to represent the strength of the spring. Its most general form, i.e., as a chain consisting of N beads connected through N–1 linearly elastic (Hookean) springs, represents the Rouse model [Rouse, 1953]. Typically, the diffusion equation is written down for the dumbbell in terms of the conformational distribution function, which, in principle, may be used to solve for the desired result. The biggest advantage of the elastic dumbbell and the Rouse chain models is that the corresponding diffusion equation for the distribution function is linear. As a consequence, an exact analytic solution is available with the distribution function represented by a Gaussian function [Lodge and Wu, 1971, p. 549; Öttinger and Rabin, 1989, Eq. (2.12); Zylka and Öttinger, 1991, Eq. (1)]. Note that in this case all the necessary information is contained within the description of the second moment, $\langle \mathbf{RR} \rangle$. This explains the value of the information contained within the second moment which, as a consequence, determines most of the structure even when a general kinetic model is used provided that the magnitude of the nonlinearities is small. Therefore, it is common to work directly with an evolution equation for $\langle \mathbf{RR} \rangle$ obtained by multiplying the diffusion equation by the dyad $\mathbf{RR}$ and by integrating over the conformation space. This reduces the theory to the conformation tensor level, at the possible expense (when the diffusion equation is nonlinear) of the introduction of a decoupling approximation which, if not enough care is paid, can lead to the exclusion of some useful physics. When this procedure is applied to the Hookean dumbbell, we arrive directly at the (exact for this model) expressions [Bird *et al.*, 1987b, pp. 63,71]

$$\overset{\nabla}{c}_{\alpha\beta} = -\frac{4K}{\zeta}c_{\alpha\beta} + \frac{4k_BT}{\zeta}\delta_{\alpha\beta}\ , \tag{8.1-46a}$$

and

$$\sigma_{\alpha\beta} = nKc_{\alpha\beta} - nk_BT\delta_{\alpha\beta}\ . \tag{8.1-46b}$$

In the above expression, ζ is a drag coefficient representing the drag from the solvent on the beads of the dumbbell. Note from (8.1-46a) that for the Hookean dumbbell model the system is assumed to be homogeneous from the outset (i.e., the non-homogeneous term associated with the material derivative of **c** is absent). Direct comparison between Eqs. (8.1-46a,b) with Eqs. (8.1-16,17), respectively, shows that the Hookean dumbbell model is equivalent to the upper-convected Maxwell model if we use the following definitions for the modulus of elasticity, G_0, and the relaxation time, λ:

$$G_0 \equiv nk_BT\ , \quad \lambda \equiv \frac{\zeta}{4K}\ . \tag{8.1-47}$$

By incorporating a viscous solvent effect into the extra stress of (8.1-46b), the Oldroyd-B model would result.

B. Finitely-extensible Nonlinear Elastic Dumbbell Models

The linearly elastic dumbbell corresponds to a Hookean approximation of the contributions of intra-chain conformations to the system free energy, as given by Eq. (8.1-2). However, as briefly mentioned in footnote 5, shortly after the development of this expression it was realized that it cannot be valid at high levels of deformation (i.e., when the magnitude of the end-to-end distance approaches the length of the fully extended chain) since in its original derivation through the "random-flight" model [Kuhn, 1934] any restrictions arising from the finite number of links in the chain have been neglected. When these restrictions are properly taken into account, Eq. (8.1-2) is modified to [Kuhn and Grün, 1942; 1946, p. 185; Peterlin, 1966, Eq. (4); Treloar, 1975, pp. 102-23][14]

$$A = -ST = k_BT\mu C + constant\ , \tag{8.1-48}$$

[14]Eqs. (8.1-48) through (8.1-51) are asymptotic expressions of the exact statistical-mechanics results for a freely-jointed chain of N segments as the number of segments $N\rightarrow\infty$. Exact results are available; however, they differ from the asymptotic approximations in an imperceptible fashion for values of N as low as 25 [Treloar, 1975, §6.5].

where C is defined as

$$C \equiv \tfrac{2}{3}L^2\left(t\beta + \ln(\frac{\beta}{\sinh\beta})\right) , \qquad (8.2\text{-}49)$$

L is the fully-extended chain length, $L \equiv Nl$, $t \equiv R/L$, β is the inverse Langevin function,

$$\beta \equiv \mathcal{L}^{-1}(t)$$

$$\underset{t\to 0}{=} 3t + \tfrac{9}{5}t^3 + \tfrac{297}{175}t^5 + \tfrac{1539}{875}t^7 + \ldots \qquad (8.1\text{-}50)$$

$$\underset{t\to 1}{=} \frac{1}{1-t} ,$$

and the Langevin function is defined as

$$\mathcal{L}(\beta) \equiv \coth(\beta) - \frac{1}{\beta} . \qquad (8.1\text{-}51)$$

All of the other parameters appearing in Eqs. (8.1-48,49) are defined as in Eq. (8.1-2). As a consequence of Eqs. (8.1-48,49), the corresponding magnitude of the elastic force per chain, F, can be calculated in a straightforward fashion as [Treloar, 1975, Eq. (6.10)]

$$F \equiv \frac{dA}{dR} = k_B T \mu \frac{dC}{dR} = \tfrac{2}{3}k_B T \mu L \mathcal{L}^{-1}(t) . \qquad (8.1\text{-}52)$$

Notice that, as it should, it reduces to the previously obtained linear relationship for Gaussian chains, Eq. (8.1-4), in the limit of small chain deformations (t<<1), as verified by using the asymptotic expression for $\mathcal{L}^{-1}(t) \approx 3t$ in Eq. (8.1-52). Also notice from Eq. (8.1-52) that the force increases more rapidly than linearly as the chains extend, becoming infinite at full extension (t=1), as verified using the asymptotic expression for $\mathcal{L}^{-1}(t)$ as $t \to 1$, this time, provided by Eq. (8.1-50). The difference from the original expression is small at small values of extension, becoming significant only at values of $t = O(1)$. Most importantly, it effectively limits the extension of the chain, t, to values smaller than unity. Consequently, if implemented in the dumbbell theory, it is anticipated to primarily influence the extensional properties of the model as seen in a comparison by Tanner and Stehrenberger [1971] of the behavior of various spring-law elastic dumbbells models (including the one provided by Eq. (8.1-52)) in an extensional flow [Tanner and Stehrenberger, 1971, Fig. 3]. Thus, nonlinear elastic dumbbell models have arisen in an effort to better take into account the internal-modes entropic contribution to the free energy.

p. 239: the number of the first equation, (8.2-49), should be (8.1-49)

B.1 The FENE-P model

To express the nonlinear elastic models in Hamiltonian form, we just need to incorporate their corresponding modifications to the elastic potential energy. For instance, the finitely-extensible, nonlinear elastic dumbbell model (FENE) has been developed in an effort to take into account the inverse Langevin expression for the force, Eq. (8.1-52), approximated in a less cumbersome way mathematically by the Warner (or FENE) force law [Warner, 1972, p. 380; Tanner, 1975, p. 54; Bird *et al.*, 1987b, §13.5]

$$F = \frac{KR}{1 - \left(\dfrac{R}{R_0}\right)^2} , \tag{8.1-53}$$

where K is a characteristic spring constant and R_0 the maximum allowed spring length, for a N bead-rod chain with no constraints $R_0=L=Nl$. Of course, more accurate approximations are still possible as, for example, obtained through the use of Padé approximants [Cohen, 1991].

When Eq. (8.1-53) is introduced into the diffusion equation for the distribution function for the end-to-end distance vector **R**, it gives rise to a nonlinear equation which, when integrated weighted with **RR**, does not lead to a closed equation for the second moment, <**RR**>, as the linear elastic law did. To obtain a closed equation, the Peterlin approximation is usually used, according to which the spring force is calculated by using a pre-average of the denominator, or, in vector form [Peterlin, 1961b, p. 259],

$$\mathbf{F} = \frac{K\mathbf{R}}{1 - \dfrac{\langle R^2 \rangle}{R_0^2}} . \tag{8.1-54}$$

Use of the above expression allows for the development of a constitutive equation expressed either directly in terms of the stress [Tanner, 1975, Eq. (65)] or indirectly in terms of the internal conformation variable <**RR**> [Wedgewood and Bird, 1988, Eq. (9); Grmela, 1989a; Beris and Edwards, 1990a, pp. 70-71; Wedgewood *et al.*, 1991, p. 120]. This model became known as the FENE-P model.

To formulate the FENE-P model within the generalized bracket formalism, it requires a change of E_ϕ to

$$E_\phi[\mathbf{c}] = -\int_\Omega \tfrac{1}{2} nKR_0^2 \ln\left(1 - \frac{\mathrm{tr}\mathbf{c}}{R_0^2}\right) \mathrm{d}^3x . \tag{8.1-55}$$

Hence

$$\frac{\delta E_\phi}{\delta c_{\alpha\beta}} = \tfrac{1}{2} nK \frac{R_0^2}{R_0^2 - \mathrm{tr}\mathbf{c}} \delta_{\alpha\beta} , \tag{8.1-56}$$

whereby

$$\sigma_{\alpha\beta} = nK\frac{R_0^2}{R_0^2 - \mathrm{tr}\,\mathbf{c}}c_{\alpha\beta} - nk_BT\delta_{\alpha\beta} \;, \tag{8.1-57a}$$

and, if we use $\boldsymbol{\Lambda}$ from (8.1-15), then we obtain the evolution equation for the conformation tensor **c** as

$$\overset{\triangledown}{c}_{\alpha\beta} = -\frac{4K}{\zeta}\frac{R_0^2}{R_0^2 - \mathrm{tr}\,\mathbf{c}}c_{\alpha\beta} + \frac{4k_BT}{\zeta}\delta_{\alpha\beta} \;. \tag{8.1-57b}$$

Eqs. (8.1-57a,b) represent the formulation of the FENE-P dumbbell model in terms of the conformation tensor. Rigorously, the evolution equation for **c** is obtained from the diffusion equation of the distribution function corresponding to the FENE model, through the usual averaging procedure mentioned above. The resulting equation is different from Eq. (8.1-57b) as it contains averages of higher moments of the end-to-end distance vector **R**. However, if the Peterlin approximation, Eq. (8.1-54), is used, or, equivalently, the higher moment average is approximated in terms of the second moment average, **c**, as

$$< \frac{R_\alpha R_\beta}{1 - \dfrac{\mathbf{R}\cdot\mathbf{R}}{R_0^2}} > \approx \frac{c_{\alpha\beta}}{1 - \dfrac{\mathrm{tr}\,\mathbf{c}}{R_0^2}} \;, \tag{8.1-58}$$

then Eq. (8.1-57b) is recovered. The fact that the same equation results, irrespective of whether the Peterlin approximation is used in the beginning (bracket approach) or at the end (kinetic theory approach) of the modeling procedure provides evidence for the consistency and the compatibility of the two approaches. Consequently, in transferring our system description from the distribution function level to the conformation tensor level, we have gained the ability to solve a simple set of equations in arbitrary circumstances at the expense of introducing a rather dubious approximation with unknown effects upon the physics of the model. It is also worth noting that the averaging procedure in and of itself restricts somewhat the physical content of the original kinetic theory model written in terms of a conformation tensor. However, for the particular case studied here, it turns out that this is not a bad approximation. Indeed, Fan [1985a] solved directly the corresponding diffusion equation without making use of the Peterlin approximation for steady simple shear flows and showed that the predictions he obtained for the viscosity and the first normal-stress coefficient matched very well those corresponding to the FENE-P model for $b \equiv KR_0^2/k_BT \geq 10$, thus justifying the use of the Peterlin approximation, at least under those conditions.

In general, as we shall see more clearly in the remainder of this subsection, in the transferral of information from the distribution function level to the conformation tensor level only certain types of terms are carried over to the macroscopic level of description. This is because of the average splitting that is required in order to express the various moments of the distribution function in terms of a second-rank tensorial representation. Specifically, in writing the dissipation associated with the conformational relaxation matrix in the form of the first integral in (8.1-5), one has explicitly set the decoupling approximation to be applied to the diffusion equation as

$$<\boldsymbol{\Lambda}:\frac{\delta H}{\delta \mathbf{c}}> \approx <\boldsymbol{\Lambda}>:<\frac{\delta H}{\delta \mathbf{c}}> . \tag{8.1-59}$$

This effectively decouples the phenomenological term associated with $\boldsymbol{\Lambda}$ from the energetic effects in the Hamiltonian. Hence a term appearing during the averaging of the diffusion equation such as

$$<\frac{R_\alpha R_\gamma R_\beta R_\gamma}{R_\varepsilon R_\varepsilon}> = <R_\alpha R_\beta> \equiv c_{\alpha\beta} , \tag{8.1-60}$$

may be reduced to the conformation tensor, whereas a similar term expressed solely through the conformation tensor,

$$\frac{c_{\alpha\gamma} c_{\beta\gamma}}{\mathrm{tr}\mathbf{c}} \neq c_{\alpha\beta} , \tag{8.1-61}$$

cannot bear the same equality. Indeed, the specific functional relationship between the two sides of the "unequation" (8.1-61) must be given by the Cayley/Hamilton theorem,

$$\mathbf{c}\cdot\mathbf{c}\cdot\mathbf{c} - I_1\mathbf{c}\cdot\mathbf{c} + I_2\mathbf{c} - I_3\boldsymbol{\delta} = \mathbf{0} , \tag{8.1-62}$$

where the I_i are the invariants of $\mathbf{c}$. Still, the same information is contained in the conformation tensor theory derived by averaging the diffusion equation as may be obtained through working solely in terms of the conformation tensor. The two types of conformation tensor theory are self-consistent; it is only in the comparison that the above problem arises as a result of the form of the transferral of information during the averaging.

B.2 The Chilcott and Rallison model

More recently, enhanced interest has been taken in the development of modified FENE-P models [Rallison and Hinch, 1988, §2.1; Chilcott and Rallison, 1988, §2]. As Chilcott and Rallison [1988, p. 385] explicitly point out, the only difference of their model, hereafter referred to as C-R, from the FENE-P described by Eqs. (8.1-55 through 8.1-57), is on the right-hand side of the evolution equation for $\mathbf{c}$, Eq. (8.1-57b). Let the nonlinear dimensionless factor $f(\mathrm{tr}\mathbf{c})=g(\mathrm{tr}\tilde{\mathbf{c}})$ be defined as

$$f(\mathrm{tr}\mathbf{c}) \equiv \frac{1}{1 - \dfrac{\mathrm{tr}\mathbf{c}}{R_0^2}} = g(\mathrm{tr}\tilde{\mathbf{c}}) \equiv \frac{1}{1 - \dfrac{\mathrm{tr}\tilde{\mathbf{c}}}{\tilde{N}}} \,, \tag{8.1-63}$$

where $\tilde{\mathbf{c}}$ is the dimensionless conformation tensor, $\tilde{\mathbf{c}} \equiv \mathbf{c}K/(k_BT) = 3\mathbf{c}/(2Nl^2)$, and $\tilde{N} \equiv 3N/2$. Then, in the C-R model, f (or g), instead of multiplying only the conformation tensor $\mathbf{c}$, as in the FENE-P model, also multiplies the unit tensor $\boldsymbol{\delta}$:

$$\overset{\triangledown}{c}_{\alpha\beta} = -\frac{4k_BT}{\zeta} g(\mathrm{tr}\tilde{\mathbf{c}})\left(\tilde{c}_{\alpha\beta} - \delta_{\alpha\beta}\right) . \qquad (C\text{-}R) \tag{8.1-64}$$

The right-hand side of Eq. (8.1-64) arises from the relaxation integral in the dissipation bracket. Thus, it would have appeared that it is the relaxation tensor $\boldsymbol{\Lambda}$ that only needs to be modified from the form that it has for the FENE-P model, Eq. (8.1-15), in order to represent the C-R model within the bracket formalism. However, Eq. (8.1-64) is incompatible with Eq. (8.1-55) regarding the value of $\tilde{\mathbf{c}}$ that they assume at equilibrium. Indeed, the value of $\tilde{\mathbf{c}}$ for which the right-hand side of Eq. (8.1-64) vanishes, $\tilde{\mathbf{c}}_e = \boldsymbol{\delta}$, is different from the value of $\tilde{\mathbf{c}}$ which minimizes the free energy defined by Eq. (8.1-55):

$$\left.\frac{\delta H_m}{\delta \mathbf{c}}\right|_e = 0 \,, \tag{8.1-65}$$

where the subscript e denotes the static equilibrium conditions under which Eq. (8.1-65) is valid. Eq. (8.1-65), in conjunction with Eqs. (8.1-14 and 8.1-56), shows us immediately that, whereas in the linear elastic (Hookean) dumbbell model,

$$\tilde{\mathbf{c}}_e = \boldsymbol{\delta} \,, \qquad (Hookean) \tag{8.1-66}$$

in the FENE-P and C-R models

$$\tilde{\mathbf{c}}_e = \frac{1}{g_e}\boldsymbol{\delta} \,, \qquad (FENE\text{-}P \;\&\; C\text{-}R) \tag{8.1-67}$$

where

$$g_e \equiv g(\mathrm{tr}\tilde{\mathbf{c}}_e) = 1 + \frac{3}{\tilde{N}} \,, \qquad (FENE\text{-}P \;\&\; C\text{-}R) \tag{8.1-68}$$

and the second equality has been evaluated by solving simultaneously Eqs. (8.1-63,68). Although for the (typically) large values of $\tilde{N}$ used, the two expressions are very close, the fact that they are not the same makes it impossible to retrieve Eq. (8.1-64) from a theory utilizing Eq. (8.1-55) as the Hamiltonian.

This incompatibility can also be seen from the calculation of the corresponding rate of energy dissipation

$$[H_m,H_m]_r \equiv -\int_\Omega \Lambda_{\alpha\beta\gamma\varepsilon}\frac{\delta H_m}{\delta c_{\alpha\beta}}\frac{\delta H_m}{\delta c_{\gamma\varepsilon}}\mathrm{d}^3x \;. \tag{8.1-69}$$

This can be obtained directly from the right-hand side of Eq. (8.1-64) and the FENE-P expression for $\delta H_m/\delta c$, calculated using Eqs. (8.1-14,56), as

$$\begin{aligned}[H_m,H_m]_r &\equiv -\int_\Omega \frac{2nKk_BT}{\zeta}g(\mathrm{tr}\tilde{c})\left(\tilde{c}_{\alpha\beta}-\delta_{\alpha\beta}\right)\\ &\qquad \left(g(\mathrm{tr}\tilde{c})\delta_{\alpha\beta}-\tilde{c}_{\alpha\beta}^{-1}\right)\mathrm{d}^3x\\ &\equiv -\int_\Omega \frac{2nKk_BT}{\zeta}g\\ &\qquad \left(g\cdot\mathrm{tr}\tilde{c}-3-3g+\mathrm{tr}\tilde{c}^{-1}\right)\mathrm{d}^3x \;.\end{aligned} \tag{8.1-70}$$

Notice that in contrast to the FENE-P model, for which use of Eq. (8.1-15) for $\boldsymbol{\Lambda}$ guarantees the satisfaction of the Onsager symmetry relations and a bilinear (quadratic) dissipation, in the C-R model, even close to equilibrium, we have an asymmetric, non-quadratic dissipation. Indeed, let the conformation tensor close to equilibrium be expressed as

$$\tilde{\mathbf{c}} = \tilde{\mathbf{c}}_e + \tilde{\boldsymbol{\epsilon}} = \frac{1}{g_e}\boldsymbol{\delta} + \tilde{\boldsymbol{\epsilon}} \;. \quad (\textit{FENE-P \& C-R}) \tag{8.1-71}$$

Then, substitution of Eq. (8.1-71) into Eq. (8.1-69) gives the following expression for the rate of relaxation dissipation:

$$\begin{aligned}[H_m,H_m]_r &\equiv -\int_\Omega \frac{2nKk_BT}{\zeta}g(\mathrm{tr}\tilde{c})g_e^2\tilde{\varepsilon}_{\alpha\beta}\\ &\qquad \times\left((1-\frac{1}{g_e})\delta_{\alpha\beta}+\tilde{\varepsilon}_{\alpha\beta}\right)\mathrm{d}^3x \;,\end{aligned} \quad (\textit{C-R}) \tag{8.1-72}$$

where, taking into account the small magnitude of $\tilde{\varepsilon}$, the following approximation for $\tilde{c}^{-1}$ has been used:

$$\tilde{\mathbf{c}}^{-1} \approx g_e\boldsymbol{\delta} - g_e^2\tilde{\boldsymbol{\epsilon}} \;. \qquad (\|\tilde{\boldsymbol{\epsilon}}\| \ll 1,\ \textit{C-R}) \tag{8.1-73}$$

As can easily be seen from Eq. (8.1-72), the only way the quadratic character for the relaxation dissipation can be restored, close to equilibrium for the C-R model, is by requiring

$$g_e = 1 \; . \tag{8.1-74}$$

To achieve this, we propose the following modification of the factor f (and g) used in the C-R model (modified Chilcott-Rallison or mC-R model)

$$f_m(\mathrm{tr}c) \equiv \frac{1-\dfrac{3K}{k_B T R_0^2}}{1-\dfrac{\mathrm{tr}c}{R_0^2}} = g_m(\mathrm{tr}\tilde{c}) \equiv \frac{1-\dfrac{3}{\tilde{N}}}{1-\dfrac{\mathrm{tr}\tilde{c}}{\tilde{N}}} \; . \tag{8.1-75}$$

Since the addition of the (very close to unity) factor $(1-3/\tilde{N})$ can be absorbed by the other parameters of the model (like z or l), the flow properties of the mC-R model are exactly the same as those of the original one. However, the proposed modification is crucial in restoring the thermodynamic admissibility of the model. Indeed, when Eq. (8.1-75) is used instead of Eq. (8.1-63), the corresponding relaxation dissipation, Eq. (8.1-69), is always non-positive. To show this, it suffices to prove that the integrand corresponding to Eqs. (8.1-69 and 8.1-75),

$$d_r(g_m,\tilde{c}) \equiv g_m \mathrm{tr}\tilde{c} - 3 - 3g_m + \mathrm{tr}\tilde{c}^{-1} \; , \tag{8.1-76}$$

is always non-negative with g_m evaluated using Eq. (8.1-75) for $\mathrm{tr}\tilde{c}$. This may be prove as follows. First, let us rewrite Eq. (8.1-76) in terms of the (positive) eigenvalues, λ_1, λ_2, and λ_3 of the symmetric, positive-definite tensor $\tilde{c}$

$$\begin{aligned} d_r(g_m,\tilde{c}) &= \sum_{i=1}^{3}\left(g_m\lambda_i - 1 - g_m + \frac{1}{\lambda_i}\right) \\ &= \sum_{i=1}^{3}\frac{1}{\lambda_i}(g_m\lambda_i - 1)(\lambda_i - 1) \; , \end{aligned} \tag{8.1-77}$$

which implies that for $g_m = 1$ and for any positive definite and symmetric $\tilde{c}$,

$$d_r(1,\tilde{c}) \geq 0 \; . \tag{8.1-78}$$

On the other hand, if, for fixed $\tilde{c}$, we calculate the partial derivative of d_r with respect to g_m, then we find

$$\left.\frac{\partial d_r}{\partial g_m}\right|_{\tilde{c}} = \mathrm{tr}\tilde{c} - 3 \; , \tag{8.1-79}$$

which is positive for $\mathrm{tr}\tilde{c} > 3$ and negative for $\mathrm{tr}\tilde{c} < 3$. Thus, given that $g_m(\mathrm{tr}\tilde{c})$, as defined through Eq. (8.1-75), is larger or smaller than unity whenever $\mathrm{tr}\bar{c}$ is larger or smaller than 3, respectively, we have that for any symmetric and positive-definite tensor $\bar{c}$

$$d_r(g_m(\mathrm{tr}\bar{c}),\bar{c}) \geq d_r(1,\bar{c}) \geq 0 \ . \tag{8.1-80}$$

Thus, it is shown that the change incorporated by Eq. (8.1-75) to the factor used in the C-R model also guarantees that the corresponding relaxation dissipation remains always non-positive. We believe that a similar correction is also necessary in order to guarantee the thermodynamic admissibility of the Rallison-Hinch model [Rallison and Hinch, 1988, Eqs. (5),(6)]. Note that since the symmetry of the relaxation dissipation for the FENE-P model is guaranteed from the form of the dissipation tensor, $\boldsymbol{\Lambda}$, a similar change to the one proposed for the C-R model for the factor f, (or g), although preferable (in order to reduce that model to the Hookean dumbbell close to equilibrium), is not thermodynamically necessary.

Finally, note that a modification proposed by Bird *et al.* [1987b, Eq. (13.5-48)] has a totally different motivation and leads to a correction which is similar to the one suggested above. The condition of Bird *et al.* is based on the self-consistency of the decoupling approximation at equilibrium and leads to a correction of the right-hand side of Eq. (8.1-57b)—as well as a corresponding correction to the stress definition, Eq. (8.1-57a)—involving the addition of the term $-4KR_0^2\varepsilon/\zeta\ \boldsymbol{\delta}$, where $\varepsilon=2/(2b+b^2)$. The resulting model—called FENE-P-B—is expected to show a closer agreement with the behavior predicted by the diffusion equation corresponding to the FENE dumbbell, than the FENE-P model (and indeed it does, albeit not by much, as demonstrated by Fan [1985a]). In contrast, our modification leads to a correction of similar form, but with a different constant $\varepsilon'=3/(3b-b^2)$. The purpose of our modification is to better approximate the original inverse Langevin spring law. Thus, we can suggest here that both modifications should be taken into account. In general, we consider it a good practice to always try to impose, as much as possible, consistency between the moment equation and its corresponding original kinetic equation. Despite the small changes that this practice resulted in for FENE-type models, this is not the case when hydrodynamic interactions are taken into account, as we shall see in the next section.

C. Dumbbell Models with Hydrodynamic Interaction

The above dumbbell models are overly simplified, not only because of the significant reduction of the degrees of freedom of the polymer chains that they aim to describe, but also because they do not take into account certain items associated with a realistic description of the physical entities represented by the dumbbells themselves. For instance, excluded volume and hydrodynamic inter-bead interactions are totally neglected despite the fact that they are present even in the dilute solution limit, for which the above models are written. Between the two, the neglect of hydrodynamic interactions is introducing by far the more significant errors. Hydrodynamic interactions occur due to the alteration in the flow field that one bead of the dumbbell feels due to the motion of the other bead. Oseen and Burgers calculated the dominant terms of the velocity disturbance induced by a sphere moving through a fluid with constant velocity [see Bird *et al.*, 1987a, pp. 28,29]. This calculation may be used (as shown below) to describe the perturbation in the velocity field of the solvent which is caused by the dumbbell beads.

The effects of the distorted flow field upon the beads of the dumbbell are conformation dependent, and are taken into account in the diffusion equation for the end-to-end distribution function, to a first-order approximation (in terms of a/R, where a is the bead radius), by modifying the ambient velocity around each bead by $K\boldsymbol{\Omega}\cdot\mathbf{R}$, where $\boldsymbol{\Omega}$ is the Oseen/ Burgers tensor [Öttinger, 1986b, p. 1669; Bird *et al.*, 1987b, §13.6]:

$$\Omega_{\alpha\beta} \equiv \frac{3h}{4\zeta}\sqrt{\frac{\pi k_B T}{K\mathbf{R}\cdot\mathbf{R}}}\left(\delta_{\alpha\beta} + \frac{R_\alpha R_\beta}{\mathbf{R}\cdot\mathbf{R}}\right) . \tag{8.1-81}$$

The dimensionless parameter h, $h \equiv (\zeta/6\eta_s)(K/\pi^3 k_B T)^{1/2}$, measures the strength of the hydrodynamic interaction and is typically believed to lie within the range $0.0 \leq h < 0.3$.[15] When $h=0.0$, the Oseen/Burgers tensor vanishes and the model reduces back to the simple Hookean dumbbell case. Note that alternative expressions to the Oseen/Burgers tensor have also been proposed in the literature, aimed to account for the finite size of the beads and/or to eliminate the singularity at $R=0$. These include the Rotne/Prager/Yamakawa tensor [Rotne and Prager, 1969; Yamakawa, 1970; Bird *et al.*, 1987b, §14.6; Öttinger, 1989b, Eq. (4b)] and various modifications of the Oseen/Burgers tensor [Fixman, 1981, Eq. (2.26); Zylka and Öttinger, 1989, Eq. (4); Öttinger, 1989b, Eq. (4)]. Although these modifications need to be used in certain applications requiring positive-definite diffusion tensors in Brownian dynamics simulations [Zylka and Öttinger, 1989, p. 475] since the Oseen/Burgers expression leads for small displacements to a non-positive diffusion tensor [Zylka and Öttinger, 1989, p. 474; Bird *et al.*, 1987b, p. 133], they lead to almost identical predictions when applied in conjunction with a simple Hookean

[15]Note that this expression is just a truncation of a more general expression in terms of the moments of **R**. The reader may refer to de Haro and Rubi [1988, p. 1249] for an expansion containing higher-order terms and additional references.

dumbbell model [Zylka and Öttinger, 1989]. Thus, for the sake of simplicity, in the following we shall only use Eq. (8.1-81), with the understanding that any one of the other expressions can be used in practice, if needed, with only minor modifications.

The major difficulty introduced in the computations when hydrodynamic interactions are taken into account, even under the simplified form of the Oseen/Burgers expression, is that they render the corresponding equation nonlinear, which, as a consequence, no longer admits a simple analytic solution. Thus, the need for approximations emerges. Zimm was the first one to use a pre-averaged form of the Oseen/Burgers expression in order to model the hydrodynamic interaction between the beads in a Rouse chain [Zimm, 1956].[16] However, since the pre-averaging was performed using the (inconsistent under flow conditions) equilibrium distribution, the final solution, although easy to calculate and still Gaussian in form, was able to correctly account for the physics only close to equilibrium (small shear rates). Moreover, it did not show a qualitative change over the predictions obtained with the Rouse chain, namely a constant viscosity and first normal-stress coefficient and a zero second normal-stress difference [Öttinger, 1987, p. 3731]. Later, Öttinger [1985; 1986a,b], Biller *et al.* [1986]; and de Haro and Rubi [1988] proposed a consistently averaged approximation of the hydrodynamic interactions for a Hookean dumbbell model. In this approach, the average value of the Oseen tensor, $\boldsymbol{\Omega}$, was consistently calculated using the (still Gaussian!—thus, fully determined from the second moment) distribution function resulting from the solution of the corresponding diffusion equation where the so approximated constant value of the Oseen tensor was used. Thus, the calculation for the average value of the Oseen tensor, Eq. (8.1-81)—performed, for example, using the formulae appearing in §E.3 of Bird *et al.* [1987b]—and the calculation for the corresponding second moment of the distribution function—as given, for example, by Eq. (4) of Öttinger [1985], for homogeneous simple shear flow—were coupled, and they had to be solved simultaneously.

The model properties resulting from the consistently averaged approximation of the hydrodynamic interactions are substantially improved over the pre-averaging one [Öttinger, 1985]. The shear properties (viscosity and normal stresses coefficients) exhibit the experimentally observed shear-thinning/shear-rate dependence characteristics and the second normal-stress is no longer predicted to be zero but rather to be a small fraction of the first normal stress. However, still the consistently averaged approximation was not perfect: the second normal

[16]A corrected version of this paper can be found in Hermans [1978, pp. 73-84]. For corrections to Zimm's paper, see also Williams [1965].

stress is predicted with the wrong sign (positive). Indeed, detailed Brownian dynamics simulation using the original diffusion equation for the distribution function showed that substantial differences still remained between the results of the Brownian dynamics simulation corresponding to a non-averaged Oseen tensor and the consistently averaged approximation [Fan, 1986 p. 6238; Fan, 1987, pp. 836-37]. The major results from the Brownian dynamics simulation were qualitatively similar (shear thinning) but quantitatively lower values for the viscosity and first normal stress than the consistently averaged approximation predictions and, most importantly, a negative second normal stress, which is more in accordance with the experimental observations [Bird *et al.*, 1987a, pp. 109-11]. The relatively poor performance of the consistently averaged approximation is attributed to not taking into consideration (when the averaging of the Oseen tensor is performed) the influence of fluctuations of the Oseen tensor on the dumbbell dynamics [Öttinger, 1989b]. This observation led naturally to the development of a more consistent "Gaussian approximation" where the overall influence of the Oseen tensor on the evolution equation for the second moment of the distribution function $<\mathbf{RR}>$—i.e., the calculation of the incurring higher moment averages—is calculated consistently based on the assumption of a Gaussian form for the distribution function [Öttinger, 1989b; Zylka and Öttinger, 1989; Wedgewood, 1989].

The predictions resulting from the Gaussian approximation (which in the case of non-averaged approximation of the Oseen tensor still involves an approximation, namely the Guassian form of the distribution function, which, in general, can have a more general functionality) compare extremely well with the predictions of a direct Brownian dynamics simulation based on the original form of the diffusion equation [Zylka and Öttinger, 1989; Wedgewood, 1989; Zylka, 1991]. Both the values of the shear thinning viscosity and first normal-stress coefficient and the sign (negative) and the magnitude of the second normal-stress coefficient agree very well over the whole range of shear-rate values [Zylka 1991, fig. 2]. In addition, Zylka and Öttinger [1989] have shown that the results depend only weakly on the specific form used to represent the hydrodynamic interactions—i.e., the reported results were found to be approximately the same when either a (modified) Oseen/ Burgers tensor or the Rotne/ Prager/Yamakava tensor was used. The importance of the last finding is that the incomplete knowledge of the exact form for the hydrodynamic interactions dynamics does not seem to affect greatly the predicted dumbbell dynamics.

These relatively recent developments are very encouraging to the modeling effort of polymer flow behavior based on internal structural parameters on two counts. First, they indicate that the use of a microscopic model does not necessarily imply the need for the solution of the corresponding Brownian dynamics problem. Indeed, the success of the consistent use of the Gaussian approximation means that a single, second-order tensor parameter (its second moment, <**RR**> or **c** as used in this work) contains all of the information that is necessary in order to accurately and faithfully reproduce the behavior of the underlying microscopic model. Thus, the Gaussian approximation (as, by the way, also all of the previous ones) leads to constitutive equations [Zylka and Öttinger, 1989, Eq. (15); Wedgewood, 1989, Eq. (12)] which, however complex, involve only <**RR**> as an internal structural parameter. Second, the success of the overall microscopic model, which, however simplistic it may be (consisting of a single connector, 2-bead Hookean elastic spring chain), has managed to provide *a priori* reasonable viscometric flow behavior, is reinforcing the belief that complicated material flow behavior can be described based on rational microscopic considerations.

What role is the generalized bracket theory anticipated to play in this modeling effort? First, notice that most of the above-mentioned models have only been used in simple, homogeneous, viscometric flows with known kinematics. To extend the usage of these models to arbitrary flows, it is useful to know more about their compatibility with continuum mechanics and irreversible thermodynamics. Thus, the role of the generalized bracket is a complimentary one to that played by microscopic kinetic theory models. The generalized bracket theory, when used in coordination with the microscopic modeling, offers at least three advantages. First, it can act as an additional safeguard for the consistency of the resulting approximative model, thermodynamical consistency this time, enforced by a non-positive rate of energy dissipation (or, equivalently, a non-negative rate of entropy production). Second, the bracket approach offers an unambiguous formulation for the stress and allows the incorporation into the model of the description of additional transport phenomena such as heat and mass transfer. Note that with microscopic models, special care needs to be exercised on which form of the stress expression is used (for example, simply the introduction of hydrodynamic interactions invalidates, already, all other stress expressions in molecular theory except the Kramers/Kirkwood expression [Bird and Curtiss, 1985; Bird *et al.*, 1987b, pp. 264-71, 275-78, 281-85, and others]. Third, by focusing the attention on the energetics of the system, one allows for the explanation of thermodynamic phenomena, such as flow-induced phase transitions or wall depletion due to entropy reduction, as discussed in more detail in chapter 9.

To illustrate the compatibility of the generalized bracket approach with the ideas exposed above for the hydrodynamic interaction dumbbell model, we analyze in the following paragraphs how this model can be described within the bracket formulation. In doing so, we could, in principle, use as a starting point the most accurate Gaussian approximation expressions for the evolution equation for the internal tensorial structural parameter **c**. However, we have chosen here, for illustration purposes, to use instead the much simpler implementation of the consistently averaged approximation. Using the Gaussian approximation would have simply led to a different (and more complex) expression for the relaxation dissipation. The "consistently averaged" expression for the Oseen/Burgers tensor can be further simplified if a decoupling approximation is used according to which this tensor is re-expressed in terms of the conformation tensor $\langle \mathbf{RR} \rangle \equiv \mathbf{c}$ simply enough as:

$$\Omega_{\alpha\beta} \simeq \frac{3h}{4\zeta}\sqrt{\frac{\pi k_B T}{K \mathrm{tr}\mathbf{c}}}\left(\delta_{\alpha\beta} + \frac{c_{\alpha\beta}}{\mathrm{tr}\mathbf{c}}\right) . \tag{8.1-82}$$

This further approximation was first proposed by Honerkamp and Öttinger [1986, Eq. (2.5)] and, later, by Wedgewood [1988, Eq. (12)].

The Oseen/Burgers tensor written in terms of the conformation tensor relates an additional relaxation mechanism, due to the flow-field perturbation, which must be included in $\boldsymbol{\Lambda}$ for the simple dumbbell models. Consequently, the relaxation matrix of (8.1-15), written in terms of the drag coefficient, ζ, becomes

$$\begin{aligned}\Lambda_{\alpha\beta\gamma\varepsilon} = \frac{2}{n\zeta}\Big[&c_{\alpha\gamma}(\delta_{\beta\varepsilon} - \zeta\Omega_{\beta\varepsilon}) + c_{\alpha\varepsilon}(\delta_{\beta\gamma} - \zeta\Omega_{\beta\gamma}) \\ &+ c_{\beta\gamma}(\delta_{\alpha\varepsilon} - \zeta\Omega_{\alpha\varepsilon}) + c_{\beta\varepsilon}(\delta_{\alpha\gamma} - \zeta\Omega_{\alpha\gamma})\Big] ,\end{aligned} \tag{8.1-83}$$

or substituting (8.1-82) into the above expression,

$$\begin{aligned}\Lambda_{\alpha\beta\gamma\varepsilon} = \frac{2}{n\zeta}&\left(1-\tfrac{3}{4}h\sqrt{\frac{\pi k_B T}{K \mathrm{tr}\mathbf{c}}}\right)\Big[c_{\alpha\gamma}\delta_{\beta\varepsilon} + c_{\alpha\varepsilon}\delta_{\beta\gamma} + c_{\beta\gamma}\delta_{\alpha\varepsilon} + \\ &c_{\beta\varepsilon}\delta_{\alpha\gamma}\Big] - \frac{3h}{n\zeta}\sqrt{\frac{\pi k_B T}{K \mathrm{tr}\mathbf{c}}}\,\frac{1}{\mathrm{tr}\mathbf{c}}\left(c_{\alpha\gamma}c_{\beta\varepsilon} + c_{\alpha\varepsilon}c_{\beta\gamma}\right) .\end{aligned} \tag{8.1-84}$$

Substitution of this latter expression for $\boldsymbol{\Lambda}$ into the general evolution equation, (8.1-7), and rewriting in terms of the extra stress tensor, gives directly the constitutive equation. The model resulting from the decoupling approximation represents the viscosity and first normal stress behavior better than the consistently averaged approximation as compared to the results obtained from the Brownian dynamics simulations (although it still predicts a positive second normal-stress coefficient and, in addition, it also predicts a small, shear thickening at small shear rates introduced by the closure approximation) [Wedgewood, 1988, p. 34;

Beris and Edwards, 1990a, pp. 73-74]. Further developments along the lines of a 2-bead chain (dumbbell) have included the combined effects of a nonlinear elastic spring law and hydrodynamic interactions [Honerkamp and Öttinger, 1986; Biller *et al.*, 1986; Diaz *et al.*, 1989]; however, at least in extensional flows, the effect of the introduction of hydrodynamic interactions is minimal [Biller *et al.*, 1986]. Other modifications, empirical in origin, that have attempted to model the effect of excluded volume in concentrated solutions and melts by including an anisotropic drag and/or Brownian force are discussed in the next section.

8.1.3 Single internal variable models for concentrated polymer solutions and melts

Because of substantially increased complexity as compared with the dilute polymer solution rheology, all available models for concentrated polymer solutions and melts, with the exception of the Doi/Edwards and Curtiss/Bird integral models [Doi and Edwards, 1978, 1979; Curtiss and Bird 1981a,b], are phenomenological in nature. Furthermore, in practice, all of them are used as a multimode expansion in order to capture the wide spectrum of relaxation times associated with the rheological behavior of concentrated polymer systems. We restrict our discussion here to the formulation of one-mode relaxation models as we defer until §8.2 the formulation of their multimode equivalent extensions. Some of the most successful models for polymer solutions and melts have already been mentioned in §8.1.1, such as, for example, the PTT and EWM models. In addition to these, other models have been developed especially for concentrated solutions and melts. Among these, three single internal variable models are the most popular. Although also phenomenological in nature, their development originated from ideas developed within the framework of network and reptation theories and the thermodynamics of rubber elasticity. As such, we felt that they deserved a special mention below, in particular in order to illustrate how the basis of these theories is related to and can be expressed through the bracket formalism. Not surprisingly at this point, we shall see several similarities with the other single internal variable models, as discussed previously.

A. The Single-mode Giesekus Model

Although the Giesekus model was originally developed within the context of a network theory of penetrating statistical continua [Giesekus, 1982], it can be more simply explained based upon an approximate kinetic theory model of a concentrated suspension of Hookean dumbbells [Bird and Wiest, 1985]. In this model, the effect on a given dumbbell due to the presence of the confining interactions with the other dumbbells is modeled by introducing an anisotropic hydrodynamic drag force under flowing conditions. Thus, a new postulate is added to the linear elastic dumbbell theory for dilute polymer solutions discussed in §8.1.2, namely, that the surrounding, flow-oriented dumbbells in a concentrated suspension would create this anisotropy in the effective molecular drag. According to this postulate, the usually constant, inverse hydrodynamic drag coefficient in kinetic theory, ζ^{-1}, Eq. (8.1-46a), which is basically inversely proportional to the relaxation time, is replaced with an anisotropic, conformation-dependent "mobility tensor," ζ^{-1} [Bird and Wiest, 1985, Eqs. (28), (29)],

$$\begin{aligned}\boldsymbol{\zeta}^{-1} &= \frac{1}{\zeta}\left((1-\alpha)\boldsymbol{\delta} + \alpha\frac{K}{k_BT}c\right) \\ &= \frac{1}{\zeta}((1-\alpha)\boldsymbol{\delta} + \alpha\tilde{c})\ ,\end{aligned} \tag{8.1-85}$$

where α is an empirical constant which lies within the range $0 \leq \alpha \leq 1$. The dependence on the conformation tensor is necessary in order for the mobility tensor to become isotropic (i.e., proportional to the identity tensor) at equilibrium (when $\tilde{\mathbf{c}}=\boldsymbol{\delta}$), since there is no preferred direction of orientation under no-flow conditions (barring external fields). Hence, replacing ζ^{-1} by $\boldsymbol{\zeta}^{-1}$ provided by Eq. (8.1-85) into Eq. (8.1-46a) results in the following constitutive relationship for the conformation tensor [Bird and Wiest, 1985, Eq. (31)]:

$$\check{c}_{\alpha\beta} = -\frac{4k_BT}{\zeta}\left((1-\alpha)\delta_{\alpha\gamma} + \alpha\tilde{c}_{\alpha\gamma}\right)\left(\tilde{c}_{\gamma\beta} - \delta_{\gamma\beta}\right)\ , \tag{8.1-86}$$

which corresponds, when the conformation tensor is eliminated between Eq. (8.1-86) and (8.1.46b), to the constitutive equation proposed by Giesekus [Giesekus, 1982, pp. 72,78]:

$$\sigma_{\alpha\beta} + \lambda\check{\sigma}_{\alpha\beta} + \frac{\alpha}{G_0}\sigma_{\alpha\gamma}\sigma_{\gamma\beta} = 2\eta A_{\alpha\beta}\ . \tag{8.1-87}$$

Note that when α is outside the suggested range, $0 \leq \alpha \leq 1$, stress magnitudes may increase upon cessation of deformation [Larson, 1988, p. 158], an aphysical result.[17]

The Giesekus model, being expressed using a conformation-dependent anisotropic drag (or, equivalently, mobility tensor), Eq. (8.1-85), is very naturally expressed in Hamiltonian form. Indeed, the same

[17]As we shall see in §8.1.4, α must lie in the above range in order to guarantee the satisfaction of the criteria imposed by thermodynamics.

p. 253: in (8.1-85), c and $\tilde{c}$ should be $\mathbf{c}$ and $\tilde{\mathbf{c}}$

Hamiltonian is used as for the Hookean dumbbell, Eqs. (8.1-12,13), and it suffices to substitute the tensorial mobility, Eq. (8.1-85), for the scalar mobility in Eq. (8.1-15)—remember $1/2(\lambda nK) = 2/(n\zeta)$—to find the corresponding phenomenological matrix $\boldsymbol{\Lambda}$ directly as

$$\Lambda_{\alpha\beta\gamma\varepsilon} = \frac{2}{n\zeta}\left[(1-\alpha)(c_{\alpha\gamma}\delta_{\beta\varepsilon} + c_{\alpha\varepsilon}\delta_{\beta\gamma} + c_{\beta\gamma}\delta_{\alpha\varepsilon} + c_{\beta\varepsilon}\delta_{\alpha\gamma}) + \alpha\frac{2nK}{G_0}(c_{\alpha\gamma}c_{\beta\varepsilon} + c_{\alpha\varepsilon}c_{\beta\gamma})\right] . \tag{8.1-88}$$

Again, we see a higher-order correction to the Hookean dumbbell relaxation matrix, (8.1-15), which has a similar form to the relaxation matrix associated with the hydrodynamic interaction. Indeed, the two expressions bear a striking resemblance to each other and a direct comparison of Eqs. (8.1-84) and (8.1-88) with respect to the various orders of the conformation tensor gives two relations for α,

$$1 - \alpha = 1 - \tfrac{3}{4}h\sqrt{\frac{\pi k_B T}{K\mathrm{tr}\mathbf{c}}} , \tag{8.1-89a}$$

and

$$\alpha = - \tfrac{3}{4}h\frac{k_B T}{K\mathrm{tr}\mathbf{c}}\sqrt{\frac{\pi k_B T}{K\mathrm{tr}\mathbf{c}}} , \tag{8.1-89b}$$

resulting in two different values for α, which differ by the dimensionless factor $-(k_B T/K\mathrm{tr}\mathbf{c})$. This discrepancy can be rationalized by considering the nature of the two models involved, i.e., the dumbbell model with hydrodynamic interaction and the Giesekus model. The elastic dumbbell model was developed for dilute solutions, and, as such, the hydrodynamic interaction force between the two beads of the dumbbell acts to increase the separation [Diaz *et al.*, 1989, p. 261; Beris and Edwards, 1990a, p. 74]. In the concentrated solutions and melts for which the Giesekus model was developed, the anisotropic dissipative force tends to push the beads together, as a result of the steric effects of the surrounding molecules [Giesekus, 1982, p. 93; Fan, 1985, p. 263].[18] The sign difference between Eqs. (8.1-89a) and (8.1-89b) accounts for the description of these opposite (higher-order in **c**) effects.

[18]Note that Fan's parameter for describing the degree of anisotropy, σ, is equivalent to $1-\alpha$ of the Giesekus model.

Later Giesekus extended his original model by allowing the mobility tensor to obey an independent relaxation equation [Giesekus, 1985b, Eq. (50)], or to be expressed as a nonlinear (either quadratic or exponential) function of the conformation tensor [Giesekus, 1988]. The first modification clearly requires an additional conformation tensor as an internal variable, and it is discussed in §8.2.2C.2. The second modification can be trivially reproduced by replacing $\boldsymbol{\zeta}^{-1}$ in the definition of the relaxation tensor $\boldsymbol{\Lambda}$, Eq. (8.1-85), accordingly. Finally, in his most recent publication [Giesekus, 1990, §7.3], combining his previous ideas of a nonlinear dependence of the mobility on the conformation with the incorporation of hydrodynamic effects into the model equation through the concept of an "effective flow field" [Schümmer and Otten, 1984], Giesekus proposed the following evolution equation for the conformation tensor **c**:

$$\overset{\nabla}{c}_{\alpha\beta} = -\frac{4k_BT}{\zeta}\Big[e^{-\zeta'(\mathrm{tr}\tilde{c}\,-\,3)}(1-\alpha)\delta_{\alpha\gamma} + \alpha\tilde{c}_{\alpha\gamma}) - \kappa'(\tilde{c}_{\alpha\gamma} - \tfrac{1}{3}\mathrm{tr}\tilde{c}\,\delta_{\alpha\gamma})\Big]\Big(\tilde{c}_{\gamma\beta} - \delta_{\gamma\beta}\Big)\,, \tag{8.1-90}$$

where ζ' and κ' are two new dimensionless coefficients with typical values of O(0.1). Note that the original Giesekus model is recovered from Eq. (8.1-90) for $\zeta'=\kappa'=0$. Through a direct comparison of Eqs. (8.1-86) and (8.1-90), the mobility tensor of the new model is evaluated as

$$\boldsymbol{\zeta}^{-1} = \frac{1}{\zeta}\Big[e^{-\zeta'(\mathrm{tr}\tilde{c}\,-\,3)}(1-\alpha)\boldsymbol{\delta} + \alpha\tilde{c}) - \kappa'(\tilde{c} - \tfrac{1}{3}\mathrm{tr}\tilde{c}\,\boldsymbol{\delta})\Big]\,, \tag{8.1-91}$$

which once again can be used to trivially reproduce this modified Giesekus model within the bracket formalism by replacing $\boldsymbol{\zeta}^{-1}$ in the definition of Eq. (8.1-85) and by following the procedure outlined immediately after that. Thus, two of the three most recent modifications of the Giesekus model essentially reduce to a trivial nonlinear generalization of the mobility tensor appearing in the original version.

Other modifications of the Giesekus model are possible. For example, Wiest [1989] has introduced into the original Giesekus formulation a finitely extensible nonlinear elastic spring force in an effort to better capture the extensional characteristics of polymer melts. His resulting equation, obtained by introducing the FENE-P expression into the kinetic theory equivalent to the Giesekus model as given by Eq. (8.1-86), can be trivially reproduced within the context of the bracket formalism if instead of using the Hookean expression for the enthalpic (internal degrees of freedom) contribution to the free energy (Hamiltonian) provided by Eq. (8.1-12) we simply use its FENE-P equivalent, Eq. (8.1-55). The

p. 255: in (8.1-91), last $\tilde{c}$ should be $\tilde{c}$

resulting equations, although similar in complexity to the ones corresponding to the original Giesekus model (simply involving terms that are cubic in the stress in addition to the quadratic ones), seem to have a substantially better capability to fit experimental data, especially extensional data [IUPAC polymer melt A; Wiest, 1989, p. 10]. However, the relaxation times needed to fit the extensional and the shear data were different. This need for different parameter values to fit different viscometric data, manifested especially for single relaxation models in the need to use different relaxation times, is common in rheology and has to be attributed to inherent model deficiencies. The major deficiencies of the single relaxation time models are discussed in section §8.2.1.

B. The Encapsulated Dumbbell Model

Bird and DeAguiar [1983] developed a model in order to account phenomenologically for the effect of the anisotropic, intermolecular interactions which occur in concentrated polymer solutions or polymer melts. This model assumes that an elastic dumbbell in a concentrated suspension, due to steric effects, is allowed a greater range of motion in the direction parallel to its connector rather than perpendicular to it. In effect, the neighboring dumbbells create a "capsule" around any particular dumbbell, which acts to constrain its motion.

Under the above hypothesis, Bird and DeAguiar phenomenologically introduced two anisotropic tensors into the usual diffusion equation of kinetic theory: one is a modified form of Stokes's law, expressed as [Bird *et al.*, 1987b, pp. 98,99]

$$\boldsymbol{\zeta}^{-1} = \frac{1}{\zeta}\left(\sigma\boldsymbol{\delta} + (1-\sigma)\frac{\mathbf{RR}}{\mathbf{R}\cdot\mathbf{R}}\right), \tag{8.1-92}$$

and the other is an anisotropic Brownian force,

$$\boldsymbol{\xi}^{-1} = \beta\boldsymbol{\delta} + (\alpha-\beta)\frac{\mathbf{RR}}{\mathbf{R}\cdot\mathbf{R}}, \tag{8.1-93}$$

with the constraint $\alpha + 2\beta = 3$. In order to preserve the physical integrity of the tensors $\boldsymbol{\zeta}^{-1}$ and $\boldsymbol{\xi}^{-1}$, it is apparent that the parameters β and σ should be restricted to the ranges $1 \le \beta \le 1.5$ and $0 \le \sigma \le 1$. σ must lie within this range in order for $\boldsymbol{\zeta}^{-1}$ to be positive definite, assuming that the motion of the dumbbell parallel to its axis is greater than the motion perpendicular to its axis. The range of β can likewise be determined by considering the anisotropic Brownian force [Bird and DeAguiar, 1983, p. 153],

$$\mathbf{F}_i = -\frac{k_B T}{\Psi}\frac{\partial}{\partial \mathbf{r}_i}\cdot\left\{\left[\alpha\frac{\mathbf{RR}}{\mathbf{R}\cdot\mathbf{R}} + \beta\left(\boldsymbol{\delta} - \frac{\mathbf{RR}}{\mathbf{R}\cdot\mathbf{R}}\right)\right]\Psi\right\}, \tag{8.1-94}$$

where $\Psi=\Psi(\mathbf{r}_1,\mathbf{r}_2,t)$ is the position-dependent distribution function and the subscript i represents bead one or two. With the anisotropy in the dumbbell motion restricting α to the range $0\leq\alpha\leq 1$, the linear relationship between α and β, $\alpha+2\beta=3$, implies the above-mentioned range on β.[19] When $\alpha=\beta=1$ and $\sigma=1$, the usual simple dumbbell model with an isotropic drag coefficient results.

By making the appropriate modifications in the diffusion equation in accordance with the anisotropic forces mentioned above, Bird and DeAguiar were able to obtain the closed-form set of equations in terms of the second moment of the distribution function in the usual fashion by averaging [Bird and DeAguiar, 1983, p. 155]:

$$\begin{aligned}\overset{\nabla}{c}_{\alpha\beta} = {} & \frac{4k_BT}{\zeta}\sigma\beta\delta_{\alpha\beta} - \frac{4}{\zeta}\left\langle R_\alpha F^c_\beta\right\rangle + \\ & \frac{4k_BT}{\zeta}(\alpha+2\beta-3\sigma\beta)\left\langle\frac{R_\alpha R_\beta}{\mathbf{R}\cdot\mathbf{R}}\right\rangle,\end{aligned} \tag{8.1-95a}$$

and

$$\begin{aligned}\sigma_{\alpha\beta} = {} & n\left\langle R_\alpha F^c_\beta\right\rangle - 2nk_BT(\alpha-\beta)\left\langle\frac{R_\alpha R_\beta}{\mathbf{R}\cdot\mathbf{R}}\right\rangle \\ & - \tfrac{1}{3}nk_BT(4\beta-\alpha)\delta_{\alpha\beta},\end{aligned} \tag{8.1-95b}$$

where $\mathbf{F}^c$ is the elastic connector force vector.[20] Shortly after the introduction of this model into the literature, Fan [1985] presented a series of calculations of the diffusion equation in simple flows, comparing his results to those of DeAguiar [1983], performed using the averaged

[19]An analysis by Phan-Thien and Atkinson [1985] indicates that this model violates the fluctuation-dissipation theorem through the introduction of the "smoothed out" Brownian force, which takes no account of any change in the Maxwellian distribution induced thereby. Thus, there is some concern as to the suitability of considering the encapsulated dumbbell model as anything more than purely phenomenological. Nevertheless, for simple flows it has been shown that, for the range of the parameters examined, there is always one base steady solution which is evolutionary [Calderer *et al.*, 1989, Table 1]—see also §8.1.6. In the following, we shall show here how arguments similar to those above may be used to derive a similar (yet slightly modified) set of model equations.

[20]Although the original model was developed in terms of the FENE-P connector force vector, in this subsection we shall use the Hookean force vector in order to keep the resulting equations analogous to those of preceding models.

equations, (8.1-95a,b). Fan found similar qualitative behavior between the two approaches (see Fan, 1985, pp. 264,65), indicating that the averaging caused no significant loss of the physical information contained within the diffusion equation *for the cases studied*.

This model is different from all of the models previously discussed in that the Hamiltonian in this case is altered due to the assumption of the anisotropic Brownian motion. Therefore, although the elastic potential, E_ϕ, may still be written as in the simple dumbbell cases, the entropic contribution to H_m is no longer based upon a spherically symmetric probability represented by ln(det$\tilde{c}$). Indeed, we must add a (phenomenological) weighting factor to the isotropic entropy expression, (8.1-13), which measures the degree of anisotropy in the permissible dumbbell conformations. Let us choose this weighting factor by considering a term which is directly proportional to the logarithm of the ratio of the arithmetic average to the geometric average of the three principal values of the conformation tensor. These principal values are identified with the squares of the molecular dimensions of the three principal axes, and as such should give a fairly accurate description of any anisotropy inherent to the system. Hence we have

$$\ln\left(\frac{\mathrm{tr}\tilde{\mathbf{c}}}{3\sqrt[3]{\det\tilde{\mathbf{c}}}}\right) = \ln\left(\tfrac{1}{3}\mathrm{tr}\tilde{\mathbf{c}}\right) - \tfrac{1}{3}\ln\left(\det\tilde{\mathbf{c}}\right) , \qquad (8.1\text{-}96)$$

so that the entropy expression becomes

$$S[\mathbf{c}] = \frac{1}{2}nk_B\int_\Omega \ln(\det\tilde{\mathbf{c}})\,\mathrm{d}^3x + (\alpha-\beta)nk_B\int_\Omega\left[\ln\left(\tfrac{1}{3}\mathrm{tr}\tilde{\mathbf{c}}\right) - \tfrac{1}{3}\ln\left(\det\tilde{\mathbf{c}}\right)\right]\mathrm{d}^3x \, . \qquad (8.1\text{-}97)$$

Therefore, the larger the value of this additional term, the greater the degree of anisotropy in the Brownian force. Note that for an isotropic system, $\tilde{\mathbf{c}}=\boldsymbol{\delta}$, and therefore this weighting factor vanishes. The functional derivative of the Hamiltonian is now

$$\frac{\delta H_m}{\delta c_{\gamma\varepsilon}} = \left(\tfrac{1}{2}nK - (\alpha-\beta)\frac{nk_BT}{\mathrm{tr}\mathbf{c}}\right)\delta_{\gamma\varepsilon} - \tfrac{1}{2}nk_BT\left(1 - \tfrac{2}{3}(\alpha-\beta)\right)c^{-1}_{\gamma\varepsilon} \, . \qquad (8.1\text{-}98)$$

Now that we have specified how the *ansatz* of anisotropic Brownian motion affects the conformational entropy of the system, we must consider the effects of the anisotropic drag upon the relaxation matrix.

Indeed, we would expect from the outset something similar to the mobility tensor of Giesekus, since the *ansatz* behind the phenomenology of the anisotropic drag is essentially the same, and this turns out to be the case:

$$\Lambda_{\alpha\beta\gamma\varepsilon} = \frac{2}{n\zeta}\left[\sigma(c_{\alpha\gamma}\delta_{\beta\varepsilon} + c_{\alpha\varepsilon}\delta_{\beta\gamma} + c_{\beta\gamma}\delta_{\alpha\varepsilon} + c_{\beta\varepsilon}\delta_{\alpha\gamma}) \right. \\ \left. + 4\frac{1-\sigma}{\mathrm{tr}\mathbf{c}}c_{\alpha\beta}c_{\gamma\varepsilon}\right] . \tag{8.1-99}$$

Hence the parameter σ is essentially equivalent to $1-\alpha$. Note, however, that the encapsulated dumbbell mobility has the additional dependence on the extension of the dumbbells, represented by $\mathrm{tr}\mathbf{c}$, over the Giesekus model.

Using relation (8.1-98) and the relaxation matrix defined by Eq. (8.1-99), the general equations reduce to the extra stress expression of (8.1-95b) and the evolution equation for the conformation tensor,

$$\overset{\nabla}{c}_{\alpha\beta} = \frac{4k_BT}{\zeta}\sigma\beta\delta_{\alpha\beta} - \frac{4K}{\zeta}c_{\alpha\beta} \\ + \frac{12k_BT}{\zeta}(1-\sigma\beta)\frac{c_{\alpha\beta}}{\mathrm{tr}\mathbf{c}} \\ + \frac{4k_BT}{\zeta}\sigma(1-\beta)\left(3\frac{c_{\alpha\beta}}{\mathrm{tr}\mathbf{c}} - \delta_{\alpha\beta}\right) , \tag{8.1-100}$$

where the constraint $\alpha+2\beta=3$ was used. The first three terms on the right-hand side of the above expression comprise exactly the evolution equation for the second moment of the encapsulated dumbbell model, (8.1-95a); however, an additional term arises in the conformation tensor description. Although at equilibrium and for slow flows, when $\mathbf{c}=\boldsymbol{\delta}$, the contribution of this term is negligible, it can play a significant role as $\mathbf{c}$ becomes arbitrarily large. In effect, this term arises in the conformation tensor theory because of the form of the averages obtained in the transferral of information between the various levels of description. Specifically, this term arises as a result of our expression of the conformational entropy of the anisotropic Brownian force. Indeed, when $\alpha=\beta=1$, the Brownian force is isotropic and the two expressions, (8.1-95a,100), are congruent. When $\alpha=\beta=\sigma=1$, the elastic dumbbell model again results. Since the two evolution equations were different, Beris and Edwards [1990b, pp. 519-22] solved both systems for steady shear and extensional flows, and found similar qualitative behavior between the two systems of equations. However, for more complex flows, no conclusions could be reached.

C. The Leonov Model

In addition to the phenomenological spring/dashpot, network theory, and the kinetic theory models discussed in the previous sections, we now wish to consider a fourth type of differential viscoelastic fluid models: those founded thoroughly in continuum mechanics and formalized through irreversible thermodynamics. This type of models is quite extensive, however, so we shall content ourselves with studying one particular example. We chose the Leonov model as our example because of its wide popularity and applicability. The Leonov model was developed after Eckart's pioneering work on the theory of the anelastic fluid [Eckart, 1948a,b]. According to Eckart's original idea, the elastic strain deformation is measured not from a fixed (constant) and equal-for-all-points (global) relaxed state, but from a variable one, both in time and space. The more the reference strain follows the actual deformation strain, the more fluid-like is the material behavior. This is determined by constitutive equations developed within the context of irreversible thermodynamics using both the total deformation and the elastic strain measure as independent variables [Eckart, 1948a, p. 379].

In the Leonov model [Leonov, 1976], the structural parameter tensor $\mathbf{c}$ is identified with the "elastic Finger strain tensor," typically denoted as $\mathbf{C}^{-1}(\mathbf{x},t)$, which changes with time only because of the recoverable "elastic deformation" of the material. For an incompressible elastic medium, the volume is preserved and hence $\det(\mathbf{C}^{-1})=1$. Leonov used this condition explicitly as a constraint for the internal variable $\mathbf{c}$ in his development. Although Leonov tried to justify his assumption against criticisms published in the literature [see, for example, Giesekus, 1982, p. 78] through arguments relating the internal variable $\mathbf{c}$ with the macromolecular stretching averaged over all the polymer molecules [Leonov, 1987, pp. 16,17], a rigorous proof for its validity based on thermodynamics, continuum mechanics, kinetic theory, or other fundamental and widely acceptable principles is still lacking. Thus, we have chosen to implement here Leonov's *ansatz* using the milder condition that the Hamiltonian is a homogeneous functional of $\mathbf{c}$, of degree of homogeneity zero, without explicitly imposing any restrictions on $\det\mathbf{c}$. It is shown that the combination of this requirement, and the assumption that $\det\mathbf{c}=1$ at equilibrium (which can always be imposed using an appropriate scaling) is equivalent to Leonov's constraint, $\det\mathbf{c}=1$, since according to the resulting evolution equations det **c** is conserved. Therefore, this formulation approach places the Leonov *ansatz* as a constitutive rather than a physical requirement, which we consider as more appropriate given the arbitrariness inherent to the physical interpretation for **c**. Furthermore, since no constraints are *a priori* placed on the tensorial parameter **c**, we do not have to worry about possible constraint-induced changes upon the Poisson bracket definition, as in §5.5.3.

The Hamiltonian functional, as a homogeneous function of zero degree with respect to **c**, can be represented as a function of the scaled first and the ratio of the second and the third invariants of **c**, I_1, and I_{-1}, respectively:

$$H_m(\mathbf{M},\mathbf{c}) \equiv \int_\Omega \frac{1}{2\rho}\mathbf{M}\cdot\mathbf{M}\,d^3x + \int_\Omega W(a_1,a_{-1})\,d^3x, \tag{8.1-101}$$

where W is the free energy (elastic potential) which depends on the scaled invariants of **c**

$$a_1 \equiv \frac{I_1}{\sqrt[3]{\det\mathbf{c}}}, \quad a_{-1} \equiv I_{-1}\sqrt[3]{\det\mathbf{c}}. \tag{8.1-102}$$

The assumption of the dependence of the elastic free energy density W on the scaled invariants only allows us to maintain the integrity of the Leonov *ansatz* without imposing explicitly the constraint det**c**=1. Note that W is never explicitly specified in this theory, although particular cases have been investigated. I_i in the above expressions are the usual invariants

$$I_1 \equiv \sum_{i=1}^{3}\lambda_i = \mathrm{tr}\mathbf{c}, \tag{8.1-103a}$$

$$I_2 \equiv \sum_{i=1}^{3}\sum_{j=1;\, j\neq i}^{3}\lambda_i\lambda_j = \frac{1}{2}\left[(\mathrm{tr}\mathbf{c})^2 - \mathrm{tr}(\mathbf{c}\cdot\mathbf{c})\right], \tag{8.1-103b}$$

$$I_3 \equiv \prod_{i=1}^{3}\lambda_i = \det\mathbf{c}, \tag{8.1-103c}$$

$$I_{-1} \equiv \sum_{i=1}^{3}\frac{1}{\lambda_i} = \mathrm{tr}\mathbf{c}^{-1} = \frac{I_2}{I_3}, \tag{8.1-103d}$$

where λ_i is the i-th eigenvalue of the tensor **c**. Because of the symmetry of **c**, we are assured that all of the λ_i are real.

In terms of the Hamiltonian defined by Eq. (8.1-101), we have

$$\begin{aligned}\frac{\delta H_m}{\delta c_{\alpha\beta}} = \frac{\partial W}{\partial a_1}\frac{1}{\sqrt[3]{\det c}}\left(\delta_{\alpha\beta} - \frac{1}{3}I_1 c^{-1}_{\alpha\beta}\right) \\ + \frac{\partial W}{\partial a_{-1}}\frac{\sqrt[3]{\det c}}{\det c}\Big(\operatorname{tr}c\,\delta_{\alpha\beta} - c_{\alpha\beta} \\ + \tfrac{1}{3}\det c\, I_{-1} c^{-1}_{\alpha\beta} \\ - \tfrac{1}{2}\left[(\operatorname{tr}c)^2 - \operatorname{tr}(\mathbf{c}\cdot\mathbf{c})\right]c^{-1}_{\alpha\beta}\Big)\ ,\end{aligned} \tag{8.1-104}$$

which vanishes at equilibrium ($\mathbf{c}=\boldsymbol{\delta}$). Therefore, following (8.1-7c), the corresponding extra stress tensor is given as

$$\begin{aligned}\sigma_{\alpha\beta} &= 2\frac{\partial W}{\partial a_1}\frac{1}{\sqrt[3]{\det c}}\left(c_{\alpha\beta} - \frac{1}{3}I_1\delta_{\alpha\beta}\right) \\ &\quad + 2\frac{\partial W}{\partial a_{-1}}\frac{\sqrt[3]{\det c}}{\det c}\Big(\operatorname{tr}c\, c_{\alpha\beta} - c_{\alpha\gamma}c_{\gamma\beta} \\ &\quad + \tfrac{1}{3}\det c\ I_{-1}\delta_{\alpha\beta} \\ &\quad - \tfrac{1}{2}\left[(\operatorname{tr}c)^2 - \operatorname{tr}(\mathbf{c}\cdot\mathbf{c})\right]\delta_{\alpha\beta}\Big) \\ &= 2\frac{\partial W}{\partial a_1}\frac{1}{\sqrt[3]{\det c}}c_{\alpha\beta} - 2\frac{\partial W}{\partial a_{-1}}\sqrt[3]{\det c}\, c^{-1}_{\alpha\beta} \\ &\quad - \frac{2}{3}\left(\frac{\partial W}{\partial a_1}\frac{I_1}{\sqrt[3]{\det c}} - \frac{\partial W}{\partial a_{-1}}\sqrt[3]{\det c}\, I_{-1}\right)\delta_{\alpha\beta}\ ,\end{aligned} \tag{8.1-105}$$

where the Cayley/Hamilton identity multiplied by $\mathbf{c}^{-1}$,

$$\mathbf{c}\cdot\mathbf{c} - I_1\mathbf{c} + I_2\boldsymbol{\delta} - I_3\mathbf{c}^{-1} = \mathbf{0}\ , \tag{8.1-106}$$

was used to simplify the original expression for Eq. (8.1-105). Note that the extra stress of Eq. (8.1-105) vanishes at equilibrium, and reduces to Leonov's stress expression within an isotropic (pressure) term when detc=1.

The corresponding evolution equation for the deformation tensor, $\mathbf{c}=\mathbf{C}^{-1}$, is then obtained by setting

$$\Lambda_{\alpha\beta\gamma\varepsilon} = \frac{2}{\lambda G_0}\left[c_{\alpha\varepsilon}c_{\gamma\beta} + c_{\alpha\gamma}c_{\beta\varepsilon} - \tfrac{2}{3}c_{\varepsilon\gamma}c_{\alpha\beta}\right], \tag{8.1-107}$$

from which

$$\lambda \overset{\nabla}{c}_{\alpha\beta} = - D^p_{\alpha\gamma}c_{\gamma\beta} - c_{\alpha\gamma}D^p_{\gamma\beta}, \tag{8.1-108}$$

where

$$\begin{aligned} G_0 D^p_{\alpha\beta} &= 2\frac{\partial W}{\partial a_1}\left(c_{\alpha\beta} - \tfrac{1}{3} I_1 \delta_{\alpha\beta}\right) \\ &\quad - 2\frac{\partial W}{\partial a_{-1}}\left(c^{-1}_{\alpha\beta} - \tfrac{1}{3} I_{-1}\delta_{\alpha\beta}\right). \end{aligned} \tag{8.1-109}$$

Note that the usually assigned constraint, det**c**=1, is not imposed on the conformation tensor explicitly. Rather, it is enforced implicitly on **c** by the evolution equation and the initial condition; i.e., at equilibrium, **c**=$\boldsymbol{\delta}$. Indeed, using the properties of the determinant [Truesdell and Noll, 1965, p. 26; Hulsen, 1990, Eq. (14)], the time evolution for det**c**, is

$$\begin{aligned} \frac{d(\det\mathbf{c})}{dt} &= \frac{d(\det\mathbf{c})}{d\mathbf{c}} : \frac{d\mathbf{c}}{dt} \\ &= \det\mathbf{c}\ \mathrm{tr}\left(\mathbf{c}^{-1}\cdot\frac{d\mathbf{c}}{dt}\right) \\ &= \det\mathbf{c}\ c^{-1}_{\beta\alpha}\frac{dc_{\alpha\beta}}{dt}, \end{aligned} \tag{8.1-110}$$

where d·/dt is the *material* or *substantial derivative* defined by (6.4-4). However, according to Eq. (8.1-108) and the definition of the upper-convected time derivative, Eq. (8.1-11), we have

$$\begin{aligned} c^{-1}_{\beta\alpha}\frac{dc_{\alpha\beta}}{dt} &= c^{-1}_{\beta\alpha}\Big(c_{\alpha\gamma}\nabla_\gamma \upsilon_\beta + c_{\gamma\beta}\nabla_\gamma \upsilon_\alpha \\ &\quad - \frac{1}{\lambda}\left(D^p_{\alpha\gamma}c_{\gamma\beta} + c_{\alpha\gamma}D^p_{\gamma\beta}\right)\Big) \\ &= 2\nabla_\gamma \upsilon_\gamma - \frac{2}{\lambda}D^p_{\gamma\gamma} = 0, \end{aligned} \tag{8.1-111}$$

since both $\boldsymbol{\nabla}\boldsymbol{\upsilon}$ and $\mathbf{D}^p$ (the latter through its definition, Eq. (8.1-109)) are traceless. Thus, as a natural consequence of the evolution equation for **c**, Eq. (8.1-108), det**c** remains unchanged with time: det**c**=1. Also note that for the simplest Leonov constitutive equation, $W = \tfrac{1}{2}G_0 I_1$, for which $\boldsymbol{\sigma} = G_0\mathbf{c}$ to within an isotropic term. For this free energy density and for α=1/2, the Leonov model reduces to the Giesekus model when only incompressible planar deformations are involved (for which $\mathrm{tr}\mathbf{c}=\mathrm{tr}\mathbf{c}^{-1}$) [Larson, 1988, p. 162]. Finally, note that viscous stress contributions, isotropic or non-isotropic (conformation dependent), can also be incorporated into the above model, if desired, simply by setting $\mathbf{Q}=\mathbf{Q}(\mathbf{c})\neq\mathbf{0}$ in (8.1-7c), where $\mathbf{Q}$ satisfies Onsager symmetry and stress symmetry criteria [Leonov, 1987, pp. 26-32].

The relaxation matrices of all of the models discussed in this and previous sections of §8.1 are presented in Table 8.1. This table may be used to appreciate the similarities and differences between the various constitutive relationships, despite the widely varying approaches used in their development. (We shall discuss this topic more thoroughly in §8.1.5.)

8.1.4 *Dissipation and admissibility criteria for single-mode viscoelastic fluids*

Within the flowing, incompressible, viscoelastic fluid, irreversible effects occur as the kinetic energy dissipates into heat through viscous friction and the conformational potential energy and entropy relax in time through rearrangements of the molecular structure. This dissipation must obey the laws of non-equilibrium thermodynamics as described in Part I. Consequently, we must see a degradation of the mechanical energy of the system, which, for uncoupled variables $\mathbf{M}$ and $\tilde{\mathbf{c}}$,[21] results in the thermodynamic admissability criteria $\eta_s \geq 0$ (see §7.2.2) and

$$-\,[H_m,H_m] = \int_\Omega \Lambda_{\alpha\beta\gamma\varepsilon}\frac{\delta A}{\delta \tilde{c}_{\alpha\beta}}\frac{\delta A}{\delta \tilde{c}_{\gamma\varepsilon}}\mathrm{d}^3x \geq 0 \;. \tag{8.1-112}$$

For a typical model, as represented by Eqs. (8.1-7b,c), this inequality can easily shown to be equivalent to the dissipative expression of Sarti and Marrucci [1973, p. 1055],

$$\boldsymbol{\sigma}:\nabla\boldsymbol{v} - \frac{\mathrm{d}A[\tilde{\mathbf{c}}]}{\mathrm{d}t} \geq 0 \;, \tag{8.1-113}$$

where $\mathrm{d}\cdot/\mathrm{d}t$ is the *material* or *substantial derivative* defined by (6.4-4). In the following, we will demonstrate the equivalence for the particular case where $\mathbf{L}$ is given by Eq. (8.1-26) (i.e., for a general viscoelastic model where a mixed-convected time derivative is utilized) since this case

[21]In this subsection, we shall find it convenient to work in terms of the dimensionless conformation tensor, $\tilde{\mathbf{c}}$. Note that this redefinition only affects the constant coefficients involved in the relaxation matrices, $\boldsymbol{\Lambda}$, corresponding to the various models.

Model	$\Lambda_{\alpha\beta\gamma\varepsilon}(\mathbf{c})$
Elastic dumbbell (UCM, Oldroyd-A, B)	$\frac{1}{2\lambda nK}\left(c_{\alpha\gamma}\delta_{\beta\varepsilon}+[perm.]\right)$
Elastic dumbbell with hydrodynamic interactions	$\frac{1}{2\lambda nK}\left(c_{\alpha\gamma}\left(\delta_{\beta\varepsilon}-\frac{3}{4}h\sqrt{\frac{\pi k_BT}{K\mathrm{tr}c}}\,(\delta_{\beta\varepsilon}+\frac{c_{\beta\varepsilon}}{\mathrm{tr}c})\right)+[perm.]\right)$
Giesekus	$\frac{1}{2\lambda nK}\left(c_{\alpha\gamma}\left((1-\alpha)\delta_{\beta\varepsilon}+\frac{\alpha K}{k_BT}c_{\beta\varepsilon}\right)+[perm.]\right)$
Phan-Thien/ Tanner (linear)	$\frac{1}{2\lambda nK}\left(1+\frac{\varepsilon}{a}\mathrm{tr}(\frac{K}{k_BT}\mathbf{c}-\boldsymbol{\delta})\right)\left(c_{\alpha\gamma}\delta_{\beta\varepsilon}+[perm.]\right)$
FENE-P	$\frac{1}{2\lambda nK}\left(c_{\alpha\gamma}\delta_{\beta\varepsilon}+[perm.]\right)$ *or any other form*
Modified encapsulated dumbbell	$\frac{1}{2\lambda nK}\left(\sigma c_{\alpha\gamma}\delta_{\beta\varepsilon}+4\frac{(1-\sigma)}{\mathrm{tr}c}c_{\alpha\beta}c_{\gamma\varepsilon}+[perm.]\right)$
Leonov	$\frac{1}{2\lambda nK}\left(c_{\alpha\varepsilon}c_{\gamma\beta}-\frac{2}{3}c_{\alpha\beta}c_{\gamma\varepsilon}+[perm.]\right)$
Extended White/ Metzner	$\frac{1}{2\lambda nK}\left(\frac{\mathrm{tr}\tilde{c}}{3}\right)^{-k}\left(c_{\alpha\gamma}\delta_{\beta\varepsilon}+[perm.]\right)$

Table 8.1: The relaxation matrices for various viscoelastic fluid models. (Adapted from [Beris and Edwards, 1990b, p. 526].)

includes all models investigated in the present work. If wished, the equivalence can be proven for the most general case in a similar fashion. In Eq. (8.1-113) let us replace $\boldsymbol{\sigma}$ by its expression for a general viscoelastic model utilizing the mixed convected derivative, as provided by Eq. (8.1-27a), and $\mathrm{d}A/\mathrm{d}t$ by $(\delta A/\delta\tilde{c}_{\alpha\beta})(\mathrm{d}\tilde{c}_{\alpha\beta}/\mathrm{d}t)$, thereby obtaining

$$\left((a+1)\,\tilde{c}_{\beta\gamma}\frac{\delta A}{\delta\tilde{c}_{\gamma\alpha}}+(a-1)\tilde{c}_{\alpha\gamma}\frac{\delta A}{\delta\tilde{c}_{\gamma\beta}}\right)\nabla_\beta\upsilon_\alpha - \frac{\delta A}{\delta\tilde{c}_{\alpha\beta}}\frac{\mathrm{d}\tilde{c}_{\alpha\beta}}{\mathrm{d}t}\geq 0\;. \tag{8.1-114}$$

Then, making the substitution implied by the general evolution equation for the conformation tensor, Eq. (8.1-7b), using (consistently with the stress substitution) the value corresponding to the mixed convected derivative for the tensor **L**, as given by Eq. (8.1-26)—see also the evolution equation for the Johnson/Segalman model, Eq. (8.1-27b)— yields

$$\begin{aligned}&\left((a+1)\tilde{c}_{\beta\gamma}\frac{\delta A}{\delta\tilde{c}_{\gamma\alpha}}+(a-1)\tilde{c}_{\alpha\gamma}\frac{\delta A}{\delta\tilde{c}_{\gamma\beta}}\right)\nabla_\beta\upsilon_\alpha\\ &-\frac{\delta A}{\delta\tilde{c}_{\alpha\beta}}\left(\frac{a+1}{2}\left(\tilde{c}_{\alpha\gamma}\nabla_\gamma\upsilon_\beta+\tilde{c}_{\beta\gamma}\nabla_\gamma\upsilon_\alpha\right)\right.\\ &\left.+\frac{a-1}{2}\left(\tilde{c}_{\alpha\gamma}\nabla_\beta\upsilon_\gamma+\tilde{c}_{\beta\gamma}\nabla_\alpha\upsilon_\gamma\right)-\Lambda_{\alpha\beta\gamma\varepsilon}\frac{\delta A}{\delta\tilde{c}_{\gamma\varepsilon}}\right)\;\geq\;0\;,\end{aligned} \tag{8.1-115}$$

or since $\tilde{\mathbf{c}}\cdot\delta A/\delta\tilde{\mathbf{c}}$ is symmetric,[22]

$$\frac{\delta A}{\delta\tilde{c}_{\alpha\beta}}\Lambda_{\alpha\beta\gamma\varepsilon}\frac{\delta A}{\delta\tilde{c}_{\gamma\varepsilon}}\geq 0\;. \tag{8.1-116}$$

A necessary and sufficient condition for (8.1-112) to be satisfied in the flow domain Ω is that the integrand is a non-negative quantity. This requirement places restrictions on the permissible forms of the fourth-

[22]This assumption is not as restrictive as it appears. It is satisfied by all models examined here. In fact, it is always true if the only anisotropic tensor present in the model is a symmetric tensor **c**: then, all second-rank tensorial quantities, by necessity, have to be represented as a linear superposition of δ, **c** and **c**$\cdot$**c**—see also Eq. (8.1-117) below— each one of which is a symmetric tensor.

rank phenomenological matrix $\boldsymbol{\Lambda}$, which, for a general form of the Helmholtz free energy, A, are quite complex. However, these can be considerably simplified in the present case where the only structural parameter tensor characterizing the material behavior is the conformation tensor $\tilde{\mathbf{c}}$. In this case, using the Cayley/Hamilton theorem, (8.1-106), the most general admissible expressions for $\delta A/\delta\tilde{\mathbf{c}} \equiv \delta H_m/\delta\tilde{\mathbf{c}}$ and $\boldsymbol{\Lambda}$ are

$$\frac{\delta H_m}{\delta \tilde{c}_{\alpha\beta}} = \omega_1 \delta_{\alpha\beta} + \omega_2 \tilde{c}_{\alpha\beta} + \omega_3 \tilde{c}_{\alpha\gamma}\tilde{c}_{\gamma\beta} \,, \tag{8.1-117}$$

and

$$\begin{aligned}
\Lambda_{\alpha\beta\gamma\varepsilon} = \;& b_1\delta_{\alpha\beta}\delta_{\gamma\varepsilon} && + b_2\delta_{\alpha\gamma}\delta_{\beta\varepsilon} && + b_3\delta_{\alpha\varepsilon}\delta_{\beta\gamma} \\
& + b_4\tilde{c}_{\alpha\beta}\delta_{\gamma\varepsilon} && + b_5\tilde{c}_{\alpha\gamma}\delta_{\beta\varepsilon} && + b_6\tilde{c}_{\alpha\varepsilon}\delta_{\beta\gamma} \\
& + b_7\delta_{\alpha\beta}\tilde{c}_{\gamma\varepsilon} && + b_8\delta_{\alpha\gamma}\tilde{c}_{\beta\varepsilon} && + b_9\delta_{\alpha\varepsilon}\tilde{c}_{\beta\gamma} \\
& + b_{10}\tilde{c}_{\alpha\beta}\tilde{c}_{\gamma\varepsilon} && + b_{11}\tilde{c}_{\alpha\gamma}\tilde{c}_{\beta\varepsilon} && + b_{12}\tilde{c}_{\alpha\varepsilon}\tilde{c}_{\beta\gamma} \\
& + b_{13}\tilde{c}_{\alpha\mu}\tilde{c}_{\mu\beta}\delta_{\gamma\varepsilon} && + b_{14}\tilde{c}_{\alpha\mu}\tilde{c}_{\mu\gamma}\delta_{\beta\varepsilon} && + b_{15}\tilde{c}_{\alpha\mu}\tilde{c}_{\mu\varepsilon}\delta_{\beta\gamma} \\
& + b_{16}\delta_{\alpha\beta}\tilde{c}_{\gamma\mu}\tilde{c}_{\mu\varepsilon} && + b_{17}\delta_{\alpha\gamma}\tilde{c}_{\beta\mu}\tilde{c}_{\mu\varepsilon} && + b_{18}\delta_{\alpha\varepsilon}\tilde{c}_{\beta\mu}\tilde{c}_{\mu\gamma} \\
& + b_{19}\tilde{c}_{\alpha\mu}\tilde{c}_{\mu\beta}\tilde{c}_{\gamma\varepsilon} && + b_{20}\tilde{c}_{\alpha\mu}\tilde{c}_{\mu\gamma}\tilde{c}_{\beta\varepsilon} && + b_{21}\tilde{c}_{\alpha\mu}\tilde{c}_{\mu\varepsilon}\tilde{c}_{\beta\gamma} \\
& + b_{22}\tilde{c}_{\alpha\beta}\tilde{c}_{\gamma\mu}\tilde{c}_{\mu\varepsilon} && + b_{23}\tilde{c}_{\alpha\gamma}\tilde{c}_{\beta\mu}\tilde{c}_{\mu\varepsilon} && + b_{24}\tilde{c}_{\alpha\varepsilon}\tilde{c}_{\beta\mu}\tilde{c}_{\mu\gamma} \\
& + b_{25}\tilde{c}_{\alpha\eta}\tilde{c}_{\eta\beta}\tilde{c}_{\gamma\mu}\tilde{c}_{\mu\varepsilon} && + b_{26}\tilde{c}_{\alpha\eta}\tilde{c}_{\eta\gamma}\tilde{c}_{\beta\mu}\tilde{c}_{\mu\varepsilon} && + b_{27}\tilde{c}_{\alpha\eta}\tilde{c}_{\eta\varepsilon}\tilde{c}_{\beta\mu}\tilde{c}_{\mu\gamma} \,,
\end{aligned} \tag{8.1-118}$$

where the coefficients ω_i and b_i are, in general, functions of the invariants of $\tilde{\mathbf{c}}$. Since the Onsager relations require that $\Lambda_{\alpha\beta\gamma\varepsilon}=\Lambda_{\gamma\varepsilon\alpha\beta}$, and since $\delta H_m/\delta\tilde{\mathbf{c}}$ is symmetric, requiring as well that $\Lambda_{\alpha\beta\gamma\varepsilon}=\Lambda_{\beta\alpha\gamma\varepsilon}=\Lambda_{\alpha\beta\varepsilon\gamma}$, it is apparent that only 12 of the 27 b_i are independent. Hence we may rewrite (8.1-118) as

$$\begin{aligned}
\Lambda_{\alpha\beta\gamma\varepsilon} = \;& a_1\delta_{\alpha\beta}\delta_{\gamma\varepsilon} \\
& + a_2(\delta_{\alpha\gamma}\delta_{\beta\varepsilon} + \delta_{\alpha\varepsilon}\delta_{\beta\gamma}) \\
& + a_3(\tilde{c}_{\alpha\beta}\delta_{\gamma\varepsilon} + \delta_{\alpha\beta}\tilde{c}_{\gamma\varepsilon}) \\
& + a_4(\tilde{c}_{\alpha\gamma}\delta_{\beta\varepsilon} + \tilde{c}_{\alpha\varepsilon}\delta_{\beta\gamma} + \delta_{\alpha\gamma}\tilde{c}_{\beta\varepsilon} + \delta_{\alpha\varepsilon}\tilde{c}_{\beta\gamma}) \\
& + a_5\tilde{c}_{\alpha\beta}\tilde{c}_{\gamma\varepsilon} \\
& + a_6(\tilde{c}_{\alpha\gamma}\tilde{c}_{\beta\varepsilon} + \tilde{c}_{\alpha\varepsilon}\tilde{c}_{\beta\gamma}) \\
& + a_7(\tilde{c}_{\alpha\mu}\tilde{c}_{\mu\beta}\delta_{\gamma\varepsilon} + \delta_{\alpha\beta}\tilde{c}_{\gamma\mu}\tilde{c}_{\mu\varepsilon}) \\
& + a_8(\tilde{c}_{\alpha\mu}\tilde{c}_{\mu\gamma}\delta_{\beta\varepsilon} + \tilde{c}_{\alpha\mu}\tilde{c}_{\mu\varepsilon}\delta_{\beta\gamma} \\
& \qquad + \delta_{\alpha\gamma}\tilde{c}_{\beta\mu}\tilde{c}_{\mu\varepsilon} + \delta_{\alpha\varepsilon}\tilde{c}_{\beta\mu}\tilde{c}_{\mu\gamma})
\end{aligned}$$

$$
\begin{aligned}
&+ a_9(\tilde{c}_{\alpha\mu}\tilde{c}_{\mu\beta}\tilde{c}_{\gamma\varepsilon} + \tilde{c}_{\alpha\beta}\tilde{c}_{\gamma\mu}\tilde{c}_{\mu\varepsilon}) \\
&+ a_{10}(\tilde{c}_{\alpha\mu}\tilde{c}_{\mu\gamma}\tilde{c}_{\beta\varepsilon} + \tilde{c}_{\alpha\mu}\tilde{c}_{\mu\varepsilon}\tilde{c}_{\beta\gamma} \\
&\qquad + \tilde{c}_{\alpha\gamma}\tilde{c}_{\beta\mu}\tilde{c}_{\mu\varepsilon} + \tilde{c}_{\alpha\varepsilon}\tilde{c}_{\beta\mu}\tilde{c}_{\mu\gamma}) \\
&+ a_{11}\tilde{c}_{\alpha\eta}\tilde{c}_{\eta\beta}\tilde{c}_{\gamma\mu}\tilde{c}_{\mu\varepsilon} \\
&+ a_{12}(\tilde{c}_{\alpha\eta}\tilde{c}_{\eta\gamma}\tilde{c}_{\beta\mu}\tilde{c}_{\mu\varepsilon} + \tilde{c}_{\alpha\eta}\tilde{c}_{\eta\varepsilon}\tilde{c}_{\beta\mu}\tilde{c}_{\mu\gamma}) \ ,
\end{aligned}
\tag{8.1-119}
$$

where the coefficients a_i constitute an independent subset of the coefficients b_i. Now if we use the expansion defined by Eq. (8.1-117) for $\delta H_m/\delta\tilde{\mathbf{c}}$, we can rewrite (8.1-116) as

$$
\frac{\delta A}{\delta\tilde{c}_{\alpha\beta}}\Lambda_{\alpha\beta\gamma\varepsilon}\frac{\delta A}{\delta\tilde{c}_{\gamma\varepsilon}} = \omega_\alpha\Phi_{\alpha\beta}\omega_\beta \geq 0 \ , \tag{8.1-120}
$$

where $\boldsymbol{\omega}=(\omega_1,\omega_2,\omega_3)^T$ and $\boldsymbol{\Phi}$ is the 3×3 symmetric matrix with components

$$
\begin{aligned}
\Phi_{11} &= \delta_{\alpha\beta}\Lambda_{\alpha\beta\gamma\varepsilon}\delta_{\gamma\varepsilon} \\
\Phi_{12} &= \Phi_{21} = \tfrac{1}{2}\left(\delta_{\alpha\beta}\Lambda_{\alpha\beta\gamma\varepsilon}\tilde{c}_{\gamma\varepsilon} + \tilde{c}_{\alpha\beta}\Lambda_{\alpha\beta\gamma\varepsilon}\delta_{\gamma\varepsilon}\right) \\
\Phi_{13} &= \Phi_{31} = \tfrac{1}{2}\left(\delta_{\alpha\beta}\Lambda_{\alpha\beta\gamma\varepsilon}\tilde{c}_{\gamma\zeta}\tilde{c}_{\zeta\varepsilon} + \tilde{c}_{\alpha\zeta}\tilde{c}_{\zeta\beta}\Lambda_{\alpha\beta\gamma\varepsilon}\delta_{\gamma\varepsilon}\right) \\
\Phi_{22} &= \tilde{c}_{\alpha\beta}\Lambda_{\alpha\beta\gamma\varepsilon}\tilde{c}_{\gamma\varepsilon} \\
\Phi_{23} &= \Phi_{32} = \tfrac{1}{2}\left(\tilde{c}_{\alpha\beta}\Lambda_{\alpha\beta\gamma\varepsilon}\tilde{c}_{\gamma\zeta}\tilde{c}_{\zeta\varepsilon} + \tilde{c}_{\alpha\zeta}\tilde{c}_{\zeta\beta}\Lambda_{\alpha\beta\gamma\varepsilon}\tilde{c}_{\gamma\varepsilon}\right) \\
\Phi_{33} &= \tilde{c}_{\alpha\zeta}\tilde{c}_{\zeta\beta}\Lambda_{\alpha\beta\gamma\varepsilon}\tilde{c}_{\gamma\zeta}\tilde{c}_{\zeta\varepsilon} \ .
\end{aligned}
\tag{8.1-121}
$$

Then a necessary and sufficient condition for the inequality (8.1-120) to be satisfied for every $\boldsymbol{\omega}$ is that $\boldsymbol{\Phi}$ is non-negative definite.

The requirement that $\boldsymbol{\Phi}$ be non-negative definite translates into a series of inequalities in terms of the coefficients a_i. These are

$$
\begin{aligned}
\Phi_{\underline{ii}} &\geq 0 \ , \quad \underline{i}=1,2,3 \ , \\
\Phi_{\underline{ii}}\Phi_{\underline{jj}} - \Phi_{\underline{ij}}\Phi_{\underline{ji}} &\geq 0 \ , \quad \underline{i},\underline{j}=1,2,3 \ , \\
\det\Phi &\geq 0 \ ,
\end{aligned}
\tag{8.1-122}
$$

where the underlining of the indices signifies that the implicit summation over repeated indices is turned off. The conditions involved in Eq. (8.1-121) when expressed in terms of the a_i are quite lengthy. Therefore, in the following, we opt to apply our analysis to the individual cases investigated in the preceding subsections and thus to define the ranges of the parameter values for which the thermodynamic criteria expressed by the above relationships are satisfied. Note that conditions (8.1-122), whereas always sufficient to guarantee a non-negative rate of energy dissipation, are only necessary if one wishes to leave the specification of the system Hamiltonian open. For specific models, it is always simpler, and one obtains less restrictive conditions, if in the calculation of the rate of energy dissipation one considers a specific form for the system Hamiltonian, as, for example, was the case in Eq. (8.1-72). In contrast, here and in the rest of this section, seeking the most general forms of admissible dissipation brackets, we have left the selection of the Hamiltonian in purpose unspecified. Nevertheless, in the cases where this general approach leaves the admissibility question unanswered, the reader should be able to check the specific model that he/she has in mind by direct substitution of the corresponding Hamiltonian without much mathematical trouble (along the lines of the discussion in Section 8.1.2B).

For all models in which the relaxation matrix, $\boldsymbol{\Lambda}$, is proportional to $[c_{\alpha\gamma}\delta_{\beta\varepsilon}$ + permutations], the dissipation bracket $[H_m,H_m]$ leads to the simpler expression

$$[H_m,H_m] \propto \int_\Omega \frac{\delta H_m}{\delta \tilde{c}_{\alpha\beta}} \tilde{c}_{\alpha\gamma} \frac{\delta H_m}{\delta \tilde{c}_{\gamma\beta}} d^3x \propto \int_\Omega \frac{\delta H_m}{\delta \tilde{c}_{\alpha\beta}} \sigma_{\alpha\beta}\, d^3x \,. \tag{8.1-123}$$

These cases correspond to the elastic dumbbell models, Maxwell model, Oldroyd-B, Phan-Thien/Tanner model, etc. For all these cases, the $\boldsymbol{\Phi}$ matrix has the following structure:

$$\Phi \propto \begin{bmatrix} \mathrm{tr}\tilde{c} & \mathrm{tr}(\tilde{c}\cdot\tilde{c}) & \mathrm{tr}(\tilde{c}\cdot\tilde{c}\cdot\tilde{c}) \\ \mathrm{tr}(\tilde{c}\cdot\tilde{c}) & \mathrm{tr}(\tilde{c}\cdot\tilde{c}\cdot\tilde{c}) & \mathrm{tr}(\tilde{c}\cdot\tilde{c}\cdot\tilde{c}\cdot\tilde{c}) \\ \mathrm{tr}(\tilde{c}\cdot\tilde{c}\cdot\tilde{c}) & \mathrm{tr}(\tilde{c}\cdot\tilde{c}\cdot\tilde{c}\cdot\tilde{c}) & \mathrm{tr}(\tilde{c}\cdot\tilde{c}\cdot\tilde{c}\cdot\tilde{c}\cdot\tilde{c}) \end{bmatrix}, \tag{8.1-124}$$

for which all components are non-negative (assuming that the tensor $\tilde{c}$ is non-negative definite). Furthermore, under the same assumption, the following invariants of $\boldsymbol{\Phi}$

$$\mathrm{tr}\Phi = \sum_{i=1}^{3} \left(\lambda_i + \lambda_i^3 + \lambda_i^5\right),$$

$$\begin{aligned} I_2(\Phi) &\equiv \tfrac{1}{2}\left((\mathrm{tr}\Phi)^2 - \mathrm{tr}(\Phi\cdot\Phi)\right) \\ &= \lambda_1\lambda_2(\lambda_1-\lambda_2)^2(1 + (\lambda_1+\lambda_2)^2 + \lambda_1^2\lambda_2^2) \\ &\quad + 2 \textit{ more terms arising from} \\ &\quad \textit{the cyclic permutation } 1\rightarrow 2\rightarrow 3 \end{aligned} \,, \tag{8.1-125}$$

$$\det\Phi = \lambda_1\lambda_2\lambda_3(\lambda_1-\lambda_2)^2(\lambda_2-\lambda_3)^2(\lambda_3-\lambda_1)^2 \,,$$

where λ_i denotes the i-th eigenvalue of $\tilde{c}$, are also non-negative. Therefore, $\boldsymbol{\Phi}$ is non-negative and, as a consequence, a relaxation matrix with components, $\Lambda_{\alpha\beta\gamma\varepsilon}$, proportional to $[\tilde{c}_{\alpha\gamma}\delta_{\beta\varepsilon}$ + permutations] always leads to thermodynamically consistent models, irrespective of the expression for the Hamiltonian, provided that $\tilde{c}$ is non-negative definite.

p. 269: all Φ's should be bold ($\boldsymbol{\Phi}$)

This last requirement has been proven for the Maxwell model [Joseph, 1990, p. 15], but, in general, we need to develop a technique for ascertaining this criterion (see §8.1.5).

For the Leonov model, the relaxation matrix is proportional to $[c_{\alpha\varepsilon}c_{\gamma\beta} - (2/3)c_{\alpha\beta}c_{\gamma\varepsilon}$ + permutations], and this translates into the following components of $\boldsymbol{\Phi}$:

$$\begin{aligned}
\Phi_{11} &= \mathrm{tr}(\mathbf{c}\cdot\mathbf{c}) - \tfrac{1}{3}(\mathrm{tr}\mathbf{c})^2 \\
\Phi_{12} = \Phi_{21} &= \mathrm{tr}(\mathbf{c}\cdot\mathbf{c}\cdot\mathbf{c}) - \tfrac{1}{3}\mathrm{tr}\mathbf{c}\,\mathrm{tr}(\mathbf{c}\cdot\mathbf{c}) \\
\Phi_{13} = \Phi_{31} &= \mathrm{tr}(\mathbf{c}\cdot\mathbf{c}\cdot\mathbf{c}\cdot\mathbf{c}) - \tfrac{1}{3}\mathrm{tr}\mathbf{c}\,\mathrm{tr}(\mathbf{c}\cdot\mathbf{c}\cdot\mathbf{c}) \\
\Phi_{22} &= \mathrm{tr}(\mathbf{c}\cdot\mathbf{c}\cdot\mathbf{c}\cdot\mathbf{c}) - \tfrac{1}{3}\big(\mathrm{tr}(\mathbf{c}\cdot\mathbf{c})\big)^2 \\
\Phi_{23} = \Phi_{32} &= \mathrm{tr}(\mathbf{c}\cdot\mathbf{c}\cdot\mathbf{c}\cdot\mathbf{c}\cdot\mathbf{c}) - \tfrac{1}{3}\mathrm{tr}(\mathbf{c}\cdot\mathbf{c})\,\mathrm{tr}(\mathbf{c}\cdot\mathbf{c}\cdot\mathbf{c}) \\
\Phi_{33} &= \mathrm{tr}(\mathbf{c}\cdot\mathbf{c}\cdot\mathbf{c}\cdot\mathbf{c}\cdot\mathbf{c}\cdot\mathbf{c}) - \tfrac{1}{3}\big(\mathrm{tr}(\mathbf{c}\cdot\mathbf{c}\cdot\mathbf{c})\big)^2 \,.
\end{aligned} \tag{8.1-126}$$

This matrix has the following invariants

$$\begin{aligned}
\mathrm{tr}\,\boldsymbol{\Phi} &= \tfrac{1}{3}\Big((\lambda_1-\lambda_2)^2 + (\lambda_1^2-\lambda_2^2)^2 + (\lambda_1^3-\lambda_2^3)^2 \\
&\qquad + \textit{cyclic permutations } 1\to2\to3\Big)\,, \\
I_2(\boldsymbol{\Phi}) &= \tfrac{1}{3}(\lambda_1-\lambda_2)^2(\lambda_2-\lambda_3)^2(\lambda_3-\lambda_1)^2\Big[1 + (\mathrm{tr}\mathbf{c})^2 \\
&\qquad + (\lambda_1\lambda_2)^2 + (\lambda_2\lambda_3)^2 + (\lambda_3\lambda_1)^2 + 2\det\mathbf{c}\,\mathrm{tr}\mathbf{c}\Big]\,, \\
\det\boldsymbol{\Phi} &= 0\,,
\end{aligned} \tag{8.1-127}$$

written in terms of the eigenvalues and the invariants of $\mathbf{c}$. The last identity in Eq. (8.1-127) may be proven by simply realizing that, via the Cayley/Hamilton theorem, the last column of $\boldsymbol{\Phi}$ can be obtained as a linear superposition of the first two columns:

$$\begin{bmatrix}\Phi_{31}\\ \Phi_{32}\\ \Phi_{33}\end{bmatrix} = I_1(\mathbf{c})\begin{bmatrix}\Phi_{21}\\ \Phi_{22}\\ \Phi_{23}\end{bmatrix} - I_2(\mathbf{c})\begin{bmatrix}\Phi_{11}\\ \Phi_{12}\\ \Phi_{13}\end{bmatrix}. \tag{8.1-128}$$

As with the Maxwell-type models, λ_i is the i-th eigenvalue of $\mathbf{c}$, and is also non-negative. Therefore, all of the invariants of $\boldsymbol{\Phi}$ are non-negative quantities, which implies that $\boldsymbol{\Phi}$ is non-negative definite. As a consequence, a relaxation matrix, $\boldsymbol{\Lambda}$, proportional to $[c_{\alpha\varepsilon}c_{\gamma\beta}-(2/3)c_{\alpha\beta}c_{\gamma\varepsilon}$ + permutations] always leads to thermodynamically consistent models, irrespective of the expression for the Hamiltonian, provided again that $\mathbf{c}$ has positive eigenvalues. Notice that the constraint det(c)=1 automatically implies that $\mathbf{c}$ be positive definite, provided that it is so initially (at rest), since an eigenvalue which is negative must pass through zero, thus voiding the unit-determinant constraint.

For the more elaborate expressions for the relaxation matrix, $\boldsymbol{\Phi}$ becomes too complex to allow a general analysis as above. Rather, we focus our further analysis on the individual model Hamiltonians covered in the preceding subsections and make use of Eq. (8.1-112) directly instead. Again, the final equations are expressed in terms of the eigenvalues and the invariants of the corresponding dimensionless conformation tensor, **c**, λ_i, and I_i, i=1,2,3, respectively.

For the Giesekus model, (8.1-112) reduces to the constraint

$$(1-3\alpha)I_1 + (1-\alpha)\frac{I_2}{I_3} + \alpha\left(I_1^2-2I_2\right) - 3(2-3\alpha) \geq 0 \ , \qquad (8.1\text{-}129)$$

which may be rewritten as

$$\sum_{i=1}^{3} \frac{(\lambda_i-1)^2}{\lambda_i}\left[(1-\alpha) + \alpha\lambda_i\right] \geq 0 \ . \qquad (8.1\text{-}130)$$

A sufficient condition for this inequality to be always satisfied is, trivially, $0 \leq \alpha \leq 1$, and thus we see that the range of permissible values for the parameter α as first suggested by Giesekus is recovered as the range for α such that the thermodynamic criteria are guaranteed to be satisfied for all flows.

For the Hookean dumbbell model with hydrodynamic interaction, the criterion corresponding to (8.1-112) is

$$\left(I_1 + \frac{I_2}{I_3} - 6\right)\left(4 - 3h\sqrt{\frac{\pi}{I_1}}\right) - \frac{3h}{I_1}\sqrt{\frac{\pi}{I_1}}\left(\sum_{i=1}^{3}(\lambda_i-1)^2\right) \geq 0 \ . \qquad (8.1\text{-}131)$$

From this expression, it is determined that the model is guaranteed to satisfy the thermodynamic criterion provided that $0 \leq h \leq (2/3)(I_1/\pi)^{1/2}$.

The multiple parameters of the encapsulated (FENE-P) dumbbell model make the corresponding constraint even more complicated:

$$\begin{aligned} &\left(\frac{1}{2}\frac{b}{b-I_1} - \frac{\alpha-\beta}{I_1}\right)^2 I_1 \\ &+ \left(1 - \tfrac{2}{3}(\alpha-\beta)\right)^2\left(\frac{1}{4}\sigma\frac{I_2}{I_3} + \frac{9}{4}\frac{1-\sigma}{I_1}\right) \qquad (8.1\text{-}132) \\ &- 3\left(1 - \tfrac{2}{3}(\alpha-\beta)\right)\left(\frac{1}{2}\frac{b}{b-I_1} - \frac{\alpha-\beta}{I_1}\right) \geq 0 \ , \end{aligned}$$

where b is the FENE constant associated with the maximum extension of the dumbbells. When $\sigma=\alpha=\beta=1$ and $b\to\infty$, this inequality, (8.1-131), reduces to the Maxwell constraint, and the system is always thermodynamically consistent. However, the complexity of Eq. (8.1-132) is only an apparent one. In the general case, one can find a sufficient set of conditions as follows. Let us devide both members of the inequality (8.1-132) by $(1-2/3(\alpha-\beta))^2$ (assumed to be non-zero; if it is zero, then (8.1-132) is trivially satisfied since only the first non-negative term survives). Then, the left hand side of the inequality (8.1-132) can be considered as a quadratic form with respect to the variable z, defined as

$$z \equiv \frac{\frac{1}{2}\frac{b}{b-I_1} - \frac{\alpha-\beta}{I_1}}{1-\frac{2}{3}(\alpha-\beta)} \; . \tag{8.1-133}$$

Therefore, a sufficient condition for the inequality to be always satisfied (for every z) is that the discriminent be non-positive, which leads to the constraint

$$9 \leq \frac{I_1 I_2}{I_3} = I_1 I_{-1} \; , \tag{8.1-134}$$

which is trivially satisfied as can be seen by expanding each one of the invariants in terms of the eigenvalues. Thus, the variation of the encapsulated dumbell model supported by the bracket formalism is always thermodynamically consistent.

8.1.5 *The non-negative character of the conformation tensor for single-variable viscoelastic fluids*

All of the conclusions on the non-negative character of the dissipation rate discussed above in §8.1.4A are contingent upon the conformation tensor being a positive-definite (more accurately, non-negative definite) entity. From a physical perspective, viewing **c** as representing the conformation of a dumbbell or the deformation of a continuous medium, this requirement is tantamount to asserting the reality of the physical quantity in question. It is hard to imagine, say, a realistic conformation for a dumbbell represented by a non-positive tensor. Hence, as a foundation for the thermodynamic criteria discussed in §8.1.4A, we made explicit use of the positive-definite property of **c**; i.e., we used as a fact that **c** is positive definite. Similarly, in numerical simulations, since the loss of positive definiteness might be due to numerical error, this is a property of the solution that should be checked, if not guarded off (from aquiring a negative eigenvalue), during the course of the computations. Computationally, the loss of positive definiteness might have particularly serious consequences on the accuracy, and in particular the stability, of the numerical calculations since the positive-definite character of **c** is often associated with the evolutionarity (and thus the well-posedness) of the governing equations—see §8.1.6. This being the case, it is particularly important to be able to ascertain that in the exact solution for a given model **c** remains positive definite (or, at least, non-negative definite) at all times. Fortunately, although few proofs were available only a few years ago, we have now several techniques which allow us to simplify the analysis considerably and to ascertain the non-negative definite character of **c** as predicted from several models, as described below.

A. Analysis Through (formal) Integration in Time

The contravariant components $\breve{c}^{\alpha\beta}$ of the conformation tensor **c** in a codeforming (with the flow) curvilinear coordinate system are connected to the components $c_{\gamma\varepsilon}$ in a fixed orthogonal Cartesian coordinate system, both at time t', as [Bird *et al.*, 1987a, p. 494]

$$\breve{c}^{\alpha\beta}(\breve{\mathbf{x}},t';t) \equiv E_{\alpha\gamma}c_{\gamma\varepsilon}(\mathbf{x},t')E_{\beta\varepsilon} \;, \tag{8.1-135}$$

where $\mathbf{x}=\mathbf{x}(\breve{\mathbf{x}},t';t)$ is the position at time t' in the fixed Cartesian coordinate frame of a particle which was at location $\breve{\mathbf{x}}$ at time t. Here and in the following we denote with a breve the quantities referring to the codeforming coordinate system. Also, especially for this subsection only, dimensionless quantiites are used with the tilde omitted for simplicity. In the above expression, **E** is the displacement gradient tensor field, for which the Cartesian components are defined as [Bird *et al.*, 1987a, p. 426]

$$E_{\alpha\beta}(x,t';t) \equiv \frac{\partial x_\alpha(\breve{x},t';t)}{\partial \breve{x}_\beta} \;. \tag{8.1-136}$$

Then, the evolution equation for the conformation tensor in the upper-convected viscoelastic fluid models, Eq. (8.1-7b) with **L=0**, can be rewritten in terms of the contravariant components as

$$\frac{\partial \breve{c}^{\alpha\beta}}{\partial t'} = -\, E_{\alpha\gamma}\Lambda_{\gamma\varepsilon\eta\nu}\frac{\partial H_m}{\partial c_{\eta\nu}}E_{\beta\varepsilon} \;, \tag{8.1-137}$$

where $\boldsymbol{\Lambda}$ denotes the dimensionless relaxation matrix, t' the dimensionless time (made so with the relaxation time λ), and use was made of the identity [Bird *et al.*, 1987a, p. 495]

$$\frac{\partial \breve{c}^{\alpha\beta}}{\partial t'} = E_{\alpha\gamma}\breve{c}_{\gamma\varepsilon}E_{\beta\varepsilon} \;, \tag{8.1-138}$$

p. 273: in (8.1-137), $\dfrac{\partial H_m}{\partial c_{\eta\nu}}$ should be $\dfrac{\delta H_m}{\delta c_{\eta\nu}}$

where the check on **c** denotes, as usual, the upper-convected time derivative.

Substituting for $\boldsymbol{\Lambda}$ the expression corresponding to the linear elastic dumbbell model (i.e., the upper-convected Maxwell model), Eq. (8.1-15), suitably non-dimensionalized, into Eq. (8.1-137) results in a linear first-order differential equation

$$\frac{\partial \check{c}^{\alpha\beta}}{\partial t'} = - \check{c}^{\alpha\beta} + E_{\alpha\gamma}E_{\beta\gamma} \ . \qquad (8.1\text{-}139)$$

Eq. (8.1-139) admits the formal solution (see, for example, Bird *et al.* [1987a, pp. 259, 60])

$$\check{c}^{\alpha\beta}(\check{x},t';t) = \int_{-\infty}^{t'} e^{-(t'-t'')} E_{\alpha\gamma}(x(\check{x},t'';t),t'';t) E_{\beta\gamma}(x(\check{x},t'';t),t'';t) dt' , \qquad (8.1\text{-}140)$$

which, since $\mathbf{E}\cdot\mathbf{E}^{\mathrm{T}}$ is non-negative definite,[23] guarantees that $\check{\mathbf{c}}$ is non-negative definite (and therefore so is **c**: this proof is along the lines of the original proof provided by Joseph *et al.* [1985, p. 228]).

For a more complex model, we may still derive an integral solution, as follows. First, let us observe that using the Cayley/Hamilton theorem and exploiting the fact that at static equilibrium **c** is proportional to the delta function, we can express the (dimensionless) dissipative relaxation term on the right-hand side of Eq. (8.1-137) as

$$\Lambda_{\alpha\beta\gamma\varepsilon}\frac{\delta H_m}{\delta c_{\gamma\varepsilon}} = \alpha_1 c_{\alpha\beta} - \alpha_0 \delta_{\alpha\beta} - \alpha_2 (c_{\alpha\gamma} - \alpha_{01}\delta_{\alpha\gamma})(c_{\gamma\beta} - \alpha_{01}\delta_{\gamma\beta}) , \qquad (8.1\text{-}141)$$

where **c** has been made dimensionless, as usual, by $K/(k_BT)$, and α_0, α_1 and α_2 are, in general, functions of the invariants of **c** at time t' (thus, functions of t') which depend on the particular model in consideration. The remaining parameter in Eq. (8.1-141), α_{01}, is a numerical constant factor equal to the ratio of α_0/α_1 at static equilibrium so that $\mathbf{c}_{eq}=\alpha_{01}\boldsymbol{\delta}$. It has been introduced in the expansion of Eq. (8.1-141) for convenience so that close to static equilibrium the last (quadratic) term can be safely neglected in favor of the linear one. Table 8.2 lists the above parameters for the various models discussed in this work.

If Eq. (8.1-141) is substituted into Eq. (8.1-137), then the components of the contravariant tensor, $\check{c}^{\alpha\beta}$, can be formally obtained as

[23]Given a square matrix **G**, the matrix $\mathbf{G}^{\mathrm{T}}\cdot\mathbf{G}$ is always non-negative definite since for any vector **x** the quadrature $\mathbf{x}^{\mathrm{T}}\cdot\mathbf{G}^{\mathrm{T}}\cdot\mathbf{G}\cdot\mathbf{x}$ is always non negative as equal to a sum of squares, $\mathbf{y}^{\mathrm{T}}\cdot\mathbf{y}$, where $\mathbf{y}\equiv\mathbf{G}\cdot\mathbf{x}$.

Model	α_1	α_0/α_1	α_2
Elastic dumbbell (UCM, Oldroyd-B)	1	1	0
Elastic dumbbell with hydrodynamic interactions	$1-\frac{3}{4}h\sqrt{\frac{\pi}{\mathrm{tr}\tilde{\mathbf{c}}}}\left(1+\frac{1}{\mathrm{tr}\tilde{\mathbf{c}}}\right)$	1	$\frac{3h\sqrt{\pi}}{4(\mathrm{tr}\tilde{\mathbf{c}})^{\frac{3}{2}}}$
Giesekus	1	1	$-\alpha$
Phan-Thien/Tanner (linear, for a=1)	$1+\varepsilon\,\mathrm{tr}(\tilde{\mathbf{c}}-\boldsymbol{\delta})$	1	0
FENE-P / No hydrodynamic interactions	$\frac{1}{1-\frac{\mathrm{tr}\tilde{\mathbf{c}}}{b}}$	$1-\frac{\mathrm{tr}\tilde{\mathbf{c}}}{b}$	0
Modified encapsulated dumbbell	$1-\frac{3}{\mathrm{tr}\tilde{\mathbf{c}}}(1+\sigma(1-2\beta))$	$\frac{1}{\frac{3}{\mathrm{tr}\tilde{\mathbf{c}}}+\frac{1-\frac{3}{\mathrm{tr}\tilde{\mathbf{c}}}}{\sigma(2\beta-1)}}$	0
Chilcott-Rallison (modified according to Eq. 8.1-75)	$\frac{1-\frac{3}{\tilde{N}}}{1-\frac{\mathrm{tr}\tilde{\mathbf{c}}}{\tilde{N}}}$	1	0
Extended White/Metzner	$\left(\frac{\mathrm{tr}\tilde{\mathbf{c}}}{3}\right)^k$	1	0

Table 8.2: Parameters for the integral equation (8.1-142) for various viscoelastic fluid models. The parameters are expressed here in terms of the dimensionless conformation tensor, $\tilde{\mathbf{c}}$, $\tilde{\mathbf{c}}\equiv\mathbf{c}K/(k_BT)$.

$$\check{c}^{\alpha\beta}(\check{x},t';t) = \int_{-\infty}^{t'} \exp\left(-\int_{t''}^{t'} \alpha_1(t^*)\,dt^*\right) E_{\alpha\gamma}(x(\check{x},t'';t),t'';t)[\alpha_0\delta_{\gamma\varepsilon}$$

$$+\alpha_2(c_{\gamma\eta}-\alpha_{01}\delta_{\gamma\eta})(c_{\eta\varepsilon}-\alpha_{01}\delta_{\eta\varepsilon})]E_{\beta\gamma}(x(\check{x},t'';t),t'';t)\,dt'' \tag{8.1-142}$$

$$= \int_{-\infty}^{t'} \exp\left(-\int_{t''}^{t'} \alpha_1(t^*)\,dt^*\right)\Big[\alpha_0 E_{\alpha\gamma}(t'';t)E_{\beta\gamma}(t'';t)$$

$$+\alpha_2 F_{\alpha\gamma}(t'';t)F_{\beta\gamma}(t'';t)\Big]dt'' ,$$

where $\mathbf{F}\equiv\mathbf{E}\cdot(\mathbf{c}-\alpha_{01}\boldsymbol{\delta})$ and the right-hand side of this expression is evaluated consistently for $\mathbf{c}$ corresponding to the contravariant components $\check{c}^{\alpha\beta}$—see Marrucci [1970, p. 326] for a similar expression obtained for non-isothermal flows. Therefore, if the integrand in Eq. (8.1-142) is always non-negative definite, $\check{\mathbf{c}}$ will be non-negative definite, and correspondingly, $\mathbf{c}$. Given the non-negative definiteness of $\mathbf{E}\cdot\mathbf{E}^T$ and $\mathbf{F}\cdot\mathbf{F}^T$ according to the previous footnote, sufficient conditions for this to happen for arbitrary kinematics are $\alpha_0\geq0$, and $\alpha_2\geq0$.

Judging from the sign of the coefficients in Table 8.2, for homogeneous flows and for the range of the parameters indicated in the preceding discussion, almost all of the models are expected to produce a non-negative conformation tensor for arbitrary flows. In particular, the variable relaxation time models (like, for example, the Phan-Thien /Tanner or the extended White/Metzner model) corresponding to a positive effective relaxation time, $\lambda_{eff}\equiv\lambda/\alpha_1$, obviously satisfy this criterion since $\alpha_2=0$ and $\alpha_0=\alpha_1>0$. For these models, Eq. (8.1-142) reduces to (written in simplified form with the dependencies on $\check{\mathbf{x}}$, t omitted)

$$\check{c}^{\alpha\beta}(t') = \int_{-\infty}^{t'} \exp\left(-\int_{t''}^{t'} \alpha_1(t^*)\,dt^*\right)\alpha_1(t'')E_{\alpha\gamma}(t'')E_{\gamma\beta}(t'')\,dt' \ . \tag{8.1-143}$$

Given the positive definiteness of $\mathbf{E}\cdot\mathbf{E}^T$, this integral is always positive as long as α_1 is positive and integrable. Of course, for $\alpha_1=1$, Eq. (8.1-143) duly reduces to Eq. (8.1-140) valid for the upper-convected Maxwell model. Based on the negative sign of α_2, questions can be raised for the Giesekus model. However, according to the discussion in the previous paragraph, this term is expected to be small, at least close to static equilibrium. Thus, at least for small enough values of the Giesekus parameter, α, the positive-definiteness of $\mathbf{c}$ should still hold. Moreover, for the Giesekus model, and for simple, homogeneous (steady shear and extensional) flows, it is known that for $0 \leq \alpha \leq 1$ the tensor **c** is non-negative [Giesekus, 1982, pp. 82,93]. For arbitrary flows, one would have to make use of the sufficient condition for positive definiteness which is described in the analysis that follows.

B. A Sufficient Condition for a Positive-definite Conformation Tensor in Differential Models

This recently developed condition [Hulsen, 1990] exploits the mathematical properties of the determinant and makes a direct usage of the evolution equation for the conformation tensor **c** in order to show that once positive definite, the conformation tensor will remain always positive definite (or at least non-negative definite) in time. Isolated application of the same principles have appeared before when they were used to prove the positive definiteness of **c** for the upper-convected Maxwell model [Dupret and Marchal, 1986a, Eq. (56)], the Johnson/Segalman model [Dupret and Marchal, 1986b, p. 158, Theorem IV], and, in a planar flow, for the Leonov and Giesekus fluid models [Hulsen, 1988]. However, Hulsen [1990] developed their most general expression in the rigorous form of a theorem. The development of the theorem, simple, yet ingenious and powerful, requires only mild conditions to be satisfied by the evolution equation, as explained below. Briefly, Hulsen examines the rate of change of the determinant of the conformation tensor, det**c**, under conditions of criticality (i.e., when one or more of the eigenvalues are zero). He then proves that the non-negative definiteness is conserved by showing that the rate of change of the determinant is always non-negative under critical conditions. His proof, adapted for the form of the evolution equations as they arise within the context of the bracket theory, and applications of his theorem to the models discussed in this section, follow below.

Within the context of the bracket theory, the general evolution equation for the conformation tensor, **c**, for a single internal variable viscoelastic model, can be written as

$$\frac{\mathrm{d}c_{\alpha\beta}}{\mathrm{d}t} = \frac{a+1}{2}\left[c_{\alpha\gamma}\nabla_{\gamma}\upsilon_{\beta} + c_{\beta\gamma}\nabla_{\gamma}\upsilon_{\alpha}\right] + \frac{a-1}{2}\left[c_{\alpha\gamma}\nabla_{\beta}\upsilon_{\gamma} + c_{\beta\gamma}\nabla_{\alpha}\upsilon_{\gamma}\right] + g_1(\mathbf{c})\delta_{\alpha\beta} + g_2(\mathbf{c})c_{\alpha\beta} + g_3(\mathbf{c})c_{\alpha\gamma}c_{\gamma\beta} \,, \tag{8.1-144}$$

where g_i, i=1,2,3, are in general functions of the invariants of **c**, which is assumed without loss of generality to be equal to the unit tensor, $\boldsymbol{\delta}$, at equilibrium. Then, Hulsen's theorem goes as follows.

THEOREM [Hulsen, 1990, p. 95]: A differential evolution equation for the conformation tensor **c** of the form given by Eq. (8.1-141) leads to a positive-definite tensor **c**(t), t>0 if $g_1(\mathbf{c}) > 0$ for any semi-positive definite tensor **c**, **c**(t=0) is positive definite, and $\nabla\mathbf{v}$ is finite.

PROOF: Use is made in the proof of the three invariants of **c**, I_1, I_2, and I_3, defined in Eq. (8.1-103) in terms of the three (non-negative) eigenvectors of **c**, λ_i, i=1,2,3. The requirement for a finite velocity gradient is there to guarantee that **c** evolves in time in a continuous fashion. As we follow the positive-definite tensor **c** in time, continuity implies that the transition to an indefinite state (requiring that at least one of the eigenvalues be zero or negative) will first occur at the boundaries of the positive quadrant ($\lambda_i \geq 0$, i=1,2,3) provided by the planes

$$P_i = (\lambda_i = 0,\ \lambda_j \geq 0,\ j=1,2,3,\ j \neq i)\ ,\ i=1,2,3\ . \tag{8.1-145}$$

Therefore, in order to show that **c** remains positive definite for t>0, it suffices to show that these planes cannot be approached from the interior of the first octant except through a layer asymptotically close to these planes where $\mathrm{d}(I_3)/\mathrm{d}t>0$. This implies that if I_3 was initially positive it will remain positive for all t. Thus, the transition of the character of **c** to indefinite can never occur.

The time derivative of the determinant, I_3, given by Eq. (8.1-110), becomes according to Eq. (8.1-144)

$$\begin{aligned}\frac{\mathrm{d}(\det\mathbf{c})}{\mathrm{d}t} &= \det\mathbf{c}\ c_{\beta\alpha}\Bigg(\frac{a+1}{2}\Big[c_{\alpha\gamma}\nabla_\gamma \upsilon_\beta + c_{\beta\gamma}\nabla_\gamma \upsilon_\alpha\Big] \\ &\quad + \frac{a-1}{2}\Big[c_{\alpha\gamma}\nabla_\beta \upsilon_\gamma + c_{\beta\gamma}\nabla_\alpha \upsilon_\gamma\Big] + g_1(\mathbf{c})\delta_{\alpha\beta} \\ &\quad + g_2(\mathbf{c})c_{\alpha\beta} + g_3(\mathbf{c})c_{\alpha\gamma}c_{\gamma\beta}\Bigg) \\ &= \det\mathbf{c}\ \Big(2a\nabla_\gamma \upsilon_\gamma + g_1(\mathbf{c})\mathrm{tr}(\mathbf{c}^{-1}) \\ &\quad + 3g_2(\mathbf{c}) + g_3(\mathbf{c})\mathrm{tr}\mathbf{c}\Big)\ .\end{aligned} \tag{8.1-146}$$

From the Cayley/Hamilton theorem, Eq. (8.1-106), we have

$$\det\mathbf{c}\ \mathrm{tr}(\mathbf{c}^{-1}) = \mathrm{tr}(\mathbf{c}^2) - (\mathrm{tr}\mathbf{c})^2 + 3I_2 = I_2\ , \tag{8.1-147}$$

which, in conjunction with Eq. (8.1-146), implies that

$$\frac{\mathrm{d}(\det\mathbf{c})}{\mathrm{d}t} = 2aI_3\nabla_\gamma \upsilon_\gamma + g_1 I_2 + (3g_2 + g_3 I_1)I_3\ . \tag{8.1-148}$$

Thus, when I_3=detc=0, and taking into account that g_1>0, we have one of the following three cases:

1. Only one of the eigenvalues λ_i, i=1,2,3, is zero (and the rest positive) $\Rightarrow I_1 > 0$, $I_2 > 0$ and d(detc)/dt > 0.
2. Only one of the eigenvalues λ_i, i=1,2,3, is positive and the rest are zero $\Rightarrow I_1 > 0$, $I_2 = 0$ and d(detc)/dt = 0, but a Taylor expansion in the neighborhood of the line boundary shows that d(detc)/dt > 0 [Hulsen, 1990, pp. 97-98] and therefore this boundary could not have been reached from the first octant.
3. All three of the eigenvalues λ_i, i=1,2,3, are zero $\Rightarrow$ from the original equation, dc/dt = $g_1\boldsymbol{\delta}$ and the path is directed toward the interior of the first octant; i.e., this point also could not have been reached from within the first octant.

Therefore, in all cases, the boundary cannot be reached from within the first octant following the evolution equation, Eq. (8.1-148), which implies that, for g_1>0, the conformation tensor **c**, once positive definite, remains positive definite at all times. QED. ■

Therefore, to apply Hulsen's theorem, it suffices to evaluate the corresponding g_1 for a given evolution equation and to examine conditions under which it is larger than zero for any positive definite **c**. As shown in Table 8.3 (see also Hulsen, 1990, §4), most of the models satisfy the conditions of the theorem within a wide range of parameter values. Note that it is important to understand that the theorem does not exclude the possibility of solutions to the steady-state equations for which **c** is not positive definite. However, what the theorem says is that these solutions cannot be reached from equilibrium. Thus, the steady-state solutions for these models which correspond to a non-positive definite conformation tensor need to be dismissed as aphysical. This implies, if no other steady solutions exist, that there are no steady-state solutions which can be reached from equilibrium, i.e., that the solution which is established has to be time-dependent. This situation is particularly evident with a class of models, like the Johnson/Segalman model, which present, for a certain range of parameters, a maximum in the shear-stress/shear-rate curve. These models exhibit a rich dynamic behavior even for flows of such simplicity as slit-die (plane Pouseuille) flow. Indeed, for certain values of the parameters, one can observe very long transients during the start-up of plane Pouseuille flow which exhibit three characteristic regions: an initial "Newtonian" phase with a parabolic velocity profile, an intermediate "latency" phase, and a final "spurt" phase during which a singular surface is developed across which a jump in shear rate takes place [Kolkka *et al.*, 1988, §3.3.1; Malkus *et al.*, 1990, §3d]. Moreover, for certain values of the parameters, a behavior similar to the

Model	$\frac{\lambda K}{k_B T} g_1(\tilde{c})$	$-\lambda g_2(\tilde{c})$	$\lambda \frac{k_B T}{K} g_3(\tilde{c})$
Elastic dumbbell (UCM, Oldroyd-A,B Johnson/ Segalman)	1	1	0
Elastic dumbbell with hydrodynamic interactions	$1-\frac{3}{4}h\sqrt{\frac{\pi}{\mathrm{tr}\tilde{c}}}$	$1-\frac{3}{4}h\sqrt{\frac{\pi}{\mathrm{tr}\tilde{c}}}\ (1-\frac{2}{\mathrm{tr}\tilde{c}})$	$\frac{3h\sqrt{\pi}}{2(\mathrm{tr}\tilde{c})^{\frac{3}{2}}}$
Giesekus	$1-\alpha$	$1-2\alpha$	$-\alpha$
Phan-Thien/ Tanner (linear)	$1+\varepsilon\,\mathrm{tr}(\tilde{c}-\boldsymbol{\delta})$	$1+\varepsilon\,\mathrm{tr}(\tilde{c}-\boldsymbol{\delta})$	0
FENE-P / No hydrodynamic interactions	1	$\frac{1}{1-\frac{\mathrm{tr}\tilde{c}}{b}}$	0
Modified encapsulated dumbbell	$\sigma(2\beta-1)$	$1-\frac{3}{\mathrm{tr}\tilde{c}}(1+\sigma(1-2\beta))$	0
Leonov	$\frac{4K}{G_0^2}\frac{\partial W}{\partial a_{-1}}$	$\frac{4}{3G_0}\left(I_1\frac{\partial W}{\partial a_1}-I_{-1}\frac{\partial W}{\partial a_{-1}}\right)$	$-\frac{4}{K}\frac{\partial W}{\partial a_1}$
Extended White/ Metzner	$\left(\frac{\mathrm{tr}\tilde{c}}{3}\right)^{-k}$	$\left(\frac{\mathrm{tr}\tilde{c}}{3}\right)^{-k}$	0

Table 8.3: Parameters for the evolution equation for the conformation tensor, Eq. (8.1-144), for various viscoelastic fluid models. The parameters, non-dimensionalized as shown, are expressed here in terms of the dimensionless conformation tensor, $\tilde{c}$, $\tilde{c} \equiv cK/(k_B T)$.

Johnson/Segalman model of the flow behavior of monodomain polymeric liquid crystals has been observed to lead to solutions for the start-up of shear flow which eventually develop into limit cycles [Edwards, 1991, ch. 5].

8.1.6 *Evolutionary character of the governing equations of single-variable viscoelastic fluids*

In section §8.1.4, we examined *thermodynamic admissibility* criteria for the governing equations imposed from (irreversible) thermodynamic considerations. We saw that, for the most part, the equations examined in this section were thermodynamically admissible, provided that the associated conformation tensor **c** is positive definite. The positive-definite character of **c** is also necessary in order to make meaningful any microscopic interpretation of **c**, so this question can be considered as addressing the *physical admissibility* of a given model. However, the ultimate goal of the mathematical modeling is to represent the system under investigation through mathematical equations, the solution of which can provide a good approximation of its evolution in time. A consideration of the necessary and/or sufficient conditions for the well-posedness of the mathematical problem gives rise to yet another set of criteria based on the *mathematical admissibility* of the model. These are associated with the proper selection of *boundary conditions* and the *mathematical stability* (i.e., the sensitivity of the equations to boundary and initial value data).

The question on the proper boundary conditions is intimately related to the mathematical *type* (i.e., elliptic, parabolic, hyperbolic, or mixed) of the differential equations. Both questions (i.e., for the type and boundary conditions) have attracted considerable interest among the mathematical research community, and since they are outside the scope of the present monograph, they are not discussed here. The interested reader can consult one of the many excellent references on the subject [Rutkevich, 1969; Joseph *et al.*, 1985; Renardy *et al.*, 1987; Renardy, 1988a,b; Van der Zanden and Hulsen, 1988; Joseph, 1990 and references therein; Cook and Schleiniger, 1991]. It suffices to say here that most problems associated with the proper type of boundary conditions are avoided in the presence of viscous dissipation (i.e., when the solvent viscosity is explicitly considered in the calculations). For this type of models (called *J-type models* by Van der Zanden and Hulsen [1988]), the boundary conditions that need to be imposed are simply those for the velocity $\boldsymbol{v}$ over the whole boundary and the conformation tensor **c** on the inflow boundary [Renardy, 1988b, p. 22; Van der Zanden and Hulsen, 1988, p. 105]. Note that although the solvent viscosity might not provide an important contribution to the rheological behavior of the medium for flows within smooth geometries, it might dominate close to geometrical singularities, such as corners.[24]

What we would like to address here is the question of mathematical stability of the equations, primarily because it has been linked to the thermodynamic and physical (positive-definiteness) criteria discussed in the previous two subsections [Rutkevich, 1972, p. 293; Dupret and Marchal, 1986b, p. 170]. The mathematical stability or instability of the equations (well-posedness or ill-posedness, respectively, of the corresponding initial value Cauchy problem) are related to the sensitivity of the solution to the initial data (this is a notion strictly weaker than hyperbolicity [Joseph and Saut, 1986, p. 117]). Following Courant and Hilbert [1962, p. 227]:

> A mathematical problem which is to correspond to physical reality should satisfy the following basic requirements:
> (1) The solution must exist.
> (2) The solution should be uniquely determined.
> (3) The solution should depend continuously on the data (requirement of stability).
> ...
> Any problem which satisfies our three requirements will be called a *properly posed* problem.

However, the same authors a few pages later recognize the physical importance of "improperly posed" problems [Courant and Hilbert, 1962, p. 230]:

> The stipulation of article 2 about existence, uniqueness and stability of solutions dominate classical mathematical physics. They are deeply inherent in the ideal of a unique, complete, and stable determination of physical events by appropriate conditions at the boundaries, at infinity, at time t =0, or in the past.

[24]The question of the behavior of the solution of the mathematical equations next to geometrical singularities is still far from being resolved for even the simplest of the viscoelastic models, the Hookean dumbbell (upper-convected Maxwell). In this case, a dominant balance using the scaling for the Newtonian solution [Dean and Montagnon, 1949; Moffatt, 1964] fails. However, recent work indicates that the Newtonian dominant balance can be restored with certain variable relaxation time models (i.e., modified upper-convected Maxwell or C-R where the viscous solvent contributions either dominate or are equally important to the viscoelastic contributions, respectively) [Brown, 1992].

> Laplace's vision of the possibility of calculating the whole future of the physical world from complete data of the present state is an extreme expression of this attitude. However, this rational ideal of causal-mathematical determination was gradually eroded by confrontation with physical reality. Nonlinear phenomena, quantum theory, and the advent of powerful numerical methods have shown that "properly posed" problems are by far not the only ones which appropriately reflect real phenomena.

The instability (ill-posedness) is manifested with the support of infinite-frequency, infinitesimal oscillations reflecting the lack of continuous dependence of the solution on the initial data (Hadamard instability [Joseph, 1990, pp. 69-70]). In consequence, the appearance of this type of instabilities effectively prohibits the convergence of any numerical calculations. Since this type of mathematical behavior is usually associated with parabolic equations (such as the heat conduction equations) when they are solved backwards in time (thus, in a non-evolutionary fashion using the future state of the system as the initial condition), the stability (or well-posedness) of the equations with respect to the initial data is said to represent the criteria of an *evolutionary* system [Dupret and Marchal, 1986, p. 143]. Evolutionarity is then connected with physical determinism and represents a property sought from the equations of a mathematically admissible model (although, as we saw above in Courant and Hilbert's comments, there is no unanimous agreement on that, since the loss of evolutionarity of the equations might simply represent the onset of a physically induced instability; see also Joseph and Saut [1986, pp. 119-21] for additional comments and a list of physical "ill-posed" problems). Nonetheless, if we accept its physical underpinnings, it is then natural to examine the relationship of the evolutionarity admissibility conditions to the thermodynamic and physical ones. For this reason, we decided to provide in the following a brief overview of the relevant theory for selected simple, single internal variable, spring/dashpot type models for which the theory is available.

A. Linear Interpolated Maxwell (Johnson/Segalman and Oldroyd A & B) Models

Rutkevich [1970] was the first to examine the evolutionarity of the equations corresponding to the upper-convected, corotational, and lower-convected Maxwell models. He developed for that purpose a method of analysis based on the examination of the support by the linearized equations of infinite-frequency waves. Then, the equations are considered evolutionary if the time frequency corresponding to infinite wavenumber is real [Rutkevich, 1970, p. 38]. Later, using the same approach, Joseph and Saut [1986], and by direct evaluation of the well-posedness of the equivalent Cauchy problem subject to data along the t=0 plane, Dupret and Marchal [1986b], formulated sufficient conditions for evolutionarity for the Johnson/Segalman model for arbitrary values of a, $-1 \leq a \leq 1$. These conditions, expressed in terms of the eigenvalues σ_1, σ_2, and σ_3, $\sigma_1 \geq \sigma_2 \geq \sigma_3$, of the extra stress tensor, $\boldsymbol{\sigma}$, are

$$\frac{\eta}{\lambda} + \frac{1}{2}a(\sigma_1 + \sigma_3) - \frac{1}{2}|\sigma_1 - \sigma_3| > 0 \; . \tag{8.1-149}$$

As recognized by Dupret and Marchal [1986a,b] and Joseph and Saut [1986], the condition (8.1-149) reduces for $a=\pm 1$ (i.e., for the upper/lower-convected Maxwell models) to the condition for the positive-definiteness of the corresponding conformation tensor. Thus, since this condition is always satisfied for a general interpolated Maxwell (Johnson/Segalman) model for flows evolving from equilibrium, as discussed in §8.1.5 above, both upper- and lower-convected models are guaranteed, under the same conditions, to stay always evolutionary.

The case for the interpolated Maxwell model for an intermediate a, $-1<a<1$, appears to be different. In this case, condition (8.1-149) is more restrictive than the requirement for a positive definite $\mathbf{c} \propto \boldsymbol{\sigma} + \eta/(a\lambda)\boldsymbol{\delta}$ —see Eq. $(8.1\text{-}29)_2$ for the appropriate definition for $\mathbf{c}$ for this fluid model—which is consistent with the literature [Joseph and Saut, 1986; Dupret and Marchal, 1986b]. Indeed, condition (8.1-149) can be violated for a positive definite $\mathbf{c}$ and therefore there is, in principle, the possibility that under the same conditions as for the upper- or lower-convected Maxwell models (i.e., for flows starting from equilibrium) the equations cease to be evolutionary. However, this is proven so far for special steady-state flows only. Dupret and Marchal [1986b, §7] and Joseph and Saut [1986, §7.4] provide the 3-dimensional sink flow as such an example. The sink flow is a particular steady-state flow, with the velocity ($\upsilon_r=-q/r^2$, $\upsilon_\phi=0$, $\upsilon_\theta=0$ in a spherical (r,ϕ,θ) coordinate system) imposed from the continuity equation, irrespective of the stress constitutive equation. Thus, the flow has a strong extensional character which increases as the origin of the sink (r=0) is approached, resulting in large (infinite as $r\to 0$) values of the difference between the values of normal stresses along the flow and normal to the flow directions. It is this large difference that causes the violation of criterion (8.1-149). Furthermore note that although the flow in a (Eulerian) frame is fixed, it is Lagrangian unsteady (in a frame moving with the flow), and the material at the inflow does exist in its equilibrium state with zero stress [Dupret and Marchal, 1986b, Eq. (7.4)]. Also note that any addition of a non-zero viscosity into the model automatically renders the equations always evolutionary (i.e., viscosity "regularizes" the solution [Rutkevich, 1970, p. 39; Joseph and Saut, 1986, p. 136; Van der Zanden and Hulsen, 1988, §4.2]). Thus, in the limit of $\lambda \to 0$, we recover the well-known result that the Navier/Stokes equations are evolutionary in time [Rutkevich, 1970, p. 39].

B. Nonlinear Maxwell-type Models

The equations corresponding to models endowed with a Newtonian (solvent) viscosity being always evolutionary (see discussion at the end of previous paragraph), the interest shifts to examine models without a direct viscous contribution to the stress (i.e., with $\mathbf{Q} \equiv \mathbf{0}$ in Eq. (8.1-7c)). The evolutionary character of a few nonlinear constitutive models other than the interpolated Maxwell model has been explicitly investigated so far, such as the White/Metzner fluid [Dupret and Marchal, 1986b; Verdier and Joseph, 1989] the Giesekus fluids [Joseph, 1990, pp. 136-37] and the encapsulated dumbbell (Bird/DeAguiar) [Calderer *et al.*, 1989, §3-5] being probably the only such cases reported in the literature. The criteria for the Giesekus model were developed only for extensional flows and they reduce (as for the upper-convected Maxwell case) to the positive definiteness of the conformation tensor [Joseph and Saut, 1986, p. 134]. The criterion for evolutionarity for the Bird/DeAguiar model reduces to the non-negative character of a nonlinear function of the stress and the velocity [Calderer *et al.*, 1989, p. 215]. Evaluating numerically this criterion, Calderer *et al.*, [1989, §5] have shown for uniform, simple shear and extensional steady flow that although there might exist multiple solutions, there exists always one and only one which is evolutionary.

The evolutionarity criteria for the White/Metzner model, Eq. (8.1-31), involve, in addition to the requirement for the positive definiteness of a corresponding "conformation tensor,"

$$\mathbf{c} = \boldsymbol{\sigma} + G_0 \boldsymbol{\delta} \ , \qquad (8.1\text{-}150)$$

the satisfaction of the following inequality [Dupret and Marchal, 1986b, Eq. (4-17)]:

$$\mathbf{e}\cdot\mathbf{c}\cdot\mathbf{e} - \frac{\lambda'(\mathrm{II_D})}{\lambda^2(\mathrm{II_D})}\left(\mathbf{e}\cdot\mathbf{A}\cdot(\mathbf{e}\times(\mathbf{e}\times(\mathbf{c}\cdot\mathbf{e})))\right) \geq 0 \ , \qquad (8.1\text{-}151)$$

for arbitrary real vectors $\mathbf{e}$, where $\mathrm{II_D} \equiv (\mathrm{II_A}/2)^2$; $\mathrm{II_A}$ is defined in Eq. (8.1-32), λ' denotes the derivative $d\lambda/d(\mathrm{II_D})$ and $\times$ denotes the cross product between two vectors. Alternatively to condition (8.1-151), Dupret and Marchal [1986b, p. 167] also offer a necessary condition for evolution, which they show is violated when the material function for the relaxation time is given either through a power-law relation similar to Eq. (8.1-34) or through a Bird/Carreau law, similar to Eq. (8.1.35) with a=2. Thus, loss of evolution is possible for the White/Metzner model. Independently, Verdier and Joseph [1989], extending their earlier stability analysis method, determined an alternative formulation to condition (8.1-151) of the form

$$\lambda^2 c_{11} + (A_{12} c_{12} + A_{13} c_{13}) \lambda'(\mathrm{II_D}) \geq 0 \,, \tag{8.1-152}$$

where a coordinate system is chosen so that c_{23}=0.

Condition (8.1-152) is similar to condition (8.1-151). The important item, though, is the realization that the same perturbation approach followed by Verdier and Joseph can be used in order to examine the evolutionarity of other nonlinear models, such as the PTT (with ξ=0) and the EWM. For these models, as well as for all models that result from a modification of the original upper-convected Maxwell model limited only to the extent of considering conformation dependent parameters, the terms in Eq. (8.1-152) that multiply the rate of deformation components are missing. This happens because, as the presence of the derivative of the relaxation time in Eq. (8.1-152)—as well as in Eq. (8.1-151)—indicates, these terms originate from the direct dependence on the rate of deformation of the relaxation time of the original White/Metzner model. Thus, when the parameters of a model involve no direct dependence on the rate of deformation or any derivatives of it or the conformation tensor, the evolutionarity conditions are automatically satisfied when **c** is positive definite [Souvaliotis and Beris, 1992, p. 249]. Therefore, both PTT (ξ=0, ε<1/3) and EWM are always evolutionary (exactly as is the upper-convected Maxwell) since the positive definiteness of their corresponding conformation tensor is always satisfied, for the range of the parameter values indicated, for flows starting from equilibrium [Hulsen, 1990, Eq. (34)]—see also §8.1.5 and Table 8.3. Similarly, extending the previous argument, based solely on the behavior introduced into the equations as a result of the use of a particular form of convected time derivative, it can be inferred that nonlinear parametric extensions with respect to the conformation tensor (or, equivalently, the stress tensor) of the Johnson/Segalman model with the parameter a strictly between –1 and 1, $-1<a<1$, such as the Phan-Thien and Tanner model with $\xi\neq 0$, can develop Hadamard instabilities, as also hinted, but not proven, by Joseph [1990, p. 78].

The application of the extended Verdier and Joseph approach is sufficient to determine the evolutionary character of a large number of the models examined in the present section. For more complex nonlinear models for which the Verdier and Joseph approach is not directly applicable, evidence suggests (like the analysis for the Giesekus model [Joseph and Saut, 1986, p. 134]) that as long as the time dependency is expressed in terms of a "pure" (upper or lower) convected time derivative and the nonlinearities involve solely the conformation tensor, for models that are developed within the bracket formalism and which correspond to a non-negative rate of entropy production, the evolutionarity condition for flows starting from equilibrium may simply reduce to the requirement for a positive-definite conformation tensor. All the models developed within the bracket formalism using **L**=**0** fall into this category. Thus, we see that a particular feature which the formalism is avoiding (the direct dependence on the rate of strain tensor) helps in reducing the evolutionarity criteria to the ones for physical admissibility (positive definiteness of **c**). Models involving an interpolated (mixed) convected time derivative (or, equivalently, for which **L**≠**0**), as explained above, might lose evolutionarity. However, all that we want to say at this point is that this feature does not necessarily invalidate them as aphysical. We view the more constrained evolutionarity conditions which are applicable for these models as minimum conditions for stability of the steady-state solutions. We do not have, at this point, physically motivated reasons to believe that the corresponding models are, by necessity, aphysical. In contrast, examination of thermodynamic criteria, through the non-positive character of relaxation dissipation—see §8.1.4—in this work, as well as through the Clausius/Duhem inequality in the previous study of the Johnson/Segalman model [Dupret and Marchal, 1986b, p. 165], indicates that as far as thermodynamics is concerned, the models are physical.

Thus, at the present, we share the view of Joseph and Saut [1986] that what the possible lack of evolutionarity of the equations signifies is that under certain conditions the corresponding solution exhibits a fundamental instability (which is what the criterion for evolutionarity is, after all: an instability to short-wavelength wave perturbations). Whether this instability is physical or aphysical we believe at this time that this has more to do with the overall success of the corresponding equation to model a specific reality rather than it arises by necessity from the formulation. Thus, the major benefit that we see coming from the study of the evolutionary character of the equations for a given flow solution is in determining whether this solution can be physically realized or not. The solution for which the equations are not evolutionary cannot be physically realized, whereas in the inverse case, stability depends on stricter linear and nonlinear stability criteria. Definitely, for the "co-rotational-like" type of models, we anticipate time-independent flows to occur less frequently, as indeed has been computationally observed to be the case [Kolkka *et al.*, 1988; Edwards, 1991, ch. 5], since the range of accessible steady solutions is decreased as a consequence of the evolutionarity/stability criterion. Certainly, models which tend to use a variable mix of the convected derivatives (as the ones developed by Hinch [1977] and Rallison and Hinch [1988]) appear to be more promising as they have additional parameters to allow further fine-tuning and to avoid entering the instability region.

In conclusion, we have seen, overall, a convergence and a tight intercorrelation among all admissibility criteria: thermodynamic, physical, and mathematical. This congruence is the strongest supportive evidence in favor of a thermodynamic approach in constructing the governing equations. The only exceptions to this rule are the mixed-convected derivative models obtained through the use of an antisymmetric component in the dissipation bracket. Although they might just correspond to more unstable models, we believe that there is a definite need for more analysis on the role of dissipation and the physical and mathematical interpretation of the antisymmetric terms which appear in the dissipation potential. These terms, despite the fact that they formally belong to the dissipation bracket, cause no additional increase in the rate of entropy production, because of their antisymmetry. Yet they can not be assimilated with the rest of the antisymmetric terms into the Poisson bracket because that would violate the Jacobi identity [Beris and Edwards, 1990a, App. B].

8.2 Incompressible Viscoelastic Fluid Models in Terms of Multiple Conformation Tensors

In the previous development, we have seen how most common rheological models for viscoelastic fluids are derived from the same general set of evolution equations, Eq. (8.1-7), simply by considering the physical form of the Hamiltonian and phenomenological matrices $\mathbf{Q}$, $\boldsymbol{\Lambda}$, and $\mathbf{L}$ for the system in question. All of the models considered were shown to involve only slight variations of these quantities, and specifically, the relaxation matrix $\boldsymbol{\Lambda}$ was generalized from the simple to the more complex models by incorporating higher-order (in $\mathbf{c}$) terms into the expression.

The single internal variable, differential constitutive equations have proven very useful in making exploratory fluid dynamics computations possible, which have given insight into the qualitative effects of viscoelasticity in a complex flow field [Bird *et al.*, 1987a, ch. 7]. Yet, a price is paid in the poor approximation to some of the material functions. The most serious defect of single internal variable differential models is their inherent inability to describe accurately the observed dynamic behavior due to the use of a single characteristic time constant, λ, to characterize the material's relaxation. As several dynamic experiments demonstrate, such as small-amplitude oscillatory shear flow, the observed behavior of any macromolecular system clearly does not correspond to the one predicted by a single relaxation time. Indeed, as more involved, multi-bead, spring/chain models demonstrate [Rouse, 1953, Eq. (31); Zimm,

1956, §7], several relaxation times (approaching a continuous spectrum in the limit $N \rightarrow \infty$) are present in the relaxation behavior of even single, isolated, polymer chains. Thus, the need arises for an extension of the previously developed models, which is the subject of the present section.

8.2.1 *Guidelines on multiple-variable viscoelastic fluid models from the perspective of linear viscoelasticity* [25]

For small deformations, the use of Boltzmann's superposition principle [Tschoegl, 1989, §2.1.2] simplifies considerably the description of the most general rheological behavior. Under these restrictions, the behavior of the material can be considered as linear; i.e., the effect of each successive deformation strain accumulates (superimposes) on the effect of the previous ones. The principles and implications of the theory dealing with small deformations constitute the subject of linear viscoelasticity. Linear viscoelasticity has grown very rapidly since the synthesis of the first polymers to become a mature science today. An enormous volume of literature has been created, culminating in the monograph by Tschoegl [1989]. We are only concerned here with the implications of linear viscoelasticity to the modeling of polymeric systems behavior through an internal variables approach. We refer the interested reader to the textbook by Ferry [1980] and to the above-mentioned monograph by Tschoegl [1989] for a description of the linear viscoelastic properties of the most common polymeric systems and their mathematical interrelationships, respectively.

The principle in which linear viscoelasticity is based is that of linear superposition: In the limit of infinitesimal deformations (from rest), the viscoelastic fluid stress behavior can be described as a linear superposition of the strain deformations suitably weighted in time. In the continuum (in time) limit, this is simply expressed mathematically through a linear integral equation in time. For example, for a pure shear flow, the classical theory of linear viscoelasticity describes the (shear) stress response of a fluid to the (shear) strain history as the integral over all previous times [Bird *et al.*, 1987a, p. 263],

$$\sigma(t) = \int_{-\infty}^{t} G(t-t')\dot{\gamma}(t')dt' , \tag{8.2-1a}$$

[25]This subsection closely follows the development of the arguments in [Beris and Edwards, 1993].

or defining $s \equiv t - t'$,

$$\sigma(t) = \int_0^\infty G(s)\dot{\gamma}(t-s)ds \ , \tag{8.2-1b}$$

where $\sigma(t)$ and $\dot{\gamma}(t-s)$ are the shear stress and shear rate at times t and $t-s$, respectively, and $G(s)$ is the *linear stress relaxation modulus* [Pipkin, 1986, p. 8; Bird *et al.*, 1987a, p. 263]. The expectation of a finite stress after the imposition of a step strain or strain rate requires $G(t)$ to be finite and integrable for all $t>0$, respectively [Pipkin, 1986, p. 47]:

$$G(t) < \infty \ , \quad \int_0^t G(s)ds < \infty \ , \quad t > 0 \ . \tag{8.2-2}$$

Moreover, the assumption of thermodynamic stability at static equilibrium requires the shear stress to follow the same sign as that of a constant step strain, or, following Eq. (8.2-1) for $\dot{\gamma}(t)=\delta(t)$

$$G(t) \geq 0 \quad t \geq 0 \ . \tag{8.2-3}$$

The only other assumption in addition to the linear superposition principle (which, by necessity, limits the validity of Eq. (8.2-1) only to strain histories involving strains of small enough magnitudes) is that of the *fading memory*. According to the fading memory principle, viscoelastic materials tend to better "remember" (and as a consequence, more violently respond) to the most recent deformations. In other words, for a given strain, the magnitude of the corresponding stress response monotonically decreases with time—for the full mathematical implications of this statement see [Beris and Edwards, 1993] and the more detailed discussion below. The fading memory principle is again expected to be valid in the limit of small deformations. Despite these and other shortcomings (as, for example, the accounting of rotations as material strains), Eq. (8.2-1) still enjoys broad use because of its simplicity and generality. In particular, it is often used in order to evaluate the dynamical behavior of a viscoelastic material since the relaxation function can be obtained relatively easily through either start-up or oscillatory shear flow rheological measurements. Eq. (8.2-1) can be generalized for a more general flow in a straightforward fashion by substituting the tensorial equivalent forms $\boldsymbol{\sigma}$ and $\mathbf{A}$, for the scalars σ and II_A($= \dot{\gamma}$ in simple shear flow), respectively.

It is instructive, at this point, to evaluate the relaxation functions corresponding to a viscoelastic fluid model with a single relaxation mode, such as the ones described in §8.1.1 through §8.1.3. Since all of the model nonlinearities are neglected, it is straightforward to see that all the single relaxation mode models correspond to the same linear viscoelastic flow behavior, namely, one where the relaxation function is provided by a single exponential,

$$G(s) = G_0 \exp(-\frac{s}{\lambda}) , \tag{8.2-4}$$

where λ is a characteristic (zero shear rate) relaxation time for the material. However, the relaxation behavior observed with real materials is distinctively different from that represented by Eq. (8.2-4) [Bird *et al.*, 1987a, p. 127; see also p. 274 for a comparison of the experimental data with the predictions obtained for a single relaxation time in small-amplitude oscillatory shear]. Not surprisingly, due to the inherent molecular complexity of real polymers as opposed to their elastic spring/dashpot or dumbbell mechanical analogues, we have a much more complex dynamic behavior. Even for the simplest possible polymer systems, that of dilute solutions, both theory and experiments indicate that instead of one exponential mode, a superposition of several of them is obtained [Ferry, 1980, ch. 9]. For example, the simplest theory of a Rouse chain, corresponding to N Hookean springs in a bead/spring chain without hydrodynamic interactions (see also §8.2.3), predicts [Rouse, 1953, Eq. (31); Ferry, 1980, pp. 187-88; Bird *et al.*, 1987b, p. 159]

$$G(s) = nk_BT \sum_{p=1}^{N} \exp(-\frac{s}{\tau_p}) , \tag{8.2-5}$$

where n is the number density of the polymer molecules, N is the number of relaxation modes equal to the number of springs per polymer chain, and τ_p is the relaxation time associated with the mode p,

$$\tau_p = \frac{\zeta}{8K} \sin^{-2}\left(\frac{p\pi}{2(N+1)}\right) , \tag{8.2-6}$$

where ζ is the friction coefficient per bead and K the elastic spring constant.

Fluid behavior like the one corresponding to Eq. (8.2-5) demonstrates the need for the use of multi-variable/multi-mode models like the ones discussed in the following sections, §8.2.2 through §8.2.5, to which corresponds the following general relaxation behavior

$$G(s) = \sum_{p=1}^{N} G_p \exp(-\frac{s}{\tau_p}) , \tag{8.2-7}$$

where G_p and τ_p are the model-dependent relaxation modulus and relaxation time for relaxation mode p, p=1,...,N. However, a question that emerges is that of adequacy: Can these models represent any possible observable behavior? Or, in other words, since their corresponding relaxation modulus has always the form given by Eq. (8.2-7), i.e., that of a series of decaying exponential functions, can this series approximate any possible material behavior, and, if yes, what are the restrictions that are imposed on G_p and τ_p?

This question becomes more relevant when we start considering the relaxation behavior of crosslinking polymer systems such as that of polydimethyl-siloxane (PDMS) model networks, for which Winter and Chambon [1986] have observed a power-law relaxation behavior

$$G(s) = Ss^{-\frac{1}{2}} , \tag{8.2-8}$$

where S is the only material parameter involved, called the "strength" of the network at gel point or the "GP-strength" [Winter and Chambon, 1986, p. 375]. Observations of similar behavior, which date as far back as the beginning of this century [Nutting, 1921, p. 681], have given rise to constitutive equations in terms of "fractional rates of deformation" obtained when we apply the kernel defined in Eq. (8.2-8)—or its generalization for an arbitrary exponent $-\alpha$—to the Finger strain measure of deformation [VanArsdale, 1985, p. 854]. Equivalently, the stress can be provided implicitly through an equation which can be conceived as a generalization of the UCM or Oldroyd-B models where a "fractional derivative" replaces the upper-convected one. Although these generalizations simplify the calculations related to small deformation flows, the fact of the matter is that they lead to complex integro-differential equations for more general flows, and the question therefore is whether they can be replaced by conventional differential models.

It is, consequently, of interest to see what the most general linear viscoelastic behavior can be, whether it can be approximated through a series of decaying exponentials as given by Eq. (8.2-7), and what are the constraints imposed upon the system by the physics and the thermodynamics. Traditionally, the only constraint which is imposed *a priori* on the relaxation function, in addition to the finiteness and non-negative definiteness conditions described by Eqs. (8.2-2) and (8.2-3), respectively, is the monotonic condition [Brennan *et al.*, 1988, p. 208; Akyildiz *et al.*, 1990, p. 482]

$$G'(t) \leq 0 \ , \quad t \geq 0 \ , \tag{8.2-9a}$$

where the apostrophe denotes differentiation with respect to t. Occasionally, as needed for the analysis, the above condition is supplemented with additional ones, like convexity [Breuer and Onat, 1962, p. 358]

$$G''(t) \geq 0 \ , \quad t \geq 0 \ , \tag{8.2-9b}$$

or that the work done on any strain path starting from equilibrium is non-negative [Akyildiz *et al.*, 1990, p. 482],

$$\omega \equiv \int_{-\infty}^{t} \sigma(\tau)\dot{\gamma}(\tau)d\tau \geq 0 \ . \qquad (8.2\text{-}10)$$

However, as recognized by Brennan *et al* [1988, p. 208] and Akyildiz *et al.* [1990], the conditions mentioned above are necessary but not sufficient to guarantee that the relaxation function fulfills the requirements necessary for an interpretation in terms of mechanical analogues. Indeed, as shown below, additional requirements are needed in order to be able to consider $G(s)$ as representing a series of parallel spring-dashpot pairs which are physically realizable (i.e., which correspond to positive spring and dashpot constants).

Starting with the linear viscoelasticity constitutive relationship, Eq. (8.2-1), in order to develop the mathematical requirements on the relaxation modulus, we use here only two additional assumptions. The first one involves the continuity of $G(s)$ and it is justified as long as we agree to subtract from G any discontinuous part (like a viscous response) for which we can examine, if we wish, the thermodynamic and fading memory restrictions separately. The second one is the fading memory concept in viscoelasticity.

Moreover, in order to define a complete set of conditions for $G(s)$, it is instructive to consider the response of the system with respect to a sequence of especially chosen strain histories. Given the integral nature of the definition for the stress provided by Eq. (8.2-1) and the assumed continuity for $G(s)$, we can extend the applicability of Eq. (8.2-1) to certain generalized functions using the theory of distributions [Richards and Youn, 1990; Friedlander, 1982]. Indeed, let us choose as special test cases the strain-rate history given by an n-th derivative of the Dirac delta function [Richards and Youn, 1990; Friedlander, 1982; Korn and Korn, 1968]

$$\dot{\gamma}_n(t) = \delta^{(n)}(t) \equiv \frac{d^n\delta}{dt^n}(t) \ , \qquad (8.2\text{-}11)$$

where the superscript (n) denotes the n-th derivative and (0) denotes the function itself. Despite the fact that these strain-rate histories are singular at t=0, the corresponding stress, as defined through Eq. (8.2-1), is well-defined for t>0—see Eq. (8.2-14) below. These strain histories are of course an idealization which can only be approximated but never exactly reproduced in practice. Nevertheless, they are very useful in mathematical manipulations and they are often used in physical applications as they serve the role of "prototype" or "primitive" functions in terms of which an arbitrary function can be represented. For example, in electromagnetism, a charge density represented by the the n-th derivative of the delta function corresponds to a 2^n-multipole—see Eqs. (8.2-17) and (8.2-18) below. In particular in linear viscoelasticity, the n-th derivative of the delta function can be considered as an idealization which is obtained from a sequence of $n+1$ constant strain steps of alternating sign, occurring at times t=0, h, $2h$, ... , nh in the limit $h \to 0$—see below.

Further, let us use the simplest possible interpretation for the fading memory requirement on the linear viscoelastic response of a material: a material possesses fading memory if its stress magnitude monotonically decays (i.e., relaxes) after the completion of the application of any strain history [Bird *et al.*, 1987a, p. 263; Larson, 1988, p. 19]. This general definition for a fading memory and the assumption of thermodynamic stability for static equilibrium allow for the determination of the stress magnitude to the strain-rate history represented by the n-th derivative of the Dirac delta function, as follows.

First, since the Dirac delta function $\delta(t)$ (and its derivatives, $\delta^{(n)}(t)$) are not functions in the regular sense (their value is not defined everywhere in $\mathbb{R}$) the definition provided by Eq. (8.2-11) needs to be interpreted in the sense of distributions [Lighthill, 1958; Richards and Youn, 1990]. In particular, despite the fact that $\delta^{(n)}(t-s)$ is not defined everywhere, the integral of $\delta^{(n)}(t-s)$ with an arbitrary C^∞ function ψ is well defined as

$$\int_0^\infty \psi(s)\delta^{(n)}(t-s)ds = \psi^{(n)}(t) \ , \quad t > 0 \ . \tag{8.2-12}$$

Eq. (8.2-12) is obtained by n consecutive integrations-by-parts using the defining property of the Dirac delta function

$$\int_{-\infty}^\infty \delta(s)\psi(s)ds \equiv \psi(0) \ , \quad \psi \in C^\infty \ , \tag{8.2-13}$$

and the fact that $\psi(s)$ is assumed to be infinitely differentiable. Thus, the fact that the integrals of $\delta^{(n)}(t-s)$ are well-defined leads, when they are used to represent the strain-rate history, to well-defined stresses (for t>0), evaluated through Eq. (8.2-2), as:

$$\begin{aligned} \sigma(t) &\equiv \int_0^\infty G(s)\dot{\gamma}(t-s)ds \\ &= \int_0^\infty G(s)\delta^{(n)}(t-s)ds = G^{(n)}(t) \end{aligned} \quad , \quad t > 0 \ . \tag{8.2-14}$$

The use of generalized functions is only a mathematical trick aimed at rendering the analysis simpler and clearer. In practice, as mentioned above, the corresponding strains cannot be realized but only approximated. In this sense, we can view each strain history represented by $\delta^{(n)}(t)$ in Eq. (8.2-11) as representing the limit of a sequence of strain histories.

For example, $\delta^{(0)}(t)$ (i.e., Dirac's delta function) can be considered as a unit "pulse" obtained (among other possibilities) at the limit $\varepsilon \to 0$ from the (positive) pulse function $\Pi(x,\varepsilon)$ defined as

$$\Pi(x,\varepsilon) \equiv \begin{cases} 0 & |x| > \varepsilon \\ \dfrac{1}{\varepsilon} & |x| \le \varepsilon \end{cases} . \tag{8.2-15}$$

Indeed, it is straightforward to show (exploiting the continuity of $\psi(s)$) that use of the definition

$$\delta(x) \equiv \lim_{\varepsilon \to 0} \Pi(x,\varepsilon) , \tag{8.2-16}$$

leads to the same generalized function as defined from Eq. (8.2-13). It is easy to show (through an integration by parts) that this strain-rate history corresponds to the application of a unit positive strain at t=0. Similarly, the generalized derivatives of $\delta(s)$, $\delta^{(n)}(s)$, n=1,2..., in Eq. (8.2-11), can be also defined as

$$\delta^{(n)}(x) \equiv \lim_{h \to 0} M_n(x,h) , \tag{8.2-17}$$

where the *2^n-pole* $M_n(x,h)$ is defined recursively as

$$M_n(x,h) \equiv \begin{cases} \delta(x) & n = 0 \\ \dfrac{1}{h}\left(M_{n-1}(x-h,h) - M_{n-1}(x,h)\right) & n > 0 \end{cases} . \tag{8.2-18}$$

Again, it is straighforward to show (through a repeated use of integration by parts), that definition (8.2-17) is equivalent to (8.2-11) leading to the same integral expressions as those shown in Eq. (8.2-12). Also note that both the delta function (or *1-pole* or *monopole* according to the above definition) and its multipole extensions have found a widespread use in physics, as for example in electromagnetism. Thus, although the exact distribution function cannot of course be realized as a strain-rate history in practice, a close approximation of it can.[26]

[26]Of course, how close we can get to $\delta(s)$ and still have a linear viscoelastic response, as indicated by Eq. (8.2-1), depends on how steep the changes we can implement are without having other effects (for example, inertial effects) come into play. However, similar restrictions are quite common with all mathematical models of reality, the continuum approximation itself being a first-rate example.

As implied from its definition, Eq. (8.2-16), as the limit of the non-negative pulse function $\Pi(t,\varepsilon)$, the unit monopole, $M_0(t)$ (i.e., the delta function, $\delta(t)$), when used to describe the shear-rate history results in a positive step change in the strain which, in order for the static equilibrium to be thermodynamically stable, results in a non-negative stress response, $\sigma(t)\geq 0$, $t\geq 0$. Moreover, by invoking the fading memory principle, the unit dipole, $M_1(t,h)$, when used to describe the shear-rate history can only be expected to generate for $t>h$ a non-negative stress response since the corresponding strain-rate history involves two equal in magnitude but opposite in sign strain actions with the positive one occurring more recently in time, as indicated from the defining Eq. (8.2-18) for $n=1$. In the same way, we can recursively show that the general 2^n-pole should result, for $t>nh$, to a non-negative stress response, since, according to the definition provided by Eq. (8.2-18), it corresponds to two equal in magnitude but opposite in sign strain-rate actions with the positive one occurring more recently. Thus, we can safely generalize and consider as a result of the fading memory of a material that its stress response for a strain-rate history represented by a 2^n-pole (for $n\geq 1$) is expected (for $t>nh$) to be non-negative. Now, if this is valid for any finite value of h, it is also valid (due to the continuity of the relaxation function) if we take the limit $h\rightarrow 0$ of the before-mentioned multipole actions. But in that limit, according to Eq. (8.2-17), the 2^n-pole (for $n\geq 1$) coincides with $(-1)^n\delta^{(n)}(s)$. Therefore, the stress response initiated by a strain-rate history represented by $\delta^{(n)}(t)$, $n\geq 1$, should be non-negative or non-positive for n even or odd, respectively. This observation together with the corresponding result for $n=0$ discussed above, leads, in conjunction with Eq. (8.2-14), to the following inequalities for the modulus $G(s)$ and its derivatives:

$$(-1)^n G^{(n)}(s) \geq 0 \ , \quad n=0,1,2,\dots \ . \tag{8.2-19}$$

Hence the complete set of constraints imposed on the relaxation modulus, $G(s)$, as implied by the conditions of a thermodynamically stable static equilibrium and the fading memory principle, is given by Eq. (8.2-19). This set completely encompasses all other conditions mentioned above (except the finiteness constraints), i.e., those described by Eqs. (8.2-9) and (8.2-10)—see Example 8.1 for a proof of the last inequality which relies on the inverse Laplace transform of $G(s)$ as defined in the next paragraph.

The conditions described by Eq. (8.2-19) are exactly those that are necessary in order to define $G(s)$ as a *completely monotonic function*, a term firstly employed by Bernstein [1928, ch. 1]—see also Pipkin [1986, p. 47]. According to Bernstein, a function is completely monotonic if and only if its inverse Laplace transform is non-negative. Thus, there is a one-to-one correspondence between a completely monotonic function $G(t)$ with the Laplace transform of a non-negative function $P(x)$, called the *relaxation spectrum*, defined through the usual relationship [Bernstein, 1928, p. 56; Pipkin, 1986, p. 47],[27]

$$G(t) = \int_0^\infty P(x)\exp(-xt)\mathrm{d}x \ , \quad P(x) \geq 0 \ . \tag{8.2-20}$$

Furthermore, $G(t)$ as a completely monotonic function can be arbitrarily closely approximated (as $N \to \infty$) by a series of exponential functions

$$G(t) \approx \sum_{i=1}^{N} G_i \exp(-\frac{t}{\lambda_i}) , \quad G_i \geq 0 , \quad \lambda_i > 0 , \quad i = 1,2,...,N \ , \tag{8.2-21}$$

where G_i and λ_i, the characteristic relaxation modulus and relaxation time of the i-th mode, are positive constants uniquely determined from the optimum approximation of $G(t)$ [Bernstein, 1928, Theorem E, p. 20]. The optimum approximation can be constructed so that, for given N, the first N derivatives at t=0 are identical and the difference between the remaining ones are bounded, alternating from below and above [Bernstein, 1928, Theorem E, p. 20].

What is especially important about the representation offered by Eq. (8.2-21) is that it corresponds to an arrangement of N spring/dashpot pairs in parallel, with the i-th spring corresponding to the elastic spring constant $K_i \equiv G_i$ connected in series with the i-th dashpot corresponding to a viscosity $\mu_i \equiv \lambda_i G_i$, i=1,2,...,N. The requirement that all the constants involved are non-negative exactly matches the conditions obtained demanding the thermodynamic consistency of the mechanical analog model, i.e., those arising from the requirement of a stable thermodynamic static equilibrium (G_i>0) and a non-negative rate of entropy production ($\mu_i \geq 0$). Thus, the description for linear viscoelasticity provided by the linear superposition of linear Maxwell modes is completely equivalent to the one corresponding to the integral expression, Eq. (8.2-1), including all the inequality constraints required by the stability of the static equilibri-

[27]Actually the finiteness conditions defined by Eq. (8.2-2) result in restricting G(s) to a subset of the completely monotonic functions for which the corresponding relaxation spectrum satisfies the integral constraints [Pipkin, 1986, p. 47]

$$\int_0^x P(y)\mathrm{d}y < 0 \ , \quad \int_0^x \frac{P(y)}{y}\mathrm{d}y < 0 \ , \quad 0 < x < \infty \ .$$

um and the fading memory principle.[28] It therefore also suggests as an obvious generalization of the single-variable models for arbitrary flow deformations their multiple-mode linear superposition. This extension is addressed in the following section.

EXAMPLE 8.1: Show that the thermodynamic inequality of (8.2-10) is satisfied automatically for a completely monotonic relaxation function. In this sense, the fading memory principle provides a stronger criterion than the thermodynamic work, at least as long as both are applied within the limitations of linear viscoelasticity.

Since we know that the relaxation modulus is given by the Laplace transform of the relaxation spectrum, $P(x)$, we can use Eqs. (8.2-1) and (8.2-20) to rewrite (8.2-10) (for a strain-rate history $\dot{\gamma}(s)=0$ when $s<0$) as

$$\omega = \int_{x=0}^{\infty} P(x) \int_{\tau=0}^{t} \int_{s=0}^{\tau} \exp(-x(\tau-s))\dot{\gamma}(s)\dot{\gamma}(\tau)\mathrm{d}s\mathrm{d}\tau\mathrm{d}x \ .$$

Then, if we approximate $\dot{\gamma}$ by a series of delta functions,

$$\dot{\gamma}(s) \approx \sum_{i=1}^{N} \Delta\dot{\gamma}_i \delta(t-t_i) \ , \quad 0<t_1<t_2<\ldots<t_N<t \ ,$$

ω becomes

[28]Regarding the related question of the appropriateness of the use of negative Maxwell modes (i.e., modes corresponding to negative relaxation moduli) within a discrete spectrum of relaxation modes, the above discussion does not imply that their use always (that is, for every deformation history) leads necessarily to a violation of the fading memory principle or to an overall negative rate of entropy production. The conclusion simply is that there exist deformations for which utilization of negative Maxwell modes can provide aphysical results. Thus, whether or not one should use negative Maxwell modes in approximating experimentally determined stress relaxation data is a question of judgement, not unlike the question of utilization of materially objective constitutive equations. In a certain limited class of applications the use of negative Maxwell modes might provide acceptable approximations. (We can also draw here the parallelism with material objectivity which is deliberately violated in linear viscoelasticity without any adverse consequences so long as the applied deformation does not involve large rotations.) However, for which class of flows are negative Maxwell modes allowed remains an open question. We believe that the only way to make sure aphysical results are avoided is to use the criteria established in the preceding paragraphs which are also consistent with an internal variable representation of a series of spring dashpot pairs in parallel.

$$\omega = \int_{x=0}^{\infty} P(x) \tfrac{1}{2} \sum_{n=1}^{N} \sum_{m=1}^{N} \exp(-x|t_n - t_m|) \Delta\dot{\gamma}_n \Delta\dot{\gamma}_m \, dx \ ,$$

which is always non-negative since

$$P(x) \geq 0 \ , \quad x \geq 0 \ ,$$

and

$$\sum_{n=1}^{N} \sum_{m=1}^{N} \exp(-x|t_n - t_m|) \Delta\dot{\gamma}_n \Delta\dot{\gamma}_m \geq 0 \ , \quad x \geq 0 \ .$$

Note that the last inequality in the above expression is due to the non-negative nature of the quadratic form, as established through a straightforward calculation of the determinants of the principal minors of the corresponding $N \times N$ square matrix:

$$\det(i\text{-}th \ principal \ minor) = \prod_{n=2}^{i} \left(1 - \exp(-2x|t_n - t_{n-1}|)\right) \ .$$

By taking the limit as $N \to \infty$, in accordance with the usual practice, the non-negative character of the continuous integral expression immediately follows. ■

8.2.2 *Phenomenological multiple-variable models*

As we saw from the previous section, most viscoelastic media exhibit not just one, but a whole spectrum of relaxation times, which may or may not be related to the polydispersity index of the molecular weight distribution since they are observed even for dilute solutions of single molecular-weight chains. As we shall see in this subsection, the use of multiple conformation tensors is a natural representation of the varied microstructure of these materials, which can fully compensate for a range of system relaxation times. Through the generalized bracket, it also becomes apparent that, unlike previous multi-relaxation modes models, some form of coupling between the various conformation tensors must be allowed for a completely general system description. This coupling leads to a new class of models based upon the concept of multiple relaxation modes, as illustrated below.

A. Decoupled Multiple Relaxation Modes Models

Typically, when modelers construct constitutive equations for materials with multiple relaxation times, they couple the momentum equation, (8.1-7a), with any given number, say N, of *independent, decoupled* modes for the extra-stress tensor [Larson, 1988, pp. 47,71,200-04]:

$$\sigma^S_{\alpha\beta} = \sum_{i=1}^{N} \sigma^i_{\alpha\beta} \ , \tag{8.2-22}$$

each one of the stress tensors $\boldsymbol{\sigma}^i$ satisfying an independent constitutive relationship

$$\lambda_i \check{\sigma}^i_{\alpha\beta} + \sigma^i_{\alpha\beta} = 2\eta_i A_{\alpha\beta} \ , \quad i=1,2,\ldots,N \ , \tag{8.2-23}$$

where we have chosen, as an example, the upper-convected Maxwell equation; however, the particular form of the constitutive equations can be any one of the forms discussed in §8.1. Note that we now have an entire spectrum of relaxation times and viscosities, one of each for all relaxation modes. Hence the above system corresponds to N (non-interacting) Maxwell models arranged in a parallel configuration.

Eqs. (8.2-23) form a system of N first-order differential equations which, although coupled implicitly through the velocity field which responds to the total extra stress tensor present in the momentum equation, have no explicit coupling between the various modes of $\boldsymbol{\sigma}^S$. Using standard mathematical techniques, the system of (8.2-23) could also be expressed as a single differential equation of N-th order.

EXAMPLE 8.2: When N=2, show how the two first-order differential equations of (8.2-23) may be combined to give a single second-order differential equation for the total extra stress.

First, we have to define the total extra stress, $\boldsymbol{\sigma}^S$, and another quantity, $\boldsymbol{\sigma}^D$, as

$$\boldsymbol{\sigma}^S \equiv \boldsymbol{\sigma}^1 + \boldsymbol{\sigma}^2 \quad \text{and} \quad \boldsymbol{\sigma}^D \equiv \boldsymbol{\sigma}^1 - \boldsymbol{\sigma}^2 \ ,$$

whereby

$$\boldsymbol{\sigma}^1 \equiv \tfrac{1}{2}(\boldsymbol{\sigma}^S + \boldsymbol{\sigma}^D) \quad \text{and} \quad \boldsymbol{\sigma}^D \equiv \tfrac{1}{2}(\boldsymbol{\sigma}^S - \boldsymbol{\sigma}^D) \ .$$

Now let us substitute the first of the above equations into (8.2-23) for i=1 and i=2, and then multiply by λ_1 and λ_2, respectively, to obtain

$$\tfrac{1}{2}\lambda_2(\boldsymbol{\sigma}^S + \boldsymbol{\sigma}^D) + \tfrac{1}{2}\lambda_1\lambda_2(\check{\boldsymbol{\sigma}}^S + \check{\boldsymbol{\sigma}}^D) = 2\eta_1\lambda_2\mathbf{A} \ ,$$

and

$$\tfrac{1}{2}\lambda_1(\boldsymbol{\sigma}^S - \boldsymbol{\sigma}^D) + \tfrac{1}{2}\lambda_1\lambda_2(\check{\boldsymbol{\sigma}}^S - \check{\boldsymbol{\sigma}}^D) = 2\eta_2\lambda_1\mathbf{A} \ .$$

Adding these two expressions together gives

$$\tfrac{1}{2}(\lambda_1+\lambda_2)\boldsymbol{\sigma}^S+\lambda_1\lambda_2\breve{\boldsymbol{\sigma}}^S-2(\eta_1\lambda_2+\eta_2\lambda_1)\mathbf{A}=\tfrac{1}{2}(\lambda_1-\lambda_2)\boldsymbol{\sigma}^D\ .$$

Substituting $\boldsymbol{\sigma}^D$ as given above into the first one of the two first-order differential equations right above it gives a second-order differential equation for $\boldsymbol{\sigma}^S$:

$$\boldsymbol{\sigma}^S+\tfrac{1}{2}(\lambda_1+\lambda_2)\breve{\boldsymbol{\sigma}}^S+\lambda_1\lambda_2\ddot{\boldsymbol{\sigma}}^S=2(\eta_1+\eta_2)\mathbf{A}+2(\eta_1\lambda_2+\eta_2\lambda_1)\breve{\mathbf{A}}\ ,$$

where by the doubledot we denote here the upper-convected derivative of the upper-convected derivative. Note that the positivity of η_1, η_2, λ_1, and λ_2 implies the positivity of the parameters appearing in the above equation. ■

The representation of the decoupled-mode model using the generalized formalism is discussed below as a particular case of the coupled multiple-modes model.

B. General Formulation of Multiple Relaxation Modes Models

The Poisson bracket for a system displaying multiple relaxation modes is expressed in terms of functionals of the momentum density vector field and the various modes of the conformation tensor: $F=F[\mathbf{M},\mathbf{c}^1,\mathbf{c}^2,\ldots,\mathbf{c}^N]$. The procedure of §5.5.2 is then easily generalized to the multiple mode case as

$$\begin{aligned}\{F,G\}=&-\int_\Omega\left(\frac{\partial F}{\partial M_\gamma}\nabla_\beta\Big(\frac{\partial G}{\partial M_\beta}M_\gamma\Big)-\frac{\partial G}{\partial M_\gamma}\nabla_\beta\Big(\frac{\partial F}{\partial M_\beta}M_\gamma\Big)\right)d^3x\\&-\sum_{i=1}^N\int_\Omega\left(\frac{\partial F}{\partial c^i_{\alpha\beta}}\nabla_\gamma\Big(c^i_{\alpha\beta}\frac{\partial G}{\partial M_\gamma}\Big)-\frac{\partial G}{\partial c^i_{\alpha\beta}}\nabla_\gamma\Big(c^i_{\alpha\beta}\frac{\partial F}{\partial M_\gamma}\Big)\right)d^3x\\&-\sum_{i=1}^N\int_\Omega c^i_{\gamma\alpha}\left(\frac{\partial G}{\partial c^i_{\alpha\beta}}\nabla_\gamma\Big(\frac{\partial F}{\partial M_\beta}\Big)-\frac{\partial F}{\partial c^i_{\alpha\beta}}\nabla_\gamma\Big(\frac{\partial G}{\partial M_\beta}\Big)\right)d^3x\\&-\sum_{i=1}^N\int_\Omega c^i_{\gamma\beta}\left(\frac{\partial G}{\partial c^i_{\alpha\beta}}\nabla_\gamma\Big(\frac{\partial F}{\partial M_\alpha}\Big)-\frac{\partial F}{\partial c^i_{\alpha\beta}}\nabla_\gamma\Big(\frac{\partial G}{\partial M_\alpha}\Big)\right)d^3x\ .\end{aligned}\tag{8.2-24}$$

Analogously to (8.1-5), the dissipation bracket (minus the entropy production terms) is defined as

$$
\begin{aligned}
[F,G] = &-\sum_{i=1}^{N}\sum_{j=1}^{N}\int_{\Omega}\Lambda^{ij}_{\alpha\beta\gamma\varepsilon}\frac{\delta F}{\delta c^{i}_{\alpha\beta}}\frac{\delta G}{\delta c^{j}_{\gamma\varepsilon}}\mathrm{d}^3x \\
&-\int_{\Omega}Q_{\alpha\beta\gamma\varepsilon}\nabla_{\alpha}\frac{\delta F}{\delta M_{\beta}}\nabla_{\gamma}\frac{\delta G}{\delta M_{\varepsilon}}\mathrm{d}^3x \\
&-\sum_{i=1}^{N}\int_{\Omega}L^{i}_{\alpha\beta\gamma\varepsilon}\left(\nabla_{\alpha}\frac{\delta F}{\delta M_{\beta}}\frac{\delta G}{\delta c^{i}_{\gamma\varepsilon}} - \nabla_{\alpha}\frac{\delta G}{\delta M_{\beta}}\frac{\delta F}{\delta c^{i}_{\gamma\varepsilon}}\right)\mathrm{d}^3x \ ,
\end{aligned}
\tag{8.2-25}
$$

where we have neglected again the inhomogeneous terms (**B**=**0** in (8.1-5)), and we have assumed that the Onsager relations require that $\boldsymbol{\Lambda}^{ij}=\boldsymbol{\Lambda}^{ji}$. Note that in general this dissipation bracket allows explicit coupling between cross modes (as proportional to $\boldsymbol{\Lambda}^{ij}$, $i\neq j$), in contrast to the usual modeling procedure of (8.2-22,23).

The Hamiltonian for multi-mode systems is a straightforward generalization of the Hamiltonian in terms of a single conformation tensor:

$$
\begin{aligned}
H_m[\mathbf{M},\mathbf{c}^1,\mathbf{c}^2,\ldots,\mathbf{c}^N] = &\int_{\Omega}\frac{\mathbf{M}\cdot\mathbf{M}}{2\rho}\mathrm{d}^3x \\
&+\tfrac{1}{2}\sum_{i=1}^{N}\int_{\Omega}\left(n_iK_i\mathrm{tr}\mathbf{c}^i - n_ik_BT\ln(\det\tilde{\mathbf{c}}^i)\right)\mathrm{d}^3x \ ,
\end{aligned}
\tag{8.2-26}
$$

where n_i and K_i are the elasticity density and the spring constant for the i-th mode, respectively. In this case, for convenience, we have assumed the simplest form of the Helmholtz free energy used in §8.1. Although the modes are coupled through the dissipation bracket, we see no coupling in the Hamiltonian between the various modes. Hence the expression of (8.2-26) is probably rather naïve, yet we shall use it in this subsection in order to keep the generalization to multi-mode systems as straightforward as possible. By writing the Hamiltonian of (8.2-26) as

$$
H_m[\mathbf{M},\mathbf{c}^1,\mathbf{c}^2,\ldots,\mathbf{c}^N] = H^0_m[\mathbf{M}]+H^1_m[\mathbf{c}^1]+H^2_m[\mathbf{c}^2]+\ldots+H^N_m[\mathbf{c}^N] \ , \tag{8.2-27}
$$

we can express the functional derivatives of H_m in terms of the functional derivatives of H^i_m evaluated as

$$
\frac{\delta H^i_m}{\delta \mathbf{c}^j} = \begin{cases} \frac{1}{2}n_iK_i\boldsymbol{\delta} - \frac{1}{2}n_ik_BT(\mathbf{c}^i)^{-1} & , \ i=j \\ 0 & , \ i\neq j \end{cases} . \tag{8.2-28}
$$

This expression is analogous to (8.1-14) for a single-mode system. The dynamical equation for our arbitrary functional F is then given by

$$\frac{dF}{dt} = \int_\Omega \left(\frac{\partial F}{\partial \mathbf{M}} \cdot \frac{\partial \mathbf{M}}{\partial t} + \sum_{i=1}^{N} \frac{\partial F}{\partial \mathbf{c}^i} \cdot \frac{\partial \mathbf{c}^i}{\partial t} \right) d^3x \equiv \{[F,H]\} \;, \tag{8.2-29}$$

so that the coupled evolution equations for the multi-mode system may now be derived as

$$\rho \frac{\partial \upsilon_\alpha}{\partial t} = -\rho \upsilon_\beta \nabla_\beta \upsilon_\alpha - \nabla_\alpha p + \nabla_\beta \sigma^S_{\alpha\beta} \;, \tag{8.2-30a}$$

and

$$\begin{aligned} \frac{\partial c^i_{\alpha\beta}}{\partial t} = &- \rho \upsilon_\gamma \nabla_\gamma c^i_{\alpha\beta} + c^i_{\alpha\gamma} \nabla_\gamma \upsilon_\beta + c^i_{\gamma\beta} \nabla_\gamma \upsilon_\alpha \\ &- \sum_{j=1}^{N} \Lambda^{ij}_{\alpha\beta\gamma\varepsilon} \frac{\delta H^j_m}{\delta c^i_{\gamma\varepsilon}} + L^i_{\alpha\beta\gamma\varepsilon} \nabla_\gamma \upsilon_\varepsilon \end{aligned} \;, \; i=1,2,\ldots,N \;, \tag{8.2-30b}$$

where

$$\begin{aligned} \sigma^S_{\alpha\beta} &\equiv Q_{\alpha\beta\gamma\varepsilon} \nabla_\gamma \upsilon_\varepsilon + \sum_{i=1}^{N} \sigma^i_{\alpha\beta} \\ &\equiv Q_{\alpha\beta\gamma\varepsilon} \nabla_\gamma \upsilon_\varepsilon + \sum_{i=1}^{N} \left(2 c^i_{\beta\gamma} \frac{\delta H^i_m}{\delta c^i_{\alpha\gamma}} + 2 L^i_{\alpha\beta\gamma\varepsilon} \frac{\delta H^i_m}{\delta c^i_{\gamma\varepsilon}} \right) . \end{aligned} \tag{8.2-30c}$$

This set of equations is completely general (for homogeneous systems), irrespective of the Hamiltonian or the phenomenological matrices chosen as a particular case. Hence it represents the generalization of the set of equations, Eq. (8.1-7), for the single-mode system to the present situation of multiple, coupled relaxation modes.

Now let us look at a specific case for the phenomenological parameters. Let us assume that the $\mathbf{L}^i$ have the simple form of (8.1-26),

$$L^i_{\alpha\beta\gamma\varepsilon} = \frac{a_i - 1}{2} (c^i_{\alpha\gamma} \delta_{\beta\varepsilon} + c^i_{\alpha\varepsilon} \delta_{\beta\gamma} + c^i_{\beta\gamma} \delta_{\alpha\varepsilon} + c^i_{\beta\varepsilon} \delta_{\alpha\gamma}) , \; i=1,2,\ldots,N \;, \tag{8.2-31}$$

and that the diagonal (in mode space) components of $\boldsymbol{\Lambda}^{ij}$ (i.e., $\boldsymbol{\Lambda}^{ii}$, i=1,2,...,N) have the form of the Giesekus mobility tensor,

$$\begin{aligned} \Lambda^{ii}_{\alpha\beta\gamma\varepsilon} = \frac{1}{2\lambda_i} \Bigg(&\frac{1-\alpha_i}{n_i K_i} (c^i_{\alpha\gamma} \delta_{\beta\varepsilon} + c^i_{\alpha\varepsilon} \delta_{\beta\gamma} + c^i_{\beta\gamma} \delta_{\alpha\varepsilon} + c^i_{\beta\varepsilon} \delta_{\alpha\gamma}) \\ &+ \frac{2\alpha_i}{G^i_0} (c^i_{\alpha\gamma} c^i_{\beta\varepsilon} + c^i_{\alpha\varepsilon} c^i_{\beta\gamma}) \Bigg) . \end{aligned} \tag{8.2-32}$$

Furthermore, although we could easily pick up a solvent viscosity of the form (8.1-19), let us assume for simplicity that **Q**=**0**. This leaves us only to specify the off-diagonal components of $\boldsymbol{\Lambda}^{ij}$ (i.e., when $i \neq j$). Rather than write a general expression for $\boldsymbol{\Lambda}^{ij}$ which gives $\boldsymbol{\Lambda}^{ii}$ as (8.2-32) when $i=j$, we opt here to consider $\boldsymbol{\Lambda}^{ij}$, $i \neq j$, explicitly in order to illustrate how the coupling between the various modes of the conformation tensor must be treated in this type of formalism. Specifically, we require that the phenomenological matrix $\boldsymbol{\Lambda}^{ij}$, $i \neq j$, maintain the integrity of the symmetries embodied in the dissipation bracket and the reciprocal relations, as (8.2-32) obviously does. With these properties in mind, we define the simplest anisotropic $\boldsymbol{\Lambda}^{ij}$, $i \neq j$, as

$$\Lambda^{ij}_{\alpha\beta\gamma\varepsilon} = \frac{1}{2k_BT}\frac{\theta_{ij}}{\sqrt{n_in_j\lambda_i\lambda_j}}\left(c^i_{\alpha\gamma}c^j_{\beta\varepsilon} + c^i_{\alpha\varepsilon}c^j_{\beta\gamma} + c^i_{\beta\gamma}c^j_{\alpha\varepsilon} + c^i_{\beta\varepsilon}c^j_{\alpha\gamma}\right),\ i \neq j\ , \quad (8.2\text{-}33)$$

where the dimensionless scalar coefficients θ_{ij} are, in general, symmetric functions of the invariants of the tensors $\mathbf{c}^i$ and $\mathbf{c}^j$. Eq. (8.2-33) explicitly illustrates the coupling that occurs between the various $\mathbf{c}^i$ for nonzero values of the coupling coefficients θ_{ij}. Note that when a_i=1, α_i=0, and $\boldsymbol{\Lambda}^{ij}$=**0**, $i \neq j$, the above system of equations reduces to the parallel Maxwell models of (8.2-22,23).

From Onsager's symmetry relations, the coupling matrix $\boldsymbol{\theta}$ (in the $N \times N$ mode space) should be symmetric. Even so, not all values of θ_{ij} are necessarily acceptable from a thermodynamic point of view, i.e., lead to a positive dissipation. To obtain the thermodynamically imposed constraints upon the θ_{ij}, the approach explained in §8.1.4 needs to be followed. This is straightforward, but very tedious, if it is to be worked out for the most general case. Consequently, let us examine here the simplest case when N=2, a_i=1, and α_i=0 for i=1,2. In this special case, called—for convenience—the two-coupled Maxwell-modes model, we have

$$\sigma^S_{\alpha\beta} = \sum_{i=1}^{2}\left(n_iK_ic^i_{\alpha\beta} - n_ik_BT\delta_{\alpha\beta}\right)\ , \quad (8.2\text{-}34a)$$

and

$$\overset{\nabla}{c}{}^1_{\alpha\beta} = -\frac{1}{\lambda_1}c^1_{\alpha\beta} + \frac{k_BT}{\lambda_1K_1}\delta_{\alpha\beta} - \frac{\theta}{2k_BT}\sqrt{\frac{n_2}{n_1}}\frac{1}{\sqrt{\lambda_1\lambda_2}} \quad (8.2\text{-}34b)$$

$$\times\left(K_2(c^1_{\alpha\gamma}c^2_{\gamma\beta} + c^2_{\alpha\gamma}c^1_{\gamma\beta}) - 2k_BTc^1_{\alpha\beta}\right)\ ,$$

where $\theta = \theta_{12} = \theta_{21}$, and the evolution equation for $\mathbf{c}^2$ is obtained from Eq. (8.2-34b) by cyclic permutation in the indices *1 and 2*.

After substitution of the expressions for $\boldsymbol{\Lambda}$ given by Eqs. (8.2-32) and (8.2-33) into (8.1-112), the following expression for the corresponding rate of energy dissipation is obtained

$$
\begin{aligned}
-\frac{[H_m, H_m]}{k_B T} &= \sum_{i=1}^{2} \frac{n_i}{2\lambda_i}\left(\mathrm{tr}\tilde{\mathbf{c}}^i - 6 + \mathrm{tr}((\tilde{\mathbf{c}}^i)^{-1})\right) \\
&\quad + \theta \sqrt{\frac{n_1 n_2}{\lambda_1 \lambda_2}} \left(\tilde{\mathbf{c}}^1 : \tilde{\mathbf{c}}^2 - \mathrm{tr}\tilde{\mathbf{c}}^1 - \mathrm{tr}\tilde{\mathbf{c}}^2 + 3\right) ,
\end{aligned}
\tag{8.2-35}
$$

where $\tilde{\mathbf{c}}$ denotes, as usual, a dimensionless conformation tensor defined as

$$
\tilde{c}^i_{\alpha\beta} \equiv \frac{K_i}{k_B T} c^i_{\alpha\beta} , \tag{8.2-36}
$$

so that it reduces to the delta function at static equilibrium, at which condition the rate of energy dissipation defined by Eq. (8.2-35) reduces (as it should) to zero.

Several of the models that we have discussed so far can be recovered as particular cases of the constitutive equations defined by (8.2-34b) and its equivalent for component 2. First, of course, in the characteristic limit θ=0, the model describing a pair of decoupled Maxwell modes is recovered. Thus, in this limit it is trivial to show[29] that the requirement of a non-negative rate of energy dissipation is duly satisfied for arbitrary values of the (positive definite) conformation tensors $\mathbf{c}^i$, i=1,2. Second, in the characteristic limit where the principal material parameters are the same (i.e., $n_1=n_2$ and $\lambda_1=\lambda_2$) the two modes obey similar evolution equations which are identical to that of the single-mode Giesekus model, Eq. (8.1-86), for $\alpha=\theta/(1+\theta)$ and $\lambda(\text{Giesekus})=\lambda/(1+\theta)$. Therefore, in this limit, we can use the inequality $0 \leq \alpha \leq 1$ as a sufficient criterion for a non-negative rate of energy dissipation—see Eq. (8.1-130)—which translates in the requirement $\theta \geq 0$ in the two coupled Maxwell modes model notation.

[29]This task can be most easily accomplished by simply expressing $\mathrm{tr}\tilde{\mathbf{c}}^i$ and $\mathrm{tr}(\tilde{\mathbf{c}}^i)^{-1}$ as a sum of the eigenvalues and the inverse of eigenvalues λ^i_j, j=1,2,3 of $\tilde{\mathbf{c}}^i$, i=1,2, respectively, and by collecting terms corresponding to the same eigenvalue:

$$
\begin{aligned}
\mathrm{tr}\mathbf{c} - 6 + \mathrm{tr}(\mathbf{c}^{-1}) &= \sum_{i=1}^{3} (\lambda_i - 2 + \frac{1}{\lambda_i}) \\
&= \sum_{i=1}^{3} \frac{1}{\lambda_i} (\lambda_i - 1)^2 \geq 0 ,
\end{aligned}
$$

where λ_i, i=1,2,3 are the (positive) eigenvalues of the posititve definite second-rank tensor $\mathbf{c}$. It can also be shown as a particular case of the first case examined in §8.1.4.

p. 305: line after (8.2-36), "delta function" should be "unit tensor"

p. 306: [Beris *et al.*, 1993] should be [Beris *et al.*, 1994] everywhere

This sufficient condition is consistent with the condition conjectured in Example 8.3, below, based on the analysis of steady simple shear flow in the limit of small shear rates:

$$|\theta| \leq 1 \quad or \quad \theta \geq \frac{\lambda_1 n_1 + \lambda_2 n_2}{\sqrt{\lambda_1 n_1 \lambda_2 n_2}} \; . \tag{8.2-37}$$

Unfortunately, the complexity of the expression for the rate of dissipation in Eq. (8.2-35) prevents us from formulating the thermodynamic admissibility criterion for general flows.

Based on the analysis of linear viscoelastic behavior, as discussed in the following paragraph, we can obtain the stricter constraint $|\theta| \leq 1$ by invoking the fading memory principle [Beris *et al.*, 1994]. Furthermore, examination of the zero shear rate value of the ratio between the second and first normal stress coefficients reveals that this remains within the experimentally established bounds (small in magnitude and negative) only when θ is small and positive. Again, in the Giesekus model limit, these θ values correspond to the typically utilized in practice values for the Giesekus α parameter, small and positive. Thus, one should expect that the regime of values of θ for which a physically realistic behavior is obtained is $0 \leq \theta \ll 1$ [Beris *et al.*, 1993]. Even with these limitations, the range of the interaction parameter values is large enough to make mode interaction effects very important. This is especially true in relation to the second normal stress and the linear viscoelastic behavior [Beris et al., 1993], which are among the few model properties that have been investigated so far. The nonlinear characteristics, can only at this stage be anticipated by analogy with the Giesekus model: to show a shear thinning viscosity and normal stress behavior- a full analysis still remains to be done.

The linear viscoelastic behavior of the two coupled Maxwell modes model appears to be especially interesting. As is shown in [Beris *et al.*, 1993], the linear stress relaxation modulus is described by two modes

$$G(t) = G_1 \exp(-\frac{t}{t_1}) + G_2 \exp(-\frac{t}{t_2}) \; , \tag{8.2-38}$$

where the parameters G_1, G_2, t_1, and t_2 are non-negative (thus, acceptable, according to the fading memory based arguments presented in §8.2.1) if and only if $|\theta| \leq 1$. Interestingly enough, the two characteristic times t_1 and t_2 are, in general, different—and on many occasions very different—from the characteristic relaxation times λ_1 and λ_2. This observation is significant, in particular when one is trying to interpret the relaxation spectrum of a complex polymeric material derived in the limit of linear viscoelasticity since it suggests that the relaxation times obtained there might not necessarily correspond to intrinsic relaxation times of structural components of the material under investigation. Instead, they might just be the result of a complicated interaction between various structural components which is especially important to have in mind when one studies multiphase systems such as polymer blends and 3-dimensional elastic networks. Of course, the two coupled Maxwell modes model, due to its inherent simplicity and arbitrary phenomenological nature on the way the interaction between the two Maxwell modes is introduced, can only provide some qualitative answers on the way mode interactions can influence the flow behavior. For more quantitative results, models based on a clearer connection with the underlying physics should be used, as the ones discussed in the following subsection. Still, we believe such that the principles are the same and since those are more clearly illustrated in the simple model introduced above we consider it useful to offer a few more details with the opportunity of addressing the range of non-negative rates of energy dissipation in the following example.

EXAMPLE 8.3: Derive a set of necessary conditions on the coupling parameter θ of the two coupled Maxwell modes model described above by imposing the constraint of a non-negative rate of energy dissipation on the steady simple shear flow for low values of shear rate.

In an orthogonal (Cartesian) system of coordinates, x_1, x_2 and x_3, let us assume the establishment of a (homogeneous) steady simple shear flow, such that $\boldsymbol{v}=(\dot{\gamma}x_2,0,0)^T$. In the limit of a vanishingly small shear rate, the stress components have the following dependence on $\dot{\gamma}$:

$$\sigma_{12} = \eta_0\dot{\gamma} \ ,$$

$$\sigma_{11} - \sigma_{22} = \Psi_1^0\dot{\gamma}^2 \ ,$$

$$\sigma_{22} - \sigma_{33} = \Psi_2^0\dot{\gamma}^2 \ ,$$

where η_0, Ψ_1^0 and Ψ_2^0 are constants representing material parameters, the zero shear-rate viscosity, first and second normal-stress coefficients, respectively [Bird *et al.*, 1987a, §3.3]. These stress dependencies are quite generic and represent the lowest order terms (up to $\dot{\gamma}^2$) in a series expansion with respect to $\dot{\gamma}$, as, for example, can be obtained from the retarded motion expansion valid for small velocity gradients [Bird *et al.*, 1987a, p. 297]. More specifically, the above expressions are those obtained using the simplest second-order fluid representation retaining up to second-order dependencies on $\dot{\gamma}$; see [Bird *et al.*, 1987, p. 297] for the next highest (third-order) correction in $\dot{\gamma}$ (applicable for the shear stress, σ_{12}) obtained using a third order fluid representation (which involves the next higher order correction in the retarded motion expansion). Note that the next correction for the normal stresses is fourth order in $\dot{\gamma}$. (In general the expansions involve only even or odd terms as a reflection of the symmetry to flow reversal).

Guided from the form of the dependency of the stress components on $\dot{\gamma}$ and given the linear dependence between the stress and conformation tensor for a Maxwellian (Hookean) free energy, Eq. (8.2-34a), the lowest order expansions (up to and including second-order terms) on $\dot{\gamma}$ for the components of the (dimensionless) conformation tensors $\tilde{\mathbf{c}}^i$, i=1,2, are:

$$\tilde{c}^i_{12} = K_i\dot{\gamma} \ , \quad i=1,2 \ ,$$

$$\tilde{c}^i_{11} - \tilde{c}^i_{22} = L_i\dot{\gamma}^2 \ , \quad i=1,2 \ ,$$

$$\tilde{c}^i_{22} - \tilde{c}^i_{33} = M_i\dot{\gamma}^2 \ , \quad i=1,2 \ ,$$

where K_i, L_i, and M_i, i=1,2, are (as yet undetermined) constants. Based on the undisturbed by the planar flow in the (1,2) plane equilibrium value for the diagonal coefficient in the third direction, $\tilde{c}^i_{33}=1$, the above expressions lead to the following dependencies for the four non-zero independent components of the (dimensionless) conformation tensor $\tilde{\mathbf{c}}^i$ (valid up to the third order in $\dot{\gamma}$):

$$\left.\begin{aligned} \tilde{c}^i_{12} &= K_i\dot{\gamma} \\ \tilde{c}^i_{11} &= 1 + (L_i + M_i)\dot{\gamma}^2 \\ \tilde{c}^i_{22} &= 1 + M_i\dot{\gamma}^2 \\ \tilde{c}^i_{33} &= 1 \end{aligned}\right\} \ , \quad i=1,2 \ .$$

To obtain the values of the coefficients K_i, L_i, and M_i, i=1,2, it suffices to substitute the above expressions into the steady-state evolution equation for $\tilde{\mathbf{c}}^1$:

$$\lambda_1(\tilde{c}^1_{\alpha\gamma}\nabla_\gamma\upsilon_\beta + \tilde{c}^1_{\gamma\beta}\nabla_\gamma\upsilon_\alpha) = \tilde{c}^1_{\alpha\beta} - \delta_{\alpha\beta} + \frac{\theta\omega}{2}\left(\tilde{c}^1_{\alpha\gamma}\tilde{c}^2_{\gamma\beta} + \tilde{c}^2_{\alpha\gamma}\tilde{c}^1_{\gamma\beta} - 2c^1_{\alpha\beta}\right) \ ,$$

and the equivalent equation for $\tilde{\mathbf{c}}^2$

$$\lambda_2(\tilde{c}^2_{\alpha\gamma}\nabla_\gamma\upsilon_\beta + \tilde{c}^2_{\gamma\beta}\nabla_\gamma\upsilon_\alpha) = \tilde{c}^2_{\alpha\beta} - \delta_{\alpha\beta} + \frac{\theta}{2\omega}\left(\tilde{c}^1_{\alpha\gamma}\tilde{c}^2_{\gamma\beta} + \tilde{c}^2_{\alpha\gamma}\tilde{c}^1_{\gamma\beta} - 2c^2_{\alpha\beta}\right) \ ,$$

where

$$\omega \equiv \sqrt{\frac{n_2\lambda_1}{n_1\lambda_2}} \ .$$

Using the fact that the only nonzero component of the velocity gradient is $\nabla_2 v_1 = \dot{\gamma}$, collecting terms of equal order in $\dot{\gamma}$, and equating to zero the lowest order in $\dot{\gamma}$ in the above equations yields the following solution for K_1, K_2

$$K_1 = \frac{\lambda_1 - \lambda_2 \theta \omega}{1 - \theta^2} ,$$

$$K_2 = \frac{\lambda_2 - \lambda_1 \dfrac{\theta}{\omega}}{1 - \theta^2} ,$$

in terms of which all the other coefficients can be expressed:

$$L_1 = 2\frac{\lambda_1 K_1 - \lambda_2 K_2 \theta \omega}{1 - \theta^2} ,$$

$$L_2 = 2\frac{\lambda_2 K_2 - \lambda_1 K_1 \dfrac{\theta}{\omega}}{1 - \theta^2} ,$$

and

$$M_1 = - \theta\omega \frac{1 - \dfrac{\theta}{\omega}}{1 - \theta^2} K_1 K_2 ,$$

$$M_2 = - \frac{\theta}{\omega} \frac{1 - \theta\omega}{1 - \theta^2} K_1 K_2 .$$

It is easy to verify that the above expressions reduce to the known zero-shear rate values for the upper-convected Maxwell and Giesekus models in the characteristic limits of $\theta=0$ and $(\lambda_1, n_1)=(\lambda_2, n_2)$, respectively.

To proceed further in the calculation of the rate of energy dissipation we need to evaluate, according to Eq. (8.2-35), $\mathrm{tr}\tilde{c}^i$, $\mathrm{tr}(\tilde{c}^i)^{-1}$, i=1,2 and $\tilde{c}^1 : \tilde{c}^2$. An asymptotic expression in $\dot{\gamma}$ for the first one is trivially obtained by summing up the expressions for the diagonal components of $\tilde{c}^i$:

$$\mathrm{tr}\tilde{c}^i = 3 + (M_i + 2L_i)\dot{\gamma}^2 + O(\dot{\gamma}^4) .$$

For the second one, the Cayley/Hamilton theorem can be used to express it in terms of the second and third invariants of $\tilde{c}^i$, Eq. (8.1-103d). The assymptotic expression for the second invariant is then obtained as

$$I_2^i \equiv \frac{1}{2}\Big((\mathrm{tr}\tilde{c}^i)^2 - \mathrm{tr}\tilde{c}^i : \mathrm{tr}\tilde{c}^i\Big) = 3 + \Big(2(L_i + 2M_i) - K_i^2\Big)\dot{\gamma}^2 + O(\dot{\gamma}^4) .$$

Moreover, the assymptotic expression for the third invariant ($\det\tilde{c}^i$) is

$$I_3^i \equiv \det\tilde{\mathbf{c}}^i = 1 + \left(L_i + 2M_i - K_i^2\right)\dot{\gamma}^2 + O(\dot{\gamma}^4) \ .$$

Thus, the asymptotic expression for $\mathrm{tr}(\tilde{\mathbf{c}}^i)^{-1}$ is

$$\mathrm{tr}(\tilde{\mathbf{c}}^i)^{-1} \equiv \frac{I_2^i}{I_3^i} = 3 + (L_i + 2M_i)\dot{\gamma}^2 + O(\dot{\gamma}^4) \ .$$

Finally, the asymptotic expression for $\tilde{\mathbf{c}}^1\!:\!\tilde{\mathbf{c}}^2$ is

$$\tilde{\mathbf{c}}^1\!:\!\tilde{\mathbf{c}}^2 = 3 + \left(L_1 + L_2 + 2(M_1 + M_2 + K_1K_2)\right)\dot{\gamma}^2 + O(\dot{\gamma}^4) \ .$$

Substituting the above expressions into Eq. (8.2-35) yields

$$-\frac{[H_m, H_m]}{k_BT\dot{\gamma}^2} = \sum_{i=1}^{2} \frac{n_i}{\lambda_i} M_i^2 + 2\theta\sqrt{\frac{n_1 n_2}{\lambda_1\lambda_2}}\, M_1 M_2 \ .$$

Finally, substituting into the above expression the value for M_1, M_2 and rearranging we get the particularly simple final expression

$$-\frac{[H_m, H_m]}{k_BT\dot{\gamma}^2} = \frac{n_1\lambda_1 + n_2\lambda_2 - 2\sqrt{n_1\lambda_1 n_2\lambda_2}\ \theta}{1 - \theta^2} \ ,$$

which can also be written in a more intuitive form as

$$-\frac{[H_m, H_m]}{\frac{1}{2}k_BT\dot{\gamma}^2} = \frac{(n\lambda)_A - (n\lambda)_G\theta}{1 - \theta^2} \ ,$$

in terms of the *arithmetic* and *geometric* means of $n_i\lambda_i$, i=1,2, $(n\lambda)_A$ and $(n\lambda)_G$, respectively, defined for a sequence m_i, i=1,N, as [Bronshtein and Semendyayev, 1985, p. 103]:

$$(m)_A \equiv \frac{1}{N}\sum_{i=1}^{N} m_i \ , \quad (m)_G \equiv \sqrt[N]{\prod_{i=1}^{N} m_i} \ .$$

Thus, it follows from the fact that for two positive m_i, i=1,2,

$$(m)_G \leq (m)_A \ ,$$

with the equality valid if and only if $m_1 = m_2$ [Bronshtein and Semenyayev, 1985, p. 103], that the sufficient and necessary conditions for the rate of the energy dissipation to be non-negative is

$$|\theta| \leq 1 \text{ or } \theta \geq \frac{(n\lambda)_A}{(n\lambda)_G} \ ,$$

which is exactly the same as the condition on θ stipulated by Eq. (8.2-37) above.

A couple of final remarks are appropriately made here. First, notice that the final condition is always duly satisfied for the upper-convected Maxwell and Giesekus models in the characteristic limits of $\theta=0$ and $(\lambda_1, n_1)=(\lambda_2, n_2)$, respectively. In both these cases, the rate of energy dissipation is provided simply as

$$-\frac{[H_m, H_m]}{\frac{1}{2}k_B T\dot{\gamma}^2} = n\lambda \ ,$$

where, for the Giesekus model, the Giesekus relaxation time λ was used related to the two coupled modes (common) relaxation time $\lambda_1=\lambda_2\equiv\lambda_c$ by $\lambda=\lambda_c/(1+\theta)$, as mentioned before. It is particularly interesting to mention here that the case of the Giesekus model corresponding to $\alpha=1/2$ emerges here (corresponding to $\theta=1$) as a structurally unstable one, at least as related to the thermodynamic admissibility of the steady simple shear flow since any small perturbation in the parameters of the two modes away from equality will lead, according to Eq. (8.2-37), to a "window" of θ values slightly greater than unity for which the steady state of the simple shear flow is lost. Whereas this observation might not be directly related with any aphysical behavior, it at least brings into question the physical meaning of any steady state solutions for simple shear flow predicted at this range of the parameters which, in consequence, should be expected to be unstable. This remark is further reinforced from the analysis of the linear viscoelasticity behavior of the two coupled Maxwell modes model which shows that the fading memory principle restricts $|\theta|$ in all cases to be less than or equal to unity [Beris *et al.*, 1993]. It is particularly interesting that this appears to take place for the Giesekus model close to $\alpha=1/2$, which is the limiting value beyond which the shear stress according to the model ceases to be a monotonically increasing function of the shear rate (in the absence of a viscous solvent contribution). Guided from this similarity, we can anticipate a similar behavior for the values of θ greater than unity in the two coupled Maxwell modes model, even those for which the steady state for simple shear flow is thermodynamically admissible. One should anticipate at least a time periodic behavior in that range of the parameter values. ■

C. Network Theory Based Models Involving Coupled Relaxation Modes

In an effort to extend the applicability of simple two-bead (dumbbell) models to more concentrated systems and, especially, in an effort to better capture a complex relaxation behavior, a few coupled relaxation

p. 311: ibid.

modes models have also been introduced. The first model [Marrucci *et al.*, 1973] involves the idea of a variable number of chain segments between entanglements which changes during the flow subject to its own balance equation between the rates of formation, destruction, and accumulation. The second one involves a generalization of the configuration-dependent anisotropic mobility model [Giesekus, 1985a] allowing for the mobility tensor to obey a separate relaxation equation. These models represent a first effort toward representing a complex multiple relaxation times behavior associated with a complicated, multiple-degrees-of-freedom structure of concentrated polymer solutions and melts through a representation in terms of a few internal variables. As Giesekus states [1985b, pp. 174-75]:

> The real state of configuration is by no means fully described by the configuration tensor **b**.[30] This is particularly so in the extremely crude approximation of the one-mode model, because this quantity only contains the second-order moment of the distribution function but no information at all on, for example, fourth-order moments or correlations between neighboring configuration vectors $\mathbf{R}_j$ and $\mathbf{R}_{j+k}$, which may also influence the kinetics of the beads. As a result it seems more realistic to consider that **b** no longer characterizes the complete momentary configuration but rather only the basic configuration associated with the large structures which convey the stresses . . . the relative mobility $\boldsymbol{\beta}$ should no longer be considered to be as a function of the present value of the configuration tensor **b** but as a functional of its history."

It is indeed with the coupled models that the idea of a representation in terms of internal variables finds its greatest usefulness. In the following sections, the Marrucci and the Giesekus two-mode models are analyzed within the perspective of the generalized bracket theory in order to show the special advantages of this systematic approach in these cases, induced from the coupling of the involved relaxation processes. Note that although the possibility (even more, the certainty) of coupling between the relaxation modes within more traditional multi-mode descriptions of viscoelasticity, such as the ones described in §8.2.2A, has also been mentioned [Giesekus, 1982, pp. 107-08], no concrete applications have been reported so far.

C.1 The Marrucci model

The earliest model involving multiple, coupled relaxation modes is the Marrucci model [Marrucci *et al.*, 1973; Acierno *et al.*, 1976a,b] which is

[30]In this work denoted by **c**.

developed in a phenomenological fashion based on the concept of a "balance of entanglements" from polymer network theory. Using the notion of a conformation tensor variable, the Marrucci model can be described through two relaxation equations. The first, as usual, describes the relaxation of the conformation tensor,

$$\check{c}_{\alpha\beta} = -\frac{1}{\lambda}c_{\alpha\beta} + \frac{k_BT}{\lambda K}\delta_{\alpha\beta} \,. \tag{8.2-39}$$

The stress equation is that corresponding to a Hookean dumbbell, Eq. (8.1-17),

$$\sigma_{\alpha\beta} = nKc_{\alpha\beta} - nk_BT\delta_{\alpha\beta} \,, \tag{8.2-40}$$

from which, if we solve for $c_{\alpha\beta}$, we obtain (8.1-18),

$$c_{\alpha\beta} = \frac{\sigma_{\alpha\beta}}{nK} + \frac{k_BT}{K}\delta_{\alpha\beta} \,. \tag{8.2-41}$$

By substituting the above expression for $c_{\alpha\beta}$ into Eq. (8.2-39), we get back the expression of Marrucci *et al.* [1973, Eq. 4],

$$\lambda\left(\frac{\check{\sigma}_{\alpha\beta}}{nk_BT}\right) + \frac{1}{nk_BT}\sigma_{\alpha\beta} = 2\lambda A_{\alpha\beta} \,, \tag{8.2-42}$$

provided that we make the substitution

$$G = nk_BT \,. \tag{8.2-43}$$

Note that in obtaining Eq. (8.2-42) we assumed (as Marrucci *et al.* did) a constant spring constant K, but we allowed for a variable (in general) number density of elastic segments, n, (and therefore G) and relaxation time, λ. So far, the development has followed closely the traditional upper-convected Maxwell model described in §8.1.1A. Where things change, though, is in the description of the parameters n and λ. Marrucci *et al.* accept in their model a variable segment density, n, by correlating it to c,[31] the number density of the macromolecules, through a number (in general variable) of entanglements per macromolecule χ (n in the notation of Marrucci *et al.* [1973]) as

$$n = c(\chi + 1) \,. \tag{8.2-44}$$

Similarly, they considered a variable relaxation time λ connected to the number of segments (strands) per macromolecule as

$$\lambda \propto (\chi + 1)^{1.4} \,. \tag{8.2-45}$$

[31]Note the difference in notation used in this sub-section C.1 as compared to the rest of chapter 8.

Finally, Marrucci *et al.* obtained their second relaxation equation for the number of strands χ by balancing the rate of formation of entanglements to the rate of destruction due to stress as

$$\frac{d\chi}{dt} = \frac{\chi_0 - \chi}{\lambda} - \frac{a}{\eta}\sqrt{\chi\chi_0}\ \mathrm{II}_{\boldsymbol{\sigma}}, \tag{8.2-46}$$

where a is a constant parameter related to the constant number of entanglements that, under equilibrium conditions, join two macromolecules, χ_0 represents the number of entanglements per macromolecule at equilibrium, and $\mathrm{II}_{\boldsymbol{\sigma}}$ is the second invariant of the stress tensor, $\mathrm{II}_{\boldsymbol{\sigma}}^2 \equiv \frac{1}{2}\boldsymbol{\sigma}:\boldsymbol{\sigma}$. Note that, as Giesekus [1985b, p. 174] points out, "the network configuration tensor is the basic variable in these (the above) equations." Thus, the Marrucci model is especially well suited to be represented within the bracket formalism.

In the Acierno *et al.* [1976a] version of the Marrucci model, the right-hand side term of the kinetic equation, Eq. (8.2-46), was modified so that to involve the first invariant (i.e, the trace) of the stress tensor instead of the second. In justifying their change the authors noted [Acierno *et al.*, 1976a, p. 132] "...it seems that the trace may represent a more physically plausible choice. It allows us to relate the destruction rate to the elastic energy content of the system." Indeed, this comment goes very well along the lines of development of the bracket formalism representation of the Marrucci model as will be seen in the following. Later, Jongschaap [1981] showed formally how the Marrucci model can be derived from transient network theory [Wiegel, 1969a,b]. In that process, he correctly identified the corresponding proper forms for the creation and annihilation rates for the entanglements. At the same time, comparison of the model predictions (the Acierno *et al.* modification) with experiments showed its deficiencies for quantitative predictions, in particular, as far as large-amplitude oscillatory shear [Tsang and Dealy, 1981, Figs. 9, 10] or stress relaxation upon the cessation of shear flow [De Cleyn and Mewis, 1981, Figs. 7-9] are concerned.

These observations have prompted further modifications of the Marrucci model. These consisted of adjustments on the right-hand side of the kinetic equation for the structure, emanating either empirically [De Cleyn and Mewis, 1981, Eq. 7] or from further considering its connection with the transient network theory [Mewis and Denn, 1983, Eq. 46]. The latter modification appeared to improve the perfomance of the Marrucci/Acierno models in large-amplitude oscilatory shear but not in shear stress growth in start-up flows [Giacomin *et al.*, 1993]. Since all of the above modifications do not change the fundamental character of the original model, they can equally well be compared with the final kinetic equation derived from the bracket formalism, Eq. (8.2-54), and therefore they do not need to be considered separately. Moreover, it should be mentioned that an attempt to change from an evolution equation for the structure to a (phenomenological) evolution equation for the stress (i.e., taking n outside the upper-convected derivative in eq. (8.2-42)) has produced results distinctively worse than the original model [Giacomin *et al.*, 1993], which provides further evidence in favor of the internal structural parameter approach.

To express the Marrucci model within the bracket formalism, we essentially need to incorporate into the formalism used for the upper-convected Maxwell the new variable χ. At the same time, given the fact that all variables entering the formalism, from physical grounds, need to be represented in the form of a physical density, and given the fact that the density of chains is no longer conserved, we need to consider $\mathbf{b}\equiv(\chi+1)\mathbf{c}$ as the new independent conformation tensor variable, in lieu of $\mathbf{c}$. Moreover, the spring coefficient cannot be considered any more as a constant. Indeed, as Eq. (8.1-3) indicates, K is inversely proportional to the number of Kuhn segments per chain, N. However, as more entanglements are created, each individual chain between two entanglements is bound to have less segments, and thus a reasonable approximation appears to be to consider K proportional to $(\chi+1)$, or $K = K_0(\chi+1)$. Furthermore, in order to represent the dynamic equilibrium which determines the number of entanglements at equilibrium, χ_0, an additional nonlinear term in χ needs to be added to the Hamiltonian. In the absence of any more specific information, we consider here a logarithmic term in χ (similar to the Flory/Huggins mixing correction term).

To a first approximation, the structural part of the Hamiltonian, $H_s\equiv H_m-K[\mathbf{M}]$, where $K[\mathbf{M}]$ is the kinetic energy defined in Eq. (8.1-6b), can then be expressed as

$$H_s[\mathbf{b},\chi] = \int_\Omega \left(\tfrac{1}{2} nK_0 \mathrm{tr}\mathbf{b} + \tfrac{1}{2} nk_BT \ln\det(\frac{K_0\mathbf{b}}{k_BT}) - \tfrac{3}{2}\chi_0 ck_BT \ln\chi \right) d^3x \,, \quad (8.2\text{-}47)$$

where n, the number density of chain segments, is related to the number of entanglements per chain, χ, through Eq. (8.2-44). Note that Eq. (8.2-47) duly satisfies the consistency requirement that at equilibrium, in the absence of any stress and when the conformation is isotropic and Gaussian, $\mathbf{b}_{\mathrm{equil}}=k_BT/K_0\boldsymbol{\delta}$, the Hamiltonian (8.2-47) is minimized for $\chi=\chi_0$, which can, in general, be considered as a function of both the concentration, c, and the temperature, T.

The Poisson bracket, (8.1-1), of course, also needs to be changed. Except that it needs to be expressed in terms of $\mathbf{b}$ instead of $\mathbf{c}$, another term has to be added in order to account for the convection of χ. This additional term has exactly the same form as the standard term responsible for the convection of mass density, as any of the ν components in the first term in Eq. (7.3-6) valid for multicomponent mixtures, with ρ_i replaced by χ. Finally, the dissipation bracket should also involve additional terms coupling χ with itself and with $\mathbf{b}$. Since χ is a non-equilibrium parameter, two relaxation terms can be added to the lowest-order (bilinear) approximation to the dissipation bracket, Eq. (8.1-5). The first, $[F,G]_\chi$, is of similar form as the relaxation for **c**:

$$[F,G]_\chi = -\int_\Omega \Lambda_\chi \frac{\delta F}{\delta \chi}\frac{\delta G}{\delta \chi} d^3x \ , \tag{8.2-48}$$

where Λ_χ is a parameter (in general dependent on χ and **b**), inversely proportional to a characteristic relaxation time for the segments, λ_χ. The second term couples the relaxation of χ and **b** together, $[F,G]_{\chi\mathbf{b}}$, and, to the lowest order of approximation, has the general form

$$[F,G]_{\chi\mathbf{b}} = -\int_\Omega \Lambda^*_{\alpha\beta}\left(\frac{\delta F}{\delta b_{\alpha\beta}}\frac{\delta G}{\delta \chi} + \frac{\delta G}{\delta b_{\alpha\beta}}\frac{\delta F}{\delta \chi}\right) d^3x \ , \tag{8.2-49}$$

where $\boldsymbol{\Lambda}^*$ is a coupling tensorial parameter (in general dependent on χ and **b**). These terms are considered in addition to all the terms appearing in Eq. (8.1-5), although, in the following, as in the upper-convected Maxwell case, we use **B**=**0**, **Q**=**L**=**0**. Thus, the question that remains to be answered is whether it is possible through a suitable selection of the parameters Λ_χ, $\boldsymbol{\Lambda}^*$, and $\boldsymbol{\Lambda}$ (of the previous form of the Poisson bracket) to derive Marrucci's equations, Eqs. (8.2-42,46).

It is useful to evaluate first the Volterra derivatives of the Hamiltonian corresponding to the expression (8.2-47), $\delta H/\delta c_{\alpha\beta}$ and $\delta H/\delta \chi$:

$$\frac{\delta H}{\delta b_{\alpha\beta}} = \frac{nK_0}{2}\delta_{\alpha\beta} - \frac{nk_BT}{2}b^{-1}_{\alpha\beta} \ , \tag{8.2-50a}$$

$$\frac{\delta H}{\delta \chi} = \tfrac{1}{2}cK_0\mathrm{tr}\mathbf{b} + \tfrac{1}{2}ck_BT\ln\det(\frac{K_0\mathbf{b}}{k_BT}) - \tfrac{3}{2}\frac{\chi_0}{\chi}ck_BT \ . \tag{8.2-50b}$$

To simplify the analysis, and based on previous experience, let us use the same relaxation parameter $\boldsymbol{\Lambda}$ as for the Maxwell fluid, Eq. (8.1-15), evaluated for **c**=**b**, $\lambda=\lambda_1$, and $K=K_0$ and a similar form for $\boldsymbol{\Lambda}^*$ and Λ_χ:

$$\Lambda_{\alpha\beta\gamma\varepsilon} = \frac{1}{2\lambda_1 nK_0}(b_{\alpha\gamma}\delta_{\beta\varepsilon} + b_{\alpha\varepsilon}\delta_{\beta\gamma} + b_{\beta\gamma}\delta_{\alpha\varepsilon} + b_{\beta\varepsilon}\delta_{\alpha\gamma}) \ , \tag{8.2-51a}$$

$$\Lambda^*_{\alpha\beta} = \tfrac{2}{3}\frac{1+\chi}{\lambda_2 nk_BT}\chi b_{\alpha\beta} \ , \tag{8.2-51b}$$

$$\Lambda_\chi = \tfrac{2}{3}\frac{1+\chi}{\lambda_3 nk_BT}\chi \ , \tag{8.2-51c}$$

where λ_i, i=1,2,3 are three, in general different, time constants. The values of λ_i are subject to certain constraints (as, for example, λ_1, λ_3 >0, and $|\lambda_2|$ is bounded) in order for the corresponding rate of energy dissipation to be non-negative (see also §8.1.4).

From the generalized bracket corresponding to the above set of variables, the following equations can be obtained:

$$\overset{\nabla}{b}_{\alpha\beta} = -\left(\frac{1}{\lambda_1} + \tfrac{1}{3}\left[\frac{K_0}{k_BT}\mathrm{tr}\mathbf{b} + \ln\det(\frac{K_0\mathbf{b}}{k_BT})\right]\frac{\chi}{\lambda_2} - \frac{\chi_0}{\lambda_2}\right)b_{\alpha\beta} \qquad (8.2\text{-}52a)$$

$$+ \frac{k_BT}{\lambda_1 K_0}\delta_{\alpha\beta} \,,$$

and

$$\sigma_{\alpha\beta} = nK_0 b_{\alpha\beta} - nk_BT\delta_{\alpha\beta} \,. \qquad (8.2\text{-}52b)$$

Now if we solve (8.2-52) for **b**,

$$b_{\alpha\beta} = \frac{\sigma_{\alpha\beta}}{nK_0} + \frac{k_BT}{K_0}\delta_{\alpha\beta} \,, \qquad (8.2\text{-}53)$$

and substitute into (8.2-52a), when no coupling is allowed (i.e., $\lambda_2=\infty$), we do indeed arrive back at the constitutive relationship for the extra stress, (8.2-42) for $\lambda=\lambda_1$. However, we believe that $\lambda_2\neq\infty$ is necessary to describe the stress-induced entanglement restructuring processes envisioned in the original model.

The corresponding evolution equation for χ is

$$\frac{d\chi}{dt} = -\frac{1}{\lambda_3}\left([\frac{K_0}{k_BT}\mathrm{tr}\mathbf{b} + \ln\det(\frac{K_0\mathbf{b}}{k_BT})]\frac{\chi}{3} - \chi_0\right) - \tfrac{1}{3}\frac{1+\chi}{\lambda_2 nk_BT}\chi\mathrm{tr}\boldsymbol{\sigma}, \qquad (8.2\text{-}54)$$

which, although not identical to the expression of Marrucci *et al.*, Eq. (8.2-46), is very close. Indeed, when $\lambda_3=\lambda_1=\lambda$, the first term on the right-hand side of (8.2-54) reduces, close to equilibrium, to the corresponding rate of formation due to thermal processes $(\chi_0-\chi)/\lambda$, whereas the second negative term is very close to Marrucci's rate of destruction by the stress, being proportional to the stress magnitude and inversely proportional to the viscosity, $\eta=nk_BT\lambda$, with the coefficient of proportionality, $\lambda_1/(3\lambda_2)(1+\chi)\chi$, depending on χ, as in Marrucci's expression, Eq. (8.2-46). The biggest drawback of Eq. (8.2-54) is that because of the dependence on the trace of the stress, the rate of entanglement destruction is predicted to be independent of the shear stress, as Marrucci envisioned in his original derivation [Marrucci *et al.*, 1973, p. 271]. This dependence can be recovered, if desired, by simply changing the relaxation parameter $\boldsymbol{\Lambda}^*$ to a more involved function of the conformation tensor **b**. However, this is beyond the scope of this simple analysis, which was to show the capability of the generalized bracket (and the concomitant corrections it offers) to rather complex viscoelastic modeling. Finally, note that the effect of a non-infinite λ_2 is simply to change the effective relaxation time for **b** from λ_1 to $\lambda_{eff}(\mathbf{b},\chi)$. As seen from Eq. (8.2-52a), a sufficient condition for λ_{eff} to be positive is that $\chi_0 \leq \lambda_2/\lambda_1$.

The original formulation of the Marrucci model, Eqs. (8.2-42,46), not only predicts quite satisfactorily most steady viscometric functions, especially when it is used in its multi-mode generalization [Acierno *et al.*, 1976b, Fig. 2], but also provides a qualitative description of start-up phenomena. In particular, it predicts stress overshoots not only in shear but also in extension [Acierno *et al.*, 1976b, Figs. 3-6]. However, due to its upper-convected formulation, it predicts a zero second normal-stress difference [Giesekus, 1985b, p. 174]. The presence of the nonlinear correction in its bracket formulation, Eq. (8.2-54), alleviates this problem. Of course, another approach would have been to use an anisotropic mobility tensor, as in the Giesekus two-mode model, described below.

C.2 The Giesekus two-mode mobility model

In extending his earlier configuration-dependent mobility model (see §8.1.3A), Giesekus proposed a model involving two coupled tensorial parameters, using, in addition to the previously introduced configuration parameter **c** (**b** in Giesekus's notation), the (in general) anisotropic mobility tensor $\boldsymbol{\zeta}^{-1} \equiv \boldsymbol{\beta}$ as a second independent internal variable [Giesekus, 1985a]. $\boldsymbol{\beta}$ was assumed to obey a relaxation-type evolution equation [Giesekus, 1985a, Eq. (17)],

$$(\boldsymbol{\beta} - \boldsymbol{\delta}) + \kappa(\check{\boldsymbol{\beta}} - \check{\boldsymbol{\delta}}) = \alpha(\tilde{\mathbf{c}} - \boldsymbol{\delta}) \; , \tag{8.2-55}$$

which contains the linear relation, Eq. (8.1-85), as its limiting case when the characteristic "retardation of the mobility" time, κ, is equal to zero. The question that thus arises is how to represent within the generalized bracket formalism Eq. (8.2-55), the corresponding evolution equation for the configuration tensor $\tilde{\mathbf{c}}$,

$$\check{\tilde{c}}_{\alpha\beta} = -\frac{1}{2\lambda}\Big(\beta_{\alpha\gamma}(\tilde{c}_{\gamma\beta} - \delta_{\gamma\beta}) + (\tilde{c}_{\alpha\gamma} - \delta_{\alpha\gamma})\beta_{\gamma\beta}\Big) \; , \tag{8.2-56}$$

and the stress equation, (8.1-17). Eq. (8.2-56) is a generalization of the single-mode Giesekus equation, (8.1-86), when the tensors $\boldsymbol{\beta}$ and **c** do not commute [Giesekus, 1985a, Eq. (10)].

To represent the two-mode Giesekus model within the bracket formalism, we need to expand the Hamiltonian description of the single-mode model presented in §8.1.3A to accommodate one additional (symmetric) tensorial structural parameter, $\boldsymbol{\beta}^* \equiv \boldsymbol{\beta} - \boldsymbol{\delta}$. It is important to decide beforehand the behavior of $\boldsymbol{\beta}^*$ under time inversion (parity). The fact that $\boldsymbol{\beta}^*$ is physically related to an effective change in the mobility of the chain, and the fact that it modifies the time derivative of **c**, Eq. (8.2-56), indicate an opposite parity from **c**, i.e., odd. Thus, it appears that $\boldsymbol{\beta}^*$ is related to the rate at which a given structure changes rather than to a static structure. The parity behavior is important because it determines the mathematical form of the $\boldsymbol{\beta}^*$-dependent terms in both the Hamiltonian and the dissipation bracket that are compatible with it. Next, the Hookean dumbbell free energy, which is also used for the single-mode Giesekus model, (8.1-12,13), needs to be expanded to accommodate changes in the free energy of the system due to deviations of $\boldsymbol{\beta}^*$ from its equilibrium value, **0**. The simplest way to incorporate a $\boldsymbol{\beta}^*$ contribution to the free energy, which is also compatible with its odd-parity interpretation, is through a quadratic term, $a_1(\boldsymbol{\beta}^*)^2$, similar to that for a kinetic energy contribution. Then, the structural part of the Hamiltonian, $H_s \equiv H_m - K$, in dimensionless form, can be written as

$$\tilde{H}_s[\tilde{\mathbf{c}},\boldsymbol{\beta}^*] = \int_\Omega \frac{1}{2}n\left(\mathrm{tr}\tilde{\mathbf{c}} + \ln(\det\tilde{\mathbf{c}}) + a_1\boldsymbol{\beta}^*\!:\!\boldsymbol{\beta}^*\right)\mathrm{d}^3x \ , \tag{8.2-57}$$

where a_1 is a positive parameter independent of $\tilde{\mathbf{c}}$ and $\boldsymbol{\beta}^*$ characterizing the inertia of the system due to the changes described by $\boldsymbol{\beta}^*$, and the Hamiltonian has been made dimensionless by k_BT. Correspondingly, the Volterra derivatives of the Hamiltonian specified by Eq. (8.2-57) are

$$\frac{\delta\tilde{H}}{\delta\tilde{\mathbf{c}}} = \frac{n}{2}(\boldsymbol{\delta} - \tilde{\mathbf{c}}^{-1}) \ , \qquad \frac{\delta\tilde{H}}{\delta\boldsymbol{\beta}^*} = na_1\boldsymbol{\beta}^* \ . \tag{8.2-58}$$

Note that they are both symmetric, given that $\tilde{\mathbf{c}}$ and $\boldsymbol{\beta}^*$ are also symmetric.

Next, the Poisson bracket, Eq. (8.1-1) needs to be expanded, which can be accomplished by simply duplicating the terms involving **c** and replacing it with $\boldsymbol{\beta}^*$. Finally, two new terms are added to the dissipation bracket, (8.1-5), in order to represent the relaxation phenomena involving the new variable $\boldsymbol{\beta}^*$, which thus acquires the general form (with the terms corresponding to **B**, **Q**, and **L** in Eq. (8.1-5) neglected)

$$\begin{aligned}[F,G] = &-\int_\Omega \Lambda_{\alpha\beta\gamma\varepsilon}\frac{\delta F}{\delta\tilde{c}_{\alpha\beta}}\frac{\delta G}{\delta\tilde{c}_{\gamma\varepsilon}}\mathrm{d}^3x - \int_\Omega N_{\alpha\beta\gamma\varepsilon}\frac{\delta F}{\delta\beta^*_{\alpha\beta}}\frac{\delta G}{\delta\beta^*_{\gamma\varepsilon}}\mathrm{d}^3x \\ &-\int_\Omega M_{\alpha\beta\gamma\varepsilon}\left(\frac{\delta F}{\delta\tilde{c}_{\alpha\beta}}\frac{\delta G}{\delta\beta^*_{\gamma\varepsilon}} - \frac{\delta G}{\delta\tilde{c}_{\alpha\beta}}\frac{\delta F}{\delta\beta^*_{\gamma\varepsilon}}\right)\mathrm{d}^3x \ ,\end{aligned} \tag{8.2-59}$$

where the relaxation matrices $\boldsymbol{\Lambda}$, **M**, and **N** are defined as

$$\Lambda_{\alpha\beta\gamma\varepsilon} = \frac{1-\alpha}{2\lambda_1 n}\left(\tilde{c}_{\alpha\gamma}\delta_{\beta\varepsilon} + \tilde{c}_{\alpha\varepsilon}\delta_{\beta\gamma} + \tilde{c}_{\beta\gamma}\delta_{\alpha\varepsilon} + \tilde{c}_{\beta\varepsilon}\delta_{\alpha\gamma}\right) \ , \tag{8.2-60a}$$

$$M_{\alpha\beta\gamma\varepsilon} = \frac{1}{4\lambda_2 a_1 n}\left(\tilde{c}_{\alpha\gamma}\delta_{\beta\varepsilon} + \tilde{c}_{\alpha\varepsilon}\delta_{\beta\gamma} + \tilde{c}_{\beta\gamma}\delta_{\alpha\varepsilon} + \tilde{c}_{\beta\varepsilon}\delta_{\alpha\gamma}\right) , \tag{8.2-60b}$$

$$N_{\alpha\beta\gamma\varepsilon} = \frac{1}{2\lambda_3 a_1 n}\left(\delta_{\alpha\gamma}\delta_{\beta\varepsilon} + \delta_{\alpha\varepsilon}\delta_{\beta\gamma}\right) , \tag{8.2-60c}$$

where λ_i, i=1,2,3 are three, in general, different relaxation time constants and special care has been taken to choose a form for the relaxation matrices compatible with Onsager reciprocity and the parity of the variables.

The resulting evolution equations are

$$\check{\tilde{c}}_{\alpha\beta} = -\frac{1-\alpha}{\lambda_1}(\tilde{c}_{\alpha\beta} - \delta_{\alpha\beta}) - \frac{1}{2\lambda_2}(\beta^*_{\alpha\gamma}\tilde{c}_{\gamma\beta} + \tilde{c}_{\alpha\gamma}\beta^*_{\gamma\beta}) , \tag{8.2-61a}$$

and

$$\check{\boldsymbol{\beta}}^* = \frac{1}{a_1\lambda_2}(\tilde{\mathbf{c}} - \boldsymbol{\delta}) - \frac{1}{\lambda_3}\boldsymbol{\beta}^* , \tag{8.2-61b}$$

which coincide with the evolution equations for the Giesekus two-mode model, (8.2-55,56) when $\lambda_1=\lambda_2=\lambda$, $a_1=\kappa/(\alpha\lambda)$, and $\lambda_3=\kappa$ at conditions close to the $\boldsymbol{\beta}^*$ equilibrium, where $\boldsymbol{\beta}^* = \alpha(\tilde{\mathbf{c}}-\boldsymbol{\delta})$. However, notice, that the stress equation needs to be modified with respect to Giesekus's equivalent [Giesekus, 1985a, Eq. (14)] as

$$\tilde{\sigma}_{\alpha\beta} = \tilde{c}_{\alpha\beta} - \delta_{\alpha\beta} + 2a_1\beta^*_{\alpha\gamma}\beta^*_{\gamma\beta} , \tag{8.2-62}$$

the stress tensor being rendered dimensionless with respect to $G_0 \equiv nk_BT$. The importance of the correction on the model predictions is, as yet, unknown. The only way to avoid this additional correction to the stress is by using a corotational derivative in the evolution equation for $\boldsymbol{\beta}^*$ (see §8.1.1C). However, the predictions of the model in that case become aphysical [Giesekus, 1985a, App.]. Finally, note also that from the non-positive character of the dissipation corresponding to a non-negative relaxation, (8.2-60a), we automatically get the requirement $0 \leq \alpha \leq 1$.

In conclusion, the two-mode Giesekus model (or, at least, a very close approximation to it) can be obtained through the bracket formalism. This model has been shown to give predictions for the steady-state, uniaxial, extensional viscosity with a maximum [Giesekus, 1985a, Figs. 4,5] and transient stress curves with overshoots [1985a, Fig. 12]. However, these last curves under high shear rates can give such intense oscillations that, in particular for the negative second normal-stress coefficient, the first minimum can reach negative values [1985a, Fig. 13]. Again, how these features will be affected from the model corrections suggested by the bracket formalism is something that remains to be seen.

8.2.3 *The bead/spring-chain model of kinetic theory*

Similar arguments apply to the bead/spring-chain model as to the multiple relaxation times model; the main difference being that the structural parameters for the bead/spring model, $\mathbf{c}^{ij}$, are now identified by two superscripted indices instead of one. In this model, N beads are connected by N–1 elastic springs, which are here assumed to be Hookean springs with equal constant K. Here, $\mathbf{c}^{ij}$ is identified as the average $\langle \mathbf{R}^i\mathbf{R}^j \rangle$, $\mathbf{R}^i$ being the connector vector for spring i. The total extra stress is derived from kinetic theory as [Bird *et al.*, 1987b, p. 156]

$$\sigma^{\mathrm{T}}_{\alpha\beta} = \sum_{i=1}^{N-1} nKc^{ii}_{\alpha\beta} - nk_BT(N-1)\delta_{\alpha\beta} \ . \tag{8.2-63}$$

Multiplying the diffusion equation by $R^i_\alpha R^j_\beta$ and integrating over d^3R^i, an evolution equation is obtained for the ij-th structural parameter [Öttinger, 1989b, p. 465],

$$\overset{\nabla}{c}{}^{ij}_{\alpha\beta} = \frac{2k_BT}{\zeta}A_{ij}\delta_{\alpha\beta} - \frac{K}{\zeta}\sum_{m=1}^{N-1}(A_{im}c^{mj}_{\alpha\beta} + A_{jm}c^{im}_{\alpha\beta}) \ , \tag{8.2-64}$$

where i,j=1,2,...,N–1 and $\mathbf{A}$ is the Rouse matrix [Bird *et al.*, 1987b, pp. 23,24],

$$A_{ij} = \begin{cases} 2, & i=j \\ -1, & |i-j|=1 \ . \\ 0, & otherwise \end{cases} \tag{8.2-65}$$

Note that not all of the coupled equations (8.2-64) are independent (see Öttinger [1989, §3]).

The Hamiltonian for the bead/spring-chain model consists of an elastic potential, similar to (8.1-12),

$$E_\phi[\mathbf{c}^{11},\mathbf{c}^{22},...,\mathbf{c}^{(N-1)(N-1)}] = \int_\Omega \tfrac{1}{2}nK\sum_{i=1}^{N-1}\mathrm{tr}\mathbf{c}^{ii}\,\mathrm{d}^3x \ , \tag{8.2-66}$$

and the entropy,

$$S[\mathbf{c}^{11},\mathbf{c}^{12},...,\mathbf{c}^{(N-1)(N-1)}] = \int_\Omega \frac{N-1}{2}nk_B\ln(\det\boldsymbol{\Gamma})\,\mathrm{d}^3x \ , \tag{8.2-67}$$

where

$$\boldsymbol{\Gamma} \equiv \frac{1}{N-1}\sum_{i=1}^{N-1}\sum_{j=1}^{N-1}C_{ij}\tilde{\mathbf{c}}^{ij} \ , \tag{8.2-68}$$

C being the Kramers matrix [Bird *et al.*, 1987b, pp. 23, 24],

$$C_{ij} = \begin{cases} i\dfrac{N-j}{N}, & i \le j \\ j\dfrac{N-i}{N}, & i > j \end{cases} . \tag{8.2-69}$$

Note that $\mathbf{C}=\mathbf{A}^{-1}$. Consequently, the functional derivative of the Hamiltonian is given as

$$\frac{\delta H_m}{\delta c^{kl}_{\gamma\beta}} = \frac{nK}{2}\delta_{\gamma\beta}\delta_{kl} - \frac{nk_BT}{2}C_{kl}\Gamma^{-1}_{\gamma\beta} . \tag{8.2-70}$$

Substitution of Eq. (8.2-70) into the stress expression obtained from the corresponding Poisson bracket, given, by direct analogy to (8.2-30c), as

$$\sigma^{\mathrm{T}}_{\alpha\beta} = 2\sum_{k=1}^{N-1}\sum_{l=1}^{N-1} c^{kl}_{\alpha\gamma}\frac{\delta H_m}{\delta c^{kl}_{\gamma\beta}} , \tag{8.2-71}$$

yields the same equation for the stress as Eq. (8.2-63) (assuming that the corresponding matrices to $\mathbf{L}^{ij}$ and $\mathbf{Q}$ in the dissipation bracket are set equal to zero).

The corresponding relaxation matrix is specified by defining $\boldsymbol{\Lambda}$ as

$$\Lambda^{ijkl}_{\alpha\beta\gamma\varepsilon} = \frac{1}{2n\zeta}\Big((A_{il}c^{kj}_{\alpha\gamma} + A_{lj}c^{ik}_{\alpha\gamma})\delta_{\beta\varepsilon} + [perm.]\Big) , \tag{8.2-72}$$

where by [*perm.*] we denote all possible permutations between the greek indices in the first term that lead to different terms when the symmetries of the tensors involved are taken into account (i.e., a permutation which only interchanges the indices of $\boldsymbol{\delta}$ is not included in [*perm.*] since it trivially leads to the same result). Substitution of Eqs. (8.2-70) and (8.2-72) into the evolution equation for $c^{ij}_{\alpha\beta}$, which is obtained from the analogous bracket to (8.2-21,22) as

$$\overset{\nabla}{c}{}^{ij}_{\alpha\beta} = -\sum_{k=1}^{N-1}\sum_{l=1}^{N-1}\Lambda^{ijkl}_{\alpha\beta\gamma\varepsilon}\frac{\delta H_m}{c^{kl}_{\gamma\varepsilon}} , \tag{8.2-73}$$

leads to the evolution equation

$$\overset{\nabla}{c}{}^{ij}_{\alpha\beta} = \frac{2k_BT}{\zeta}c^{ij}_{\alpha\gamma}\Gamma^{-1}_{\gamma\beta} - \frac{K}{\zeta}\sum_{m=1}^{N-1}(A_{im}c^{mj}_{\alpha\beta} + A_{jm}c^{im}_{\alpha\beta}) . \tag{8.2-74}$$

Note that the coupling resulting from the first term on the left-hand side of (8.2-74) is different from that of (8.2-64); however, both expressions reduce to the Hookean dumbbell model when N=2. This returns us to the issue of the interpretation of the nature of the structural tensor—see below. Again, it does not appear that the behavior of the model is changed significantly by the discrepancy between Eqs. (8.2-74) and (8.2-64). These equations can be written to account for hydrodynamic interactions by redefining **A** in Eq. (8.2-65) to incorporate the Oseen/Burgers tensor [Öttinger, 1989b, p. 464].

EXAMPLE 8.4: Derive the entropy expression in terms of the conformation tensors, $\mathbf{c}^{lm}$, for the bead/spring-chain model.

In the determination of the proper entropy for the bead/spring system, the following arguments are used. The entropy of a collection of N beads is taken to be proportional to the number of configurations, $\tilde{N}$, available to them which are consistent with their distribution function,

$$S = \frac{N-1}{2} k_B \ln \tilde{N} ,$$

where the coefficient of proportionality reflects the fact that there are N–1 degrees of freedom with respect to the center of mass. The number of available configurations, $\tilde{N}$, can be estimated from the determinant of the variance tensor for the position of a bead with respect to the center of mass:

$$\tilde{N} = \det\left[\frac{\sum_{k=1}^{N} \langle \mathbf{Q}^k \mathbf{Q}^k \rangle}{N-1}\right] ,$$

where $\mathbf{Q}^k$ is the position of the k-th bead from the center of mass,

$$\mathbf{Q}^k \equiv \mathbf{r}_k - \mathbf{r}_c = \sum_{l=1}^{N-1} B_{kl} \mathbf{R}^l ,$$

with $\mathbf{r}_k$ and $\mathbf{r}_c$ the position vector for the k-th bead and the center of mass, respectively, $\mathbf{R}^l$ the displacement vector between the l and l+1 beads, and the matrix element B_{kl} defined as

$$B_{kl} = \begin{cases} \dfrac{l}{N}, & l < k \\ -\dfrac{N-l}{N}, & l \geq k \end{cases} .$$

Using the definition of $\mathbf{Q}^k$, the expression for the sum of averages $\langle \mathbf{Q}^k \mathbf{Q}^k \rangle$becomes

$$\sum_{k=1}^{N} <\mathbf{Q}^k\mathbf{Q}^k> = \sum_{k=1}^{N}\sum_{l=1}^{N-1}\sum_{m=1}^{N-1} B_{kl}B_{km}<\mathbf{R}^l\mathbf{R}^m> = \sum_{l=1}^{N-1}\sum_{m=1}^{N-1} C_{lm}\mathbf{c}^{lm} ,$$

where we have used the identity $\Sigma^k B_{ki}B_{kj}=C_{ij}$ [Bird *et al.*, 1987b, p. 23]. Using the above relation, the original expression for the entropy of N beads becomes

$$S = \frac{N-1}{2} nk_B \ln(\det\boldsymbol{\Gamma}) ,$$

where the tensor $\boldsymbol{\Gamma}$ is defined in Eq. (8.2-68), which exactly corresponds to the expression given for the entropy functional by Eq. (8.2-67). ■

The major drawback of the model represented by Eq. (8.2-74) is that it introduces, for $N \neq 2$, to the evolution equation for $c^{ij}_{\alpha\beta}$ dependencies on the other spatial components, $c^{ij}_{\gamma\varepsilon}$, $\gamma \neq \alpha$ and/or $\varepsilon \neq \beta$. Thus, a major simplification of the original equations, (8.2-63, 64), cannot take place. The simplification is affected through an orthogonal transformation of the original variables

$$d^{ij}_{\alpha\beta} = \sum_{k=1}^{N-1}\sum_{l=1}^{N-1} \Omega_{ik} c^{kl}_{\alpha\beta} \Omega^{\mathrm{T}}_{lj} , \tag{8.2-75}$$

where $\boldsymbol{\Omega}$ is the orthogonal matrix (of $(N-1)\times(N-1)$ dimensions) which diagonalizes the Rouse matrix $\mathbf{A}$, and $\mathbf{d}^{ij}$ are the transformed internal variables. Pre- and post-multiplying Eq. (8.2-64) by $\boldsymbol{\Omega}$ and $\boldsymbol{\Omega}^{\mathrm{T}}$, respectively (considering, for the same set of α,β varying indices i,j, $i,j=1,..,N-1$), results in a set of equivalent evolution equations for $\mathbf{d}^{ij}$ of which the corresponding to the diagonal (in $(N-1)\times(N-1)$ space) components are totally decoupled [Bird *et al.*, 1987b, pp. 158, 159]:

$$\check{d}^{ii}_{\alpha\beta} = -\frac{1}{\lambda_i} d^{ii}_{\alpha\beta} + \frac{k_B T}{\lambda_i K}\delta_{\alpha\beta} , \tag{8.2-76}$$

where the relaxation times λ_i, $i=1,..,N-1$ are defined in terms of the eigenvalues of the Rouse matrix $\mathbf{A}$, $a_i \equiv 4\sin^2(i\pi/(2N))$ [Bird *et al.*, 1987b, Eq. (11.6.9)], $i=1,..,N-1$ as [*Bird et al.*, 1987b, p. 159]

$$\lambda_i = \frac{\zeta}{2Ka_i} . \tag{8.2-77}$$

Interestingly enough, the equation for the stress, Eq. (8.2-63), if expressed in terms of the transformed variables, remains unchanged [*Bird et al.*, 1987b, Eq. (15.3-14)]:

$$\sigma^{\mathrm{T}}_{\alpha\beta} = \sum_{i=1}^{N-1} nKd^{ii}_{\alpha\beta} - n(N-1)k_BT\delta_{\alpha\beta} \ . \tag{8.2-78}$$

Thus, since Eqs. (8.2-77, 78) constitute a complete set of equations for the stress, this transformation (called the transformation to a normal form [Bird *et al.*, 1987b, §15.3a] results in a significant reduction of the number of internal variables, from $(N-1)(N-2)/2$ to $(N-1)$.

Although the transformation to the normal form does not work with the model equations corresponding to the generalized bracket analog of the Rouse chain model, (8.2-74), it is interesting to note that the normal form equations, (8.2-77, 78), can be trivially represented through the bracket formalism as outlined in §8.2.2B. However, it is more interesting to examine what happens when a more realistic nonlinear spring-chain model is considered, as, for example, the FENE-P chain. In this case, the coefficients **A** under the summation symbol on the right-hand side of the corresponding evolution equation for the second moments, (8.2-64), are multiplied by a factor Z_m which, under the standard Peterlin approximation (see §8.1.2b), depends explicitly on the conformation tensors $\mathbf{c}^{ij}$:

$$Z_m = \frac{1}{1 - \dfrac{\mathrm{tr}\mathbf{c}^{mm}}{R_0^2}} \ . \tag{8.2-79}$$

As a consequence of this dependence, the normal form transformation in this case does not lead to a reduction of the number of the variables [Wedgewood *et al.*, 1991, p. 121]. However, a recently proposed modification of the Peterlin approximation [Wedgewood *et al.*, 1991, Eq. (5)],

$$Z_m = Z \equiv \frac{1}{1 - \dfrac{\sum_{j=1}^{N-1} \mathrm{tr}\mathbf{c}^{jj}}{R_0^2}} \ , \tag{8.2-80}$$

circumvents this problem yet still leads to a decoupled system of equations involving only the diagonal components of the normal variables $\mathbf{d}^{ii}$, i=1,..,N–1. The resulting equations are identical with Eqs. (8.2.76, 78), with the $\mathbf{d}^{ii}$ appearing on the right-hand side weighted by the factor Z, which can be expressed as a function of $\mathbf{d}^{ii}$, i=1,..,N–1, by Eq. (8.2-80) simply by replacing $\mathbf{c}^{jj}$ with $\mathbf{d}^{jj}$ [Wedgewood *et al.*, 1991, pp. 124-25]. The resulting (still coupled) system of equations can be represented trivially within the bracket formalism, using the same Poisson and dissipation bracket as in the case of the Rouse chain discussed above, with the new enthalpic contribution to the Hamiltonian

$$E_\phi[\mathbf{d}^{11},\mathbf{d}^{22},\ldots,\mathbf{d}^{(N-1)(N-1)}] = \int_\Omega \frac{N-1}{2} nKR_0^2 \ln\Big(1-\frac{\sum_{i=1}^{N-1} \mathrm{tr}\mathbf{d}^{ii}}{(N-1)R_0^2}\Big) d^3x \,, \tag{8.2-81}$$

replacing that corresponding to the Hookean-springs chain. Of course, in the limit $R_0 \to \infty$, the Rouse chain behavior is recovered.

In conclusion, we see that although complex (detailed) kinetic theory models might not be easily developed within the generalized bracket formalism,[32] we can always find close analogs which can. In addition, as the number of variables are reduced (through the normal-form transformation, for example) this task of reformulation becomes considerably easier. We hope that the above examples show well the close connection between microscopic modeling and continuum approximation through the generalized bracket. We believe that there are significant advantages in reformulating kinetic models under the bracket point of view. Among other things, we can then associate a free energy into the system, which is absolutely necessary if more complex phenomena are to be described (for example, diffusion or wall-induced migration—see chapter 9). On the other hand, there is considerable need (indeed even more so now than before) in basing the application of the bracket formalism on the results of a physically sound and considerably detailed microscopic analysis, such as the one which is provided through the application of kinetic theory outlined in such standard texts on the subject, as Bird *et al.* [1987b]. We are confident that future work along these lines will show in practice the benefits that can be achieved.

[32]We cannot say here "not possible" because definitely we have not performed an exhaustive enough search for alternative Hamiltonian and dissipation bracket expressions to be able, in any rigorous way, to prove the non-existence of equivalent bracket expressions; however, what we can say here is that if one exists, for example, for the original variables Rouse chain model, Eqs. (8.2-61, 64), then the corresponding expressions for the Hamiltonian and the dissipation bracket must be involved and not intuitively obvious. For example, the reasoning behind the selection of the entropy (see Example 8.4) is difficult to repeat for another expression.

9
Transport Phenomena in Viscoelastic Fluids

The job of theorists...is to suggest new experiments. A good theory makes not only predictions, but surprising predictions that then turn out to be true....Let a theorist produce just one theory of the type sketched above and the world will jump to the conclusion (not always true) that he has special insight into difficult problems.

—*Francis Harry Compton Crick*

In chapter 8, we applied the bracket formalism to the relatively old problem of incompressible and isothermal viscoelastic fluids. In addition to these assumptions concerning the state of the fluid medium, we therein assumed that the polymer concentration (in the case of polymer solutions) was constant. Although even in this case we were able to find some new results, it is through new applications altogether that the major advantages of this technique will be applied to the fullest extent. Thus, in this chapter we wish to study three new applications of the generalized bracket to outstanding problems concerning viscoelastic fluids. The first section of this chapter is concerned with the complications induced in viscoelastic-fluid modeling by considering compressible and non-isothermal systems. In the second section, we present the analysis of simultaneous concentration and deformation changes associated with the bulk flow of dilute polymer solutions in a form also suitable for the description of flow-induced phase separation. In the third and final section we focus our attention on the solid-surface/polymer interactions which may lead to an apparent "slip velocity" or "adsorption layer" at the interface.

We consider §9.3 as the culmination of chapters 8 and 9, and therefore we present it last despite the fact that the natural order for chapter 9 would have been the reverse of what is given below, going from the simplest to the most complex. This last section is the perfect example of our theme of consistently abstracting microscopic information to the macroscopic level of description. Because of the abundance and variety of thought on the issue of flow-induced polymer migration, §9.2 is very inconclusive at this point in time. Its presence here is solely to stimulate additional thought upon this issue.

9.1 Compressible and Non-Isothermal Viscoelastic Fluid Models

In industrial applications involving polymers, rarely does the engineer deal with an isothermal, and, consequently, incompressible fluid. Most processes are performed at extremely high temperatures, and much heating and cooling design goes into the successful process. Indeed, even if industrial processes were performed at constant temperature, one would still need to handle non-isothermalities since polymers produce large degrees of viscous heating during flow.

As a result of the above arguments, the modeling of actual industrial processes usually requires the incorporation of the non-isothermalities into the constitutive equations. So far, however, this has only been achieved in a satisfactory fashion for constitutive equations represented either by an abstract (generic) functional formalism or by a linear with respect to the strain (thus limited to small deformations) integral relationship—see Crochet [1975], Pearson and McIntire [1979] and references therein. When one considers the difficulties arising when modeling incompressible, isothermal viscoelastic fluids by differential models, it is not surprising that the more complex, non-isothermal problem has received little attention in the literature and industrial practice.

Another potential future application of compressible constitutive equations arises from the result that most simple viscoelastic constitutive relationships for the extra stress result in a fully hyperbolic system of equations [Edwards and Beris, 1990, p. 417]. This is of importance in both the mathematical analysis and the numerical solutions to these equations because it allows the implementation of hyperbolic solution algorithms without having to worry about the parabolic region in the computational space. An attempt at such a device was made recently by Phelan *et al.* [1989] by applying a standard pressure/density constitutive relationship based upon a Taylor expansion of the density.

The goal of this section is to demonstrate that non-isothermal, viscoelastic fluids may be consistently modeled through the bracket description of transport phenomena. Indeed, the simplicity (i.e., symmetry) inherent to the bracket allows a straightforward extension of previous cases to this more complex situation. So far, in comparison to the numerous studies for isothermal polymer systems (see chapter 8), little work has been devoted to the study of non-isothermal polymer flows [Bird and Öttinger, 1992, p. 397]. Dutta and Mashelkar [1987] have summarized the experimental techniques, data, and theory up to 1983. In particular for polymer solutions, previous work has primarily focused on the measurement and the correlation of the solution's conductivity with the polymer concentration [Dutta and Mashelkar, 1987, §6.1]. The only work addressing the effects of flow on heat conduction in polymers is the more recent contribution of van der Brule [1990] based on the

Hookean dumbbell model for polymer solutions. We comment more on this work further below.

Let us define the problem so as to be as simple as possible. When one considers a compressible polymeric solution, it is necessary to consider the polymer density and solvent density as independent dynamic variables. Although this can easily be incorporated into the analysis, we disregard it here to avoid confusion, and consider a polymer melt which contains only entangled polymer molecules. Furthermore, most polymers are, in reality, polydisperse in molecular weight; i.e., they contain molecules of different sizes. Any truly accurate theory of polymeric materials should incorporate this effect, possibly by considering the densities of different molecular-weight components as independent dynamic variables, so that $F=F[s,\mathbf{M},\mathbf{C},\rho_1,\rho_2,...]$. Again, this leads to unnecessary complication, and we shall consider only a monodisperse polymer melt in this section.

In the scenario described above, we again have the dynamical variables of §5.5, ρ, **M**, s, and **C**, with the operating space of (5.5-5) (with the no-slip condition on $\partial\Omega$ substituted for the no-penetration condition). The Hamiltonian can be simply written as in Eq. (5.5-12), without external fields:

$$H[\rho,\mathbf{M},s,\mathbf{C}] = \int_\Omega [e_k(\rho,\mathbf{M}) + u(\rho,s,\mathbf{C})]\,\mathrm{d}^3x \ . \tag{9.1-1}$$

Note that we are no longer dealing only with the mechanical contribution to the total energy of the system, as in chapter 8. The Poisson bracket is given by (5.5-11), and the dissipation bracket is derived as [Edwards and Beris, 1991b, p. 2477]

$$\begin{aligned}
[F,G] = &-\int_\Omega \Lambda_{\alpha\beta\gamma\varepsilon}\frac{\delta F}{\delta C_{\alpha\beta}}\frac{\delta G}{\delta C_{\gamma\varepsilon}}\,\mathrm{d}^3x \\
&-\int_\Omega \mathrm{B}_{\alpha\beta\gamma\varepsilon\eta\nu}\nabla_\gamma\frac{\delta F}{\delta C_{\alpha\beta}}\nabla_\nu\frac{\delta G}{\delta C_{\varepsilon\eta}}\,\mathrm{d}^3x \\
&-\int_\Omega Q_{\alpha\beta\gamma\varepsilon}\nabla_\alpha\frac{\delta F}{\delta M_\beta}\nabla_\gamma\frac{\delta G}{\delta M_\varepsilon}\,\mathrm{d}^3x \\
&-\int_\Omega L_{\alpha\beta\gamma\varepsilon}\left(\nabla_\alpha\frac{\delta F}{\delta M_\beta}\frac{\delta G}{\delta C_{\gamma\varepsilon}} - \nabla_\alpha\frac{\delta G}{\delta M_\beta}\frac{\delta F}{\delta C_{\gamma\varepsilon}}\right)\mathrm{d}^3x \\
&-\int_\Omega \alpha_{\alpha\beta}\nabla_\alpha\frac{\delta F}{\delta s}\nabla_\beta\frac{\delta G}{\delta s}\,\mathrm{d}^3x
\end{aligned}$$

$$+\int_\Omega \frac{1}{T}\frac{\delta F}{\delta s}\Lambda_{\alpha\beta\gamma\varepsilon}\frac{\delta G}{\delta C_{\alpha\beta}}\frac{\delta G}{\delta C_{\gamma\varepsilon}}\mathrm{d}^3x$$

$$+\int_\Omega \frac{1}{T}\frac{\delta F}{\delta s}\mathrm{B}_{\alpha\beta\gamma\varepsilon\eta\nu}\nabla_\gamma\frac{\delta G}{\delta C_{\alpha\beta}}\nabla_\nu\frac{\delta G}{\delta C_{\varepsilon\eta}}\mathrm{d}^3x \tag{9.1-2}$$

$$+\int_\Omega \frac{1}{T}\frac{\delta F}{\delta s}Q_{\alpha\beta\gamma\varepsilon}\nabla_\alpha\frac{\delta G}{\delta M_\beta}\nabla_\gamma\frac{\delta G}{\delta M_\varepsilon}\mathrm{d}^3x$$

$$+\int_\Omega \frac{1}{T}\frac{\delta F}{\delta s}\alpha_{\alpha\beta}\nabla_\alpha\frac{\delta G}{\delta s}\nabla_\beta\frac{\delta G}{\delta s}\mathrm{d}^3x \ .$$

Compared with Eq. (8.1-5) the above expression has two major differences. First, all the terms are included in the bracket (including the entropic ones) and second, since $\mathbf{C}\equiv\rho\mathbf{c}$ is used here as the independent variable instead of $\mathbf{c}$, all the phenomenological tensors that multiply 1 or 2 Volterra derivatives with respect to $\mathbf{C}$ are different from the ones defined in chapter 8 by a factor ρ^n where n is 1 or 2, respectively. The new parameter $\boldsymbol{\alpha}$ represents the, in general, anisotropic thermal conductivity matrix, $\boldsymbol{\alpha}=\boldsymbol{\alpha}_{00}$ in Eq. (7.2-1). The phenomenological parameters are not totally independent, but they obey constraints imposed from Onsager symmetry relationships. In general, they may be functions of the dynamic variables; however, typically, a dependence on the temperature rather than the entropy density appears. As an example, the relaxation time is typically considered empirically as an exponential function of the temperature,

$$\lambda = \lambda_0 \exp(\frac{A_0}{k_B T}) \ , \tag{9.1-3}$$

where A_0 is a (constant) activation energy.[1]

The above brackets result in the compressible, non-isothermal transport equations in terms of the dynamic variables [Edwards and Beris, 1991b, p. 2478],

$$\frac{\partial\rho}{\partial t} = -\ \nabla_\beta(\upsilon_\beta\rho) \ , \tag{9.1-4a}$$

[1]Note that an explicit temperature dependence implies that T is a dynamic variable, as we shall see shortly.

$$\rho\frac{\partial \upsilon_\alpha}{\partial t} = -\rho\upsilon_\beta\nabla_\beta\upsilon_\alpha - \nabla_\alpha p + \nabla_\beta\sigma_{\alpha\beta} , \tag{9.1-4b}$$

$$\begin{aligned}\frac{\partial s}{\partial t} &= -\nabla_\beta(\upsilon_\beta s) + \frac{1}{T}Q_{\alpha\beta\gamma\varepsilon}\nabla_\alpha\upsilon_\beta\nabla_\gamma\upsilon_\varepsilon + \frac{1}{T}\nabla_\alpha(\alpha_{\alpha\beta}T\nabla_\beta T) \\ &+ \frac{1}{T}\Lambda_{\alpha\beta\gamma\varepsilon}\frac{\delta H}{\delta C_{\alpha\beta}}\frac{\delta H}{\delta C_{\gamma\varepsilon}} + \frac{1}{T}B_{\alpha\beta\gamma\varepsilon\zeta\eta}\nabla_\gamma(\frac{\delta H}{\delta C_{\alpha\beta}})\nabla_\eta(\frac{\delta H}{\delta C_{\varepsilon\zeta}}) ,\end{aligned} \tag{9.1-4c}$$

and

$$\begin{aligned}\frac{\partial C_{\alpha\beta}}{\partial t} &= -\nabla_\gamma(\upsilon_\gamma C_{\alpha\beta}) + C_{\alpha\gamma}\nabla_\gamma\upsilon_\beta + C_{\gamma\beta}\nabla_\gamma\upsilon_\alpha \\ &- \Lambda_{\alpha\beta\gamma\varepsilon}\frac{\delta H}{\delta C_{\gamma\varepsilon}} + L_{\alpha\beta\gamma\varepsilon}\nabla_\gamma\upsilon_\varepsilon + \nabla_\gamma(B_{\alpha\beta\gamma\varepsilon\zeta\eta}\nabla_\eta\frac{\delta H}{\delta C_{\varepsilon\zeta}}) ,\end{aligned} \tag{9.1-4d}$$

where the pressure is as defined in (5.5-13e) and the extra stress is defined as

$$\sigma_{\alpha\beta} = Q_{\alpha\beta\gamma\varepsilon}\nabla_\gamma\upsilon_\varepsilon + 2c_{\beta\gamma}\frac{\delta H}{\delta c_{\alpha\gamma}} + 2L_{\alpha\beta\gamma\varepsilon}\frac{\delta H}{\delta c_{\gamma\varepsilon}} . \tag{9.1-4e}$$

As before, the corresponding energy equations may be written down as well [Edwards and Beris, 1991b, p. 2478].

The above system of equations represents a fully general description of the non-inertial, polymeric system. The various phenomenological matrices can take any of the forms described in chapter 8, consistently generalizing the incompressible viscoelastic fluid models to non-isothermal situations.[2] The only new phenomenological matrix is $\boldsymbol{\alpha}$, the thermal conductivity matrix, which we should no longer expect to be isotropic.

Since the polymeric fluid is anisotropic, at least under flowing conditions, we hypothesize that the structure of the fluid affects the thermal conductivity to the extent dictated by the Cayley/Hamilton theorem,

$$\alpha_{\alpha\beta} = a_1\delta_{\alpha\beta} + a_2C_{\alpha\beta} + a_3C_{\alpha\gamma}C_{\gamma\beta} , \tag{9.1-5}$$

where the scalar coefficients, a_i, i=1,2,3, are, in general, functions of the invariants of **C** and the other dynamic variables. By checking the entropy inequality, assuming that the appropriate term is completely independent of all the others, we find that [Edwards and Beris, 1992, Eq. (3-19)][3]

[2]One must realize, however, that the compressible nature of the fluid might oblige additional terms in the phenomenological matrices which are not present for the incompressible case—see §10.1.5.

[3]These conditions were actually obtained for the constrained conformation tensor theory with unit trace. The result is probably generalizable to the non-constrained case, however.

$$a_1 \geq 0 , \quad \text{and} \quad a_2 + a_3 \geq 0 . \tag{9.1-6}$$

Eq. (9.1-5) is all that can be gleaned from continuum mechanics; to derive a more specific form, the use of a microscopic model is necessary. As a particular case, for an elastic dumbbell model, the 0-th order approximation of van der Brule [1990, p. 419] suggests that a_1 is the thermal conductivity of the solvent (for a polymer solution), a_3=0, and

$$a_2 = \tfrac{3}{2}\alpha\rho\frac{\zeta}{m}k_B , \tag{9.1-7}$$

where α is the mass fraction of the polymer, ζ is the friction coefficient, and m is the mass of a bead. Clearly, however, this subject is still in its infancy and much work is necessary to elaborate on the forms for the thermal conductivity matrix for various models. Nevertheless, the bracket description dictates the form of the equations with which future non-isothermal models must conform.

In order to derive explicit equations for particular models, it is necessary to specify not only the phenomenological matrices but the Hamiltonian as well. The obvious thing to do at this preliminary stage is just to use the Hamiltonians and matrices of chapter 8, although written in terms of the appropriate dynamic variables for this more general situation. The former presents a problem, however, in that to generalize the various Hamiltonians of chapter 8 we need to work in terms of the Helmholtz free energy because the temperature, and not the entropy, appears as a natural variable. Hence we wish to transform $H[\rho,s,\mathbf{M},\mathbf{C}]\rightarrow\tilde{A}[\rho,T,\mathbf{M},\mathbf{C}]$ in the generalized bracket.

Due to the above arguments, we write an extended free energy, $\tilde{A}$, as

$$\tilde{A}[\rho,\mathbf{M},s,\mathbf{C}] = \int_\Omega [e_k(\rho,\mathbf{M}) + a(\rho,T,\mathbf{C})]\,\mathrm{d}^3x , \tag{9.1-8}$$

and hence

$$\tilde{A} = K[\rho,\mathbf{M}] + A[\rho,T,\mathbf{C}] . \tag{9.1-9}$$

Note that $a = u - Ts$, from §4.3.3, and that

$$\tilde{A} = H - TS = H - \int_\Omega Ts\,\mathrm{d}^3x . \tag{9.1-10}$$

Now the functional derivatives needed for the generalized bracket may be calculated as

$$\left.\frac{\delta H}{\delta \mathbf{C}}\right|_{\rho,\mathbf{M},s} = \left.\frac{\delta \tilde{A}}{\delta \mathbf{C}}\right|_{\rho,\mathbf{M},s} + s\left.\frac{\partial T}{\partial \mathbf{C}}\right|_{\rho,\mathbf{M},s}$$

$$= \left.\frac{\delta \tilde{A}}{\delta \mathbf{C}}\right|_{\rho,\mathbf{M},T} + \left.\frac{\delta \tilde{A}}{\delta T}\right|_{\rho,\mathbf{M},\mathbf{C}}\left.\frac{\partial T}{\partial \mathbf{C}}\right|_{\rho,\mathbf{M},s} + s\left.\frac{\partial T}{\partial \mathbf{C}}\right|_{\rho,\mathbf{M},s}$$

$$= \left.\frac{\delta \tilde{A}}{\delta \mathbf{C}}\right|_{\rho,\mathbf{M},T} ,$$

$$\left.\frac{\delta H}{\delta \rho}\right|_{s,\mathbf{M},\mathbf{C}} = \left.\frac{\delta \tilde{A}}{\delta \rho}\right|_{T,\mathbf{M},\mathbf{C}} , \tag{9.1-11}$$

$$\left.\frac{\delta H}{\delta \mathbf{M}}\right|_{s,\rho,\mathbf{C}} = \left.\frac{\delta \tilde{A}}{\delta \mathbf{M}}\right|_{T,\rho,\mathbf{C}} ,$$

$$\left.\frac{\delta H}{\delta s}\right|_{\rho,\mathbf{M},\mathbf{C}} = \left.\frac{\delta \tilde{A}}{\delta s}\right|_{\rho,\mathbf{M},\mathbf{C}} + \left.\frac{\partial (Ts)}{\partial s}\right|_{\rho,\mathbf{M},\mathbf{C}}$$

$$= \left.\frac{\delta \tilde{A}}{\delta T}\right|_{\rho,\mathbf{M},\mathbf{C}}\left.\frac{\partial T}{\partial s}\right|_{\rho,\mathbf{M},\mathbf{C}} + T + s\left.\frac{\partial T}{\partial s}\right|_{\rho,\mathbf{M},\mathbf{C}}$$

$$= T .$$

EXAMPLE 9.1: Given the definition of the pressure in terms of the internal energy density, (5.5-13e), find an equivalent definition in terms of the Helmholtz free energy density.

Eq. (5.5-13e) states that

$$p = -u + \rho\frac{\partial u}{\partial \rho} + s\frac{\partial u}{\partial s} + C_{\alpha\beta}\frac{\partial u}{\partial C_{\alpha\beta}} .$$

Substituting $u = a + Ts$ into this expression, we find that

$$
\begin{aligned}
p &= -a - Ts + \rho \left.\frac{\partial(a+Ts)}{\partial \rho}\right|_{\mathbf{C},s} + s \left.\frac{\partial(a+Ts)}{\partial s}\right|_{\mathbf{C},\rho} + \mathbf{C}:\left.\frac{\partial(a+Ts)}{\partial \mathbf{C}}\right|_{\rho,s} \\
&= -a - Ts + \rho \left.\frac{\partial a}{\partial \rho}\right|_{\mathbf{C},s} + \rho s \left.\frac{\partial T}{\partial \rho}\right|_{\mathbf{C},s} + s \left.\frac{\partial a}{\partial s}\right|_{\mathbf{C},\rho} + sT + s^2 \left.\frac{\partial T}{\partial s}\right|_{\mathbf{C},\rho} \\
&\quad + \mathbf{C}:\left.\frac{\partial a}{\partial \mathbf{C}}\right|_{\rho,s} + s\mathbf{C}:\left.\frac{\partial T}{\partial \mathbf{C}}\right|_{\rho,s} \\
&= -a + \rho \left.\frac{\partial a}{\partial \rho}\right|_{\mathbf{C},T} + \rho \left.\frac{\partial a}{\partial T}\right|_{\mathbf{C},\rho} \left.\frac{\partial T}{\partial \rho}\right|_{\mathbf{C},s} + \rho s \left.\frac{\partial T}{\partial \rho}\right|_{\mathbf{C},s} + s \left.\frac{\partial a}{\partial T}\right|_{\mathbf{C},\rho} \left.\frac{\partial T}{\partial s}\right|_{\mathbf{C},\rho} \\
&\quad + s^2 \left.\frac{\partial T}{\partial s}\right|_{\mathbf{C},\rho} + \mathbf{C}:\left.\frac{\partial a}{\partial \mathbf{C}}\right|_{\rho,T} + \left.\frac{\partial a}{\partial T}\right|_{\mathbf{C},\rho} \mathbf{C}:\left.\frac{\partial T}{\partial \mathbf{C}}\right|_{\rho,s} + s\mathbf{C}:\left.\frac{\partial T}{\partial \mathbf{C}}\right|_{\rho,s} \\
&= -a + \rho \left.\frac{\partial a}{\partial \rho}\right|_{\mathbf{C},T} + \mathbf{C}:\left.\frac{\partial a}{\partial \mathbf{C}}\right|_{\rho,T} .
\end{aligned}
\tag{9.1-12}
$$

The final equality defines the pressure in terms of the Helmholtz free energy density.

Let us check to see if this definition is consistent with the pressure as defined through the absolute Helmholtz free energy. This is given by (4.2-4b) as

$$
p = -\left.\frac{\partial A}{\partial V}\right|_{T,\mathbf{C}} ,
$$

so that, defining $\hat{\mathbf{C}} \equiv \mathbf{C}/\rho$, we have

$$
\begin{aligned}
p &= -\left.\frac{\partial \hat{A}}{\partial(\frac{1}{\rho})}\right|_{\hat{\mathbf{C}},T} = -\left.\frac{\partial(\frac{a}{\rho})}{\partial(\frac{1}{\rho})}\right|_{\hat{\mathbf{C}},T} = -a - \frac{1}{\rho}\left.\frac{\partial a}{\partial(\frac{1}{\rho})}\right|_{\hat{\mathbf{C}},T} \\
&= -a - \frac{1}{\rho}\left(\left.\frac{\partial a}{\partial(\frac{1}{\rho})}\right|_{\mathbf{C},T} + \left.\frac{\partial a}{\partial \mathbf{C}}\right|_{T,\rho} : \left.\frac{\partial \mathbf{C}}{\partial(\frac{1}{\rho})}\right|_{\hat{\mathbf{C}},T} \right) \\
&= -a + \rho \left.\frac{\partial a}{\partial \rho}\right|_{\mathbf{C},T} + \mathbf{C}:\left.\frac{\partial a}{\partial \mathbf{C}}\right|_{\rho,T} .
\end{aligned}
$$

Hence we have arrived independently at the same expression. ■

Upon substitution of (9.1-11) into the generalized bracket, we arrive at a coupled set of evolution equations in terms of the extended Helmholtz free energy: Eq. (9.1-4a), (9.1-4b) with the pressure defined by (9.1-12), and (9.1-4c) through (9.1-4e) with $\delta\tilde{A}/\delta\mathbf{C}$ replacing $\delta H/\delta\mathbf{C}$. Consequently, any of the Hamiltonians of chapter 8 may now be re-expressed in the more general formalism. For example, taking the Hamiltonian for the Maxwell model, (8.1-12,13), we see that

$$a = a_0(\rho,T) + \tfrac{1}{2}\alpha K \operatorname{tr}\mathbf{C} - \tfrac{1}{2}\alpha\rho k_B T \ln\det\left(\frac{\mathbf{C}K}{\rho k_B T}\right) . \tag{9.1-13}$$

In this expression, $a_0(\rho,T)$ represents the typical Helmholtz free energy density (i.e., not depending on $\mathbf{C}$) of a fluid. The form of this term must be specified by a constitutive relationship, yet it will only appear in the pressure. The important point, however, is that the various quantities appearing in the free energy may now have an explicit dependence upon the temperature, such as the spring constant (a misnomer, in this case), K. Future work should focus on determining the functional forms for these parameters from microscopic theories, as well as specific relationships for the phenomenological matrices.

9.2 Modeling of the Rheology and Flow-Induced Concentration Changes in Polymer Solutions

In this section,[4] the systematic development of two models for the hydrodynamics of polymer solutions is presented using the generalized bracket formulation. This new development avoids the ambiguities or arbitrary assumptions introduced in previous models [Helfand and Fredrickson, 1989; Onuki, 1989; 1990]. As regards the coupling of the elastic stress and the concentration and the modeling of interphases, the same phenomenology is described as in the previous models; however, the equations are quantitatively different. In addition, more general descriptions of polymer dynamics and thermodynamics can be accounted for.

In subsection §9.2.1, a single-fluid model is presented where the polymer solution is treated as a coherent single continuum medium, similar to the picture used in chapter 8 and §9.1. Thus, a minimum of assumptions is used, mostly based on the description of the transport phenomena through a bilinear dissipation form. A drawback of this approach is the proliferation of material parameters.

[4]This section was written in collaboration with V. G. Mavrantzas and follows the ideas outlined in [Mavrantzas and Beris, 1992a].

One of the advantages of the bracket formulation is that it allows the elucidation of the role of the assumptions necessary for the construction of a particular model. This is achieved by separating the mathematics involved in the development of the model from its underlying physical principles. By eliminating any arbitrariness in the mathematical derivation, the validity of the underlying principles is reflected in the validity of the resulting models. Thus, it allows for a direct re-examination of various assumptions through a comparison of the corresponding models. One consequence of this, and a caveat to the user of the theory, is that the use of the bracket theory in conjunction with a poor description of the physics of the system can lead to poor results, despite the correct use of the mathematics.

With the above remarks in mind, we embark in §9.2.2 to present the development of an alternative theory for the description of hydrodynamic interactions in polymer solutions, based on a two-fluid model of the underlying physics. The adoption of a two-fluid model is equivalent to the notion of two interpenetrating continua. The interactions between the two components can vary from very vigorous—leading to the complete equilibration of momentum and energy as it is associated with the single-fluid/two-component model—to very weak—leading to separate momenta and thermal energies as the ones often associated with the electrons and the neutral and ionized species in a dilute plasma [Bataille and Kestin, 1977, p. 49]; see also chapter 12. The case of dilute polymer solutions can be considered as an intermediate one, where complete thermal equilibration but only partial momentum equilibration is established between the macromolecules and the solvent. Two-fluid model descriptions in terms of hydrodynamic equations of motion have been applied quite successfully in the past to systems with a condensed phase, such as charged and neutral superfluids, and dielectric and magnetic crystals [Enz, 1974].

9.2.1 *Single-fluid model of polymer solution hydrodynamics allowing for concentration variations*

The phase separation behavior of dilute and semidilute polymer solutions under shear flow has attracted considerable attention in the literature. Rangel-Nafaile *et al.* [1984] examined stress-induced phase separations by resorting to the dependence of the free energy on the chain conformation and hence on the imposed deformation rate. More recently, Helfand and Fredrickson [1989] and Onuki [1989; 1990] have proposed three different phenomenological theories of the dynamic behavior of polymer solutions under flow. The theories attempted to couple the internal deformation state and the number density of the polymeric molecules in the description of the rheology and mass transfer. As early as in 1977, Tirrell and Malone [1977] had considered diffusion of macromolecules driven by gradients of an entropic potential arising from distortion of molecular conformation by deformation as well as by chain concentration gradients. However, there are differences and ambiguities among the proposed models. The bracket formalism is used here as an alternative approach for the derivation of a thermodynamically consistent set of governing equations which elucidates and limits the underlying assumptions.

As in all previous models, the starting point in the modeling development is the definition of the governing Hamiltonian (extended Helmholtz free energy) of the system, H. In the present formulation, the Hamiltonian is defined as a functional involving the sum of the kinetic, h_k, (external) potential, h_v, and internal free energy, h_e, densities of the system:

$$H \equiv \int_{\Omega} (h_k + h_v + h_e)\, \mathrm{d}^3x \ . \tag{9.2-1}$$

Each one of the energy densities is then defined as a function of the external (thermodynamic and dynamic, i.e., densities, temperature, etc.) and internal (as, for example, elastic deformation) variables of the system. The governing equations are then derived as the balance and constitutive equations, involving the primary variables and the partial derivatives of the energy densities of the system.

Where the present approach is different from the previous ones is in basing the development of the governing equations on the generalized bracket formalism. For simplicity, we focus our attention here on the description of an isothermal system close to equilibrium. Thus, we forgo the entropy correction term from the description of the dissipation bracket, keeping in mind that the remaining bracket no longer conserves the Hamiltonian (rather, the Hamiltonian of a closed system decreases monotonically corresponding to the energy loss due to the dissipation).

The primary variables are the two mass densities, ρ_1, ρ_2, for the polymer and the solvent, respectively, the momentum density, $\mathbf{M}\equiv\rho\mathbf{u}$, where ρ is the total density, $\rho_1+\rho_2$, and $\mathbf{u}$ the velocity, and the conformation density tensor, $\mathbf{C}$. Notice that within the generalized bracket theory for compressible viscoelastic fluids, the appropriate internal variable to describe viscoelasticity is proportional to the polymer density, i.e., $\mathbf{C}=\rho_1\mathbf{c}$, and not just the conformation tensor $\mathbf{c}$, where $\mathbf{c}$, for the Rouse model, can be identified with the second moment of the polymer end-to-end distance vector, $\langle\mathbf{RR}\rangle$.[5] This is also of importance in comparing our final expressions with those of Onuki [1990] and Milner [1991] ($\mathbf{c}\equiv\mathbf{w}$, in their notation). The incompressibility constraint, corresponding to a constant

[5]The fact that in chapter 8 the conformation tensor $\mathbf{c}$ was used as the structural variable is not inconsistent with the above requirement, since, in the cases investigated in chapter 8, ρ_1 is assumed always to be constant and therefore its inclusion in $\mathbf{C}$ does not change any of the results.

mass density, ρ, is also introduced at this stage, as an additional approximation. Thus, the independent primary variables are ρ_1, $\mathbf{M}$ and $\mathbf{C}$.

The corresponding Hamiltonian involves only the kinetic energy density defined as

$$h_k = \frac{1}{2}\frac{M^2}{\rho} , \tag{9.2-2}$$

and a general internal free energy density, $h_e(\rho_1,\mathbf{C})$. The Poisson bracket is

$$\begin{aligned}\{F,G\} = &-\int_\Omega \left(\frac{\partial F}{\partial \rho_1}\nabla_\beta(\frac{\partial G}{\partial M_\beta}\rho_1) - \frac{\partial G}{\partial \rho_1}\nabla_\beta(\frac{\partial F}{\partial M_\beta}\rho_1) \right) d^3x \\ &-\int_\Omega \left(\frac{\partial F}{\partial M_\gamma}\nabla_\beta(\frac{\partial G}{\partial M_\beta}M_\gamma) - \frac{\partial G}{\partial M_\gamma}\nabla_\beta(\frac{\partial F}{\partial M_\beta}M_\gamma) \right) d^3x \\ &-\int_\Omega \left(\frac{\partial F}{\partial C_{\alpha\beta}}\nabla_\gamma(C_{\alpha\beta}\frac{\partial G}{\partial M_\gamma}) - \frac{\partial G}{\partial C_{\alpha\beta}}\nabla_\gamma(C_{\alpha\beta}\frac{\partial F}{\partial M_\gamma}) \right) d^3x \\ &-\int_\Omega C_{\gamma\alpha}\left(\frac{\partial G}{\partial C_{\alpha\beta}}\nabla_\gamma(\frac{\partial F}{\partial M_\beta}) - \frac{\partial F}{\partial C_{\alpha\beta}}\nabla_\gamma(\frac{\partial G}{\partial M_\beta}) \right) d^3x \\ &-\int_\Omega C_{\gamma\beta}\left(\frac{\partial G}{\partial C_{\alpha\beta}}\nabla_\gamma(\frac{\partial F}{\partial M_\alpha}) - \frac{\partial F}{\partial C_{\alpha\beta}}\nabla_\gamma(\frac{\partial G}{\partial M_\alpha}) \right) d^3x \, .\end{aligned} \tag{9.2-3}$$

The first term represents a modification over the Poisson bracket corresponding to an incompressible viscoelastic fluid, Eq. (8.1-1), since the polymer density is now allowed to vary.

The dissipation terms are introduced through the use of the following expression for the dissipation bracket

$$\begin{aligned}[F,G] = &-\int_\Omega \frac{\eta_s}{2}\left(\nabla_\alpha\frac{\delta F}{\delta M_\beta} + \nabla_\beta\frac{\delta F}{\delta M_\alpha}\right)\left(\nabla_\alpha\frac{\delta G}{\delta M_\beta} + \nabla_\beta\frac{\delta G}{\delta M_\alpha}\right) d^3x \\ &-\int_\Omega \Lambda_{\alpha\beta\gamma\varepsilon}\frac{\delta F}{\delta C_{\alpha\beta}}\frac{\delta G}{\delta C_{\gamma\varepsilon}} d^3x \\ &-\int_\Omega D_{\alpha\beta}\nabla_\alpha\frac{\delta F}{\delta \rho_1}\nabla_\beta\frac{\delta G}{\delta \rho_1} d^3x\end{aligned}$$

$$- \int_\Omega E_{\alpha\beta\gamma\varepsilon} \left(\nabla_\alpha (C_{\beta\lambda} \frac{\delta F}{\delta C_{\lambda\gamma}}) \nabla_\varepsilon \frac{\delta G}{\delta \rho_1} + \nabla_\alpha (C_{\beta\lambda} \frac{\delta G}{\delta C_{\lambda\gamma}}) \nabla_\varepsilon \frac{\delta F}{\delta \rho_1} \right) d^3x$$

$$- \int_\Omega \mathrm{B}_{\alpha\beta\gamma\varepsilon\eta\nu} \nabla_\alpha (C_{\beta\lambda} \frac{\delta F}{\delta C_{\lambda\gamma}}) \nabla_\varepsilon (C_{\zeta\mu} \frac{\delta G}{\delta C_{\mu\eta}}) d^3x \ . \tag{9.2-4}$$

In Eq. (9.2-4), η_s represents the solution viscosity and $\boldsymbol{\Lambda}$ the relaxation tensor—see §8.1 for various available expressions depending on the rheological effects taken into consideration. **D** is the, in general, anisotropic diffusion coefficient for the polymer species. The third term represents stress-induced concentration diffusion and the fourth one dissipation driven by large gradients in the conformation of the polymer structure. Although Eq. (9.2-4) might appear utterly complex, still it does not represent the most general expression that can be written for the dissipation, but rather a first-order approximation. As explained in §7.1, the dissipation bracket $[F,G]$ is a nonlinear functional of G. However, in most applications, retaining the linear terms has proven to be sufficient. Whether this is the case in the present application remains to be seen.

The transport coefficients η_s, **D**, **E**, and **B** are, in general, functions of the dynamic variables. The use of Onsager symmetry relationships and of the thermodynamic inequalities similar to the ones expressed in chapter 7, emanating from the non-negative character of the corresponding entropy production, places certain constraints on them. Still, many degrees of freedom are left to be useful, so we refrain from presenting the resulting thermodynamic constraints in their most general form. Instead, they are used further below, in §9.2.3, where the use of a microscopic theory and/or some additional isotropicity assumptions substantially decreases their complexity.

The large number of degrees of freedom included in the general transport coefficients also prohibits their direct determination experimentally. This leaves their determination through a microscopic theory as the only viable alternative. Previous works based on the polymer kinetic theory have primarily focused their attention on the determination of the diffusion tensor **D** (see the recent review by Bird and Öttinger [1992, p. 395] for a listing of the most important work). Even there, the earlier work is limited either next to equilibrium or for homogeneous flows (steady shear). The most important conclusion there is the anisotropicity of the diffusion tensor under flow conditions (defined with respect to the concentration gradients as the driving force) and its strong dependence on the flow rate [Öttinger, 1989a, pp. 283-84; Prakash and Mashelkar, 1991, pp. 3746-47]. Moreover, the latter work presents a form for the diffusivity as a function of the polymer conformation, which can, in principle, be used in the bracket formulation. However, no information was provided for the remaining two transport coefficients. This has been accomplished (to a first approximation) through a recent attempt to extend these works to non-homogeneous systems [Bhave *et al.*, 1991]. Alternatively, to the same degree of approximation, information on all the transport coefficients can be obtained through a two-fluid modeling approach which considers the momentum of the macromolecules separately from that of the solvent. This approach is followed in detail in the next subsection.

Finally, note that incorporation of surface tension effects between phases can be achieved by including a dependence of the free energy density h_e on the gradient of the polymer mass density ρ_1 as well, $h_e=h_e(\rho_1, \nabla\rho_1, \mathbf{C})$. Thus, in the most general case considered here, h_e is assumed to involve three types of contributions: First, a Flory/Huggins term accounting for entropic and enthalpic contributions due to mixing of polymer with the solvent; second, an intra-molecular free energy term accounting for the deviation of the polymer conformation from the equilibrium one; and third, a term depending on polymer concentration gradient accounting for the "free-surface" type of interactions. A typical expression for h_e, corresponding to the simplest model for the polymer chains (Hookean dumbbell), is

$$H = \int_\Omega \frac{1}{2}\rho u^2 \mathrm{d}^3x + \int_\Omega k_B T(n\ln\phi + n_s \ln\phi_s)\mathrm{d}^3x$$

$$+\int_\Omega \frac{1}{2}K\mathrm{tr}\mathbf{C}\,\mathrm{d}^3x - \int_\Omega \frac{1}{2}nk_B T\ln\det(\frac{K\mathbf{C}}{nk_B T})\mathrm{d}^3x \qquad (9.2\text{-}5)$$

$$+\int_\Omega \frac{1}{2}K_1(\nabla\phi)^2\mathrm{d}^3x \ .$$

In Eq. (9.2-5), ϕ and ϕ_s are the volume fraction of the polymer and solvent, respectively [Flory, 1953, p. 502], n the polymer number density and n_s the solvent number density.

The resulting evolution equations consist of the continuity of the polymer species

$$\frac{\partial\rho_1}{\partial t} = -u_\beta\nabla_\beta\rho_1 + \nabla_\alpha(D_{\alpha\beta}\nabla_\beta\frac{\delta H}{\delta\rho_1}) + \nabla_\varepsilon(\tfrac{1}{2}E_{\alpha\beta\gamma\varepsilon}\nabla_\alpha\sigma_{\beta\gamma}) \ , \qquad (9.2\text{-}6)$$

the momentum equation

$$\frac{\partial M_a}{\partial t} = -u_\beta\nabla_\beta M_\alpha - \nabla_\alpha p - \nabla_\beta(\frac{\partial h_e}{\partial(\nabla_\beta\rho_1)}\nabla_\alpha\rho_1) \qquad (9.2\text{-}7)$$

$$+\nabla_\beta\sigma_{\alpha\beta} + \nabla_\beta[\eta(\nabla_\beta v_\alpha + \nabla_\alpha v_\beta)] \ ,$$

and the evolution equation for the structural parameter **C**

$$\overset{\triangledown}{C}_{\alpha\beta} = -\Lambda_{\alpha\beta\gamma\varepsilon}\frac{\delta H}{\delta C_{\varepsilon\gamma}} + C_{\alpha\gamma}\nabla_{\varepsilon}\left(E_{\varepsilon\gamma\beta\lambda}\nabla_{\lambda}\frac{\delta H}{\delta \rho_1}\right)$$
$$+ C_{\alpha\gamma}\nabla_{\varepsilon}\left(B_{\varepsilon\gamma\beta\lambda\zeta\eta}\nabla_{\lambda}(C_{\zeta\mu}\frac{\delta H}{\delta C_{\mu\eta}})\right), \tag{9.2-8}$$

where, as usual, the polymeric contribution to the stress, $\boldsymbol{\sigma}$, is defined as

$$\sigma_{\alpha\beta} = 2C_{\beta\gamma}\frac{\delta H}{\delta C_{\gamma\alpha}} \ . \tag{9.2-9}$$

As Eq. (9.2-6) shows, the flux in polymer solutions is, in general, expected to depend on both the gradient of the chemical potential, $\boldsymbol{\nabla}(\delta H/\delta\rho_1)$ and the gradient of the extra stress, $\boldsymbol{\nabla}\cdot\boldsymbol{\sigma}$. Similarly, additional terms appear in the evolution equation for the conformation tensor, Eq. (9.2-8). However, the exact magnitude of their contributions cannot be decided from the formalism alone. In §9.2.3 this issue is further investigated through a direct comparison of the above expression against the two-fluid model described below, in §9.2.2, and various models which have appeared in the literature. There, the significance of the additional terms will be assessed.

9.2.2 *Two-fluid model for polymer solutions*

In a recent Letter, Milner [1991] developed a two-fluid model based on the projection of the Langevin equations for a Rouse chain, thus offering a theoretical justification of a similar model developed previously by Onuki [1990]. However, he did not take into account sharp changes in the concentration as the changes encountered in the vicinity of interphases and his approach is difficult to carry out to describe a more complex polymer thermodynamic and/or dynamic behavior. It is the purpose of the present section to apply the generalized bracket formalism in the derivation of the governing equations corresponding to a two-fluid model description of polymer dynamics.

As in the one-fluid model, the starting point in the modeling development is again the definition of the governing Hamiltonian (extended Helmholtz free energy) of the system, H, provided by Eq. (9.2-1). For clarity, we start with the description of a dissipationless, structureless, homogeneous medium, and we work our way out toward more complex systems, considering one extra feature at a time.

The compressible, ideal (dissipationless), two-fluid mixture is considered first, as the basic case. The dynamic variables are the two mass densities, ρ_1, ρ_2, and the two momentum densities, $\mathbf{M}_1 \equiv \rho_1 \mathbf{u}_1$ and $\mathbf{M}_2 \equiv \rho_2 \mathbf{u}_2$, of the system, where $\mathbf{u}_1$, $\mathbf{u}_2$ are the corresponding velocities. The corresponding Hamiltonian involves only the kinetic energy density defined as

$$h_k = \tfrac{1}{2}\left(\frac{M_1^2}{\rho_1} + \frac{M_2^2}{\rho_2}\right) , \qquad (9.2\text{-}10)$$

and a general internal free energy density, $h_e(\rho_1,\rho_2)$. The corresponding Poisson bracket involves the direct sum of the terms corresponding to a single compressible fluid, Eqs. (5.3-30), written for the individual mass and momentum density pairs, $(\rho_1, \mathbf{M}_1)$ and $(\rho_2, \mathbf{M}_2)$, respectively. The resulting evolution equations consist then of the continuity and momentum balances of the individual species,

$$\begin{aligned} \frac{\partial \rho_i}{\partial t} &= -\nabla_\beta(u_{i\beta}\rho_i) , \quad i=1,2 , \\ \frac{\partial M_{i\alpha}}{\partial t} &= -\nabla_\beta(u_{i\beta}M_{i\alpha}) - \rho_i \nabla_\alpha\left(\frac{\partial h_e}{\partial \rho_i}\right) , \quad i=1,2 , \end{aligned} \qquad (9.2\text{-}11)$$

where, as usual, Greek subscripts denote spatial coordinates, Latin subscripts denote species and the Einstein summation convention is assumed over repeated Greek indices only. Equation (9.2-11) represents the mathematical description of non-interacting, interpenetrating continua. For a further discussion on the subject with applications to plasma glow discharge, see chapter 12.

In order to incorporate the incompressibility constraint corresponding to a constant total mass density, as it is customarily assumed [Onuki, 1990], it is first necessary to switch to a new set of variables:

$$\begin{aligned} \rho_+ &= \rho_1 + \rho_2 , \qquad & \mathbf{M}_+ &= \mathbf{M}_1 + \mathbf{M}_2 \\ \rho_- &= \rho_1 , \qquad & \mathbf{M}_- &= \frac{\rho_2}{\rho_1+\rho_2}\mathbf{M}_1 - \frac{\rho_1}{\rho_1+\rho_2}\mathbf{M}_2 . \end{aligned} \qquad (9.2\text{-}12)$$

The choice of the variables is dictated by the nature of the incompressibility constraint

$$\boldsymbol{\nabla}\cdot\mathbf{M}_+ = 0 \;\Rightarrow\; \boldsymbol{\nabla}\cdot\mathbf{v} = 0 , \qquad (9.2\text{-}13)$$

where $\mathbf{v}$ is the mass average velocity of the mixture,

$$\mathbf{v} = \frac{\rho_1}{\rho_1+\rho_2}\mathbf{u}_1 + \frac{\rho_2}{\rho_1+\rho_2}\mathbf{u}_2 \ , \tag{9.2-14}$$

so that the Volterra derivative of the Hamiltonian with respect to $\mathbf{M}_+$ in the new set of variables is also divergence free ($\delta H/\delta \mathbf{M}_+ = \mathbf{v}$). Note that in this variables framework, the Volterra derivative of the Hamiltonian with respect to $\mathbf{M}_-$ also has a simple form, $\delta H/\delta \mathbf{M}_- = \boldsymbol{\Delta}\mathbf{v}$, where $\boldsymbol{\Delta}\mathbf{v} \equiv \mathbf{u}_1 - \mathbf{u}_2$.

Re-expressing the generalized bracket in terms of the new variables and exploiting the incompressibility constraint imposed by the assumption of constant total mass density, ρ_+=constant=ρ_0, we are led to the following set of conservation equations

$$\begin{aligned}\frac{\partial \rho_-}{\partial t} &= -\nabla_\beta(u_{1\beta}\rho_1) \\ &= -v_\beta\nabla_\beta\rho_- - \nabla_\beta[\rho_-(1-\phi)\Delta v_\beta] \ ,\end{aligned} \tag{9.2-15}$$

$$\begin{aligned}\frac{\partial M_{+\alpha}}{\partial t} = &-\nabla_\beta(u_{1\beta}M_{1\alpha} + u_{2\beta}M_{2\alpha}) - M_{1\beta}\nabla_\alpha u_{1\beta} - M_{2\beta}\nabla_\alpha u_{2\beta} \\ &- \rho_-\nabla_\alpha\left(\frac{\partial h_e}{\partial \rho_-}\right) - \nabla_\alpha p \ ,\end{aligned} \tag{9.2-16}$$

$$\begin{aligned}\frac{\partial M_{-\alpha}}{\partial t} = &-(1-\phi)\nabla_\beta(u_{1\beta}M_{1\alpha}) + \phi\nabla_\beta(u_{2\beta}M_{2\alpha}) - (1-\phi)M_{1\beta}\nabla_\alpha u_{1\beta} \\ &+ \phi M_{2\beta}\nabla_\alpha u_{2\beta} - (1-\phi)\rho_-\nabla_\alpha\left(\frac{\partial h_e}{\partial \rho_-}\right) \ ,\end{aligned} \tag{9.2-17}$$

where $\phi \equiv \rho_1/\rho_0$, and the pressure p is evaluated by solving a Poisson equation obtained by taking the divergence of Eq. (9.2-16).

By using the inverse of the transformation shown in Eq. (9.2-12), we obtain the following species momentum density conservation equations

$$\begin{aligned}\frac{\partial M_{1\alpha}}{\partial t} &= \frac{\partial M_{-\alpha}}{\partial t} + \phi\frac{\partial M_{+\alpha}}{\partial t} + M_{+\alpha}\frac{\partial \phi}{\partial t} \\ &= -\nabla_\beta(u_{1\beta}M_{1\alpha}) - M_{1\beta}\nabla_\alpha u_{1\beta} \\ &\quad - \rho_-\nabla_\alpha\left(\frac{\partial h_e}{\partial \rho_-}\right) - \phi\nabla_\alpha p - v_\alpha\nabla_\beta(\rho_1 u_{1\beta}) \ ,\end{aligned} \tag{9.2-18}$$

$$\frac{\partial M_{2\alpha}}{\partial t} = \frac{\partial M_{+\alpha}}{\partial t} - \frac{\partial M_{1\alpha}}{\partial t}$$

$$= -\nabla_\beta(u_{2\beta}M_{2\alpha}) - M_{2\beta}\nabla_\alpha u_{2\beta} \tag{9.2-19}$$

$$-(1-\phi)\nabla_\alpha p + v_\alpha \nabla_\beta(\rho_1 u_{1\beta}) \ .$$

So far the two components of the mixture have been assumed to be structureless. If one of them, for example, component 1, is viscoelastic, additional contributions to the momentum equation arise. In order to consistently calculate such contributions, it suffices to introduce as an additional internal variable the conformation density tensor, **C**, in a fashion similar to the description of compressible viscoelasticity (see §9.1). Thus, the free energy density becomes a function of both ρ_- and **C**, $h_e = h_e(\rho_-, \mathbf{C})$.

By comparison with the traditional description of a one-fluid viscoelastic model (see chapter 8), the velocity which has to be used in the corresponding two-fluid model is the velocity of the polymer phase, $\mathbf{u}_1$. This arises from the connection of **C** with the Finger strain tensor in the elastic phase as revealed by the development of the Poisson bracket (see §5.5). If the new terms in the generalized bracket are expressed in the transformed variables framework, see Eq. (9.2-12), it is found that they lead to two additional contributions in Eq. (9.2-18) (the continuity equation and the momentum equation for component 2 remain unchanged)

$$\frac{\partial M_{1\alpha}}{\partial t} = -\nabla_\beta(u_{1\beta}M_{1\alpha}) - M_{1\beta}\nabla_\alpha u_{1\beta}$$

$$-\rho_-\nabla_\alpha\left(\frac{\partial h_e}{\partial \rho_-}\right) - \phi\nabla_\alpha p - v_\alpha\nabla_\beta(\rho_1 u_{1\beta}) \tag{9.2-20}$$

$$-C_{\gamma\beta}\nabla_\alpha\left(\frac{\partial h_e}{\partial C_{\gamma\beta}}\right) + \nabla_\gamma\left(2C_{\gamma\beta}\frac{\partial h_e}{\partial C_{\alpha\beta}}\right) ,$$

while the conformation tensor obeys the following equation

$$\frac{\partial C_{\alpha\beta}}{\partial t} = -\nabla_\gamma(u_{1\gamma}C_{\alpha\beta}) + C_{\gamma\beta}\nabla_\gamma(u_{1\alpha})$$

$$+ C_{\alpha\gamma}\nabla_\gamma(u_{1\beta}) - \Lambda_{\alpha\beta\gamma\varepsilon}\frac{\partial h_e}{\partial C_{\varepsilon\gamma}} , \tag{9.2-21}$$

where $\boldsymbol{\Lambda}$ is a fourth-rank tensor which varies inversely in proportion to the polymer relaxation time (see Table 8.1 for various expressions of $\boldsymbol{\Lambda}$, depending on the viscoelastic model assumed).

Additional dissipation (viscous) terms can be introduced through the use of the following expression for the dissipation bracket

$$\begin{aligned}[F,G] = &-\int_\Omega \frac{\eta_s}{2}[\nabla_\alpha \frac{\delta F}{\delta M_{+\beta}} + \nabla_\beta \frac{\delta F}{\delta M_{+\alpha}}][\nabla_\alpha \frac{\delta G}{\delta M_{+\beta}} + \nabla_\beta \frac{\delta G}{\delta M_{+\alpha}}]d^3x \\ &- \int_\Omega Z_{\alpha\beta} \frac{\delta F}{\delta M_{-\alpha}} \frac{\delta G}{\delta M_{-\beta}} d^3x \ ,\end{aligned} \tag{9.2-22}$$

where η_s represents the solvent viscosity, which is constrained to be non-negative, and $\mathbf{Z}$ is the drag coefficient tensor, assumed to be symmetric and non-negative definite. Note that the description of the interaction between the two media through the drag tensor is the simplest one possible. Whether it is an adequate one to represent the complex interactions between macromolecules and the surrounding solvent is still an issue which deserves further study. Certainly, we anticipate the tensor $\mathbf{Z}$ to be a function of the conformation tensor $\mathbf{C}$ and, in general, anisotropic, at least far away from the static equilibrium. Furthermore, incorporation of surface tension effects between phases can be achieved by including a dependence of the free energy density h_e on the gradient of the polymer mass density ρ_- as well. Moreover, any non-local effects (such as, for example, those introduced due to the adjacency of an interface—see §9.3 below) can be described through terms in the Hamiltonian which have a direct spatial dependence. Thus, the dependence of the free energy density on the primary variables shall now be assumed as $h_e = h_e(\mathbf{x}, \rho_-, \boldsymbol{\nabla}\rho_-, \mathbf{C})$. Then, the final equations valid for a polymer solution modeled as an incompressible, two-fluid system with component 1 viscoelastic are found to have the following form:

$$\begin{aligned}\frac{\partial M_{1\alpha}}{\partial t} = &-\nabla_\beta(u_{1\beta}M_{1\alpha}) - M_{1\beta}\nabla_\alpha u_{1\beta} - v_\alpha \nabla_\beta(\rho_1 u_{1\beta}) \\ &- \nabla_\alpha \Pi + \nabla_\beta(2C_{\beta\gamma}\frac{\partial h_e}{\partial C_{\gamma\alpha}}) - \frac{\partial h_e}{\partial x_\alpha} - Z_{\alpha\beta}\Delta v_\beta \\ &+ \nabla_\beta\left(\rho_- \nabla_\alpha \frac{\partial h_e}{\partial(\nabla_\beta \rho_-)}\right) - \phi\nabla_\alpha p + \phi\nabla_\beta\left(\eta_s(\nabla_\beta v_\alpha + \nabla_\alpha v_\beta)\right) \ ,\end{aligned} \tag{9.2-23}$$

$$\frac{\partial M_{2\alpha}}{\partial t} = -\nabla_\beta(u_{2\beta}M_{2\alpha}) - M_{2\beta}\nabla_\alpha u_{2\beta}$$
$$+ v_\alpha \nabla_\beta(\rho_1 u_{1\beta}) - (1-\phi)\nabla_\alpha p \tag{9.2-24}$$
$$+ (1-\phi)\nabla_\beta[\eta_s(\nabla_\alpha v_\beta + \nabla_\beta v_\alpha)] + Z_{\alpha\beta}\Delta v_\beta \ ,$$

where Π is the osmotic pressure defined as

$$\Pi \equiv \rho_- \frac{\partial h_e}{\partial \rho_-} + (\nabla_\beta \rho_-)\frac{\partial h_e}{\partial(\nabla_\beta \rho_-)} + C_{\gamma\beta}\frac{\partial h_e}{\partial C_{\gamma\beta}} - h_e \ . \tag{9.2-25}$$

The equations for the total and reduced momentum densities are

$$\frac{\partial M_{+\alpha}}{\partial t} = -\nabla_\beta(u_{1\beta}M_{1\alpha} + u_{2\beta}M_{2\alpha}) - M_{1\beta}\nabla_\alpha u_{1\beta} - M_{2\beta}\nabla_\alpha u_{2\beta}$$
$$-\nabla_\alpha \Pi + \nabla_\beta(\rho_- \nabla_\alpha \frac{\partial h_e}{\partial(\nabla_\beta \rho_-)}) - \frac{\partial h_e}{\partial x_\alpha} \tag{9.2-26}$$
$$-\nabla_\alpha p + \nabla_\beta(2C_{\beta\gamma}\frac{\delta H}{\delta C_{\gamma\alpha}}) + \nabla_\beta[\eta_s(\nabla_\beta v_\alpha + \nabla_\alpha v_\beta)] \ ,$$

$$\frac{\partial M_{-\alpha}}{\partial t} = -(1-\phi)\nabla_\beta(u_{1\beta}M_{1\alpha}) + \phi\nabla_\beta(u_{2\beta}M_{2\alpha}) - (1-\phi)M_{1\beta}\nabla_\alpha u_{1\beta}$$
$$+ \phi M_{2\beta}\nabla_\alpha u_{2\beta} - Z_{\alpha\beta}\Delta v_\beta \tag{9.2-27}$$
$$-(1-\phi)\left(\nabla_\alpha \Pi + \frac{\partial h_e}{\partial x_\alpha} - \nabla_\beta(\rho_- \nabla_\alpha \frac{\partial h_e}{\partial(\nabla_\beta \rho_-)} + 2C_{\beta\gamma}\frac{\delta H}{\delta C_{\gamma\alpha}})\right) .$$

The term on the left-hand side and the first four terms on the right-hand side of Eq. (9.2-27) represent the inertial effects in the equation for the reduced momentum. An order of magnitude analysis shows the ratio of these inertial terms to the viscous interaction term $\mathbf{Z}\cdot\boldsymbol{\Delta}\mathbf{v}$ to be proportional to $Re_L\,(\xi/L)^2$, where Re_L is the (macroscopic) Reynolds number and ξ and L are characteristic lengths of the polymer molecules and the flow, respectively. Thus, for small Re_L flows and/or flow scales much larger than the characteristic dimension of the polymer molecules, these terms are unimportant and can safely be neglected. Under those conditions, Eq. (9.2-27) reduces to a linear algebraic relationship for $\boldsymbol{\Delta}\mathbf{v}$ from which the following expression for Δv_α can be obtained:

$$\Delta v_\alpha = -Z_{\alpha\beta}^{-1}(1-\phi)\Bigg(\nabla_\beta \Pi + \frac{\partial h_e}{\partial x_\beta} \\ - \nabla_\gamma (2C_{\gamma\varepsilon}\frac{\delta H}{\delta C_{\varepsilon\beta}} + \rho_- \nabla_\beta \frac{\partial h_e}{\partial(\nabla_\gamma \rho_-)})\Bigg) . \tag{9.2-28}$$

Then, by substituting Eq. (9.2-28) into Eq. (9.2-15), a one-fluid model approximation is obtained, represented by Eq. (9.2-26) and a diffusion equation for ρ_-:

$$\frac{\partial \rho_-}{\partial t} = -\nabla_\beta(v_\beta \rho_-) + \nabla_\varepsilon \Bigg(Y_{\varepsilon\beta}\Bigg[\nabla_\beta \Pi + \frac{\partial h_e}{\partial x_\beta} \\ - \nabla_\gamma (2C_{\gamma\alpha}\frac{\delta H}{\delta C_{\alpha\beta}} + \rho_- \nabla_\beta \frac{\partial h_e}{\partial(\nabla_\gamma \rho_-)})\Bigg]\Bigg) , \tag{9.2-29}$$

where $\mathbf{Y}$ is the diffusivity tensor defined as

$$\mathbf{Y} \equiv \rho_1 (1-\phi)^2 \mathbf{Z}^{-1} . \tag{9.2-30}$$

The term $\partial h_e/\partial x_\alpha$ on the right-hand side of Eq. (9.2-29) expresses the partial derivative of the internal part of the free energy density with respect to the spatial coordinates obtained by keeping the other variables (i.e., ρ_1, $\nabla\rho_1$ and $\mathbf{C}$) constant. Thus, this term arises only when there is an explicit dependence of h_e on the spatial coordinates, as for example, the one obtained from consideration of surface/polymer interactions (see §9.3). When only bulk effects are considered, this term is absent and can be omitted from the equation, as it appeared in [Mavrantzas and Beris, 1992b, Eq.(11)].

Equation (9.2-29) concludes the development of the two-fluid model. In the next subsection, §9.2.3, Eqs. (9.2-21), (9.2-25), (9.2-26), (9.2-29), and (9.2-30) are compared against the corresponding equations of the one-fluid model and those from the literature.

9.2.3 *Comparison of various theories for polymer solution hydrodynamics*

In the previous two subsections, two different models for polymer solution hydrodynamics were derived. In the literature, at least five different models have already been reported over the last three years [Helfand and Fredrickson, 1989; Onuki, 1989; 1990; Doi, 1990; Milner, 1991; Bhave *et al.*, 1991] on the same subject—see also Larson [1992] for a recent review. In view of this very recent increased interest in the field, we feel that it is important to devote this subsection to some comparisons among the different models. At the same time, we need to caution the reader that the conclusions are preliminary in this rapidly evolving new area of research.

As a general remark, it is rather surprising and reassuring to find that there are more similarities than differences among the resulting governing equations, despite the sometimes very different character of the approaches taken and the assumptions made. As is discussed in more detail below, all the models lead to a diffusion-like equation for the polymer concentration with both density and stress gradients acting as a forcing term. However, there are still some unanswered questions related to the magnitude of the predicted effects for the concentration, even for such simple viscometric flows as realized in the cone-and-plate and parallel-plate configurations [Doi, 1990]. The fact that similar predictions are obtained irrespective of the model (or, equivalently, the approach) used signifies the fact that if there is a problem, it needs to be sought at a higher level in the assumption hierarchy. The representation of the interactions between polymers and solvent through a single term in the dissipation bracket has already been mentioned as a weak link in the thought process. However, similar results are obtained with the kinetic theory employing the elastic dumbbell model. Again, we do not feel that we have an answer at this early stage of the investigation; our objective of mentioning these preliminary analyses is simply to raise the awareness about them.

First, both the single- and two-fluid models, developed using the generalized bracket formalism, lead (at least under steady-state conditions) to similar equations. Indeed, a similar equation for $\rho_1(=\rho_-)$ is obtained from both the one-fluid bracket formulation, Eq. (9.2-6), and the two-fluid model under the steady-state approximation, Eq. (9.2-29). The major difference is that in the one-fluid model the driving force for the diffusion involves gradients of the chemical potential, $\delta H/\delta\rho_1$, as opposed to gradients of the osmotic pressure, Π, included in the two-fluid model. This implies, according to the definition of the osmotic pressure, Π, Eq. (9.2-25), and the fact that the elastic part of h_e is a linear function of the polymer density ρ_1, that only the non-elastic part of the free energy density h_e appears in the driving force for the two-fluid model. This is in agreement with the work of Helfand and Fredrickson [1989]. In addition, from Eq. (9.2-29), the diffusivity tensor $\mathbf{Y}$, corresponding to the two-fluid model, is proportional to the polymer concentration, in agreement with all other works. Of course, in the single-fluid model, the diffusion coefficients $\mathbf{B}$, $\mathbf{D}$, and $\mathbf{E}$, being phenomenological ones, are left to be specified by the user.

As regards the "stress-diffusion" term in the single-fluid constitutive equation (last term in equation (9.2-8) involving the sixth-rank tensor $\mathbf{B}$), this is associated with the development of boundary layers of microscopic dimensions next to solid boundaries where significant changes in molecular conformations occur [Bhave *et al.*, 1991]. However, its presence necessitates the imposition of boundary conditions for the conformation tensor **C** (or, alternatively, for the elastic stress $\boldsymbol{\sigma}$), which, in general, are not known *a-priori*. Therefore, for computational simplicity, it is preferable to omit it from macroscopic simulations. In this case, a word of caution needs to be called for. As it turns out, based on a consideration of the rate of dissipation corresponding to the one-fluid model, Eq. (9.2-4), a minimum, positive-definite value for each one of **B** and **D** is necessary in order to balance the effect of the coupling terms introduced by a non-zero value of the tensorial parameter **E**. In fact, the tensors **B**, **D**, and **E** are related to each other; each one can be considered as part of a larger matrix coupling the effects on the sytem dissipation of the gradients in the chemical potential ($\equiv\delta H/\delta\rho_1$) and the conformation potential ($\equiv\delta H/\delta\mathbf{C}$). In fact, in the particular case where a kinetic theory microscopic model is used, it can be shown that all matrices are proportional to the same microscopic parameter $1/\zeta$, where ζ is the bead friction coefficient [Bhave *et al.*, 1991]—see also Eqs. (9.2-31) and (9.2-32) below. Thus, it should not be surprising if thermodynamic consistency requires for a non-zero **E** matrix (which can be considered the off-diagonal component of the larger coupling matrix) also non-zero **B** and **D** (representing the diagonal components).

Moreover, as it is revealed through an order of magnitude analysis of the terms introduced from the governing equations (9.2-6)-(9.2-8), whenever the coupling introduced through **E** becomes important, the stress-diffusion terms introduced through **B** are also equally important (in general) and cannot be neglected. Fortunately, for the parallel-plate flow only the coupling terms proportional to **E** are necessary for the determination of the concentration changes, and the secondary flow which depends on both **E** and **B** is identically zero [Mavrantzas *et al.*, 1993]. However, for a more general flow, such as the cone-and-plate flow, this is not any longer true and omittance of the **B** term has been found to lead to both inaccuracies in the solution and numerical instabilities [Mavrantzas *et al.*, 1993]. Note that due to the different form of the dissipation bracket for the two-fluid model, Eq. (9.2-22), a term similar to the stress-diffusion term which is proportional to the tensor **B**, although allowed, is not necessary to be added in order to preserve the positive rate of entropy production when the term responsible for flow-induced migration phenomena (the term proportional to the tensor **Z**) is present. In the single-fluid model though, in order to produce flow-induced migration phenomena, a non-zero value for the tensor **E** is necessary. Thus, the tensors **B**, **D** need to be introduced also with non-zero, positive-definite values. We believe that this is the single most important drawback of the single-fluid theory which places serious doubts on its utility in the future, at least in relation to macroscopic numerical simulations.

In the absence of dependencies of h_e on the polymer concentration gradient, $\nabla\rho_1$, our two-fluid model concentration and momentum equations under the steady-state approximation, Eqs. (9.2-26) and (9.2-29), are astonishingly similar to those of Helfand and Fredrickson [1989, Eqs. (1), (2)], Onuki [1990, Eqs. (5)-(7)], Doi [1990, Eqs. (3.6), (3.7)], and Milner [1991, Eq. (11)]. With respect to the stress constitutive equation, the bracket formalism is the only one capable of evaluating corrections resulting from the presence of compressibility and a changing concentration. In addition, our work explains the choice of reference velocity made by Doi [1990] as it is imposed by the structure of the Poisson bracket.

Where our results qualitatively differentiate from all previous works is in the evaluation of the effects of terms in the analysis that relate to surface tension contributions in phase-separated systems. Both the single- and two-fluid models show that such interphase-related interactions manifest themselves in two ways: they act as an additional stress in the momentum equation (third term on the right-hand side of Eq. (9.2-7) or eighth term on the right-hand side of Eq. (9.2-23)), and they modify the equation for the osmotic pressure (second term on the right-hand side of Eq. (9.2-25)). Notice that the last contribution is preserving the symmetry of the other terms. Both contributions of the gradients are different from the previous ones proposed by Helfand and Fredrickson [1989] and Onuki [1990]. The assessment of the correct form of these contributions is very important in the study of systems that may lead to separations of phases under the application of a shear stress.

The two previous microscopic approaches to the problem [Milner, 1991; Bhave *et al.*, 1991] are both based on the Rouse chain model. Milner employed a projection of the Fokker/Planck equations for the distribution function of microscopic phase-space variables down to the hydrodynamic variables in order to capture the coupling between concentration and stress fields. Using kinetic theory principles, Bhave *et al.* [1991] have reported governing equations for the Hookean dumbbell model in the presence of non-homogeneous stress fields. By resorting to a microscopic (kinetic theory) model and allowing the configurational distribution function to depend on spatial location, they were able to derive specific expressions for the transport coefficients entering the constitutive equations for the mass flux vector and the structure tensor in terms of the microscopic parameters of their elastic dumbbell model. It is interesting to note that both microscopic approaches lead to equations that are similar to the ones corresponding to the single- and two-fluid models presented here. In particular, their expressions for **D** and **E**,

$$D_{\alpha\beta} = \frac{1}{2\zeta}\rho_1\delta_{\alpha\beta}$$
$$E_{\alpha\beta\gamma\varepsilon} = -\frac{1}{\zeta}\delta_{\alpha\beta}\delta_{\gamma\varepsilon} \,, \tag{9.2-31}$$

are the same as the predictions obtained with the steady-state approximation of the two-fluid model (within a factor of $(1-\phi)^2$) if the drag coefficient tensor, **Z**, is assumed isotropic, as a direct comparison of Eq. (9.2-6) with Eq. (9.2-29) demonstrates. This similarity suggests that their modeling assumptions are very similar to that of interpenetrating continua. Indeed, all the elasticity is assumed to be confined within the chain or dumbbell. However, the equations derived by Bhave *et al.* [1991] are cast in the single- rather than the two-fluid formalism. As such, they have all the disadvantages associated with this formalism, as were discussed above. In particular, they necessitate the use of a non-zero, non-negative-definite tensor **B**. In remarkable agreement with this thermodynamic requirement, according to Eqs. (35) and (43) of Bhave *et al.*, the transport coefficient **B** is given as

$$B_{\alpha\beta\gamma\varepsilon\zeta\eta} = \frac{1}{\rho_1\zeta}\delta_{\alpha\varepsilon}\delta_{\beta\zeta}\delta_{\gamma\eta} \,. \tag{9.2-32}$$

Two additional points are worth mentioning about the work of Bhave *et al.* [1991]. First, there are some subtle differences from the bracket equations: the constitutive equation for the structure tensor lacks a term which arises from the coupling of stresses with polymer concentration, the second term on the right-hand side of Eq. (9.2-8). Moreover, their mass flux equation, Eq. (35) of Bhave *et al.* [1991], involves an additional "inertial" term, $-\,m/2(\nabla\!\cdot\!\mathbf{c})\cdot\nabla\mathbf{v}$, which is absent in our formulation. Note that this term is also absent from the equations developed by Öttinger [1992, Eq. (18)] through a body tensor formulation of continuum mechanics [Lodge, 1974]. Second, Bhave *et al.* [1991] have applied their model to the study of the rheology of dilute polymer solutions next to a solid wall. Their constitutive model for the conformation tensor, Eq. (43) of [Bhave *et al.*, 1991], contains a "stress-diffusion" term equivalent to the term weighted by the **B** tensor in our phenomenological one-fluid model, Eq. (9.2-8), with **B** given as in Eq. (9.2-32) above. Since this term, as mentioned before, takes into account rapid changes within a scale of length commensurate with that of polymer chain, it can lead to predictions which appear reasonable (i.e., boundary layers having the correct scale of length) when the equations are applied in situations where there is an incompatibility between the imposed boundary condition for the conformation (for example, at a solid wall surface) and the conformation established in response to an external driving force in the bulk (for example, in the bulk of a viscometric flow). Although macroscopically this approach leads to the correct (qualitatively) prediction of an apparent slip at the wall, it is deficient as far as describing any details of the changes occuring within the boundary layer in the flow or in the concentration. The reason for that is simply the inadequate description of polymer/wall effects which are of thermodynamic origin.

Indeed, as shown in the next section, §9.3, it is the free energy increase near the wall, due to the loss of conformational entropy caused by the presence of the solid barrier, that is primarily responsible for the polymer/wall effects. These eventually result in phenomena such as the apparent slip in the shear flow of the polymer solution and the depletion of the interfacial region in polymer molecules. This belief is reinforced by the very good, quantitative agreement with experimental data for the slip velocity and the depletion phenomena as a function of the imposed shear stress and the geometrical parameters, presented in §9.3.5. The approach followed by Bhave *et al.* [1991] inherently "smears-out" those effects that take place within a length scale of the radius of gyration. The principal reason for the inadequacy of this approach to catch any details in the transition region is that, although the stress-diffusion term is adequate to phenomenologically describe a sharp response taking place within a length scale of the radius of gyration due to incompatible conditions on the conformation, it is totally inadequate to describe the cause for this incompatibility. This is manifested in the arbitrariness of the boundary condition which needs to be used in conjunction with stress-diffusion models, as already recognized by the previous work of El-Kareh and Leal [1989]. Accompanied by a procedure for a self-consistent determination of the missing boundary condition for the conformation, this approach could have led to a first-order approximation of the polymer/wall effects, at least for phenomena manifested within a length scale of a chain length.

The model equations derived by Helfand and Fredrickson [1989] have been applied by them to the study of concentration fluctuations in polymer solutions under shear. Using linear analysis, they were able to show that scattering is greatly enhanced by the coupling of polymer concentration and shear flow through the dependence of the viscosity and the normal stress coefficients on the polymer concentration. According to their analysis, for strong flows, fluctuations are enhanced along $\beta=45°$, where β is the angle that the scattering wave vector forms with the flow direction. This is in very good agreement with elastic light scattering experimental data ($\beta \approx 40°$) for semidilute polymer solutions under a uniform laminar shear flow [Wu *et al.*, 1991].

The works of Helfand and Fredrickson [1989] and Bhave *et al.* [1991] constitute the first attempts to use the aforementioned models in order to give quantitative answers to problems of great importance in polymer rheology. In contrast, in their papers, Doi [1990] and Onuki [1990] restricted their analyses to a qualitative assessment for the possible implications of these models to viscometric flows. Although both of them concluded that the presence of inhomogeneities in the flow field may lead to inhomogeneities in the concentration, neither gave a quantitative answer about the magnitude of these gradients. However, recently, Osaki and Doi [1991] estimated the concentration gradient in a cone-and-plate viscometer based on a general thermodynamic argument. More recently, Mavrantzas *et al.* [1993] conducted a direct numerical simulation based on the full set of governing equations for the concentration, velocity, and conformation tensor outlined in this section. Both the one- and two-fluid models were used. Very briefly, the major results from their analysis can be summarized as follows.

For the parallel-plate flow problem and the two-fluid model the analytical solution reported by Brunn [1984] was recovered. This solution describes a polymer migration towards the centerline which depends solely on the Weissenberg number $We \equiv RDe/h$, where R is the radius of the plate and h the gap thickness. The results for the single-fluid model were qualitatively similar; however, they also depended on additional molecular parameters. Variations in the geometry, modeled as a cone-and-plate flow with a small inclination angle and a large gap thickness, were found to initiate a small secondary recirculation that resulted in a significant decrease of the concentration gradients. When the centerline gap thickness became less than about 10^{-3}, the ideal cone-and-plate flow behavior dominated for which no axial, but substantial radial, variations in the concentration were observed, mostly around the centerline. This behavior is consistent with the experimental observations of Shafer *et al.* [1974] and Dill and Zimm [1979] with dilute DNA solutions, as well as with the predictions of the analytic results by Aubert *et al.* [1980] and Brunn [1984], derived from kinetic theory arguments under the assumption of a purely azimuthal flow and the estimate of Osaki and Doi [1991].

9.3 Surface Effects on the Microstructure and Concentration in Incompressible and Isothermal Viscoelastic Fluid Flows

The generalized bracket is used in this section[6] to develop a consistent set of evolution equations to describe the rheological behavior of high molecular weight, dilute polymer solutions near planar, smooth, solid surfaces. As we shall see, this complex theoretical problem can be made tractable through the bracket description, and, indeed, it is not obvious

[6]This section was written in collaboration with V. G. Mavrantzas and follows the analysis presented in [Mavrantzas and Beris, 1992a].

how one could have obtained a consistent macroscopic set of equations valid under arbitrary flow conditions otherwise. Although the investigation reported in this section has just begun, it has already produced illuminating results concerning the surface/polymer interaction for surfaces which interact with the polymer molecules non-energetically, as described below. For completion purposes, in the last subsection, §9.3.6, we venture into exploring how energetic interactions can be taken into account and how they might affect the polymer concentration and conformation. However, we caution the reader that at this time (1994) this is a much less understood subject, and chances are that developments in the near future will make the material presented in §9.3.6 obsolete. It should still be of some value though, at least as another example for the use of the bracket formulation in effectively transferring information from the microscopic to the macroscopic scale in the modeling of an inherently complex structural fluid.

9.3.1 Introduction

The interfacial properties of polymers, both in the condensed, amorphous state and in solution, are important in a broad range of technical areas. Their knowledge is essential to understanding the mechanical properties of particulate and fiber-reinforced polymeric composites and semi-crystalline polymers, and in manufacturing mechanical components with well-characterized properties. Their role is significant in processes such as colloidal stability and flocculation, gel chromatography, separation by flow, tertiary oil recovery, and resin transfer molding. In addition, surface-related dynamic phenomena seem to profoundly affect not only the quality of products but also the rate of production in polymer-processing operations such as extrusion and film blowing. For example, breakdown of adhesion at the polymer/metal interface in the die land region, at stresses above a critical value, appears to be the primary factor responsible for the initiation of slip and the onset of surface irregularities in blown-film fabrication processes (i.e., surface melt fracture).

From the modeling point of view, the very steep stress gradients that are predicted to develop in the neighborhood of a bounding surface during the flow of a Maxwell fluid, leading to anomalous behavior (relative to experience based on the flow of Newtonian liquids), have been a prevailing theme in viscoelastic fluid-mechanics calculations. Thus the polymer/wall interaction has long been suspected to be the reason for the convergence failure of numerical simulations of viscoelastic flows at stresses above a critical value. The implication that a faulty description of the physics near the wall is partly behind such numerical difficulties seems to be consistent with Ramamurthy's [1986a, p. 353] claim that, at such high stresses, failure of adhesion leads to the onset of instabilities and causes extrudate distortions.

Although considerable work has been done to study the polymer behavior at interfaces under quiescent conditions, much less work has been undertaken in order to study the effects associated with a flow field. However, most polymer processes involve flow fields above the adsorbing surface. The main reasons for the scarcity of work on the problem of the flow/surface interaction seem to be the complexity of the problem and the unavailability of a strict mathematical formalism that would facilitate the incorporation of all the physically important phenomena that take place near the solid surface into a model.

Ramamurthy [1986a,b] investigated slip phenomena and the influence of materials of construction on the observed extrudate irregularities during the flow of high-viscosity molten polymers. He experimented mainly with linear low-density polyethylene (LLDPE) (but also with other polymers) in both capillary and blown film dies. His work has shown that in flows of such high-viscosity molten polymers, slip in the die land region invariably accompanies the observed extrudate irregularities. Moreover, by proper choice of the materials of construction and the addition of adhesion promoters into the resin, the rate-limiting effects of melt fracture with these polymers can be virtually eliminated.

Previously, Cohen and Metzner [1985] studied the flow behavior of aqueous and organic polymer solutions in laminar Poiseuille flow and measured experimental flow rates that were much higher than those predicted based on viscometric cone-and-plate data. This abnormal flow enhancement was quantified by an effective slip velocity. The slip velocity was found to increase with increasing shear rate and with decreasing tube diameter. The investigators speculated, at that time, that the apparent slip phenomenon occurs in the flow of high-molecular-weight polymer solutions in the presence of inhomogeneous stress fields. Dutta and Mashelkar [1984] solved numerically the concentration and momentum equations formulated by Cohen and Metzner for the steady-state, fully developed flow through a capillary, verifying the experimentally observed flow enhancement.

More recent experimental data by Kalika and Denn [1987, pp. 818-20] on LLDPE demonstrated the existence of four regimes in the flow curve depending on the applied wall shear stress. The first corresponds to a stable flow curve where extrudates with smooth, glossy surfaces are observed. The second region, called the "sharkskin region," corresponds to extrudates with high-frequency variations on the surface. Immediately after the sharkskin region, a stick/slip region is observed where extrudates that alternate in a periodic fashion, between patterns with rough, sharkskinned surfaces and smooth, glossy ones, are seen. Finally, at very high shear rates, a wavy region is established where the extrudates have much more severe irregularities. In this region, the fracture depth is believed to be on the order of the extrudate diameter. A similar succession of no-slip, alternate sticking and slipping, and slip flow (as the flow rate was increased) was reported by Lim and Schowalter [1989, pp. 1360,1373-75,1381] for a series of polybutadienes when extruded through slit capillary dies. Recently, Hatzikiriakos and Dealy [1991; 1992] have reported wall slip of molten, high-density polyethylene through sliding plate and capillary rheometer studies and have advanced a semi-phenomenological model for the slip velocity based on a model of adhesion originally developed by Lau and Showalter [1986].

Direct experimental evidence of wall slip is also available for materials ranging from polymer solutions and gels to elastomers. Müller-Mohnssen *et al.* [1990], using a laser-differential anemometer and a total-reflection-microscope anemometer, measured velocity profiles in ducted flows of aqueous polyacrylamide solutions up to distances of 0.15μm from the wall. Solutions of various concentrations were used. At the lowest concentration (0.005% wt.), these authors were able to show a sharp transition in the velocity profile at very small distances from the wall (on the order of the radius of gyration of the polymer chains, 0.1μm) from zero to a finite value. This is the first instance where the apparent character of the slip phenomena was directly verified experimentally. Extrapolation of the velocity profiles led to finite velocity values at the surface (i.e., an apparent slip velocity, υ_s), while the ratio of the slip velocity to the wall shear stress, i.e., the slip coefficient, was found to be constant for constant polymer concentration over all of the region of shear stresses examined. Further measurements of the average viscosity in the interfacial region indicated the existence of a depletion layer.

Some years earlier, Ausserré *et al.* [1986] and Rondelez *et al.* [1987], using the newly developed (at that time) evanescent-wave-induced-fluorescence technique (EWIF), obtained the first accurate measurements of the monomer concentration as a function of the distance from the wall for a dilute aqueous solution of xanthan (a stiff polysaccharide of high molecular weight). The experiments demonstrated the existence of a depletion layer next to a non-adsorbing, fused-silica surface. Ausserré *et al.* [1991] have used the same technique recently to study the polymer concentration profiles of polymer solutions under shear. They have found that the shear rate increases the "surface excess," a quantity that measures the extent of the depletion, relative to that exhibited under equilibrium conditions. Very recently, Migler *et al.* [1993] combined the EWIF technique with fringe pattern fluorescence recovery after photobleaching (FPFRAP) in measuring for the first time directly the local velocity of a sheared polymer melt within the first 100nm from the solid/liquid surface (polydimethylosiloxane, PDMS, on a silanated silica surface). A transition from weak to strong slip was observed at high enough shear rates. When conditions of a strong interaction between the surface and the polymer were established the slip was strongly reduced. These results seem compatible with a theoretical model proposed by Brochard-Wyart *et al.* [1992].

Although the experimental work has advanced considerably and many of the major characteristics of the polymer/wall interaction problem have been investigated, a theoretical explanation of the rheological properties of such systems still lags behind and is only in the first stages of its development. With homogeneous polymer melts, loss of adhesion might predominantly account for slip phenomena. This explanation is supported by the fact that slip phenomena seem to appear above a critical value of the shear stress [Kalika and Denn, 1987, p. 818]. The situation for inhomogeneous polymer melts and polymer solutions is more complex. There, thermodynamic phenomena seem to control the behavior of the system. These ideas are further addressed in §9.3.3.

A polymer molecule is a very complex mechanical system with a large number of degrees of freedom, capable of realizing many conformations. The presence of the solid wall reduces considerably the number of conformations that are available to a single chain in a way that depends strongly on the distance from the wall. Indeed, quite early the principal cause of the apparent slip phenomena in polymer solutions was identified with the entropy changes caused by the deformation of the macromolecules [Garner and Nissan, 1946, p. 634], although initially the cause for the deformation was sought exclusively in the stress field. Therefore, in the early theoretical work following Garner and Nissan, there was an attempt to attribute the apparent slip to concentration changes produced by inhomogeneous flow fields [Metzner *et al.*, 1979, §3].

The first work to account for the influence of the wall on the macromolecular conformations was due to Brunn [1976]. Brunn used the kinetic theory of non-interacting, elastic (Hookean) dumbbells, and obtained an approximate expression for the configurational distribution function around the known result for a uniform shear flow [1976, §2]. Thus, his early results were only valid at large distances from the wall. The usefulness of this approach is severely limited, particularly from a computational standpoint, but even more so from physical considerations. Aubert and Tirrell [1982] followed a similar procedure, but used a perturbation in terms of the shear rate rather than the wall influence. Consequently, this approach was only valid for small shear rates. As such, their results compared very favorably with Chauveteau's data [1982] in the limit of zero shear rate. Later, Brunn [1985], Brunn and Grisafi [1987], and Grisafi and Brunn [1989] obtained solutions valid for arbitrary values of shear rates and a broad range of parameters (narrow, intermediate, and wide channels) for both steady simple shear and Poiseuille flow, by refining their analytical and numerical techniques.

In all of these works, the presence of a boundary layer in the number density of the dumbbells, and, as a consequence, in the velocity and in the shear stress components, has been demonstrated. The boundary layer extends over one or two radii of gyration [Grisafi and Brunn, 1989]; however, its effects on the average rheological quantities are felt using channels with widths up to ten times as much.

In parallel to the elastic dumbbell theories, a Brownian dynamics method with excluded-volume interactions has been used recently [Duering and Rabin, 1990] to simulate the behavior of polymer chains in a dilute solution in simple shear flow between impenetrable walls. The major assumption in these simulations was the existence of a constant velocity gradient both in the bulk and in the interfacial region. The simulations provided information for the projections of the polymer radius of gyration and end-to-end distance along and perpendicular to the flow direction and for the surface excess as a function of the distance from the wall. Contrary to the previously mentioned evanescent-wave-induced-fluorescence studies (EWIF) [Ausserré *et al.*, 1991], these simulations have shown the depletion phenomena near the wall to decrease with the shear rate. More recently, de Pablo *et al.* [1992] developed a stochastic algorithm to solve the time evolution of the probability density function for a dilute solution of rigid rods next to a neutral wall. Their results confirmed the experimental data at high shear rates only. Starting from equilibrium, the depletion layer was found first to decrease and then, after passing through a minimum value, to increase with increasing shear rate [de Pablo *et al.*, 1992, Fig. 9].

In this section, we follow a different way of modeling slip phenomena for dilute polymer solutions. For polymer molecules, such parameters are "averaged" quantities related with the size of the chains. The formulation is based on the free energy of the system expressed appropriately in terms of the dynamic variables, which involve the velocity and parameters reflecting the internal microstructure. All that is then needed in order to apply the formulation to viscoelastic flow problems is the specification of the extended free energy (Hamiltonian) of the system (as described in chapter 8), taking into account the effects of the wall on the conformational entropy of the system. Stress-induced migration effects as analyzed in the previous section are believed to be of secondary importance for these phenomena, as the very good comparison of the predictions of the present approach with available experimental data demonstrates (see §9.3.5).

9.3.2 *Derivation of the governing equations in terms of the Hamiltonian*

As before, it is first necessary to set up the problem by determining the appropriate variables needed for an adequate description of the system dynamics. In the previous chapter, we considered only homogeneous viscoelastic fluids, where the properties of the system were more or less independent of the spatial coordinates in that no variations were allowed in the polymer solution concentration nor directly in the microstructure from point to point (**B**=**0**). Hence the macroscopic state of the system could be adequately described with only the two dynamic variables **M** and **c**.[7] By incorporating wall effects into the system, however, we have explicitly introduced a spatial inhomogeneity into the system, and the above-mentioned effects are therefore expected to play an important role in the physics of the situation. As such, a new dynamic variable (over chapter 8) is now needed which accounts for spatial variations in the polymer concentration. The obvious choice is ρ, which we divide into two contributions,

$$\rho = \rho_p + \rho_s \,, \tag{9.3-1}$$

where ρ_p is the density of the polymer and ρ_s is the density of the solvent. Since, in this case, the polymeric solution is required to be dilute, we assume that ρ=constant, and that the overall system is still incompressible (as well as isothermal). This assumption is not exactly correct, but we shall use it for convenience (as is always done in the literature). ρ_p is allowed to vary, but as this quantity is assumed to be quite small, ρ_s is assumed to change negligibly (i.e., $\rho \approx \rho_s$). With this assumption, the proper dynamic variables for the description of this system are $\mathbf{M}=\rho_s \boldsymbol{v}$, $\rho_p = n/\alpha$, and $\mathbf{C}=\rho_p \mathbf{c}$, where we now see that as $\rho_p \to 0$, $\mathbf{C} \to \mathbf{0}$, and the effects of the microstructure vanish. Note that α represents the (constant) measure of the degree of elasticity per mass.

The operating space for this problem is the same as that of chapter 8, except that the variable ρ_p is added to the description. The Poisson bracket is simply that of (5.5-11), with ρ replaced with ρ_p and the entropy density neglected, as before. This leaves only the dissipation bracket to be specified, which is given by Eq. (8.1-5), expressed now in terms of Volterra derivatives with respect to **C** instead of **c**, plus the additional integral[8]

[7]Once again, whether or not one is allowed to make such an assumption depends critically upon the time and length scales of the experiment or process.

[8]Note that the inclusion of this term necessitates the presence of a parallel term involving the chemical potential of the solvent, in accordance with the conservation of mass principle—see §7.3. Yet, in the limit of dilute solutions, the influence of the additional term on the equations can be safely neglected. Also, we neglect for simplicity in this preliminary analysis any coupling between $\boldsymbol{\nabla}(\delta F/\delta\rho_p)$ and the other dynamic variables, although these effects might conceivably play a role in the system description—see §9.2 for a separate description of their effect in flow-induced concentration and conformation changes. Moreover, since the dissipation involves derivatives with respect to $\mathbf{C} \equiv \rho_p \mathbf{c}$ rather than **c**, this implies that the corresponding tensorial parameters now differ from those of (8.1-5) by factors of ρ_p.

$$-\int_\Omega \kappa_{\alpha\beta}\nabla_\alpha\left(\frac{\delta F}{\delta\rho_p}\right)\nabla_\beta\left(\frac{\delta G}{\delta\rho_p}\right)d^3x \,. \tag{9.3-2}$$

Since we are considering an isothermal and incompressible fluid, as in chapter 8, we shall only be concerned with the mechanical contribution to the total energy (Hamiltonian) of the system, and only the flux terms in the dissipative bracket (and hence $dH_m/dt<0$ again).[9]

The general evolution equations for the dynamic variables of this system are now easily found to be[10]

$$\rho\frac{\partial v_\alpha}{\partial t} = -\rho v_\beta\nabla_\beta v_\alpha - \nabla_\alpha p + \nabla_\beta\sigma_{\alpha\beta} \,, \tag{9.3-3a}$$

$$\begin{aligned}\frac{\partial C_{\alpha\beta}}{\partial t} &= -\nabla_\gamma(v_\gamma C_{\alpha\beta}) + C_{\alpha\gamma}\nabla_\gamma v_\beta + C_{\gamma\beta}\nabla_\gamma v_\alpha \\ &\quad - \Lambda_{\alpha\beta\gamma\varepsilon}\frac{\delta H_m}{\delta C_{\gamma\varepsilon}} + L_{\alpha\beta\gamma\varepsilon}\nabla_\gamma v_\varepsilon + \nabla_\gamma\left(B_{\alpha\beta\gamma\varepsilon\zeta\eta}\nabla_\eta\frac{\delta H_m}{\delta C_{\varepsilon\zeta}}\right) \,,\end{aligned} \tag{9.3-3b}$$

and

$$\frac{\partial\rho_p}{\partial t} = -v_\beta\nabla_\beta(\rho_p) + \nabla_\alpha\left(\kappa_{\alpha\beta}\nabla_\beta\frac{\delta H_m}{\delta\rho_p}\right) \,, \tag{9.3-3c}$$

where the extra stress is defined as

$$\sigma_{\alpha\beta} = Q_{\alpha\beta\gamma\varepsilon}\nabla_\gamma v_\varepsilon + 2C_{\beta\gamma}\frac{\delta H_m}{\delta C_{\alpha\gamma}} + 2L_{\alpha\beta\gamma\varepsilon}\frac{\delta H_m}{\delta C_{\gamma\varepsilon}} \,. \tag{9.3-3d}$$

Again, even before we specify the Hamiltonian of the system, we shall make a number of simplifying assumptions. Although the effects of inhomogeneities in the **C** field are certainly important in this case, we shall once again neglect **B**, as this sixth-rank tensor unnecessarily complicates our simplified model at this stage of the analysis. Also, non-affine effects should be particularly important for systems with polymer/surface interactions, but we set **L**=0 as well. As a first approximation and for simplicity in the following preliminary calculations, the phenomenological matrix associated with the diffusivity, $\boldsymbol{\kappa}$, is assumed

[9]Note that for a non-reacting system, there are no relaxational affinities associated with $\delta H_m/\delta\rho_p$; i.e., we have only the gradient terms with respect to ρ_p in (7.1-24), such as $\boldsymbol{\nabla}(\delta H_m/\delta\rho_p)$.

[10]Note that the manipulation of (8.1-9) is no longer explicitly correct, but must be amended to incorporate the additional dependence on the density, $A=A[\mathbf{C},\rho_p]\Rightarrow a=a(\mathbf{C},\rho_p)$. The vanishing of the two appropriate integrals in the Poisson bracket then follows directly.

to be constant and isotropic, $\boldsymbol{\kappa} \equiv \kappa \boldsymbol{\delta}$. Although, in general, variations in the diffusivity and/or deviations from isotropicity are anticipated at close proximity to a solid surface [Mansfield and Theodorou, 1989] and under flow [Prakash and Mashelkar, 1991], these are not usually taken into account in the existing model calculations because of the added complexity and uncertainty introduced into the equations. Finally, we take $\mathbf{Q}$ to be (8.1-19), since we desire that the evolution equations in the bulk domain (i.e., away from the walls) reduce to the Oldroyd-B constitutive equation, which is a reasonably well-suited model for performing calculations of viscoelastic fluid flows.

In order to expedite the following analysis, and to keep it in harmony with the existing literature on the subject, we shall now make use of the definition $\rho_p \equiv n/\alpha$ in order to rewrite $H_m[\mathbf{C},\rho_p] \rightarrow H_m[\mathbf{C},n]$ and (9.3-3c) as

$$\frac{\partial n}{\partial t} = - \upsilon_\beta \nabla_\beta n + \nabla_\alpha (\kappa \nabla_\alpha \frac{\delta H_m}{\delta n}) , \tag{9.3-4}$$

since α is assumed to be a constant. Similarly, it is understood that the diffusivity κ associated with Eq. (9.3-4) is different from the one of (9.3-3c) by a factor of α^2. Similar changes are implemented to all the other equations in (9.3-3). This will allow us to deal directly with volume fractions in the subsequent discussion, as we shall soon see. The functional derivative $\delta H_m/\delta n = \partial h/\partial n$ (since there is no dependence of the free energy density h on the gradient of n in this case) is defined as μ^e, an effective chemical potential.

9.3.3 *Determination of the Hamiltonian functional*

The above system of equations, (9.3-3,4), is completely general for our simple system in that it describes the dynamics of our medium in the bulk flow domain as well as near the bounding interface(s). It is thus in the Hamiltonian, or Helmholtz free energy, where the confining aspects of the walls will limit the configurational contribution to the entropy functional. In the following, we proceed to the determination of the Hamiltonian for a non-adsorbing polymer solution, in two stages, corresponding to the incorporation of the bulk and the surface interactions, respectively. Although not explicitly apparent, the key underlying assumption for the derivation of the Hamiltonian is a Gaussian distribution for the end-to-end polymer distance. This assumption is certainly valid in the bulk. However, next to a surface it appears to be more characteristic of a semi-flexible chain than a flexible one. This observation might explain why the comparison against experimental data obtained with semirigid xanthan macromolecules turns out to be as good as it is—see §9.3.5.

A. Contributions of bulk viscoelasticity

Let us first specify the Hamiltonian in the absence of the walls, so that we may see how the results of chapter 8 are generalized to include the additional dynamic variable n. Again, we identify the Hamiltonian as the (extended) Helmholtz free energy defined as

$$A[\mathbf{C},n] \equiv E_\phi[\mathbf{C}] - TS[\mathbf{C},n] \,, \tag{9.3-5}$$

which, for the Maxwell model, we may write as

$$E_\phi[\mathbf{C}] = \int_\Omega \tfrac{1}{2}K\mathrm{tr}\mathbf{C}\,\mathrm{d}^3x \,, \tag{9.3-6}$$

and

$$S[\mathbf{C},n] = -\int_\Omega k_B(n\ln\phi_1 + n_s\ln\phi_2)\,\mathrm{d}^3x + \int_\Omega \tfrac{1}{2}nk_B\ln\det\left(\frac{K\mathbf{C}}{nk_BT}\right)\mathrm{d}^3x \,. \tag{9.3-7}$$

In these expressions, ϕ_1 is the volume fraction of polymer, ϕ_2 is the volume fraction of solvent, and n_s is the solvent number density. The first integral in (9.3-7) is the standard Flory/Huggins entropy of random mixing [Flory, 1953, p. 509; Billmeyer, 1984, p. 162] where, for simplicity, theta-solvent conditions are assumed. If x is defined as the degree of polymerization of the polymer chains (assumed constant), then

$$\phi_1 = \frac{nx}{nx+n_s} \quad \text{and} \quad \phi_2 = \frac{n_s}{nx+n_s} \,. \tag{9.3-8}$$

For a very dilute polymer solution, $nx \ll n_s$, so that $\phi_1 \approx nx/n_s$ and $\phi_2 \approx 1$. This allows us to write the entropy functional in the bulk flow domain as

$$S[\mathbf{C},n] = -\int_\Omega nk_B\ln\left(\frac{nx}{n_s}\right)\mathrm{d}^3x + \int_\Omega \tfrac{1}{2}nk_B\ln\det\left(\frac{K\mathbf{C}}{nk_BT}\right)\mathrm{d}^3x \,. \tag{9.3-9}$$

In the limit of constant n, Eqs. (9.3-3) and (9.3-5) through (9.3-9) reduce to the system of equations for the Oldroyd-B model studied in §8.1.1. Moreover, Eq. (9.3-4) reduces to the same diffusion equation used by Tirrell and Malone [1977, Eq. 8] in the limit of dilute polymer solutions. However, the ratio k_BT/n of the Fickian to the "stress-induced" diffusion is obtained automatically within the framework of our theory.

EXAMPLE 9.2: In the absence of flow and surface interactions, find the equilibrium configuration and concentration of the polymer throughout the system.

The conditions for equilibrium, when $\mathbf{M}=0$, amount to the minimization of the free energy from (9.3-3b),

$$\frac{\delta A}{\delta C_{\alpha\beta}} = 0 \ ,$$

and to the zero-flux condition from (9.3-4),

$$\boldsymbol{\nabla}\left(\frac{\delta A}{\delta n}\right) = \mathbf{0} \quad \Rightarrow \quad \frac{\delta A}{\delta n} = \mathit{constant} \ .$$

The solutions to these equations are

$$\mathbf{C}^e = \frac{n_0 k_B T}{K}\boldsymbol{\delta} \quad \text{and} \quad n = n_0 \quad \Rightarrow \quad \tilde{n} \equiv \frac{n}{n_0} = 1 \ ,$$

where $\tilde{n}$ is defined as the reduced chain number density (concentration) relative to that in the bulk, n_0. Therefore, at static equilibrium, the polymer concentration is uniform and the conformation tensor is isotropic; i.e., the polymer molecules are in spherical configurations. ■

EXAMPLE 9.3: In the absence of surface interactions, determine the solution properties in a uniform shear flow.

The velocity profile in this system is required to have the following functional dependencies: $\upsilon_x=\upsilon_x(y)$, $\upsilon_y=\upsilon_z=0$, where x is the flow direction and y the direction of shear. Under these kinematical conditions, the continuity equation is automatically satisfied, $\mathbf{C}=\mathbf{C}(y)$, $\boldsymbol{\sigma}=\boldsymbol{\sigma}(y)$, and the momentum equation, in the absence of a pressure gradient and external forces, reduces to σ_{xy}=constant. Although the polymer chains assume a spherical, random-coil configuration at equilibrium, they are deformed and oriented under shear flow into ellipsoids with a shear-dependent angle, $\hat{a}$, with respect to the flow direction—see Figure 9.1.

Let us adopt the scaling conventions ($G_0 \equiv n_0 k_B T$)

$$\tilde{y} \equiv \sqrt{\frac{K}{k_B T}}\, y; \quad \tilde{n} \equiv \frac{n}{n_0}; \quad \tilde{\mathbf{C}} \equiv \frac{K}{G_0}\mathbf{C} = \tilde{n}\,\tilde{\mathbf{c}}; \quad \tilde{\dot{\gamma}} \equiv \lambda\dot{\gamma} = \lambda\frac{d\upsilon_x}{dy}; \quad \tilde{\sigma}_{xy} \equiv \frac{\sigma_{xy}}{G_0} \ .$$

For simplicity, from here on till the end of this chapter, the tilde is omitted from the notation of the dimensionless quantities. Thus, all quantities are dimensionless, unless otherwise indicated. Since the flow is uniform, n=1 is constant throughout the flow domain. Also, since the flow is planar, the system is invariant in the z direction, and the evolution equation for $\mathbf{C}$ reduces to three algebraic equations for the components c_{xx}, c_{yy}, and c_{xy} of the conformation tensor ($c_{xz}=c_{yz}=0$, $c_{zz}=1$):

$$2\dot{\gamma}c_{xy} - c_{xx} + 1 = 0 \ , \quad c_{yy} = 0 \ , \quad \text{and} \quad \dot{\gamma}c_{yy} - c_{xy} = 0 \ .$$

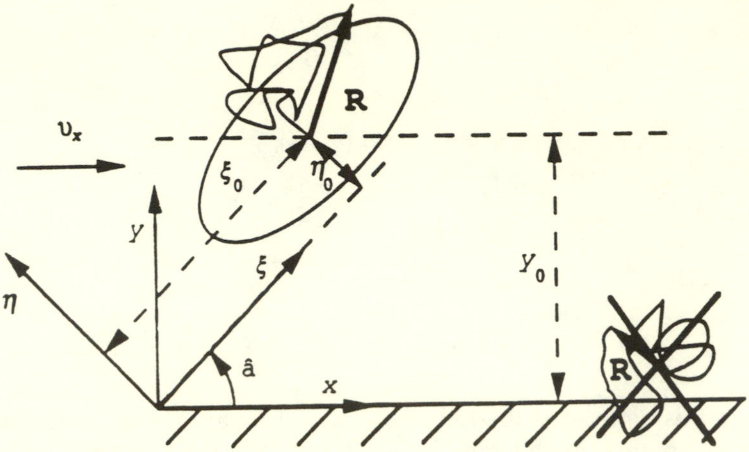

Figure 9.1: Schematic of macromolecular conformations. Two instances of chain conformations are shown as particular realizations of random walk statistical procedures corresponding to specific end-to-end distance vectors **R** and specific spatial locations. Note that the conformation on the lower right is not realizable as having segments crossing the wall surface. The average extent of the distribution function for the end-to-end vector **R** is shown schematically here for the upper left chain by an ellipse with major axes along the direction of the eigenvectors and magnitude proportional to the corresponding eigenvalues of the conformation tensor **c**. Thus, the major axis of the ellipse is parallel to the ξ-axis at an angle $\hat{a}$ with the x-axis. The location of the chain end, of the upper left chain, is also indicated in both the fixed, (x,y), and the eigenvector, (ξ,η), coordinate frame.

From the stress constitutive equation, (9.3-3d), we also know that

$$c_{xy} + \mu\dot{\gamma} = \sigma_{xy} \; ,$$

where μ is the dimensionless solvent viscosity, $\mu \equiv \eta_s/\eta_p$. This system may be solved analytically to obtain

$$\dot{\gamma} = \sigma_{xy}/(\mu+1) \; , \quad c_{xx} = 1 + 2\dot{\gamma}^2 \; , \quad c_{yy} = 1 \; , \quad \text{and} \quad c_{xy} = \dot{\gamma} \; ,$$

so that in the absence of surface interactions, all of the variables of interest are constant throughout the flow domain.

The c_{xx} component of the conformation tensor which measures the projection of the end-to-end distance vector along the flow direction is found to increase due to the applied shear stress: the higher the value of the shear stress, the larger the degree of stretching. Also, the shear stress determines the angle of orientation, $\hat{a}$—measured effectively by c_{xy}—of the ellipsoid relative to the flow direction. The higher the value of the shear stress, the smaller the size of $\hat{a}$. The c_{yy} component is not affected by the shear field. This is an inherent property of the Oldroyd-B model, indicating a zero second normal-stress difference. For dilute polymer solutions this property has been verified experimentally through small-angle neutron scattering techniques [Lindner and Oberthür, 1988]. ■

B. Contributions of entropic surface interactions for non-absorbing surfaces

Now we wish to determine what additional terms over those of (9.3-6,9) will be required in the Helmholtz free energy once we consider polymer/surface interactions. The view of the polymer molecule we now adopt, which is complementary to our previous description, is that of a random-flight chain, which in the absence of surface interactions reduces to the elastic dumbbell model. Since the pioneering work of John Kirkwood in the 1940s and 1950s [Kirkwood, 1967], the random-flight model has played a major role in describing the conformational properties of polymeric molecules.

The random-flight model, in its most general form, considers a particle that undergoes a series of displacements, $\mathbf{r}_1,\mathbf{r}_2,\ldots,\mathbf{r}_N$, in which the magnitude and direction of each is independent of those which preceded it. If the probability that the i-th displacement lies between $\mathbf{r}_i$ and $\mathbf{r}_i+\mathrm{d}\mathbf{r}_i$ is related through a distribution function, prob.$=f_i(\mathbf{r}_i)\mathrm{d}^3r_i$, then we wish to know the probability that after N displacements the coordinates of the particle will lie between $\mathbf{R}$ and $\mathbf{R}+\mathrm{d}\mathbf{R}$. In other words, we are seeking the cumulative probability distribution function for the resultant vector $\mathbf{R}$ of the N displacements, $\mathbf{r}_i$, $i=1,2,\ldots,N$, which all follow the corresponding displacement probability function, $f_i(\mathbf{r}_i)$.

In applying the random-flight model to the study of the statistical properties of flexible polymers, the displacements $\mathbf{r}_i$ correspond to the bond vectors, while the number of displacements, N, corresponds to the number of links, or monomers, of which the chain consists. In this case,

$$\mathbf{r}_i = \mathbf{R}_i - \mathbf{R}_{i-1} \ , \quad i = 1,2,\ldots,N \ , \tag{9.3-10}$$

where $\mathbf{R}_0,\mathbf{R}_1,\ldots,\mathbf{R}_N$ represent the position vectors of the N+1 joints. The size of the random walk corresponds to the size of the polymer chain, which is characterized by the end-to-end vector, $\mathbf{R}$, such that

$$\mathbf{R} \equiv \mathbf{R}_N - \mathbf{R}_0 \ . \tag{9.3-11}$$

The problem of the random flight is then translated into determining the probability $W(\mathbf{R})\mathrm{d}^3R$ that, after N displacements, the end-to-end vector of the polymer chain lies between $\mathbf{R}$ and $\mathbf{R}+\mathrm{d}\mathbf{R}$.

A common approach over the years to solve the above problem has been the *a priori* assignment of the segment probability distribution function, $f_i(\mathbf{r}_i)$, which governs the magnitude and direction of each displacement. As examples, we mention the freely jointed chain and the Gaussian one. Although certain assumptions for the form of the distribution function that describes the displacements $\mathbf{r}_i$ may be justifiable for static systems, there is little that may be safely used when the polymer chains are imposed on a non-vanishing flow field. In that case, we are faced with the problem of random flights in its most general form, which was first worked out by Chandrasekhar [1943, ch. 1].

Two assumptions are generally associated with the random-flight model: that a random flight consists of a large number of displacements ($N \gg 1$), and that all of the segment probability functions are the same. The first of these amounts to the consideration of macromolecules of high molecular weight. In the limit of $N \to \infty$, the cumulative probability function is fully determined by the first and second moments of the common segment probability distribution function, f, [Papoulis, 1984, p. 191]. Let us denote the (as yet) unknown first and second moments of f as $<x>$, $<y>$, $<z>$, $<xy>$, $<xz>$, etc., and let $<\xi^2>$, $<\eta^2>$, and $<\zeta^2>$ be the eigenvalues of the symmetric matrix formed by the second moments,

$$S.M. = \begin{bmatrix} <xx> & <xy> & <xz> \\ <yx> & <yy> & <yz> \\ <zx> & <zy> & <zz> \end{bmatrix} . \tag{9.3-12}$$

The three eigenvalues of the above matrix form an orthogonal coordinate system, which may be denoted as (ξ,η,ζ). Also let $\mathbf{R}=(\Xi,\mathrm{H},\mathrm{Z})^T$ be the end-to-end vector in this coordinate system.

Chandrasekhar [1943, pp. 8-11], using Markoff's method, proved that (as $N \to \infty$) the probability that the end-to-end vector of the chain lies between $\mathbf{R}$ and $\mathbf{R}+d\mathbf{R}$ is given by the product of three Gaussian functions,

$$W(\mathbf{R}) = \frac{1}{\sqrt{8\pi^3 N^3 <\xi^2><\eta^2><\zeta^2>}} \times \exp\left(-\frac{(\Xi - N<\xi>)^2}{2N<\xi^2>} - \frac{(\mathrm{H} - N<\eta>)^2}{2N<\eta^2>} - \frac{(\mathrm{Z} - N<\zeta>)^2}{2N<\zeta^2>} \right) . \tag{9.3-13}$$

If we adopt the notation introduced earlier, then this expression becomes

$$W(\mathbf{R}) = \frac{1}{\sqrt{8\pi^3 c_{\xi\xi} c_{\eta\eta} c_{\zeta\zeta}}} \times \exp\left(-\frac{(\Xi - N<\xi>)^2}{2c_{\xi\xi}} - \frac{(\mathrm{H} - N<\eta>)^2}{2c_{\eta\eta}} - \frac{(\mathrm{Z} - N<\zeta>)^2}{2c_{\zeta\zeta}} \right) , \tag{9.3-14}$$

and according to this expression the probability distribution of the end-to-end vector, after suffering a large number of displacements (each one governed by f), is an ellipsoidal distribution centered at $(N<\xi>, N<\eta>, N<\zeta>)$; in other words, the particle undergoes an average, systematic, net displacement of amount $(N<\xi>, N<\eta>, N<\zeta>)$ and, superimposed upon this, a completely random one governed by a general Gaussian distribution. If we shift the origin of the coordinate axes to $(N\langle\xi\rangle, N\langle\eta\rangle, N\langle\zeta\rangle)$, then the final form of the solution to the general problem of random flights may be written as

$$W(\mathbf{R}) = \frac{1}{\sqrt{8\pi^3 c_{\xi\xi} c_{\eta\eta} c_{\zeta\zeta}}} \exp\left(-\frac{\Xi^2}{2c_{\xi\xi}} - \frac{H^2}{2c_{\eta\eta}} - \frac{Z^2}{2c_{\zeta\zeta}} \right). \tag{9.3-15}$$

The above equation, which at equilibrium reduces to the random coil configuration, $c_{\xi\xi}=c_{\eta\eta}=c_{\zeta\zeta}=1$, accommodates the expected behavior for the effect of the velocity field on a flexible polymer chain. As revealed in Example 9.2, the steady-state conformation of a polymer in a planar shear flow is that of an ellipsoid with its major axis oriented at a finite, shear-dependent angle with respect to the flow direction. Now we must put our effort into quantifying the additional effects that exclusion of certain polymer conformations resulting from solid-surface interactions imposes on such a regular structure.

When a solid surface is imposed as a solid barrier at a certain distance from the origin of the chain, many of the polymer conformations are no longer available to the chain. The problem then is reduced to calculating the number of conformations that survive the imposition of the wall. In this work, the solid surface is treated as a neutral barrier whose functionality is just to lower the conformations that can be realized by a single polymer chain. All of the analysis in this section is valid for planar flows which reduce the dimensionality of the problem to two, x and y, the probability along the z-direction being unaffected. Assuming that the wall is a perfect barrier, the problem can be formulated as follows: what is the probability Wd^3R that, after N steps, the chain will find its end-to-end vector between $\mathbf{R}$ and $\mathbf{R}+\mathrm{d}\mathbf{R}$ without ever having touched or crossed the solid wall?

As a first approximation, the conformations may be calculated based on the same probability function f for each segment; i.e., by ignoring chain end effects. The solution to this problem when the solid barrier is parallel to the x-axis, which in the (ξ,η)-coordinate system is represented by the line $\eta=-\tan(\hat{a})\xi$ (see Figure 9.1), is given by

$$W(\mathbf{R}) = \frac{1}{\sqrt{4\pi^2 c_{\xi\xi} c_{\eta\eta}}} \left[\exp\left(-\frac{(\xi-\xi_0)^2}{2c_{\xi\xi}} - \frac{(\eta-\eta_0)^2}{2c_{\eta\eta}} \right) - \exp\left(-\frac{(\xi-\xi_2)^2}{2c_{\xi\xi}} - \frac{(\eta-\eta_2)^2}{2c_{\eta\eta}} \right) \right], \tag{9.3-16}$$

where

$$\xi_2 = \frac{(1-s^2)\xi_0 + 2s\sqrt{\dfrac{c_{\xi\xi}}{c_{\eta\eta}}}\,\eta_0}{s^2+1},$$

$$\eta_2 = \frac{(s^2-1)\eta_0 + 2s\sqrt{\dfrac{c_{\eta\eta}}{c_{\xi\xi}}}\,\xi_0}{s^2+1}, \qquad (9.3\text{-}17)$$

$$s = -\tan(\hat{a})\sqrt{\frac{c_{\xi\xi}}{c_{\eta\eta}}}\ .$$

(The above calculation is illustrated in Appendix D.1.) In the above expressions, (ξ_0,η_0) is the origin of the chain in the (ξ,η) space, (ξ_2,η_2) a skewed image (in a rescaled space) of the origin with respect to the line $\eta=-\tan(\hat{a})\xi$, and s measures the relative magnitude of the two principal axes of the ellipse. According to (9.3-16), W is given by the difference of the products of two Gaussian distribution functions and expresses the probability (or the statistical weight) that a chain whose origin is at point (ξ_0,η_0) arrives at point (ξ,η) after N steps, without ever having touched or crossed the solid wall. Clearly, as $(\xi_0,\eta_0)\to\infty$ (i.e., far away from the wall), W assumes its bulk, Gaussian value, corresponding to the case where no chains contact the wall.

The probability function W is relatable to both the excess partition function of the system, Θ, and the excess (negative) entropy, ΔS, which characterizes the macromolecular chains in the interfacial region near the wall due to the excluded conformations. The partition is provided by the integration of W over all possible points (ξ,η) [Doi and Edwards, 1986, p. 17] as

$$\Theta = \Theta(\xi_0,\eta_0) = \int W d\xi d\eta\ . \qquad (9.3\text{-}18)$$

In the bulk, $\Theta=1$ since the presence of the surface has no effect on the number of conformations that a chain may realize. However, in the interfacial region, $\Theta<1$ because many of the conformations are no longer available. At the surface, where no conformations are possible, $\Theta=0$.

The integral Θ, for all distances and orientations of the solid barrier, is given simply enough by (see Appendix D.2)

$$\Theta = \operatorname{erf}\left(\frac{y_0}{\sqrt{2c_{yy}}}\right), \qquad (9.3\text{-}19)$$

where erf(·) denotes the error function. This expression is valid for both static and flowing conditions, and shows that the wall affects only the component of the end-to-end vector that is perpendicular to the surface.

The difference in the Helmholtz free energy between a chain with an origin at a distance y_0 from the wall and a reference chain at infinity is then found by considering the change in the corresponding conformational entropy as

$$\frac{\Delta A}{chain} = -\ln\Theta \,, \tag{9.3-20}$$

where the Hemholtz free energy, A, is made dimensionless by dividing by k_BT. Assuming that the solution is dilute enough to neglect interactions among the chains and that we are at the θ-solvent condition,[11] we can write

$$\Delta A = -\int n \ln\Theta \, \mathrm{d}^3x \,, \tag{9.3-21}$$

for the whole system. Thus the complete Helmholtz free energy for this system is

$$A = \int \tfrac{1}{2}\mathrm{tr}\mathbf{C}\,\mathrm{d}^3x + \int n \ln(\frac{nx}{n_s})\,\mathrm{d}^3x - \int \tfrac{1}{2} n \ln\det(\frac{\mathbf{C}}{n})\,\mathrm{d}^3x - \int n \ln\Theta\,\mathrm{d}^3x \,. \tag{9.3-22}$$

This is the form of the Helmholtz free energy which we shall use in the next subsection to divine specific results for our system.

9.3.4 *Calculations with the resulting model equations*

Now let us inspect the behavior of the above system of equations in a couple of special circumstances. First we shall consider the static system in order to determine what effects the conformational loss has upon the equilibrium polymer configuration and distribution in space. Subsequently, we shall return to the investigation of steady-shear flow, which we began in Example 9.3.

A. Static conditions

In the absence of flow, from symmetry considerations, **c** is diagonal. Moreover, the same conditions on **C** and n apply as in Example 9.2: the Hamiltonian is a minimum with respect to **C** and the chemical potential of the polymer is constant. These conditions result in $c_{xx}=c_{zz}=1$,

$$-1 + c_{yy} - 2c_{yy}\frac{\partial \ln\Theta}{\partial c_{yy}} = 0 \,, \tag{9.3-23}$$

[11]The θ-solvent condition implies that the partial molar free energy due to polymer/solvent interactions is zero so that deviations from ideal solution behavior vanish [Billmeyer, 1984, p. 165].

and

$$\frac{\delta A}{\delta n} = \ln n - \tfrac{1}{2}\ln c_{yy} + \tfrac{5}{2} - \ln\Theta - n\frac{\partial \ln\Theta}{\partial n}$$
$$= constant\ (bulk\ value) \equiv \frac{5}{2}\ . \tag{9.3-24}$$

Manipulation of (9.3-24) and the use of (9.3-23) result in the following expression for the dimensionless concentration in the interfacial region:

$$n = \Theta\sqrt{c_{yy}e^{1-c_{yy}}}\ . \tag{9.3-25}$$

Thus the physical picture that the above relation conveys is that the increase of the chemical potential due to conformational loss in the interfacial region is compensated by an analogous decrease in the number of chains which occupy this region. Consequently, if a solid barrier is suddenly immersed in a dilute polymer solution, then there should occur a diffusive flux of polymer molecules away from the surface until equilibrium is re-established with the above concentration profile.

Eqs. (9.3-19,23,25) provide us with a lot of new information for the static system in consideration. Θ gives us the relative number of configurations which survive the imposition of the solid barrier, n gives us the concentration profile, and (9.3-23) gives us the spatial resolution of c_{yy}. Figures 9.2 and 9.3 reveal that, starting from their bulk values, as the surface is approached both c_{yy} and n decrease monotonically, becoming zero exactly on the wall. Because of entropic constraints, the chains are effectively "squeezed" in the direction perpendicular to the wall, while simultaneously diffusing farther away from it. The length scale of these phenomena is about three times the unperturbed, equilibrium, root-mean-square, end-to-end distance, R_e, of the chains. These results are in excellent quantitative agreement with the Monte Carlo results of Fitzgibbon and McCullough [1989, pp. 664-69] and the molecular dynamics simulations of Bitsanis and Hadziioannou [1990, §3.B].

Although the wall affects significantly the component perpendicular to the surface, the above analysis predicts that the component parallel to the wall, c_{xx}, is everywhere identical to its bulk value. This is rather obvious because the mean-square displacement of a random flight parallel to the wall should not be affected by what is happening in the direction normal to the wall. The Gaussian assumption leaves the two components of the end-to-end distance of the chain uncoupled because no excluded volume effects are taken into account. This is most definitely an aphysical assumption, but quite possibly an accurate one as the previous calculations [Fitzgibbon and McCullough, 1989, pp. 664-69; Bitsanis and Hadziioannou, 1990, §3.B] would seem to indicate, particularly for

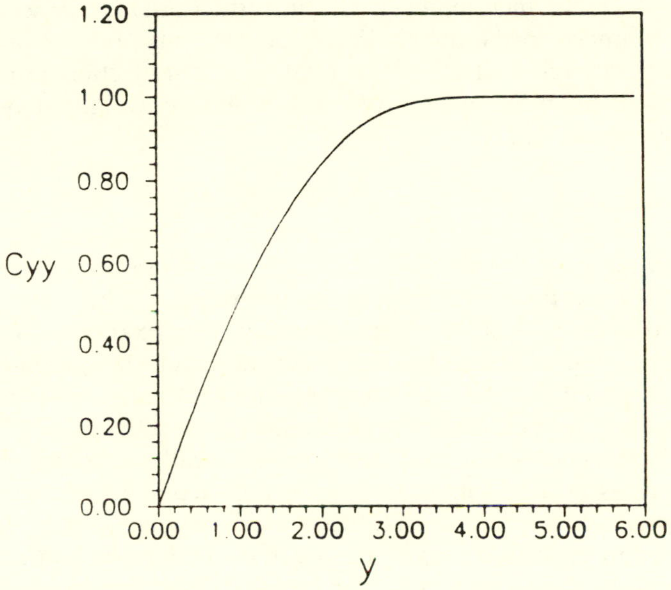

Figure 9.2: Dimensionless equilibrium component c_{yy} versus dimensionless distance from the wall. (Reprinted with permission from Mavrantzas and Beris [1992a, p. 190].)

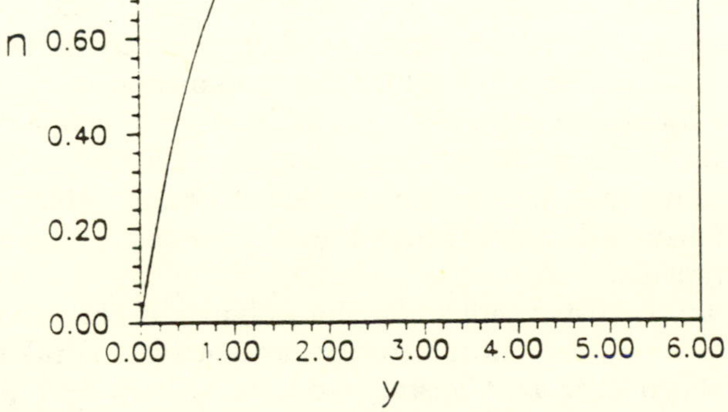

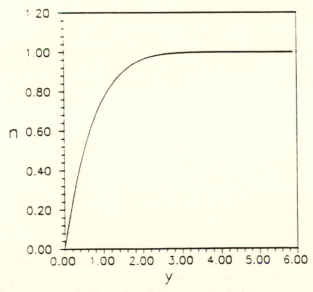

Figure 9.3: Dimensionless equilibrium concentration n versus dimensionless distance from the wall. (Reprinted with permission from Mavrantzas and Beris [1992a, p. 191].)

polymers of high molecular weight. In addition to the experiments by Ausserré *et al.* [1986] mentioned in the introduction to this section, depletion phenomena in the interfacial region have also been reported experimentally through a study using total-internal-reflection infrared measurements in mixtures of two different molecular weight polymer solutions [Pangelinan, 1991].

B. Steady-shear flow

Now we wish to apply the above theory to the problem of isothermal, planar, steady-shear flow of a high-molecular-weight, dilute polymer solution above a neutral solid surface. We recall from Example 9.3 that, with no polymer/surface interactions, a constant velocity gradient is predicted throughout the flow domain. Again, let x and y denote the directions parallel and perpendicular to the flow, respectively.

In viscoelastic steady-shear flow next to a solid wall, we perceive two time scales: one associated with the flow near the barrier, and the second which characterizes the migration/diffusion of polymer chains from regions of low to regions of high entropy (i.e., regions without constraints). Since macromolecules have very small diffusivities in solution, we expect that this second time scale might be very large. For example, based on their analysis on the developing Poiseuille flow, Tirrell and Malone [1977, p. 1579] suggest that length-to-diameter ratios of the order of 5000 may be necessary in order to achieve about 63% full development of the concentration profile. This entrance effect has been recently fully documented in the studies of Pérez-González *et al.* [1992]. An estimate of the time that polymer chains in the interfacial region spend in the channel is provided by the ratio of the length of the channel and the slip velocity, $t_F=L/\upsilon_s$. For a channel length of 10cm and for a slip velocity of 10μm/s [Müller-Mohnssen *et al.*, 1990, p. 231], t_F is found to be on the order of 10^4s. A simple expression for the diffusion coefficient is given by Einstein's equation, $D=k_BT/6\pi\eta_sR_e$. For some typical values of these parameters, one can find that the translational diffusion coefficient is very small, on the order of 10^{-11}m^2/s, which, for a depletion layer of about 0.15μm [Müller-Mohnssen *et al.*, 1990, p. 233], corresponds to a diffusion time scale of the order of 10^{-3}s. This time is much shorter than the characteristic convective time in the interfacial region, which demonstrates that diffusion is definitely present during the performance of even very fast experiments.

However, the diffusive motion of the polymer chains under flow and/or near a solid wall is inherently anisotropic [Prakash and Mashelkar, 1991, pp. 3746-47; Mansfield and Theodorou, 1989, pp. 3149, 3152], and this leads to diffusion constants which, in the direction perpendicular to the wall, are several orders of magnitude less than the corresponding ones in the bulk. The diffusion constants become even lower when entanglement effects become dominant, as is the case in polymer melts or in systems where polymer segregation phenomena are investigated. In such complex systems, the above time scale increases substantially, and can be as high as 10^5s [Pangelinan, 1991], thus justifying the consideration of both a short and a long time behavior. In general, we can define a dimensionless Peclet number, $Pe \equiv t_D/t_F$ [Tirrell and Malone, 1977, p. 1579], which can be used to characterize the two limits that we are interested in: the limit $Pe=0$ corresponds to complete migration of the polymer molecules (long flow-time analysis), while the limit $Pe=\infty$ corresponds to very fast flows where diffusion is not given the time to set in (short flow-time analysis). In reality, for intermediate values of the Pe number, the concentration profile changes continuously along the flow direction and in time. For simplicity, in the following we shall restrict ourselves only to the two limits mentioned above.

B.1 Short-time analysis ($Pe=\infty$)

In this limit of the analysis, diffusional effects are neglected so that, under steady-state conditions, we have the establishment of a one-dimensional profile, and the kinematics, extra stress, and concentration profile are those of Example 9.3. Also, the constitutive equation for the Oldroyd-B model reduces again to three algebraic equations for the effect of flow upon the three components c_{xx}, c_{yy}, and c_{xy} of the conformation tensor. Moreover, the same scaling as that used in Example 9.3 is also adopted here.

An immediate consequence of the above scaling is that velocities in our model scale with the ratio R_e/λ. This scaling involves characteristic dimensional parameters that are associated with the structure and idiosyncrasy of the polymer chains. The relaxation time λ measures the "elastic" character of the molecule, the higher the value of λ, the more pronounced the "elastic" character of the molecule, with λ going to infinity for a perfectly elastic medium. The relaxation time and the parameters associated with the macromolecule mobility determine the transport behavior of the fluid under flow conditions. The longest relaxation time provides an estimate of the rate at which chains will lose memory of their original conformational characteristics, and hence of the viscoelastic effects to be expected when they are subjected to external deformation at a rate commensurate with the intrinsic relaxation time [Mansfield and Theodorou, 1989, p. 3151]. On the other hand, the rate of diffusion of a chain's center of mass is proportional to the fluidity, or inverse viscosity, of the polymer. Characteristic length and time scales that quantify this transfer are provided by the equilibrium (bulk) size of the chains, and by their relaxation times. This ratio provides a measure for the rate at which chains self-diffuse under equilibrium conditions.

The description presented above reduces the problem of planar shear flow to a system of four algebraic, nonlinear equations with respect to four unknowns, c_{xx}, c_{yy}, c_{xy}, $\dot{\gamma}$; however, since Θ depends only on c_{yy}, proper manipulation of these equations shows that one of them, the one for c_{yy}, is decoupled from the imposed shear stress and obeys the same conservation equation as the one under equilibrium conditions, (9.3-23). All other unknowns, c_{xx}, c_{xy}, and $\dot{\gamma}$ are then given as functions of c_{yy}, σ_{xy}, and $\mu=\eta_s/\eta_p$,

$$\dot{\gamma} = \frac{\sigma_{xy}}{\mu + c_{yy}} , \tag{9.3-26}$$

$$c_{xy} = \sigma_{xy} \frac{c_{yy}^2}{\mu + c_{yy}} , \tag{9.3-27}$$

and

$$c_{xx} = 1 + 2\left(\sigma_{xy} \frac{c_{yy}}{\mu + c_{yy}}\right)^2 . \tag{9.3-28}$$

Since c_{yy} does not depend on the imposed shear stress, σ_{xy}, (9.3-26) demonstrates that the velocity gradient scales with the shear stress. In addition, since c_{yy} decreases near the wall (see Figure 9.2), the equation predicts a velocity gradient that should increase from its bulk value to the Newtonian solvent limit, σ_{xy}/μ, as the solid surface is approached. Such a velocity gradient then gives rise to a velocity profile which exhibits a sudden increase from its zero value on the wall to its bulk characteristics within a distance commensurate with the unperturbed size of the polymer chains. Extrapolation of the bulk velocity profile to intersect the wall then defines an apparent slip velocity. This slip velocity is then strictly a linear function of the imposed shear stress with a dimensionless slip coefficient, k, depending only on the solvent/polymer viscosity ratio:

$$\upsilon_s = k\sigma_{xy} . \tag{9.3-29}$$

All of this theoretical analysis is depicted in Figures 9.4 and 9.5.

Figure 9.4 is a typical plot of the velocity profile. The velocity increases rapidly (much steeper than linearly) within a distance approximately equal to unity (dimensional distance equal to R_e), reaching asymptotically the linear profile characteristic of the bulk behavior. This figure demonstrates also the strong dependence of the velocity profile on the solvent/polymer viscosity ratio. Extrapolation of the straight line to intersect the wall defines the (apparent) slip velocity.

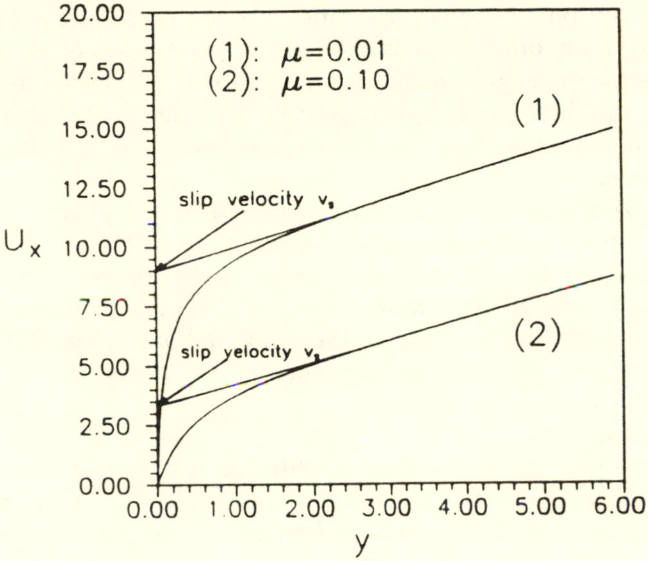

Figure 9.4: Dimensionless velocity versus dimensionless distance from the wall for different values of μ, for σ_{xy}=1.0, and for the high *Pe* analysis. (Reprinted with permission from Mavrantzas and Beris [1992a, p. 194].)

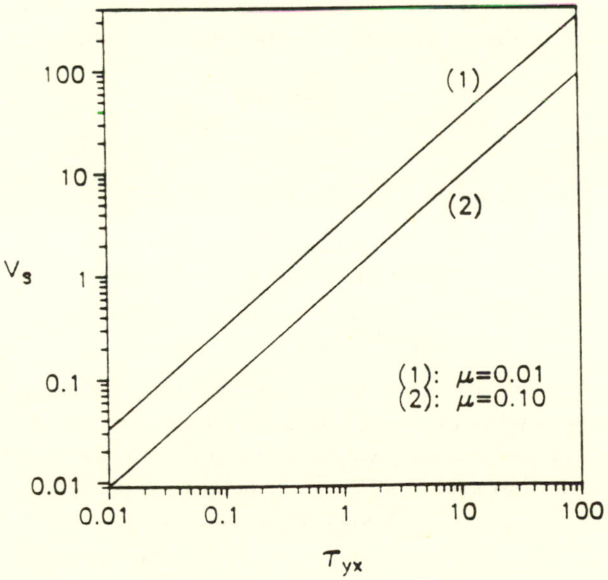

Figure 9.5: Dimensionless slip velocity versus dimensionless shear stress $\tau_{yx}\equiv\sigma_{xy}$ for different values of μ and for the high *Pe* analysis. (Reprinted with permission from Mavrantzas and Beris [1992a, p. 195].)

Like the velocity gradient, the dimensionless slip velocity depends on both the solvent/polymer viscosity ratio and the shear stress (which, since it is constant in our system, is also the wall shear stress). Figure 9.5 demonstrates the linear dependence of the slip velocity on the imposed shear stress. In addition, in Figure 9.5, it is seen that by increasing the solvent/polymer viscosity ratio, we can reduce, and ultimately eliminate, slip phenomena.

Eq. (9.3-29) has been used on a heuristic basis in the past by several investigators to describe and explain experimental data. The work presented here provides a theoretical explanation and justification. In the next subsection, we shall show how diffusion affects this picture, and we shall present results for the effects of the solid wall on the concentration of the solution, and upon the conformations of the chains.

B.2 Long-time analysis (Pe=0)

Assuming steady-state, isothermal conditions, and constant solution density ρ, we start by treating the number density of polymer chains as a variable and allowing for diffusion in the direction perpendicular to the solid surface. The diffusion equation, (9.3-4), then takes the form (assuming that κ is constant)

$$\nabla\cdot\nabla\left(\frac{\delta A}{\delta n}\right) = 0 \;\Rightarrow\; \nabla_y\left(\frac{\delta A}{\delta n}\right) = constant = 0 \,, \tag{9.3-30}$$

the last equality being due to the no-flux boundary condition on the wall. This demands for the Helmholtz free energy the condition

$$\frac{\delta A}{\delta n} = constant \,, \tag{9.3-31}$$

and therefore

$$\begin{aligned}\frac{\delta A}{\delta n} &= \ln n - \tfrac{1}{2}\ln\det\mathbf{c} + \tfrac{5}{2} - \ln\Theta - n\frac{\partial\ln\Theta}{\partial n} \\ &= constant\ (bulk\ value) \\ &= \tfrac{5}{2} - \tfrac{1}{2}\ln\det\mathbf{c}_b \,,\end{aligned} \tag{9.3-32}$$

where $\mathbf{c}_b$ is the conformation tensor in the bulk, obtained from the previous analysis without taking into account any surface effects. Proper manipulation of this equality results in the following nonlinear equation for the concentration profile in the interfacial region:

$$n = \Theta\sqrt{c_{yy}e^{1-c_{yy}}}\sqrt{\frac{1+(2-c_{yy})c_{yy}^2\left(\dfrac{\sigma_{xy}}{\mu+nc_{yy}}\right)^2}{1+\left(\dfrac{\sigma_{xy}}{\mu+1}\right)^2}} \,. \tag{9.3-33}$$

This expression has the same form as (9.3-25) at static equilibrium, but includes a correction term that takes into account flow-field effects. At any distance from the wall, this equation has to be solved numerically with certain data: the imposed shear stress, σ_{xy}, the viscosity ratio, μ, and the corresponding value of c_{yy}. Of course, the value of n also depends on the imposed shear stress.

According to (9.3-33), depletion of the interfacial region of polymer molecules is expected, and the extent of this depletion depends now upon the imposed shear stress. The behavior of n for different values of the shear stress may be seen in Figure 9.6. We see there that the effect of the shear stress is to shift the equilibrium profile to the left; the higher the imposed shear stress, the greater the shift. This means that the shear stress reduces depletion phenomena induced by the interaction with the solid wall. A similar behavior has been reported by Duering and Rabin [1990, §4,5] through Brownian dynamics simulations. The maximum value that the concentration profile can attain is evaluated for an "infinite" shear stress. It is found then that for very small values of the parameter μ, n approaches asymptotically the curve defined by the equation

$$n = \sqrt[4]{(2-c_{yy})c_{yy}e^{1-c_{yy}}} = \sqrt[4]{2-c_{yy}}\sqrt{n_e} \ , \tag{9.3-34}$$

where n_e is the equilibrium (σ_{xy}=0) concentration profile, (9.3-25).

A measure of the effect of the shear stress on the concentration profile near the wall can be provided through the surface excess function, Γ, which is the integrated polymer density:

$$\Gamma \equiv \int_0^\infty (n-1)\,dy \ , \tag{9.3-35}$$

where the distance y has been made dimensionless with R_e. Typical plots of Γ for some representative values of μ may be viewed in Figure 9.7. Since n is less than unity, Γ is always negative. Moreover, since the shear stress shifts the equilibrium profile towards the left, the absolute value of Γ decreases as the shear stress increases, approaching very rapidly its asymptotic value. Figure 9.7 also shows that, apart from the imposed shear stress, depletion phenomena are strongly affected by the relative size of the solvent and polymer viscosities that enter the equations through the parameter μ. As the ratio μ decreases, the absolute value of the surface excess Γ becomes smaller for the same value of the shear stress. The major advantage of reporting the surface excess (or its negative, i.e., the wall depletion) is that it has been studied experimentally using the EWIF technique [Ausserré *et al.*, 1991].

Since the polymer contribution to the viscosity is proportional to the polymer concentration (or, equivalently, to the chain number density), the momentum equation, (9.3-3a), in our model assumes the form

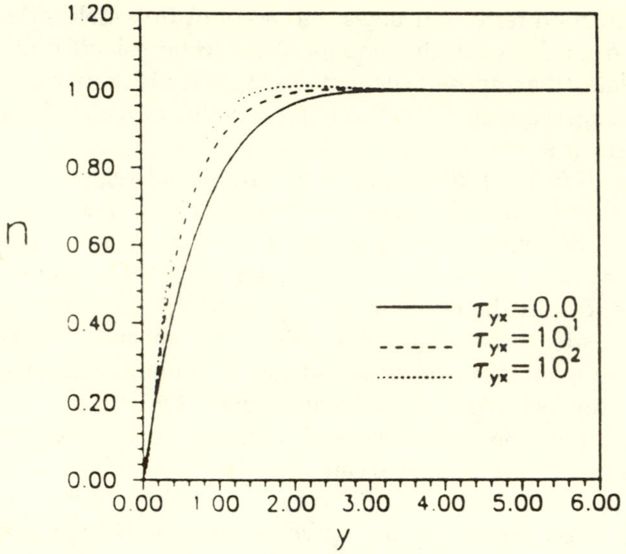

Figure 9.6: Dimensionless concentration versus dimensionless distance from the wall for $\mu=0.1$ and various values of dimensionless shear stress $\tau_{yx}\equiv\sigma_{xy}$. (Reprinted with permission from Mavrantzas and Beris [1992a, p. 197].)

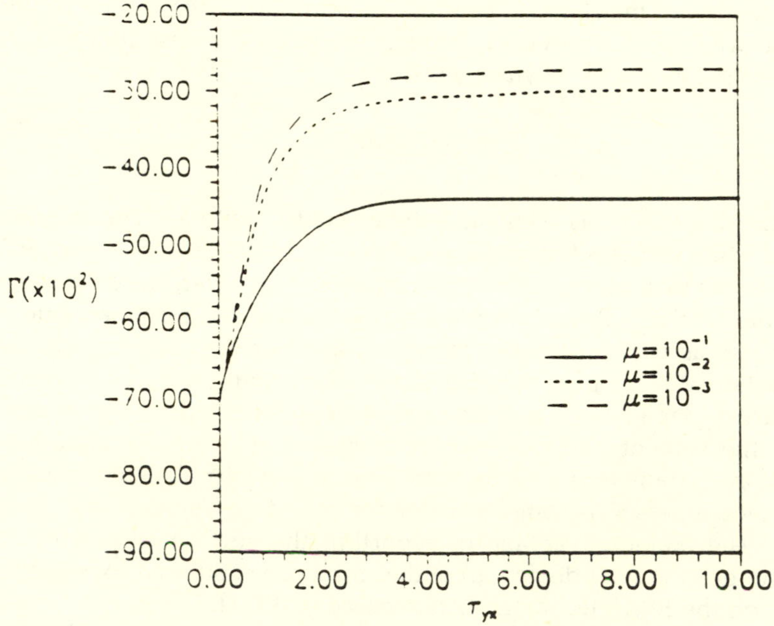

Figure 9.7: Surface excess versus dimensionless shear stress $\tau_{yx}\equiv\sigma_{xy}$ for different values of μ. (Reprinted with permission from Mavrantzas and Beris [1992a, p. 198].)

$$\dot{\gamma} = \frac{\sigma_{xy}}{\mu + n c_{yy}} , \qquad (9.3\text{-}36)$$

This equation can be compared with (9.3-26) in the absence of any concentration gradients in the flow domain. Allowance for diffusion of the chains from regions of higher to regions of lower entropic constraints leads to higher values of the velocity gradient. Eqs. (9.3-27,28) are now transformed into

$$c_{xy} = \sigma_{xy} \frac{c_{yy}^2}{\mu + n c_{yy}} , \qquad (9.3\text{-}37)$$

and

$$c_{xx} = 1 + 2\left(\sigma_{xy} \frac{c_{yy}}{\mu + n c_{yy}}\right)^2 . \qquad (9.3\text{-}38)$$

According to (9.3-36), diffusion increases the drastic changes which the velocity gradient undergoes in the interfacial region. This results in calculated apparent slip velocities that are greater than the previous ones; however, their dependence on the (wall) shear stress approximately retains its linear character (see Figures 9.8 and 9.9).

In Figure 9.9, the slip coefficient k has been plotted against μ for the two time scales of analysis and, for the low *Pe* analysis, for two different values of the shear stress. It can be seen that for moderate or high values of μ, the slip coefficient is independent of the shear stress. This can be attributed to the quantitatively small effects of the shear stress on the concentration profile near the wall. It is only for very small values of μ that the slip coefficient changes, depending on the shear stress; however, the changes are still small. As indicated in Figure 9.9, a change in shear stress by four orders of magnitude, for μ=0.01, changes k less than 50%.

Figure 9.10 shows that the normal projection of the conformation tensor along the flow direction, c_{xx}, initially exhibits a slight decrease, then increases suddenly as the chain approaches the solid wall; it reaches a peak, and finally drops sharply to assume exactly the value of unity on the wall. (In Figure 9.10, the value of c_{xx} has been scaled with that in the bulk, $c_{xx,b}$.) This picture is also in excellent agreement with the Brownian dynamics results of Duering and Rabin [1990, §4,5]. Increases in shear stress change only the quantitative aspects of this picture. Exactly on the wall there is no flow, and therefore at this location, the polymer chain should exhibit a value of c_{xx} similar to that in the bulk. This explains why c_{xx} becomes unity on the solid surface. Although this also proves that the model is consistent, no experiments have been undertaken to test its physical validity.

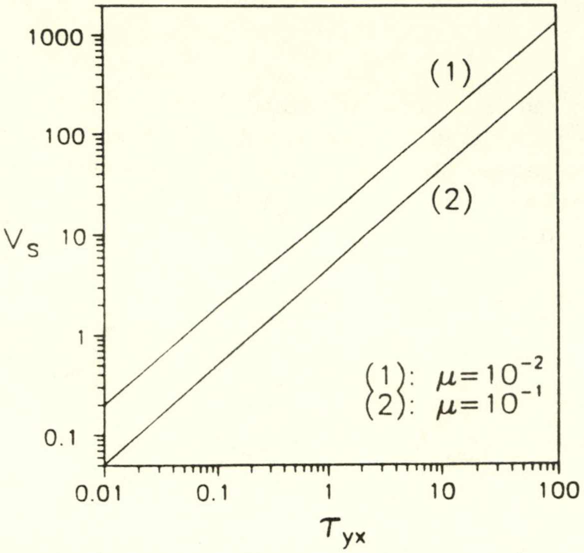

Figure 9.8: Dimensionless slip velocity versus dimensionless shear stress $\tau_{yx} \equiv \sigma_{xy}$ for different values of μ and for the low *Pe* analysis. (Reprinted with permission from Mavrantzas and Beris [1992a, p. 200].)

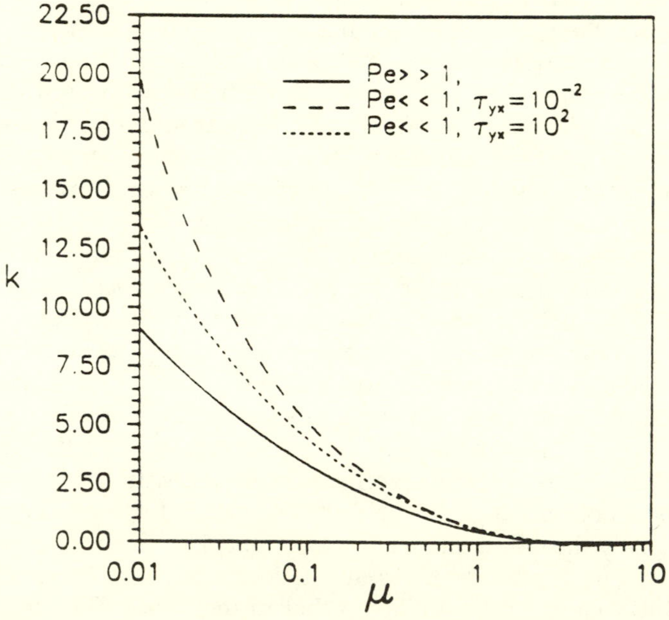

Figure 9.9: Dimensionless slip coefficient *k* versus μ for the two time scales of our analysis: the low and the high *Pe* analysis. Since for a low *Pe* and small values of μ the slip velocity v_s is not exactly a linear function of $\tau_{yx} \equiv \sigma_{xy}$, the slip coefficient *k* depends also on the imposed shear stress. (Reprinted with permission from Mavrantzas and Beris [1992a, p. 201].)

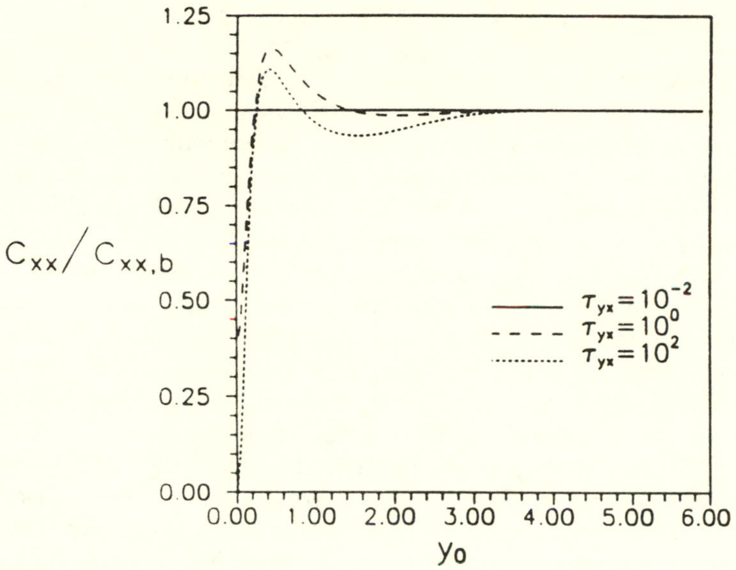

Figure 9.10: Dimensionless c_{xx} component versus dimensionless distance from the wall for μ=0.1 and various values of $\tau_{yx} \equiv \sigma_{xy}$ in the low *Pe* analysis. (Reprinted with permission from Mavrantzas and Beris [1992a, p. 202].)

In examining how quantities related to the conformation of chains evolve as the solid wall is approached, we have to realize the simultaneous effects of both the shear and the surface. At distances considerably larger than the chain size, shear-induced conformation changes prevail, while at distances of the order of the chain size, wall effects become dominant. Clearly, the increase of c_{xx} is the result of the wall- and the shear-induced flattening of the chains which occurs there; however, since on the wall there is no flow, and because of the absence of excluded-volume phenomena in the model, chains there should assume their equilibrium characteristics. This explains the sharp decrease of c_{xx} near the wall.

In addition to the velocity profile, the concentration profile, and the molecular structure, our model provides information for the behavior of the normal stresses near the solid wall. These can be calculated by use of (9.3-3d). For all the cases examined, the second normal-stress difference was found to be identically zero. In fact, (9.3-23) for c_{yy} is also the equation for a zero second normal-stress difference everywhere in the flow domain. On the contrary, the first normal-stress difference was found to change drastically in the interfacial region relative to the bulk, following closely the behavior of c_{xx}. In our model $\sigma_{xx} - \sigma_{yy}$ is proportional to $n(c_{xx} - 1)$.

9.3.5 Comparison with available experimental data and previous theoretical work

The most significant prediction of this investigation was the linear dependence of the slip velocity on the wall shear stress, which has been inferred indirectly from a large number of experiments in the past. The linear dependence of the slip velocity on the wall shear stress has been verified experimentally by Müller-Mohnssen *et al.* [1990, p. 235] during the flow of an electrolyte-free, aqueous solution of a high molecular-weight, anionic polyacrylamide. Quantitative comparison of their results with ours, however, is not feasible at the present time because of energetic interactions and of a shear-thinning behavior, both of which have not been taken into account in the present work. A rough comparison has nevertheless been attempted where the parameters (data) entering our model have been taken from Müller-Mohnssen *et al.* [1990, Table 1]. For the lowest polymer concentration (c=0.005% wt.) examined in the experiments, the following data were used: solvent viscosity, 0.96cp; bulk polymer solution viscosity, 4.1cp (this was calculated so that for a dimensionless shear stress $\sigma_{xy}=1$, the bulk velocity gradient predicted by our model is the same as the one measured experimentally); and a root-mean-square, equilibrium, unperturbed, end-to-end distance of 0.12μm. For this concentration, the experimentally measured slip coefficient, k, was found to be 4×10^{-3} cm^3/dyn/s. The model prediction for the same concentration and for a fully developed concentration profile is 1×10^{-3} cm^3/dyn/s.

This rough comparison demonstrates that our model, without the use of any adjustable parameters, can capture the correct order of magnitude for the slip coefficient. Moreover, the predicted value for the slip coefficient, and subsequently for the corresponding slip velocity, is slightly lower than the experimental one. We believe that by taking into account energetic interactions of the flowing polymer molecules with the confining solid surface and by assessing the correct dependence of the solution viscosity on the conformation state of the polymer chains, we should be able to improve our predictions even further. We should mention here that, although in the experiments specific care was taken to eliminate any energetic interactions of the polymer solution with the solid wall, such interactions seemed to be present. As is claimed in the paper, these interactions were believed to be of anionic origin and repulsive [Müller-Mohnssen *et al.*, 1990, p. 241].

In addition to the linear relation between the slip velocity and the wall shear stress, this investigation led to very specific predictions with respect to the microscopic structure (molecular orientation), concentration (number density) profiles, and stress behavior. Few explicit predictions for the orientation of macromolecules next to a solid surface are available in the literature; most of the previous works examined the effects of the wall on the concentration or the stress. We have definitely verified the alignment of the molecules along the wall, which is commonly used as a boundary condition. In addition, Duering and Rabin [1990, §4,5] reported results of computer simulations of polymers with excluded-volume interactions in simple shear flow between impenetrable walls. Their results for the spatial dependence of the components of the end-to-end vector of the chains perpendicular and parallel to the surface are in very good agreement with ours, especially for the parallel projection, c_{xx}. The simulations showed the same peak at an intermediate distance from the solid surface and then the same sharp decrease towards a finite value on the wall as this theory. With respect to the effect of the solid wall on the concentration, the theory confirms Duering and Rabin's Brownian dynamics simulations in that the flow field increases the polymer chain concentration near the wall relative to the bulk.

The model also compares favorably with the earlier model developed by Aubert and Tirrell [1982] in the prediction of the effective viscosity of dilute polymer solutions for flows through micropore membranes. Table 9.1 shows the comparisons with the zero-shear-rate effective viscosity (i.e., apparent zero-shear-rate viscosity, dimensionalized with the pure solvent viscosity) predictions obtained by Aubert and Tirrell [1982, pp. 559-60], those of this investigation, and the corresponding experimental data of Chauveteau [1982]. The comparison is very favorable for all the cases shown. Since we were interested only in the zero-shear-rate predictions, the reported values were simply calculated by taking into account only the effects of the wall on the polymer concentration using the equilibrium solution for a planar wall, Eqs. (9.3-23,25). The resulting location-dependent viscosity was then used in the calculation of the effective viscosity following Eqs. (42) and (43) in Aubert and Tirrell. For the evaluation of the corresponding Hookean elasticity parameter, $K \equiv 3k_BT/b^2$, the reported end-to-end distance b=0.82μm [Chauveteau, 1982, p. 121] was used for xanthan, the calculation thus involving no adjustable parameters. Note that although close, our predictions are different from Aubert and Tirrell's, even when we use the same value as theirs, b=0.9. The differences can be attributed to the different way of accounting for the polymer concentration (to be consistent with our model, we used a viscosity which was linearly dependent on concentration), as well as (for low concentrations) to the different way of evaluating the wall-induced entropy reduction—our expression correcting for the local width of the distribution function. Of course, due to the assumption of a dilute solution, the good agreement at higher concentrations could be considered as fortuitous.

Experimental Parameters: D=average pore diameter c_b=bulk concentration, % wt.	Experimental Results (Chauveteau, 1981)	Aubert and Tirrell (1982) b=0.9μm	This Work b=0.82μm (b=0.9μm)
D=10.9μm, c_b=400ppm	3.18	3.25	3.08 (3.02)
D=6.60μm, c_b=400ppm	2.89	3.00	2.72 (2.66)
D=4.80μm, c_b=400ppm	2.46	2.58	2.49 (2.43)
D=3.60μm, c_b=200ppm	1.63	1.47	1.61 (1.58)
D=3.60μm, c_b=400ppm	2.31	2.12	2.29 (2.23)

TABLE 9.1: Comparison between the experimental data and the theoretical predictions for the zero-shear-rate effective viscosity. Note that the first three data were obtained using a Nuclepore membrane and the last two a silicon carbide pack. (Reprinted with permission from Mavrantzas and Beris [1992a, p. 204].)

The model predictions compare very well also with the more recent experimental data by Omari *et al.* [1989a,b] for the effect of the polymer concentration on the effective viscosity of dilute solutions both for flexible (polyacrylamide, PMA) and semirigid (xanthan) polymers. In both cases, the only parameter that is needed as input into our model (in addition to the bulk concentration and zero-shear-rate intrinsic viscosity) is the value of the root-mean-square equilibrium, end-to-end distance, R_e, of the polymer chain. For the PMA solution, this was calculated from the reported value of the polymer radius of gyration, R_g=0.19μm, estimated through the use of the Flory/Fox equation [Omari, 1989b, p. 521]. However, analysis of the experimental data suggested that the estimated value of R_g should have been corrected by a factor 1.24/0.9 in order to account for excluded-volume effects existing in a good solvent (this was the case in the experiments). Such a factor corrects for deviations from Gaussian statistics that is valid only in θ-solvent conditions. Our predictions have been obtained using this "corrected" radius of gyration corresponding to a value of R_e equal to $\sqrt{6}$*0.19*1.24/0.9=0.26μm. For the two samples of the semirigid xanthan solution, the values of R_e were calculated through the relation $R_e=A\sqrt{N}$, where N is the number of statistical Kuhn segments of the polymer chain and A is their length

(A=2q, where q is the so-called "persistence length"). Both of the values of N and q are reported by Omari *et al.* [1989a, p. 4].

The comparison between our predictions and the experimental data can be seen in Table 9.2. This shows the theoretical predictions for the ratio η_b/η_p, where η_b is the bulk and η_p the effective pore viscosity for the flow through a micropore membrane, respectively. These are considered as functions of the dimensionless parameter $c_b[\eta]_0$ for different average pore diameters and in the limit of zero shear rate, together with the experimental data. Considering viscosity properties, nonlinear effects appear for a critical value of about $c_b[\eta]_0$=3.0 for flexible and of $c_b[\eta]_0$=1.2 for semirigid polymer solutions, respectively [Omari *et al.*, 1989a, Figs. 1,2]. Therefore, the predicted values are in very good agreement with the experimental data in both cases, but only in the dilute region. In the semi-dilute region where nonlinear phenomena are becoming important, our predictions based on a dilute solution theory, not surprisingly, deviate significantly. Moreover, the same table shows that in the limit of zero concentration the model predictions for the rod-like polymer solution are in better agreement with the data obtained with the semirigid (xanthan) solutions. This can be expected from the equal-probability-distribution assumption for all segments, which might be regarded as more appropriate for polymer chains characterized by a semirigid structure.

Grisafi and Brunn [1989, Table 1] have reported values for the boundary layer thicknesses for the material functions in simple shear and pressure-driven flow. The boundary layer thickness (BLT) was defined as the distance from the surface that a quantity achieves 99.5% of its ultimate value. For the simple shear flow, and for the material functions of interest (polymer density and first normal-stress difference), the BLT was found to be 1.633×R_e, the same in both cases, while for the averaged flow properties it was found to be one order of magnitude higher. In our work, the BLT for the polymer density is 2.5×R_e and for the first normal-stress difference, 3.5×R_e. Notice that because of the coupling of the equations that describe the profiles of the concentration and the conformation tensor components (which define the normal stresses in the system), not only one, but various length scales can be detected in our analysis.

The main advantage of this theory is that it leads to a general description for flows where polymers are involved which is valid both in the bulk and in the interfacial region, under both steady and unsteady conditions. In the bulk, the model reduces to the well-known Oldroyd-B fluid that has been used extensively in viscoelastic fluid mechanics calculations. Near the surface, the effects of polymer/wall interaction have been built explicitly into the equations, and this assures that the

$c_b[\eta]_0$	R_p= 12.25	R_p=8.68	R_p=5.27	R_p=6.86	R_p= 3.70
0.	1.00 (1.00)	1.00 (1.00)	1.00 (1.00)	1.00 (1.00)	1.00 (1.00)
0.5	1.08 (1.05)	1.11 (1.07)	1.15 (1.11)	1.13 (1.13)	1.19 (1.17)
1.0	1.14 (1.10)	1.18 (1.14)	1.27 (1.21)	1.22 (1.26)	1.35 (1.33)
1.5	1.18 (1.15)	1.25 (1.21)	1.37 (1.32)	1.30 (1.39)	1.48 (1.50)
2.0	1.22 (1.20)	1.30 (1.28)	1.45 (1.42)	1.36 (1.52)	1.58 (1.66)
2.5	1.25 (1.25)	1.34 (1.35)	1.52 (1.53)	1.42 (1.65)	1.68 (1.83)
3.0	1.28 (1.30)	1.38 (1.35)	1.58 (1.63)	1.47 (1.78)	1.76 (1.98)
3.5	1.31 (1.34)				
4.0	1.33 (1.38)				
4.5	1.35 (1.41)				
5.0	1.37 (1.43)				

TABLE 9.2: Comparison between the experimental data and the theoretical predictions for the zero-shear-rate ratio η_{rb}/η_{rp} as a function of the parameter $c_b[\eta]_0$. The first three columns refer to the flexible (PMA) polymer solutions and the next two to semirigid (xanthan) solutions. In every cell, the top number represents the model prediction and the bottom (inside parentheses) one the experimentally measured value. R_p denotes the pore diameter in each experiment in units of R_e. (Reprinted with permission from Mavrantzas and Beris [1992a, p. 206].)

underlying physics has been taken into account consistently. In fact, this model extends the Oldroyd-B constitutive equation to account for polymer/neutral-wall interactions. This is very important because in this form the model shows great potential for attacking problems where the role of polymer interactions with a solid surface becomes a dominant factor, such as flows through microporous media, coating processes, problems involving geometrical singularities (i.e., flow past a sudden contraction), and in elucidating many of the salient features of surface-induced instabilities. The main limitation of the above theoretical development is that adsorption effects are not taken into account. Incorporation of absorption phenomena into these models necessitates a more powerful description of the physics involved, able to distinguish changes taking place at the length scale of the monomer length, instead of the chain length which was the limiting scale for the validity of the previous analysis. Currently (1994), this is still the subject of work in progress [Mavrantzas and Beris, 1994a-c]. As an example of the anticipated effects, we have included some results from a preliminary study of the static equilibrium problem in the next section, 9.3.6.

9.3.6 *Effects of surface adsorption*

The problem of adsorption of polymers on solid surfaces has been an outstanding issue in polymer science for many years, mainly because of its great significance in areas of such technological applications as the stabilization and flocculation of colloidal suspensions, adhesion and lubrication processes, and flows through membranes and porous media. The properties of these systems are determined by the details of particle interactions mediated by adsorbed layers of polymer molecules. An understanding of these interactions requires the knowledge of the configurations of the adsorbed molecules, and this has been the subject of all the theories of polymer adsorption. So far, most of the research has focused on the adsorption under quiescent conditions. Unfortunately, the more interesting problem of the combined flow and polymer adsorption has received very little consideration. This is so because most of the works on the adsorption under static conditions have been based on lattice models, the extension of which to non-equlibrium (flow) conditions is very difficult. Similarly, a direct extension of the model developed in §9.3.1-5 for non-interacting surfaces is not possible, since the Gaussian approximation for the distribution function of the end-to-end chain distance, on which this model was based, inherently limits its validity to distances larger than or equal to the average chain length (or radius of gyration). Polymer adsorption takes place within a much smaller length scale, of the order of the monomer size from the adsorbing wall. Thus, a much finer representation is necessary.

In the present section we present the basic conceptual elements of an approach that we have recently taken in order to study the problem on a finer scale. The approach is illustrated here for the static equilibrium case, where the state of the adsorbed polymer is that of the minimum free energy. A significant effort is put therefore in identifying the relevant parameters that define the proper formulation for the free energy of a polymer solution next to an interactive (adsorbing) surface. Such an important parameter, for example, is the density of polymer chain ends, which can be used to quantify the density of the polymer component on a segment-length scale. In contrast to previous attempts which relied on a discrete description of the problem (lattice gas theories), the present approach can be easily extended to non-equilibrium conditions simply by extending the free energy to account for polymer deformation due to flow (in a similar fashion as was done in the previous chapters) and by replacing the minimization procedure with the satisfaction of the relevant dynamic equations, as they emerge from the application of the generalized bracket. This second step represents work currently in progress (1994), and it is not going to be addressed here. Instead, we will limit our discussion to a few representative results from the analysis of the static equilibrium problem. In consequence, we focus on the description of the microscopic (random flight) problem—see 9.3.3B—as it should be modified to take into account the polymer adsorption on the end-to-end distance distribution function. To achieve that, the Gaussian approximation needs to be relaxed.

Next to a solid surface, the interaction of the polymers with the confining boundary results in a change of the free energy of the system: as in the non-adsorbing case, the molecules lose entropy due to confinement (loss of conformations). However, in this case, they can also exchange enthalpy by contacting the surface. Both of these corrections are incorporated into the Hamiltonian through the last term in Eq. (9.3-22) which involves the contribution of a correction partition function, Z. This partition function is again evaluated through the use of a microscopic model, which, however, is different from the one described for the non-adsorbing case, as outlined below. Moreover, as the chains start to accumulate near the wall, excluded-volume phenomena become significant. In such a case, a mean-field theory needs to be implemented to account for this increase in polymer concentration above the value for chain overlap.

Most of the earliest theories (e.g., [Hesselink, 1969; 1971]) considered the adsorption of isolated macromolecules which has little practical relevance; despite this, they were able to demonstrate the significance of the formation of loops and tails by the adsorbed molecules on the properties of such systems. Consideration of only configurational entropy changes near the wall, however, implies that existence of tails is favored over that of loops for chains of high molecular weight [Kawaguchi and Takahashi, 1980, p. 2075], which does not agree with experimental data.

Through the use of lattice models, the Scheutjens/Fleer theory [1979; 1980] has been the most comprehensive and successful one for describing the random adsorption of homopolymers with no *a priori* assumptions about the configurations of the adsorbed polymer layer and in very good quantitative agreement with many experimental data. Lattice model theories were recently generalized by Theodorou [1988a,b] to describe not only homopolymer but also copolymer adsorption over surfaces of various adsorbing strengths.

Alternatively, a continuum approach to describe macromolecular conformations has been developed which requires the solution of a diffusion-like equation for the end-to-end distance distribution function,

$$W = W(\mathbf{r},n;\mathbf{r}_0) \ , \tag{9.3-39}$$

which expresses the probability of a segment chain of length n to reach $\mathbf{r}$ having started from $\mathbf{r}_0$ [Edwards, 1965; de Gennes, 1979, ch. 9]. This continuum framework was employed by Ploehn *et al.*, [1988] to study the static equilibrium adsorption of homopolymers next to a solid surface. Using self-consistent, mean-field arguments together with an inner/outer layer analysis of the diffusion equation that describes polymer configurations near the wall, Ploehn *et al.*, [1988] derived the following boundary condition

$$We^{\beta U} = \frac{1}{2}[\exp\left(-\beta\,(\chi_s - U)\right) + 1](W - \frac{\partial W}{\partial n}) + \frac{1}{4}\frac{\partial W}{\partial y} \ , \tag{9.3-40}$$

where χ_s is the energy of adsorption for a single monomer, $U = U(\phi_s,\phi_b,\chi_F,N)$ is the self consistent free-energy field, ϕ_s is the polymer volume fraction on the surface, ϕ_b is the polymer volume fraction in the bulk, χ_F is the Flory/Huggins parameter and N is the degree of polymerization. Note that in Eq. (9.3-40), the distance from the adsorbing surface, y, is made dimensionless by the segment length, l.

This boundary condition is similar to the one reported by de Gennes [1969, p. 198; 1979, p. 252; 1981, p. 1639], who exploited an energy balance near the surface through the use of energy functionals:

$$\frac{1}{W}\frac{dW}{dy} = \frac{\gamma_1}{L} = -K \ , \tag{9.3-41}$$

where L is a dimensionless length scale and γ_1 the second term in the expansion of the polymer interfacial tension in powers of ϕ_s:

$$\gamma = \gamma_d(\phi_s) + I(\phi_s, \phi_b) \cong \gamma_0 + \gamma_1 \phi_s \; . \tag{9.3-42}$$

The assumption under which boundary condition (9.3-41) is valid is that of the weak-coupling limit, $\gamma_1 < k_B T/l^2$, which ensures that $\phi_s << 1$; however, this is compatible with strong adsorption because of the large number of segments in a polymer chain.

To solve the full diffusion equation (i.e., excluded volume phenomena also included), Ploehn *et al.* (1988) and Ploehn and Russel (1989) expanded the probability density function in the eigenfunctions of the linear operator of the diffusion equation, and kept only the lowest order term (ground state solution). This was matched with an asymptotic expansion for the configurational probability in the outer region describing non-adsorbed chains, to provide a uniformly valid approximation. Calculated profiles for the segment volume fraction, the surface concentration, the bound fraction, and the hydrodynamic layer thicknesses were found to agree qualitatively with experimental data and with the previous lattice model predictions of Scheutjens and Fleer [1979].

From a conceptual point of view, the form of the equations for the probability density function in the work of Ploehn *et al.* [1988] can be shown to represent the continuum analogue of the discrete equations of the lattice gas approach which govern the probability of movement of a polymer segment along the lattice sites [Ploehn, 1988, p. 51]. Scheutjens and Fleer's analysis contained a free energy minimization stage where, starting from the canonical partition function, an expression for the free energy of the system was derived, which was then minimized with respect to the number of polymer molecules at a particular conformation. In contrast, Ploehn *et al.* [1988] used a form of a potential that directly rendered the counting of conformations self-consistent. The form for this self-consistent potential was derived by Helfand [1975a,b; 1976]. In this way, the stage for the free energy minimization was avoided and the final system consisted of two fundamental equations: the first is the self-consistent diffusion equation for the polymer conformations, and the second is the equation relating the polymer segment volume fraction to the integrated probability function.

An alternative approach, also a continuum model, can be formulated by considering the density of polymer chain ends as an additional variable. Then, using the expression for the transition probabilities that arises from the solution of the diffusion equation, the density of polymer segments can be calculated. The chain-ends density, together with the segment density can then be used to evaluate the free energy of the system (by generalizing, for example, the Flory/Huggins expression to non-homogeneous systems). In this way, one can express the free energy of the system solely in terms of one unknown, the chain-ends density. The equilibrium state of the system can then be evaluated through a minimization procedure. This is exactly the approach that is currently followed by Mavrantzas and Beris [1994a-c]. The advantage of this approach is that it can be formally extended to flow conditions through the use of the generalized bracket which represents work currently in progress. In the remaining paragraphs of this section, we will provide a selection of results obtained with this approach at static equilibrium conditions.

First, as regards the governing equations, these are astonishingly similar to their lattice model analogues of Scheutjens and Fleer [1979], and to the continuum ones of Ploehn *et al.* [1988]. Second, the solution for the self-consistent field (segment density) is found numerically without resorting to an eigenfunction expansion. In particular, at θ-solvent conditions, an analytical solution for the probability function W and its integral is utilized. This solution actually corresponds to a simpler form of the boundary conditions defined in Eq. (9.3-40), obtained by neglecting the term $\partial W/\partial n$. However, numerical solution utilizing the correct boundary conditions have shown this approximation to be valid provided N>10. At non-θ-solvent conditions, a full numerical solution of the diffusion equation is necessary. In all cases, the solution profiles for the quantities of interest (amount of polymer adsorbed, conformation characteristics, and layer thicknesses) are found to be in very good quantitative agreement with the Scheutjens/Fleer lattice model predictions. This agreement is illustrated in the following figures.

Figure 9.11 shows typical profiles for the adsorbed amount, Γ, versus bulk polymer volume fraction (adsorption isotherms), for both a θ- (χ_F=0.5) and an athermal (χ_F=0.0) solvent, and for two molecular weights (N=100 and N=1000). Γ is defined by

$$\Gamma \equiv \int_0^\infty (\phi(y) - \phi_f(y))\,dy \ , \tag{9.3-43}$$

where ϕ_f represents the segment density corresponding to free chains and the distance from the adsorbing surface, y, is made dimensionless using the segment length, l. The lines represent the predictions of the current approach while the points have been taken from the paper of Scheutjens/Fleer for an hexagonal lattice. According to Figure 9.11, as ϕ_b increases, Γ also increases: however, for ϕ_b less than about 10^{-3}, Γ increases very little, while for ϕ_b greater than 10^{-3}, this increase is quite substantial. This is so for both the θ- and the athermal solvent. In addition, Figure 9.11 shows that as the quality of the solvent increases (χ_F decreases), Γ decreases significantly. This happens because a better solvent tends to decrease the affinity of the polymer segment toward the surface, i.e., the polymer segments prefer being in the solution rather than in contact with the wall. All of these tendencies are in excellent qualitative and quantitative agreement with the Scheutjens/Fleer lattice model predic-

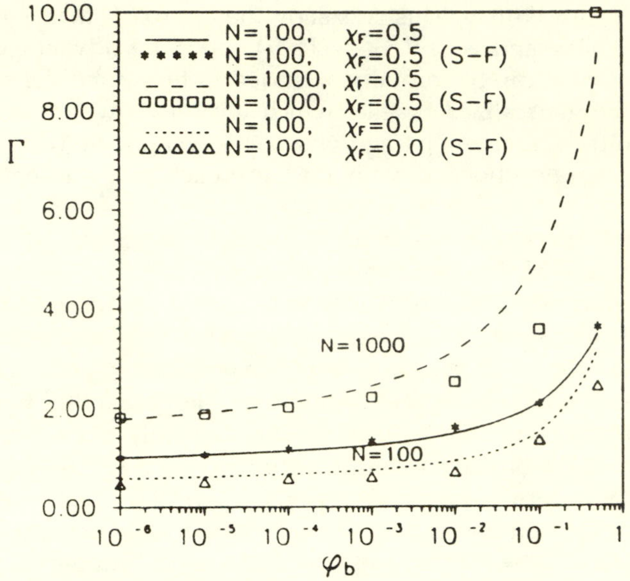

Figure 9.11: The adsorbed amount Γ as a function of the bulk volume fraction ϕ_b for various values of chain length, N, and of the solvent quality, χ_F. The strength of adsorption, χ_s, is χ_s=1. The lines represent the model predictions, and the points the results of Scheutjens and Fleer [1979, p. 1629] for an hexagonal lattice.

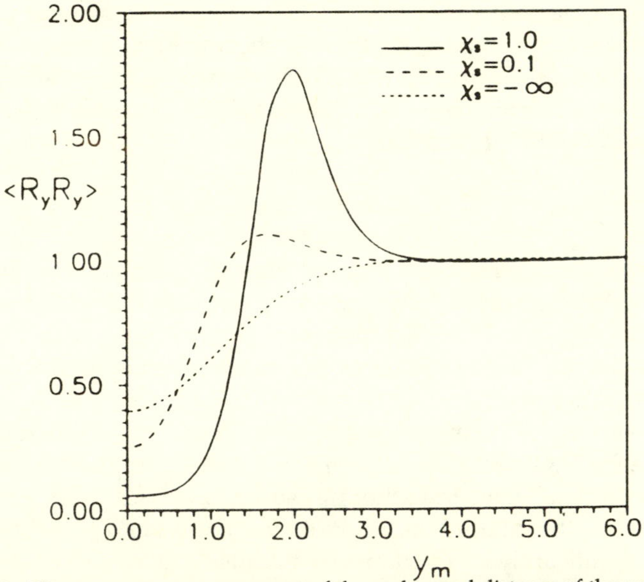

Figure 9.12: The mean square component of the end-to-end distance of the polymer chains in the direction perpendicular to the wall, c_{yy}, as a function of the dimensionless distance of the middle point of the chain from the wall, y_m, for various values of the adsorbing potential, χ_s (bulk polymer volume fraction $\phi_b=10^{-3}$, degree of polymerization $N=10^3$, χ_F=0.5). The value $\chi_s= -\infty$ corresponds to a non-interacting (purely impermeable) wall.

tions, which alludes to the similarity of our continuum-form equations to their discrete-form ones.

Figure 9.12 shows typical profiles for the mean square of the normal component of the end-to-end distance of the polymer chains perpendicular to the wall, $<R_yR_y>$, as function of the distance of the middle point of the chain from the surface for various values of the surface adsorption strength. The value of χ_s is indicative of the strength of the adsorption: $\chi_s=-\infty$ corresponds to a non-adsorbing (neutral) surface, $\chi_s=0.1$ to a weakly, and $\chi_s=1.0$ to a strongly adsorbing surface, respectively. The value of $<R_yR_y>$ is a measure of the spatial extension of the chains in the direction normal to the surface boundary. According to Figure 9.12, for a purely neutral surface the width of the distribution function exhibits a continuous decrease as the middle point of the chain is forced to lie closer to the wall, and, in fact, exactly on the surface, the value of $<R_yR_y>$ is just about four tenths its value in the bulk.

The picture of a continuously decreasing $<R_yR_y>$ changes if the surface is able to exert even a small attraction on the polymer segments (i.e., the curve corresponding to an adsorbing strength, $\chi_s=0.10$ in k_BT units). In this case, starting from the value of unity in the bulk, $<R_yR_y>$ first increases and then, after going through a maximum value, larger than unity, decreases continuously attaining its minimum (which is lower than the one corresponding to the neutral surface) exactly on the surface. This trend is characteristic for all adsorbing strengths, the initial increase, maximum value and the subsequent minimum value on the surface being more pronounced the stronger the strength of the adsorption is—compare the curves for $\chi_s=0.10$ and 1.0. In addition, the occurrence of the maximum shifts further away from the surface with increasing strength of adsorption. These findings are consistent with the prevailing picture of the underlying physics for the problem which visualizes the macromolecules to change from a predominantly perpendicular-to-the-surface conformation to a parallel one as the surface is approached and a transition takes place between chains which have at most one surface-contact point and chains with multiple segments lying on the surface. This transition becomes more pronounced as the strength of the adsorption increases.

In Figure 9.13 we attempt to compare the predictions of the new, more rigorous approach (which uses the form of the distribution function as derived from the solution of the diffusion equation subject to the appropriate boundary conditions) with the results of the more simplistic approach followed in §9.3.4 (where the polymer statistics were described by a Gaussian distribution), in the particular case of a neutral surface. The figure clearly shows that the Gaussian assumption is quantitatively valid only for length scales commensurate with the bulk size of the polymer coil, i.e., the equilibrium end-to-end chain distance, R_e.

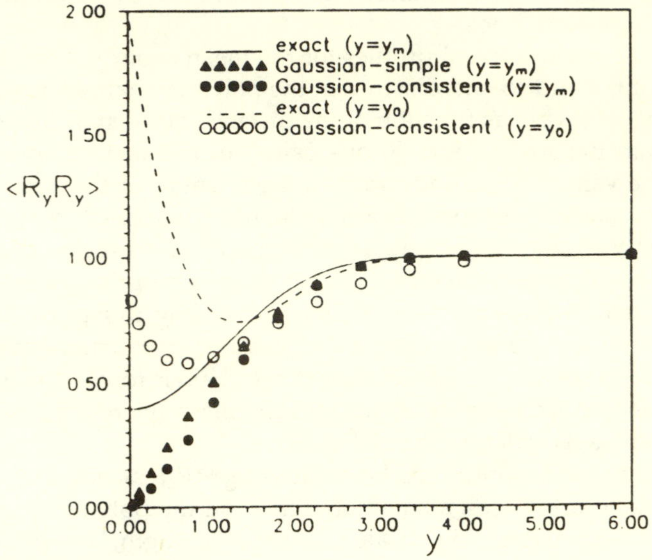

Figure 9.13: The mean square component of the end-to-end distance of the polymer chains in the direction perpendicular to the wall, $\langle R_y R_y \rangle$, as a function of the distance of either the middle or the end point of the chain from the surface, y_m or y_0, respectively, for a non-interacting (purely impermeable) wall. The lines represent the results of the new model and the points the results of the (approximate) Gaussian approach discussed in §9.3.4. The results of the latter pertaining to the middle of the chain have been calculated in two different ways: either directly as c_{yy} (labeled in the graph as Gaussian-simple) or from the distribution defined in Eq. (9.3-16) (labelled as Gaussian-consistent).

The trends in the curves are, however, qualitatively the same: at large distances from the wall, $\langle R_y R_y \rangle$ decreases monotonically as the surface is approached, and it decreases even more if the reference point is taken to be the end point rather than the middle of the chain. This trend is reversed for distances closer than about $1.5R_e$, from which point on $\langle R_y R_y \rangle$ continues to decrease if the reference point of the chain is the middle, but it starts increasing if the reference point is the end of the chain.

This change reflects the fact that if the middle of a chain is constrained to lie close to the surface, chances are that this chain will be found preferably in a configuration parallel to the wall. In contrast, if the reference point is the end, conformations which bring the chain far into the bulk dominate; these tend to favor a larger chain extension in the direction perpendicular to the wall. In fact, according to the more rigorous analysis, if the end touches the wall, the square of the chain's most probable extension is twice as big as that encountered for the same direction in the bulk. The same trends are reflected in both approaches, but the actual distribution function, becoming more asymmetric as the surface is approached, cannot possibly be represented well by the symmetric Gaussian approximation. Thus, for y approximately less than 1.5 (scaled with R_e), the results of the Gaussian approximation exhibit significant quantitative differences from those corresponding to the more rigorous approach. For even smaller distances, the Gaussian approximation appears to break down, as it is unable to capture the fine changes that happen in length scales of the order of the segment size. For example, it predicts a zero value for the width of the chains whose *middle* point is exactly on the wall (instead of 0.4), and it fails to capture the tendency for the rest of a chain whose *end* point is brought close to the wall to escape away from it. Nevertheless, it is worthwhile to mention here that the results never become aphysical and that the trends are similar; this, together with the depletion in polymer chains of the region close to the wall surface, where the differences are more pronounced, explains the good agreement with the experimental results reported in §9.3.5.

Finally a word of explanation is necessary on how the results reported in Figure 9.13 pertaining to the Gaussian approximation have been calculated (for the results corresponding to the more rigorous approach, the interested reader is referred to [Mavrantzas and Beris, 1994a-c]). As mentioned in §9.3.3B, in the Gaussian approximation the distribution function for any N-segment chain (ignoring end effects) is given as the product of three Gaussian distributions (in the space of the eigenvectors of tensor $\mathbf{c}$) defined by Eq. (9.3-15). Within the theory of random flights, this is also true for the end-to-end distribution function of any k-segment part of the chain, with the variances $c_{\xi\xi}$, $c_{\eta\eta}$, and $c_{\zeta\zeta}$ scaled accordingly (i.e., by k/N). For a constant tensor $\mathbf{c}$, these two definitions are equivalent. However, if $\mathbf{c}$ varies, an approximation is introduced, since the results of the analysis for a constant segment distribution function are used to describe multi-segment chain conformations. This is primarily the reason why the analysis reported in §9.3.3 is unable to capture changes taking place within length scales smaller than the order of the chain length. (This is referred to as Gaussian approximation in Fig. 9.13.)

The Gaussian approximation becomes worse the further away the reference point (i.e., the location to which the values of the tensor $\mathbf{c}$ refer to) is taken. Thus, for the distribution function of an N-segment chain, the best approximation is obviously the one for which the reference point is the middle one. This is exactly why we considered the results obtained with the Gaussian approximation (namely the chain density, n, and the conformation tensor, $\mathbf{c}$) as referring to the middle of the chain rather than any other point. Technically, one could refer the results to any point within the chain, since, anyhow, as for example seen in Figure 9.13, they fail to capture changes occurring within a length scale less than the end-to-end chain distance; however, based on the above discussion, the error is minimized when the reference point is taken to be the middle of the chain. Now, if in addition to that assumption, we neglect the corrections to the end-to-end distribution function introduced by the presence of the surface, $<R_y R_y>$ can be calculated simply as c_{yy}, and this represents the way the data labeled in Figure 9.3 as Gaussian-simple have been calculated. However, this approach takes also into account (as mentioned before in 9.3.3B) chain conformations which are crossing the wall. Thus, a description more consistent with the physical realities of the problem (and with the way the Hamiltonian was calculated in §9.3.3) is to exclude these conformations from calculations for the polymer statistics. This can be accomplished within the Gaussian approximation approach simply by utilizing the corrected distribution function of Eq. (9.3-16) instead of the Gaussian one, Eq. (9.3-15). When this is applied to each half of the chain, with the variance suitably adjusted so that reference to the middle of the chain is made (in accordance with the previous discussion), the results reported in Fig. 9.13 as Gaussian-consistent are obtained. For further details, see [Mavrantzas and Beris, 1994a-c].

A final point that it is worth mentioning is the close agreement between the Gaussian-simple and the Gaussian-consistent data as seen in Figure 9.13 (corresponding to the filled triangles and the filled circles, respectively). This indicates the internal consistency of the approach, despite its shortcomings in revealing the microscopic changes in quantitiative detail.

10
Non-Conventional Transport Phenomena

The only justification for our concepts is that they serve to represent the complex of our experiences; beyond this they have no legitimacy. I am convinced that the philosophers had a harmful effect upon the progress of scientific thinking in removing certain fundamental concepts, from the domain of empiricism, where they are under control, to the intangible heights of the a priori*—the universe of ideas is just as little independent of the nature of our experiences as cloths are of the form of the human body.*

—*Albert Einstein*

Après que son excellence se fut couchée, et que le secrétaire se fut approché de son visage: Il faut avouer, dit Micromégas, que la nature est bien variée.
Oui, dit le Saturnien, la nature est comme un parterre dont les fleurs...
Ah! dit l'autre, laissez là votre parterre.
Elle est, reprit le secrétaire, comme une assemblée de blondes et de brunes, dont les parures...
Eh! qu'ai-je à faire de vos brunes? dit l'autre.
Elle est donc comme une galerie de peintures dont les traits...
Eh non! dit le voyageur, encore une fois la nature est comme la nature. Pourquoi lui chercher des comparaisons?
Pour vous plaire, répondit le secrétaire.
Je ne veux point qu'on me plaise, répondit le voyageur; je veux qu'on m'instruise: commencez d'abord par me dire combien...

—*Voltaire: Micromégas*

In this chapter, we wish to exploit the availability of the bracket formalism in the description of complex, non-conventional transport phenomena. In the first section, §10.1, we analyze relaxational phenomena in heat and mass transfer. The next section, §10.2, includes the description of phase transitions in inhomogeneous media. The last section, §10.3, contains a first effort to describe inertial effects in viscoelasticity. These problems have rarely been considered in the past, and when they have it has always been from a phenomenological perspective. We explore the availability of the bracket formalism here to provide a more systematic basis for these systems than has heretofore been available, and hence we characterize the models in this chapter as semi-phenomenological. The basic approach that we use is to first establish an appropriate internal variable for the system in consideration, and then to divine an appropriate Hamil-tonian which does, in some limits, produce available phenomenological models. (The latter step indicates why we characterize the models deve-loped in this chapter as "semi-phenomenological.") As we shall see, describing the models on this more fundamental basis clears up a number of inconsistencies, as well as extending their range of validity without unduly sacrificing their simplicity.

10.1 Relaxational Phenomena in Heat and Mass Transfer

In most engineering applications[1] of heat and mass transfer, the simple linear constitutive relations of (6.4-12) are adequate in order to describe the respective transport processes. A couple of very simple examples are the heat flux, when the affinity is the temperature gradient (giving Fourier's law of heat conduction), and the mass diffusion flux, when the affinity is the chemical potential (giving Fick's law of mass diffusion). The importance of such relationships in engineering practice cannot be overestimated.

The validity of the linearized equations is generally established by *steady-state* experiments, so the question that naturally arises is whether or not the same constitutive relationship will hold for *transient* phenomena. This question cannot be answered as long as only steady-state experiments are performed.

From physical considerations alone, it is obvious that the linearized constitutive relationships cannot be complete, in and of themselves. For instance, if we consider a material initially at equilibrium and at some time we impose a finite value of the affinity, then the flux also jumps instantaneously from its equilibrium value (zero) to its steady-state value. Also, the parabolic equations resulting from constitutive relations of this form have the aphysical property that a sudden change of the affinity at some point within the material will be felt instantaneously everywhere (although with exponentially small amplitudes at distant points). Hence the parabolic equations imply infinite speeds of propagation, violating the physical principal of action at a distance, which is currently a well-accepted maxim of physics. Although this aphysical behavior does not affect most engineering applications, even transient ones, there are cases

[1]This section was written in collaboration with N. S. Kalospiros and follows closely the analysis presented in Kalospiros *et al.* [1993b].

where finite speeds of propagation are experimentally observed, such as anomalous diffusion in polymers [Crank, 1953; Neogi, 1983a,b; Camera-Roda and Sarti, 1990].

If one considers relaxational phenomena as well as those associated with transport, then both problems mentioned in the preceding paragraph are resolved. Such phenomena are associated with relaxation times which, in a loose sense, may be correlated with the time it takes a flux to respond to changes in the affinity. Furthermore, they allow for the development of discontinuities in the potential, giving rise to finite propagation velocities for temperature and concentration fronts.

In his celebrated paper of 1867, Maxwell became the first scholar to predict a heat-flux relaxation,[2] although he immediately cast out the heat flux time derivative, noting that it "...may be neglected, as the rate of conduction will rapidly establish itself [Maxwell, 1867, p. 86]." The matter has been taken up by a fair number of researchers in the last forty years, and excellent reviews of this work have appeared recently by Joseph and Preziosi [1989; 1990].

In idealized solids, the underlying physics of the concept of heat flux relaxation is related to the mechanisms by which heat is transported (conducted). In this case, thermal energy is transported by quantized electronic excitations, called free electrons, and quanta of lattice vibrations, called phonons. These quanta undergo collisions of a dissipative nature; the communication time between these collisions amounts to a relaxation time. Indeed, there is a whole spectrum of relaxation times, and, in order to explain the phenomena involved, one must consider relaxation times corresponding to modes carrying the most heat rather than a mean relaxation time [Joseph and Preziosi, 1989, p. 41].

As far as mass transfer is concerned, the presence of relaxational phenomena in the equation for mass diffusion can be justified through the basic ideas proposed by Truesdell [1962], and implicitly by Maxwell himself [1867] in his work on the kinetic theory of gases. This matter was later studied again by Müller and Villaggio [1976, §2], and a fundamental formulation of the momentum balances for interpenetrating continua also shows that the inertia of the diffusing species leads to the equivalent of a relaxation time in the mass flux equation [Aifantis, 1980, §3].

[2]According to Maxwell [1867, p. 71]: "Let Q be the value of the quantity for any particular molecule, and $<Q>$ the mean value of Q for all molecules... The quantity $<Q>$ may vary from two causes. The molecules within the element may by their mutual action or by the action of external forces produce an alteration of $<Q>$, or molecules may pass into the element and out of it, and so cause an increase or diminution of the value of $<Q>$ within it." It is this "mutual action" which in Maxwell's terms causes relaxation phenomena; see also §10.1.4.

The introduction of relaxational effects in the mass-flux relation to describe diffusion in polymers was explicitly proposed by Neogi [1983a,b] and many researchers have since elaborated upon this. However, the central idea dates back to Crank [1953], who was the first to relate mass-flux relaxation to the morphology of the system under consideration: a polymer being penetrated by a gas. It was assumed that the diffusion coefficient was concentration-history dependent, with two basic modes considered as the ones dominating the diffusion. Specifically, a penetrant molecule may jump into a "hole" which opens up in its vicinity inside of the polymer's structure, and this will, in general, leave the neighboring molecules in a state of strain. An instantaneous response to this change is attributed to the movements of individual molecular groups and small segments of the polymer chains, while the retarded response is associated with the relatively slow uncoiling and displacement of large segments of the polymer chains.

With the exception of the simple ideal gas models, such as that of Maxwell, the development of constitutive equations describing relaxational phenomena associated with heat and mass transfer is usually based upon empirical phenomenological relationships governing the corresponding fluxes. In this section, we wish to reveal the underlying common basis for relaxational phenomena in heat conduction and mass diffusion by consideration of the relaxational processes which occur in the structure of the physical system. We do this by introducing two internal variables, which, in a loose sense at least, characterize the heat and mass flux, although they do not have to be identified necessarily as such. Hence a thermodynamically consistent theoretical framework is developed for the description of transport processes influenced by relaxational changes in the internal microstructure. It is also shown herein that the bracket description provides an explanation for the observed behavior, which is consistent with the only case where a microscopic analysis is available (Maxwellian gases). In the last subsection, the potential of this method is demonstrated for bridging the gap between macroscopic models and microscopic ones in explaining anomalous mass transfer behavior in polymeric systems.

10.1.1 Flux-relaxation models

The most general linear form for heat and mass flux relaxational phenomena may be obtained through an integral equation which represents a weighted average of the corresponding affinity (here taken to be the gradient of a potential function, Φ) over the time history of the material,

$$J_\alpha = \int_{-\infty}^{t} \psi(t-t')\, \nabla_\alpha \Phi(\mathbf{x},t')\, dt' \,, \tag{10.1-1}$$

where **x** is the spatial coordinate. The kernel $\psi(s)$ in (10.1-1) is a positive, monotonically decreasing relaxation function that tends to zero as $s \rightarrow \infty$. For anisotropic materials, ψ has, in general, a tensorial character, so that (10.1-1) becomes

$$J_\alpha = \int_{-\infty}^{t} \psi_{\alpha\beta}(t-t')\, \nabla_\beta \Phi(\mathbf{x},t')\, dt' \,. \tag{10.1-2}$$

Depending upon the specific form of the kernel, one can recover various flux-relaxation models. As an example, by assuming a single, exponentially decreasing expression in time for $\psi(s)$,

$$\psi(s) = \frac{k}{\lambda} \exp\left(-\frac{s}{\lambda}\right) , \tag{10.1-3}$$

we obtain upon differentiation the equivalent differential model

$$J_\alpha + \lambda \frac{\partial J_\alpha}{\partial t} = -\, k \nabla_\alpha \Phi \,, \tag{10.1-4}$$

where λ is the relaxation time and k the corresponding diffusivity or conductivity. This was the form for the equation originally proposed by Cattaneo [1948] for heat transfer. An analogous equation was assumed by Kalospiros *et al.* [1991; 1993a] in a model describing sorption and transport phenomena in solid polymers. Pao and Banerjee [1973] have remarked that a natural generalization of (10.1-4) to anisotropic media is

$$J_\alpha + \lambda_{\alpha\beta} \frac{\partial J_\beta}{\partial t} = -\, k_{\alpha\beta} \nabla_\beta \Phi \,, \tag{10.1-5}$$

where the material parameters $\boldsymbol{\lambda}$ and $\mathbf{k}$ are positive semi-definite tensors which may depend on Φ and the morphology of the material.

When the kernel in (10.1-1) is defined as the superposition of an exponentially decreasing term and a delta function,

$$\psi(s) = k_1 \delta(s) + \frac{k_2}{\lambda} \exp\left(-\frac{s}{\lambda}\right) , \tag{10.1-6}$$

then an equation arises which is similar to the well-known model of Jeffreys [1924, §14.421-14.423][3] for the stress and strain rate in liquids:

$$J_\alpha + \lambda \frac{\partial J_\alpha}{\partial t} = -\, k \nabla_\alpha \Phi - \lambda k_1 \frac{\partial (\nabla_\alpha \Phi)}{\partial t} \,. \tag{10.1-7}$$

Here, k is the total diffusivity (or conductivity) at infinitesimally slow processes and k_1 can be interpreted as an effective diffusivity/conduct-

[3]In chapter 8, this type of model was referred to as the Oldroyd models since Oldroyd generalized Jeffreys's model to a convected coordinate system.

ivity corresponding to an instantaneous response to changes in the affinity. The relationship between k and k_1 is given by

$$k = k_1 + k_2 \ , \tag{10.1-8}$$

where k_2 is an elastic diffusivity/conductivity due to the slower relaxation modes. The Pao and Banerjee generalization is also applicable to Eqs. (10.1-7) and (10.1-8). A discussion of the physical ideas leading to (10.1-7) is given by Joseph and Preziosi [1989; 1990] for the heat transfer case. The analogous model for mass transfer was proposed by Neogi [1983a,b], and used by Camera-Roda and Sarti [1990] in their model for diffusion in polymers. The dependence of the diffusion coefficient upon the concentration history considered by Crank [1953] also results in a similar mass flux model. Inclusion of an effective diffusivity is meant to represent rapidly decaying modes in practical applications involving much longer time scales where the parabolic nature of the resulting equations and, consequently, the infinitely fast wave propagation associated with (10.1-7), is of no concern.

At this point, some comments are in order regarding the similarities between the above relaxation models for heat and mass transfer and the viscoelastic ones for the extra stress which were described in detail in chapter 8. The analogy between (10.1-4) and the Maxwell model of viscoelasticity, (8.1-10), is fairly evident. Hence the Cattaneo-type models are easily recognized as the analogues of Maxwell's equation with the conductivity k_2 corresponding to the polymer viscosity, η. As already mentioned, the Jeffreys form, (10.1-7), allows for the incorporation of an instantaneous response, similar to the Newtonian solvent viscosity of the Oldroyd models. Also notice that for the kernel of (10.1-6), the steady-state diffusivity/conductivity is $k = k_1 + k_2$. Typically, the relaxation times associated with heat and mass transfer in simple systems are so small (on the order of 10^{-10}s) that relaxational phenomena can safely be neglected.[4] On the other hand, in the case of diffusion in polymeric systems, exceedingly large relaxation times are realized, of the same or larger magnitude than the typical experimental time scale (i.e., on the order of seconds). This class of materials represents the primary motivation for the analysis of the models of relaxation phenomena presented in this section.

In addition to the relaxation models discussed above for heat conduction, there is another category of relaxation models in heat transfer originating from an effort to describe a different phenomenon, namely the second sound in dielectric crystals [Joseph and Preziosi, 1989]. These

[4]There are, however, some important exceptions, such as sintering phenomena, [Joseph and Preziosi, 1990, p. 380] where relaxational phenomena are relevant because the time scale of the process is equally small.

models have never formally been theoretically connected with the ones of the Cattaneo-type and involve a qualitatively different equation for the heat flux which does not reduce to Fourier's law even under steady-state conditions. The original equation, proposed for the description of heat waves in dielectric crystals at low temperatures, was obtained by Guyer and Krumhansl [1966, §II] through a solution of the linearized Boltzmann equation for the pure phonon field in terms of the normal-process collision operator. An interpretation of their results in terms of fluxes led to the system of two equations [Joseph and Preziosi, 1989, p. 48]

$$C_v \frac{\partial T}{\partial t} + \nabla_\alpha J_\alpha = 0 \,, \tag{10.1-9}$$

$$J_\alpha + \tau_R \frac{\partial J_\alpha}{\partial t} = -\tau_R \frac{c^2}{3} \nabla_\alpha T + \tau_R \tau_N \frac{c^2}{5} [\nabla^2 J_\alpha + 2\nabla_\alpha (\nabla_\varepsilon J_\varepsilon)] \,, \tag{10.1-10}$$

where τ_R is a relaxation time for the momentum-nonconserving processes (in which momentum is lost from the phonon system), c is the average sound speed of the phonons, and τ_N is a relaxation time for normal processes that preserve the phonon momentum. It is interesting to note that the original equations were expressed in terms of internal variables α_0 and α_1 rather than the heat flux [Guyer and Krumhansl, 1966, p. 768]. These variables characterize the non-equilibrium modification of the distribution function of the phonons. In the limit of low temperature, under the assumption of a dispersionless and isotropic phonon spectrum, these components are proportional to the local thermal-energy density (i.e., the temperature) and the heat current, respectively. Also note that under these conditions, the heat current is proportional to the momentum of the phonon gas. Finally, note that (10.1-10) does not reduce to Fourier's law under steady-state conditions and that one can have according to (10.1-10) heat flux which is not necessarily down the temperature gradient. This new type of heat transport was subsequently found experimentally in Helium IV crystals [Mezhov-Deglin, 1964, p. 1297; 1966, §5,6]. Corresponding phenomena for mass transfer, although entirely possible from a theoretical point of view, have not been reported in the literature.

10.1.2 The bracket formulation of relaxational phenomena in heat conduction

In this subsection, we are concerned with the description through the bracket formulation of relaxational phenomena associated with heat conduction only; relaxational phenomena associated with mass diffusion will be discussed separately in the next subsection. As alluded to in the introduction to this section, an internal variable, $\boldsymbol{\phi}$, is introduced in addition to the usual thermodynamic variables characterizing the system at equilibrium, i.e., the mass and entropy densities, ρ and s, respectively. If so desired, a macroscopic flow field can also be introduced following the approach detailed in chapters 5 and 7, but the analysis of its effect will be postponed until §10.1.5 in order to keep the introduction of the new ideas in this section as simple as possible. For simplicity, we also assume that the medium in consideration is incompressible, thus removing the density from the set of the dynamic variables. Hence the dynamic variables for this system are just the entropy density, s, and the internal variable, $\boldsymbol{\phi}$.

As an example of the physical interpretation of the internal variable, in the case of an idealized solid, $\boldsymbol{\phi}$ can be associated with a vector parameter describing the anisotropic deviation of the density distribution function of phonons from equilibrium [Guyer and Krumhansl, 1966, §II]. This is proportional to the momentum flux of the phonon gas. In §10.1.4, we shall consider the nature of $\boldsymbol{\phi}$ in the Maxwellian gas.

From a thermodynamic viewpoint, the internal state variable requires an equilibrium value, $\boldsymbol{\phi}^{eq}$, depending only on s; i.e., $\boldsymbol{\phi}^{eq}=\boldsymbol{\phi}^{eq}(s)$. Furthermore, $\boldsymbol{\phi}$ must have odd parity under time inversion, i.e., $\boldsymbol{\phi}(t) = -\boldsymbol{\phi}(-t)$, which arises from the physical character of the internal variable. As discussed previously, for example, in modeling the second-sound phenomena in dielectric crystals, $\boldsymbol{\phi}$ may be identified with the momentum flux of the phonons, and, therefore, it must be odd under time inversion, as is the velocity.

The general form of the Hamiltonian of systems such as these is given simply by the internal energy of the medium,

$$H[s,\boldsymbol{\phi}] = \int_\Omega u(s,\boldsymbol{\phi}) \, d^3x \, , \tag{10.1-11}$$

since we are not concerned with flow or external field effects. Since $\mathbf{M}$ is assumed to be zero, we have as an immediate consequence that $\{F,H\}=0$, which implies that $dF/dt=[F,H]$. Following the detailed guidelines presented in chapter 7, it is a simple matter to divine the most general form of the dissipation bracket which is compatible (close to equilibrium) with the variables $\boldsymbol{\phi}$ and s as

$$[F,G] = -\int_\Omega K_{\alpha\beta} \frac{\delta F}{\delta \phi_\alpha} \frac{\delta G}{\delta \phi_\beta} d^3x + \int_\Omega K_{\alpha\beta} \frac{1}{T} \frac{\delta F}{\delta s} \frac{\delta G}{\delta \phi_\alpha} \frac{\delta G}{\delta \phi_\beta} d^3x$$

$$- \int_\Omega N_{\alpha\beta\gamma\varepsilon} \nabla_\alpha \left(\frac{\delta F}{\delta \phi_\beta}\right) \nabla_\gamma \left(\frac{\delta G}{\delta \phi_\varepsilon}\right) d^3x$$

$$
\begin{aligned}
&+\int_\Omega N_{\alpha\beta\gamma\varepsilon}\frac{1}{T}\frac{\delta F}{\delta s}\nabla_\alpha(\frac{\delta G}{\delta\phi_\beta})\nabla_\gamma(\frac{\delta G}{\delta\phi_\varepsilon})\mathrm{d}^3x \\
&-\int_\Omega \Lambda_{\alpha\beta}\left[\nabla_\alpha(\frac{\delta F}{\delta s})(\frac{\delta G}{\delta\phi_\beta})-\nabla_\alpha(\frac{\delta G}{\delta s})(\frac{\delta F}{\delta\phi_\beta})\right]\mathrm{d}^3x \\
&-\int_\Omega P_{\alpha\beta}\nabla_\alpha(\frac{\delta F}{\delta s})\nabla_\beta(\frac{\delta G}{\delta s})\mathrm{d}^3x \\
&+\int_\Omega P_{\alpha\beta}\frac{1}{T}\frac{\delta F}{\delta s}\nabla_\alpha(\frac{\delta G}{\delta s})\nabla_\beta(\frac{\delta G}{\delta s})\mathrm{d}^3x \quad .
\end{aligned}
\tag{10.1-12}
$$

The phenomenological matrix **K** in the above expression is inversely proportional to a characteristic relaxation time for the internal dynamic variable $\boldsymbol{\phi}$, and $\boldsymbol{\Lambda}$ and **P** are related to the elastic and effective conductivities, respectively. In order to preserve the finite speed of propagation of heat waves, **P** should always be taken as identically zero; however (see also [Joseph and Preziosi, 1989]), in practical applications, the fast relaxation modes may be approximated using an effective (and, generally, anisotropic) conductivity, which characterizes all of the modes that have decayed within the time scale of the observer (e.g., phonon/phonon interactions in idealized solids). The elastic conductivity accounts for the slow relaxation modes, such as free-electron/phonon interactions. The terms in Eq. (10.1-12) which are proportional to **N** represent and describe the inhomogeneous structural effects, and are usually neglected as corresponding to higher-order corrections to the dissipation; however, they become important in systems where the primary dissipation is absent (**P**=**0**).

Note that the dissipation bracket defined by (10.1-12) consists of four parts, each one represented by a bilinear form in F and G and its corresponding entropy production. Also notice that the second pair of terms is antisymmetric due to the assumed parity of $\boldsymbol{\phi}$ (odd). The first three parts can be considered together as coupling the vector affinities present in the system, $\delta H/\delta\phi$ and $\nabla(\delta H/\delta s)$. Thus, the phenomenological coefficient matrices **K** and **M** should be symmetric to satisfy the Onsager relations (see §6.5). Similarly, the fourth-rank tensor **N** should also satisfy corresponding symmetry conditions.

In view of the dissipation bracket of (10.1-12), the evolution equations for the dynamic variables $\boldsymbol{\phi}$ and s are simply

$$\frac{\partial \phi_\alpha}{\partial t} = - K_{\alpha\beta} \frac{\delta H}{\delta \phi_\beta} + \Lambda_{\alpha\beta} \nabla_\beta T + \nabla_\beta [N_{\beta\alpha\gamma\varepsilon} \nabla_\gamma (\frac{\delta H}{\delta \phi_\varepsilon})] \ , \tag{10.1-13a}$$

$$\begin{aligned} \frac{\partial s}{\partial t} &= \nabla_\alpha (\Lambda_{\alpha\beta} \frac{\delta H}{\delta \phi_\beta}) + \nabla_\alpha (P_{\alpha\beta} \nabla_\beta T) + \frac{1}{T} [K_{\alpha\beta} \frac{\delta H}{\delta \phi_\alpha} \frac{\delta H}{\delta \phi_\beta} \\ &\quad + P_{\alpha\beta} (\nabla_\alpha T)(\nabla_\beta T) + N_{\alpha\beta\gamma\varepsilon} \nabla_\alpha (\frac{\delta H}{\delta \phi_\beta}) \nabla_\gamma (\frac{\delta H}{\delta \phi_\varepsilon})] \ . \end{aligned} \tag{10.1-13b}$$

By comparing (10.1-13b) with the macroscopic equation derived for the entropy density, (6.4-10), we realize that the entropy flux is simply

$$J_\alpha = - \left[\Lambda_{\alpha\beta} \frac{\delta H}{\delta \phi_\beta} + P_{\alpha\beta} \nabla_\beta T \right] , \tag{10.1-14}$$

and the local rate of entropy production is

$$\begin{aligned} \sigma = \frac{1}{T} \Bigg[& K_{\alpha\beta} \frac{\delta H}{\delta \phi_\alpha} \frac{\delta H}{\delta \phi_\beta} + P_{\alpha\beta} (\nabla_\alpha T)(\nabla_\alpha T) \\ & + N_{\alpha\beta\gamma\varepsilon} \nabla_\alpha (\frac{\delta H}{\delta \phi_\beta}) \nabla_\gamma (\frac{\delta H}{\delta \phi_\varepsilon}) \Bigg] \ . \end{aligned} \tag{10.1-15}$$

The evolution equation for the internal energy density is calculated simply from

$$\frac{\partial u}{\partial t} = \frac{\partial u}{\partial \boldsymbol{\phi}} \cdot \frac{\partial \boldsymbol{\phi}}{\partial t} + \frac{\partial u}{\partial s} \frac{\partial s}{\partial t} = \frac{\delta H}{\delta \boldsymbol{\phi}} \cdot \frac{\partial \boldsymbol{\phi}}{\partial t} + T \frac{\partial s}{\partial t} \ , \tag{10.1-16a}$$

$$\begin{aligned} \frac{\partial u}{\partial t} = \nabla_\alpha \Bigg[& \Lambda_{\alpha\beta} T \frac{\delta H}{\delta \phi_\beta} + P_{\alpha\beta} T (\nabla_\beta T) \\ & + N_{\alpha\beta\gamma\varepsilon} (\frac{\delta H}{\delta \phi_\beta}) \nabla_\gamma (\frac{\delta H}{\delta \phi_\varepsilon}) \Bigg] \ . \end{aligned} \tag{10.1-16b}$$

Another useful relation which we may derive is the one for the change in the heat of the system. As a consequence of the local equilibrium assumption for an incompressible, closed system, and in the absence of mass transfer,

$$\frac{dQ}{dt} = T\frac{\partial s}{\partial t} = -\,\boldsymbol{\nabla}\cdot\mathbf{J}^h + \sigma^h\ , \tag{10.1-17}$$

where $\mathbf{J}^h$ and σ^h are the heat flux and rate of heat production, respectively. These quantities may be evaluated by substituting (10.1-13b) in (10.1-17):

$$J_\alpha^h = -\,\Lambda_{\alpha\beta}T\frac{\delta H}{\delta\phi_\beta} - P_{\alpha\beta}T\nabla_\beta T\ , \tag{10.1-18a}$$

$$\sigma = K_{\alpha\beta}\frac{\delta H}{\delta\phi_\alpha}\frac{\delta H}{\delta\phi_\beta} + \Lambda_{\alpha\beta}(\nabla_\alpha T)\frac{\delta H}{\delta\phi_\beta} + N_{\alpha\beta\gamma\varepsilon}\nabla_\alpha\left(\frac{\delta H}{\delta\phi_\beta}\right)\nabla_\gamma\left(\frac{\delta H}{\delta\phi_\varepsilon}\right)\ . \tag{10.1-18b}$$

In order to see how the models for heat transfer discussed in the previous subsection arise through the above system of equations, we shall have to make a few assumptions about the form of the Hamiltonian. Let us assume, as is typical, that the Helmholtz free energy, $A[T,\boldsymbol{\phi}]$, takes on a Taylor expansion in terms of an equilibrium configuration, $\boldsymbol{\phi}^e = \boldsymbol{\phi}^e(T)$, and truncate after the quadratic term:

$$\boldsymbol{\phi}^e \approx \boldsymbol{\phi}^0 + \mathbf{F}(T - T^0)\ , \tag{10.1-19}$$

$$A[T,\boldsymbol{\phi}] \approx A^e[T] + \int_\Omega \tfrac{1}{2}\mathbf{Q}(T):[\boldsymbol{\phi}-\boldsymbol{\phi}^e][\boldsymbol{\phi}-\boldsymbol{\phi}^e]\ d^3x\ . \tag{10.1-20}$$

Note that $\mathbf{Q}$ is a shorthand notation for the symmetric matrix $\partial^2 a^e/\partial\boldsymbol{\phi}\partial\,\boldsymbol{\phi}$, where, as usual, $A^e = \int a^e d^3x$, and $\mathbf{F}$ is a suitable function $(\partial\boldsymbol{\phi}^{eq}/\partial T)$ with the units of $\boldsymbol{\phi}/T$. Also, note that the linear term in (10.1-20) vanishes since the free energy at equilibrium is minimized with respect to $\boldsymbol{\phi}$. Without a loss of generality, we may assume that $\boldsymbol{\phi}^0 = \mathbf{0}$ and $T^0 = 0$, so that

$$A[T,\boldsymbol{\phi}] \approx A^e[T] + \int_\Omega \tfrac{1}{2}\mathbf{Q}:[\boldsymbol{\phi}-T\mathbf{F}][\boldsymbol{\phi}-T\mathbf{F}]\ d^3x\ , \tag{10.1-21}$$

so that via the thermodynamic relations of §9.1,

$$\left.\frac{\delta H}{\delta\boldsymbol{\phi}}\right|_s = \left.\frac{\delta A}{\delta\boldsymbol{\phi}}\right|_T = \mathbf{Q}\cdot[\boldsymbol{\phi}-T\mathbf{F}]\ . \tag{10.1-22}$$

p. 407: in (10.1-18b), σ should be σ^h

Furthermore, let us assume for simplicity that **F=0**.[5]

The next step in the analysis is to make assumptions about the phenomenological matrices, in view of the fact that we have no microscopic theories to guide in their determination.[6] It is easiest to assume that the medium is isotropic, so that the most general expressions for the phenomenological matrices which are compatible with the symmetry conditions are given as

$$K_{\alpha\beta} = K\delta_{\alpha\beta}, \quad \Lambda_{\alpha\beta}, = \Lambda\delta_{\alpha\beta}, \quad P_{\alpha\beta} = P\delta_{\alpha\beta},$$
$$N_{\alpha\beta\gamma\varepsilon} = N_1(\delta_{\alpha\gamma}\delta_{\beta\varepsilon} + \delta_{\alpha\varepsilon}\delta_{\beta\gamma}) + N_2(\delta_{\alpha\beta}\delta_{\gamma\varepsilon}) \ . \qquad (10.1\text{-}23)$$

Information concerning the scalar coefficients is provided by the entropy inequality, $dS/dt = \{S,H\} \geq 0$. In general, the entropy inequality under the above assumptions is

$$K\frac{\delta H}{\delta\phi_\alpha}\frac{\delta H}{\delta\phi_\alpha} + 2N_1 H_{\alpha,\beta}H_{\alpha,\beta} + N_2 H_{\alpha,\alpha}H_{\beta,\beta} + P(\nabla_\alpha T)(\nabla_\alpha T) \geq 0 \ , \qquad (10.1\text{-}24)$$

where in the above expression we have defined a symmetric inhomogeneous term

$$H_{\alpha,\beta} \equiv \frac{1}{2}\left[\nabla_\alpha\frac{\delta H}{\delta\phi_\beta} + \nabla_\beta\frac{\delta H}{\delta\phi_\alpha}\right] . \qquad (10.1\text{-}25)$$

Since the variations in T and $\boldsymbol{\phi}$ are assumed to be independent, the second term in the inequality (10.1-24) and the rest of the terms have to be independently non-negative. Thus, we get directly the constraint $P \geq 0$ and the inequality

$$K\frac{\delta H}{\delta\phi_\alpha}\frac{\delta H}{\delta\phi_\alpha} + 2N_1 H_{\alpha,\beta}H_{\alpha,\beta} + N_2 H_{\alpha,\alpha}H_{\beta,\beta} \geq 0 \ . \qquad (10.1\text{-}26)$$

In the event that the relaxational or inhomogeneous effects vanish, inequality (10.1-26) implies that $K \geq 0$, $N_1 \geq 0$, and $2N_1 + 3N_2 \geq 0$. The last two inequalities are derived by considering separately the contributions of the traceless and isotropic parts of **H**—see (7.2-20,21).

By incorporating the above arguments into Eqs. (10.1-13a) and (10.1-18a), respectively, we obtain

$$\frac{\partial\phi_\alpha}{\partial t} = -KQ_{\alpha\beta}\phi_\beta + \Lambda\nabla_\alpha T$$
$$+ N_1\nabla^2(Q_{\alpha\beta}\phi_\beta) + (N_1 + N_2)\nabla_\alpha[\nabla_\varepsilon(Q_{\varepsilon\beta}\phi_\beta)] \ , \qquad (10.1\text{-}27)$$

[5] As we shall see in §10.1.4, **F**=**0** for a Maxwellian gas.

[6] Again, in §10.1.4 we shall derive these matrices for the Maxwellian gas.

$$J^h_\alpha = -\Lambda T Q_{\alpha\beta}\phi_\beta - PT(\nabla_\alpha T) \ . \tag{10.1-28}$$

Now if we assume that $N_1=N_2=0$, then we may solve (10.1-28) for $\boldsymbol{\phi}$,

$$\phi_\alpha = -\frac{1}{\Lambda T}Q^{-1}_{\alpha\beta}J^h_\beta + \frac{P}{\Lambda}Q^{-1}_{\alpha\beta}\nabla_\beta T \ , \tag{10.1-29}$$

and substitute this into (10.1-27) to yield, assuming that $\Lambda T\,\mathbf{Q}$ and PT are constant,

$$J^h_\alpha + \frac{1}{K}Q^{-1}_{\alpha\beta}\frac{\partial J^h_\beta}{\partial t} = -\ (P + \frac{\Lambda^2}{K})T\nabla_\alpha T - \frac{P}{K}TQ^{-1}_{\alpha\beta}\frac{\partial(\nabla_\beta T)}{\partial t} \ . \tag{10.1-30}$$

If we now assume that $\mathbf{Q}=Q\,\boldsymbol{\delta}$ (with no justification whatsoever), where Q is the trace of $\mathbf{Q}$, then (10.1-30) takes the form of Jeffreys's equation, (10.1-7). The relationships among the parameters of the two models can be determined as

$$\lambda = \frac{Q^{-1}}{K} \ , \quad k = (P + \frac{\Lambda^2}{K})T \ , \quad \text{and}$$

$$k_1 = PT \ \Rightarrow\ k_2 = \frac{\Lambda^2 T}{K} \ . \tag{10.1-31}$$

Alternatively, if we choose to set $P = 0$ and $N_1=N_2=N$, then, assuming that ΛT and $\mathbf{Q}$ are constant, we find that

$$J^h_\alpha + \frac{1}{K}Q^{-1}_{\alpha\beta}\frac{\partial J^h_\beta}{\partial t} = -\ \frac{\Lambda^2}{K}T\nabla_\alpha T + \frac{N}{K}[\nabla^2 J^h_\alpha + 2\nabla_\alpha(\nabla_\varepsilon J^h_\varepsilon)] \ , \tag{10.1-32}$$

which, again with $\mathbf{Q}=Q\,\boldsymbol{\delta}$, gives the model of Guyer and Krumhansl, (10.1-10) with the parameters determined as

$$\tau_N = \frac{5N}{3\Lambda^2 T} \ , \quad \tau_R = \frac{Q^{-1}}{K} \ , \quad \text{and} \quad c^2 = \frac{3\Lambda^2 T}{Q^{-1}} \ . \tag{10.1-33}$$

Consequently, we see that the two most often used models for relaxational heat transfer effects appear as limiting cases of the general evolution equations, (10.1-13).

10.1.3 The bracket description of relaxational phenomena in mass diffusion

Now let us consider what effects relaxational mechanisms have upon the system behavior during mass diffusion. In the previous subsection dealing with heat transfer, we assumed that the density was constant so that diffusional effects were non-existent. Similarly, in this subsection we assume that the temperature is constant so that we need not worry about any phenomena associated with heat conduction. In reality, of course, any physical transport process will involve both of these effects.

The dynamic variables for this system are the mass densities of each component, $\rho_1,\rho_2,\ldots,\rho_\nu$, the entropy density, s, and the internal variable, $\boldsymbol{\theta}$. The last of these variables characterizes the diffusional process, and may be associated with the flux of the diffusing species or, more intuitively, with the rate of change of the microscopic structure of the system. As an internal state variable, its equilibrium value depends only on the other thermodynamic variables; $\boldsymbol{\theta}^{eq}=\boldsymbol{\theta}^{eq}(s,\rho_1,\ldots,\rho_\nu)$. As was the case with $\boldsymbol{\phi}$, $\boldsymbol{\theta}$ had odd parity under time inversion; i.e., $\boldsymbol{\theta}(t)=-\boldsymbol{\theta}(-t)$.

The Hamiltonian, in this case, is generally

$$H[s,\boldsymbol{\theta},\rho_1,\ldots,\rho_\nu] = \int_\Omega u(s,\boldsymbol{\theta},\rho_1,\ldots,\rho_\nu)\, d^3x \,, \tag{10.1-34}$$

and the dissipation bracket, constructed in an analogous fashion to the heat transfer case, (10.1-12), is

$$\begin{aligned}
[F,G] = &-\int_\Omega \Phi_{\alpha\beta}\frac{\delta F}{\delta\theta_\alpha}\frac{\delta G}{\delta\theta_\beta}d^3x + \int_\Omega \Phi_{\alpha\beta}\frac{1}{T}\frac{\delta F}{\delta s}\frac{\delta G}{\delta\theta_\alpha}\frac{\delta G}{\delta\theta_\beta}d^3x \\
&-\int_\Omega \Omega_{\alpha\beta\gamma\varepsilon}\nabla_\alpha\Big(\frac{\delta F}{\delta\theta_\beta}\Big)\nabla_\gamma\Big(\frac{\delta G}{\delta\theta_\varepsilon}\Big)d^3x \\
&+\int_\Omega \Omega_{\alpha\beta\gamma\varepsilon}\frac{1}{T}\frac{\delta F}{\delta s}\nabla_\alpha\Big(\frac{\delta G}{\delta\theta_\beta}\Big)\nabla_\gamma\Big(\frac{\delta G}{\delta\theta_\varepsilon}\Big)d^3x \\
&-\sum_{i=0}^{\nu}\int_\Omega X^i_{\alpha\beta}\left[\nabla_\alpha\Big(\frac{\delta F}{\delta\rho_i}\Big)\Big(\frac{\delta G}{\delta\theta_\beta}\Big)-\nabla_\alpha\Big(\frac{\delta G}{\delta\rho_i}\Big)\Big(\frac{\delta F}{\delta\theta_\beta}\Big)\right]d^3x \\
&-\sum_{i,j=0}^{\nu}\int_\Omega \Psi^{ij}_{\alpha\beta}\nabla_\alpha\Big(\frac{\delta F}{\delta\rho_i}\Big)\nabla_\beta\Big(\frac{\delta G}{\delta\rho_j}\Big)d^3x \\
&+\sum_{i,j=0}^{\nu}\int_\Omega \Psi^{ij}_{\alpha\beta}\frac{1}{T}\frac{\delta F}{\delta s}\nabla_\alpha\Big(\frac{\delta G}{\delta\rho_i}\Big)\nabla_\beta\Big(\frac{\delta G}{\delta\rho_j}\Big)d^3x \,.
\end{aligned} \tag{10.1-35}$$

Note that since the temperature is assumed to be constant, $\boldsymbol{\nabla}(\delta H/\delta s)\equiv \boldsymbol{\nabla}T=0$. The matrix $\boldsymbol{\Phi}$ is inversely proportional to a charateristic relaxation time of the evolution of the internal state variable, $\boldsymbol{\theta}$, and $\boldsymbol{X}^i$ and $\boldsymbol{\Psi}^{ij}$, $i,j=1,2,..,\nu$, are related to the elastic and effective diffusivities, respectively. Analogously to the case of heat conduction, the $\boldsymbol{\Psi}^{ij}$ should be taken as identically zero in order to preserve the finite speed of propagation of mass waves. In practice, the $\boldsymbol{\Psi}^{ij}$ are non-zero as they are taken to represent diffusional processes with corresponding relaxation times much smaller than the observation time. Thus, for diffusion in polymers, the effective diffusivity, following Crank's arguments [Crank, 1953], is meant to represent the (almost) instantaneous response to changes in the affinity attributed to movements of individual molecular groups and small segments of chains, while the elastic diffusivity takes into account slow modes associated with the relatively slow uncoiling and displacement of large segments of the polymer chains. Moreover, not all the material parameters $\boldsymbol{X}^i$, $\boldsymbol{\Psi}^{ij}$ are independent of each other. Indeed, as is evident from the conservation of mass principle, they have to satisfy linear relationships analogous to those of §7.3:

$$\sum_{i=1}^{\nu} \boldsymbol{X}^i = 0 \quad \text{and} \quad \sum_{i=1}^{\nu} \boldsymbol{\Psi}^{ij} = 0 \;, \quad j=0,1,\ldots,\nu \;. \tag{10.1-36}$$

The fourth couple of terms in (10.1-35) represent non-homogeneous structural effects and will be neglected in the following as corresponding to higher-order corrections to the dissipation; however, they might become important in systems where the primary dissipation is absent ($\boldsymbol{\Psi}^{ij}=0$; see also a similar discussion in the previous section). Finally, also similarly to the heat flux case, the coefficients $\boldsymbol{X}^i$, $\boldsymbol{\Psi}^{ij}$, and $\boldsymbol{\Omega}$ have to satisfy suitable symmetry relations.

The evolution equations for the dynamic variables may now be easily derived:

$$\frac{\partial \theta_\alpha}{\partial t} = -\Phi_{\alpha\beta}\frac{\delta H}{\delta \theta_\beta} + \sum_{i=1}^{\nu} X^i_{\alpha\beta}\nabla_\beta \mu_i + \nabla_\beta[\Omega_{\beta\alpha\gamma\varepsilon}\nabla_\gamma(\frac{\delta H}{\delta\theta_\varepsilon})] \;, \tag{10.1-37a}$$

$$\begin{aligned}\frac{\partial s}{\partial t} = \nabla_\alpha(X^0_{\alpha\beta}\frac{\delta H}{\delta\phi_\beta}) + \sum_{j=1}^{\nu}\nabla_\alpha(\Psi^{0j}_{\alpha\beta}\nabla_\beta\mu_j) + \frac{1}{T}[\Phi_{\alpha\beta}\frac{\delta H}{\delta\theta_\alpha}\frac{\delta H}{\delta\theta_\beta} \\ + \sum_{i,j=1}^{\nu}\Psi^{ij}_{\alpha\beta}(\nabla_\alpha\mu_i)(\nabla_\beta\mu_j) + \Omega_{\alpha\beta\gamma\varepsilon}\nabla_\alpha(\frac{\delta H}{\delta\theta_\beta})\nabla_\gamma(\frac{\delta H}{\delta\theta_\varepsilon})] \;,\end{aligned} \tag{10.1-37b}$$

$$\frac{\partial \rho_i}{\partial t} = \nabla_\alpha(X^i_{\alpha\beta}\frac{\delta H}{\delta\theta_\beta}) + \sum_{j=1}^{\nu}\nabla_\alpha(\Psi^{ij}_{\alpha\beta}\nabla_\beta\mu_j) \;, \quad i=1,2,\ldots,\nu \;. \tag{10.1-37c}$$

This system of equations fully describes the mass transport within the medium with a constant temperature throughout.

Let us now assume for simplicity that the system in consideration is restricted to two components. Under this restriction, the linear relationships of (10.1-36) lead us to the equalities

$$\begin{aligned}\boldsymbol{X}^1 = -\boldsymbol{X}^2 \equiv \boldsymbol{X} \;, \quad \boldsymbol{\Psi}^{01} = -\boldsymbol{\Psi}^{02} \equiv \boldsymbol{\Psi}^0 \;, \\ \boldsymbol{\Psi}^{11} = -\boldsymbol{\Psi}^{12} = -\boldsymbol{\Psi}^{21} = \boldsymbol{\Psi}^{22} \equiv -\boldsymbol{\Psi} \;.\end{aligned} \tag{10.1-38}$$

Note that in the latter of these equalities the reciprocal relations have been employed. The dissipation bracket of (10.1-35) may now be written concisely as

$$
\begin{aligned}
[F,G] = & - \int_\Omega \Phi_{\alpha\beta} \frac{\delta F}{\delta\theta_\alpha} \frac{\delta G}{\delta\theta_\beta} \, d^3x + \int_\Omega \Phi_{\alpha\beta} \frac{1}{T} \frac{\delta F}{\delta s} \frac{\delta G}{\delta\theta_\alpha} \frac{\delta G}{\delta\theta_\beta} \, d^3x \\
& - \int_\Omega \Omega_{\alpha\beta\gamma\varepsilon} \nabla_\alpha (\frac{\delta F}{\delta\theta_\beta}) \nabla_\gamma (\frac{\delta G}{\delta\theta_\varepsilon}) \, d^3x \\
& + \int_\Omega \Omega_{\alpha\beta\gamma\varepsilon} \frac{1}{T} \frac{\delta F}{\delta s} \nabla_\alpha (\frac{\delta G}{\delta\theta_\beta}) \nabla_\gamma (\frac{\delta G}{\delta\theta_\varepsilon}) \, d^3x \\
& - \int_\Omega X_{\alpha\beta} \left[\nabla_\alpha (\frac{\delta F}{\delta\rho_1} - \frac{\delta F}{\delta\rho_2}) (\frac{\delta G}{\delta\theta_\beta}) - \nabla_\alpha (\frac{\delta G}{\delta\rho_1} - \frac{\delta G}{\delta\rho_2}) (\frac{\delta F}{\delta\theta_\beta}) \right] d^3x \\
& - \int_\Omega \Psi_{\alpha\beta} \nabla_\alpha (\frac{\delta F}{\delta\rho_1} - \frac{\delta F}{\delta\rho_2}) \nabla_\beta (\frac{\delta G}{\delta\rho_1} - \frac{\delta G}{\delta\rho_2}) \, d^3x \\
& + \int_\Omega \Psi_{\alpha\beta} \frac{1}{T} \frac{\delta F}{\delta s} \nabla_\alpha (\frac{\delta G}{\delta\rho_1} - \frac{\delta G}{\delta\rho_2}) \nabla_\beta (\frac{\delta G}{\delta\rho_1} - \frac{\delta G}{\delta\rho_2}) \, d^3x \\
& - \int_\Omega X^0_{\alpha\beta} \nabla_\alpha (\frac{\delta F}{\delta s}) \frac{\delta G}{\delta\theta_\beta} \, d^3x \\
& + \int_\Omega \Psi^0_{\alpha\beta} \nabla_\alpha (\frac{\delta F}{\delta s}) \nabla_\beta (\frac{\delta G}{\delta\rho_1} - \frac{\delta G}{\delta\rho_2}) \, d^3x \, .
\end{aligned}
\tag{10.1-39}
$$

Consequently, for the two-component mixture we find that the system of evolution equations, (10.1-37), reduces to

$$\frac{\partial\theta_\alpha}{\partial t} = -\Phi_{\alpha\beta} \frac{\delta H}{\delta\theta_\beta} + X_{\alpha\beta} \nabla_\beta (\Delta\mu) + \nabla_\beta [\Omega_{\beta\alpha\gamma\varepsilon} \nabla_\gamma (\frac{\delta H}{\delta\theta_\varepsilon})] \, , \tag{10.1-40a}$$

$$\frac{\partial\rho_1}{\partial t} = \nabla_\alpha (X_{\alpha\beta} \frac{\delta H}{\delta\theta_\beta}) + \nabla_\alpha [\Psi_{\alpha\beta} \nabla_\beta (\Delta\mu)] \, , \tag{10.1-40b}$$

$$\frac{\partial\rho_2}{\partial t} = - \nabla_\alpha (X_{\alpha\beta} \frac{\delta H}{\delta\theta_\beta}) - \nabla_\alpha [\Psi_{\alpha\beta} \nabla_\beta (\Delta\mu)] \, , \tag{10.1-40c}$$

$$\frac{\partial s}{\partial t} = \nabla_\alpha \Big(X^0_{\alpha\beta}\frac{\delta H}{\delta \phi_\beta}\Big) + \nabla_\alpha [\Psi^0_{\alpha\beta}\nabla_\beta(\Delta\mu)] + \frac{1}{T}\Big[\Phi_{\alpha\beta}\frac{\delta H}{\delta\theta_\alpha}\frac{\delta H}{\delta\theta_\beta}$$

$$+ \Psi_{\alpha\beta}\nabla_\alpha(\Delta\mu)\nabla_\beta(\Delta\mu) + \Omega_{\alpha\beta\gamma\varepsilon}\nabla_\alpha\Big(\frac{\delta H}{\delta\theta_\beta}\Big)\nabla_\gamma\Big(\frac{\delta H}{\delta\theta_\varepsilon}\Big)\Big] , \tag{10.1-40d}$$

where $\Delta\mu \equiv \mu_1 - \mu_2$. Eqs. (10.1-40b,c) imply that the mass fluxes are simply

$$J^1_\alpha = - J^2_\alpha \equiv J^m_\alpha = - X_{\alpha\beta}\frac{\delta H}{\delta\theta_\beta} - \Psi_{\alpha\beta}\nabla_\beta(\Delta\mu) . \tag{10.1-41}$$

As in the preceding subsection, let us again assume a simple phenomenological expression for the Helmholtz free energy of the system. We therefore take a Taylor expansion of $A=A[T,\boldsymbol{\theta},\rho_1,\rho_2]$ about equilibrium, where $\boldsymbol{\theta}^e=\boldsymbol{\theta}^e(T,\rho_1,\rho_2)$:

$$\boldsymbol{\theta}^e \approx \boldsymbol{\theta}^0 + \tfrac{1}{2}\mathbf{F}(\Delta\rho - \Delta\rho^0) , \tag{10.1-42}$$

$$A[T,\rho_1,\rho_2,\boldsymbol{\theta}] \approx A^e[T,\rho_1,\rho_2] + \int_\Omega \tfrac{1}{2}\mathbf{Q}:(\boldsymbol{\theta}-\boldsymbol{\theta}^e)(\boldsymbol{\theta}-\boldsymbol{\theta}^e)\,d^3x , \tag{10.1-43}$$

where $\Delta\rho\equiv\rho_1-\rho_2$, $\mathbf{F}\equiv\partial\boldsymbol{\theta}^{eq}/\partial\rho$, and $\mathbf{Q}\equiv\partial^2 a^e/\partial\boldsymbol{\theta}\partial\boldsymbol{\theta}$ is a symmetric matrix. By setting $\boldsymbol{\theta}^0=\mathbf{0}$ and $\Delta\rho^0=0$, we obtain

$$A[T,\rho_1,\rho_2,\boldsymbol{\theta}] = A^e[T,\rho_1,\rho_2] + \int_\Omega \tfrac{1}{2}\mathbf{Q}:(\boldsymbol{\theta} - \tfrac{1}{2}\mathbf{F}\Delta\rho)(\boldsymbol{\theta} - \tfrac{1}{2}\mathbf{F}\Delta\rho)\,d^3x . \tag{10.1-44}$$

Therefore,

$$\frac{\delta H}{\delta\boldsymbol{\theta}}\bigg|_{s,\rho_1,\rho_2} = \frac{\delta A}{\delta\boldsymbol{\theta}}\bigg|_{T,\rho_1,\rho_2} = \mathbf{Q}\cdot(\boldsymbol{\theta} - \tfrac{1}{2}\mathbf{F}\Delta\rho) . \tag{10.1-45}$$

As before, we now set $\mathbf{F}=\mathbf{0}$ and assume isotropic matrices, which, subject to the appropriate symmetry conditions, take a form similar to those of (10.1-23),

$$\Phi_{\alpha\beta} = \Phi\delta_{\alpha\beta} ,\quad X_{\alpha\beta} = X\delta_{\alpha\beta} ,\quad \Psi_{\alpha\beta} = \Psi\delta_{\alpha\beta} ,$$

$$\Omega_{\alpha\beta\gamma\varepsilon} = \Omega_1(\delta_{\alpha\gamma}\delta_{\beta\varepsilon} + \delta_{\alpha\varepsilon}\delta_{\beta\gamma}) + \Omega_2\delta_{\alpha\beta}\delta_{\gamma\varepsilon} , \tag{10.1-46}$$

with similar inequalities applying to the (constant) scalar coefficients. Hence (10.1-40a,41) become, respectively,

$$\frac{\partial\theta_\alpha}{\partial t} = - \Phi Q_{\alpha\beta}\theta_\beta + X\nabla_\alpha(\Delta\mu) + \Omega_1\nabla^2(Q_{\alpha\beta}\theta_\beta)$$

$$+ (\Omega_1 + \Omega_2)\nabla_\alpha[\nabla_\varepsilon(Q_{\varepsilon\beta}\theta_\beta)] , \tag{10.1-47}$$

$$J^m_\alpha = - XQ_{\alpha\beta}\theta_\beta - \Psi\nabla_\alpha(\Delta\mu)\,. \tag{10.1-48}$$

Note the complete analogy of these expressions with those of the heat transfer case of the preceding subsection.

As a special case of the general system of equations given above, if we set $\Omega_1 = \Omega_2 = 0$, and assume that $X\mathbf{Q}$ and Ψ are constant, we obtain

$$\begin{aligned} J^m_\alpha + \frac{1}{\Phi}Q^{-1}_{\alpha\beta}\frac{\partial J^m_\beta}{\partial t} &= -\left(\Psi + \frac{X^2}{\Phi}\right)\nabla_\alpha(\Delta\mu) \\ &\quad - \frac{\Psi}{\Phi}Q^{-1}_{\alpha\beta}\frac{\partial\nabla_\beta(\Delta\mu)}{\partial t}\,, \end{aligned} \tag{10.1-49}$$

which, for $\mathbf{Q}=Q\boldsymbol{\delta}$, reduces to the Jeffreys equation, with similar relationships between the parameters to those of (10.1-31). If $\mathrm{tr}\,\mathbf{Q}\to\infty$, or equivalently, if $\mathrm{tr}\,\mathbf{Q}^{-1}\to 0$ (which implies that $\boldsymbol{\theta}=\boldsymbol{\theta}^{eq}$ at all times), then Fick's law is recovered.

Alternatively, setting $\Psi=0$, $\Omega_1=\Omega_2=\Omega$ and assuming both X and $\mathbf{Q}$ to be constant, we obtain

$$J^m_\alpha + \frac{1}{\Phi}Q^{-1}_{\alpha\beta}\frac{\partial J^m_\beta}{\partial t} = -\frac{X^2}{\Phi}\nabla_\alpha(\Delta\mu) + \frac{\Omega}{\Phi}[\nabla^2 J^m_\alpha + 2\nabla_\alpha(\nabla_\varepsilon J^m_\varepsilon)]\,, \tag{10.1-50}$$

which, for $\mathbf{Q}=Q\boldsymbol{\delta}$, corresponds exactly to the mass transfer equivalent of the Guyer/Krumhansl model, (10.1-10). As far as we know, this expression has yet to find any applications in mass transfer problems.

10.1.4 *Heat and mass flux relaxation in the kinetic theory of gases*

Maxwell's classical kinetic theory of gases was one of the first (and greatest) attempts to develop a microscopic theory for describing transport processes and properties. As for relaxational effects, we have already remarked that Maxwell was the first investigator to consider relaxational effects in heat conduction, although he immediately disregarded them as occurring too rapidly to affect the essential problem. Natanson [1896, §11] demonstrated that the kinetic theory could also account for relaxational phenomena in mass diffusion. More elaborate and sophisticated treatments of the kinetic theory have since appeared; for examples, we cite those of Grad [1958] and Truesdell and Muncaster [1980].

In this subsection, we wish to present the original results obtained by Maxwell in his paper of 1867 regarding stress and heat flux relaxation. We shall also discuss how arguments similar to Maxwell's may be used in the derivation of analogous properties in the mass transfer case, and the ramifications thereof.

As in Maxwell's analysis, let us consider a system of N particles, consisting of N_1 particles of species 1, N_2 particles of species 2, etc., so that

$$N = \sum_{i=1}^{\nu} N_i \,, \tag{10.1-51}$$

where ν is the number of components in the system. The position of an individual particle, independent of species, is denoted by $\mathbf{x}$, and its velocity by $\boldsymbol{\xi}$.

According to Maxwell [1867, p. 51]:

> In the present paper I propose to consider the molecules of a gas, not as elastic spheres of definite radius, but as small bodies or groups of smaller molecules repelling one another with a force whose direction always passes very nearly through the centres of gravity of the molecules, and whose magnitude is represented very nearly by some function of the distance of the centres of gravity.

Consequently, the force, $\mathbf{X}_{jk}$, between two particles, j and k, is repulsive along the line (vector) connecting the two particles, $\mathbf{r} \equiv \mathbf{x}_j - \mathbf{x}_k$, and varies inversely as a reasonably large power of the magnitude of the distance, r, between them; i.e.,

$$\mathbf{X}_{jk} = \frac{K_{jk}\mathbf{r}}{r^{s+1}} \,, \quad s>3, \tag{10.1-52}$$

where K_{jk} is the corresponding intermolecular force constant. Obviously, $K_{jk} = -K_{kj}$. The case of the so-called "Maxwellian molecules" corresponds to $s = 5$, since as originally discovered by Maxwell, this case simplifies the analysis [1867, p. 60].[7]

Maxwell's aim was to determine the mean values of the first three moments of the mass distribution function for every species [1867, p. 55]. The 0-th moment is simply the species mass density, ρ_i, defined intuitively as

[7]As it turned out, the assumption of the repulsive force between the molecules was essential in order to make the calculation tractable. Concurrently with the work of Maxwell, Ludwig Boltzmann was attempting to solve a similar problem assuming perfectly elastic collisions through brute force calculations, a very difficult problem due to the singularities induced at the moment of impact. The introduction of the repulsive force allowed Maxwell to sidestep this issue, since he believed that the properties of a gas should be largely independent of the particular law which governed the collision of two molecules. Boltzmann later remarked on the artistic merit of Maxwell's assumption [Thomson, 1931].

$$\rho_i(\mathbf{x},t) \equiv \int f_i(\mathbf{x},\boldsymbol{\xi},t)\ d^3\xi\ , \tag{10.1-53}$$

where f_i is the mass distribution function for the i-th species. This function is basically interpreted as the probability density of finding a particle of species i at location $\mathbf{x}$ with velocity $\boldsymbol{\xi}$ at time t, multiplied by the mass of the i-th species. Hence the total number of molecules of species i is given by

$$N_i = \left(\frac{1}{m_i}\right) \int f_i(\mathbf{x},\boldsymbol{\xi},t)\ d^3\xi\ d^3x\ . \tag{10.1-54}$$

The first moment is the mass flux density for species i,

$$\mathbf{J}_i^m \equiv \rho_i \mathbf{u}_i\ , \tag{10.1-55}$$

where $\mathbf{u}_i$ is the mean velocity of the i-th species, defined as

$$\mathbf{u}_i(\mathbf{x},t) \equiv \left(\frac{1}{\rho_i}\right) \int \boldsymbol{\xi} f_i(\mathbf{x},\boldsymbol{\xi},t)\ d^3\xi\ . \tag{10.1-56}$$

The total stress tensor, $\mathbf{T}$, is defined by the second moment of the mass distribution function,[8]

$$\mathbf{T} \equiv \int \mathbf{c}\mathbf{c} f\ d^3\xi\ , \tag{10.1-57}$$

where $\mathbf{c}$ is the velocity of a particle relative to the mean, $\mathbf{c} \equiv \boldsymbol{\xi} - \mathbf{u}$. Finally, the heat flux vector, $\mathbf{q}$, is defined using the third moment of f,

$$S_{\alpha\beta\gamma} \equiv \int c_\alpha c_\beta c_\gamma f\ d^3\xi\ , \tag{10.1-58}$$

as half the contracted third moment:

$$q_\alpha \equiv \tfrac{1}{2} S_{\alpha\beta\beta} = \int \tfrac{1}{2} c_\alpha c_\beta c_\beta f\, d^3\xi = \tfrac{1}{2} \int c_\alpha c^2 f\ d^3\xi\ , \tag{10.1-59}$$

c being the magnitude of $\mathbf{c}$.

In general, knowledge of the mass distribution function f (or f_i) is required in order to evaluate the moments defined above. The distribution function can, in principle, be evaluated by solving a complex set of nonlinear integro-differential equations, the *Maxwell/Boltzmann equations*. However, analytical solutions to the Maxwell/Boltzmann equations are only available for special cases. For example, at equilibrium the corresponding solution is given by a Gaussian function, f^0, as

[8]At this point, we wish to focus on heat conduction only, so we consider a one-component system and drop the subscripts on the various quantities. We shall return later to the multi-component system.

$$f^0 = \frac{m}{(2\pi k_B T)^{3/2}} \exp\left[\frac{-mc^2}{2k_B T}\right] . \tag{10.1-60}$$

An alternative approach involves the direct evaluation of the moments from the solutions of partial-differential equations (in space) which can be obtained, as an example, from a partial integration of the Maxwell/Boltzmann equations over the velocity space. At some stage in the development of the equations, however, higher moments are always approximated as functions of the lower ones.[9] Consequently, this method always leads to approximate equations with a limited region of validity, restricted most often to conditions close to equilibrium. Nevertheless, the savings in computational effort are generally so great that they diminish the attention paid to this shortcoming.

Over the years, several approximations of the Maxwell/Boltzmann equations have been developed, a number of which are discussed by Grad [1958, §27]. The one that we use here, based upon an expansion about a locally Maxwellian distribution, is the one developed originally by Maxwell himself since it has the advantage of not excluding relaxational phenomena *a priori*.

In regards to the relaxation of the system stress, Maxwell [1867, p. 82], as well as Grad [1958, p. 272] and Truesdell [1984, p. 453], showed that if the velocity gradient is set to zero (i.e., a stationary system) for a Maxwellian gas, then the stress tensor decays exponentially in time toward its equilibrium value. The relaxation time, λ_s, characterizing this decay (i.e., appearing as the constant in the exponential rate expression) was given as a function of the mass of a particle, the density of the system, and the force constant, $K=K_{11}$, which characterizes the interaction between particles:

$$\lambda_s = \frac{1}{3}\sqrt{\frac{2m}{K}}\ \frac{m}{\rho A_2} , \tag{10.1-61}$$

where A_2 is a numerical constant equal to 1.37 (to three significant digits). A more convenient and well-known expression for λ_s is given in terms of the viscosity of the gas, μ, and its pressure,

$$\lambda_s = \frac{\mu}{p} . \tag{10.1-62}$$

This latter expression allows an easy calculation of order-of-magnitude estimates of the relaxation time. For example, Maxwell estimated a value

[9]This is similar to the practice of using decoupling approximations, as discussed in chapter 8.

of 2×10^{-10} s for λ_s for air at standard conditions (1 atm. and 17°C) [1867, p. 83].

Although the above relation is amongst the most celebrated of Maxwell's results, he was also the first researcher to derive a Cattaneo-type equation for the heat flux [1867, Eq. (143)]. Actually, the cited equation only becomes of the Cattaneo type if the density of the system is constant and use is made of the fact that Maxwellian gases obey the ideal gas law,

$$p = \frac{\rho R T}{M_w} \quad . \tag{10.1-63}$$

The corresponding heat flux relaxation time, λ_h, is of the same form as λ_s, so $\lambda_h = \lambda_s$, in Maxwell's theory. The more sophisticated formulations have shown that $\lambda_h = (3/2)\lambda_s$ [Truesdell, 1984, p. 458]. Furthermore, regarding the latter relationship, Ikenberry and Truesdell [1956, ch. VI], by proving the so-called *theorem of the trend to equilibrium*, showed that the relaxation time for any process in a Maxwellian one-component system is at most as large as the one for heat flux.

The thermal conductivity coefficient obtained by considering Maxwell's Eq. (143) is [1867, p. 87]

$$k_M = \sqrt{\frac{2m}{K}} \; \frac{p^2}{A_2 \rho^2 T (\gamma - 1)} \; , \tag{10.1-64}$$

where γ is the ratio of the heat capacity at constant pressure to the one at constant volume. Using (10.1-61) and the ideal gas relationship, the above expression becomes

$$k_M = \frac{3 \lambda_h \rho R^2}{M_w^2 (\gamma - 1)} T \; , \tag{10.1-65}$$

which states that at constant density the conductivity is proportional to the temperature.

Next we shall use some more of the fundamental results of Maxwell to demonstrate the existence of relaxational phenomena associated with the case of mass transfer in a two-component system. Natanson, as far back as 1896, was the first to show the existence of a mass flux relaxation time for a Maxwellian gas in his extension of the kinetic theory to include dissipation. However, this treatment of mass flux relaxation remained largely ignored until about eighty-five years after its original publication [Guminski, 1980; Sieniutycz, 1981]. An independent development was proposed by Sandler and Dahler [1964], who studied the special case of equimolar counter-diffusion in a mixture of two very similar chemical species. Consequently, since the kinetic theory description of relaxational phenomena in mass transfer does not seem to be widely known, we shall try to provide a more elaborate analysis of it than the one presented above for the more commonly known case of relaxational phenomena in heat transfer. Some of this analysis is similar to that of Sandler and Dahler [1964] and serves as an example of how mass flux relaxational phenomena may be described, even using the original Maxwell equations.

In order to avoid complications extraneous to the central issue, we constrain ourselves to the case of a two-component mixture of Maxwellian molecules. The two species behave very similar chemically, with equivalent molecular weights, M_w, and with intermolecular forces so nearly the same that the physical properties of the system are essentially independent of composition. Furthermore, we restrict our attention to the special circumstance of equimolar counter-diffusion; i.e.,

$$\rho \mathbf{u} = \rho_1 \mathbf{u}_1 + \rho_2 \mathbf{u}_2 = \mathbf{0} , \tag{10.1-66}$$

where the densities and velocities are defined according to Eqs. (10.1-53) and (10.1.56), respectively. The continuity equation for species 1 has the form

$$\frac{\partial \rho_1}{\partial t} + \boldsymbol{\nabla} \cdot (\rho_1 \mathbf{u}_1) = 0 , \tag{10.1-67}$$

with a similar equation for species 2. The addition of these two equations under the realization of (10.1-66) demonstrates that ρ is independent of time, depending only upon $\mathbf{x}$.

The momentum balance for species 1 (with an analogous equation for species 2) is given as [Bearman and Kirkwood, 1958, p. 143; Bataille and Kestin, 1977, p. 51]

$$\rho_1 \frac{\mathrm{d}_1 \mathbf{u}_1}{\mathrm{d}_1 t} + \boldsymbol{\nabla} \cdot [\mathbf{T}_1 - \rho_1 (\mathbf{u}_1 - \mathbf{u})(\mathbf{u}_1 - \mathbf{u})] = \rho_1 \mathbf{F} + \mathbf{f}_1 , \tag{10.1-68}$$

where $\mathrm{d}_1 \cdot / \mathrm{d}_1 t$ denotes the material derivative with respect to $\mathbf{u}_1$,

$$\frac{\mathrm{d}_1 \cdot}{\mathrm{d}_1 t} = \frac{\partial \cdot}{\partial t} + \mathbf{u}_1 \cdot \frac{\partial \cdot}{\partial \mathbf{x}} , \tag{10.1-69}$$

$\mathbf{T}_1$ is a partial stress tensor, $\mathbf{F}_1$ characterizes forces which are external to the system (per unit mass), and $\mathbf{f}_1$ represents the force (per unit volume) due to the motion resistance presented by molecules of species 2. Note that the total stress tensor is given as [Bataille and Kestin, 1977, p. 51]

$$\mathbf{T} = \mathbf{T}_1 + \mathbf{T}_2 + \rho_1 (\mathbf{u}_1 - \mathbf{u})(\mathbf{u}_1 - \mathbf{u}) + \rho_2 (\mathbf{u}_2 - \mathbf{u})(\mathbf{u}_2 - \mathbf{u}). \tag{10.1-70}$$

In the case of a binary mixture, the $\mathbf{f}_i$ are given as [Maxwell, 1867, p. 73]

$$\mathbf{f}_1 = k A_1 \rho_1 \rho_2 (\mathbf{u}_2 - \mathbf{u}_1) = - \mathbf{f}_2 , \tag{10.1-71}$$

where A_1 is a numerical constant equal to 2.66, and k is given as

$$k = \sqrt{\frac{K^{12}}{m_1 m_2 (m_1 + m_2)}} \ . \tag{10.1-72}$$

For the case at hand, $m_1=m_2=m$ and $K^{12}=K^{11}=K^{22}\equiv K$, so that (10.1-72) becomes

$$k = \frac{1}{m}\sqrt{\frac{K}{2m}} \ . \tag{10.1-73}$$

In addition, under the special circumstances of the case at hand, it is reasonable to let $\mathbf{T}_i = p_i \boldsymbol{\delta}$, where p_i is the partial pressure of species i. Consequently, in the absence of external forces and through using the definition of the diffusive flux of species i, (10.1-55), along with (10.1-68,71), we find that

$$\frac{\partial \mathbf{J}_1^m}{\partial t} + \boldsymbol{\nabla} p_1 = -\,(k A_1 \rho)\mathbf{J}_1^m \ , \tag{10.1-74}$$

with a similar expression for species 2.

Since $\mathbf{J}_1^m = -\mathbf{J}_2^m$, we can multiply (10.1-74) for each species by $1/\rho_1$ and $1/\rho_2$, respectively, and thereby obtain by subtraction and the use of the ideal gas relationship for the chemical potential,

$$d\mu_1 = \frac{M_w}{\rho_1} dp_1 \ , \tag{10.1-75}$$

the expression

$$\mathbf{J}_1^m + \lambda_m \frac{\partial \mathbf{J}_1^m}{\partial t} = -\,D\,\boldsymbol{\nabla}(\Delta\mu) \ , \tag{10.1-76}$$

with

$$\lambda_m \equiv \frac{1}{k A_1 \rho} \quad \text{and} \quad D \equiv \left(\frac{\lambda_m}{RTM_w}\right) \rho \chi_1 \chi_2 \ . \tag{10.1-77}$$

The mole fraction of species i is χ_i, λ_m is the mass flux relaxation time, and D is the diffusion coefficient using $\Delta\mu$ is the appropriate driving force. Note that (10.1-76) is of the same type as the Cattaneo version of (10.1-49) with Ψ=0. In addition, note that the expression for the diffusion coefficient changes if the density of species 1 is used as the driving force rather than the difference in the chemical potentials. In this case, (10.1-74) becomes

$$\mathbf{J}_1^m + \lambda_m \frac{\partial \mathbf{J}_1^m}{\partial t} = -\,D^*\,\boldsymbol{\nabla}\rho_1 \ , \tag{10.1-78}$$

where the corresponding diffusion coefficient is given by

$$D^* \equiv \frac{RT\lambda_m}{M_w} \tag{10.1-79}$$

This is the same as the one reported by Natanson [1896, p. 403] and Sandler and Dahler [1964, p. 1744].

The declaration of (10.1-79) allows for estimations of the mass flux relaxation times in various systems. For example, Cussler [1984, p. 106] lists a value of 0.779 cm^2/s for the diffusion coefficient of H_2 in N_2 at 24.2°C and 1 atm. Through the definition of D^*, we calculate a value of about 10^{-11}s for λ_m, using the ideal gas law to evaluate the density of the mixture. As expected, mass-flux relaxation times in gases with approximately Maxwellian behavior are very small.

All of the above results were obtained by a direct consideration of the original Maxwell kinetic theory of gases. In all cases, the relaxation terms in the equations for the fluxes are due to the collisions between particles of the various species, which result in an exchange of momentum and energy. Ergo, a correspondence between the macroscopic fluxes and internal structural variables (in the spirit of §10.1.2,3) may be established in a straightforward fashion. Table 10.1 summarizes the results of this subsection.

An interesting comment can be made regarding the relation of each one of these phenomena to a *different structural variable* of the Maxwellian gas. The fundamental molecular mechanism for relaxation is associated in all cases with the effect of the collision process on the evolution of the corresponding structural variables. These *different* rates of evolution are characterized then by *different* relaxation times. The latter may possibly be related to each other through the underlying collision process. Regarding the value of the above-mentioned analysis, it allows us to physically interpret and directly relate the relevant phenomenological coefficients of the macroscopic models developed in §10.1.2 and 10.1.3 to the diffusivities and relaxation times obtained through the arguments of kinetic theory.

Of course, in the case of gases, relaxation phenomena are very fast and unimportant to (most) engineering applications. However, within more complex systems, such as polymeric solids, relaxation times can be determined on the order of several seconds or larger. Unfortunately, a microscopic theory, analogous to the kinetic theory presented above, which would have allowed us to obtain a microscopic physical picture of the macroscopic phenomena, is still missing. In its absence, in the next subsection an overview of the relevant experimental data is presented as well as a discussion of the possible microscopic causes in an effort to generate enough interest for a thorough analysis, in a similar vein to that presented above. The macroscopic system is described below.

Type of flux, J	Character of flux	Corresponding structural variable, Λ	Proportionality constant, A, $J = A\Lambda$	Relaxation time from Maxwell's theory
Stress, $\mathbf{T}$	tensor	$\int \mathbf{c}\mathbf{c} f d^3\xi$	1	$\frac{1}{3}\sqrt{\frac{2m}{K}}\frac{m}{\rho A_2}$
Heat flux, $\mathbf{q}$	vector	$\int \mathbf{c} c^2 f d^3\xi$	$\frac{1}{2}$	$\frac{1}{3}\sqrt{\frac{2m}{K}}\frac{m}{\rho A_2}$ †
Mass flux, $\mathbf{J}_1^m$	vector	$\int \xi f_1 d^3\xi$	1	$\sqrt{\frac{2m}{K}}\frac{m}{\rho A_1}$

†Note that the relaxation time for heat flux has a coefficient of ½ instead of ⅓ in the more sophisticated formulations [Truesdell, 1984, p. 458].

TABLE 10.1: Characteristic parameters for relaxational effects in transport phenomena for Maxwellian gases.

10.1.5 Relaxational mass transfer in polymeric systems

In this subsection, theoretical models are contrasted against experimental evidence where relaxational phenomena in mass transfer are detected with relaxation times on the order of seconds or more. In contrast to mass transfer, all known relaxation times for heat transfer are several orders of magnitude smaller and, consequently, are much less important to engineering applications. The mass transfer results involve the diffusion of small molecules through polymeric solids, and frequently (but not always) are accompanied by other structural phenomena such as swelling, crystallization, and mechanical deformation of the polymer matrix.

Relaxational phenomena in glassy polymers being penetrated by low molecular-weight compounds manifest themselves in the experimentally observed, anomalous kinetic behavior of the weight, M, of a solute absorbed into a sample. The polymer sample is in the form of a rectangular slab of thickness X, and it is exposed at both surfaces at time zero to a gas or liquid phase containing the low molecular-weight solute at a specific (constant) fugacity. Consideration of Fick's law as the appropriate constitutive equation for mass flux and of a constant diffusivity leads to two theorems concerning the type of experiment described above:

1. Plots of $M(t)/M_\infty$ versus t/X^2 for different sample sizes should superimpose. Here $M_\infty \equiv \lim_{t\to\infty}[M(t)]$ is a measure of the solubility at the assigned outside fugacity;

2. $dM/dt \to \infty$ as $t \to 0$, corresponding to $M \propto t^{1/2}$ in the neighborhood of time zero.

Instead, with many solid polymers various anomalous types of diffusion behavior are observed, i.e., behavior different from the one predicted by classical diffusion theory. Here, we focus attention on three major types of such experimental results: two-stage sorption, Case II transport, and sorption overshoot. Of these, we will be concerned mainly with the last one. A schematic representation of the corresponding weight sorption curves, along with the one corresponding to Fickian diffusion, is shown in Figure 10.1.

In two-stage sorption, as seen in Figure 10.1b, the weight uptake is characterized by an initial Fickian-like (classical) behavior until an apparent equilibrium is reached, only to subsequently grow at a much smaller rate to a significantly larger final equilibrium solute content [Bagley and Long, 1955, pp. 2173,74]. It is usually observed in sequential

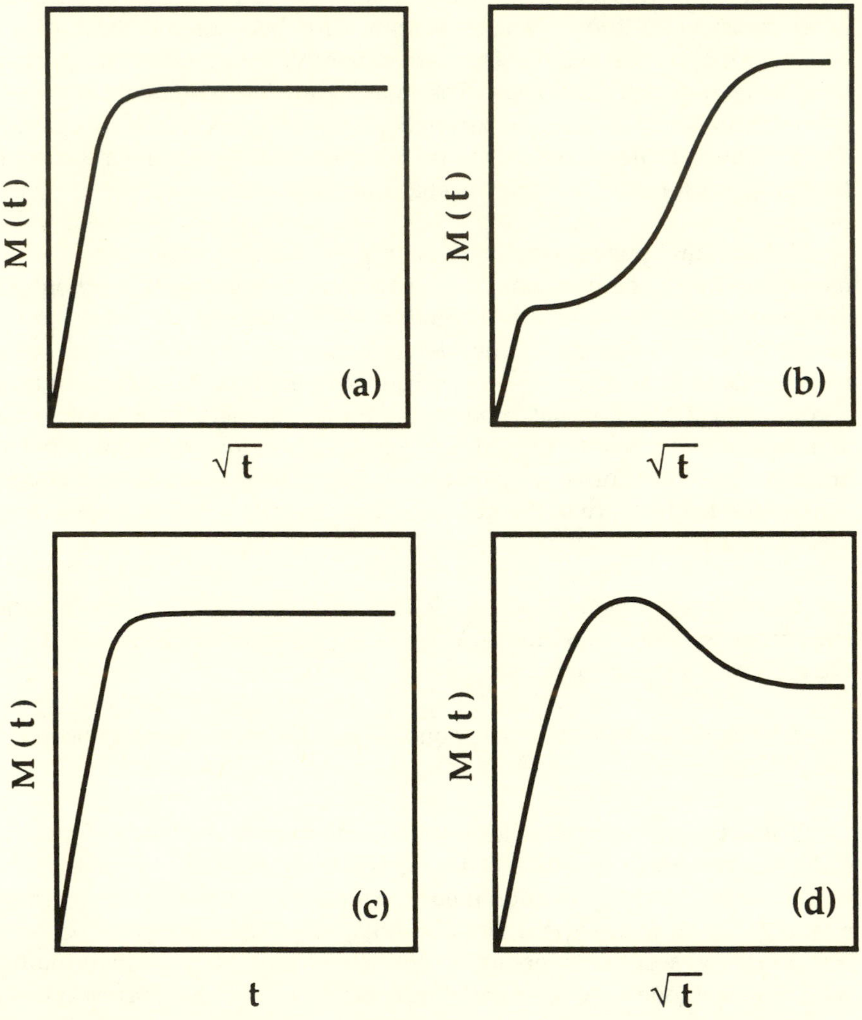

Figure 10.1: Sorption experiments with polymers. Different types for weight sorption curves are presented here schematically for a) Fickian diffusion, b) two stage sorption, c) Case II transport, and d) sorption overshoot. Note that the ordinate of graph c is different than that of the others.

experiments, where the gas-phase fugacity of the solute is increased through a series of constant values. As shown in Figure 10.1c, it is also possible that solute transport exhibits what is often called Case II transport, in which dM/dt in a neighborhood of time zero has been observed to be finite, so that M is proportional to t, often up to the time of complete saturation of the sample [Thomas and Windle, 1978, pp. 257,58]. In the same case, concentration fronts moving with an essentially constant velocity through the polymeric sample have been experimentally observed [Thomas and Windle, 1978, p. 257]. As sample size increases, the behavior predicted by classical diffusion theory is more closely approached.

The most striking experimental observation is "overshoot:" the amount of solute absorbed reaches a maximum and subsequently decreases toward its final equilibrium value (see Figure 10.1d). In some instances, this effect has been attributed to crystallization occurring at time scales larger than the time scale of the experiment [e.g., Titow *et al.*, 1974, pp. 876,79,80,83,85; Ware *et al.*, 1981, pp. 2975,77,85,87]. However, in well-documented cases [Faulkner *et al.*, 1977, p. 1134; Vrentas *et al.*, 1984, pp. 400-06; Smith and Peppas, 1985, p. 573; Peppas and Urdahl, 1988, p. 15], overshoots are observed in the absence of crystallization. Although some of these data may be attributable to irreversible morphological changes other than crystallization, this is not the case with the data reported by Vrentas *et al.* [1984, p. 405]. These authors observed overshoot in a repeat experiment on the very same sample, showing that the kinetic behavior associated with the overshoot has to be sought in some other type of mechanism.

In the same paper, Vrentas *et al.* [1984, p. 403] also propose a possible explanation, based on slow relaxation processes in the polymer/solute system: solute is absorbed into the polymeric matrix before the chains have a chance to completely relax. The structural rearrangements experienced by the sample as the polymer chains eventually reorient themselves can lead to rejection of some of the solute out of the slab. The exclusion of penetrant will become even more pronounced in instances when the diffusion rate, characterized by the usual diffusion time, is faster than the rate of chain relaxation. The latter is characterized by a relaxation time, a measure of which is given, according to Vrentas *et al.* [1984, p. 403], by the mechanical relaxation time, λ_m, defined as

$$\lambda_m \equiv \int_0^\infty t\,G(t)\,dt \;/\; \int_0^\infty G(t)\,dt \;, \tag{10.1-80}$$

where $G(t)$ is the shear relaxation modulus.

In general, λ_m depends upon concentration. When the concentration "window" of the experiment is suitably small, an average value of λ_m can be safely used. The ratio of λ_m to the diffusion time was called the diffusion Deborah number by Vrentas *et al.* [1984, p. 402]. For the system poly(ethyl-methacrylate)/ethylbenzene at 120°C, Vrentas *et al.* [1984, p. 404] reported values of the Deborah number in the range of 0.1 to 1000. The larger the value of this number, the more pronounced the overshoot phenomenon. Values of the Deborah number were also reported for the system polystyrene/ethylbenzene at 160°C, and were in the range of 10^7 to 10^{-2} [Vrentas *et al.*, 1984, p. 404]. No overshoots were reported for this system. In this case, no relaxation phenomena are actually observed and the system follows typical Fickian behavior. Large values of the Deborah number may also correspond to Case II transport behavior, suggesting that the latter can also be attributed to a mechanism closely related to the one characterizing sorption overshoots.

Here, we stress that in our view the mechanical relaxation time is only a measure of the corresponding mass-flux relaxation time, with a large value of the former probably implying a relatively large value for the latter. As discussed in §10.1.4, the two times, although related through a fundamental molecular relaxation mechanism, most probably characterize the evolution of *different* internal variables.

Vrentas *et al.'s* idea is one out of two that presently can be used to explain the observed overshoots (the other one being mass transfer relaxation, discussed in §10.1.3). Very simply, it works as follows. For thin samples (where the overshoot phenomena have been observed), it is assumed that the diffusion time t_D (i.e., the time necessary for a uniform concentration to be established throughout the sample) is much smaller than the characteristic time for the one relevant to polymer structure relaxation, λ_m; i.e., the diffusion Deborah number is assumed to be high. Thus, as a first-order approximation, the concentration changes uniformly within the sample in such a way that the corresponding chemical potential for the solute, μ, within the sample is always at equilibrium with the experimental, externally imposed value, μ_{ex}. In general, for a slowly relaxing structure, the chemical potential μ can be considered as a function of both thermodynamic equilibrium variables (the solute density, ρ, for example) and dynamic variables characterizing the relaxation process. Let us, for simplicity, use one non-equilibrium variable, θ, in addition to the solute density ρ so that $\mu \equiv \mu(\rho,\theta)$. Then, in order to observe an overshoot upon, say, a step increase of the chemical potential outside the sample, from μ_1 to μ_2, it suffices to have such a constitutive relation, $\mu=\mu(\rho,\theta)$, or equivalently, $\rho=\rho(\mu,\theta)$, that the concentration established within the sample at short times, $t \approx t_D \ll \lambda$, $\rho_D=\rho(\mu_2,\theta_1)$, is larger than the concentration observed at long times, $t \gg \lambda$, $\rho_2=\rho(\mu_2,\theta_2)$. This is certainly possible in a variety of ways. However, it ultimately will have to rely on an (as yet unverified) arbitrary hypothesis concerning the specific functional dependence of the chemical potential, μ, on the density, ρ, and the non-equilibrium parameter, θ.

p. 426: 7 lines down, 10^7 should be 10^{-7}

EXAMPLE 10.1: Show how it is possible using Vrentas *et al.'s* idea to explain the overshoot observed in sorption experiments involving thin samples of rubber in a consistent fashion with the available (at equilibrium) experimental results. For example, Kamiya *et al.* [1989, p. 882] have observed that the equilibrium uptake of CO_2 in PEMA rubber is proportional to the external pressure, i.e., a Henry's law type of behavior

$$\text{concentration} = Hp \ ,$$

is observed, where H is the Henry's constant. Since the relation of p to f, where f is the fugacity, is observed to be almost linear for similar systems under similar conditions [Berens and Huvard, 1987, Fig. 11], we can safely use as a first approximation a similar (linear) relationship between the concentration and f, and thus between ρ and f

$$\rho = H'f \ ,$$

where H' is another (positive) constant.

To achieve this goal it suffices to assume a simple relaxation model for θ:

$$\frac{d\theta}{dt} = \frac{[\theta_e(\rho) - \theta]}{\lambda} \ ,$$

where θ_e represents the equilibrium value for the structural parameter θ, together with two constitutive relations for f and θ_e

$$\theta_e = B\rho^\beta \ ,$$

$$f(\rho,\theta) = C\rho^\gamma\theta^\varepsilon \ ,$$

where B, C, β, γ, ε are, as yet, undetermined coefficients. The fugacity, rather than the chemical potential, is used here solely as a matter of convenience, the two being related in a monotonic fashion

$$\mu = RT \ln f + \mu_0 \ ,$$

where μ_0 is a reference chemical potential, constant under the conditions of the experiments. The condition at equilibrium provides one relation between the exponents β, γ, and ε,

$$1 = \gamma + \beta\varepsilon \ ,$$

whereas the solution for the intermediate density, ρ_D,

$$\frac{\rho_D}{\rho_2} = \left(\frac{\rho_2}{\rho_1}\right)^{\frac{\beta\varepsilon}{\gamma}} \ ,$$

which is obtained by equating the fugacities before and after the structure equilibration, requires as a necessary condition to observe an overshoot after a step increase on the external concentration $\beta\varepsilon/\gamma > 0$. ■

Yet, there are insufficient experimental data available to test further the above model. The most important prediction is that under conditions when an overshoot is observed after step increases of concentration ($\beta\varepsilon/\gamma > 0$), an undershoot is predicted as well for step decreases. However, if necessary, the original model can be modified (for example, by assuming a concentration-dependent relaxation time so that the step decrease process is not a high Deborah number process) to eliminate undershoots, at the expense of the model's simplicity.

Alternatively, macroscopic models have been developed that can account for the kinetic behaviors related to the relaxational phenomena mentioned above [Camera-Roda and Sarti, 1990, pp. 852-54; Kalospiros *et al.*, 1991, pp. 852-54]. The main feature of both models is the use of viscoelastic constitutive equations for the mass flux (a Cattaneo-type equation, Eq. (4a), in Kalospiros *et al.* and a Jeffreys-type one, Eq. (6), in Camera-Roda and Sarti). When the relaxation times characterizing the evolution of the mass flux are comparable to or larger than the experimental time scales—i.e., when the appropriate Deborah number is large—both Case II transport as well as sorption overshoot are observed. Inclusion of relaxation phenomena in the mass flux equation also results in the prediction of sharp concentration fronts moving inside the polymeric film. Specifically, in the model of Kalospiros *et al.*, concentration discontinuities are obtained due to the hyperbolicity of the resulting equations. Jeffreys-type equations, such as the one considered by Camera-Roda and Sarti, yield parabolic systems of partial differential equations, which do not admit discontinuous solutions. With appropriate choice of the parameters, however, sharp concentration profiles can be obtained. A sample of the predictions obtained with the model of Kalospiros *et al.* is shown in Figure 10.2 for the Case II transport and sorption overshoot.

Despite the success of the models referred to above, the choice of the appropriate mass-flux equation was empirically based on the viscoelastic nature of the polymer/penetrant systems studied. In this context, these models are phenomenological ones, with the relevant parameters masking the coupling of the elementary phenomena involved. No relationship between the relaxation phenomena included in their formulation and the (possibly complicated) morphology of the systems they wish to describe was established. A useful approach that sets the context for revealing such relationships and assigns specific physical meaning to the model parameters is the bracket formulation described in §10.1.3.

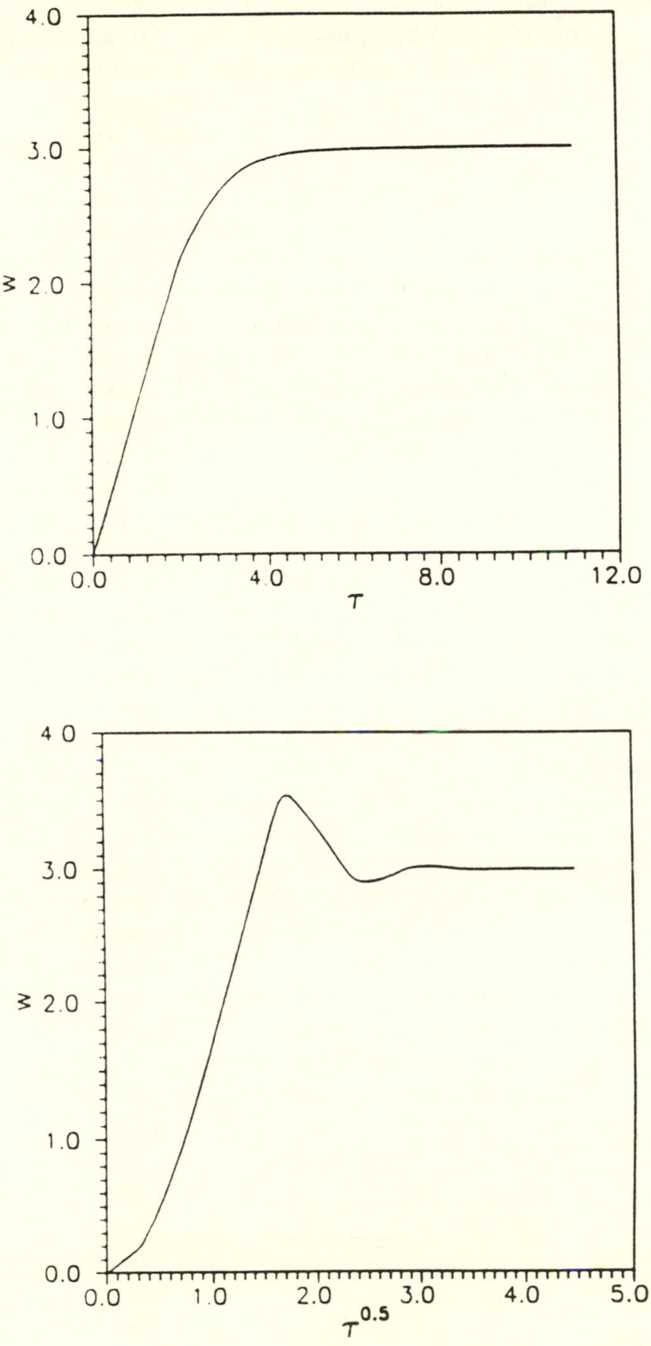

Figure 10.2: Predictions obtained with the model of Kalospiros *et al.* [1991].

Additional relaxational mechanisms that are pertinent to the viscoelastic nature of polymers have been proposed. A prominent one is the mechanism proposed by Crank [1953, §II], alluded to in the introduction. Crank considered a time-dependent diffusion coefficient, D, in Fick's law, which is given by

$$\left.\frac{\partial D}{\partial t}\right|_x = \frac{\partial D_i}{\partial c}\left.\frac{\partial c}{\partial t}\right|_x + \alpha(D_e - D) \ , \tag{10.1-81}$$

with D_i being the part of D which [Crank, 1953, p. 152] "can change instantaneously, is a function of concentration (c) only, and changes when concentration changes but not otherwise." The equilibrium diffusion coefficient is denoted by D_e and α is the rate parameter (the inverse of a relaxation time) which controls the approach to equilibrium. The above expression is equivalent to considering a Jeffreys-type equation for mass flux, with

$$\lambda = \alpha^{-1} \ , \quad k_1 = D \ , \quad \text{and} \quad k_2 = \left(\frac{1}{\alpha}\right)\left(\frac{\partial D}{\partial t}\right)_x \ . \tag{10.1-82}$$

Using this model, Crank was able to obtain Case II transport behavior.

The physical rationale Crank provides in support of the above model is as follows. A penetrant molecule may jump into a "hole" that opens up in its vicinity in the polymer structure, thus leaving the neighboring molecules in a state of strain. The slow relieving of this strain by the uncoiling of the polymer segments as the polymer structure seeks an equilibrium configuration of minimum energy, consistent with the presence of penetrant molecules, can cause a slow drift of the diffusion coefficient, described by the second term on the right-hand side of (10.1-81). The instantaneous response in the same equation is attributed to movements of individual molecular groups and small segments of chains. Note that Crank implicitly considered the diffusion coefficient itself as an internal variable, characterizing the local morphology of the system.

Indeed, following Crank's suggestion, it is possible to obtain his model provided by (10.1-81) within the bracket formalism informally as follows. Let us assume as an internal scalar variable the "elastic" part of diffusivity, D_p, which relaxes according to the equation

$$\frac{\partial D_p}{\partial t} = \alpha[D_e(c) - D_i(c) - D_p] \ , \tag{10.1-83}$$

where D_e, D_i, and α have been defined through (10.1-81). Then, it is straightforward to see that (10.1-81) arises from a dissipation bracket of the form

$$[F,G] = -\int_\Omega [D_i(c) + D_p]\left[\nabla_\alpha\left(\frac{\delta F}{\delta\rho_1} - \frac{\delta F}{\delta\rho_2}\right)\nabla_\beta\left(\frac{\delta G}{\delta\rho_1} - \frac{\delta G}{\delta\rho_2}\right)\right] d^3x$$
$$- \int_\Omega \alpha \frac{\delta F}{\delta D_p}\frac{\delta G}{\delta D_p}\, d^3x + \text{entropy production terms.} \tag{10.1-84}$$

Application of the usual rules and use of an effective Hamiltonian involving the quadratic term $(\alpha/2)[D_e(c)-D_i(c)-D_p]^2$ provides (10.1-83) as the time evolution equation for the internal variable, D_p, and an expression for the mass flux, $\mathbf{J}^m$,

$$J^m = [D_i(c) + D_p]\,\boldsymbol{\nabla}(\Lambda\mu) \quad , \tag{10.1-85}$$

corresponding to the total diffusivity $D=D_i+D_p$, in accordance with (10.1-81).

Kalospiros *et al.* [1991] were also able to model two-stage sorption by introducing an additional internal variable, the degree of swelling. In our view, the latter is related to the internal viscous stress in the model of Durning *et al.* [1985] and Durning and Tabor [1986] (see discussion below), which also accounts for two-stage sorption. In both cases, the relevant internal variables (degree of swelling and internal viscous stress) are assumed to follow first-order kinetics (i.e., Maxwell-type relationships are used to describe their evolution).

The effect of crystallization kinetics on the weight uptake kinetic behavior of polymer/diluent systems was studied by introducing another internal variable, the degree of crystallization [Kalospiros *et al.*, 1993a]. This type of relaxation phenomena, as well as the one related to the degree of swelling [Kalospiros *et al.*, 1991], can be accounted for and the thermodynamic consistency of the constitutive equations assumed for the rate of change of the internal variables can be proven by considering such additional variables in the bracket formulation and introducing the relevant terms in the respective bracket.

Several other models, relying on the coupling between the diffusion and other transport phenomena, have also been proposed. Durning *et al.* [1985, pp. 172,73] and Durning and Tabor [1986, pp. 2221-25] considered the coupling of mass flux to the internal viscous stress of the polymer/penetrant system. When stress relaxation is assumed (i.e., a Maxwell-type equation is considered), two-stage sorption behavior is recovered. In the case of Newtonian behavior for the stress, their model coincides with the one proposed by Thomas and Windle, [1982, pp. 531-32], which accounts for Case II transport. Notice that, in both models, no mass flux relaxation is taken into consideration and Fick's law is utilized.

The type of models described in the preceding paragraph have recently been integrated with the relaxational ones using ideas from extended irreversible thermodynamics (EIT)[10] by Jou *et al.* [1991, §2]. The main result of this work was the development of a set of three equations coupling the mass flux, the viscous pressure, p^v, and the traceless, viscous-stress tensor, $\mathbf{T}^v$. These equations are [Jou *et al.*, 1991, p. 3598][11]

$$\lambda_1 \frac{d\mathbf{J}_1^m}{dt} + \mathbf{J}_1^m = -D\,\boldsymbol{\nabla}(\Delta\mu) - DT\,\boldsymbol{\nabla}\cdot(\beta_2 \mathbf{T}^v) - DT\,\boldsymbol{\nabla}(\beta_0 p^v)\,, \tag{10.1-86a}$$

$$\lambda_0 \frac{dp^v}{dt} + p^v = \kappa\,\boldsymbol{\nabla}\cdot\mathbf{v} - \kappa T\beta_0\,\boldsymbol{\nabla}\cdot\mathbf{J}_1^m\,, \tag{10.1-86b}$$

$$\lambda_2 \frac{d\mathbf{T}^v}{dt} + \mathbf{T}^v = 2\mu <\boldsymbol{\nabla}\mathbf{v}> - 2\mu T\beta_2 <\boldsymbol{\nabla}\mathbf{J}_1^m>\,. \tag{10.1-86c}$$

In the above expressions, $<\mathbf{a}>$ denotes the symmetric and traceless component of $\mathbf{a}$, as defined by $\mathbf{a}^z$ in §6.4.3. Also, D is a (positive) coefficient related to the diffusivity, κ and μ are the bulk and shear viscosities, β_0 and β_2 are scalar weighting coefficients for the corresponding entropy flux (associated with the viscous pressure and stress tensor), and λ_0, λ_1, and λ_2 are (positive) relaxation times. Notice that if the viscous pressure and viscous-stress tensor are expressed in terms of the total-stress tensor (which, in this subsection, does not include the thermodynamic pressure),

$$p^v = \tfrac{1}{3}\mathrm{tr}\mathbf{T} \quad \text{and} \quad \mathbf{T}^v = <\mathbf{T}>\,, \tag{10.1-87}$$

then (10.1-86) may be written solely in terms of $\mathbf{J}_1^m$ and $\mathbf{T}$:

$$\lambda_1 \frac{d\mathbf{J}_1^m}{dt} + \mathbf{J}_1^m = -D\,\boldsymbol{\nabla}(\Delta\mu) - DT\,\boldsymbol{\nabla}\cdot(\mathbf{Z}:\mathbf{T})\,, \tag{10.1-88a}$$

$$\mathbf{W}:\frac{d\mathbf{T}}{dt} + \mathbf{T} = \mathbf{N}:<\boldsymbol{\nabla}\mathbf{v}> - T\mathbf{N}:(\mathbf{Z}:\boldsymbol{\nabla}\mathbf{J}_1^m)\,. \tag{10.1-88b}$$

The coefficient matrices in the above expressions are defined as

[10]Note that the theory expounded upon in this book is completely divorced from what is known as "extended irreversible thermodynamics" [Jou *et al.*, 1988]. We pass no judgements upon EIT at this point in time.

[11]Note that our sign convention on the stress is opposite to the one used by these authors.

$$Z_{\alpha\beta\gamma\varepsilon} = \beta_2\left[\tfrac{1}{2}(\delta_{\alpha\gamma}\delta_{\beta\varepsilon} + \delta_{\beta\gamma}\delta_{\alpha\varepsilon}) - \tfrac{1}{3}\delta_{\alpha\beta}\delta_{\gamma\varepsilon}\right] + \beta_0\tfrac{1}{3}\delta_{\alpha\beta}\delta_{\gamma\varepsilon} , \tag{10.1-89a}$$

$$W_{\alpha\beta\gamma\varepsilon} = \lambda_2\left[\tfrac{1}{2}(\delta_{\alpha\gamma}\delta_{\beta\varepsilon} + \delta_{\beta\gamma}\delta_{\alpha\varepsilon}) - \tfrac{1}{3}\delta_{\alpha\beta}\delta_{\gamma\varepsilon}\right] + \lambda_0\tfrac{1}{3}\delta_{\alpha\beta}\delta_{\gamma\varepsilon} , \tag{10.1-89b}$$

$$N_{\alpha\beta\gamma\varepsilon} = 2\mu\left[\tfrac{1}{2}(\delta_{\alpha\gamma}\delta_{\beta\varepsilon} + \delta_{\beta\gamma}\delta_{\alpha\varepsilon}) - \tfrac{1}{3}\delta_{\alpha\beta}\delta_{\gamma\varepsilon}\right] + \kappa\delta_{\alpha\beta}\delta_{\gamma\varepsilon} , \tag{10.1-89c}$$

which have the most general forms which are symmetric with respect to the indices α and β.

In the bracket description of relaxational effects in polymeric systems, we must specify the dynamic variables as ρ_1, the polymer-substrate density, ρ_2, the penetrant density, s, $\mathbf{M}=\rho\mathbf{v}$, $\mathbf{C}=\rho_1\mathbf{c}$, and $\boldsymbol{\theta}$: $F= F[\rho_1, \rho_2, s, \mathbf{M}, \mathbf{C}, \boldsymbol{\theta}]$. In terms of these variables, the Poisson bracket is

$$\begin{aligned}
\{F,G\} = &-\sum_{i=1}^{2}\int_\Omega \left[\frac{\delta F}{\delta \rho_i}\nabla_\beta\left(\frac{\delta G}{\delta M_\beta}\rho_i\right) - \frac{\delta G}{\delta \rho_i}\nabla_\beta\left(\frac{\delta F}{\delta M_\beta}\rho_i\right)\right] d^3x \\
&-\int_\Omega \left[\frac{\delta F}{\delta M_\alpha}\nabla_\beta\left(\frac{\delta G}{\delta M_\beta}M_\alpha\right) - \frac{\delta G}{\delta M_\alpha}\nabla_\beta\left(\frac{\delta F}{\delta M_\beta}M_\alpha\right)\right] d^3x \\
&-\int_\Omega \left[\frac{\delta F}{\delta \theta_\alpha}\nabla_\beta\left(\frac{\delta G}{\delta M_\beta}\theta_\alpha\right) - \frac{\delta G}{\delta \theta_\alpha}\nabla_\beta\left(\frac{\delta F}{\delta M_\beta}\theta_\alpha\right)\right] d^3x \\
&-\int_\Omega \left[\frac{\delta F}{\delta s}\nabla_\beta\left(\frac{\delta G}{\delta M_\beta}s\right) - \frac{\delta G}{\delta s}\nabla_\beta\left(\frac{\delta F}{\delta M_\beta}s\right)\right] d^3x \\
&-\int_\Omega \left[\frac{\delta F}{\delta C_{\alpha\beta}}\nabla_\gamma\left(\frac{\delta G}{\delta M_\gamma}C_{\alpha\beta}\right) - \frac{\delta G}{\delta C_{\alpha\beta}}\nabla_\gamma\left(\frac{\delta F}{\delta M_\beta}C_{\alpha\beta}\right)\right] d^3x \\
&-\int_\Omega C_{\alpha\beta}\left[\frac{\delta G}{\delta C_{\gamma\beta}}\nabla_\alpha\left(\frac{\delta F}{\delta M_\gamma}\right) - \frac{\delta F}{\delta C_{\gamma\beta}}\nabla_\alpha\left(\frac{\delta G}{\delta M_\gamma}\right)\right] d^3x \\
&-\int_\Omega C_{\alpha\beta}\left[\frac{\delta G}{\delta C_{\gamma\alpha}}\nabla_\beta\left(\frac{\delta F}{\delta M_\gamma}\right) - \frac{\delta F}{\delta C_{\gamma\alpha}}\nabla_\beta\left(\frac{\delta G}{\delta M_\gamma}\right)\right] d^3x ,
\end{aligned} \tag{10.1-90}$$

where the new (third) integral appears to describe the convection of the internal variable $\boldsymbol{\theta}$. Note that this variable is "inertially objective," rather than materially objective like **C**, analogously to the momentum density. The dissipation bracket, again assuming a homogeneous system temperature, is given likewise as

$$
\begin{aligned}
[F,G] = & -\int_\Omega \Phi_{\alpha\beta} \frac{\delta F}{\delta \theta_\alpha} \frac{\delta G}{\delta \theta_\beta} d^3x + \int_\Omega \Phi_{\alpha\beta} \frac{1}{T} \frac{\delta F}{\delta s} \frac{\delta G}{\delta \theta_\alpha} \frac{\delta G}{\delta \theta_\beta} \mathrm{d}^3x \\
& - \int_\Omega R_{\alpha\beta\gamma\varepsilon} \nabla_\alpha (\frac{\delta F}{\delta M_\beta}) \nabla_\gamma (\frac{\delta G}{\delta M_\varepsilon}) \mathrm{d}^3x \\
& + \int_\Omega R_{\alpha\beta\gamma\varepsilon} \frac{1}{T} \frac{\delta F}{\delta s} \nabla_\alpha (\frac{\delta G}{\delta M_\beta}) \nabla_\gamma (\frac{\delta G}{\delta M_\varepsilon}) \mathrm{d}^3x \\
& - \int_\Omega X_{\alpha\beta} \left[\nabla_\alpha (\frac{\delta F}{\delta \rho_1} - \frac{\delta F}{\delta \rho_2}) (\frac{\delta G}{\delta \theta_\beta}) - \nabla_\alpha (\frac{\delta G}{\delta \rho_1} - \frac{\delta G}{\delta \rho_2}) (\frac{\delta F}{\delta \theta_\beta}) \right] \mathrm{d}^3x \\
& - \int_\Omega \Psi_{\alpha\beta} \nabla_\alpha (\frac{\delta F}{\delta \rho_1} - \frac{\delta F}{\delta \rho_2}) \nabla_\beta (\frac{\delta G}{\delta \rho_1} - \frac{\delta G}{\delta \rho_2}) \mathrm{d}^3x \\
& + \int_\Omega \Psi_{\alpha\beta} \frac{1}{T} \frac{\delta F}{\delta s} \nabla_\alpha (\frac{\delta G}{\delta \rho_1} - \frac{\delta G}{\delta \rho_2}) \nabla_\beta (\frac{\delta G}{\delta \rho_1} - \frac{\delta G}{\delta \rho_2}) \mathrm{d}^3x \\
& - \int_\Omega X^0_{\alpha\beta} \nabla_\alpha (\frac{\delta F}{\delta s}) \frac{\delta G}{\delta \theta_\beta} \mathrm{d}^3x \\
& + \int_\Omega \Psi^0_{\alpha\beta} \nabla_\alpha (\frac{\delta F}{\delta s}) \nabla_\beta (\frac{\delta G}{\delta \rho_1} - \frac{\delta G}{\delta \rho_2}) \mathrm{d}^3x \\
& - \int_\Omega \Lambda_{\alpha\beta\gamma\varepsilon} \frac{\delta F}{\delta C_{\alpha\beta}} \frac{\delta F}{\delta C_{\alpha\beta}} \mathrm{d}^3x \\
& + \int_\Omega \Lambda_{\alpha\beta\gamma\varepsilon} \frac{1}{T} \frac{\delta F}{\delta s} \frac{\delta G}{\delta C_{\alpha\beta}} \frac{\delta F}{\delta C_{\alpha\beta}} \mathrm{d}^3x \\
& - \int_\Omega \Theta_{\alpha\beta\gamma\varepsilon} \left[\nabla_\alpha (\frac{\delta F}{\delta \theta_\beta}) \frac{\delta G}{\delta C_{\gamma\varepsilon}} - \nabla_\alpha (\frac{\delta G}{\delta \theta_\beta}) \frac{\delta F}{\delta C_{\gamma\varepsilon}} \right] \mathrm{d}^3x \; .
\end{aligned}
\tag{10.1-91}
$$

As usual, the phenomenological matrices appearing in the above integrals are not known *a priori*, in the absence of a detailed microscopic theory as discussed in the previous subsection (under which condition this analysis offers nothing of value, unless one is interested in reducing the number of degrees of freedom to a computationally tractable total). Consequently, we shall later have to make some assumptions regarding these coefficients.[12]

The above brackets lead directly to the coupled set of evolution equations (with $H=H[\rho_1,\rho_2,s,\mathbf{M},\mathbf{C},\boldsymbol{\theta}]$ and no external fields)

$$\frac{\partial\theta_\alpha}{\partial t} = -\nabla_\beta(v_\beta\theta_\alpha) - \Phi_{\alpha\beta}\frac{\delta H}{\delta\theta_\beta} + X_{\beta\alpha}\nabla_\beta(\Delta\mu) - \nabla_\beta\left(\Theta_{\beta\alpha\gamma\varepsilon}\frac{\delta H}{\delta C_{\gamma\varepsilon}}\right), \tag{10.1-92a}$$

$$\frac{\partial\rho_1}{\partial t} = -\nabla_\beta(v_\beta\rho_1) + \nabla_\alpha(X_{\alpha\beta}\frac{\delta H}{\delta\theta_\beta}) + \nabla_\alpha[\Psi_{\alpha\beta}\nabla_\beta(\Delta\mu)], \tag{10.1-92b}$$

$$\frac{\partial\rho_2}{\partial t} = -\nabla_\beta(v_\beta\rho_2) - \nabla_\alpha(X_{\alpha\beta}\frac{\delta H}{\delta\theta_\beta}) - \nabla_\alpha[\Psi_{\alpha\beta}\nabla_\beta(\Delta\mu)], \tag{10.1-92c}$$

$$\rho(\frac{\partial v_\alpha}{\partial t} + v_\beta\nabla_\beta v_\alpha) = -\nabla_\alpha p + 2\nabla_\beta(C_{\beta\gamma}\frac{\delta H}{\delta C_{\gamma\alpha}}) + \nabla_\beta(R_{\alpha\beta\gamma\varepsilon}\nabla_\gamma v_\varepsilon), \tag{10.1-92d}$$

$$\frac{\partial C_{\alpha\beta}}{\partial t} = -\nabla_\gamma(v_\gamma C_{\alpha\beta}) + C_{\gamma\beta}\nabla_\gamma v_\alpha + C_{\alpha\gamma}\nabla_\gamma v_\beta - \Lambda_{\alpha\beta\gamma\varepsilon}\frac{\delta H}{\delta C_{\gamma\varepsilon}} - \Theta_{\gamma\varepsilon\alpha\beta}\nabla_\gamma\frac{\delta H}{\delta\theta_\varepsilon}, \tag{10.1-92e}$$

[12]The attentive reader will have noticed by now that we do not always incorporate all of the dependencies required by (7.1-19) into the dissipation bracket. For isotropic media, the coupling in the dissipation bracket is restricted by the Curie principle, but, in general, this restriction is not evident. Ergo, sometimes we exclude certain terms from the dissipation bracket which we feel do not contribute to the physics of the situation. Whether or not we are justified in doing so is a matter which posterity will have to judge, as we shall continue doing it for simplicity, where applicable, in the remainder of this book.

$$\frac{\partial s}{\partial t} = -\nabla_\beta(v_\beta s) + \frac{1}{T}\left[\Phi_{\alpha\beta}\frac{\delta H}{\delta\theta_\alpha}\frac{\delta H}{\delta\theta_\beta} + \Psi_{\alpha\beta}\nabla_\alpha(\Delta\mu)\nabla_\beta(\Delta\mu)\right.$$

$$\left. + \Lambda_{\alpha\beta\gamma\varepsilon}\frac{\delta H}{\delta C_{\alpha\beta}}\frac{\delta H}{\delta C_{\gamma\varepsilon}} + R_{\alpha\beta\gamma\varepsilon}(\nabla_\alpha v_\beta)(\nabla_\gamma v_\varepsilon)\right] \tag{10.1-92f}$$

$$+ \nabla_\alpha(X^0_{\alpha\beta}\frac{\delta H}{\delta\theta_\beta}) + \nabla_\alpha[\Psi^0_{\alpha\beta}\nabla_\beta(\Delta\mu)] \ .$$

Note that the thermodynamic pressure is defined in the usual fashion as

$$p = -u + \sum_{i=1}^{2}\rho_i\frac{\partial u}{\partial \rho_i} + s\frac{\partial u}{\partial s} + \boldsymbol{\theta}\cdot\frac{\partial u}{\partial \boldsymbol{\theta}} + \mathbf{C}:\frac{\partial u}{\partial \mathbf{C}} \ . \tag{10.1-93}$$

Also note that Eqs. (10.1-92b,c) imply that the mass fluxes are the same as given by (10.1-41).

As before, for the sake of similar arguments, let us make use of the thermodynamic relationships

$$\left.\frac{\partial u}{\partial \mathbf{C}}\right|_{s,\rho_1,\rho_2,\boldsymbol{\theta}} = \left.\frac{\partial a}{\partial \mathbf{C}}\right|_{T,\rho_1,\rho_2,\boldsymbol{\theta}} , \tag{10.1-94a}$$

$$\left.\frac{\partial u}{\partial \boldsymbol{\theta}}\right|_{s,\rho_1,\rho_2,\mathbf{C}} = \left.\frac{\partial a}{\partial \boldsymbol{\theta}}\right|_{T,\rho_1,\rho_2,\mathbf{C}} , \tag{10.1-94b}$$

and let us assume that the Helmholtz free energy density may be written as the sum of two terms,

$$a(T,\rho_1,\rho_2,\boldsymbol{\theta},\mathbf{C}) = a_v(T,\rho_1,\rho_2,\mathbf{C}) + \tfrac{1}{2}\mathbf{Q}:\boldsymbol{\theta}\boldsymbol{\theta} \ , \tag{10.1-95}$$

with the first term being, in the first approximation, the free energy density in the absence of $\boldsymbol{\theta}$ as a dynamic variable, and the second term being the Taylor expansion, as in §10.1.2 and §10.1.3. As a particular example, let us consider the case of the Helmholtz free energy density corresponding to the Maxwell model,

$$a_v = \frac{\alpha K}{2}\mathrm{tr}\mathbf{C} - \frac{\alpha\rho_1 k_B T}{2}\ln\det(\frac{K\mathbf{C}}{\rho_1 k_B T}) + a_0(\rho_1,\rho_2,T) \ . \tag{10.1-96}$$

In the above expression, remember that K is the Hookean spring constant, which may be temperature dependent, and α is the measure of the elasticity per mass. The third term on the right-hand side of this equation represents the usual free energy density in the absence of elastic effects. Given the above relationship, we may calculate immediately

$$\frac{\partial a}{\partial \mathbf{C}} = \frac{\alpha K}{2}\boldsymbol{\delta} - \frac{\alpha \rho_1 k_B T}{2}\mathbf{C}^{-1} \quad \text{and} \quad \frac{\partial a}{\partial \boldsymbol{\theta}} = \mathbf{Q}\cdot\boldsymbol{\theta}\,. \tag{10.1-97}$$

In the following, we shall again set $\mathbf{Q}=Q\boldsymbol{\delta}$ where Q is a positive constant.

As a first approximation of the phenomenological matrices, let us assume that they are isotropic and that there is no direct (Fickian) diffusion or viscous (Newtonian) stress contribution; i.e.,

$$\boldsymbol{\Phi} = \Phi\boldsymbol{\delta}\,,\ \ \mathbf{X} = X\boldsymbol{\delta}\,,\ \ \mathbf{X}^0 = X^0\boldsymbol{\delta}\,,\ \ \boldsymbol{\Psi}=0\,,\ \ \boldsymbol{\Psi}^0=0\,,\ \ \mathbf{R}=0\,, \tag{10.1-98}$$

where $\boldsymbol{\Phi}$ (like Q) is a positive constant (X can be either positive or negative). As for the remaining two matrices, $\boldsymbol{\Lambda}$ and $\boldsymbol{\Theta}$, we choose $\boldsymbol{\Lambda}$ as the corresponding matrix for the Maxwell model generalized for a compressible medium,

$$\Lambda_{\alpha\beta\gamma\varepsilon} = \frac{1}{2\alpha K\lambda_2^c}(C_{\alpha\gamma}\delta_{\beta\varepsilon}+C_{\alpha\varepsilon}\delta_{\beta\gamma}+C_{\beta\gamma}\delta_{\alpha\varepsilon}+C_{\beta\varepsilon}\delta_{\gamma\alpha}-\tfrac{4}{3}\delta_{\alpha\beta}C_{\gamma\varepsilon}) + \frac{2}{3\alpha K\lambda_0^c}\delta_{\alpha\beta}C_{\gamma\varepsilon}\,, \tag{10.1-99}$$

where λ_0^c, λ_2^c are two relaxation times corresponding to the relaxation of the trace and the traceless part of the conformation tensor (i.e., with respect to the first two indices), respectively.[13] The matrix $\boldsymbol{\Theta}$ we choose as the analogous expression to (10.1-99),

$$\Theta_{\alpha\beta\gamma\varepsilon} = \Theta_2(C_{\alpha\gamma}\delta_{\beta\varepsilon} + C_{\alpha\varepsilon}\delta_{\beta\gamma} + C_{\beta\gamma}\delta_{\alpha\varepsilon} + C_{\beta\varepsilon}\delta_{\gamma\alpha} - \tfrac{4}{3}\delta_{\alpha\beta}C_{\gamma\varepsilon}) + \Theta_0\delta_{\alpha\beta}C_{\gamma\varepsilon}\,. \tag{10.1-100}$$

In light of the above simplifications, we observe that

$$\theta_\alpha = -\frac{J_\alpha^m}{XQ}\,, \tag{10.1-101}$$

so that, if we choose a suitable scaling such that

[13]From a mathematical perspective, it is apparent that the matrix $\boldsymbol{\Lambda}$ of chapter 8 might also have an associated second relaxation time. As it turned out, however, all of the models therein considered had $\lambda_0^c \equiv 0$. λ_0^c should probably appear in the $\boldsymbol{\Lambda}$ when used in §9.1, depending upon the particular system in consideration.

$$X = -\frac{1}{Q}\,, \tag{10.1-102}$$

we can identify θ_α with the mass flux J^m_a. Substitution of Eqs. (10.1-94) through (10.1-102) into (10.1-92) results in:

$$\frac{\partial J^1_\alpha}{\partial t} = -\nabla_\beta(v_\beta J^1_\alpha) - \Phi Q J^1_\alpha - \frac{1}{Q}\nabla_\alpha(\Delta\mu) - \nabla_\beta\left[\Theta_{\beta\alpha\gamma\varepsilon}\frac{\delta H}{\delta C_{\gamma\varepsilon}}\right], \tag{10.1-103a}$$

$$\frac{\partial C_{\alpha\beta}}{\partial t} = -\nabla_\gamma(v_\gamma C_{\alpha\beta}) + C_{\gamma\beta}\nabla_\gamma v_\alpha + C_{\alpha\gamma}\nabla_\gamma v_\beta - \Theta_{\gamma\varepsilon\alpha\beta}\nabla_\gamma(QJ^1_\varepsilon) - \frac{1}{\lambda_2}(C_{\alpha\beta} - \tfrac{1}{3}C_{\gamma\gamma}\delta_{\alpha\beta}) - \frac{1}{3\lambda_0}C_{\gamma\gamma}\delta_{\alpha\beta} + \frac{\rho_1 k_B T}{K\lambda_0}\delta_{\alpha\beta}\,, \tag{10.1-103b}$$

and if we substitute for the internal variable **C** its equivalent expression

$$C_{\alpha\beta} = \frac{1}{\alpha K}T_{\alpha\beta} + \frac{\rho_1 k_B T}{K}\delta_{\alpha\beta}\,, \tag{10.1-104}$$

then we get (for isothermal conditions)

$$\frac{1}{\Phi Q}\left(\frac{DJ^1_\alpha}{Dt} + J^1_\alpha\nabla_\beta v_\beta\right) + J^1_\alpha = -\frac{1}{\Phi Q^2}\nabla_\alpha(\Delta\mu) - \frac{T}{\Phi Q^2}\nabla_\beta\left[\frac{2Q\Theta_2}{T}(\tau_{\alpha\beta} - \tfrac{1}{3}\tau_{\gamma\gamma}\delta_{\alpha\beta}) + \frac{2Q\Theta_0}{3T}\delta_{\alpha\beta}\tau_{\gamma\gamma}\right], \tag{10.1-105a}$$

$$\frac{\mathcal{D}\tau_{\alpha\beta}}{\mathcal{D}t} + \frac{1}{\lambda_2}(\tau_{\alpha\beta} - \tfrac{1}{3}\tau_{\gamma\gamma}\delta_{\alpha\beta}) + \frac{1}{3\lambda_0}\tau_{\gamma\gamma}\delta_{\alpha\beta} = \rho_1\nu k_B T(\nabla_\alpha v_\beta + \nabla_\beta v_\alpha) - \nu K\Theta_{\gamma\varepsilon\alpha\beta}\nabla_\gamma(QJ^1_\varepsilon) + \nu k_B T\,\delta_{\alpha\beta}\nabla_\alpha J^1_\alpha\,, \tag{10.1-105b}$$

where

$$\frac{\mathcal{D}\tau_{\alpha\beta}}{\mathcal{D}t} \equiv \frac{\partial\tau_{\alpha\beta}}{\partial t} + \nabla_\gamma(v_\gamma\tau_{\alpha\beta}) + \tau_{\gamma\beta}\nabla_\gamma v_\alpha + \tau_{\alpha\gamma}\nabla_\gamma v_\beta\,, \tag{10.1-106}$$

denotes the Oldroyd derivative [1950] of the stress tensor.[14] By premultiplying (10.1-105b) with the fourth-rank tensor **Z**, defined by (10.1-89a), it is straightforward to show that the system of Eq. (10.1-105) can be brought in the form proposed by Jou *et al.* (10.1-88), at the limit close to equilibrium, where

$$C_{\alpha\beta} = \left(\frac{\rho_1 k_\beta T}{K}\right) \delta_{\alpha\beta} \quad , \tag{10.1-107}$$

with the exception of two additional terms arising in our formulation: a density correction to the mass flux equation (the second term on the left-hand side of Eq. (10.1-105a)), and a correction to the stress equation (last term on the right-hand side of Eq. (10.1-105b)).

By direct comparison of the resulting system of equations from Eqs. (10.1-105) with the equations proposed by Jou *et al.* [1991], (10.1-88), their parameters are expressed in terms of the coefficients introduced in the bracket formulation as

$$\lambda_1 = \frac{1}{\Phi Q} \quad , \quad D = \frac{1}{\Phi Q^2} \quad , \tag{10.1-108a}$$

$$\beta_2 = -\frac{\Theta_2 Q}{T} \quad , \quad \beta_0 = -\frac{2\Theta_0 Q}{T} \quad , \tag{10.1-108b}$$

$$\mu = \rho_1 \nu k_B T \lambda_2 \quad , \quad \kappa = 2\rho_1 \nu k_B T \lambda_0 \quad . \tag{10.1-108c}$$

Note that the above relations imply that the ratio of the viscosities κ/μ is equal to twice the ratio of the corresponding relaxation times, λ_0/λ_2. Although this correlation has been derived by using a specific viscoelastic model (the elastic dumbbell), it is believed that it holds even when a more complicated viscoelastic model is used, as long as only two relaxation times are involved. Thus, we can see a crucial advantage of the bracket approach versus the extended irreversible thermodynamics used by Jou *et al.*, enabling the establishment of correlations and thus achieving a reduction of the number of arbitrary parameters necessarily introduced for the description of the problem. This is a feature that is seen on many occasions, as, for example, in the scalar/vector description

[14]This derivative has also been called the "Truesdell derivative" [Eringen, 1962, pp. 253,60] as utilized by Truesdell in the description of anelastic, compressible solids. However, it is clearly a particular case of the general codeformational time-derivative of Oldroyd [1950].

of liquid crystals in §11.6.2. Moreover, since the model equations proposed by Jou *et al.* [1991] are a generalization of the previous models that coupled viscoelastic stress relaxation and mass flux [Thomas and Windle, 1978, 1982; Durning *et al.*, 1985; Durning and Tabor, 1986], the latter can also be considered as particular cases of the general equations, (10.1-92), derived from the bracket formalism.

An important point concerning the comparison of the results from the above analysis (as well as its simpler no-flow subcase considered in §10.1.3) with those of the Crank and Vrentas/Duda models, discussed previously, is that although two-stage sorption and Case II transport can be predicted from Crank's model and that sorption overshoot can be described with the model presented in Example 10.1, both phenomena can be captured by the same model only when direct mass-flux relaxational phenomena are considered—see Figure 10.2. The common mechanism between Case II transport and sorption overshoot, revealed by the Deborah number measurements of Vrentas *et al.* [1984, p. 404], has also been shown to be related exactly to these phenomena. Indeed, the models containing mass-flux relaxation times are the only ones predicting both Case II transport and sorption overshoot for large Deborah numbers.

In fairness, we need to caution the reader that there are still unanswered questions related to whether or not it is appropriate to use mass-flux relaxational models with large relaxation times in explaining the diffusion of solutes in polymers. More specifically, the fact that the internal variables which are necessarily introduced (for example, $\boldsymbol{\theta}$) are inertial in nature with an odd parity raises the question of how it is possible to associate a large relaxation time with them. This question is really related to the overall problem of introducing inertial phenomena in polymer fluid dynamics. It is only recently that awareness of this issue has arisen (see, for example, discussion on the Hinch and Giesekus 2-mode models, §8.1.1.C and §8.2.2.C.2, respectively). A first attempt to describe the importance of inertial phenomena in the development and change of the internal structure within a viscoelastic medium during flow is offered in §10.3. Clearly, however, more work is warranted in order to clarify all the issues related to this subject.

In conclusion, the most significant contribution of the above theoretical development is the unification of all relaxation processes associated with transport phenomena. Granted, the bracket formalism cannot distinguish *a priori* which one of the many available mechanisms for transport is more important and dominates a particular application. However, it offers the flexibility to accommodate the mechanics under a common theoretical framework. By modeling relaxation phenomena associated with heat and mass transfer using an internal-variables approach, which is extensively used in describing viscoelastic stress-relaxation behavior, we begin to see a common framework for the description of all these cases. In addition, this approach allows for the straightforward description of coupling between various transport processes through their dependence on the same relaxing internal variable. Similar coupling between mass transfer and stress, consistent with the above analysis, was also seen in the analysis of concentration changes in polymer solutions, §9.2. Finally, the validity of the internal-variables approach has been reinforced by showing its consistency with the kinetic theory of gases.

10.2 Phase Transitions in Inhomogeneous Media

The next system which we wish to consider (although not in such detail) is the very important area of physics concerned with the development and evolution of inhomogeneous (in space) phase transitions in fluid systems. This problem is characterized by the onset of instabilities (spinodal decomposition) when a fluid phase becomes thermodynamically unstable and gradually (or rapidly, depending on the time scale of the observer) changes into another. An example of industrial interest in such a model stems from co-polymer systems pushed beyond the onset of immiscibility of the two components, possibly due to stresses induced in the material upon application of flow.[15] The most important previous theoretical work in this field has been the development of a series of simple phenomenological models in order to describe the phase change and diffusion of the components. These are typically referred to generically as "Cahn/Hilliard equations," since the initial work on this subject was done by these authors [1958]. Today, there are three different models which are commonly used to deal with this subject, each being, more or less, a phenomenological meshing of various effects. We shall discuss each in turn, then proceed to our discussion of their development and formulation in terms of the bracket theory. Several inconsistencies and incongruities will be ironed out in the process, leading to a better overall model for describing such an interesting physical phenomenon.

10.2.1 The kinetics of phase transitions

In the simplest form, the kinetics of phase transitions are phenomenologically described through an evolution equation for a scalar "order parameter," ϕ, which, for example, can represent the volume fraction of one component in a two-component mixture. This evolution equation is

[15]None of the models discussed in this section has ever considered flow effects, so we shall neglect them entirely (i.e., **M** is not a dynamic variable), although we could, if desired, easily extend the following development to include flow effects.

provided by a partial-differential equation involving derivatives in both space and time. The starting point in these model derivations is usually the specification of a generalized Helmholtz free energy functional, more or less along the lines detailed previously. The key characteristic of A, however, is now its dependence, in addition to ϕ itself, on the gradient of ϕ as well,

$$A[\phi] = \int_{\Omega} a(\phi, \boldsymbol{\nabla}\phi)\, \mathrm{d}^3x \; . \tag{10.2-1}$$

In the most general case, the free energy density is typically defined as [e.g., Cahn and Hilliard, 1958, p. 259; Langer, 1986, p. 171]

$$a(\phi, \boldsymbol{\nabla}\phi) \equiv a_h(\phi) + \frac{1}{2}\kappa |\boldsymbol{\nabla}\phi|^2 \; , \tag{10.2-2}$$

where $a_h(\phi)$ is the homogeneous (i.e., not depending on $\boldsymbol{\nabla}\phi$) contribution to the free energy density and κ is a positive constant (which might depend on the temperature).

Although the free energy, as given above, is fairly standardized among these models, the exact form of the evolution equation for the order parameter generally takes one of three forms.[16] In the first, which we shall call "Model A" in keeping with tradition, the evolution equation for ϕ is a relaxational (i.e., in terms of a relaxational affinity) expression of the form

$$\frac{\partial\phi}{\partial t} = -K\frac{\delta A}{\delta\phi} \; , \tag{10.2-3}$$

where K is a positive coefficient which might depend on ϕ and the temperature [Halperin *et al.*, 1974, p. 141]. This model is proposed for systems where the internal order parameter ϕ characterizes a non-equilibrium structure which eventually (as $t \rightarrow \infty$) approaches an equilibrium value obtained by minimizing the free energy function.

In cases where the order parameter represents an equilibrium thermodynamic variable (such as, for example, the concentration), "Model B" is proposed:

$$\frac{\partial\phi}{\partial t} = \boldsymbol{\nabla}\cdot\left(M\, \boldsymbol{\nabla}\frac{\delta A}{\delta\phi}\right) \; , \tag{10.2-4}$$

where M is a positive coefficient related to diffusivity. This expression was introduced by Cahn as a model for spinodal decomposition [1961, p. 799].

[16]An interesting review article discussing these three models and their applications has been written by Hohenberg and Halperin [1977].

Sometimes, the order parameter is insufficient to describe the local state of the system, and additional fields need to be incorporated into coupled model equations. The most popular model for this case, called the "phase-field" or "Model C," involves, in addition to ϕ, the temperature T written as [Halperin *et al.*, 1974, p. 142]

$$\alpha\xi^2 \frac{\partial\phi}{\partial t} = \xi^2\nabla^2\phi + g(\phi) - f(T) , \tag{10.2-5a}$$

$$\frac{\partial f}{\partial t} - \lambda \frac{\partial\phi}{\partial t} = \nabla^2 f , \tag{10.2-5b}$$

where f is a strictly increasing function of the temperature (which may, in fact, just be associated with the dimensionless temperature), g is a function with the form of a negative, double-well function, and λ, ξ, and α are positive constants. It is apparent that the description of Model C is more specific than the previous two cases. Indeed, although introduced phenomenologically by Halperin *et al.* [1974, p. 142], a recent attempt was made by Penrose and Fife [1990] to place this model upon a more fundamental foundation. These authors re-expressed (10.2-5) in terms of an entropy functional as [Penrose and Fife, 1990, p. 50]

$$\frac{\partial\phi}{\partial t} = K\frac{\delta S}{\delta\phi} , \tag{10.2-6a}$$

$$\frac{\partial u}{\partial t} = -\,\boldsymbol{\nabla}\cdot\left(M\,\boldsymbol{\nabla}\frac{1}{T}\right) , \tag{10.2-6b}$$

where K and M are positive constants, similar to those above, and u is the (homogeneous) energy density, defined as [Penrose and Fife, 1990, p. 51]

$$u \equiv f(T)\nu(\phi) + e_\phi(\phi) = \frac{\mathrm{d}\left(\dfrac{a_h(T,\phi)}{T}\right)}{\mathrm{d}\left(\dfrac{1}{T}\right)} , \tag{10.2-7}$$

with ν and e_ϕ defined as the number of degrees of freedom per unit volume and the potential energy, respectively. The entropy functional is defined by Penrose and Fife [1990, p. 49] as

$$S = \int_\Omega \left[s(u,\phi) - \tfrac{1}{2}\kappa^*|\boldsymbol{\nabla}\phi|^2\right] \mathrm{d}^3x , \tag{10.2-8}$$

with $\kappa^* \equiv \kappa / T$ and s representing the homogenous entropy density, defined in the usual sense through the thermodynamic relation

$$s = \frac{u}{T} - \frac{a_h(T,\phi)}{T} . \tag{10.2-9}$$

Note that in applying the standard thermodynamic relations above, only the homogeneous densities were used by Penrose and Fife [1990, p. 48].

10.2.2 *The bracket description of the kinetics of phase transitions*

Although one of the key advantages of using the bracket description is the straightforward incorporation of flow phenomena into the system derivation, in order to keep the discussion as simple as possible (and also to enable comparisons with previous work), we shall restrict ourselves here to the consideration of purely dissipative phenomena. Of course, the incorporation of flow phenomena is essentially trivial along the lines discussed earlier in conjunction with the Poisson bracket, but we focus our attention now upon the definition of the dissipation bracket.

Models A and B may be trivially recovered through the use of a single internal variable, the order parameter ϕ, the mechanical Hamiltonian of (10.2-1,2),[17] and the dissipation brackets

$$[F,G] = -\int_\Omega K \frac{\delta F}{\delta \phi} \frac{\delta G}{\delta \phi} \, d^3x , \tag{10.2-10a}$$

$$[F,G] = -\int_\Omega M \, \boldsymbol{\nabla}(\frac{\delta F}{\delta \phi}) \cdot \boldsymbol{\nabla}(\frac{\delta G}{\delta \phi}) \, d^3x , \tag{10.2-10b}$$

respectively. The first bracket just involves intra-subsystem relaxational effects and the second bracket inter-subsystem diffusive effects. Each has a range of validity within a certain window on the time scale of the observer, and an obvious generalization would be that when the time scales of both phenomena are comparable to that of the experimenter, both effects would be present. The second law of thermodynamics again requires the non-negativity of the phenomenological coefficients, and since the system is considered to be spatially homogeneous, it is apparent that M should really be an anisotropic tensor so that

$$[F,G] = -\int_\Omega M_{\alpha\beta} \nabla_\alpha (\frac{\delta F}{\delta \phi}) \nabla_\beta (\frac{\delta G}{\delta \phi}) \, d^3x . \tag{10.2-11}$$

[17]These models are implicitly restricted to isothermal and incompressible fluid systems, so that the Hamiltonian and Helmholtz free energy are essentially equivalent for $\mathbf{M}=\mathbf{0}$—see (8.1-3).

By far the most interesting phenomena are observed when one considers the coupling between the phase transitions and various other transport processes. Here, for simplicity, we shall consider only the coupling with heat transfer, as in Model C. In this case, we have the total Hamiltonian (system energy) as a functional of ϕ and s. As we saw in §9.2, however, the functional derivatives of H may be directly related to those of $A(\phi,T)$, so that we may use the free energy density of (10.2-2) without loss of rigor. In this approach, therefore, the thermodynamic definitions apply directly to the energy densities,

$$u = \frac{d(\frac{a}{T})}{d(\frac{1}{T})} , \tag{10.2-12}$$

$$s = \frac{u}{T} - \frac{a(T,\phi)}{T} , \tag{10.2-13}$$

in contrast to (10.2-7,9).

Now the general expression for the system's dissipation bracket is

$$\begin{aligned}
[F,G] = &- \int_\Omega K \frac{\delta F}{\delta \phi} \frac{\delta G}{\delta \phi} \, d^3x + \int_\Omega K \frac{1}{T} \frac{\delta F}{\delta s} \frac{\delta G}{\delta \phi} \frac{\delta G}{\delta \phi} \, d^3x \\
&- \int_\Omega M_{11} \nabla_\gamma \left(\frac{\delta F}{\delta \phi}\right) \nabla_\gamma \left(\frac{\delta G}{\delta \phi}\right) d^3x \\
&+ \int_\Omega M_{11} \frac{1}{T} \frac{\delta F}{\delta s} \nabla_\gamma \left(\frac{\delta G}{\delta \phi}\right) \nabla_\gamma \left(\frac{\delta G}{\delta \phi}\right) d^3x \\
&- \int_\Omega M_{21} \nabla_\gamma \left(\frac{\delta F}{\delta s}\right) \nabla_\gamma \left(\frac{\delta G}{\delta \phi}\right) d^3x \\
&+ \int_\Omega M_{21} \frac{1}{T} \frac{\delta F}{\delta s} \nabla_\gamma \left(\frac{\delta G}{\delta s}\right) \nabla_\gamma \left(\frac{\delta G}{\delta \phi}\right) d^3x \\
&- \int_\Omega M_{22} \nabla_\gamma \left(\frac{\delta F}{\delta s}\right) \nabla_\gamma \left(\frac{\delta G}{\delta s}\right) d^3x \\
&+ \int_\Omega M_{22} \frac{1}{T} \frac{\delta F}{\delta s} \nabla_\gamma \left(\frac{\delta G}{\delta s}\right) \nabla_\gamma \left(\frac{\delta G}{\delta s}\right) d^3x ,
\end{aligned} \tag{10.2-14}$$

where **M** is a symmetric (assuming that the reciprocal relations are valid), non-negative-definite matrix. This bracket leads to the coupled evolution equations

$$\frac{\partial\phi}{\partial t} = - K\frac{\delta A}{\delta\phi} + \nabla_\gamma(M_{11}\nabla_\gamma\frac{\delta A}{\delta\phi}) + \nabla_\gamma(M_{12}\nabla_\gamma T) \,, \qquad (10.2\text{-}15a)$$

$$\frac{\partial s}{\partial t} = - \nabla_\gamma J_s^\gamma + \sigma \,, \qquad (10.2\text{-}15b)$$

where, for convenience, the entropy flux, $\mathbf{J}_s$, and the entropy production rate, σ, are defined as

$$J_s^\gamma = - M_{21}\nabla_\gamma(\frac{\delta A}{\delta\phi}) - M_{22}\nabla_\gamma T \,, \qquad (10.2\text{-}16a)$$

$$\begin{aligned}\sigma = \frac{1}{T}\Big[K\frac{\delta A}{\delta\phi}\frac{\delta A}{\delta\phi} &+ M_{11}\nabla_\gamma(\frac{\delta A}{\delta\phi})\nabla_\gamma(\frac{\delta A}{\delta\phi}) \\ &+ 2M_{21}\nabla_\gamma(\frac{\delta A}{\delta\phi})\nabla_\gamma T + M_{22}(\nabla_\gamma T)(\nabla_\gamma T)\Big] \,.\end{aligned} \qquad (10.2\text{-}16b)$$

Therefore, the corresponding rate of change for the energy density is

$$\begin{aligned}\frac{\partial u}{\partial t}(\phi, \boldsymbol{\nabla}\phi, s) &= \frac{\partial u}{\partial s}\frac{\partial s}{\partial t} + \frac{\partial u}{\partial\phi}\frac{\partial\phi}{\partial t} + \frac{\partial u}{\partial(\nabla_\gamma\phi)}\nabla_\gamma(\frac{\partial\phi}{\partial t}) \\ &= - T\nabla_\gamma J_s^\alpha + T\sigma + \left[\frac{\partial a(T,\phi,\boldsymbol{\nabla}\phi)}{\partial\phi} + \frac{\partial a(T,\phi,\boldsymbol{\nabla}\phi)}{\partial(\nabla_\gamma\phi)}\nabla_\gamma\right] \\ &\times\left[- K\frac{\delta A}{\delta\phi} + \nabla_\beta(M_{11}\nabla_\beta\frac{\delta A}{\delta\phi}) + \nabla_\beta(M_{12}\nabla_\beta T)\right] \,.\end{aligned}$$

(10.2-17)

Comparison of Eqs. (10.2-16,17) with Model C, as defined by (10.2-5) or (10.2-6), reveals that these later equations are not complete. More specifically, (10.2-5a) can be recovered if we set

$$M_{11} = M_{12} = 0 \,,\ \kappa K = \frac{1}{\alpha} \,,\ Ka_h(T,\phi) = -\frac{1}{\alpha\xi^2}\Big(\int g(\phi)\mathrm{d}\phi - f(T)\phi\Big) \,; \qquad (10.2\text{-}18)$$

however, the corresponding (consistent) equation for the energy density then works out to be (with the assumption of a constant heat capacity), with the substitution of (10.2-7),

$$\frac{\partial f}{\partial t} - \lambda \frac{\partial \phi}{\partial t} = \nabla^2 f + \omega_1 [K(\frac{\delta A}{\delta \phi})^2 + M_{22}(\boldsymbol{\nabla} T)^2] \ . \tag{10.2-19}$$

In this expression, ω_1 is a dimensionalization constant and λ is the linear differential operator

$$\lambda = \omega_2 [g(\phi) - u - \kappa (\nabla_\gamma \phi) \nabla_\gamma] \ , \tag{10.2.20}$$

with ω_2 being another dimensionalization constant. By comparison of (10.2-19) with (10.2-5b), it is evident that the internally consistent set above reduces to Model C only when the additional terms in (10.2-19) are negligible with respect to the others. The validity of this assumption, as well as the validity of the other assumptions in this model, is still a matter to be investigated. Still, it is easy to verify the nature of the additional terms in (10.2-19). The last two terms represent entropy production associated with relaxation of the order parameter and the heat conduction, respectively. The additional differential operator in λ, (10.2-20), is due to the dependence of the energy density on the spatial gradients of the order parameter. From the above discussion, it is evident that Eqs. (10.2-6) are also incomplete, with similar terms as discussed above missing from the expression. However, the system of equations derived above is, in principle, no more difficult to solve computationally than the original one.

10.3 The Inertial Description of Incompressible Viscoelastic Fluids

In this section, we discuss briefly the bracket theory applied to isothermal and incompressible inertial viscoelastic fluids. We do this at this point because, in essence, the development for inertial fluids is still not well developed (i.e., semi-phenomenological), and because it provides the natural lead-in to liquid-crystalline theory (chapter 11), which is well developed, although from the perspective of continuum mechanics.

We shall consider a viscoelastic material which is described not only by the momentum density and the conformation tensor, but also by an additional second-rank tensor, **w**, which represents the conjugate variable to **c**. Hence we define this new variable as $\mathbf{w} \equiv \sigma \dot{\mathbf{c}}$, where σ is an inertial constant and $\dot{\mathbf{c}}$ is the material derivative of the conformation tensor (see (6.4-4)). This additional variable should allow us to describe a viscoelastic system in which inertial effects are important. Although it is generally believed that inertial forces play no role in most circumstances, it is possible that such effects may become important in instances where the local acceleration of the fluid is very large, e.g., at the corner of a four-to-one contraction. Using these three dynamic variables, we can write our arbitrary functional as

$$F[\mathbf{M},\mathbf{c},\mathbf{w}] = \int_\Omega f(\mathbf{M},\mathbf{c},\mathbf{w})\ d^3x\ , \tag{10.3-1}$$

so that the dynamical equation for the system is again given as (7.1-1).

When one deals with inertial systems whose dynamic variables include not only certain characteristic variables, but their time derivatives as well, a fuzzy region of physics is entered in which not much work has been done. Indeed, the only work that we are aware of on this subject is that of Machlup and Onsager [1953], who attacked this problem from the point of view of establishing an extended set of reciprocal relations, with dubious results.[18] This being the case, we opt below to write the dissipation bracket in its most restricted form; i.e., we write it so that it gives directly the non-inertial equations (chapter 8), in which we are fairly confident, in the limit when $\sigma \to 0$. The full generality on the inertial level still applies, however, although we again have a restriction on the form of the information passed between various levels of description (Figure 8.1). The fully general inertial system remains an outstanding problem awaiting investigation. Consequently, the discussion in this section should not be taken as gospel, but only as a first step in the process of describing inertial systems. Nevertheless, we shall show in the next chapter that the material presented in this section does indeed give us seemingly realistic results, at least in some limits.

The Hamiltonian of the inertial system is a functional of the same three dynamic variables as F. It is here expressed as

$$H_m[\mathbf{M},\mathbf{c},\mathbf{w}] = \int_\Omega \left(\frac{1}{2\rho}\mathbf{M}\cdot\mathbf{M} + \frac{1}{2\sigma}\mathbf{w}\cdot\mathbf{w}\right) d^3x + A[\mathbf{c}]\ , \tag{10.3-2}$$

where $A[\mathbf{c}]$ is the Helmholtz free energy of chapter 8, and the kinetic energy has been expanded to include an additional term associated with the rotational kinetic energy of the microstructure of the medium. In the simple case of a dumbbell, or a rigid rod, it represents the energetic contribution due to the rotation of the physical entity suspended in the

[18]Note that in defining the material derivative of the conformation tensor as a dynamic variable (i.e., $F[\mathbf{M},\mathbf{c},\mathbf{w}]=\int f(\mathbf{M},\mathbf{c},\mathbf{w})d^3x$) we are still able to use the variational derivative of (5.2-6). On the other hand, if we had chosen to incorporate this dependence into $\mathbf{c}$, as we did for $\nabla\mathbf{c}$ (i.e., $F[\mathbf{M},\mathbf{c}] = \int f(\mathbf{M},\mathbf{c},\nabla\mathbf{c})d^3x$), then we would have had to redefine the Volterra derivative of (5.2-6) appropriately. This is trivial. In this case, we feel that using $\mathbf{w}$ as a dynamic variable is more appropriate; a feeling which will be borne out in the next chapter. Note, however, that it is along the latter line of reasoning that Machlup and Onsager [1953, p. 1513] proceed.

solvent. Note that we make no claims as to the effect of inertia on the Helmholtz free energy, $A \neq A[\mathbf{c},\mathbf{w}]$.

In the inertial description of viscoelastic fluids, the Poisson bracket is slightly different from the one of (8.1-1) in that we have the additional dynamic variable, **w**, which is related to the time derivative of **c**. Indeed, the relationship between **w** and **c** is implied directly by the canonical form of Hamilton's equations,

$$\dot{c}_{\alpha\beta} = \frac{\delta H_m}{\delta w_{\alpha\beta}} = c_{\alpha\gamma} \nabla_\gamma v_\beta + c_{\gamma\beta} \nabla_\gamma v_\alpha \, , \tag{10.3-3}$$

with the last equality stating the material objectivity requirement. If we now replace **v** in (10.3-3) with $\delta H_m/\delta \mathbf{M}$, and express the result in terms of arbitrary F and G (which implies that F and G are no longer completely arbitrary, but are forced to depend upon **M** and **w** only through the kinetic energy of (10.3-2)), then the last two integrals in (8.1-1) must be reexpressed in the inertial description as a single integral which bears a close resemblance to the canonical form in the material description of fluid flow (see §5.3). After adding the additional term for the material derivative of **w**, the Poisson bracket becomes

$$\begin{aligned}
\{F,G\} = &- \int_\Omega \left[\frac{\delta F}{\delta M_\gamma} \nabla_\beta \left(\frac{\delta G}{\delta M_\beta} M_\gamma \right) - \frac{\delta G}{\delta M_\gamma} \nabla_\beta \left(\frac{\delta F}{\delta M_\beta} M_\gamma \right) \right] d^3x \\
&- \int_\Omega \left[\frac{\delta F}{\delta c_{\alpha\beta}} \nabla_\gamma \left(\frac{\delta G}{\delta M_\gamma} c_{\alpha\beta} \right) - \frac{\delta G}{\delta c_{\alpha\beta}} \nabla_\gamma \left(\frac{\delta F}{\delta M_\gamma} c_{\alpha\beta} \right) \right] d^3x \\
&- \int_\Omega \left[\frac{\delta F}{\delta w_{\alpha\beta}} \frac{\delta G}{\delta c_{\alpha\beta}} - \frac{\delta G}{\delta w_{\alpha\beta}} \frac{\delta F}{\delta c_{\alpha\beta}} \right] d^3x \\
&- \int_\Omega \left[\frac{\delta F}{\delta w_{\alpha\beta}} \nabla_\gamma \left(\frac{\delta G}{\delta M_\gamma} w_{\alpha\beta} \right) - \frac{\delta G}{\delta w_{\alpha\beta}} \nabla_\gamma \left(\frac{\delta F}{\delta M_\gamma} w_{\alpha\beta} \right) \right] d^3x \, .
\end{aligned} \tag{10.3-4}$$

Note that in the limit of negligible inertial effects, $\sigma \rightarrow 0$ implies that $\mathbf{w} \rightarrow \mathbf{0}$, and the integral in (10.3-4) vanishes. However, $\delta F/\delta \mathbf{w}$ does not necessarily vanish, and indeed one must make the back substitution implied by (10.3-3), at which point the bracket (8.1-1) is reinstated as the proper system descriptor.[19]

[19]The reader will note the phenomenological manner in which we arrived at the above

The general form of the dissipation bracket for this incompressible system is analogous to (8.1-2), except that we must recognize that including $\mathbf{w}=\sigma\dot{\mathbf{c}}$ as a dynamic variable in the problem formulation has certain implications concerning the form of the functional of (7.1-19). In this case, when considering inertial effects, we have effectively reduced the time scale of our observations to the point where $\mathbf{c}$, other than being convected with the fluid, is a frozen variable with respect to the dissipation. In other words, the effects associated with the relaxation of the microstructure are so much slower than the dissipative inertial effects that they are excluded from the problem formulation. Specifically, this constraint is expressed by the definition of $\mathbf{w}$ itself as being linearly proportional to $\dot{\mathbf{c}}$. Hence the dissipation bracket for the inertial description is simply

$$\begin{aligned}[F,G] = &-\int_\Omega \Theta_{\alpha\beta\gamma\varepsilon}\frac{\delta F}{\delta w_{\alpha\beta}}\frac{\delta G}{\delta w_{\gamma\varepsilon}}\,d^3x \\ &-\int_\Omega Q^I_{\alpha\beta\gamma\varepsilon}\nabla_\alpha\Big(\frac{\delta F}{\delta M_\beta}\Big)\nabla_\gamma\Big(\frac{\delta G}{\delta M_\varepsilon}\Big)\,d^3x \\ &-\int_\Omega Z_{\alpha\beta\gamma\varepsilon}\left[\nabla_\alpha\Big(\frac{\delta F}{\delta M_\beta}\Big)\frac{\delta G}{\delta w_{\gamma\varepsilon}} + \nabla_\alpha\Big(\frac{\delta G}{\delta M_\beta}\Big)\frac{\delta F}{\delta w_{\gamma\varepsilon}}\right]d^3x\ ,\end{aligned} \tag{10.3-5}$$

where the inhomogeneous term (analogous to $\mathbf{B}$ in (8.1-2)) has been neglected for simplicity. $\boldsymbol{\Theta}$, $\mathbf{Q}^I$, and $\mathbf{Z}$ are again phenomenological matrices, which may in general depend upon the dynamic variables. Note that since $\mathbf{w}$ is proportional to $\dot{\mathbf{c}}$, $\delta H_m/\delta\mathbf{w}$ has the opposite parity to $\delta H_m/\delta\mathbf{c}$, and therefore the minus sign in the fourth integral of (8.1-2) is a plus sign in the third integral of the above expression. This term now contributes to the overall entropy production rate within the system.

With the Hamiltonian of (10.3-2), and the brackets of (10.3-4,5), the evolution equations for our three dynamic variables are

$$\rho\frac{\partial v_\alpha}{\partial t} = -\rho v_\beta\nabla_\beta v_\alpha - \nabla_\alpha p + \nabla_\beta\sigma^I_{\alpha\beta}\ , \tag{10.3-6a}$$

Poisson bracket. The strict derivation of this bracket from first principles has only recently been carried out [Edwards and McHugh, 1994, §4].

$$\frac{\partial c_{\alpha\beta}}{\partial t} = - v_\gamma \nabla_\gamma c_{\alpha\beta} + \dot{c}_{\alpha\beta} \ , \tag{10.3-6b}$$

$$\frac{\partial w_{\alpha\beta}}{\partial t} = - v_\gamma \nabla_\gamma w_{\alpha\beta} - \Theta_{\alpha\beta\gamma\varepsilon} \dot{c}_{\gamma\varepsilon} - \frac{\delta A}{\delta c_{\alpha\beta}} - Z_{\alpha\beta\gamma\varepsilon} \nabla_\gamma v_\varepsilon \ , \tag{10.3-6c}$$

where

$$\sigma^I_{\alpha\beta} = Q^I_{\alpha\beta\gamma\varepsilon} \nabla_\gamma v_\varepsilon + Z_{\alpha\beta\gamma\varepsilon} \dot{c}_{\gamma\varepsilon} \ . \tag{10.3-6d}$$

Note that (10.3-6b) just defines the material derivative of the conformation tensor. Now if we require that these equations reduce to those of the non-inertial theory, (8.1-6,7), then we obtain the constraints on the phenomenological matrices

$$\begin{aligned} \Theta^{-1}_{\alpha\beta\gamma\varepsilon} &= \Lambda_{\alpha\beta\gamma\varepsilon} \ , \quad Q_{\alpha\beta\gamma\varepsilon} = Q^I_{\alpha\beta\gamma\varepsilon} - Z_{\alpha\beta\eta\nu} \Theta^{-1}_{\eta\nu\xi\rho} Z_{\xi\rho\gamma\varepsilon} \ , \\ - \Theta^{-1}_{\alpha\beta\eta\nu} Z_{\eta\nu\gamma\varepsilon} &= L_{\alpha\beta\gamma\varepsilon} + c_{\alpha\gamma} \delta_{\varepsilon\beta} + c_{\beta\gamma} \delta_{\varepsilon\alpha} \ , \\ - Z_{\alpha\beta\eta\nu} \Theta^{-1}_{\eta\nu\gamma\varepsilon} &= L_{\alpha\beta\gamma\varepsilon} + c_{\beta\gamma} \delta_{\varepsilon\alpha} + c_{\beta\varepsilon} \delta_{\alpha\gamma} \ . \end{aligned} \tag{10.3-7}$$

Note that the symmetry of the product of $\boldsymbol{\Theta}$ and $\mathbf{Z}$ is split in the last two relationships of the above equation. Although we have assumed that the reciprocal relations hold, which implies the symmetry of the individual matrices, there is no theorem in linear algebra which states that the product of two symmetric matrices must also be symmetric. In fact, simple counter examples abound. Hence the above relations do not surprise us, and, as we shall see in the next chapter, they are borne out in practice. Also due to the above relations, one could show that the inequalities associated with the entropy production rate for both the inertial and the non-inertial dissipation brackets result in equivalent admissibility criteria for the problem parameters. We shall see this more clearly in the following chapter as well.

11
The Dynamical Theory of Liquid Crystals

The essence of rhythm is the fusion of sameness and novelty... A crystal lacks rhythm from excess of pattern, while a fog is unrhythmic in that it exhibits a patternless confusion of detail.

—*Alfred North Whitehead*

Liquid crystals (LCs) present a state of matter with properties—as the name suggests—intermediate between those of liquids and crystalline solids. Liquid-crystalline materials, as all liquids, cannot support shear stresses at static equilibrium. Their molecules are characterized by an anisotropy in the shape and/or intermolecular forces. Thus, there is the potential for the formation of a separate phase(s), called a "mesophase(s)," where a partial order arises in the molecular orientation and/or location, which extends over macroscopic distances. This partial long-range molecular order, reminiscent of (but not equivalent to) the perfect order of solid crystals, in addition to the material fluidity, is primarily responsible for the many properties which are inherent characteristics of liquid-crystalline phases, such as a rapid response to electric and magnetic fields, anisotropic optical and rheological properties, etc.—see, for examples, the reviews by Stephen and Straley [1974] and Jackson and Shaw [1991], the monographs by de Gennes [1974], Chandrasekhar [1977], and Vertogen and de Jeu [1988], and the edited volumes by Ciferri *et al.* [1982] and Ciferri [1991].

The variety of the liquid-crystalline macroscopic properties is such that trying to derive a theory capable of describing the principal liquid-crystalline dynamic characteristics can be a very frustrating task if one does not approach the issue in a systematic fashion. Characteristically, the main two theories that have been advanced over the last thirty years for the description of the liquid-crystalline flow behavior—the Leslie/Ericksen (LE) theory and the Doi theory—are essentially models developed from a set theoretical frame work—continuum mechanics and molecular theory, respectively. Nevertheless, each one of these theories has a limited domain of application. The description of the dynamic liquid-crystalline behavior through the bracket formalism, as seen in this chapter, leads naturally to a single conformation tensor theory with an extended domain of validity. This conformation theory consistently generalizes both previous theories, which can be recovered from it as particular cases. This offers additional evidence that the wealth of inherent information in LCs can only be appropriately handled when pursued in a systematic, fundamental manner.

In this chapter, we focus our attention on the dynamics of nematic liquid crystals. Other types of liquid-crystalline states can be investigated in a similar fashion. Moreover, since most of the current interest on the subject is directed toward the understanding of the flow behavior of polymeric liquid crystals (PLCs), we focus our attention on the theories that are the most suitable ones for this particular class of liquid crystals. Nevertheless, since the most notable difference between typical, low molecular-weight liquid crystals (LMWLCs) and PLCs is a quantitative rather than qualitative one (the size of the molecules), the models described herein are, in general, applicable to all liquid-crystalline systems. The bracket formalism used to provide a description of the dynamical equations is provided in a form that allows the general interaction between the liquid crystals and external fields. Our emphasis is in the modeling. Applications arising from specific interactions, such as electrohydrodynamic, magnetohydrodynamic, and optical effects, or effects induced by walls and/or propagation of defects (discontinuities) in the nematic structure are not discussed here; the interested reader is referred to the many excellent references for these topics that are available in the literature [Helfrich, 1969; de Gennes, 1971, §3.2; Wolff *et al.*, 1973; Berreman, 1973; de Gennes, 1974, chs. 3,4; Stephen and Straley, 1974, §VII, §XVII; Kléman, 1975; Ericksen, 1976, pp. 273-84; Goossens, 1978; Leslie, 1979, pp. 46-75; Lonberg *et al.*, 1984; Martin and Stupp, 1987; Moore and Stupp, 1987; Kléman, 1991; Rey, 1991].

We also restrict ourselves to discussing lyotropic (defined below) LCs (Doi and LE theories) for three reasons. First, all of the flow models previously available (although, possibly, not initially intended to) are more applicable to lyotropic LCs since they are concerned with incompressible fluids. The description of thermotropic (also defined below) LCs should allow for compressibility since these systems have, in general, an anisotropic temperature profile and, in consequence, density changes cannot be neglected since the density is, in general, a sensitive function of the temperature. Second, most test systems where experimental rheological data are available for a variety of well-characterized flow kinematics are (due to experimental convenience) lyotropic systems. Third, analysis of the thermodynamics introduced by non-isothermality is more crucial for thermotropic LCs and should be the subject of a thorough, independent investigation; the interested reader is referred to the monograph by Vertogen and de Jeu [1988] for an introduction to the physics of thermotropic liquid crystals. From the above arguments, it is easily deduced that any attempt to describe both types of systems in an overview chapter such as this would have probably led to utter confusion on the part of the reader, especially considering all of the technicalities involved in using concentration as an additional variable in the generalized bracket. Note that a first attempt to apply the bracket theory for the description of a general, dynamical theory for thermotropic LCs, including anisotropic flow and heat transfer, has already been made and may be found in a recent article [Edwards and Beris, 1992] which will not be discussed here.

Thus, the main emphasis of the present chapter is the systematic development of the general equations governing the isothermal dynamic behavior of lyotropic, nematic liquid crystals. We start in §11.1 with a brief introduction to liquid crystals, where terms such as "lyotropic" and "thermotropic," mentioned above, are defined. In the next section, §11.2, an overview is presented of the available descriptions for the Hamiltonian (free energy) of liquid crystals under static conditions. In section §11.3, the two main (classical) theories (LE and Doi) are presented. As a thermodynamically consistent continuum model, the LE theory can be exactly reproduced within the bracket formalism, as shown in §11.4. In the next section, §11.5, the generalized bracket formalism is used to develop a generalization of the LE theory, based on a conformation-tensor description. It is demonstrated in §11.6 that the conformation theory consistently reduces, in the uniaxial nematic limit, to the classical LE theory. In addition, in the same section, it is shown how the conformation theory can accommodate a generalization of the Doi conformation-tensor theory, which can consistently account for tumbling, viscous, and (macroscopically) elastic phenomena. In creating a single conformation theory which can accommodate the previous ones as particular cases, obvious deficiencies present in each one of the two classical theories are rectified. Thus a much wider range of liquid-crystalline behaviors can be accommodated easily without having to necessarily resort to microscopic (distribution function) descriptions. We achieve this by building heavily upon our experience working with the bracket formalism, as exhibited in the previous chapters. Indeed, LC dynamics are the perfect test of the thesis of this book, and, in a sense, a mild climax, since in order to describe LCs we shall have to use nearly everything that we have learned thus far.

Because of the complexity of this subject, we shall not be as explicit in the presentation of the mathematics as we have been in the previous chapters. However, by now, most of it should be apparent anyway, and what is not may probably be found in the dissertation of Edwards [1991], upon which this chapter is largely based. Furthermore, several papers on this subject have been written which present most of the material from a different viewpoint [Edwards and Beris, 1989a,b; Edwards *et al.*, 1990a,b; 1991; Edwards and Beris, 1992], which also might prove helpful to the interested reader. Nevertheless, this chapter, in its present form, should provide a good introduction to the recent LC continuum theoretical efforts and their possible implications to the understanding of the macroscopic, nonlinear, dynamic, LC behavior.

11.1 Introduction to Liquid Crystals

As the name implies, liquid crystals appear as phases which possess properties intermediate to those of crystalline solids and amorphous liquids, giving them the most peculiar and most interesting properties to be found in condensed matter physics. Specifically, LC phases possess a certain degree of the long-range order of anisotropic, crystalline solids, but deform continuously under the application of a tangential stress, as do fluids. Hence one might expect that such a partially oriented, anisotropic fluid phase would show somewhat unusual behavior. Over the years, the term "mesomorphic phase" or, in short, "mesophase," has come to denote the LC phase, indicating the intermediate nature of the properties (between liquids and crystalline solids) exhibited by these materials [de Gennes, 1974, p. 1; Chandrasekhar, 1977, p. 1; Vertogen and de Jeu, 1988, §1.1; Jackson and Shaw, 1991, pp. 4-6].

Systems which form these anisotropic fluid phases are composed of molecules which, by necessity, have themselves a degree of structural anisotropy, such as inflexible rod-like or disk-like molecules with an inherent direction of orientation. Generally, highly symmetric molecular systems such as those composed of neutral, spherical molecules exhibit a direct transition from a highly ordered, isotropic crystalline state to a disordered, but also isotropic, liquid phase. Therefore, one expects to find LC phases associated with only certain materials, with the vast majority of systems showing no mesophase whatsoever. One may refer to many reviews of liquid-crystalline materials in the literature for careful discussions concerning specific systems which form mesophases [Ekwall, 1975; Gray, 1976; Samulski and DuPré, 1979; Varshney, 1986; Wright and Mermin, 1989; Lauprêtre and Noël, 1991; Jackson and Shaw 1991].

Figure 11.1 illustrates schematically how a mesophase differs from a crystalline solid and an isotropic fluid for a material composed of rod-like molecules. In a crystalline solid, the molecules are arranged on a regular lattice with a specific distance between nearest neighbors in each of a molecule's two characteristic directions. This structure is rigidly maintained due to electrostatic forces and will deform only under the application of an intrinsically large stress, and then it will break across some characteristic plane or defect rather than deform perse. The isotropic fluid is composed of a random jumble of the rod-like molecules, with an average orientation of zero. The material flows as a fluid would

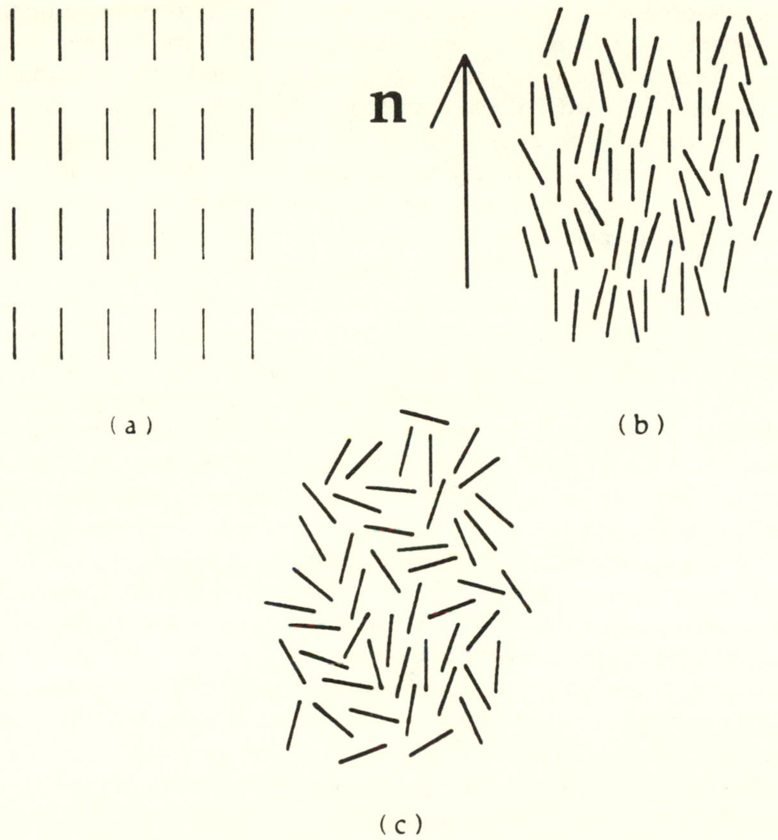

Figure 11.1: Schematic illustration of various possible phases for condensed matter: a.) crystalline solid; b.) liquid crystal; c.) isotropic fluid. **n** represents the local orientation of the liquid crystal.

under the application of a tangential stress, however small, with a viscosity determined by the steric effects and friction due to molecular interactions. The mesophase exhibits an intermediate behavior between the two extremes above in that it possesses a long-range orientation, specified by the so-called "director," **n**; yet still deforms continuously under the application of a tangential stress. Indeed, due to the partial alignment of the molecules, a LC actually has a lower viscosity than the jumbled, isotropic fluid since fewer molecular interactions occur in the mesophase. The alignment of the anisotropic molecules in LCs is also responsible for their optical activity, i.e., their birefringence observed when polarized light is transmitted through a given sample [e.g., de Gennes, 1971, pp. 193,94; Wissbrun, 1981, p. 654; Srinivasarao and Berry, 1991, p. 381; Threefoot, 1991, §5.3.3.2].

A mesophase is known to occur in either of two widely different circumstances. The first circumstance is obtained by heating a solid crystal up to a critical temperature, at which point the mesophase is formed via a first-order phase transition. This occurs at the threshold where the thermal energy acquired by the molecules of the material offsets the electrostatics of the crystalline lattice enough to seriously affect the positional state of the molecules. Thus, the molecules become free to move relative to each other, yet still maintain some characteristic axis of orientation due to steric effects. Mesophases of this type are termed "thermotropic" or "nonamphiphilic" LCs [Gray, 1976, p. 1] and are primarily associated with LMWLCs [Chandrasekhar, 1977, §1.2], although they are encountered in PLCs as well [Lauprêtre and Noël, 1991, §3]. Obviously, any further increase in temperature will push the material beyond another threshold, at which point the molecules will have acquired enough thermal energy to overcome the steric effects which maintain the average orientation, diffusing finally into a randomly oriented isotropic fluid.

The second circumstance occurs in solutions of rigid, high molecular-weight molecules in various solvents, with concentration (rather than temperature) as the driving force behind the mesophase formation. Thus, it is primarily associated with PLCs, such as those formed by solutions of synthetic polypeptides—as, for example, poly-γ-benzyl-L-glutamate (PBLG)—in organic liquids when the concentration exceeds a certain critical value [Chandrasekhar, 1977, §1.1.2; Lauprêtre and Noël, 1991, §4]. If one gradually dissolves inflexible molecules in a solvent, initially an isotropic fluid will be formed with no average orientation. However, at some point any further increase in the number of molecules in a given volume of solvent will force some type of partial alignment due to steric effects alone. Hence "lyotropic" LCs form at a critical concentration which depends not only on the nature of the solvent, but on the length and diameter of the polymer molecules as well. Lyotropic LCs are also called amphiphilic since, in order to facilitate the liquid-crystalline phase formation, individual molecules typically, but not always, contain two distinct regions, one non-polar (hydrophobic) and the other polar (hydrophilic), from which property we deduce the close relationship between lyotropic LCs and surfactants. However, since the amphiphilic character of the molecule is not always present in lyotropic systems (for example, in polymeric LCs), it is better to consider the two terms lyotropic and amphiphilic distinct [de Gennes, 1974, pp. 5-7]. Moreover, the distinction between lyotropic and thermotropic systems can become diffuse when we consider cases where either the lyotropic system is sensitive to the temperature or the thermotropic one also contains some solvent while retaining its liquid-crystalline characteristics [Gray, 1976, p. 2].

As one passes through the isotropic-to-LC phase transition in lyotropic systems, one usually sees a dramatic decrease in the viscosity of the material [Hermans, 1962; Kwolek *et al.*, 1977; Papkov *et al.*, 1974, pp. 1758,59,64,65; Wissbrun, 1981, pp. 636,39] because of the greater freedom of the molecules which are less sterically confined in the oriented state. Of course, both of the above cases are somewhat simplified in that several different mesophase types may exist, a number of which can succeed each other in the transition from the solid to the isotropic liquid or upon an increase of concentration [Wright and Mermin, 1989, §II.A]. However, a similar dramatic decrease in viscosity has also been observed in the single transition from the isotropic to nematic phase, upon increase in the concentration, as, for example, seen with racemic solutions of Poly(benzyl-glutamate) (PBG) [Kiss and Porter, 1978, pp. 196-99], and with solutions of poly(1,4-phenylene-2,6-benzobisthiazole) (PBT), in methane sulfonic acid (MSA), containing 3% by weight chlorosulfonic acid [Einaga *et al.*, 1985, Fig. 16].

The liquid-crystalline mesophases are classified broadly into three types: *nematic*, *cholesteric*, and *smectic*. In nematic LCs, there exists a long range orientational order of the molecules—along the direction of the director, **n**—but no positional order—see Figure 11.1b. Cholesteric liquid crystals characterize a mesophase which is very similar to that of the nematics, except that in addition to the nematic order at small distances (comparable to the molecular length) it also exhibits a supermolecular structure corresponding to a helical rotation of the director, along a neutral direction, of an intermediate length scale [de Gennes, 1974, p. 9; Freizdon and Shibaev, 1993, p. 253]. The spiral arrangement of the molecules that results, when the wavelength of the oscillations falls within the visible spectrum of light, is responsible for the unique optical properties associated with the LMWLCs used in optical displays. The cholesteric mesophase is only formed by chiral (i.e., with no mirror symmetry, thus, optically active) molecules [de Gennes, 1974, p. 9; Freizdon and Shibaev, 1993, p. 253]. Moreover, the addition of a minute quantity of an optically active substance, even non-mesomorphic, is usually sufficient at equilibrium to change a liquid-crystalline structure from nematic to cholesteric [Chandrasekhar, 1977, p. 3]. However, especially with helical polypeptide solutions such as the PBLG solutions mentioned above, the cholesteric structure is quite weak and does not persist in a strong enough magnetic field or flow field, where it typically changes to nematic [Kiss and Porter, 1978, p. 209; Moldenaers and Mewis, 1990, p. 361]. It is interesting to note that use of a racemic mixture, i.e., of equal amounts of both enantiomorphic forms of an optically active molecule typically corresponding to a cholesteric phase, also results in a nematic mesophase, as, for example, observed with the racemic mixture of PBLG and PBDG [Kiss and Porter, 1978, p. 193]. The third category, smectics, is quite distinct from the previous two, since it corresponds to a mesophase which exhibits positional as well as orientational molecular order. Several categories of smectics exist, each one corresponding to a different (partial) positional and orientational ordering of the molecules. In general, the molecules are aligned in layers, vertical to the director, of thickness on the order of the length of the free molecules [Chandrasekhar, 1977, pp. 4-8]. In the following, we will limit our discussion to nematic LCs since they represent, by far, the most studied mesophase of PLCs.

Mesogenic polymers can exist in three different types: *backbone, comb,* and *plate* [Warner *et al.*, 1985, p. 3007; Lauprêtre and Noël, 1991, §3 and §4.1]. The backbone (or mainchain) PLCs owe their liquid-crystalline character to a global or mainchain (as opposed to local or sidechain) rigidity. This is typically induced due to the presence of an intrinsically stiff mesogenic unit at the main chain, such as benzene rings interlinked at para-positions. However, on other occasions (such as encountered with PBG), the global molecular rigidity arises because the molecule assumes a helical conformation in some solvents. Correspondingly, comb (or sidechain) PLCs have rigid units in the side chains. The final category, plate PLCs, is not as common as the previous two and corresponds to polymers with a special planar rigid structure. In the following, we shall consider theories appropriate for the most common backbone PLCs only.

The interest in PLCs as well as in their gross macroscopic analogues, the short-fiber suspensions, derives primarily from the processing industry where one is always on the lookout for materials combining good processability with superior structural performance. Additional interest in PLCs arises because of biomedical applications related to the liquid crystallinity of biopolymers and membranes [Varshney, 1986, p. 554]. As already discussed in chapter 8, polymeric materials with an orientable internal microstructure are very important in the processing industry because of their low viscosity in the liquid-crystalline state (implying a potential for easy processing) combined with the good mechanical properties of the finished product, particularly on a by-weight basis. This is especially true for materials consisting of rigid, orientable units, such as polymeric liquid crystals, which after processing and solidification lead to products which show exceptional heat resistance, dimensional stability, and mechanical strength [Gabriele, 1990]. Indeed, as quoted from a recent review by Dobb and McIntyre [1984, p. 63],

> A major impetus was given to work, both academic and industrial, in the field of lyotropic systems by the development by duPont of commercial fibres having exceptionally high tensile strength and modulus through the use of nematic anisotropic

> solutions of relatively rigid-chain aromatic polyamides. The earliest product to appear, Fibre B, was based upon poly (p-benzamide), but was replaced by the fully commercial product, Kevlar, based upon poly (p-phenylene terephthalamide). Arenka, from Akzo, also has the latter chemical repeating unit.

These fibers are the primary reinforcing agents in polymer composites [Jackson and Shaw, 1991, p. 168]. Moreover, promise for eventual use of the combined electrical and mechanical properties of PLCs still exists [Jackson and Shaw, 1991, pp. 180-84]. Thus, the main issue in processing PLCs or fiber-filled composites is to know how the molecules or fibers orient during the processing so that one obtains the desired final condition.

The description of the thermodynamics of phase transitions in liquid crystals requires the introduction (in addition to the director, **n**, which characterizes the *direction* of the molecular alignment) of at least one more internal structural parameter in order to characterize the *degree* of the alignment. The first idea would have been to use the average, $<\cos\theta>$, of the cosine of the angle between the direction of a particular molecule, **a**, and the director, **n**, for uniaxial (i.e., oriented along a single direction) nematics. However, this vanishes identically for nematics given the fact the directions **n** and -**n** are equivalent. Thus a higher moment of the uniaxial molecular orientation distribution needs to be used. The lowest-order one giving a non-trivial answer is the *order parameter*, *S*, defined as [de Gennes, 1974, p. 24; McMillan, 1975, p. 103]

$$S \equiv \frac{1}{2} <3\cos^2\theta - 1> , \qquad (11.1\text{-}1)$$

where the angular brackets denote an ensemble average over all possible molecular conformations.

For more general (than uniaxial) molecular orientations, the tensorial equivalent form of the scalar order parameter, *S*, is used, **S**, defined in terms of the distribution of the orientation unit vector, **a**, of the molecules as [Stephen and Straley, 1974, p. 620]

$$\mathbf{S} \equiv <\mathbf{a}\mathbf{a} - \tfrac{1}{3}\boldsymbol{\delta}> . \qquad (11.1\text{-}2)$$

Note that, by definition, **S** corresponds to a symmetric and traceless tensor. Alternatively, for experimental purposes, the order parameter can be defined in terms of an anisotropic macroscopic property, such as the anisotropic part of the magnetic susceptibility [de Gennes, 1969b; 1974, p. 31]. Since this is directly related to the microscopic definition above, we shall, for simplicity, limit our discussion to the definition of the order parameter provided by Eqs. (11.1-1,2).

As mentioned above, the scalar order parameter is used when the liquid-crystalline state can be approximated as a uniaxial nematic corresponding to a single preferred direction **n**. Under this assumption, (11.1-2) simplifies to [Stephen and Straley, 1974, p. 621]

$$\mathbf{S} = S(\mathbf{nn} - \tfrac{1}{3}\boldsymbol{\delta}) \quad . \tag{11.1-3}$$

The most important thermodynamic descriptions of LCs at static equilibrium, due to Onsager [1949], Maier and Saupe [1958; Stephen and Straley, 1974, pp. 622-24], Flory [1956], and Flory and Ronca [1979] are reviewed in the next section, §11.2. They typically lead to expressions for the Helmholtz free energy which can be approximated by a low-order polynomial expression in S or **S** (called a "Landau/de Gennes expansion" [de Gennes, 1969b; 1974, p. 48]) as discussed in §11.2.1.A.

The thermodynamic description of LCs at equilibrium typically assumes a homogeneous system characterized by a uniform molecular orientation—see §11.2.A. However, even under static conditions, the structure which is often realized exhibits variations in orientation. These typically arise either from the presence of structural *defects*, called *disclinations*,[1] which are similar to the dislocations encountered in non-equilibrium crystalline structures, or from incompatible boundary conditions [de Gennes, 1974, ch. 4]. Given the much weaker (as compared to solid crystals) elastic forces present in liquid-crystalline solutions, much more numerous disclinations are expected (and, indeed, experimentally observed [Hudson *et al.*, 1987, Fig. 4]), to the point of giving rise to a "multidomain" or "textured" structure at a "mesoscopic" scale (on the order of a μm) [Srinivasarao and Berry, 1991, p. 380; Burghardt and Fuller, 1991, p. 2550]—see discussion below. In general, disclinations can only be avoided through the use of an especially strong external field (magnetic, electric, or flow). Irrespective of its origin, spatial inhomogeneity implies a greater free energy of the system. This can be accounted for by a correction to the Helmholtz free energy density which resembles that which accounts for the free energy of a macroscopic elasticity—the "Oseen/Frank distortion energy" [Oseen, 1933; Frank, 1958; Chandrasekhar, 1977, §3.2]. To a first approximation, the Oseen/Frank distortion energy, W, is represented by a lower-order polynomial in terms of the gradients of **n** and S, or **S**, as shown in §11.2B. Obviously, both the homogeneous (equilibrium) and the non-homogeneous (Frank elasticity) components of the Helmholtz free energy are needed for a systematic study of LCs, under static as well as dynamic (flow) conditions.

[1]These are acutally called "disinclinations" in the older literature [Frank, 1958, p. 26].

In the past, other researchers have studied the dynamics of anisotropic fluids, and it has been more than thirty years since the first dynamical theories for liquid-crystalline materials appeared in the literature. In the early 1960s, Ericksen wrote down the conservation laws for LCs and provided the general framework for the continuum description of liquid-crystalline dynamics, based upon a unit-vector description of the local orientation (i.e., the director of Figure 11.1) [Ericksen, 1960a,b; 1961]. Toward the end of the same decade, Leslie completed the development of the dynamical theory by writing down the final form of the conservation laws as well as the entropy inequality for the system [Leslie, 1968]. An informative review of this dynamical theory (the LE theory) was published by Leslie a decade later [Leslie, 1979]—see also §11.3.1. Alternatively, as a thermodynamically consistent continuum model of dynamic behavior, the LE model can be developed using the bracket formalism, as shown in §11.4.

In general, the LE model was remarkably successful in describing the rheological behavior of low molecular-weight LCs, especially at low deformation rates. A particularly important feature of this model was the incorporation of the elastic energy arising due to spatial inhomogeneities in the orientation—the Oseen/Frank distortion energy, W—not to mention the inclusion of the effects of director rotation. However, the theory still suffered from several drawbacks, most of which were associated with the fact that the internal parameter, $\mathbf{n}$, used in the theory carries no information on the deviations from a perfect, uniaxial mesophase. Thus, it was only applicable in the well-oriented, liquid-crystalline regime. Any realistic model of LCs has to account for changes in the degree of orientation, as, for example, communicated through the order parameter S or $\mathbf{S}$ defined above, in order to be able to predict phase transitions. Moreover, one cannot describe a biaxial LC (i.e., a LC with two characteristic directions of orientation) in terms of a single unit vector. Thus, the use of the unit vector to describe the orientation excludes from the problem formulation interesting physics and experimentally observed behaviors which occur in LCs.

In the mid 1970s, Hess introduced an alternative microscopic description which was especially applicable for the prediction of the dynamic behavior of LCs. Hess developed a Fokker/Plank equation for the orientational distribution function where the interparticle correlation was taken into account through a molecular field, which, for a uniaxial alignment, becomes equivalent to the Maier/Saupe interaction [Hess, 1976a]. In the same work, Hess also developed a series of "hierarchy" equations for the time evolution of the second and fourth moments of the orientation distribution functions. This was accomplished by neglecting contributions arising from higher-rank tensors than fourth, and by replacing the expression thus obtained for the fourth-rank tensor in the evolution equation for the second-rank tensor with a single constitutive equation for the second moment of the distribution function [Hess, 1976a, Eq. (3)]. The key advantage of this microscopic approach is that the interaction of the orientation dynamics with the flow is solely dictated by two microscopic parameters, the values of which can be inferred (in principle) from hydrodynamics. For example, in the absence of intermolecular interactions, these coefficients have been determined in the analysis of the flow orientation in dilute colloidal suspensions [Pokrovskii, 1972; Hess, 1974] following the pioneering work of Jeffery in the analysis of the flow orientation of rigid ellipsoids [Jeffery, 1922]. The values of these coefficients are critical in defining a "tumbling" or "flow-aligning" behavior in simple shear flow [Hess, 1976a, p. 1036], as discussed in more detail in §11.3.4. This theory was subsequently used by Hess to describe changes in the isotropic/nematic transition induced by the flow [Hess, 1976b]. Later, Hess also added a non-homogeneous component to the free energy arising from gradients in the molecular orientation which led to an estimation of the Frank elasticity coefficients [Hess and Pardowitz, 1981; Pardowitz and Hess, 1982]. Although Hess's microscopic theory correctly captured the influence of the flow on the molecular orientation, his elaboration of the inverse effect (i.e., the influence of the molecular orientation on the flow) was less successful. A thermodynamically consistent theory was not developed until a few years later by Doi [1981].

The starting point of Doi's theory was similar to Hess's; i.e. he used kinetic theory for describing the dynamics of lyotropic, polymeric liquid crystals by writing a diffusion equation for the distribution function of rigid, rod-like molecules [Doi, 1981; and Doi and Edwards, 1986, chs. 8-10]. However, in addition to Hess's development, Doi used more powerful thermodynamic arguments based on the calculation of the work of deformation in order to obtain an expression for the stresses in terms of the molecular orientation—see also §11.3.2 for a more detailed description. Similarly to Hess, in order to simplify the theory, primarily for applications to non-homogeneous flows, Doi also suggested a continuum approximation based on an internal tensor parameter representing the second moment of the distribution function. However, to develop a closed set of equations he had to restrict the excess information on the kinetic theory level by using one of the decoupling approximations mentioned in chapter 8. As compared to the LE theory discussed previously, this restricted continuum theory, in terms of a conformation tensor, also had successes: it was applicable in a broader flow regime where the degree of orientation of the LC system can vary, and it could describe rudimentary phase-transition behavior. More specifically, Doi's theory could explain the decrease (roughly by a factor of two) in the viscosity typically observed between the isotropic and the nematic phases at the isotropic/nematic phase transition [Wissbrun, 1981, p. 637]. It also seemed to correctly predict the dependence of the viscosity (in the Newtonian region II—see second paragraph below) on the molecular weight for lyotropic PLC's. Moreover, the restricted Doi theory has been applied as a phenomenological model with two adjustable parameters with success in fitting the steady shear behavior in regions II and III (see the definitions below) and the steady extensional behavior observed with lyotropic (Hydroxypropylcellulose, HPC) PLCs [Metzner and Prilutski, 1986].

The success of the predictions for the first normal stress is partial. For the Newtonian region II of liquid-crystalline behavior, the Doi theory predicts a linear dependence of the first normal-stress difference with respect to the shear rate, which appears to qualitatively agree (except, occasionally, for the observation of a negative sign) with the experimental data [Kiss and Porter, 1978, Fig. 16; Moldenaers and Mewis, 1990, Fig. 1]. However, the predictions are quantitatively off by at least a factor of two [Moldenaers and Mewis, 1990, p. 362]. Doi theory's primary liabilities are due to the fact that it cannot describe the influence of spatial inhomogeneities in the orientation (i.e., the disclinations) on the free energy of the system, and, to a lesser effect for PLCs (which are typically highly viscous), inertial effects associated with molecular rotation. Debate has also arisen as to the suitability of the decoupling approximation employed by Doi for describing liquid-crystalline rigid-rod systems because of the inability of the original form of the restricted theory to describe phenomena associated with director tumbling [Marrucci, 1985, p. 1547; Marrucci and Maffettone, 1989; Larson, 1990]. Tumbling phenomena in PLCs have recently been verified through both direct optical measurements (on a 20% by weight PBG solution [Burghardt and Fuller, 1991, p. 2550]) and mechanical experiments (on a uniformly aligned sample through the action of an electric field of poly(n-hexyl isocyanate) in p-xylene [Yang, 1992]).

A more complex behavior, also indicating that the macromolecules are not homogeneously aligned with the flow, has been observed through conoscopic measurements on a liquid-crystalline solution of PBT by Srinivasarao and Berry [1991]. Ironically, the potential for tumbling was preserved (and duly mentioned) in the evolution equation for the second moment of the orientation distri-bution function proposed several years before Doi by Hess [Hess, 1976a] which was also based on a rotational diffusion (Fokker/Planck) equation, as explained above in the introduction. Indeed, as seen in §11.6.5, the description of a tumbling behavior can be relatively easily reinstated into the second-rank conformation tensor theory through an extension which arises naturally when we examine the reformulation of that theory under the perspective of the bracket approach. These drawbacks of the restricted Doi theory are also thought to be primarily responsible for the inability of the theory to correctly account for the oscillatory transient phenomena [Moldenaers and Mewis, 1986; 1990] which are typically observed and negative first normal stress differences [Kiss and Porter, 1978, pp. 202-04; Wissburn, 1981, pp. 643-46; Moldenaers and Mewis, 1990, Fig. 1]—see also §11.3.3.

The behavior of the viscosity at variable shear rates is of importance to the processing of PLCs. In contrast with the viscoelastic, shear-flow behavior corresponding to solutions and melts of flexible polymers, three regions of flow behavior are thought to occur with PLCs—see Figure 11.2. At very low shear rates, PLCs (in the liquid-crystalline mesophase) show generally a shear-thinning behavior, followed by a constant viscosity (Newtonian) region for low to intermediate shear rate values, and a final shear-thinning region at intermediate to high shear rates [Onogi and Asada, 1980, p. 129; Wissbrun, 1981, §IV; Einaga *et al.*, 1985, Fig. 9; Yanase and Asada, 1987, pp. 281-83; Muir and Porter, 1989, §3].

A number of PLC solutions exhibit behavior showing all three regions, albeit some show only region I or regions II and III [Wissbrun, 1981, Fig. 8]. Sometimes the data are in conflict; for example, Yanase and Asada [1987, p. 284] report a three-region behavior for 20 wt% solution of PBG in m-cresol, whereas Moldenaers and Mewis [1986, p. 571] saw only a region II and III behavior for a 12% solution of PBLG in the same solvent. However, it is believed that most PLCs do indeed show a shear-thinning behavior at very low shear rates [Muir and Porter, 1989, p. 85]. This behavior has sometimes been described as viscoplastic, characterized by a yield value [Wissbrun, 1981, p. 648], but at least in the 40 wt% solution of lyotropic HPC in glacial acetic acid, investigated by Metzner and Prilutski [1986, p. 679], this was definitely not the case since these authors reported the observation of small rising bubbles through the sample. This is also consistent with the observations of Einaga *et al.* [1985] on nematic solutions of PBT and PTTA which exhibited a fluid-like behavior, even at the lowest shear rates. Moreover, the slope of the function $-d(\ln\eta)/d(\ln\dot{\gamma})$ was not as large as unity, as would have been expected for a material with a yield stress [Berry, 1988, p. 353].

Rheo-optical measurements suggest that the rheological behavior in region I is associated with the motion of "domains" within the sample which are regions of more or less uniform orientation separated by abrupt changes in the orientation (disclinations) [Onogi and Asada, 1980, p. 129; Wissbrun, 1981, p. 650 and 1985, p. 161; Muir and Porter, 1989, p. 85; Burghardt and Fuller, 1991]—see also Figure 11.3. It is believed that with increasing shear rate the domains become smaller in size and more oriented [Muir and Porter, 1989, p. 85], as recently verified directly through rheo-optical measurements [Burghardt and Fuller, 1991, p. 2550]. This decrease in the characteristic size of the domains with increasing intensity of the external field (a role herein played by the flow and quantified by the shear rate) is also consistent with both observations and

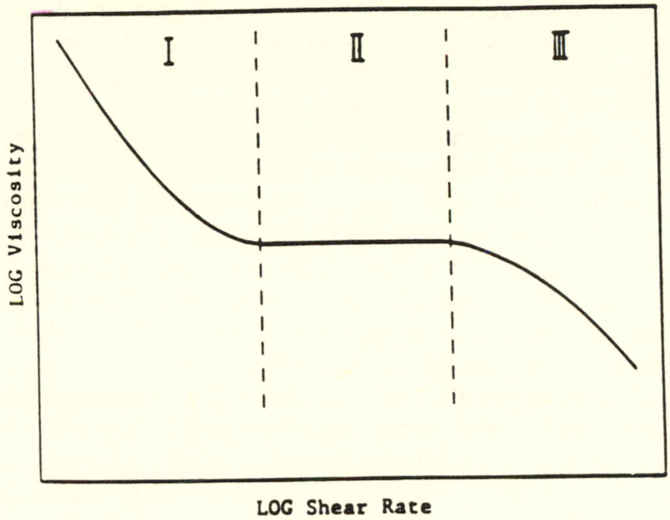

Figure 11.2: Schematic representation of the shear rate dependence of viscosity for liquid-crystalline polymers. (Reprinted with permission from [Onogi and Asada, 1980, Fig. 3].)

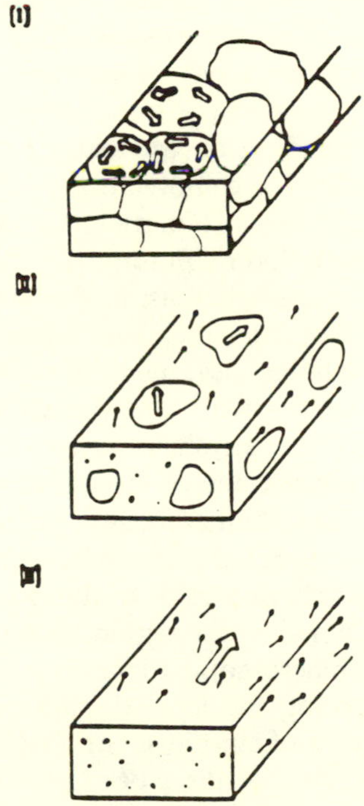

Figure 11.3: Schematic representation of the bulk structure of polymer liquid crystals after Onogi and Asada: [I] Piled polydomain system, [II] dispersed polydomain system, and [III] monodomain continuous phase. (Reprinted with permission from [Onogi and Asada, 1980, Fig. 8].)

experiments concerning similar changes pertaining to the size of other field-induced periodic structures in nematic liquid crystals, such as the twist/bend Fréedericksz distortion observed after the sudden application of intense magnetic fields [Lonberg *et al.*, 1984, Fig. 4].

Several attempts have been made to model the dynamic behavior of PLCs as induced by the presence of domains [Marrucci, 1985; Wissbrun, 1985; Marrucci and Maffettone, 1990a,b; Larson, 1990; Burghardt and Fuller, 1990; Larson and Doi, 1991]. The first two approaches offered an order-of-magnitude estimation of the correlation between the relevant para-meters of the problem by equating the Oseen/Frank elastic forces to the shear ones for an average domain [Marrucci, 1985, p. 1549; Wissbrun, 1985, p. 162]. The second model, utilizing an "ansatz" for the viscosity originated from non-equilibrium molecular dynamics for dense fluids, led to specific predictions for the shear-thinning behavior of the viscosity which are consistent with the observed behavior in region I. Later, Larson and Doi [1991] improved this phenomenological model by also considering the liquid-crystalline contributions to the stress through a "mesoscopic averaging" approximation of the LE equations, thus allowing them to qualitatively represent the transient experimental data of Moldenaers *et al.* [1989, Fig. 4]; see also Larson and Doi [1991, Figs. 5,6].

A more rigorous approach has been followed by Burghardt and Fuller [1990]. These authors have solved numerically the LE model equations for transient, two-dimensional, simple shear flow between two parallel plates. Depending on the sign of the ratio of the LE viscosity coefficients, α_3/α_2, they have shown that either an aligning or a tumbling regime is established—see also pertinent discussion in §11.3.4.A.2. In the first regime, the orientation monotonically changes from the initial to the final (steady-state) value, whereas, in the second, the orientation decays to its steady-state value after the observation of oscillations which decay in amplitude. In the latter case, the total rotation of the director depends on the distance from the wall and the value of the Ericksen number, $Er \equiv \tau/(K_1/d^2)$, where τ is the shear stress, K_1 is the Oseen/Frank elastic constant for splay deformation of the director field, and d is the distance between the plates. Based on an order of magnitude analysis, Burghardt and Fuller have demonstrated that the *local* Ericksen number, determined based on a distance d characteristic of the experimentally observed defect structure is, typically (i.e., in the region II), much higher than the Ericksen number based on a macroscopic distance for the flow, and it is expected to be the one governing the local dynamics. Moreover, following the transient data (start up of simple shear flow) of Moldenaers *et al.* [1989, Fig. 4], which indicate a scaling of the oscillations with strain and a superposition of the normalized results with shear rate, they hypothesized the existence of a limiting value for the Ericksen number, Er_{sat}, estimated to be in the range 50-200 [Burghardt and Fuller, 1990, p. 991].

Burghardt and Fuller were also able to explain the constancy of the recoverable strain for tumbling nematics observed in constrained recoil experiments on PBLG and HPC solutions by Larson and Mead [1989, Figs. 6,7] based on their numerical results for the tumbling regime [Burghardt and Fuller, 1990, Fig. 7a], and the assumption of a limiting value for the Ericksen number [p. 986]. Thus, they have verified, indirectly, the suspected occurrence of director tumbling for these lyotropic PLC systems. Moreover, the same assumption for the existence of Er_{sat} leads to the scaling of the relaxation time upon cessation of shear flow with the previously established shear rate [p. 987], again in agreement with the experimental observations on PBLG [Larson and Mead, 1989, Fig. 6; Moldenaers and Mewis, 1990, p. 367] and HPC [Larson and Mead, 1989, Fig. 7]. This time scale for the relaxation phenomena is also consistent with experimental observations of structural behavior, as, for example, the ones related to band formation upon cessation of shear flow [Burghardt and Fuller, 1990, p. 988]. Later, Burghardt and Fuller conducted a detailed mechanical and rheo-optical study on the flow reversal and constrained recoil of a 20 wt% PBG solution which led to results consistent with their numerical predictions [Burghardt and Fuller, 1991].

It is also of interest to note here the remark of Burghardt and Fuller [1990, p. 989] concerning the similarities and the differences between their model and the previous ones by Marrucci and Wissbrun. Although the hypothesis for a limiting value of the Ericksen number leads naturally to a scaling law for the characteristic defect structure, similar to the one derived in the Marrucci and Wissbrun models, it is a qualitatively different prediction in that Marrucci and Wissbrun assumed the scaling law to be applicable for the shear-thinning region I, whereas Burghardt and Fuller considered it more appropriate for the intermediate (Newtonian) region II, instead. This restriction results from the requirement that the magnitude of the shear rate is low enough so that the viscoelastic polymer effects can be safely neglected, but high enough so that the corresponding characteristic length is smaller than the macroscopic one established by the flow geometry. Thus, it is believed that the shear-thinning flow behavior in regime I is a result of the confining walls interacting with the flow defects, as also advocated by Berry [1988, pp. 353-54; 1991, p. 976]. This interpretation is consistent with very recent experimental results which showed that the viscosity measured at intermediate times (100-500s) in a previously aligned unidomain sample (as verified by conoscopic measurements) can be significantly lower than the viscosity observed with a polydomain sample [Yang, 1992]. We could actually draw a rough analogy between a defect-ridden nematic and a foam with the Frank elasticity and the domain boundaries playing the role of the surface tension and liquid bridges. In this case, a shear-thinning yield-type behavior is expected, as experimentally observed

[Khan *et al.*, 1988, Fig. 6] and successfully modeled theoretically [Princen, 1983; Khan and Armstrong, 1986; 1987; Kraynik *et al.*, 1991] in foams and concentrated liquid emulsions.

The other polydomain theories [Marrucci and Maffettone, 1990; Larson, 1990] are based on an analysis of the molecular theory of Doi in its original formulation, in terms of a distribution function. Domain interaction effects have been neglected, or, at best, taken into account in an "ad hoc" phenomenological fashion. Marrucci and Maffettone [1990] estimated the rheological response of a polydomain nematic material by simply calculating the average either with respect to time (intermediate shear rates) or with respect to a uniform distribution of initial rod orientations (high shear rates), based on the response calculated in the single domain situation. They evaluated the latter by solving a two-dimensional approximation—i.e., restricting the orientations of the rods to the shear plane—of the Smoluchovski (diffusion) equation proposed by Doi and based on the mean-field approximation for the rod interactions represented by a Maier/Saupe potential (see §11.2.1.B). The use of the molecular theory of Doi allowed the predictions of the flow behavior at intermediate as well as at high shear rates, where a nonlinear, shear-thinning rheological behavior is observed. Indeed, both the steady shear viscosity and first normal-stress difference predictions are consistent with typical experimental observations, such as the one obtained with PBLG [Marrucci and Maffettone, 1990, p. 1233]. The neglect of the interdomain interactions seems not to have much influence on the steady-state predictions, especially the ones corresponding to high shear rates, as expected since the stress tensor is dominated there by the liquid-crystalline contributions within the domains. Of course, the shear-thinning response corresponding to regime I (very low shear rates) cannot be predicted. In contrast, the success of the averaging procedure in determining the transient response of the system is limited to, at best, a qualitative agreement of the predictions with the experimental data [Marrucci and Maffettone, 1990, pp. 1229-30,1242-43]. Moreover, in the case of the tumbling nematic regime at intermediate shear rates (region II), to obtain even this qualitative agreement in the transient material behavior it was necessary to introduce a phenomenological correction (in the form of a pseudo-diffusivity) to the Smoluchovski equation in order to take into account the interdomain interactions. Of course, the interactions arise because of sudden changes in the director orientation and are governed by the Oseen/Frank elasticity terms, which are absent from the original Doi formulation. They appear to be of importance in determining the transient material behavior.

The work of Larson [1990] progressed along parallel lines to that of Marrucci and Maffettone with the use of the fully three-dimensional

Smoluchovski equation and a different potential (of the Onsager form, see §11.2.1.C) as primary differences. Because of the use of the more accurate (however, also more computationally involved) three-dimensional formulation, the steady-state predictions are closer to the experimental observations [Larson, 1990, Figs. 11,15]. In addition, a semi-quantitative prediction of the first and second normal-stress differences measured for a 14 wt% PBLG solution has been demonstrated at intermediate and high shear rates [Magda *et al.*, 1991, Fig. 15]. The agreement becomes even more noteworthy given the fact that an experimentally measured abnormal (for viscoelastic standards) region corresponding to negative first and positive second normal-stress difference values was correctly predicted by the model. Moreover, the predicted transient behavior is richer than the one obtained by Marrucci and Maffettone. A regime where the director "wags" around a fixed direction, resulting in sustained or even damped—for higher shear rates—oscillations in time, is also predicted [Larson, 1990, p. 3989] in addition to the tumbling observed at low shear rates [p. 3988]. However, the lack of accounting for inter-domain elastic interactions again limits, as in the Marrucci and Maffetone work, the validity of the predictions to intermediate and high shear rates, especially for the transient behavior. Of course, when Doi's molecular theory is directly compared against experimental results, the fact that most of the available PLCs consist of macromolecules which are neither monodisperse (i.e., of the same molecular weight) nor perfectly rigid, as the theory assumes, needs to be taken into account. Although considerable theoretical effort has been devoted to taking into account these two effects in the thermodynamic description of PLCs at conditions of static equilibrium (see §11.2.1), little work has been devoted to address the same issues in the theoretical description of the flow behavior. Indeed, we know of only two such studies so far: Marrucci and Grizzuti [1984] have adapted Doi's theory for a concentrated solution of polydisperse rodlike polymers; Semenov [1987] has studied the dynamical behavior of an athermal nematic solution of semiflexible, persistent macromolecules for slow flows. In the following, for simplicity, we shall neglect these two effects in the theoretical development.

Even so, from the review of the analyses available thus far of the dynamical behavior, it is obvious that a theory able to accommodate, even qualitatively, the observed behavior in all regions is still lacking. The LE model provides a good accounting for the low to intermediate shear rate dynamic flow behavior, whereas the Doi model, in its original distribution-function formulation (with the drawback of a substantial increase in the computational requirements for the solution of the more detailed equations involved in this microscopic formulation) produces results which are in qualitative agreement with experiments at intermediate and high shear rates. The situation is even worse if we limit our selection to continuum models, although, still, we can find that each one of the previous models is reasonably successful within a certain limited domain of application: the LE theory is applicable for slow flows whereas the (restricted) Doi theory seems appropriate for the description of high shear rate flows when the contributions from the elastic forces induced by structural non-homogeneities are not important. Therefore, it makes sense to require from any continuum model, with an extended domain of validity, the reduction to the previous theories under the above-mentioned special circumstances.

This was the set-up of the problem that we required our dynamical equations to describe from the outset. Can the advantages of both previous theories be incorporated into a single model without simultaneously carrying along most of their disadvantages and limitations? Moreover, is it possible to transfer to a simpler continuum formalism the richer dynamical behavior corresponding to the original (distribution function) Doi formalism? We believe that the development of the conformation tensor theory in §11.5 answers both these questions affirmatively, as its comparison with the previous theories shows in §11.6. A significant component of this modeling task is carried out by a consistent and systematic use of the bracket formalism through various levels of description. As an example of the benefits gained, it suffices to mention here that, as shown by Edwards [1991] and further explained in §11.6.3, the deficiency of the restricted Doi model in reproducing director tumbling phenomena can be understood through its inertial bracket representation (see §10.3) and can be easily rectified though a modification of the objective time derivative of the order parameter tensor.

Concluding this brief overview of the most important characteristics of flowing LCs and PLCs, let us summarize our understanding of their very different levels of description. When the molecules of a liquid-crystalline mesophase are visualized as rigid rods, they can be described by a single orientation (unit) vector, $\mathbf{n}$, on an individual basis. Collectively, a probability distribution function Ψ is necessary for the most general description, of which internal parameter models use the various moments. In Figure 11.4, we depict schematically the levels of description on which LC theories may be based from a continuum perspective. Of course, the most exact method for modeling LC dynamics is through the higher-order phenomenological theories; however, as before, these are still too complicated, involving far too many parameters to be of any practical use. The most general method of description which does allow solutions to be obtained for homogeneous flow problems is the one based on the probability distribution function Ψ. However, to evaluate the solution to general non-homogeneous flow problems, a further abstraction to the next description level as shown in Figure 11.3 is necessary, reducing the unknown variables to a single conformation tensor field, $\mathbf{m}$,

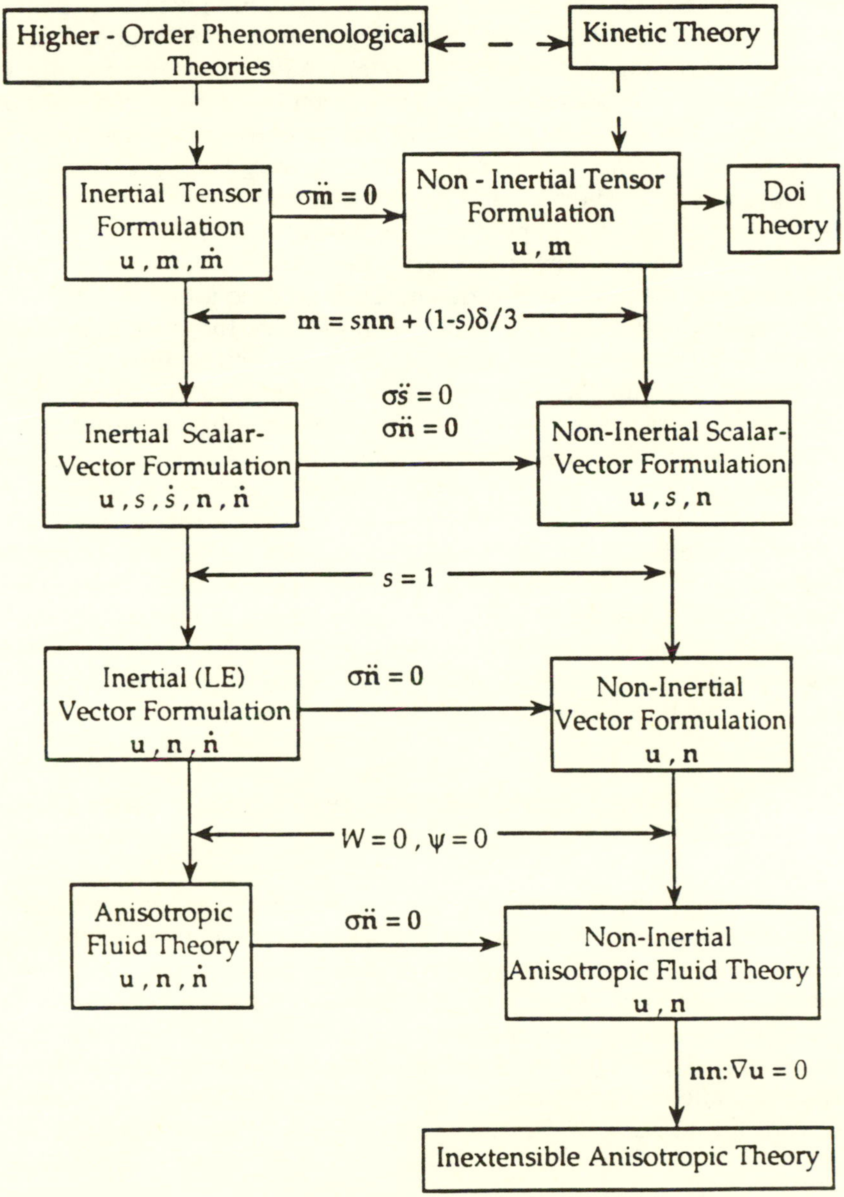

Figure 11.4: Levels of description for liquid-crystalline systems.

and the velocity vector field, **v**. As special cases, we get all of the lower-level theories represented by the scalar-vector[2] and vector levels when **m** reduces by definition to $\mathbf{m} \equiv s\mathbf{nn}+(1-s)\boldsymbol{\delta}/3$ and $\mathbf{m} \equiv \mathbf{nn}$, respectively, where s is a scalar order parameter describing the spread of the distribution around the director, similar to S but not necessarily defined by (11.1-1). In addition, the Doi theory may also be obtained under suitable restrictions of the general conformation tensor theory. As we reduce the level of description, we preclude more and more physics from the ensuing development. Hopefully, the second-rank tensor theory in terms of **m** represents the best compromise between the higher-order (but very complicated) theories and the simple (but often oversimplified) vector description. The important point, however, as we shall see, is that the Hamiltonian form dictated by the generalized bracket is preserved as we transcend the cascade of different levels of description.

In most situations, it is generally believed that the inertia associated with molecular rotation is negligible; however, it conceivably plays a role in some circumstances. Therefore, it is important to incorporate these effects into the modeling as well. In fact, one can split the modeling into two categories: one where the molecular inertia of rotation is included explicitly (inertial theory—see §10.3); and one in which it is neglected (non-inertial theory). These categories are also reflected in Figure 11.4, where the inertial theory treats the material derivative of the structure descriptor (i.e., $\dot{\mathbf{m}}$, $\dot{\mathbf{n}}$, or $\dot{s}$) as a dynamic variable as well. Hence it is crucial for the consistency of the modeling approach that the equations developed in this chapter for the inertial theory reduce to the equations developed for the non-inertial theory when inertial effects are neglected; i.e., when $\sigma \to 0$ (σ being an inertial constant as in §10.3).

11.2 Thermodynamics of Liquid Crystals under Static Conditions

Before we begin discussion of the theories describing the dynamic liquid-crystalline behavior, it is helpful to review in the present section the most important theories pertaining to the modeling of the thermodynamics of liquid crystals under static conditions. We can further distinguish between the description of the homogeneous thermodynamics of LCs, presented in §11.2.1, and the description of the non-homogeneous thermodynamics, allowing for spatial variations of the state variables (primarily the director, **n**, but also the order parameter, S), discussed in §11.2.2.

[2]Ericksen [1991] has recently attempted to extend the LE theory to cases where the degree of orientation varies by working on a scalar-vector level. This theory will be discussed when appropriate, later in this chapter—see §11.3.2.

11.2.1 *Equilibrium thermodynamics of homogeneous liquid crystals*

Several macroscopic (continuum) and microscopic (through statistical mechanics) approaches have been followed for the description of homogeneous liquid-crystalline thermodynamics. We shall review here the ones which are of importance to the description of the dynamic behavior of PLCs, placing our emphasis on the formulation of the contribution of the liquid-crystalline structure to the system (Helmholtz) free energy density, a. Although no theory can claim, as yet, quantitative agreement with experiments conducted with real liquid crystals (especially with PLCs) most of them provide a quite adequate qualitative explanation of the phenomena. Moreover, these theoretical results (or, at least, good approximations) can be obtained by using only a small number of parameters in order to characterize the internal structure of the medium under investigation.

A. Landau/de Gennes theory

The first approach, named after Landau and de Gennes, is a phenomenological one and involves the approximation of the system (Helmholtz) free energy density, a, by a truncated polynomial with respect to the invariants of **S** [Landau and Lifshitz, 1958, ch. XIV; de Gennes, 1969b; Fan and Stephen, 1970, p. 500; Stephen and Straley, 1974, p. 621; de Gennes, 1974, p. 48; Mada, 1984, Eqs. (1),(3); Gramsbergen *et al.*, 1986, Eq. (2.2)][3]

$$\begin{aligned} a = a_{\mathrm{iso}} &+ \tfrac{1}{2} A S_{\alpha\beta} S_{\alpha\beta} - \tfrac{1}{3} B S_{\alpha\beta} S_{\beta\gamma} S_{\gamma\alpha} \\ &+ \tfrac{1}{4} C S_{\alpha\beta} S_{\alpha\beta} S_{\gamma\delta} S_{\gamma\delta} - \tfrac{1}{3} \Delta\chi H_\alpha S_{\alpha\beta} H_\beta \\ &- \tfrac{1}{3} \frac{1}{4\pi} \Delta\alpha E_\alpha S_{\alpha\beta} E_\beta \, , \end{aligned} \tag{11.2-1}$$

where a_{iso} is the Helmholtz free energy density of the isotropic ($\mathbf{S} = \mathbf{0}$) state, in general dependent on the thermodynamic variables p and T and the external electric and magnetic field intensities, **E** and **H**, respectively. A, B, and C are phenomenological coefficients, in general dependent on

[3]The original form proposed by de Gennes [1969b] and Fan and Stephen [1970] for the Helmholtz free energy density also contained two extra terms involving the gradients of **S**. These are not included in (11.2-1) since they are relevant to the modeling of nonhomogeneous phenomena only, the discussion of which, for simplicity in presentation, is deferred to §11.2.2.

the pressure and the temperature, and $\Delta\chi \equiv n(\chi_\parallel - \chi_\perp)$ and $\Delta\alpha \equiv \varepsilon_\parallel - \varepsilon_\perp$ are the anisotropy in the magnetic susceptibilities per unit volume and the anisotropy in dielectric constants, respectively, where $\chi_\parallel(\chi_\perp)$ and $\varepsilon_\parallel(\varepsilon_\perp)$ are the magnetic susceptibility of a single molecule and the dielectric constant of a fully aligned nematic, respectively, with the property measured in a direction parallel (normal) to the molecular long axis [Sheng and Priestley, 1974, p. 162].

Note that although the first four terms in expression (11.2-1) arise as the most general expansion for the free energy in terms of the independent rotational invariants of **S**, up to the fourth order, and are, in consequence, valid only for small magnitudes of **S**, the last two terms expressing the influence of the magnetic and electric fields are valid for arbitrary values of the order parameter. Thus, these terms can be used to represent the influence of an external magnetic and electric field together with any one of the following descriptions for the nematic free energy, from which, for simplicity, they will be omitted in the following. For a symmetric and traceless tensor (such as **S**), there are no other independent moments of **S** of order greater than or equal to four [Stephen and Straley, 1974, p. 621]. The alternate quartic invariant $S_{\alpha\beta}S_{\beta\gamma}S_{\gamma\varepsilon}S_{\varepsilon\alpha}$ is not independent of the quartic invariant appearing in (11.2-1). Rather, as shown by Wright and Mermin [1989, App. B] and Edwards *et al.* [1990b, App. A]—see also Theorem 11.4 and Example 11.3—we have

$$S_{\alpha\beta}S_{\beta\gamma}S_{\gamma\varepsilon}S_{\varepsilon\alpha} = \frac{1}{2} S_{\alpha\beta}S_{\alpha\beta}S_{\gamma\delta}S_{\gamma\delta} \,. \quad (\mathbf{S} \text{ symmetric and traceless}) \quad (11.2\text{-}2)$$

It should be mentioned here that an essential feature of Landau's theory, on which the first four terms in expression (11.2-1) are based, is the assumption that all thermodynamic functions admit regular series expansions with respect to the thermodynamic state variables. Although next to transition points theoretical work on soluble models—such as the two-dimensional Ising model—have shown this assumption to be invalid, Landau's general description of the thermodynamic consequences due to symmetry changes and the kind of phase transitions to be expected remain valid and have been successfully used in many applications [Ter Haar, 1969, p. 7]. In particular, it can be shown[4] that the coefficient B of the third-order term in **S** in (11.2-1) is never zero (except, possibly, at an

[4]The existence of non-zero odd terms in **S** in the free energy expansion is due to the fact that a change in the sign of **S** has a non-trivial physical (microscopic) consequence: a change in the orientational conformation of the molecules from axial to planar. Since there is no reason why the free energy corresponding to these two physically distinct cases should be the same, the change of sign in **S** should, in general, induce a change in the free energy, thus, implying the existence of non-zero contributions to the free energy from odd terms [de Gennes, 1974, p. 48; Stephen and Straley, 1974, p. 622; Sheng and Priestley, 1974, p. 155].

isolated point), thus guaranteeing that the isotropic ($\mathbf{S}=\mathbf{0}$) to nematic ($\mathbf{S}\neq\mathbf{0}$) transition is first order (except, possibly, at an isolated point) [Landau, 1937, p. 215]. However, it must be emphasized that the discontinuities in the variables (for example, in the density) accompanying the isotropic to nematic phase transition are small: the transition is "weakly first order" which implies that the coefficient B is relatively small [de Gennes, 1974, p. 49].

For uniaxial nematics, substitution of $\mathbf{S}$ from (11.1-3) into (11.2-1) allows us to express the free energy as a function of S only as [Stephen and Straley, 1974, p. 622]

$$a = a_{\text{iso}} + \frac{1}{3}AS^2 - \frac{2}{27}BS^3 + \frac{1}{9}CS^4 , \tag{11.2-3}$$

in the absence of an external field. Minimization of the expression (11.2-3) with respect to S provides one to three solutions for S representing the one to three equilibrium steady states—see also Figures 11.7 and 11.8, which are found in §11.5. Typically, above a critical temperature (or, below a critical concentration, depending upon whether the system is thermotropic or lyotropic) the parameter A is positive, thus guaranteeing the existence of the isotropic ($S=0$) state at least as a local minimum. Below that characteristic temperature (respectively, above the characteristic concentration) the isotropic state becomes unstable and is replaced by two anisotropic states, of which the nematic one ($S>0$) is physically realizable. Other quantities of thermodynamic interest (like the value of S at and beyond the phase transition, etc.) can also be provided from (11.2-3) in a straightforward way [Stephen and Straley, 1974, p. 622]. Because of the neglect of contributions higher than fourth-order in S into the free energy, the Landau expression, provided by (11.2-1) or (11.2-3), is expected to be quantitatively valid only for low values of S (isotropic state), typically up to $S\approx0.5$, below the isotropic/nematic phase transition [de Gennes, 1971, p. 197; Stephen and Straley, 1974, p. 621]. However, its predictions remain qualitatively valid in the whole region.

The parameters A, B, and C entering the Landau/de Gennes expression shall be considered as functions of the temperature and the concentration (density). Close to the isotropic/nematic phase transition, specific phenomenological expansions can be used in terms of the critical temperature and/or critical concentration in order to determine the influence of these parameters on the onset and appearance of the phase transition [Stephen and Straley, 1974, Eq. (2.2)]. Alternatively, in order to reduce the number of arbitrary parameters involved in the expressions, it is necessary to evaluate the Landau coefficients A, B, and C in terms of physical parameters within the context of an appropriate molecular/statistical model. One approach is to use one of the other theories presented below, for example, the Maier/Saupe theory (see (11.2-7) below). Alternatively, a more specific model to the system under investigation can be used, such as the persistent model of semi-flexible (worm-like) macromolecules of Jähnig [1979], with the Landau coefficients determined through a formal expansion in terms of the order parameter [ten Bosch *et al.*, 1983, §3;[5] Rusakov and Shliomis, 1985]. Note that in this particular case, Rusakov and Shliomis [1985, p. L-942] report that near the critical point for the phase transition this expansion retains all the essential features of the original molecular/statistical model. This observation reinforces the viability of the Landau/de Gennes approach in simplifying the free energy expression for LCs and PLCs; its biggest advantage is the need of only one macroscopic structural parameter, the order parameter (scalar or tensor), for its evaluation. Yet another advantage of the Landau/de Gennes theory is that it can be extended by incorporating dependencies on other order parameters, so that it can describe other liquid-crystalline phases, such as the smectic A [de Gennes, 1972, § 3.1] and smectic C [Chu and McMillan, 1977].

B. Maier/Saupe theory

An alternative to Landau/de Gennes theory has been provided by Maier and Saupe [1958], primarily for LMWLCs. It is based on three assumptions [Stephen and Straley, 1974]:

a) an attractive, orientation-dependent van der Waals interaction between the molecules,
b) the configuration of the centers of mass is not affected by the orientational-dependent interaction, and
c) the mean field approximation.

The Maier/Saupe theory leads to the following expression for the Helmholtz free energy density for a uniaxial liquid crystal [Stephen and Straley, 1974, p. 623]:

$$a(T,S) = -\tfrac{1}{2} n A_0 S^2 + n k_\beta T \frac{\ln C}{4\pi} , \qquad (11.2\text{-}4)$$

where n is the number density of molecules, A_0 is a quantity independent of the molecular orientation and dependent on interparticle spacing, and C is a function of S according to the normalization condition

$$C^{-1} \equiv 2\pi \int_0^\pi \exp\left(-\frac{V(\theta,S)}{k_B T}\right) \sin\theta \, d\theta , \qquad (11.2\text{-}5a)$$

where the intermolecular potential $V(\theta,S)$ is defined, in the uniaxial approximation, as

[5]Later corrected by Rusakov and Shliomis [1985, p. L-936].

$$V(\theta,S) = -A_0 S(\frac{3}{2}\cos^2\theta - \frac{1}{2}) . \tag{11.2-5b}$$

Minimization of $a(S,T)$ at constant T yields the order parameter at the equilibrium state

$$S \equiv 2\pi C \int_0^{\pi} (\frac{3}{2}\cos^2\theta - \frac{1}{2}) \exp(-\frac{V(\theta,S)}{k_B T}) \sin\theta \, d\theta , \tag{11.2-6}$$

which provides a self-consistent evaluation for S, according to its original definition, (11.1-1), based on the distribution function, f

$$f(\theta,S) = C \exp(-\frac{V(\theta,S)}{k_B T}) . \tag{11.2-7}$$

Eq. (11.2-6) has always the isotropic phase (S=0) as a solution. However, this becomes unstable at the critical temperature T_c such that $A_0/(k_B T_c)$= 4.55, in which case a phase transition to a nematic phase, $S=S_c \equiv 0.43$, is predicted to take place. In reality, there is a small discontinuous change in density at T_c, and the proper procedure for the calculation of the phase transition requires us to equate the chemical potential at the isotropic and the nematic phases. However, this results in only a weak dependence of $A_0/(k_B T_c)$ and S_c on the density [Stephen and Straley, 1974, p. 623].

The parameter A_0 in the original theory was meant to represent only the attractive part of the intermolecular forces and, as a consequence, its calculation from the van der Waals interactions alone was found to provide values for the temperature corresponding to the liquid phase transition which were too small [Stephen and Straley, 1974, p. 623]. Thus, the Maier/Saupe theory is best viewed now as a semi-phenomenological model with the parameter A_0 including contributions from both attractive and repulsive intermolecular forces with quadrupole symmetry. Even so, the value for the order parameter S at the isotropic/nematic phase transition ($S \equiv S_c$=0.43 [Stephen and Straley, 1974, Eq. (2.22)]) is too low compared to the one which is experimentally observed, especially with PLCs. Thus, forces of other symmetries (such as near-neighbor interactions) and/or excluded-volume effects are important and need to be taken into account. An account of the efforts to include near-neighbor effects, which are more important for LMWLCs, can be found in Chandrasekhar [1977, §2.5] and are not discussed here. Excluded-volume interactions, of special importance to PLC's, are addressed below. Finally, note that the Maier/Saupe theory has been generalized to two-component mixtures [Palffy-Muhoray *et al.*, 1985].

In macroscopic (continuum) calculations, the Maier/Saupe theory is often easier to use by first adapting its predictions to a Landau/de Gennes form which is obtained by expanding the free energy, (11.2-5), in powers of S. This results in (11.2-3), with the following identifications [Stephen and Straley, 1974, p. 624]

$$A = \tfrac{3}{2} n A_0 \left(1 - \frac{A_0}{5k_B T}\right) , \tag{11.2-8a}$$

$$B = \tfrac{9}{70} \frac{n A_0^3}{(k_B T)^2} , \tag{11.2-8b}$$

$$C = \tfrac{9}{700} \frac{n A_0^4}{(k_B T)^3} . \tag{11.2-8c}$$

In the isotropic phase (even in the presence of external fields), where S is small, these relations provide a very accurate description of the free energy.

C. Onsager theory

A third theory was provided by Onsager [1949] (see also Straley [1973b, §3.1] and Stephen and Straley [1974, p. 624]). This is a microscopic theory based on statistical mechanics, obtained by application of the virial theorem [Chandler, 1987, p. 205] to a collection of orientable particles interacting pair-wise through some potential which depends on both position and orientation. In its most simple form advocated by Onsager [1949], it expresses the free energy of the system in terms of an integral over the distribution function involving the second virial coefficient, $B(\mathbf{m},\mathbf{n})$ [Stephen and Straley, 1974, Eq. (2.24)]:

$$\frac{a}{nk_B T} = \frac{a_0}{nk_B T} + \int f(\mathbf{n}) \ln f(\mathbf{n}) \, \mathrm{d}^3 n + \ln n + \frac{n}{2} \int \int f(\mathbf{m}) f(\mathbf{n}) \, B(\mathbf{m},\mathbf{n}) \, \mathrm{d}^3 m \, \mathrm{d}^3 n . \tag{11.2-9}$$

For the particular case of hard-wall interactions, $B(\mathbf{m},\mathbf{n})$ is just the volume which a particle of orientation $\mathbf{m}$ may not enter due to the presence of another particle of orientation $\mathbf{n}$. In the particular case of cylindrical particles with spherical caps [Stephen and Straley, 1974, Eq. (2.27)],

$$B(\mathbf{m},\mathbf{n}) = 2\pi L d^2 + \frac{4\pi d^3}{3} + 2L^2 d \left|\sin\gamma\right| , \tag{11.2-10}$$

where γ is the angle between $\mathbf{m}$ and $\mathbf{n}$, and L and d are the length and the diameter of the rods. Exact expressions for the second virial coefficient are also available for cylinders and prolate ellipsoids of revolution [Straley, 1973b, App. I]. Since in the limit of long rods, $L/d \gg 1$, only the last term in (11.2-10) is important; it is the only term that must be used to obtain a qualitative answer [Straley, 1973c, p. 10]. Moreover, if $|\sin\gamma|$ is replaced in (11.2-10) by $\sin^2\gamma$, then the equilibrium distribution function f can be evaluated exactly [Isihara, 1951, p. 1146]. It turns out, through comparison with the results of numerical calculations using the exact form, that this approximation results in qualitatively similar predictions. However, there are some quantitative differences: the transition densities and the order parameter are about 20% smaller [Straley, 1973b, p. 341].

To evaluate the equilibrium value for the distribution function f, it suffices (in a first-order approximation) to minimize the free energy with respect to f subject to the normalization condition

$$\int f(\mathbf{n})\, d^3n = 1 . \tag{11.2-11}$$

The solution leads to the integral equation [Stephen and Straley, 1974, p. 624]

$$f(\mathbf{n}) = C \exp\left[-n \int B(\mathbf{n},\mathbf{m}) f(\mathbf{m})\, d^3m\right] , \tag{11.2-12}$$

where C is a constant determined from the normalization condition, (11.2-11). A limiting form of the solution at the anisotropic (nematic) limit has been provided by Onsager [1949], who used a variational method with the trial function for the distribution function f for the orientation of a test rod [Odijk, 1986, p. 2314]

$$f(\mathbf{n}) = \frac{\alpha}{4\pi \sinh\alpha} \cosh(\alpha\cos\theta) , \tag{11.2-13}$$

where θ is the angle between the director and the test rod and α is the variational parameter. As noted by Odijk [1986, p. 2314], the selection of (11.2-13) for the form of the distribution function is a particularly fortunate one because it can accommodate as particular cases all cases of importance, for instances, isotropic ($f = 1/(4\pi)$ for $\alpha = 0$), uniaxial nematic ($f = c_1 + c_2\cos^2\theta$, $\alpha = O(1)$), and highly ordered states (f = Gaussian, $\alpha \gg 1$). Use of this expression for $f(\mathbf{n})$ and minimization of the free energy leads to a description of the isotropic/nematic transition as a first-order transition corresponding (in the limit $L/d \to \infty$) to the following values for the order parameter S (in the nematic phase) and the rod volume fractions ϕ_n and ϕ_i in the nematic and isotropic phase, respectively [de Gennes, 1974, p. 37; Samulski and DuPré, 1979, p. 127; Khokhlov, 1991, p. 97]:

$$S \approx 0.84 , \quad \phi_n \approx 4.5 \frac{d}{L} , \quad \phi_i \approx 3.3 \frac{d}{L} , \tag{11.2-14}$$

where the volume fraction, ϕ, is defined in terms of the number density, n, as $\phi \equiv n\pi d^2 L/4$.

The integral equation (11.2-12) has been solved numerically by Lasher [1970, p. 4144] through an expansion of the distribution function in terms of Legendre polynomials followed by the minimization of the free energy functional. A more accurate solution, together with a formal analysis of the bifurcation (first-order transition), has been presented later by Kayser and Raveché [1978] and Lekkerkerker *et al.* [1984, Table I] which led (again, for $L/d \to \infty$) to the values [Odijk, 1986, p. 2315; Khokhlov, 1991, p. 100]

$$S = 0.792 \ , \quad \phi_n = 4.191 \frac{d}{L} \ , \quad \phi_i = 3.290 \frac{d}{L} \ . \tag{11.2-15}$$

A comparison of Eqs. (11.2-14) and (11.2-15) shows that the simple trial function utilized by Onsager gives a quite reasonable approximation.

The truncation of the virial expansion at the second-order term restricts the validity of the Onsager theory to dilute solutions, $\phi \equiv nd^2L \ll 1$, although the exact limit of validity of the theory has been the matter of considerable debate [Khokhlov, 1991, p. 99]. For a suspension of rigid rods, simple estimates of the second and the next higher (third) virial coefficients at the isotropic/nematic transition (occurring, according to Eqs. (11.2-14) and (11.2-15), at $\phi \equiv \phi_{crit} \propto d/L$) have shown that the contribution from the third virial coefficient is less than 10% that of the second only as long as the particles have a very high aspect ratio, $L/d > 100$ [Straley, 1973c, p. 17]. Moreover, arguments based on estimations of the co-volume per particle (i.e., the volume per particle in which interaction takes place) as between L^3 and L^2d, place doubts on the quantitative validity of Onsager's theory for any aspect ratio L/d as long as $\phi \geq d/L$ [Flory and Ronca, 1979a, pp. 306,07; Ballauff, 1989, pp. 211,12]. The fact that the changes in the concentration and the orientation introduced from the incorporation of the third virial coefficient in the analysis, albeit small, do not seem to decrease as $L/d \to \infty$ [Odijk, 1986, Table I] also points towards the same conclusion. However, the theory certainly remains correct qualitatively for L/d as low as 10 [Straley, 1973b, p. 339; Straley, 1973c, p. 17; Stephen and Straley, 1974, p. 624].

The constraint $L/d > 10$ is certainly difficult to satisfy with LMWLCs but not with PLCs, which explains the popularity of Onsager's theory with the latter. A comparison of the critical volume fraction predictions against various other theories and experimental data obtained with lyotropic PLCs has indeed shown the Onsager approximation predictions to be qualitatively correct when applied to dilute and semi-dilute solutions of long, rigid-rod macromolecules [Straley, 1973b, Fig. 8; Ciferri, 1991, p. 224]. A comparison of the predictions of the values of the order parameter for PBLG and PBA even shows quantitative agreement [Ciferri, 1991, Fig. 11], although this is probably the result of a fortunate cancellation of errors. Onsager's theory eventually leads to results which are qualitatively similar to the Maier/Saupe theory with the principal difference being that the Onsager theory, being most appropriately applied to dilute solutions (lyotropic systems) for which the free energy change with density is relatively small, tends to predict large changes in density at the transition in contrast to the Maier/Saupe theory which is most usually applied to thermotropic liquids with minimal density changes [Stephen and Straley, 1974, p. 624].

Various attempts have been made to extend Onsager's theory for systems at higher concentrations. The most obvious one involves inclusion of the contribution from the third virial coefficient, $C(\mathbf{l},\mathbf{m},\mathbf{n})$, in the virial expansion provided by (11.2-9), as well as consideration of the full expression, (11.2-10), for $B(\mathbf{m},\mathbf{n})$ [Odijk, 1986, §III]. However, the extended theory that results never introduces more than a perturbative effect on the original Onsager theory, without significantly extending its domain of validity: the order parameter is systematically shifted upward, whereas the change in the volume fraction from the isotropic to the nematic phase increases from 30% to 80%, with both ϕ_i and ϕ_n decreasing for $L/d = 20$, and ϕ_i almost unchanged and ϕ_n increasing for $L/d = 50$ [Odijk, 1986, Table 1]. Among the other approaches, the most successful one is undoubtedly the use of a decoupling approximation by Parsons [1979] (see also Khokhlov and Semenov [1985, §3] and Khokhlov [1991, §3]). Starting from a mean-field approximation of the rigorous evaluation of the steric interactions using the pair-correlation function, it reduces in its simplest form to a multiplicative correction term for the second virial coefficient, the last term in (11.2-9). It involves the term $J(\phi)/\phi$, with $J(\phi)$ defined as [Parsons, 1979, p. 1227]

$$J(\phi) \equiv \int_0^{\phi} \alpha(\phi')\, d\phi' \,, \tag{11.2-16}$$

where $\alpha(\phi)$ is a coefficient which depends on the density, $\alpha \to 1$ as $\phi \to 0$, and is usually determined empirically through an equation of state. The advantage of Parsons's approach is that it allows an extension of the Onsager theory in a consistent fashion for concentrated solutions of rigid rods.

Finally, other works have extended Onsager's approach in order to account for polydispersity [Lekkerkkerker *et al.*, 1984, §III,IV; Odijk and Lekkerkerker, 1985] and a partial flexibility of the macromolecular backbone corresponding to a microscopic (molecular) picture of either a chain consisting of freely jointed, rigid segments [Khokhlov, 1978] or to a persistent chain where the flexibility is distributed homogeneously along the chain contour [Khokhlov and Semenov, 1981, 1982]. The most interesting finding from the latest work is the fact the semi-flexibility drastically reduces the value of the order parameter at the isotropic/

nematic transition point, a fact which is still at odds with the available experimental data—the predictions ($S \approx 0.4$-0.42) are approximately half the values measured experimentally. For example, using infrared spectroscopy, $S = 0.76$ for a solution of polyparabenzamide (PBA) in dimethylacetamide (DMA), while $S = 0.94$-0.98 for a solution of PBG in dichloromethane (DCM) and trifluoroacetic acid (TFA) [Semenov and Khokhlov, 1988, §3.5]. Clearly, the aligning effects of polydispersity, higher-order excluded-volume, and/or thermal effects (attractive intermolecular interactions) has to be taken into account. Indeed, some recent works incorporating Parsons's decoupling approximation for the description of excluded volume effects [DuPré and Yang, 1991] and a Maier/Saupe term for the description of thermal effects [Khokhlov, 1991, §10] have already given encouraging results. However, the field is still under development. Thus, we refrain from discussing these theories any further, and instead refer the interested reader to the above-mentioned references, as well as to the reviews by Odijk [1986], Semenov and Khokhlov [1988], and Khokhlov [1991].

D. Flory theory

Flory's lattice theory [Flory, 1956], as later modified by Flory and Ronca [1979a] in order to remove certain shortcomings of the original theory, involves a totally different approach. Here we follow closely the development of the theory offered by Semenov and Khokhlov [1988, §2.2], which makes its comparison with the previous theories easier. Every rod is modeled by a discrete sequence of cells in a cubic lattice. The unit lattice is assumed to be a cube of characteristic dimension d, so that the length of the sequence x plays the role of the aspect ratio L/d (see Figure 11.5a). To describe an inclined rod, Flory used y shorter segments of length x/y each (see Figure 11.5b). If the x axis is taken to coincide with the average orientation of the rods (i.e., with the orientation of the director $\mathbf{n}$ in a uniaxial nematic), then the average value of y can be obtained from the orientation distribution function, $f(\mathbf{u})$, as [Flory and Ronca, 1979a, Eq. (36); Khokhlov and Semenov, 1988, Eq. (2.14)]

$$\bar{y} = \frac{4x}{\pi}\int \sin\theta f(\mathbf{u})\, d^2u \;, \tag{11.2-17}$$

where $\mathbf{u}$ is a unit vector along the direction of the rod and θ is the angle between that direction and the director $\mathbf{n}$.

The free energy density, a, of the system is then considered as the sum of two components

$$a = a_{\text{orient}} + a_{\text{comb}} \;, \tag{11.2-18}$$

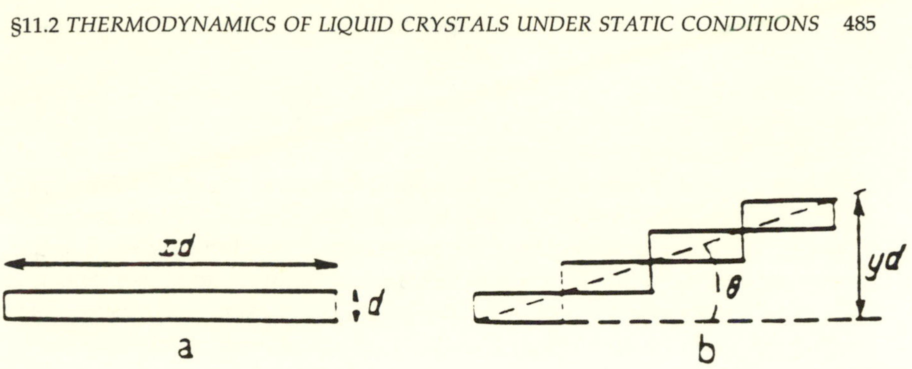

Figure 11.5: Arrangement of rods on a lattice in the theory of Flory. a—Rod oriented along the axis of ordering. b—Inclined rod at an angle θ. (Reprinted with permission from [Semenov and Khokhlov, 1988, Fig. 1].)

where a_{orient} is the (usual) part due to the molecular-orientation distribution,

$$a_{\text{orient}} = nk_BT \int f(\mathbf{n}) \ln[4\pi f(\mathbf{n})] \, d^2n \ , \tag{11.2-19}$$

and a_{comb} is a combinatory (steric) part related to the possible number of ways in which a particular conformation can be realized in the lattice. It is calculated (in the mean-field approximation) as [Flory and Ronca, 1979a, Eq. (7); Khokhlov and Semenov, 1988, Eq. (2.21)]

$$\begin{aligned} \frac{a_{\text{comb}}}{k_BT} &= n_0 \ln(1-\phi) + n\ln\left(\frac{n}{n_T}\right) \\ &- (n_0 + \bar{y}n)\ln\left[1-\phi\left(1-\frac{\bar{y}}{x}\right)\right] + n(\bar{y}-1) \ , \end{aligned} \tag{11.2-20}$$

where n_T is the number density of lattice cells, $n_0 \equiv n_T - nx$ is the number density of the free cells, and $\phi \equiv nx/n_T$ is the volume fraction of rods on the lattice.

Minimization of the free energy leads to the solution of the problem. Driven from probability arguments, consistent with the development of the free energy, Flory and Ronca [1979a] arrived at the following form for the orientation distribution function at equilibrium [Eq. (12)]:

$$f = f_1 \exp(-\alpha \sin\theta) \ , \tag{12.2-21}$$

where

$$\alpha = -\frac{4}{\pi} x \ln[1 - \phi(1 - \frac{\bar{y}}{x})] , \qquad (11.2\text{-}22)$$

and f_1 is a normalization constant. Using this form for the distribution function, Flory and Ronca [1979a] calculated the isotropic/nematic transition as a first-order transition corresponding (in the limit $L/d \to \infty$) to the following values for the order parameter S (in the nematic phase) and the rod volume fractions ϕ_n and ϕ_i in the nematic and isotropic phase, respectively [Flory and Ronca, 1979a, p. 307; Semenov and Khokhlov, 1988, Eq. (2.22)]:

$$S \approx 0.92 , \quad \phi_n \approx 11.57 \frac{d}{L} , \quad \phi_i \approx 7.89 \frac{d}{L} . \qquad (11.2\text{-}23)$$

Comparison between Eqs. (11.2-14) or (11.2-15) and Eq. (11.2-23) reveals that the Flory theory predicts a more oriented nematic state and (more than twice) higher volume fractions at the isotropic/nematic transition than the Onsager approach. The discrepancy between the two approaches may be traced to either a failure in convergence of the virial expansion (see comments by Flory and Ronca [1979a, p. 308]) and/or the artificial character (i.e., errors due to the lattice discretization) of the lattice model [Semenov and Khokhlov, 1988, p. 991]. For lyotropic PLCs, Flory's theory seems to provide better predictions for the critical polymer volume fraction than Onsager's [Ciferri, 1991, p. 224]. A closer inspection of Flory's, Onsager's, and the extension of Onsager's theory to higher concentrations by Parsons or Khokhlov and Semenov [1985, §3], reveals that the three theories are quite similar, only differing in the estimation of the steric contributions to the free energy [Ballauff, 1989, p. 222]. Actually, it can be shown that the Flory lattice theory reduces to Parsons's theory for semi-dilute systems and both Flory's and Parsons's theories reduce to Onsager's theory at low concentrations [Ballauff, 1989].

Flory's theory has also been extended to describe several additional effects, such as polydispersity—the earlier (approximate) theory of Flory [1956] was used; see [Flory, 1982] and references therein—orientation-dependent interactions that are operative between rods [Flory and Ronca, 1979b], and, more recently, a partial-chain flexibility [Flory, 1989; Yoon and Flory, 1989]. The accounting of partial-chain flexibility (semi-flexibility) offers more difficulties using Flory's approach than Onsager's [Abe and Ballauff, 1991, p. 164]. However, in contrast to Onsager's theory, the Flory theory is not limited to low concentrations. This seems to be the most critical factor and explains why, in comparisons with experimental data obtained with lyotropic PLCs, the Flory theory offers better agreement than the other theories [Ciferri, 1991, p. 224]. The observed semi-quantitative agreement between the predictions and the experimental results is encouraging given the limitations of the modeling assumptions (rigid rods) and the discretization errors introduced from the lattice approximation.

11.2.2 The Oseen/Frank description of non-homogeneous nematic liquid crystals

Oseen's theory [1933], as later refounded on a more secure basis by Frank [1958], attempts to take into consideration changes in the free energy of uniaxial liquid crystals as a result of spatial non-homogeneities. Spatial inhomogeneities can arise in liquid-crystalline systems for a variety of reasons. For instance, liquid-crystalline molecules in the vicinity of a solid surface tend to align in a specific direction, parallel or perpendicular to it [Berreman, 1973; Wolff *et al.* , 1973; Mada, 1979; Martin *et al.* , 1986]. Thus, if a thin sample lies between two (similar) surfaces, then the director assumes a uniform orientation compatible with the boundary conditions. A magnetic field or a flow field can disturb a uniformly aligned state in such a way that the direction changes continuously until a restoring force of an elastic nature holds the applied force in equilibrium [Zocher, 1933, p. 945; de Gennes, 1974, §3; Rey, 1991], resulting in a spatially inhomogeneous director orientation. Alternatively, if the two confining solid surfaces are dissimilar so that the imposed orientation of the molecules in their vicinity is incompatible with one another, these boundary conditions can be accommodated only through a change (usually smooth) in the orientation of the director orientation within the bulk of the sample [de Gennes, 1974, Fig. 3.1]. In addition, as already mentioned in §11.1, structural defects in the molecular orientation are almost always present in liquid crystals, implying sudden changes in the director orientation in their vicinity [de Gennes, 1974, §4; Chandrasekhar, 1977, §3.5; Kléman, 1983]. These spatial changes in the orientation of the director imply the presence of a non-zero curvature, and, consequently, a higher free energy. Oseen [1933] and Frank [1958] described this inhomogeneity-induced increase of the free energy through a macroscopic, phenomenological theory analogous to that of solid elasticity. Then the static equilibrium configurations that correspond to various defect structures should, at least locally, minimize the bulk free energy that is subject to the imposed boundary conditions [Cohen *et al.* , 1987; Lin, 1989].

As lucidly explained by Frank [1958, p. 20], the Oseen/Frank theory is analogous, but not identical, to that governing solid elasticity:

> In the latter theory, when we calculate equilibrium curvatures in bending, we treat the material as having undergone homogeneous strains in small elements: restoring forces are considered to oppose the change of distance between neighboring points in the

> material. In a liquid, there are no permanent forces opposing the change of distance between points: in a bent liquid crystal, we must look for restoring torques which directly oppose the curvature. We may refer to these as torque-stresses, and assume an equivalent of Hooke's law, making them proportional to the curvature-strains, appropriately defined, when these are sufficiently small. It is an equivalent procedure to assume that the free-energy density is a quadratic function of the curvature-strains, in which the analogues of elastic moduli appear as coefficients: this is the procedure we shall actually adopt.

Frank goes even further in drawing the difference between the liquid-crystalline, "torque-induced" elasticity and the solid elasticity in estimating the former for elastic media as well. However, an order-of-magnitude analysis shows that it is negligible in magnitude, even for beams of thickness down to about 1μm [Frank, 1958, p. 28].

In the following, we shall present two alternative formulations of the Oseen/Frank elasticity for liquid crystals. The first is described in terms of spatial gradients of the director, **n**, while the second uses an analogous description in terms of spatial gradients of the order parameter tensor, **S**.

A. Description of the Oseen/Frank elasticity in terms of the director

With respect to a reference orientation at the origin, **n**(**0**), aligned along the z axis, Frank [1958] identified three possible independent modes according to which the director can vary in the neighborhood of the origin, as represented in Figure 11.6. In the first one, the "splay" mode, the curvature is represented by the components $\partial n_x/\partial x$ and $\partial n_y/\partial y$, in the second, the "twist" mode, the curvature is represented by $-\partial n_y/\partial x$ and $\partial n_x/\partial y$, and in the third, the "bend" mode, by $\partial n_x/\partial z$ and $\partial n_y/\partial z$. Based on this physical picture, and using material symmetry (i.e., corresponding to the absence of polarity and enantiomorphy) and material objectivity (i.e., corresponding to invariance with respect to rotation around the z-axis), Frank showed that, within the quadratic limit, the curvature elasticity can be represented for nematic liquid crystals in a coordinate-free notation as [Frank, 1958, Eq. (25)]

$$\begin{aligned}\Delta a = {} & \tfrac{1}{2}k_{11}(\boldsymbol{\nabla}\cdot\mathbf{n})^2 + \tfrac{1}{2}k_{22}(\mathbf{n}\cdot\boldsymbol{\nabla}\times\mathbf{n})^2 + \tfrac{1}{2}k_{33}\left|(\mathbf{n}\cdot\boldsymbol{\nabla})\mathbf{n}\right|^2 \\ & - \tfrac{1}{2}(k_{22}+k_{24})\left[(\boldsymbol{\nabla}\cdot\mathbf{n})^2 + \left|\boldsymbol{\nabla}\times\mathbf{n}\right|^2 - \boldsymbol{\nabla}\mathbf{n}:\boldsymbol{\nabla}\mathbf{n}\right],\end{aligned} \qquad (11.2\text{-}24a)$$

or, equivalently [Ericksen, 1966, Eq. (5); 1976, Eq. (10)],

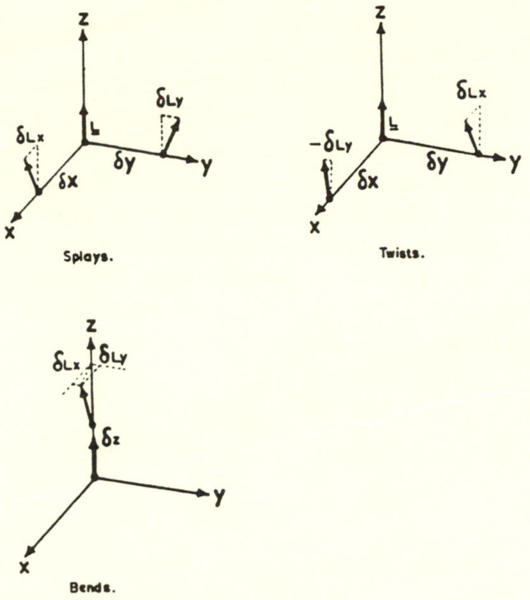

Figure 11.6: Schematic representation of the three independent modes of director variations in uniaxial liquid crystals. a) splays, b) twists, and c) bends. (Reprinted with permission from [Frank, 1958, Fig. 1].)

$$\begin{aligned}\Delta a &= \tfrac{1}{2}k_{11}(\boldsymbol{\nabla}\cdot\mathbf{n})^2 + \tfrac{1}{2}k_{22}(\mathbf{n}\cdot\boldsymbol{\nabla}\times\mathbf{n})^2 + \tfrac{1}{2}k_{33}\,|\mathbf{n}\times(\boldsymbol{\nabla}\times\mathbf{n})|^2 \\ &\quad + \tfrac{1}{2}(k_{22}+k_{24})\left[\mathrm{tr}(\boldsymbol{\nabla}\mathbf{n}\cdot\boldsymbol{\nabla}\mathbf{n}) - (\boldsymbol{\nabla}\cdot\mathbf{n})^2\right],\end{aligned} \tag{11.2-24b}$$

where k_{11}, k_{22}, k_{33}, and k_{24} are four, in general non-vanishing, elastic moduli. In the more general case of enantiomorphic materials (such as, cholesterics) three more terms need to be added to (11.2-24) [Frank, 1958, §3.3].

As discussed by Ericksen [1966, Eq. (18); 1976, Eq. (11)], the elastic moduli are subject to the inequalities

$$k_{11} \geq 0\,, \quad k_{22} \geq |k_{24}|\,, \quad k_{33} \geq 0\,, \quad 2k_{11} \geq k_{22} + k_{24}\,, \tag{11.2-25}$$

so that the homogeneous state corresponds to a minimum of the free energy. As the last term in (11.2-24) represents a total differential, it can be accounted for by a surface integral [Ericksen, 1962, p. 191]. This implies that the participation of the last term in the bulk free-energy density—and, as a consequence, the corresponding coefficient, k_{24}—is ambiguous, given that, without any change of the total free energy, an arbitrary multiple of it can be added provided that it is suitably compensated by the addition of a suitably chosen surface term. This observation certainly explains the difficulty in measuring k_{24} as a bulk material property; it is definitely associated with the surface forces at the interface and, in consequence, with the boundary conditions [Ericksen, 1962, p. 191]. It makes sense, then, to consider the contribution of this term in the analysis separately. For example, we can rewrite the inequalities on the moduli of (11.2-25) in two installments: first, we can write three inequalities involving solely k_{11}, k_{22}, and k_{33}, as

$$k_{11} \geq 0 \ , \quad k_{22} \geq 0 \ , \quad k_{33} \geq 0 \ , \tag{11.2-26a}$$

and, second, we can write two inequalities involving k_{24}:

$$-\,k_{22} \leq k_{24} \leq \min\left(2k_{11} - k_{22} ,\, k_{22}\right) \ . \tag{11.2-26b}$$

For the description of the bulk free energy, only inequality (11.2-26a) is important. Once k_{11}, k_{22}, and k_{33} are determined, inequality (11.2-26b) can be considered as restraining, due to stability reasons, the remaining coefficient k_{24}, or, equivalently, the surface free-energy term. Similarly, without loss of generality, we can neglect the fourth term from the right-hand side of (11.2-24), reducing the free-energy density expression to

$$\Delta a = \tfrac{1}{2}k_{11}(\boldsymbol{\nabla}\cdot\mathbf{n})^2 + \tfrac{1}{2}k_{22}(\mathbf{n}\cdot\boldsymbol{\nabla}\times\mathbf{n})^2 + \tfrac{1}{2}k_{33}\left|\mathbf{n}\times(\boldsymbol{\nabla}\times\mathbf{n})\right|^2 \ . \tag{11.2-27}$$

Again, the contribution of the omitted term to the free energy can be taken into account separately as a surface-density term.

Thus, if we are concerned only with the description of the bulk phenomena (as we are in this work), we need only use (11.2-27) for the definition of the free-energy density, involving the three moduli, k_{11}, k_{22}, and k_{33}, which are subject to the inequalities (11.2-26a). This is usually the choice made in the literature [Straley, 1973a, Eq. (7); Poniewierski and Stecki, 1979, Eq. (3.3); Mada, 1984, Eq. (8)]. Of course, by adding multiples of the exact differential, other equivalent expressions can be derived, as, for example [Edwards and Beris, 1989a, Eq. (1)],

$$\begin{aligned} \Delta a = {} & \tfrac{1}{2}k_{11}n_{i,i}n_{j,j} + \tfrac{1}{2}k_{22}\left(n_{i,j}n_{i,j} - n_{i,i}n_{j,j} - n_i n_j n_{k,i} n_{k,j}\right) \\ & + \tfrac{1}{2}k_{33}n_i n_j n_{k,i} n_{k,j} \ , \end{aligned} \tag{11.2-28}$$

which, for simplicity, is written in component form, and the shorthand notation $n_{i,j} \equiv \partial n_i/\partial x_j$ is used. Consistent with the above analysis is the observation [Ericksen, 1962, p. 191] that an exact differential in the free energy does not contribute to the stress and evolution equations, and, therefore, this apparently arbitrary choice in defining the free energy density has no physical consequences, at least locally.

B. Description of the Oseen/Frank elasticity in terms of the conformation tensor

When the order parameter tensor, **S**, is used in order to characterize the nematic liquid-crystalline structure, the contribution of inhomogeneities (Oseen/Frank elasticity) to the Helmholtz free energy is described to the lowest order in **S** as [de Gennes, 1969b, Eq. (2); Mada, 1984, Eq. (2); Wright and Mermin, 1989, Eq. (4.14) with $q_0 = 0$]

$$\Delta a = \tfrac{1}{2} L_1 (\nabla_\alpha S_{\beta\gamma})(\nabla_\alpha S_{\beta\gamma}) + \tfrac{1}{2} L_2 (\nabla_\alpha S_{\alpha\gamma})(\nabla_\beta S_{\beta\gamma}) \, , \qquad (11.2\text{-}29)$$

where the elastic moduli L_1 and L_2, in the uniaxial limit, can be correlated to the Frank elastic moduli k_{11}, k_{22}, and k_{33} as [Edwards and Beris, 1989a, p. 1191][6]

$$\begin{aligned} k_{11} &= (2L_1 + L_2) S^2 \, , \\ k_{22} &= 2L_1 S^2 \, , \\ k_{33} &= (2L_1 + L_2) S^2 \, . \end{aligned} \qquad (11.2\text{-}30)$$

Eqs. (11.2-30) have been developed by simply equating (within an exact differential) expressions (11.2-28) and (11.2-29) in the limit of uniaxial orientation, $\mathbf{S} = S(\mathbf{nn} - \tfrac{1}{3}\boldsymbol{\delta})$, and by exploiting the normalization condition on the director, $\mathbf{n} \cdot \mathbf{n} = 1$.

From stability considerations (i.e., requiring the free energy to be a minimum at homogeneous equilibrium), two independent inequality constraints can be deduced for the elastic moduli, L_1 and L_2 [de Gennes, 1971, p. 201; Wright and Mermin, 1989, p. 401]:

$$L_1 \geq 0 \, , \quad L_1 + \tfrac{2}{3} L_2 \geq 0 \, . \qquad (11.2\text{-}31)$$

Note that the above inequalities, together with the defining equalities (11.2-30), imply

[6] In the similar transformation relations reported by Mada [1984, Eq. (9)], there is a term $(9/2)L_1$ missing from the expression for L_1'.

$$k_{22} \geq 0 \ , \quad k_{11}, k_{33} \geq \tfrac{1}{4} k_{22} \ . \tag{11.2-32}$$

This set of inequalities imposes slightly stricter restrictions on the Frank moduli than the Ericksen inequalities (11.2-26a). These additional constraints arise from the fact that the elastic energy expression (11.2-29) involves additional modes as compared with the Frank form, (11.2-27). The additional modes involve gradients of the order parameter S, as explained in the following paragraph. As a consequence, in order to guarantee that the free energy is minimum at equilibrium with respect to these additional modes, additional constraints are necessary.

The free-energy expression provided by (11.2-29) is, in a sense, more general than (11.2-28), even in the uniaxial limit, since it involves, in addition to gradients in the director orientation, gradients of the (scalar) order parameter. Indeed, the difference in the free energy between expressions (11.2-28) and (11.2-29), Δa_S, can be expressed—in the limit of the uniaxial approximation and provided the relationships between the moduli (11.2-30) hold—as[7]

$$\begin{aligned} \Delta a_S = \tfrac{1}{3} L_1 (\nabla_\alpha S)^2 + \tfrac{1}{2} L_2 \Big[& \tfrac{1}{3}(n_\alpha n_\beta + \tfrac{1}{3}\delta_{\alpha\beta})(\nabla_\alpha S)(\nabla_\beta S) \\ & + \tfrac{2}{3}(2 n_\alpha \nabla_\beta n_\beta - n_\beta \nabla_\beta n_\alpha) S (\nabla_\alpha S) \Big] \ . \end{aligned} \tag{11.2-33}$$

However, looked upon in a different way, expression (11.2-29) is also more restricted than (11.2-28), since it forces a particular relation between the Frank elastic moduli,

$$k_{11} = k_{33} \ , \tag{11.2-34}$$

as easily deduced from relations (11.2-30a) and (11.2-30c).

This relation arises because of the restriction of (11.2-29) to the lowest-order contributions of **S**. This restriction provides a reasonable approximation for small values of S. Indeed, the constraint imposed by (11.2-34) has been independently derived in the limit $\langle P_4 \rangle \ll \langle P_2 \rangle \equiv S$, where $\langle P_i \rangle$ represents the orientation average of the Legendre polynomial, P_i [Priest, 1973, p. 724]. This last inequality is expected to be valid in the limit of small S, consistent with the derivation of (11.2-34). However, (11.2-34) is not expected to be valid in general (especially for the important nematic phase, $S > 0.5$) and indeed, experimentally this relationship does not hold [Straley, 1973, p. 2183]. To remove this restriction from the conformation tensor theory, it suffices to include higher-order terms in S and/or involve higher moments of the orientation distribution function [Marrucci

[7] In the similar expression of Mada [1984, Eq. (8)], the value for L_1'' needs to be corrected by removing the factor 1/3.

and Greco, 1991]. However, the final expression is no longer unique. One of the simplest possible extensions involves a cubic term in S [Edwards and Beris, 1989a, Eq. (7)]:

$$\Delta a = \tfrac{1}{2}L_1(\nabla_\alpha S_{\beta\gamma})(\nabla_\alpha S_{\beta\gamma}) + \tfrac{1}{2}L_2(\nabla_\alpha S_{\alpha\gamma})(\nabla_\beta S_{\beta\gamma}) + \tfrac{1}{2}L_3 S_{\alpha\beta}(\nabla_\alpha S_{\gamma\varepsilon})(\nabla_\beta S_{\gamma\varepsilon}) , \tag{11.2-35}$$

incorporating one additional elastic modulus L_3.

The advantage of the extended expression is that now a one-to-one correspondence between the Frank and the conformation-tensor elastic moduli can be established [Edwards and Beris, 1989a, p. 1191]:

$$\begin{aligned} k_{11} &= (2L_1 + L_2)S^2 - \tfrac{2}{3}L_3 S^3 , \\ k_{22} &= 2L_1 S^2 - \tfrac{2}{3}L_3 S^3 , \\ k_{33} &= (2L_1 + L_2)S^2 + \tfrac{4}{3}L_3 S^3 , \end{aligned} \tag{11.2-36}$$

and, inversely,

$$L_1 = \frac{3k_{22} - k_{11} + k_{33}}{6S^2} , \quad L_2 = \frac{k_{11} - k_{22}}{S^2} , \quad L_3 = \frac{k_{33} - k_{11}}{2S^3} , \tag{11.2-37}$$

following the same procedure as before (equating the expressions in the uniaxial limit). However, we should reiterate here the involved arbitrariness, since higher-order terms in S, can, in principle, also be present. Clearly, the question of the most suitable form can only be answered by resorting to a microscopic (molecular) level analysis, which we offer in the next subsection. Of course, as was the case with the director description, the order-parameter description of the free energy can also be extended to cholesteric LCs by including an additional second-order term [de Gennes, 1971, §5.1; Wright and Mermin, 1989, Eq. (4.14)].

C. Evaluation of the Oseen/Frank elasticity coefficients from molecular theories

Albeit the Frank elasticity correction to the free energy appears purely phenomenological, considerable work since its introduction has shown that it can be based on molecular theories. In addition to providing a microscopic explanation, this work has helped to elucidate the dependence of the Frank moduli on the parameters characterizing the internal microstructure of LCs. Most (if not all) of the analyses are restricted to comparing the simplest form, (11.2-27), with the results of a microscopic analysis which is based on a mean-field approximation of the free energy of a system with long-wavelength inhomogeneities. Only the final results are listed here.

One of the first evaluations of the elastic moduli was through the use of the Maier/Saupe theory [Priest, 1972; Ten Bosch and Sixou, 1987, Eq. (6)]:

$$k_{11} = k_{22} = \kappa k_{33} \propto S^2 , \qquad (11.2\text{-}38a)$$

where κ, i.e., the ratio of k_{33} to k_{11}, is equal to one if the intermolecular interaction between two molecules, *i* and *j*, is assumed to be of the Maier /Saupe-type

$$V_{ij} = - A(\mathbf{r}_{ij}) P_2(\cos\theta_{ij}) , \qquad (11.2\text{-}38b)$$

with the interaction function $A(\mathbf{r})$ spherically symmetric and if the spatial distribution of the molecules is assumed to be isotropic [Marrucci and Greco, 1991]. As it has been shown by Priest [1972], assuming a non-isotropic center-of-mass correlation can result in a higher value for k_{33} than $k_{11}=k_{22}$ [Priest, 1972]. Moreover, the fact that the first two moduli are predicted to be equal was shown later to arise because of an evaluation of the intermolecular interactions using an average calculated over a spherical neighborhood of the "test" particle [Marrucci and Greco, 1991]. Thus, assuming (in addition to spatial anisotropicity) that the intermolecular interactions are calculated using as a reference an extended (rod-like) structure leads to predictions involving three different moduli, $k_{22}<k_{11}<k_{33}$ as experimentally observed with molecules that possess a rod-like shape, i.e., where the liquid-crystalline structure is induced primarily because of geometrical reasons rather than enthalpic intermolecular interactions. In this sense, although the Maier/Saupe theory was primarily developed for LMWLCs, as based on intermolecular interactions—see also the relevant discussion for the Maier/Saupe theory in §11.2.1B—it can also have some limited value for PLCs. Indeed, in one approach where the Maier/Saupe theory is used, ten Bosch and Sixou [1987] offered the accounting of the screening effect of the polymer chain as the basis for developing a Maier/Saupe correction to the free energy of PLCs.

Using Onsager's theory, Priest [1973] has shown that the Frank elastic constants may be expressed as a series of the orientation distribution average of the even-order Legendre polynomials, $<P_i>$, i=2,4,... as [Priest, 1973, p. 720]

$$\frac{k_{11} - \bar{k}}{\bar{k}} = C - 3C' \frac{<P_4>}{<P_2>} + \dots ,$$

$$\frac{k_{22}-\bar{k}}{\bar{k}} = -2C - C'\frac{<P_4>}{<P_2>} + \dots ,$$
$$\frac{k_{33}-\bar{k}}{\bar{k}} = C + 4C'\frac{<P_4>}{<P_2>} + \dots , \tag{11.2-39}$$

where

$$\bar{k} \equiv \tfrac{1}{3}(k_{11} + k_{22} + k_{33}) , \tag{11.2-40}$$

and C, C' are constants which depend on the details of the system. For the case of hard rods with a sphero-cylindrical shape, use of the excluded-volume, second virial coefficient theory of Onsager [1949] gives [Priest, 1973, Eq. (39)]

$$C = \frac{2(\frac{L}{d})^2 - 2}{7(\frac{L}{d})^2 + 20} , \quad C' = 27\frac{\frac{1}{16}(\frac{L}{d})^2 - \frac{1}{6}}{7(\frac{L}{d})^2 + 20} . \tag{11.2-41}$$

Later, Poniewierski and Stecki [1979, p. 1937], using a slightly different approach, arrived at the same expressions as shown in (11.2-39), but corresponding to slightly different coefficients:

$$C = \frac{2(\frac{L}{d})^2 - 3}{7(\frac{L}{d})^2 + 21} , \quad C' = 54\frac{\frac{1}{16}(\frac{L}{d})^2 - \frac{1}{6}}{7(\frac{L}{d})^2 + 21} . \tag{11.2-42}$$

These coefficients assume values very similar to those corresponding to Priest's, (11.2-41), except for a factor of two in C'. Of course, the relations presented by (11.2-39), in the limit of $<P_i> = 0$, $i = 4, 6, \dots$, which is expected to hold when $<P_2> = S \ll 1$, are fully equivalent to the lowest order in S (S^2) predictions in (11.2-36) for

$$\bar{k} = (2L_1 + \tfrac{2}{3}L_2)\,S^2 , \quad C = \frac{L_2}{6L_1 + 2L_2} . \tag{11.2-43}$$

Straley [1973] has calculated the Frank elastic moduli corresponding to (11.2-27), also using the Onsager theory, but avoiding the use of spherical-harmonic expansions, which limit the validity of the results to small S values. Instead, he used a simplified form of the distribution of rod orientations which reproduces very well the exact homogeneous solution. Later, Ponierwierski and Stecki [1979] and Lee and Meyer [1986] repeated the calculations using a higher-order approximation of the distribution function in terms of spherical harmonics and a numerical solution to Onsager's integral equation, respectively. For a specific rod aspect ratio, *L/d*, and number density, *n* (corresponding to a specific orientation distribution, Ψ, and order parameter, *S*), each of the k_{ii} (no sum) can be represented in the form [Straley, 1973, Table 1;[8] Ponierwierski and Stecki, 1979, Table 2]

$$k_{ii} = n^2 L^4 d k_B T C_{ii}[F(\Psi)] \ , \qquad (\text{ No sum }) \qquad (11.2\text{-}44)$$

where $C_{ii}[F(\Psi)]$ is a numerical factor, depending on the orientation distribution of the rods, Ψ, with typical values about 0.01-0.22. For the reported values of $S = 0.71 - 0.84$, $C_{11} \approx 0.03 = 3C_{22}$, and C_{22} are practically constant while C_{33} increases from 0.12 for $S = 0.71$ to 0.22 for $S = 0.84$. The numerical results of Lee and Meyer [1986] have essentially verified the previous data (again, after correcting Straley's data for the missing factor 1/2), exhibiting only minor quantitative differences, especially with respect to the Ponierwierski and Stecki data [1979, Table 2]. Moreover, they considerably extended the range of the calculations, covering values of *S* from 0.6 to 0.95 [Lee and Meyer, 1986, Table III and Fig. 2]. Their results verified the previously mentioned trend for the elastic moduli: as the concentration increases—and, therefore, the sample becomes more oriented ($S \rightarrow 1$)—whereas C_{ii}, for i=1,2 remain almost constant, C_{33} diverges to infinity at perfect order. Since this is in agreement with intuition, based on the physics represented by the bend modulus k_{33}, it is not believed to be an artifact of the mean-field approximation [Lee and Meyer, 1986, p. 3446]. More recently, these results have also been verified by Lee and Meyer [1986, Fig. 2].

De Gennes [1977] discussed the influence of a partial flexibility of polymeric chains on the Oseen/Frank elastic moduli of PLCs. He argued that the principal factor dictating the magnitude of the splay constant, k_{11}, is the intermolecular strain free energy, rather than the relative compression and dilation in the spatial distribution of the chain ends, which reduces their entropy. He concluded that for partly flexible chains, the ratio k_{11}/k_{33} can be much higher than for rigid chains, which (in the limit of small *S*) assumes the value of unity. His theory was improved by Meyer [1982] by considering interdigitation of the polymer chains with one another; he concluded that k_{11} should increase, but not as fast as predicted originally by de Gennes: proportionally to *L* instead of L^2 [Meyer, 1982, p. 138; Lee and Meyer, 1991, pp. 347,348]. Later, Grosberg

[8]The results reported in this table are larger than the correct values by a factor of 2 due to a missing factor of ½ from Eq. (3) in Straley [1973], as later pointed out by Ponierwierski and Stecki [1979, p. 1938].

and Zhestkov [1986] examined the influence of the partial flexibility of the chain on the magnitude of the elastic moduli in a more quantitative fashion using a general, self-consistent statistical description of the conformation of long chains under the influence of an external field, originally developed by Lifshitz [1969] and later introduced into the study of semi-flexible PLCs by Khokhlov and Semenov [1981; 1982].

The results of Grosberg and Zhestkov [1986] depend on an additional quantity, the persistence length, ℓ, which is defined as the length of the effective Kuhn segment [Flory, 1953, pp. 411-413]. To apply their theory, the persistence length ℓ is always assumed to be much larger than the average monomer diameter, d,

$$\ell \gg d \ . \tag{11.2-45}$$

Then, by analogy with the corresponding description of the liquid-crystalline state, the predictions for the elastic moduli depend on whether the persistence length, ℓ, is much smaller, of equal magnitude, or much larger than the total macromolecular contour length, L. The results can be summarized as follows.

A. $\ell \ll L$ This represents the limiting case of long, semi-flexible macromolecules, which is examined here in the limit of persistent chains (i.e., the flexibility is homogeneously distributed along the chain contour) and for an athermal solution (i.e., where the excluded volume is the only interaction between the molecules) [Grosberg and Zhestkov, 1986, p. 100]:

$$\begin{aligned} k_{11}^{\infty} &\approx \frac{k_B T}{4d} x^{\frac{1}{3}} \exp(3.82\, x^{\frac{1}{2}} - 7.1\, x^{\frac{1}{6}}) \ , \\ k_{22}^{\infty} &\approx \frac{k_B T}{d} x \left(\frac{1-S}{12} - \frac{\pi}{4} \int (1+u_1)^2 f_0^2 \, du_1 \right) , \\ k_{33}^{\infty} &\approx \frac{k_B T}{2d} x S \ , \end{aligned} \tag{11.2-46}$$

where the superscript ∞ denotes the requirement $L \gg \ell$ and x represents a dimensionless number density defined as

$$x \equiv n L \ell d \ . \tag{11.2-47}$$

Grosberg and Zhestov argued that the large increase in the splay constant, k_{11}, is due to the presence of "hairpin" defects produced by splay distortions. More recently, Vroege and Odijk [1988, Eq. (IX.7)] have provided a more accurate estimation of the k_{11} elastic modulus, based on a numerical solution to the corresponding integro-differential equation for the distribution function:

$$k_{11}^{\infty} \approx 0.257 \frac{k_B T}{d} c^{\frac{1}{3}} \exp(1.824 c^{\frac{2}{3}} - 10.18 c^{-\frac{1}{3}}) , \qquad (11.2\text{-}48)$$

where c, the number of persistence-length segments present in a volume $(\pi/4)\ell^2 d$, is defined as

$$c \equiv \frac{\pi}{4} \ell^2 d \frac{L}{\ell} n = \frac{\pi}{4} \ell d L n \; . \qquad (11.2\text{-}49)$$

In addition, they have defined more accurately the domain of validity of the infinite-chain analysis, based on the "global persistence length," g, rather than the persistence length, ℓ: $L \gg g$. Since the global persistence length, physically corresponding to the average distance between two "hairpin" defects in the chain can be typically many times the scale of the persistence length, ℓ [Vroege and Odijk, 1988, Table III], it is not surprising that no experimental data are presently available against which the theory can be tested in this characteristic limit.

From (11.2-46), it is clear that as the order parameter S increases to unity, the twist modulus k_{22} decreases to zero, whereas the bending modulus k_{33} increases, reaching asymptotically a constant value. Moreover, from Eqs. (11.2-46) and (11.2-48), we can show that the cross-wise flexing modulus k_{11} increases exponentially as the concentration increases. This is attributed to a sudden energetic deficiency of chain folding, and it is to be contrasted with the calculation of k_{11} by de Gennes [1973], where $k_{11} \propto L^2$, which does not assume chain folding [Grosberg and Zhestov, 1986, p. 101].

B. $\ell \approx L$ This is the most difficult case for the analysis. By analogy with the susceptibility to a dipolar field, the behavior of k_{11} is inferred to be [Grosberg, and Zhestov, 1986, p. 101]

$$k_{11} \approx k_{11}^{\infty} \left(1 - \frac{1 - \exp\left(- \dfrac{k_B T L^2 n}{2 k_{11}^{\infty}} \right)}{\left(\dfrac{k_B T L^2 n}{2 k_{11}^{\infty}} \right)} \right) . \qquad (11.2\text{-}50)$$

As this formula shows, the value of k_{11} abruptly decreases from an exponentially high value to unity as the length of the molecule decreases from ∞ to about unity. In contrast, the values of the other moduli remain practically unchanged from their $L/\ell = \infty$ values.

C. $\ell \gg L$ This case corresponds to the rigid-rod limit for which the results obtained with the Onsager theory (as discussed above) are expected to be valid. Indeed, one obtains $k_{22} = k_{11}/3$ always, and, in the limit of a well-oriented sample ($S \to 1$), the following asymptotic expressions [Grosberg and Zhestov, 1986, p. 102] are obtained:

$$k_{11} = \frac{k_B T n L^2}{4} , \quad k_{33} = k_{11} \frac{1.1}{1-S} , \tag{11.2-51}$$

where $x \equiv nL\ell d$ is assumed to be inversely proportional to $(1-S)^{1/2}$ in the limit of $S \to 1$. These results are consistent with the Onsager expressions, (11.2-44), if the dimensionless number density, $nL^2 d$, is assumed to have a (constant) value (≈ 10).

Thus, whereas the effect of rigidity is to increase k_{33}/k_{22} with increased aspect ratio, the effect of semi-flexibility is to increase the ratio k_{11}/k_{22}. Indeed, experimental measurements with PBG reported that both of these ratios were high and increased with molecular weight [Lee and Meyer, 1991, Table 5]. Several investigations have addressed additional effects on the elastic moduli arising, for example, from attractive long-range interactions [Vertogen, 1982], high concentration [Lee and Meyer, 1991, p. 347], screening of the intermolecular potential due to density correlations along the polymer chain [ten Bosch and Sixou, 1987], director fluctuations [Vroege and Odijk, 1988, p. 2856], a small wavelength to molecular-length ratio [Lo and Pelcovits, 1990], etc. However, all these effects have proven, under typical conditions, to change slightly and only quantitatively the previously described predictions and are not further addressed here.

11.3 The LE and Doi Models for Flowing Liquid-Crystalline Systems

In setting up the stage for the application of the bracket formalism to the modeling of the dynamical behavior of liquid crystals, it is worth some time and effort, at this point, to review briefly the previous dynamical theories for liquid-crystalline materials. This will allow us to understand the shortcomings of each, as well as the mathematical rationale behind them. It will also aid us in §11.6 when we compare the general conformation tensor theory to these models. We need to keep in mind that consideration of the dynamics adds substantially to the complexity of an already difficult problem. This explains why, when compared to the theories for the static behavior of LCs described in the previous sections, the theories for the flow behavior risk being considered primitive.

There are two primary theories: the Leslie/Ericksen theory, based on continuum-mechanics considerations (presented first in §11.3.1), and the Doi theory, based on a microscopic rotational diffusion model for a suspension of rigid rods (considered in §11.3.3). An intermediate, scalar-vector theory developed recently by Ericksen [1991], which attempts to bridge the gap between the other two theories, is also discussed for comparison purposes in §11.3.2. Finally, in §11.3.4 the significance of the values of the parameters of the Leslie/Ericksen theory in controlling the dynamic flow behavior of LCs is discussed through an overview of the analysis of simple shear flow.

11.3.1 The Leslie/Ericksen theory

The Leslie/Ericksen (LE) theory was originally developed in an inertial form; however, it is straightforward to convert it into the non-inertial form. In this section, we first present this theory in its original form [Ericksen, 1960a,b; Leslie, 1966; 1968] and then transform it into its counterpart in the representation where the moment of inertia $\sigma \rightarrow 0$.

A. The inertial form

In the LE theory for incompressible, liquid-crystalline systems, the short-range order of the system is described by the vector **n**, called the director (see Figure 11.1), which is defined to be of unit length and, as such, subject to the constraint

$$\mathbf{n} \cdot \mathbf{n} = 1 \; . \tag{11.3-1}$$

The states of **n** and $-\mathbf{n}$ are physically indistinguishable from each other due to the invariance of the molecular orientation under a rotation of 180 degrees.

For flowing systems, two constitutive relations describe the kinematics of the director:

$$\rho \dot{v}_\alpha = F_\alpha - p_{,\alpha} - \left(\frac{\partial W}{\partial n_{\gamma,\beta}} n_{\gamma,\alpha} \right)_{,\beta} + t_{\alpha\beta,\beta} \; , \tag{11.3-2}$$

$$\sigma \ddot{n}_\alpha = G_\alpha + \gamma n_\alpha - \frac{\partial W}{\partial n_\alpha} + \left(\frac{\partial W}{\partial n_{\alpha,\beta}} \right)_{,\beta} + g_\alpha \; , \tag{11.3-3}$$

where a superscripted dot denotes the material derivative, i.e.,

$$\dot{n}_\alpha \equiv \frac{\partial n_\alpha}{\partial t} + v_\beta n_{\alpha,\beta} \; , \tag{11.3-4}$$

and $\ddot{\mathbf{n}}$ denotes the material derivative of $\dot{\mathbf{n}}$.

These expressions were derived following standard practices in continuum mechanical modeling, yet it will take a little bit of time for us to explain what each equation (and each term in each equation) represents. The first expression, (11.3-2), represents the usual balance of linear or translational momentum for the particular fluid particle at position **x** at time t. Hence, the velocity, **v**, is once again a function of position and time, $\mathbf{v} = \mathbf{v}(\mathbf{x},t)$. The second expression is a balance of angular momentum as in §10.3, with the director also being a function of position and time, $\mathbf{n} = \mathbf{n}(\mathbf{x},t)$. Therefore, both the velocity vector field and the director field are distinct parameters of the problem, and each is highly dependent upon the other as explicitly stated through (11.3-2,3). Let us now consider each particular momentum balance in turn.

As already stated, (11.3-2) reflects the law of conservation of linear momentum. In this expression, ρ is the constant density of the system, which requires that **v** is subject to the incompressibility constraint, as in §5.4,

$$\operatorname{div} \mathbf{v} = 0 \,, \tag{11.3-5}$$

and **F** is a body force vector due to an external force field. Here we treat this external field as a magnetic field, because of the significance of these fields in the physics of LCs. Many studies have been concerned with the effects of magnetic fields upon LCs and may be found in the literature mentioned in the introduction. For an external magnetic field, **H**, which induces in the material a magnetization, $\boldsymbol{\Phi}$,

$$F_\alpha = \Phi_\beta H_{\beta,\alpha} \,, \tag{11.3-6}$$

where

$$\Phi_\alpha \equiv \chi_\perp H_\alpha + (\chi_\parallel - \chi_\perp) n_\beta H_\beta n_\alpha \,, \tag{11.3-7}$$

and $\chi_\perp$ and $\chi_\parallel$ are the magnetic susceptibilities perpendicular and parallel to the director, respectively.

The last two terms on the right-hand side of (11.3-2) represent the effects of the extra stress. The first term is due to the distortion stress, which reflects the elastic stress of the system due to gradients in the director field, arising from the Frank distortion energy [Frank, 1958, §2]. As explained in §11.2.2.A, this is given by

$$2W = k_{11}(\operatorname{div}\mathbf{n})^2 + k_{22}(\mathbf{n}\cdot\operatorname{curl}\mathbf{n})^2 + k_{33}[(\mathbf{n}\cdot\boldsymbol{\nabla})\mathbf{n}]^2 \,, \tag{11.3-8}$$

where each one of the elastic constants in (11.3-8), k_{ii}, is associated with a characteristic deformation which the molecular orientation may assume. The deformations for k_{11}, k_{22}, and k_{33} are, respectively, the splay, twist, and bend modes (see also Figure 11.5). $t_{\alpha\beta}$ is the dissipative portion of the stress, which must satisfy thermodynamic criteria. In its most general form, which is linear in the velocity gradient for an anisotropic medium subject to material frame indifference, it is given as

$$t_{\alpha\beta} = \alpha_1 n_\gamma n_\varepsilon A_{\gamma\varepsilon} n_\alpha n_\beta + \alpha_2 N_\alpha n_\beta + \alpha_3 N_\beta n_\alpha + \alpha_4 A_{\alpha\beta} + \alpha_5 A_{\alpha\gamma} n_\gamma n_\beta + \alpha_6 A_{\beta\gamma} n_\gamma n_\alpha \,, \tag{11.3-9}$$

where the α_i represent the anisotropic (constant) viscosities, which for an isotropic system are all (except α_4) identically zero. These α_i must satisfy all of the thermodynamic criteria presented by Leslie [1979, p. 13], based upon the requirement that the rate of entropy production must be non-negative—see Example 11.1 for a discussion of these criteria—as well as the relation of Parodi [1970, p. 581]:

$$\alpha_2 + \alpha_3 = \alpha_6 - \alpha_5 \,, \tag{11.3-10}$$

which arises as a consequence of the Onsager/Casimir reciprocal relations of irreversible thermodynamics. (Later, we shall see how this relation arises through the bracket description.)

In (11.3-9), the rotational derivative of the director is defined using the antisymmetric, velocity-gradient tensor:

$$N_\alpha = \dot{n}_\alpha - \Omega_{\alpha\beta} n_\beta \,. \tag{11.3-11}$$

Now that we have specified all of the terms in (11.3-2), it is easily recognizable as the standard linear momentum balance of fluid mechanics, although applied to an anisotropic system. For an isotropic system, with no average orientation ($\mathbf{n} = \mathbf{0}$), we recover the Navier/Stokes equations for the flow of a constant-viscosity (Newtonian) fluid.

The remaining evolution equation, (11.3-3), is the balance of angular momentum. In this expression, σ is an inertial constant (i.e., a moment of inertia, as in §10.3), $\ddot{\mathbf{n}}$ is the material derivative of $\dot{\mathbf{n}}$, and $\mathbf{G}$ is a director body force associated with the external magnetic field, expressed by the relation

$$G_\alpha = \chi_a n_\beta H_\beta H_\alpha \,, \tag{11.3-12}$$

with χ_a being the difference in the magnetic susceptibilities, $\chi_\parallel - \chi_\perp$. Dissipation is included in this expression through the "couple stress," g_α, which, as the name implies, couples the director evolution equation with the velocity gradient:

$$g_\alpha = -\gamma_1 N_\alpha - \gamma_2 A_{\alpha\beta} n_\beta \;;\; \gamma_1 = \alpha_3 - \alpha_2 \quad \text{and} \quad \gamma_2 = \alpha_6 - \alpha_5 \,. \tag{11.3-13}$$

The presence of γ in (11.3-3) is required in order to ensure that $\mathbf{n}$ remains a unit vector; i.e., dotting (11.3-3) with $\mathbf{n}$ must give a result that vanishes, and hence γ must be defined as

$$\gamma = -G_\beta n_\beta + \frac{\partial W}{\partial n_\beta} n_\beta - n_\beta \left(\frac{\partial W}{\partial n_{\beta,\gamma}} \right)_{,\gamma} \tag{11.3-14}$$
$$+ (\alpha_2 + \alpha_3) v_{\beta,\gamma} n_\gamma n_\beta - \sigma \dot{n}_\beta \dot{n}_\beta \ .$$

The angular momentum balance for an isotropic fluid, $\mathbf{n} = \mathbf{0}$, obviously vanishes, leaving only the linear momentum equation expressing the dynamics of the medium.

The above system of equations forms a closed set, which, in principle, may be solved simultaneously for arbitrary kinematics. Note that (11.3-1,4) imply the additional constraint

$$\dot{\mathbf{n}} \cdot \mathbf{n} = 0 \ . \tag{11.3-15}$$

We shall see later what effects this constraint will have upon the system.

Example 11.1: Find the inequalities which the Leslie coefficients (the α_i) must satisfy for thermodynamic consistency.

For an incompressible fluid, **A** is both symmetric and traceless. Thus, it can be expressed, in terms of an unconstrained, second-order tensor $\hat{\mathbf{A}}$, as:

$$A_{\alpha\beta} = \tfrac{1}{2}(\hat{A}_{\alpha\beta} + \hat{A}_{\beta\alpha}) - \tfrac{1}{3}\hat{A}_{\gamma\gamma}\delta_{\alpha\beta} \ . \tag{11.E-1}$$

Also, **N** must satisfy the constraint

$$N_\alpha n_\alpha = 0 \ , \tag{11.E-2}$$

as can be easily evaluated from the constraint (11.3.15) using (11.3.11) and the fact that $\boldsymbol{\Omega}$ is antisymmetric. Similarly with **A**, **N** can be expressed in terms of an unconstrained vector $\hat{\mathbf{N}}$ as

$$N_\alpha = \hat{N}_\alpha - \hat{N}_\beta n_\beta n_\alpha \ .$$

From the principle of non-negative entropy production, as calculated through the bracket representation of the LE model, (11.4-13), we have, substituting for **t** and **g**, using Eqs. (11.3-9) and (11.3-13), respectively

$$\alpha_1 (n_\alpha n_\beta A_{\alpha\beta})^2 + \alpha_4 A_{\alpha\beta} A_{\alpha\beta} + \eta N_\alpha A_{\alpha\beta} n_\beta \tag{11.E-4}$$
$$+ \beta A_{\alpha\beta} n_\beta A_{\alpha\gamma} n_\gamma + \gamma_1 N_\alpha N_\alpha \geq 0 \ ,$$

where $\beta = \alpha_5 + \alpha_6$, and $\eta = \alpha_2 + \alpha_3 + \alpha_6 - \alpha_5$. Since the coordinate system is completely arbitrary, we can choose our Cartesian coordinates so that the x_1 direction lies in the direction of n_1, so that $\mathbf{n}$ may be written as

$$\mathbf{n} = (1,0,0)^T . \tag{11.E-5}$$

Substituting (11.E-1,3,5) into inequality (11.E-4) yields a 9×9 symmetric matrix, $\mathbf{B}$, satisfying the inequality

$$\mathbf{V}^T \cdot \mathbf{B} \cdot \mathbf{V} \geq 0 , \tag{11.E-6}$$

where all the components of vector $\mathbf{V}$,

$$\mathbf{V} \equiv (\hat{A}_{11}, \hat{A}_{22}, \hat{A}_{33}, \hat{A}_{12}, \hat{A}_{23}, \hat{A}_{31}, \hat{N}_1, \hat{N}_2, \hat{N}_3)^T ,$$

can be independently varied. Thus, in order for the inequality (E.6) to be valid, $\mathbf{B}$ must be non-negative definite. The non-zero components of the matrix $\mathbf{B}$, are

$$\begin{aligned}
B_{11} &= \tfrac{4}{9}(\alpha_1 + \beta) + \tfrac{2}{3}\alpha_4 , \\
B_{22} &= B_{33} = \tfrac{1}{9}(\alpha_1 + \beta) + \tfrac{2}{3}\alpha_4 , \\
B_{44} &= B_{66} = 2\alpha_4 + \beta , \\
B_{55} &= 2\alpha_4 , \\
B_{88} &= B_{99} = \gamma_1 , \\
B_{12} &= B_{21} = B_{13} = B_{31} = -\tfrac{2}{9}(\alpha_1 + \beta) - \tfrac{1}{3}\alpha_4 , \\
B_{23} &= B_{32} = \tfrac{1}{9}(\alpha_1 + \beta) - \tfrac{1}{3}\alpha_4 , \\
B_{84} &= B_{48} = B_{96} = B_{69} = \tfrac{1}{2}\eta .
\end{aligned} \tag{11.E-7}$$

Since $\mathbf{B}$ must be non-negative-definite, all the diagonal components must be greater than or equal to zero, as well as the determinants of all of the principal minors. This easily results in a number of conditions on the α_i, five of which are not superfluous:

$$\begin{aligned}
\alpha_4 &\geq 0 , \\
2\alpha_4 + \beta &\geq 0 , \\
\gamma_1 &\geq 0 , \\
2\alpha_1 + 3\alpha_4 + 2\beta &\geq 0 , \\
4\gamma_1(2\alpha_4 + \beta) - \eta^2 &\geq 0 .
\end{aligned} \tag{11.E-8}$$

These are the conditions of Leslie [1979, Eq. (35)]. ■

B. The non-inertial form

A "quasi-steady state" theory can easily be generated from the preceding system of equations for applications in which the inertial constant, σ, is very small. For very viscous materials, in which rotation of the director is largely inhibited, this approximation is probably not overly restrictive. Therefore, in this limit, we may rewrite the system of equations neglecting the director inertia. Hence the term $\sigma \ddot{n}_\alpha$ vanishes from (11.3-3), and we wish to write coupled evolution equations for **v** and **n** (rather than **v**, **n**, and $\dot{\mathbf{n}}$).

The evolution equation for **n** may be obtained from (11.3-3) by solving for $\dot{\mathbf{n}}$, which appears in this expression through the couple stress via Eqs. (11.3-11,13). Performing this calculation, one obtains

$$\dot{n}_\alpha = \frac{\alpha_2}{(\alpha_2-\alpha_3)} v_{\alpha,\beta} n_\beta + \frac{\alpha_3}{(\alpha_2-\alpha_3)} v_{\beta,\alpha} n_\beta - \frac{1}{(\alpha_2-\alpha_3)} G_\alpha$$
$$- \frac{1}{(\alpha_2-\alpha_3)} \gamma n_\alpha + \frac{1}{(\alpha_2-\alpha_3)} \left[\frac{\partial W}{\partial n_\alpha} - \left(\frac{\partial W}{\partial n_{\alpha,\beta}} \right)_{,\beta} \right], \tag{11.3-16}$$

where we have used the Parodi relation of (11.3-10).

Now in order to obtain the linear momentum balance, one must substitute (11.3-16) into (11.3-2) via the extra stress of (11.3-9,11). This results in a new expression for the extra stress tensor,

$$t^n_{\alpha\beta} = \beta_1 A_{\gamma\varepsilon} n_\gamma n_\varepsilon n_\alpha n_\beta + \beta_2 A_{\alpha\beta} + \beta_3 [A_{\alpha\gamma} n_\gamma n_\beta + A_{\beta\gamma} n_\gamma n_\alpha]$$
$$- \frac{(\alpha_2+\alpha_3)}{(\alpha_2-\alpha_3)} \gamma n_\alpha n_\beta - \frac{1}{(\alpha_2-\alpha_3)} [\alpha_2 G_\alpha n_\beta + \alpha_3 G_\beta n_\alpha]$$
$$+ \frac{1}{(\alpha_2-\alpha_3)} \left[\alpha_2 \frac{\partial W}{\partial n_\alpha} n_\beta + \alpha_3 \frac{\partial W}{\partial n_\beta} n_\alpha \right] \tag{11.3-17}$$
$$- \frac{1}{(\alpha_2-\alpha_3)} \left[\alpha_2 \left(\frac{\partial W}{\partial n_{\alpha,\gamma}} \right)_{,\gamma} n_\beta + \alpha_3 \left(\frac{\partial W}{\partial n_{\beta,\gamma}} \right)_{,\gamma} n_\alpha \right],$$

where

$$\beta_1 = \alpha_1 , \quad \beta_2 = \alpha_4 , \quad \text{and} \quad \beta_3 = \frac{\alpha_6 \alpha_2 - \alpha_5 \alpha_3}{(\alpha_2 - \alpha_3)} . \tag{11.3-18}$$

Thus we arrive at the non-inertial version of the LE theory as given by Eqs. (11.3-2,16,17).

If we were to set $\mathbf{G} = \mathbf{0}$ and $W = 0$ (i.e., no external field nor distortion energy) in the above equations, then the equations should reduce to Ericksen's transversely isotropic fluid [Ericksen, 1960b, §3]. In this case, if we incorporate γ into the first three terms of (11.3-16), we get the new phenomenological coefficients

$$\beta_1^n = \alpha_1 - \frac{(\alpha_2 + \alpha_3)^2}{(\alpha_2 - \alpha_3)} , \quad \beta_2^n = \alpha_4 , \quad \text{and} \quad \beta_3^n = \frac{\alpha_6\alpha_2 - \alpha_5\alpha_3}{(\alpha_2 - \alpha_3)} . \tag{11.3-19}$$

We shall use only the parameters of (11.3-19) in the remainder of this dissertation, since it will always be possible to incorporate γ into the other terms in this way.

11.3.2 The scalar-vector theory

Ericksen [1991] has recently tried to extend the director-based LE theory to a scalar-vector theory, with little intrinsic gain. We shall not discuss this theory in any detail, since it bears such a close resemblance to the LE theory, but we shall content ourselves with providing a very brief overview of the non-inertial form of this theory so that we may see how easily it derives as a special case of the conformation tensor theory in §11.6.2.

The non-inertial, scalar-vector theory involves two parameters: a scalar order parameter, s, describing the spread or distribution of the orientation, and a unit vector, the director $\mathbf{n}$, which is the sole device in the LE theory. For $s = 1$, perfect alignment is assumed, with random alignment corresponding to $s = 0$. An evolution equation is provided for each parameter, s and $\mathbf{n}$, by solving the corresponding equations for $\ddot{s}$ and $\ddot{\mathbf{n}}$ in the limit where $\sigma \to 0$ [Ericksen, 1991]:[9]

$$\dot{s} = -\frac{1}{\hat{\beta}_2(s)}\frac{\delta H_m}{\delta s} - \frac{\hat{\beta}_1(s)}{\hat{\beta}_2(s)}\mathbf{n}^{\mathrm{T}}\cdot\mathbf{A}\cdot\mathbf{n} , \tag{11.3-20}$$

$$\dot{\mathbf{n}} = \boldsymbol{\Omega}\cdot\mathbf{n} + \frac{\hat{\gamma}_2(s)}{\hat{\gamma}_1(s)}[\mathbf{n}(\mathbf{n}^{\mathrm{T}}\cdot\mathbf{A}\cdot\mathbf{n}) - \mathbf{A}\cdot\mathbf{n}] - \frac{1}{\hat{\gamma}_1(s)}\frac{\delta H_m}{\delta\mathbf{n}} . \tag{11.3-21}$$

Here, H_m is the mechanical energy of the system (or Hamiltonian), which Ericksen never specifies. He simply writes the energy as a general function of the system variables, $H_m = H_m[s, \boldsymbol{\nabla}s, \mathbf{n}, \boldsymbol{\nabla}\mathbf{n}]$.

[9]The $\hat{\beta}_i$ and $\hat{\gamma}_i$ are once again viscosity coefficients, similar to the α_i of LE theory, but which may depend on the scalar order parameter in general.

The linear momentum balance assumes the standard form from fluid mechanics, with the total-stress tensor field, $\mathbf{T}'$, given by the expression

$$T'_{\alpha\beta} = -p\delta_{\alpha\beta} + \hat{\alpha}_1(s) n_\alpha n_\beta n_\gamma n_\varepsilon A_{\gamma\varepsilon} + \hat{\alpha}_2(s)[\dot{n}_\alpha n_\beta - \Omega_{\alpha\gamma} n_\gamma n_\beta] + \hat{\alpha}_3(s)[\dot{n}_\beta n_\alpha - \Omega_{\beta\gamma} n_\gamma n_\alpha] + \hat{\alpha}_4(s) A_{\alpha\beta} + \hat{\alpha}_5(s) n_\beta n_\gamma A_{\gamma\alpha} + \hat{\alpha}_6(s) n_\alpha n_\gamma A_{\gamma\beta} + \hat{\beta}_1(s)\dot{s} n_\alpha n_\beta - n_{\gamma,\alpha}\frac{\partial W'}{\partial n_{\gamma,\beta}} - s_{,\alpha}\frac{\partial W'}{\partial s_{,\beta}}\,, \tag{11.3-22}$$

where $\hat{\gamma}_1 = \hat{\alpha}_3 - \hat{\alpha}_2$ and $\hat{\gamma}_2 = \hat{\alpha}_6 - \hat{\alpha}_5 = \hat{\alpha}_2 + \hat{\alpha}_3$. W' represents the unspecified distortion energy which now depends also upon s, $\boldsymbol{\nabla} s$ as well as $\mathbf{n}$, $\boldsymbol{\nabla}\mathbf{n}$. When s=1, W' should be identical to W.

The above theory involves seven phenomenological parameters (excluding H_m), as general functions of s, which we can also see by rewriting the stress equation, (11.3-22), using (11.3-20,21) to substitute for $\dot{\mathbf{n}}$ and $\dot{s}$:

$$T'_{\alpha\beta} = -p\delta_{\alpha\beta} + \hat{\alpha}_4 A_{\alpha\beta} + \left(\hat{\alpha}_1 + \frac{\hat{\gamma}_2^2}{\hat{\gamma}_1} - \frac{\hat{\beta}_1^2}{\hat{\beta}_2}\right) n_\alpha n_\beta n_\gamma n_\varepsilon A_{\gamma\varepsilon} + \left(\hat{\alpha}_5 - \hat{\alpha}_2\frac{\hat{\gamma}_2}{\hat{\gamma}_1}\right)[n_\alpha n_\gamma A_{\gamma\beta} + n_\beta n_\gamma A_{\gamma\alpha}] - \frac{\hat{\beta}_1}{\hat{\beta}_2} n_\alpha n_\beta \frac{\delta H_m}{\delta s} - s_{,\alpha}\frac{\partial W'}{\partial s_{,\beta}} - n_{\gamma,\alpha}\frac{\partial W'}{\partial n_{\gamma,\beta}} - \frac{1}{2}\frac{\hat{\gamma}_2}{\hat{\gamma}_1}\left(n_\alpha \frac{\delta H_m}{\delta n_\beta} + n_\beta \frac{\delta H_m}{\delta n_\alpha}\right) - \frac{1}{2}\left(n_\alpha \frac{\delta H_m}{\delta n_\beta} - n_\beta \frac{\delta H_m}{\delta n_\alpha}\right). \tag{11.3-23}$$

From Eqs. (11.3-20,21,23), we see that the seven independent parameters are $\hat{\beta}_1$, $\hat{\beta}_2$, $\hat{\gamma}_1$, $\hat{\gamma}_2$, $\hat{\alpha}_4$, $\hat{\alpha}_1$, and $\hat{\alpha}_5$.

11.3.3 The Doi theory

Of far more importance to the description of LCs than the theory of the previous subsection is the Doi theory of rigid, rod-like molecules [Doi, 1981, pp. 230-36; and Doi and Edwards, 1986, chs. 8-10]. This theory was developed on the level of the distribution function, then approximated in terms of a second-rank tensor, $\mathbf{S}$, in order to be able to use the theory in calculations of complex flows. The original distribution function theory lacked the capability of describing the elastic stress accompanying spatial distortions in the orientation, and so, relatively recently, Doi *et al.* [1988, §IIA] have tried to extend this molecular theory to incorporate such effects. In principle, they were successful; however, practically speaking, only a small gain was realized since the resulting equations in terms of the distribution function are impossible to solve, except for a few cases (such as the spinodal decomposition [Shimada *et al.*, 1988]), even numerically by making use of the most sophisticated of today's computers. Other problems with the Doi theory of solutions of rigid rod-like molecules will be discussed when appropriate. In this section, we shall briefly describe each of these theories as background material for use in §11.6.

A. The original Doi theory

The original molecular theory of Doi [1981] considers the dynamics of inflexible molecules modeled as rigid rods of length L and diameter d. This non-inertial theory describes the rotational motion of the rods for a given flow field through a kinetic (or diffusion) equation for the orientational distribution function, f. This distribution function depends only on time and the unit orientation vector, $\mathbf{q}$.[10] Hence $f(\mathbf{q}, t)d\Omega$ is the probability that a given rod lies inside a solid angle of differential size, $d\Omega$, at time t.

The kinetic equation for the distribution function is derived as [Doi, 1981, p. 232]

$$\frac{\partial f}{\partial t} = \boldsymbol{\nabla}^s \cdot D_r \left(\boldsymbol{\nabla}^s f + \frac{f}{k_B T} \boldsymbol{\nabla}^s V \right) - \boldsymbol{\nabla}^s \cdot (\bar{\mathbf{q}} f) \,, \tag{11.3-24}$$

where $D_r = D_r(\mathbf{q}, f)$ is a rotational diffusion coefficient, $V = V(\mathbf{q}, f)$ is the interaction potential responsible for the alignment of the rods in the liquid-crystalline phase, and $\bar{\mathbf{q}}$ is the local affine rate of change of the unit vector $\mathbf{q}$ due to the macroscopic velocity field,

$$\bar{\mathbf{q}} = (\boldsymbol{\nabla}\mathbf{v})^{\mathrm{T}} \cdot \mathbf{q} - [\mathbf{q}^{\mathrm{T}} \cdot (\boldsymbol{\nabla}\mathbf{v})^{\mathrm{T}} \cdot \mathbf{q}]\mathbf{q} \,. \tag{11.3-25a}$$

Furthermore, the $\boldsymbol{\nabla}^s$ in (11.3-24) represents the "gradient operator" on the unit sphere, with the gradient of a scalar, a, defined in a spherical coordinate system as

$$\boldsymbol{\nabla}^s a \equiv \mathbf{e}_\theta \frac{\partial a}{\partial \theta} + \frac{1}{\sin\theta} \mathbf{e}_\phi \frac{\partial a}{\partial \phi} \,, \tag{11.3-25b}$$

and the divergence of a vector, $\mathbf{w}$, defined as

[10]We use $\mathbf{q}$ here, rather than $\mathbf{n}$, in order to keep the formulation distinct from the LE theory.

$$\boldsymbol{\nabla}^s \cdot \mathbf{w} \equiv \frac{1}{\sin\theta}\left[\frac{\partial(w_\theta \sin\theta)}{\partial\theta} + \frac{\partial w_\phi}{\partial\phi}\right]. \tag{11.3-25c}$$

The expression (11.3-24) is strictly valid only for homogeneous systems (i.e., for cases where f does not change in space) since no mechanism is included for such effects, as through a translational diffusivity or the material derivative. Implicitly, however, f may depend on $\mathbf{x}$ through the velocity gradient tensor of (11.3-25), since $\mathbf{v} = \mathbf{v}(\mathbf{x},t)$, which is evaluated from the linear momentum equation with the extra stress of (11.3-34) and appropriate boundary conditions. The solution to this coupled problem is extremely impractical. Therefore, for all intents and purposes, it is implicit that $\boldsymbol{\nabla}\mathbf{v}$ may not change in space, leaving f unchanged as well. Thus, in the case of a homogeneous flow, although the kinematics affect the orientational state of the system, the orientational state does not affect the kinematics, which remain homogeneous.

What now remains is to specify the interaction (or mean-field) potential, as well as the rotational diffusivity, which appear in (11.3-24). The mean-field potential is taken as arising from the free-energy expression of Onsager [1949, p. 643]—see also §11.2.1.C, and is written as the integral expression

$$V(\mathbf{q},f) = 2ndL^2 k_B T \int f(\mathbf{q}',t)\,\mathbf{q}'\times\mathbf{q}\; d^2q' , \tag{11.3-26}$$

where n is the number density of rods, $\mathbf{q}'$ and $\mathbf{q}$ represent (unit) orientation vectors and d^2q indicates integration over the unit sphere.[11] The rotational diffusivity is also represented by an integral expression [Doi, 1981, p. 231]:

$$D_r(\mathbf{q},f) = \nu_1 D_{r0}(nL^3)^{-2}\left[\frac{4}{\pi}\int f(\mathbf{q}',t)\,\mathbf{q}'\times\mathbf{q}\; d^2q'\right]^{-2} , \tag{11.3-27}$$

where D_{r0} is the rotational diffusivity of a dilute solution of rods, and ν_1 is a numerical constant of order unity.

Substitution of Eqs. (11.3-25) through (11.3-27) into (11.3-24) results in an integro-differential equation which is extremely difficult to solve, even using the advanced supercomputers. Hence, in order to make use of the theory, a series of approximations is made which eventually reduces the model to the conformation tensor level. First, a more convenient form of the potential is derived as a first-order approximation of the integral in (11.3-26) [Doi, 1981, p. 232],

[11]These types of arguments are now quite standard; however, a thorough discussion may be found in Doi and Edwards [1986].

$$V(\mathbf{q}) \approx -\tfrac{3}{2}Uk_BT(q_\alpha q_\beta - \tfrac{1}{3}\delta_{\alpha\beta})S_{\alpha\beta} \; . \tag{11.3-28}$$

The intensity of this potential is characterized by the parameter U, which is a dimensionless concentration:

$$U \equiv \nu_2 ndL^2 \; , \tag{11.3-29}$$

with ν_2 being another unspecified numerical constant of order unity. The average orientation of the liquid-crystalline material influences V through the order parameter tensor, $\mathbf{S}$, which is defined through the conformation tensor $\mathbf{m}$ as

$$S_{\alpha\beta} \equiv m_{\alpha\beta} - \tfrac{1}{3}\delta_{\alpha\beta} \; , \tag{11.3-30}$$

where

$$m_{\alpha\beta} \equiv \langle q_\alpha q_\beta \rangle \; , \tag{11.3-31}$$

with $\langle\cdot\rangle$ denoting the average over the distribution function. Note that the trace of $\mathbf{m}$, as defined by (11.3-31), must be unity, which implies that $\mathbf{S}$ is traceless.

Likewise, the rotational diffusivity of (11.3-27) can be replaced by an average diffusivity, which upon further approximation is written as [Doi, 1981, p. 235]

$$D_r \approx \nu_1^* D_{r0}(nL^3)^{-2}\left(1-\frac{3}{2}S_{\mu\nu}S_{\mu\nu}\right)^{-2} \equiv D_r^*\left(1-\frac{3}{2}S_{\mu\nu}S_{\mu\nu}\right)^{-2} \; , \tag{11.3-32}$$

with ν_1^* being yet another numerical constant (related to ν_1 and ν_2), since this is sufficient to ensure that D_r increases with an increasing degree of rod alignment. Eq. (11.3-32) thus completes a closed system of equations, (11.3-24,25,28,30-32), which may be solved iteratively for the distribution function after specifying the kinematics and the value of U.

Notice that when $\mathbf{v}=\mathbf{0}$ (no flow), $f=1/(4\pi)$ is always one solution to the steady-state problem, corresponding to the isotropic phase. However, for $\mathbf{v}=\mathbf{0}$ this solution becomes unstable for $U\geq 3$, corresponding to the transition from the isotropic to the liquid-crystalline phase. For non-zero values of $\nabla\mathbf{v}$, the corresponding "isotropic" and "liquid-crystalline" equivalent solutions may be evaluated by continuation, their stability depending upon both U and $\nabla\mathbf{v}$. For $U\geq 3$, the "isotropic" solution is always unstable.

The complexity of the approximate diffusion equation, as given by (11.3-24,25,28,32), is still such that it is quite difficult to solve in other than homogeneous flows situations [Edwards and Beris, 1989b; Larson, 1990]. Consequently, further simplifications are introduced by Doi [1981], starting with multiplying (11.3-24) by $(q_\alpha q_\beta - \tfrac{1}{3}\delta_{\alpha\beta})$ and then integrating over all possible orientations, which yields[12]

$$\frac{\partial S_{\alpha\beta}}{\partial t} = -6D_r S_{\alpha\beta} + 6D_r U\left(S_{\alpha\gamma} \langle q_\gamma q_\beta \rangle - \langle q_\alpha q_\beta q_\gamma q_\varepsilon \rangle S_{\gamma\varepsilon}\right) \tag{11.3-33}$$
$$+ \langle q_\gamma q_\beta \rangle \nabla_\gamma v_\alpha + \langle q_\alpha q_\gamma \rangle \nabla_\gamma v_\beta - 2\langle q_\alpha q_\beta q_\gamma q_\varepsilon \rangle \nabla_\varepsilon v_\gamma \; .$$

The extra stress may also be calculated in terms of the order parameter tensor and the fourth-order averages of (11.3-33), and is found to contain three terms:

$$\boldsymbol{\sigma} = \boldsymbol{\sigma}^e + \boldsymbol{\sigma}^v + \boldsymbol{\sigma}^s \; , \tag{11.3-34}$$

where we use $\boldsymbol{\sigma}$ to represent the extra stress in the Doi model rather than the $\mathbf{t}$ of LE theory. Here, $\boldsymbol{\sigma}^e$ is the contribution due to elasticity, $\boldsymbol{\sigma}^v$ is the viscous contribution, and $\boldsymbol{\sigma}^s$ is the solvent contribution, which are given by the following expressions:

$$\sigma^e_{\alpha\beta} = 3nk_BT\left(S_{\alpha\beta} - U[S_{\alpha\gamma}(S_{\beta\gamma} + \tfrac{1}{3}\delta_{\beta\gamma}) - \langle q_\alpha q_\beta q_\gamma q_\varepsilon \rangle S_{\gamma\varepsilon}]\right) \; , \tag{11.3-35}$$

$$\sigma^v_{\alpha\beta} = n\zeta_r \langle q_\alpha q_\beta q_\gamma q_\varepsilon \rangle \nabla_\varepsilon v_\gamma \; , \tag{11.3-36}$$

$$\sigma^s_{\alpha\beta} = \eta_s(\nabla_\alpha v_\beta + \nabla_\beta v_\alpha) \; , \tag{11.3-37}$$

where ζ_r is a rotational drag coefficient and η_s is the solvent viscosity. In the limit of low deformation rates, only $\boldsymbol{\sigma}^e$ is expected to contribute to $\boldsymbol{\sigma}$.

Yet Eqs. (11.3-33) do not form a closed set, since one still needs to evaluate f in order to calculate the averages appearing in this expression. To be sure, the second-order averages may be replaced by $\mathbf{S}$ through (11.3-30,31); but one is handicapped by the fourth-order averages, regardless. At this point, Doi [1981, p. 236] introduces the decoupling approximation,

$$\langle q_\alpha q_\beta q_\gamma q_\varepsilon \rangle \approx m_{\alpha\beta} m_{\gamma\varepsilon} \; , \tag{11.3-38}$$

in order to arrive at the closed set of equations

[12]Note the dependence of this equation on the fourth moment of the distribution function.

$$\frac{\partial S_{\alpha\beta}}{\partial t} = -\, 6D_r[(1-\frac{U}{3})S_{\alpha\beta} - U(S_{\alpha\mu}S_{\beta\mu} - \tfrac{1}{3}\delta_{\alpha\beta}S_{\mu\nu}S_{\mu\nu})$$
$$+ US_{\alpha\beta}S_{\mu\nu}S_{\mu\nu}] + \tfrac{1}{3}(\nabla_\alpha v_\beta + \nabla_\beta v_\alpha) + S_{\mu\beta}\nabla_\mu v_\alpha \qquad (11.3\text{-}39)$$
$$+ S_{\mu\alpha}\nabla_\mu v_\beta - \tfrac{2}{3}\delta_{\alpha\beta}S_{\mu\nu}\nabla_\nu v_\mu - 2S_{\alpha\beta}S_{\mu\nu}\nabla_\nu v_\mu \, .$$

The same decoupling approximation, (11.3-38), may be applied to the elastic stress component as well, obtaining

$$\sigma^e_{\alpha\beta} = 3nk_BT\left[(1-\frac{U}{3})S_{\alpha\beta} - U(S_{\alpha\mu}S_{\beta\mu} - \tfrac{1}{3}\delta_{\alpha\beta}S_{\mu\nu}S_{\mu\nu}) + US_{\alpha\beta}S_{\mu\nu}S_{\mu\nu}\right] . \qquad (11.3\text{-}40)$$

For specific kinematics, the system of equations (11.3-39) may now be solved directly in terms of **S**. These much simpler calculations have appeared in a number of places for a variety of kinematics. Still, one must wonder exactly what information was lost through the integration and decoupling approximation. Indeed, a minor controversy has arisen over the value of the averaged equations. The results of Edwards and Beris [1989b][13] (as also discussed in chapter 3 of Edwards [1991]) have shown that they constitute a good approximation of the distribution function equations, (11.3-24,25,28,32b), as least as far the *steady-state* solutions of simple homogeneous flows (simple-shear or elongational) are concerned. However, as other researchers have shown [Kuzuu and Doi, 1983; Marrucci and Maffettone, 1989; Larson, 1990], the introduction of the decoupling approximation, (11.3-38), stabilizes the equations and thus excludes some interesting physics (specifically, the tumbling of the rods in shear flow—see §11.3.4) from the system of equations, (11.3-39). However, as shown in §11.6.5, it is possible to fix this deficiency by simply replacing the upper-convected derivative in the evolution equation for the order parameter, (11.3-39), with a mixed-convected one. This has also been verified by detailed calculations of "tumbling" in simple shear flow (for example, the order-parameter, time-periodic behavior shown by

[13]Part of the controversy was due to the report of an approximate solution of Eqs. (11.2-24,25,28,32) by Edwards and Beris [1989b, §III], termed there the "first approach," according to which the diffusion equation, (11.3-24), was solved with the potential V calculated based on the order parameter evaluated from the solution of the decoupled equations. However, this technique was only used in parts A and B of the results section, §IV, and not in the most crucial part C on which the evaluation of the closure approximation was based [p. 550]: "By incorporating the definition of $S_{\alpha\beta}$ into the Newton's iteration loop and solving for converged values of $S_{\alpha\beta}$ as well as the distribution, a consistent solution to the full set of equations...was obtained."

Edwards [1991, Fig. 5.97]—reproduced here as Figure 11.9, below—is almost identical to that shown by Larson [1990, Fig. 7].) This change naturally emerges from the discussion of the conformation tensor (order parameter) representation of the inertialess Doi/LE theory, as §11.4 demonstrates.

B. The extended Doi theory

Recently, Doi *et al.* [1988, §IIA] extended the original theory of Doi to incorporate the effects of spatial inhomogeneities in the system behavior. In order to accomplish this, they phenomenologically introduced a new term on the right-hand side of the diffusion equation, (11.3-24), which we rewrite here in component form as

$$\frac{\partial}{\partial x_\gamma}\left[[(D_\parallel - D_\perp) q_\gamma q_\varepsilon + D_\perp \delta_{\gamma\varepsilon}] f \frac{\partial}{\partial x_\varepsilon}(\ln f + W) \right] . \qquad (11.3\text{-}41)$$

In the above expression, $D_\parallel$ and $D_\perp$ are the diffusivities of a rod parallel and perpendicular to the rod axis, and W is the generalization of V from (11.3-26) to include inhomogeneities in the orientation:

$$W(\mathbf{x},\mathbf{q},t) \equiv \int\!\!\int V^*(\mathbf{x}-\mathbf{x}',\mathbf{q},\mathbf{q}') f(\mathbf{x}',\mathbf{q}',t)\, d^2q'\, d^3x' \; . \qquad (11.3\text{-}42)$$

Note that the distribution function now depends on $\mathbf{x}$ as well as $\mathbf{q}$ and t: $f = f(\mathbf{x},\mathbf{q},t)$. A general form for the function V^* is not specified, due to the complexity of the above system of equations. Consequently, the extended theory is of even less use than the original theory in practical applications. However, one interesting calculation has been performed by Shimada *et al.* [1988, §III]; i.e., the calculation of the initial stages of the spinodal decomposition of an isotropic material to a liquid-crystalline material under the driving force of a concentration fluctuation. This treatment will be very helpful in §11.6.4, and we shall discuss it more thoroughly at that time.

11.3.4 The Leslie coefficients: Implications for flow behavior and molecular determination

The values of the Leslie coefficients are of importance in that they determine the steady and dynamic LC flow behavior, irrespective of the underlying physical model, at least for slow flows. Qualitatively different behavior can be observed due to subtle changes in the values of these coefficients, as discussed in subsection A, below. As such, their experimental evaluation and/or *a priori* microscopic determination in terms of molecular parameters has generated significant interest both as a means of understanding the corresponding liquid-crystalline flow behavior and as a means of comparison and validation of various theories and approximations. Any of the flows mentioned below can be used to measure the Leslie coefficients. Alternatively, their theoretical evaluation from the Doi theory is presented in subsection B.

A. LE theory predictions for simple flows

Even after using the Parodi relation, (11.3-10), the LE theory still involves five viscosity coefficients, in addition to the Frank elasticity moduli. These are usually determined experimentally, although microscopic theories can also be used to express them in terms of molecular parameters, as shown in the next subsection. Thus, there is interest in studying the prediction of the LE theory for simple flows which are easily realized experimentally. In the following, we distinguish between flows where the molecular orientation can be forced into a specific direction through the application of an external field (§A.1), and cases where the molecular orientation is dictated by the flow field (§A.2). Only a few illustrative examples will be discussed in each case.

A.1 Controlled orientation flows

One of the simplest experiments involves evaluation of the shear viscosity for a fully oriented sample. The sample orientation is usually obtained from the application of a magnetic field, although an electric field can also be used for materials which have a positive dielectric anisotropy. If the field is strong enough that a possible misalignment due to the orienting effects of the solid walls can be safely ignored [de Gennes, 1974, p. 163], but not so excessively strong as to induce flow instabilities, a uniformly aligned sample in the field direction is obtained. Then the measured viscosity depends on the orientation of the sample director with respect to the shear plane and flow direction [de Gennes, 1974, p. 164]. There are three independent outcomes of the measurements, η_a, η_b, and η_c, corresponding to director orientations which are perpendicular to the plane of shear, along the flow direction, and along the direction of the velocity gradient, respectively.

Miesowicz [1935, 1946] was the first one to use this technique, employing a slowly oscillating thin plate immersed in the fluid investigated [1983, p. 2]—thus, the name "Miesowicz viscosities" for η_a, η_b, and η_c. The reported values for p-azoxyanisole ranged from 0.024 poise for the smallest viscosity, η_b, (along the flow direction) to 0.034 for η_a and

0.092 for the largest, η_c (along the direction of shear) [de Gennes, 1974, p. 164]. These variations are typical for LMWLCs. However, with PLCs, as the aspect ratio *L/d* increases, changes of up to two orders of magnitude between the Miesowicz viscosities are common [Lee and Meyer, 1991, Tables 6,7]. Ignoring surface effects, the expression (11.3.9) for the stress tensor in the LE theory can be used directly in order to correlate the Miesowicz viscosities to the Leslie coefficients as [de Gennes, 1974, p. 165]

$$\eta_\alpha = \tfrac{1}{2}\alpha_4 \ , \quad \eta_\beta = \tfrac{1}{2}(\alpha_3 + \alpha_4 + \alpha_6) \ , \quad \eta_c = \alpha_4 + \alpha_5 - \alpha_2 \ . \tag{11.3-43}$$

Thus, from measurements of the viscosity of well-oriented samples in shear flow, we can obtain three relations between the Leslie coefficients.

The predictions of the LE theory for steady uniaxial elongational flow are also available, although the usefulness of the elongational flow for the experimental determination of the Leslie coefficients is limited due to the usual experimental difficulties for its realization under controlled conditions. In any event, the predictions for the elongational viscosity, $\bar{\eta}$, are [Kuzuu and Doi, 1984, p. 1036]

$$\bar{\eta} \equiv \frac{\sigma_{zz} - \sigma_{xx}}{\dot{\varepsilon}} = \alpha_1 + \tfrac{3}{2}\alpha_4 + \alpha_5 + \alpha_6 \ , \tag{11.3-44}$$

where $\sigma_{zz} - \sigma_{xx}$ denotes the first normal-stress difference (z is the direction of the flow) and $\dot{\varepsilon}$ is the extension rate.

Other experiments can be used to evaluate additional combinations of the Leslie coefficients. The effective viscosities associated with the three types of director distortions

$$\eta_{\text{twist}} = \alpha_3 - \alpha_2 \equiv \gamma_1 \ , \quad \eta_{\text{bend}} = \gamma_1 - \frac{\alpha_3^2}{\eta_b} \ , \quad \eta_{\text{splay}} = \gamma_1 - \frac{\alpha_2^2}{\eta_c} \ , \tag{11.3-45}$$

are measured from inelastic light-scattering experiments [de Gennes, 1974, §5.2.5]. Ultrasound measurements can be used to evaluate the "ultrasonic viscosities" $\tilde{\eta}_a$, $\tilde{\eta}_b$, and $\tilde{\eta}_c$ [de Gennes, 1974, §5.2.2], where

$$\tilde{\eta}_a = \eta_a \ , \quad \tilde{\eta}_b = \eta_b - \frac{\alpha_3^2}{\gamma_1} \ , \quad \tilde{\eta}_c = \tilde{\eta}_b \ . \tag{11.3-46}$$

Finally, observations on the dynamics of the onset of instabilities after the sudden application of an external electric or magnetic field can also be used for the experimental determination of the LE model parameters. For example, in the study of the Fréedericksz transition between two orienting glass plates, the magnetic field intensity is increased abruptly from zero to a certain value, H_0 (larger than a critical value H_c necessary for the transition to take place), and the dynamics are observed, for example, with an optical method. Depending on the arrangement, various combinations of the Leslie parameters can be measured in this way [de Gennes, 1974., §5.2.4.3]. de Gennes supplies additional information on those and other experiments, as well as supplying in tabular form a compilation of the data obtained with various methods for MBBA [Table 5.1]. Lee and Meyer [1991, Tables 6,7] provide similar information applicable to PLCs, with data for PBG.

A.2 Simple shear flow

In the previous subsection, by fixing the molecular orientation we only needed to take into consideration the anisotropic viscous behavior of the liquid crystals characterized by multiple viscosity coefficients, as in the general theory for an anisotropic viscous fluid [Hand, 1962]—the Leslie coefficients, as discussed above. However, under general flow conditions, elastic as well as viscous effects need to be taken into account (neglecting inertial effects, which are usually much less important, especially for highly viscous PLCs). According to (11.3-2), elastic stresses develop in liquid-crystalline systems when the director orientation changes and these are proportional to the Oseen/Frank elastic moduli, $F_{elas} \approx K/\overline{L}^2$, where K is a characteristic elastic modulus and $\overline{L}$ a characteristic length over which changes in the director orientation take place. Typically, $\overline{L}$ scales with the apparatus, albeit under flow conditions it can often be much smaller, characterizing internal rearrangement in the director orientation [Marrucci, 1991, p. 4177]. On the other hand, the viscous stresses, according to (11.3-9), scale as $F_{visc} \approx \eta\dot{\gamma}$, where η is a characteristic viscosity of the system. Of course, due to the anisotropicity of the liquid crystals, both K and η depend on the application; sometimes a geometric average of the relevant elastic moduli and Leslie coefficients are used [Pieranski and Guyon, 1973, p. 436]. Then the importance of the viscous effects versus the elastic effects can be judged from their ratio, called the Ericksen number, Er, defined as [Cladis and Torza, 1975, p. 1286]

$$Er \equiv \frac{\eta\dot{\gamma}\overline{L}^2}{K} . \qquad (11.3\text{-}47)$$

This definition for the Ericksen number coincides with the inverse of Leslie's parameter, ζ [Leslie, 1968, p. 279], and with Pikin's parameter R/λ_{11} [Pikin, 1974, p. 1246]. It is also essentially the same as the Ericksen number originally proposed by de Gennes, defined as the ratio of convective to diffusive transport of the director orientation [Cladis and Torza, 1975, p. 1286]. The use of the Ericksen number as a characteristic dimensionless parameter can considerably simplify and/or elucidate the mathematical analysis. For example, for shear flows, as the shear rate increases, the Ericksen number increases so that eventually, at high enough Ericksen numbers—or, equivalently, small enough ζ—the elastic effects are confined to thin boundary layers next to the walls [Leslie, 1968, p. 281], and they can therefore be neglected. Ignoring the elastic effects might result in some loss of information. However, this is usually compensated by the insight to be gained by pondering the analytic form of a limiting case. Thus, in the following, we shall first examine the simple shear flow behavior neglecting elastic effects before considering how the results of the analysis affect the combined (elastic plus viscous) problem.

Let us assume that the simple shear flow is established between two infinite parallel plates, placed parallel to the *x-y* plane of a Cartesian coordinate system, at a distance $\overline{L}$ apart, with the top plate moving with a constant velocity, V, along the x direction with respect to the other. As customary, we assume the establishment of a linear velocity profile,

$$v_x = \dot{\gamma} z \,, \quad v_y = v_z = 0 \,, \tag{11.3-48}$$

and we neglect (director) inertial effects. If, in addition, we neglect the elastic forces, the governing equations corresponding to the inertialess LE theory, (11.3-16,17), simplify considerably, reducing to the much simpler equations for transversely isotropic fluids [Ericksen, 1960b, Eqs.(19),(20)]. Using, for simplicity, a spherical coordinate system where the components of the unit vector, n, representing the director are given as

$$n_x = \cos\theta\cos\phi \,, \quad n_y = \cos\theta\sin\phi \,, \quad n_z = \sin\theta \,. \tag{11.3-49}$$

The angular momentum balance equations become [Leslie, 1981, §3; 1987, §4]

$$\gamma_1 \frac{d\theta}{dt} + \dot{\gamma}\,(\alpha_3\cos^2\theta - \alpha_2\sin^2\theta)\cos\phi = 0 \,, \tag{11.3-50a}$$

$$\gamma_1\cos\theta\,\frac{d\phi}{dt} - \dot{\gamma}\,\alpha_2\sin\theta\,\sin\phi = 0 \,, \tag{11.3-50b}$$

where $\gamma_1 \equiv \alpha_3 - \alpha_2$. Eq. (11.3-50) is equivalent to Eq. (24) of Ericksen [1960b], with the parameter λ identified as

$$\lambda = \frac{1+\varepsilon}{1-\varepsilon} \;\Rightarrow\; \varepsilon = \frac{\lambda - 1}{\lambda + 1} \,, \tag{11.3-51a}$$

where ε is defined as the ratio of the Leslie coefficients α_3 and α_2

$$\varepsilon \equiv \frac{\alpha_3}{\alpha_2} \,. \tag{11.3-51b}$$

From the form of Eqs. (11.3-50), the steady solutions admitted can be immediately recovered [Ericksen, 1960b, §3-5]. From (11.3-50b), a steady solution can be realized if and only if $\sin\theta=0$ or $\sin\phi=0$. The first case is always a solution, corresponding to an alignment normal to the plane of shear ($\theta=0$, $\phi=\pi/2$). This is unstable for $\varepsilon>0$ [Ericksen, 1960b, p. 120] and neutrally stable for $\varepsilon<0$. From (11.3-50a), we can see that the second case ($\sin\phi=0$) leads to the steady (homogeneous) solution [Leslie, 1966, p. 366]

$$\tan^2\theta = \varepsilon \equiv \frac{\alpha_3}{\alpha_2} , \tag{11.3-52a}$$

or, equivalently,

$$\cos(2\theta) = \frac{1}{\lambda} , \tag{11.3-52b}$$

if and only if (11.3-52) admits a real solution, i.e., only when $\varepsilon\geq 0$ or, equivalently, $|\lambda| \geq 1$, implying that α_2 and α_3 have the same sign. In this case, the angle θ that is realized is (taking into account the fact that γ_1 has to be positive—see Example 11.1) [Leslie, 1987, p. 243]

$$\text{a) when } \alpha_2,\alpha_3 < 0 \;\Rightarrow\; \alpha_2 < \alpha_3 < 0 \;\Rightarrow\; 0 < \theta < \frac{\pi}{4} , \tag{11.3-52a}$$

$$\text{b) when } \alpha_2,\alpha_3 > 0 \;\Rightarrow\; 0 < \alpha_2 < \alpha_3 \;\Rightarrow \frac{\pi}{4} < \theta < \frac{\pi}{4} . \tag{11.3-52b}$$

This flow regime, realized when $\varepsilon \geq 0$ or, equivalently, $|\lambda| \geq 1$, is called the "flow aligning" case.

The most interesting case, however, is when α_2 and α_3 have an opposite sign ($\varepsilon < 0$ or $|\lambda| < 1$). In this case, (11.3-52) does not admit any real solutions and therefore can not be physically satisfied. Furthermore, stability analysis shows that the steady, out-of-plane solution is always (neutrally) unstable [Leslie, 1981, p. 117], as can also be seen from the integral curves of the time-dependent solution of (11.3-50) [Ericksen, 1960b, Fig. 3]. Thus, we reach the conclusion that *when* $\alpha_2\alpha_3 < 0$, *no stable steady solution to the simple shear flow exists for the LE theory*.[14] The solution under these circumstances can be found through a direct integration of (11.3-50) in time [Ericksen, 1960b, §5]. It leads to a perpetual oscillatory motion where the director continuously undergoes a "tumbling" motion, thus giving the name "tumbling regime" to this case. Actually, for $\varepsilon < 0$, (11.3-50) can be shown [Ericksen, 1960b, §5] to be the same as the equations derived by Jeffery [1922, Eqs.(45)-(47)] for the motion of an ellipsoid

[14]This occurrence is due to the fact that the bulk free energy of the type discussed in §11.2.1 is missing from the LE theory, being incompatible with the unit vector description of the LE microstructure. The presence of a bulk free energy in the analysis was shown by Edwards [1991, p. 152] to stabilize the corresponding solution for small shear rates in the conformation tensor theory.

p. 518: the numbers of the second pair of equations (11.3-52a) and (11.3-52b) should be (11.3-53a) and (11.3-53b), respectively of revolution in a Newtonian fluid. This assumes that the director **n** is taken to be the axis of revolution, and that [Ericksen, 1960b, §5]

$$\lambda = \frac{a^2 - b^2}{a^2 + b^2} \Rightarrow \varepsilon = -\left(\frac{b}{a}\right)^2 , \qquad (11.3\text{-}54)$$

where a is the length of the unequal (major) axis and b is the common length of the other two axes. Thus, the solutions for the director orientation are identical to Jeffery's orbits evaluated for the motion of rigid ellipsoidal particles in dilute suspensions under shear. Moreover, according to (11.3-54), the higher the aspect ratio a/b (the longer the particles are), the closer ε is to zero (approaching it always from below).

The above analysis suggests that a drastically different dynamic behavior is realized depending on whether α_2 and α_3 have the same or different signs. Such differences are indeed observed experimentally. However, care needs to be exercised in extrapolating the simple picture presented above, where the elastic effects were neglected, to real liquid-crystalline systems, especially in the tumbling regime. The reason that even qualitative differences might be observed is because the tumbling solution mentioned above is a singular limit; i.e., the introduction of any amount of non-zero Oseen/Frank elasticity introduces a boundary-orienting effect which significantly changes the simplified picture presented above. In addition, the observations depend critically on the nature of the director orientation next to the confining solid surfaces. The situation is similar to the singular character of the inviscid approximation to the Navier/Stokes equations.

The changes for $\varepsilon < 0$, outlined in more detail below, can be briefly summarized here as follows. At small enough shear rates, a steady solution always exist. At shear rates above a threshold value, a transition to time-dependent flow may take place, depending on the boundary conditions. The location of the transition and the nature of the resulting solution are a strong function of the Ericksen number and the boundary (anchoring) conditions specifying the director orientation at the solid walls.

When elastic effects are taken into account, the Ericksen number is the natural dimensionless number to use as a parameter (as mentioned above) in order to characterize the relative importance of the viscous and the elastic effects. For simple shear flow realized between two parallel plates at a distance $\overline{L}$ from each other and a relative tangential velocity V it is defined as [Pieranski *et al.* , 1976, Eq. (5)]

$$Er \equiv \frac{|\alpha_2| V \overline{L}}{k_{11}} . \qquad (11.3\text{-}55)$$

In addition, since taking into account elastic effects in the equations introduces second-order spatial derivatives into the governing equations,

Eq. (11.3-2), additional boundary conditions are necessary in order to have a mathematically well-posed problem. These are supplied through the specification of the director orientation at the solid boundaries. The director orientation next to a solid surface depends on the surface/molecule interactions, and can be experimentally controlled through a chemical and/or mechanical treatment (for example, by introducing grooves in a particular direction [Berreman, 1973]) of the confining solid surfaces.

For planar shear flow, when the molecules are aligned parallel to the solid surfaces at the boundaries and along the flow direction, and under the simplifying assumptions, $\alpha_1 = 0$ and $|\varepsilon| \ll 1$ (which is typically satisfied with PLCs), the following results are reported [Pikin, 1974]. The structure of the steady-state solution critically depends on both the parameter ε and the magnitude of the Ericksen number. In the flow-aligning case, and for large Ericksen numbers, $Er \gg \varepsilon^{-1/2} \gg 1$, the Leslie solution, (11.3-53), is realized everywhere, except at narrow layers next to the wall, with thicknesses on the order of

$$\frac{\Delta z}{L} \approx (2\varepsilon)^{\frac{1}{4}} Er^{-\frac{1}{2}} , \tag{11.3-56}$$

in which the orientation of the director changes exponentially from the value $\theta \approx \varepsilon^{1/2}$ in the bulk to $\theta = 0$ at the solid boundary [Pikin, 1974]. Note that, in general, there are multiple stable steady-state solutions, of which one can be singled out as the least dissipative [Currie and Macsithigh, 1979]. This last solution falls into one of two categories, depending on the angle of the molecular orientation at the boundaries (for a general molecular alignment in the plane of shear at a specific angle with respect to the surface normal) [Currie and Macsithigh, 1979].

In the tumbling regime and for large Ericksen numbers, i.e, when $Er \gg |\varepsilon|^{-1/2} \gg 1$, in contrast to the zero-elasticity analysis presented above, there is a steady solution with the molecules oriented in the shear plane:

$$\theta(z) = 3\,(2\,|\varepsilon|)^{\frac{1}{2}} \wp\!\left(\frac{z + C}{\Delta z}; -\tfrac{2}{3}; g_3\right) , \tag{11.3-57}$$

where $\wp$ is the Weierstrass's $\wp$ function [Abramowitz and Stegun, 1964, §18.1.2] and C and g_3 are determined from the boundary conditions. In this case, the dependence of $\theta(z)$ on z is periodic with a period on the order of Δz.

Finally, for both the flow-aligning and tumbling regimes and for intermediate Ericksen numbers, $|\varepsilon|^{-1/2} \gg Er \gg 1$, there is a steady-state solution which takes the approximate form

$$\theta(z) \approx -\frac{\varepsilon}{2} Er \frac{z}{L}\left(\frac{z}{L} - 1\right) . \tag{11.3-58}$$

What is equally (if not more) interesting than the presentation of the various steady-state solutions, is the analysis of their stability. In the flow-aligning case (i.e., when $\varepsilon > 0$), the steady solution is always stable against infinitesimally small perturbations [Pikin, 1974, p. 1249]. Therefore, any flow instability in Couette flow has to involve finite amplitude perturbations, exactly as in the case for Newtonian isotropic fluids. In contrast, in the tumbling nematic regime (i.e., when $\varepsilon < 0$), the stationary motion characterized by the single minimum, (11.3-55), is unstable for values of the Ericksen number larger than a critical value [Pikin, 1974, p. 1249]

$$Er_c \approx 10\,|\varepsilon|^{-\frac{1}{4}} \; ; \quad -10^{-4} < \varepsilon < 0 \; . \qquad (11.3\text{-}59)$$

In the narrow limit of validity of this case, $0 < |\varepsilon| < 10^{-4}$, Er_c varies from infinity to a value of about 100. For $\varepsilon < -10^{-4}$ and $Er \ll |\varepsilon|^{-1/2}$, the stationary flow is stable. In the same ε region when $Er \gg |\varepsilon|^{-1/2}$, the flow is also unstable with a threshold value, Er_c, the critical value where the single minimum solution vanishes [Pikin, 1974, p. 1250]:

$$Er_c \approx 5\,|\varepsilon|^{-\frac{1}{2}} \; ; \quad \varepsilon < -10^{-4} \; . \qquad (11.3\text{-}60a)$$

Later Pieranski *et al.* [1976] refined this limit through numerical integration of the equations:

$$Er_c = 4.8\,|\varepsilon|^{-\frac{1}{2}} \; ; \quad \varepsilon < -10^{-4} \; , \qquad (11.3\text{-}60b)$$

a result which compared favorably with the experimentally determined value for p-n hexyloxybenzilidène p′-aminobenzonitrile at T=77°C, $4.5|\varepsilon|^{-1/2}$ [Pieranski *et al.* , p. C1-5]. These results have also been verified through a direct numerical integration in time of the relevant two-dimensional (in plane) equations [Manneville, 1981; Carlsson, 1984]. This numerical work also showed the form of the instability: as the velocity of the upper plate increases, the director "tumbles" to another steady solution corresponding to higher values of the director orientation [Manneville, 1981, §III]. This exchange of solutions happens in regions where a solution multiplicity is obtained, implying hysteresis phenomena typically associated with phase transitions [Carlsson, 1984]. It is interesting to note that after successive "tumblings," the region where the director crosses the $\theta=\pi/2$ value is encountered closer and closer to the wall. However, these higher-order solutions are unlikely to be realized experimentally since the director is expected to escape in the transversal (out-of-plane) direction in order to minimize the elastic energy [Manneville, 1981, p. 240] (see also the discussion on out-of-plane stability below).

The in-plane solution for the planar flow with the director aligned with the flow can also become unstable towards an out-of-plane perturbation where the molecules, for sufficiently high shear rates, leave their plane of symmetry and orient themselves along a direction which is simultaneously perpendicular to the direction of the velocity and the velocity gradient [Pieranski *et al.* , 1976, §2.3]. Stability calculations indicate that the critical *Er* corresponding to this mode of instability, for $\varepsilon<0$, is [Pieranski and Guyon, 1973, Eq. (3); Pieranski *et al.* , 1976, Eq. (10)]

$$Er_c \approx 9.5 \sqrt{\frac{k_{22}}{k_{11}}} |\varepsilon|^{-\frac{1}{2}} ; \quad \varepsilon < 0 . \tag{11.3-61}$$

Thus, depending on the ratio k_{22}/k_{11}, this instability can set in before or after the in-plane instability discussed above. In any case, it has been experimentally observed that the state established after the onset of instability for the in-plane solution does appear to involve the director rotating in an out-of-plane configuration [Pieranski *et al.* , 1976, p. C1-6]. Moreover, the out-of-plane configuration seems to be energetically more favorable than the tumbling motion calculated when the Frank elasticity is neglected. These results are supported from recent detailed theoretical calculations for out-of-plane instabilities by Zúñiga and Leslie [1989]. These authors have also shown that, in general, the out-of-plane instability occurs very soon, allowing, at a maximum, only one tumble of the director to take place. However, since when $\varepsilon<0$ there is no stationary, out-of-plane solution for the director, subject to boundary conditions corresponding to a planar orientation (along the solid boundaries), the flow that is finally established cannot be steady [Pieranski *et al.* , 1976, p. C1-6]. In contrast, when the boundary conditions at the solid surfaces impose a perpendicular orientation to the plane of shear, then the behavior for $\varepsilon<0$ or $\varepsilon>0$ is reversed: for $\varepsilon>0$ the steady, out-of-plane solution becomes unstable for *Er* higher than a critical value, Er_c (evaluated approximately by a relationship identical to (11.3-61) [Pieranski *et al.* , 1976, p. C1-6] (see Manneville and Dubois-Violette [1976]; Leslie [1976; 1979, p. 35] for a more accurate description). In contrast, for $\varepsilon<0$, the uniform, out-of-plane configuration is unconditionally stable [Pieranski *et al.* , 1976, §1.2].

Thus, even when the influence of the boundaries is taken into account, the previous conclusion of no existence of a stable solution remains valid, provided that the orientation of the director imposed at the boundaries is in the plane of shear and that the Ericksen number is higher than some critical value. However, the orientation that is realized is more complex than the Jeffery's orbit suggested from the analysis, which neglected the Frank elasticity. Note that similar conclusions with respect to the liquid-crystalline flow dynamics are reached when $\varepsilon<0$ for other flows as well, such as Poiseuille flow [Manneville and Dubois-Violette, 1976b], oscillatory shear [Clark *et al.* , 1981], under the presence of an electric field [Skarp *et al.* , 1981], etc. This "tumbling" flow behavior generates much more complex structures than the flow-aligning one. It is observed with both LMWLCs [Pieranski and Guyon, 1974; Pieranski *et al.* , 1976; Skarp *et al.* , 1981] and PLCs [Burghardt and Fuller, 1991, p. 2550], although it is more common with the latter. It is of importance because it can be used to explain abnormal macroscopic observations, such as oscillatory transients in the start-up and relaxation of shear flows [Moldenaers *et al.* , 1989; Moldenaers and Mewis, 1990], negative first normal stresses [Kiss and Porter, 1978], etc.—see also the relevant discussion at the end of §11.1. Thus, it is important to preserve these features in the flow constitutive equations, which essentially amounts to the consideration of models for which the corresponding Leslie parameter ε is negative (or, equivalently, $|\lambda|<1$). The predictions from the Doi theory and its modifications are outlined in the next subsection.

B. Theoretical evaluation of the Leslie coefficients

In the limit of slow flows, any theory of liquid-crystalline flow behavior ought to reduce to the phenomenological, continuum description of the LE theory. This generality of the LE description, arising from its thermodynamic foundation, allows the fitting of its phenomenological parameters against any other available theoretical model. The Doi theory, as the only (so far) available microscopic theory describing the nonlinear, liquid-crystalline flow behavior, has been used almost exclusively for the *a priori* determination of the LE parameters. Both the restricted and the kinetic theory (with the pre-averaging approximation for the rotational diffusivity) have been used over the years, with the results summarized in Table 11.1 [Lee and Meyer, 1991, Table 2]. In addition to the scalar order parameter, S, the other three parameters which appear in Table 11.1 are the fourth moment R,

$$R \equiv \langle P_4(\theta)\rangle \; , \tag{11.3-62}$$

where P_4 is the fourth-order Legendre polynomial and $\langle\cdot\rangle$ represents the average over the orientational distribution, the parameter Γ, defined as

$$\Gamma \equiv 2S \left\langle \frac{d\overline{V}}{d\theta} g \cos\theta \right\rangle^{-1} ; \quad \overline{V} \equiv \frac{V}{k_B T} \; , \tag{11.3-63}$$

with $V=V(\theta)$ the self-consistent field potential defined by (11.3-26), $g(\theta)$ the solution to

$$\frac{1}{\sin\theta}\frac{d}{d\theta}\left(\sin\theta\frac{dg}{d\theta}\right) - \frac{g}{\sin^2\theta} - \frac{d\overline{V}}{d\theta}\frac{dg}{d\theta} = -\frac{d\overline{V}}{d\theta}, \qquad (11.3\text{-}64)$$

and the parameter ζ the dimensionless rotational drag coefficient,

$$\zeta \equiv 2\frac{\zeta_r D_r}{nk_B T}. \qquad (11.3\text{-}65)$$

As Table 11.1 shows, in the characteristic limit of an isotropic state (S=R=0), all the coefficients from both expressions vanish, except the isotropic viscosity, α_4, as they should. Furthermore, the Parodi relationship, (11.3-10), and the inequalities for the Leslie coefficients (see Example 11.1) are duly satisfied. In addition, $\alpha_2+\alpha_3=\alpha_6-\alpha_5=-2S$ in both models. However, comparing further the results obtained from the restricted (decoupled) and the distribution function theories, we notice that, although there are qualitative similarities (e.g., in the isotropic limit, S=R=1, and for Γ=1, the two expressions are almost identical), there are significant quantitative differences. What are more important, in terms of dictating the dynamic liquid-crystalline flow behavior, are the predictions for the ratio ε of α_3 over α_2. Whereas the restricted Doi theory predicts unequivocally a positive ε, the predictions of the microscopic theory depend upon whether the microscopically determined coefficient Γ is larger or smaller than 1:

$$\lambda = \Gamma. \quad \text{(microscopic Doi theory)} \qquad (11.3\text{-}66)$$

As it turns out, the results of both approximate [Semenov, 1983, Fig. 2] and detailed numerical calculations [Kuzuu and Doi, 1984, Fig. 1] indicate that within the region of the order parameter values of interest, $0.5<S<1$, Γ is smaller than one (equivalently, ε is negative), implying through relationship (11.3-66) a tumbling dynamic behavior. Clearly, in view of the drastic differences in behavior between the tumbling and flow-aligning cases, as illustrated in the previous subsection, this discrepancy represents a major defect of the restricted Doi theory, a defect usually attributed to the decoupling approximation, (11.3-38). However, as is shown in the next section, this defect can be easily rectified within the bracket formalism.

A few attempts were also made toward improving the predictions of the Doi theory by taking into account concentration effects on the rotational diffusivity [Kuzuu and Doi, 1984] and semi-flexibility [Semenov, 1987] effects. The effects of diffusion arise through the scaling of the Leslie viscosities and so do not affect the ε (or λ) predictions. The effect of the semi-flexibility (not surprisingly) is to increase the value of λ which, now, for a sufficiently flexible macromolecule, is predicted to be larger than unity, therefore corresponding to a flow-aligning behavior [Semenov, 1987, Fig. 5]. Experimentally, the theoretically predicted trends

Parameter[†]	Restricted Doi theory[‡]	Microscopic Doi theory using the preaveraged diffusivity
α_1	$-2S^2$	$-(2-\overline{\zeta})R$
α_2	$-2S\dfrac{1+2S}{2+S}$	$-(1+\dfrac{1}{\Gamma})S$
α_3	$-2S\dfrac{1-S}{2+S}$	$-(1-\dfrac{1}{\Gamma})S$
α_4	$\frac{2}{3}(1-S)$	$\frac{2}{105}\left(7(3+\overline{\zeta})-5(3+2\overline{\zeta})S-3(2-\overline{\zeta})R\right)$
α_5	$2S$	$\frac{2}{7}\left((5+\overline{\zeta})S+(2-\overline{\zeta})R\right)$
α_6	0	$-\frac{2}{7}(2-\overline{\zeta})(S-R)$
References	Marrucci, 1982; Doi, 1983.	Semenov, 1983; Kuzuu and Doi, 1983, 1984; Lee and Meyer, 1986; Lee, 1988.

† The parameters are scaled with respect to $1/2\ nk_BT\ D_r$.

‡ The decoupling approximation, (11.3-38), was used.

Table 11.1: The Leslie coefficients, α_i, for a nematic LC of rigid rods. (See text for an explanation of the symbols.)

seem to have been verified. There are certain systems of rigid macromolecules, such as PBG, for which substantial experimental evidence has been accumulated, both optical and mechanical, for a tumbling behavior [Burghardt and Fuller, 1991, p. 2550]. Similarly, other systems, less rigid, like PBT, seem to show a flow-aligning behavior [Berry, 1991, p. 968].

In conclusion, this long diversion has had the purpose of demonstrating the significance of the physics which is conveyed through a model in order to have the flexibility to accommodate the correct ε ratio. This is very important if the model is to represent, even qualitatively, the correct dynamic flow behavior.

11.4 The Bracket Description of the LE Theory

In this section, we derive the LE equations presented in the preceding section for both the inertial and non-inertial forms using the generalized bracket. This exercise allows us to rewrite the LE equations in an alternate form; i.e., in Hamiltonian form. Once we realize the underlying structure of this system of equations embodied in its Hamiltonian form, we begin to appreciate how this form is preserved through the various levels of description for LCs.

11.4.1 The inertial form

As we have repeatedly seen in Part II, the first step in developing a set of equations for a given system is to choose consistently the parameters which are necessary to describe uniquely the state of the system. For an incompressible LC in the inertial form, let us assume that the state of the system at any time may be completely described by three vector fields, each a function of position, $\mathbf{x}$, and time, t: the momentum density, $\mathbf{M}(\mathbf{x},t)$, the director, $\mathbf{n}(\mathbf{x},t)$, and the conjugate orientation "momentum" $\boldsymbol{\omega}(\mathbf{x},t)$. Since the fluid is incompressible, for convenience we treat ρ as a constant of value unity, so that the velocity field, $\mathbf{v}(\mathbf{x},t)$, becomes the necessary parameter rather than $\mathbf{M}$, under the realization that the units of $\mathbf{v}$ are the units of $\mathbf{M}$. The director is required to be a unit vector from the outset, subject to the constraint $\mathbf{n}\cdot\mathbf{n}=1$. The last parameter, $\boldsymbol{\omega}$, will later be identified as the material derivative of the director, multiplied by an inertial constant, $\sigma\dot{\mathbf{n}}$. Here we treat it as an independent parameter in order to illustrate that applying the constraint $\mathbf{n}\cdot\mathbf{n}=1$ to the generalized bracket automatically insures that the constraint $\dot{\mathbf{n}}\cdot\mathbf{n}=0$ is satisfied without the necessity of imposing it explicitly.

The operating space contains all the functions which satisfy the appropriate boundary and initial conditions on the variables $\mathbf{v}$, $\mathbf{n}$, and $\boldsymbol{\omega}$, as well as the imposed constraints due to the physics of the situation. Thus, for our incompressible LC, we define the operating space as

$$P \equiv \begin{cases} \mathbf{v}(\mathbf{x},t) \in \mathbb{R}^3; & \boldsymbol{\nabla}\cdot\mathbf{v} = 0 \text{ in } \Omega \ , \quad \mathbf{v}(\mathbf{x},t) = \mathbf{0} \quad \text{on } \partial\Omega \\ \mathbf{n}(\mathbf{x},t) \in \mathbb{R}^3; & \mathbf{n}\cdot\mathbf{n} = 1 \quad \text{in } \Omega \\ \boldsymbol{\omega}(\mathbf{x},t) \in \mathbb{R}^3 & \end{cases} , \tag{11.4-1}$$

with appropriate initial conditions on all three vector fields.

Now that we have chosen the proper variables to describe the liquid-crystalline material and have defined the proper operating space P, we need to specify the total energy of the system. In this case, however, we are dealing with an incompressible medium, and hence we only need to specify the mechanical contribution to the total energy, which we shall denote again as H_m, the Hamiltonian.

The Hamiltonian is written as a functional of the above-mentioned parameters which we chose for describing the system. As a consequence, $H_m = H_m[\mathbf{v},\mathbf{n},\boldsymbol{\omega}]$, and we can write the Hamiltonian explicitly for this system as

$$H_m[\mathbf{v},\mathbf{n},\boldsymbol{\omega}] \equiv \int_\Omega \left(\frac{1}{2}\rho\mathbf{v}\cdot\mathbf{v} + \frac{1}{2\sigma}\boldsymbol{\omega}\cdot\boldsymbol{\omega} + W + \psi_m\right) \mathrm{d}^3x \ . \tag{11.4-2}$$

In this expression, the first two terms represent the translational and rotational kinetic energy, respectively, as in §10.3. In the following, the units of mass and length are chosen so as to make the density, ρ, equal to unity (for convenience, as discussed above). W is the Oseen/Frank distortion energy, given by (11.3-8), which represents the effects of spatial variations on the director field. Note that W depends not only on the director but on the gradients of $\mathbf{n}$ as well, i.e., $W = W(\mathbf{n}, \boldsymbol{\nabla}\mathbf{n})$. The last term, ψ_m, incorporates the effects of an external magnetic field into the system of equations. It is given as

$$\psi_m \equiv -\tfrac{1}{2}\left(\chi_a(\mathbf{n}\cdot\mathbf{H})^2 + \chi_\perp\mathbf{H}\cdot\mathbf{H}\right) . \tag{11.4-3}$$

Note the difference in notation between the mechanical energy, H_m (a global scalar), and the magnetic field, $\mathbf{H}$ (a local vector).

The next step in our procedure is to introduce the generalized bracket which describes the dynamics of an arbitrary functional, F, which depends on the same parameters as H_m; i.e., $F = F[\mathbf{v},\mathbf{n},\boldsymbol{\omega}]$:

$$F[\mathbf{v},\mathbf{n},\boldsymbol{\omega}] \equiv \int_\Omega f(\mathbf{v},\mathbf{n},\boldsymbol{\nabla}\mathbf{n},\boldsymbol{\omega}) \ \mathrm{d}^3x \ . \tag{11.4-4}$$

The total time derivative of this functional again takes the form of (5.2-7),

$$\frac{\mathrm{d}F}{\mathrm{d}t} = \int_\Omega \left[\frac{\delta F}{\delta \mathbf{v}}\cdot\frac{\partial \mathbf{v}}{\partial t} + \frac{\delta F}{\delta \mathbf{n}}\cdot\frac{\partial \mathbf{n}}{\partial t} + \frac{\delta F}{\delta \boldsymbol{\omega}}\cdot\frac{\partial \boldsymbol{\omega}}{\partial t}\right] \mathrm{d}^3x \ , \tag{11.4-5}$$

where the variational derivatives are defined as

$$\frac{\delta F}{\delta v_\alpha} \equiv I_\alpha(\frac{\partial f}{\partial \mathbf{v}}) \ , \quad \frac{\delta F}{\delta \omega_\alpha} \equiv \frac{\partial f}{\partial \omega_\alpha} \ ,$$
$$\frac{\delta F}{\delta n_\alpha} \equiv \frac{\partial f}{\partial n_\alpha} - \frac{\partial f}{\partial n_\beta} n_\beta n_\alpha - \left(\frac{\partial f}{\partial n_{\alpha,\beta}}\right)_{,\beta} + \left(\frac{\partial f}{\partial n_{\gamma,\beta}}\right)_{,\beta} n_\gamma n_\alpha \ . \tag{11.4-6}$$

Note that $\mathbf{n}\cdot[\delta F/\delta \mathbf{n}]=0$, which implies that $\delta F/\delta \mathbf{n} \in P$.

With the above definitions of the Volterra derivatives, the dynamics of F are completely described in terms of the evolution equations for the dynamic variables by (11.4-5). Inversely, we can use (11.4-5) as a means to define the evolution equations by postulating that F also obeys the dynamical relation of (7.1-1,2)

$$\frac{\mathrm{d}F}{\mathrm{d}t} = \{F,H_m\} + [F,H_m] \ . \tag{11.4-7}$$

The inertial form of the Poisson bracket for two arbitrary functionals, $F[\mathbf{v},\tilde{\mathbf{n}},\boldsymbol{\omega}]$ and $G[\mathbf{v},\tilde{\mathbf{n}},\boldsymbol{\omega}]$, where $\tilde{\mathbf{n}}$ is the unconstrained direction vector (i.e., not necessarily a unit vector), can be written as in §10.3:

$$\begin{aligned}\{F,G\} = &- \int_\Omega v_\alpha \left[\frac{\delta F}{\delta v_\gamma}\nabla_\gamma \frac{\delta G}{\delta v_\alpha} - \frac{\delta G}{\delta v_\gamma}\nabla_\gamma \frac{\delta F}{\delta v_\alpha}\right] \mathrm{d}^3x \\ &- \int_\Omega \tilde{n}_\alpha \left[\frac{\delta F}{\delta v_\gamma}\nabla_\gamma \frac{\delta G}{\delta \tilde{n}_\alpha} - \frac{\delta G}{\delta v_\gamma}\nabla_\gamma \frac{\delta F}{\delta \tilde{n}_\alpha}\right] \mathrm{d}^3x \\ &- \int_\Omega \left[\frac{\delta F}{\delta \omega_\alpha}\frac{\delta G}{\delta \tilde{n}_\alpha} - \frac{\delta G}{\delta \omega_\alpha}\frac{\delta F}{\delta \tilde{n}_\alpha}\right] \mathrm{d}^3x \\ &- \int_\Omega \omega_\alpha \left[\frac{\delta F}{\delta v_\gamma}\nabla_\gamma \frac{\delta G}{\delta \omega_\alpha} - \frac{\delta G}{\delta v_\gamma}\nabla_\gamma \frac{\delta F}{\delta \omega_\alpha}\right] \mathrm{d}^3x \ .\end{aligned} \tag{11.4-8}$$

This expression satisfies all of the properties of a Poisson bracket, yet when $\tilde{\mathbf{n}}$ is constrained to be a unit vector, the Poisson bracket of (11.4-8) is not necessarily valid, as in §5.5.3. In order to find the appropriate expression for the constrained case, we need to construct a projection mapping which transforms the bracket (11.4-8) into the properly constrained bracket. Consequently, we define this mapping, R, as

$$R(\tilde{\mathbf{n}}) \rightarrow \frac{\tilde{\mathbf{n}}}{(\tilde{n}_\beta \tilde{n}_\beta)^{1/2}} \equiv \mathbf{n} \ , \tag{11.4-9}$$

i.e., we restrict our functionals F and G to those which depend on $\tilde{\mathbf{n}}$ only through their dependence on $\mathbf{n}$, the constrained vector. Thus we may use the following theorem to transform the Volterra derivative in the unconstrained space, $\delta F/\delta\tilde{\mathbf{n}}$, to the equivalent derivative in the constrained space, $\delta F/\delta\mathbf{n}$.[15]

Theorem 11.1:

$$\frac{\delta F}{\delta \tilde{n}_\alpha} = \frac{1}{(\tilde{n}_\gamma \tilde{n}_\gamma)^{1/2}} (\delta_{\alpha\beta} - n_\beta n_\alpha) \frac{\delta F}{\delta n_\beta} \ . \tag{11.4-10}$$

We note the happy circumstance that the second term in (11.4-10) is zero, since $\mathbf{n}\cdot[\delta F/\delta\mathbf{n}]=0$, and hence the Poisson bracket does not change under the substitution of Theorem 11.1 into (11.4-8), except that $\tilde{\mathbf{n}}$ is replaced with $\mathbf{n}$.[16] This bracket must then satisfy the properties of a Poisson bracket, mentioned earlier, even for constrained $\mathbf{n}$, by construction.

The dissipation bracket is a phenomenological form which was derived by postulating the most general expression possible, and then restricting it to the bilinear (close-to-equilibrium) limit. Drawing from past experience again, we can define this bracket as

$$\begin{aligned}[F,G] \equiv &- \int_\Omega Q_{\alpha\beta\gamma\varepsilon} \nabla_\alpha\left(\frac{\delta F}{\delta v_\beta}\right) \nabla_\gamma\left(\frac{\delta G}{\delta v_\varepsilon}\right) d^3x \\ &+ \int_\Omega \alpha_2 \left[\frac{\delta F}{\delta \omega_\alpha} - \nabla_\beta\left(\frac{\delta F}{\delta v_\alpha}\right) n_\beta\right]\left[\frac{\delta G}{\delta \omega_\alpha} - \nabla_\gamma\left(\frac{\delta G}{\delta v_\alpha}\right) n_\gamma\right] d^3x \\ &- \int_\Omega \alpha_3 \left[\frac{\delta F}{\delta \omega_\alpha} + \nabla_\alpha\left(\frac{\delta F}{\delta v_\beta}\right) n_\beta\right]\left[\frac{\delta G}{\delta \omega_\alpha} + \nabla_\alpha\left(\frac{\delta G}{\delta v_\gamma}\right) n_\gamma\right] d^3x \ ,\end{aligned} \tag{11.4-11}$$

where the phenomenological matrix, $\mathbf{Q}$, is given by

[15]The proof of Theorem 11.1 is analogous to the one given in §5.5.3, and so it is not presented here; however, the interested reader may find it in Appendix C of Edwards [1991].

[16]We shall not be so lucky, however, when we turn to the conformation tensor theory.

$$\begin{aligned} Q_{\alpha\beta\gamma\varepsilon} \equiv\ & \alpha_1 n_\alpha n_\beta n_\gamma n_\varepsilon + \tfrac{1}{2}\alpha_4(\delta_{\alpha\gamma}\delta_{\beta\varepsilon} + \delta_{\beta\gamma}\delta_{\alpha\varepsilon}) \\ & + \tfrac{1}{2}(\alpha_2 + \alpha_5)(\delta_{\beta\varepsilon} n_\alpha n_\gamma + \delta_{\beta\gamma} n_\alpha n_\varepsilon) \\ & + \tfrac{1}{2}(\alpha_6 - \alpha_3)(\delta_{\alpha\varepsilon} n_\beta n_\gamma + \delta_{\alpha\gamma} n_\beta n_\varepsilon)\ . \end{aligned} \tag{11.4-12}$$

Although on first inspection this bracket looks different from those discussed in previous chapters, it is not. We have simply rearranged it somewhat in order to illustrate the exact nature of the dissipation in this inertial system. Note that the second and third terms in the above expression, (11.4-11), represent the dissipative effects of rotational motion, and, ergo, possess the intrinsic forms of the upper- and lower-convected derivatives, respectively. Note that this bracket couples the variables $\boldsymbol{\omega}$ and $\mathbf{v}$ with themselves, and with each other. $\delta F/\delta\mathbf{n}$ does not appear in the dissipation bracket since $\mathbf{n}$ is a conserved (frozen) quantity when the time scale of the problem is such that the inertial effects are present (and thus we do not need to worry about Theorem 11.1 for this bracket). Because of the nature of $\ddot{\mathbf{n}}$, $\dot{\mathbf{n}}$ must satisfy (11.3-4), and the dissipation shows itself in the equation for angular momentum. Note that we have labeled the constants (the α_i) in this expression so as to be consistent with the LE theory, with no loss of generality whatsoever.

The above dissipation bracket provides for the Hamiltonian, defined by (11.4-2), the same dissipation (provided that $\boldsymbol{\omega}$ is set equal to $\sigma\dot{\mathbf{n}}$) as that which was evaluated by Leslie [1979, p. 13]:

$$\frac{\mathrm{d}H_m}{\mathrm{d}t} = -\,[H_m, H_m] = -\int_\Omega t_{\alpha\beta} A_{\alpha\beta}\, \mathrm{d}^3x + \int_\Omega g_\alpha N_\alpha\, \mathrm{d}^3x \leq 0\ , \tag{11.4-13}$$

where $t_{\alpha\beta}$ and g_α are defined by Eqs. (11.3-9,13), respectively, since the Parodi relationship, (11.3-10), is assumed to hold.[17] As a consequence, the non-negative definite nature of the dissipation bracket imposes the same restrictions on the parameters α_i as appear in the LE theory [Leslie, 1979, Eq. (35)].

Now we may calculate the functional derivatives of the Hamiltonian, which, for the case of (11.4-2), are found to be

$$\frac{\delta H_m}{\delta v_\alpha} = v_\alpha\ , \tag{11.4-14}$$

[17]In fact, it can immediately be seen now that this relation is implied from (11.3-12): since $\mathbf{Q}$ is a phenomenological dissipative parameter, it satisfies the reciprocal relations as well as material frame invariance; $Q_{\alpha\beta\gamma\varepsilon}=Q_{\gamma\varepsilon\alpha\beta}=Q_{\beta\alpha\gamma\varepsilon}=Q_{\alpha\beta\varepsilon\gamma}$.

$$\frac{\delta H_m}{\delta n_\alpha} = -\chi_a n_\beta H_\beta H_\alpha + \chi_a n_\beta n_\gamma H_\beta H_\gamma n_\alpha + \frac{\partial W}{\partial n_\alpha}$$

$$-\left(\frac{\partial W}{\partial n_{\alpha,\gamma}}\right)_{,\gamma} - \frac{\partial W}{\partial n_\beta} n_\beta n_\alpha + \left(\frac{\partial W}{\partial n_{\beta,\gamma}}\right)_{,\gamma} n_\beta n_\alpha \,, \tag{11.4-15}$$

$$\frac{\delta H_m}{\delta \omega_\alpha} = \frac{1}{\sigma}\omega_\alpha \,. \tag{11.4-16}$$

Then, the two forms of the dynamical equation for F, (11.4-5,7), may be directly compared to yield the following evolution equations:

$$\dot{v}_\alpha = F_\alpha - p_{,\alpha} - \left(\frac{\partial W}{\partial n_{\beta,\gamma}} n_{\beta,\alpha}\right)_{,\gamma} + t'_{\alpha\gamma,\gamma} \,, \tag{11.4-17}$$

$$\frac{\partial n_\alpha}{\partial t} = -v_\gamma n_{\alpha,\gamma} + \frac{1}{\sigma}(\omega_\alpha - \omega_\beta n_\beta n_\alpha) \,, \tag{11.4-18}$$

$$\frac{\partial \omega_\alpha}{\partial t} = -v_\gamma \omega_{\alpha,\gamma} - \frac{\partial W}{\partial n_\alpha} + \frac{\partial W}{\partial n_\beta} n_\beta n_\alpha$$

$$+\left(\frac{\partial W}{\partial n_{\alpha,\beta}}\right)_{,\beta} - \left(\frac{\partial W}{\partial n_{\gamma,\beta}}\right)_{,\beta} n_\gamma n_\alpha + G_\alpha - G_\beta n_\beta n_\alpha \tag{11.4-19}$$

$$+\alpha_2\left[\frac{1}{\sigma}\omega_\alpha - v_{\alpha,\gamma} n_\gamma\right] - \alpha_3\left[\frac{1}{\sigma}\omega_\alpha + v_{\gamma,\alpha} n_\gamma\right] ,$$

where $t'_{\alpha\gamma}$ is defined by (11.3-9) with $\boldsymbol{\omega}/\sigma$ replacing $\dot{\mathbf{n}}$. From (11.4-18), we get

$$\sigma\dot{n}_\alpha = \omega_\alpha - \omega_\beta n_\beta n_\alpha \;\rightarrow\; \omega_\alpha = \sigma\dot{n}_\alpha + \varepsilon n_\alpha \,, \tag{11.4-20}$$

with the parameter $\varepsilon = \omega_\beta n_\beta$ as yet unspecified. In order to get $\sigma\ddot{n}_\alpha$, we take the material derivative of (11.4-20):

$$\sigma\ddot{n}_\alpha = \dot{\omega}_\alpha - \dot{\omega}_\beta n_\beta n_\alpha - \omega_\beta \dot{n}_\beta n_\alpha - \varepsilon\dot{n}_\alpha \,. \tag{11.4-21}$$

Substitution of (11.4-19) into the above expression then yields

$$\sigma \ddot{n}_\alpha = G_\alpha + \gamma n_\alpha - \frac{\partial W}{\partial n_\alpha} + \left(\frac{\partial W}{\partial n_{\alpha,\beta}} \right)_{,\beta} + g_\alpha - \varepsilon \dot{n}_\alpha , \qquad (11.4\text{-}22)$$

where the parameter γ is evaluated as (11.3-14). From the definition of the Hamiltonian provided by (11.4-2) and $\dot{\mathbf{n}} \equiv \delta H_m / \delta \boldsymbol{\omega} = \boldsymbol{\omega}/\sigma$, we can now specify the parameter $\varepsilon = \dot{n}_\beta n_\beta = 0$ from (11.4-20). For this value of ε, Eqs. (11.4-17,22) reduce exactly to the LE equations, (11.3-2,3). Hence we have shown that the LE equations in inertial form may be derived directly from the generalized bracket, with no *a priori* knowledge about the system except (a) a proper choice of the descriptive variables characterizing the state, (b) the system Hamiltonian which is implied from equilibrium thermodynamic information, and (c) a knowledge of the structure and symmetry inherent to the Poisson and dissipation brackets.

11.4.2 *The non-inertial form*

It is also possible to use the bracket formalism in order to develop the LE equations under the quasi-steady state approximation—Eqs. (11.3-2,16,17). For this purpose, the Poisson bracket and dissipation expressions developed in the last subsection need to be modified accordingly, as discussed briefly in §10.3.

We shall now assume that the time scale of the problem is such that the director inertia is negligible; i.e., $\sigma \to 0 \Rightarrow \boldsymbol{\omega} = \mathbf{0}$. As such, the number of dynamic variables in our problem is reduced from three to two:

$$F[\mathbf{v},\mathbf{n}] \equiv \int_\Omega f(\mathbf{v},\mathbf{n},\boldsymbol{\nabla}\mathbf{n}) \, d^3x , \qquad (11.4\text{-}23)$$

and, consequently, the Hamiltonian is written solely in terms of **v** and **n** as

$$H_m[\mathbf{v},\mathbf{n}] \equiv \int_\Omega (\tfrac{1}{2}\mathbf{v}\cdot\mathbf{v} + W + \psi_m) \, d^3x . \qquad (11.4\text{-}24)$$

From (11.4-8), we know that the inertial form of the Poisson bracket is given as

$$\{F,G\} = -\int_\Omega v_\alpha \left[\frac{\delta F}{\delta v_\gamma} \nabla_\gamma \frac{\delta G}{\delta v_\alpha} - \frac{\delta G}{\delta v_\gamma} \nabla_\gamma \frac{\delta F}{\delta v_\alpha} \right] d^3x$$

$$- \int_\Omega n_\alpha \left[\frac{\delta F}{\delta v_\gamma} \nabla_\gamma \frac{\delta G}{\delta n_\alpha} - \frac{\delta G}{\delta v_\gamma} \nabla_\gamma \frac{\delta F}{\delta n_\alpha} \right] d^3x$$

$$- \int_\Omega \left[\frac{\delta F}{\delta(\sigma \dot{n}_\alpha)} \frac{\delta G}{\delta n_\alpha} - \frac{\delta G}{\delta(\sigma \dot{n}_\alpha)} \frac{\delta F}{\delta n_\alpha} \right] d^3x$$

$$- \int_\Omega \sigma \dot{n}_\alpha \left[\frac{\delta F}{\delta v_\gamma} \nabla_\gamma \frac{\delta G}{\delta(\sigma \dot{n}_\alpha)} - \frac{\delta G}{\delta v_\gamma} \nabla_\gamma \frac{\delta F}{\delta(\sigma \dot{n}_\alpha)} \right] d^3x \ , \tag{11.4-25}$$

where we have replaced ω_α with $\sigma \dot{n}_\alpha$. As $\sigma \rightarrow 0$, the fourth integral in this expression vanishes; however, we still have the functional derivatives with respect to $\sigma \dot{\mathbf{n}}$ in the third integral, which must be reduced in terms of $\mathbf{v}$ and $\mathbf{n}$. In order to accomplish this task, let us invoke the equation of continuity for the vector $\mathbf{n}$:

$$\dot{n}_\alpha - n_\gamma v_{\alpha,\gamma} = 0 \ , \tag{11.4-26}$$

where we have used the upper-convected derivative for a vectorial quantity to take into account the codeformation with the medium. Next, by definition,

$$\frac{\delta H_m}{\delta(\sigma \dot{n}_\alpha)} = \dot{n}_\alpha = n_\gamma \frac{\partial}{\partial x_\gamma} \frac{\delta H_m}{\delta v_\alpha} \ , \tag{11.4-27}$$

where (11.4-26) has been used to substitute for $\dot{\mathbf{n}}$. Substituting this expression, written for general functionals $F{=}H_m$ and $G{=}H_m$, into the bracket, (11.4-25), yields the Poisson bracket for the functionals $F[\mathbf{v},\mathbf{n}]$ and $G[\mathbf{v},\mathbf{n}]$:

$$\{F,G\} = - \int_\Omega v_\alpha \left[\frac{\delta F}{\delta v_\gamma} \nabla_\gamma \frac{\delta G}{\delta v_\alpha} - \frac{\delta G}{\delta v_\gamma} \nabla_\gamma \frac{\delta F}{\delta v_\alpha} \right] d^3x$$

$$- \int_\Omega n_\alpha \left[\frac{\delta F}{\delta v_\gamma} \nabla_\gamma \frac{\delta G}{\delta n_\alpha} - \frac{\delta G}{\delta v_\gamma} \nabla_\gamma \frac{\delta F}{\delta n_\alpha} \right] d^3x \tag{11.4-28}$$

$$- \int_\Omega n_\gamma \left[\frac{\delta G}{\delta n_\alpha} \nabla_\gamma \frac{\delta F}{\delta v_\alpha} - \frac{\delta F}{\delta n_\alpha} \nabla_\gamma \frac{\delta G}{\delta v_\alpha} \right] d^3x \ .$$

Thus we arrive at the materially objective Poisson bracket for a liquid-crystalline system in terms of functionals of the dynamic variables $\mathbf{v}$ and $\mathbf{n}$. Note that this bracket applies whether or not $\mathbf{n}$ is constrained to be a unit vector, since Theorem 11.1 still applies.

To develop the proper dissipation bracket, we consider the changes which occur in the problem as we adjust the time scale to allow the neglect of the system inertia. The concept of a time scale is very important in determining which variables couple in the dissipation bracket. For example, if the time scale is particularly short, any phenomena occurring over a longer time scale are "fixed" in the system, and may be ignored. Alternatively, any phenomena occurring on a much shorter time scale will happen so rapidly as to reveal an "average" value, which also appears fixed. Thus, as we adjust our time scale to a value where the inertia can be neglected safely, the dissipation manifests itself in a different way. In terms of the dissipation bracket, the variables **v** and **n** now couple with themselves and with each other. Close to equilibrium, in a manner similar to the previous case, the dissipation can then be expressed as a quadratic form in terms of the affinities $\delta G/\delta \mathbf{n}$ and $\boldsymbol{\nabla}(\delta G/\delta \mathbf{v})$:

$$
\begin{aligned}
[F,G] = &- \int_\Omega Q^n_{\alpha\beta\gamma\varepsilon} \nabla_\alpha\Big(\frac{\delta F}{\delta v_\beta}\Big) \nabla_\gamma\Big(\frac{\delta G}{\delta v_\varepsilon}\Big)\, d^3x \\
&- \int_\Omega P_{\alpha\beta\gamma\varepsilon} \frac{\delta F}{\delta n_\alpha} n_\beta \frac{\delta G}{\delta n_\gamma} n_\varepsilon \, d^3x \\
&- \int_\Omega L_{\alpha\beta\gamma\varepsilon} \left[\nabla_\alpha\Big(\frac{\delta F}{\delta v_\beta}\Big) n_\gamma \frac{\delta G}{\delta n_\varepsilon} - \nabla_\alpha\Big(\frac{\delta G}{\delta v_\beta}\Big) n_\gamma \frac{\delta F}{\delta n_\varepsilon} \right] d^3x \,,
\end{aligned}
\tag{11.4-29}
$$

where, in order to produce a form compatible with the inertial expression of the equations in the limit $\sigma \to 0$, the coefficient matrices in (11.4-29) should be

$$
\begin{aligned}
Q^n_{\alpha\beta\gamma\varepsilon} \equiv\ & \beta^n_1 n_\alpha n_\beta n_\gamma n_\varepsilon + \tfrac{1}{2}\beta^n_2 (\delta_{\alpha\gamma}\delta_{\beta\varepsilon} + \delta_{\beta\gamma}\delta_{\alpha\varepsilon}) \\
& + \tfrac{1}{2}\beta^n_3 (\delta_{\beta\varepsilon} n_\alpha n_\gamma + \delta_{\beta\gamma} n_\alpha n_\varepsilon + \delta_{\alpha\varepsilon} n_\beta n_\gamma + \delta_{\alpha\gamma} n_\beta n_\varepsilon) \,,
\end{aligned}
\tag{11.4-30}
$$

$$
P_{\alpha\beta\gamma\varepsilon} \equiv \frac{1}{\alpha_3 - \alpha_2} (\delta_{\alpha\gamma}\delta_{\beta\varepsilon} + \delta_{\beta\gamma}\delta_{\alpha\varepsilon}) \,,
\tag{11.4-31}
$$

$$
L_{\alpha\beta\gamma\varepsilon} \equiv \frac{\alpha_3}{\alpha_2 - \alpha_3} (\delta_{\alpha\gamma}\delta_{\beta\varepsilon} + \delta_{\beta\gamma}\delta_{\alpha\varepsilon}) \,.
\tag{11.4-32}
$$

Previously, no dissipation was allowed in the $\dot{\mathbf{n}}$ equation; however, as we alter the time scale to neglect inertia, the dissipation must appear here. Thus the second term in the dissipation bracket represents relaxational phenomena, while the third allows for any non-affine coupling between the director and the velocity gradient. This dissipation bracket has exactly the same form as the previous bracket, (11.4-11), although at first glance this may not appear to be the case. Of course, the phenomenological coefficients are not the same as in the previous bracket, but this is to be expected. Again, we set the constant coefficients so as to be consistent with the LE theory equations.

Through the identity $[H_m, H_m] \geq 0$, and the procedure of Example 11.1, we can arrive at the following conditions on the phenomenological coefficients for the non-inertial theory:

$$\alpha_3 - \alpha_2 \geq 0 \ , \tag{11.4-33}$$

$$\beta_2^n \geq 0 \ , \tag{11.4-34}$$

$$\beta_3^n + \beta_2^n \geq 0 \ , \tag{11.4-35}$$

$$2(\beta_1^n + 2\beta_3^n) + 3\beta_2^n \geq 0 \ . \tag{11.4-36}$$

Upon substitution of (11.3-19) into these inequalities, they become the equivalent inequalities of Leslie [1979, p. 13], which were derived in the inertial form. Hence we see that in the limit as $\sigma \to 0$ the dissipation is completely equivalent to that corresponding to the inertial form.

Using the Poisson bracket, Eq. (11.4-28), and the above dissipation bracket, (11.4-29), the dynamical equation, (11.4-7), can easily be shown to be equivalent to the non-inertial LE equations, (11.3-2,16,17). Thus we have arrived at a bracket formalism for each particular case of the LE theory, and in so doing, we have realized the underlying structure and symmetry common to each case.

11.5 The Conformation Tensor Theory

Now that we know the underlying structure embodied in the generalized bracket for a liquid-crystalline system described by a unit vector, we may also see how it is preserved when we describe the system in terms of the conformation tensor, **m** (or, equivalently, **S**). Thus, we may develop the general tensorial theory for LCs, and feel confident that we have consistently incorporated all the applicable mathematical nuances into the governing system of equations. To be sure, we could do the same for the scalar-vector theory, but it seems rather pointless to do so since we may obtain this theory simply by restricting the tensorial theory to a special case. Hence we present in this section the development of the tensorial equations in both the inertial and non-inertial forms following the discussion of the preceding section.

As already mentioned, the tensorial parameter better reflects the symmetries of LCs: for instance, a turn of the director by 180 degrees (i.e., a change in sign of the director) does not influence the state of the material. The only way in which this insensitivity of the material to changes in sign of the director can be fully reflected into the equations is if we require them to depend on the tensor structural parameter, **m**,

$$\mathbf{m} \equiv \mathbf{nn} \ , \tag{11.5-1}$$

and its time derivative(s). However, the above equation for **m** is overly restrictive, assuming a perfect alignment of the molecules along the director axis **n**.

In general, a distribution of different orientations is established, in which case (11.5-1) should be replaced by an average of **nn** over the distribution function. In this case, only the more general constraint

$$\mathrm{tr}(\mathbf{m}) = 1 \ , \tag{11.5-2}$$

is applicable. In addition to allowing the correct representation of the molecular symmetries in the equations, a reformulation in terms of a general, symmetric tensor parameter, **m**, subject to the constraint (11.5-2), allows the description of a wider variety of liquid-crystalline structures: nematic, cholesteric, and blue phases, as well as the states with limited (statistical) order as encountered in polymeric liquid crystals. In addition, this formulation allows the description of various liquid-crystalline states at equilibrium and during the transitions between these states. This is only feasible through a description of the material's free energy in terms of the various moments of **m** and its spatial gradients. Note that this is not possible if (11.5-1) is valid (or, which is equivalent, if we still consider the problem in terms of a unit vector description of the material), since in that case all the moments are equal to unity. Indeed, this has been performed before by many authors by incorporating a Landau/de Gennes expansion [de Gennes, 1971, p. 197] of the bulk free energy into the system energy of the form

$$H_b \equiv \int_\Omega [a_2 m_{\alpha\beta} m_{\beta\alpha} + a_3 m_{\alpha\beta} m_{\beta\gamma} m_{\gamma\alpha} + a_4 m_{\alpha\beta} m_{\beta\gamma} m_{\gamma\varepsilon} m_{\varepsilon\alpha}] \, d^3x \ , \tag{11.5-3}$$

where the a_i are coefficients which depend on the system temperature and concentration. This expression has been used many times in the past to describe phase transitions in liquid-crystalline systems for static problems; however, no such expression is possible in terms of **n** since the unit vector constraint will reduce an equivalent of (11.5-3) to a constant.

When **m** can be represented as **nn**, it corresponds to a uniaxial structure; however, a more general representation requires two independent, orthogonal directions, $\boldsymbol{\nu}\boldsymbol{\nu} + \mathbf{nn}$, corresponding to a biaxial structure.

Furthermore, it is well known that even in the nematic liquid-crystalline state the uniaxial state can be unstable relative to the biaxial one [Wright and Mermin, 1989, §II.C], although the second component is much smaller in magnitude and usually can be neglected.

With these general remarks behind us, let us now go on to the derivation of the equations of motion for the conformation tensor theory of LCs. Again, let us not get bogged down in the mathematics so that we may be able to view the final system of equations and the method of their development from an uninterrupted perspective.

11.5.1 *The inertial form*

Now we shall consider a liquid-crystalline material which is described by the velocity vector, **v**, and two tensor fields, **m** and **w**. As before, we shall later set $w_{\alpha\beta}=\sigma^m \dot{m}_{\alpha\beta}$, the material derivative of the structure tensor, $m_{\alpha\beta}$, multiplied by a different inertial constant, σ^m. These variables belong to a similar space P as (11.4-1), except that **n** and $\boldsymbol{\omega}$ are replaced with **m** and **w**. We can therefore write an arbitrary functional of these three variable fields as

$$F[\mathbf{v},\mathbf{m},\mathbf{w}] \equiv \int_\Omega f(\mathbf{v},\mathbf{m},\boldsymbol{\nabla}\mathbf{m},\mathbf{w})\ \mathrm{d}^3x\ . \tag{11.5-4}$$

Once again, the dynamical equation for this functional is

$$\begin{aligned}\frac{\mathrm{d}F}{\mathrm{d}t} &= \int_\Omega \left[\frac{\delta F}{\delta \mathbf{v}}\cdot\frac{\partial \mathbf{v}}{\partial t} + \frac{\delta F}{\delta \mathbf{m}}:\frac{\partial \mathbf{m}}{\partial t} + \frac{\delta F}{\delta \mathbf{w}}:\frac{\partial \mathbf{w}}{\partial t}\right]\mathrm{d}^3x \\ &= \{[F,H_m]\} \equiv \{F,H_m\} + [F,H_m]\ ,\end{aligned} \tag{11.5-5}$$

where the Hamiltonian is given by

$$H_m[\mathbf{v},\mathbf{m},\mathbf{w}] \equiv \int_\Omega [\frac{1}{2}\rho\mathbf{v}\cdot\mathbf{v} + \frac{1}{2\sigma^m}\mathbf{w}:\mathbf{w} + W + \psi_m]\ \mathrm{d}^3x + H_b[\mathbf{m}]\ . \tag{11.5-6}$$

In this Hamiltonian, ρ is again taken to be unity, using suitable units for mass and length, in order to simplify the final expressions. Similarly, ψ_m and W are now expressed in terms of **m** as

$$\psi_m \equiv -\tfrac{1}{2}\,[\chi_a\,\mathbf{HH}:\mathbf{m} + \chi_\perp\,\mathbf{H}\cdot\mathbf{H}]\ , \tag{11.5-7}$$

$$W \equiv \tfrac{1}{2}\,b_1 m_{\alpha\beta,\gamma}m_{\alpha\beta,\gamma} + \tfrac{1}{2}\,b_2 m_{\alpha\beta,\alpha}m_{\gamma\beta,\gamma}\ . \tag{11.5-8}$$

Note that W is a truncated series including only the lowest-order terms in this expansion, which are generally believed to be enough to describe adequately all of the necessary physics associated with distortional effects in most cases. This being the case, it is unnecessary to carry over any additional terms in the expansion which would imply additional parameters (and additional complications). In the case where $\mathbf{m}=\mathbf{nn}$, W reduces to the Frank distortion energy of (11.3-8), provided that we set $k_1=k_3=2b_1+b_2$ and $k_2=2b_1$, which is trivial to show.

There is still one term left in the Hamiltonian (11.5-6) to discuss: the bulk free energy, H_b, which reduces to a constant under the assumption that $\mathbf{m}=\mathbf{nn}$. As already mentioned, it is this term which allows us to incorporate the physics of the phase transition into our model, and therefore we wish to be very careful about the functional form which we give it. This term has the form of (11.5-3), yet, following standard practice, we shall write it in terms of the traceless tensor $\mathbf{S}$ rather than $\mathbf{m}$, with no loss of generality:

$$H_b[\mathbf{S}] \equiv \int_\Omega [a_1 S_{\alpha\beta}S_{\beta\alpha} + a_2 S_{\alpha\beta}S_{\beta\gamma}S_{\gamma\alpha} + a_3 S_{\alpha\beta}S_{\beta\gamma}S_{\gamma\epsilon}S_{\epsilon\alpha}]\, d^3x \ . \quad (11.5\text{-}9)$$

Thus we use the Landau/de Gennes expansion in terms of the moments of $\mathbf{S}$, truncated after the first three terms. All higher-order moments may be reduced to the above three via the Cayley/Hamilton theorem, hence only affecting slightly the definitions of the parameters a_i (which are arbitrary anyway). However, products of the various moments may be included in (11.5-9) as well, but we do not do so here for the sake of simplicity. The expression as given by (11.5-9) is thus restricted to uniaxial systems (at equilibrium), but by including these additional terms, one can extend this free energy to biaxial systems as well [Allender and Lee, 1984, p. 335; Gramsbergen *et al.*, 1986, §2.3]. Thus this section is restricted to the study of uniaxial phase transitions, but the inclusion of biaxial effects is extremely trivial and can be done simply by including two more terms in (11.5-9). Still, however, once one imposes a flow field upon the system, the system of equations may display biaxial characteristics anyway.

The specification of H_b is not yet complete since we have not determined the values of the parameters a_i. These parameters have units of *energy/volume*, and we pick them up in their standard forms from the Landau/de Gennes theory of phase transitions for static systems [Gramsbergen *et al.*, 1986, §2.2], with the numerical constants chosen for later convenience (see below):

$$a_1 = \tfrac{3}{4}\nu_1(1-U) \ , \quad a_2 = -\tfrac{3}{2}\nu_2 U \ , \quad a_3 = \tfrac{9}{4}\nu_3 U \ . \quad (11.5\text{-}10)$$

Here, the ν_i are numerical constants, which, for simplicity in later calculations, we shall set equal to each other (i.e., $\nu_1=\nu_2=\nu_3=\nu$), and U is the dimensionless concentration mentioned in §11.3.3.[18] With these parameters, H_b is uniquely determined for any value of the dimensionless concentration.[19]

For the sake of later discussion, we show a plot of H_b/ν versus s, the scalar order parameter, in Figure 11.7. This figure was obtained using the uniaxial approximation for **S**:

$$S_{\alpha\beta} = s\,(q_\alpha q_\beta - \tfrac{1}{3}\delta_{\alpha\beta})\,, \tag{11.5-11}$$

whereupon (11.5-9) becomes

$$H_b = \tfrac{1}{2}\nu s^2(1-U) - \tfrac{1}{3}\nu s^3 U + \tfrac{1}{2}\nu s^4 U\,. \tag{11.5-12}$$

As is evident from the figure, for a dimensionless concentration, U, smaller than the fraction 8/9, the isotropic phase is globally stable, representing the only minimum in the free-energy curve. For $U>1$, the liquid-crystalline state becomes globally stable, yet in between these two values there can exist either a superconcentrated isotropic state, or (possibly) a biphasic region.

The behavior of the critical points can be examined by taking the derivative of H_b with respect to s:

$$\frac{\partial H_b}{\partial s} = 0 = \nu s(1-U) - \nu s^2 U + 2\nu s^3 U\,, \tag{11.5-13}$$

so that the solutions are given by

$$s = 0\,,\quad \frac{1}{4} \pm \frac{3}{4}\left(1 - \frac{8}{9U}\right)^{1/2}\,. \tag{11.5-14}$$

These are plotted in Figure 11.8. Here, the straight line corresponds to the $s=0$ solution, the upper branch of the parabola to the liquid-crystalline state, and the lower branch to the mathematical bifurcation between the two states.

It is also possible to examine the stability of the solutions via a numerical recipe, i.e., through the Jacobian arising through Newton's method. The Jacobian of this system is given by $-\partial\cdot/\partial s$ of (11.5-13). Consequently, driven from the relaxation equation for the system close to equilibrium, $ds/dt = -\lambda(\partial h_b/\partial s)$, $\lambda>0$, when all of the eigenvalues (in this case, there is only one eigenvalue) of the Jacobian are negative, the system is stable with respect to arbitrary, small perturbations, otherwise it is unstable.

[18]Here, U is treated as a parameter of the problem. In general, however, U may change as the orientation changes, and should be considered an additional dynamic variable.

[19]Of course, only positive values of U make sense physically.

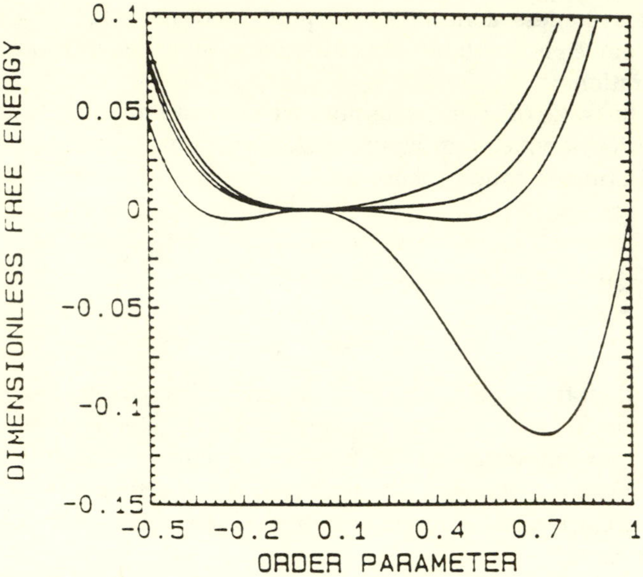

Figure 11.7: Dimensionless bulk free energy versus the scalar order parameter for various U in descending order: U=0.75, 0.87, 0.96, 1.5. (Reprinted with permission from Edwards [1991, p. 91].)

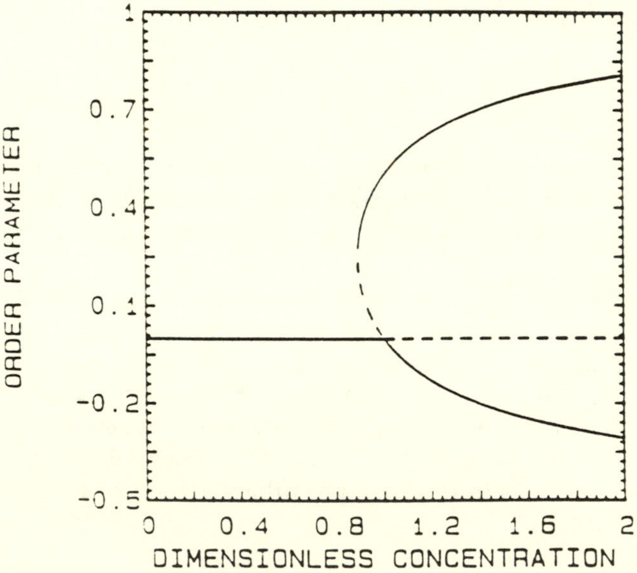

Figure 11.8: Order parameter versus dimensionless concentration. (Reprinted with permission from Edwards [1991, p. 93].)

Analyzing the solutions of (11.5-14), we see that the upper branch of the parabola is stable down to U=8/9, and the lower branch down to U=1. The isotropic curve is stable up to U=1, and then unstable thereafter. For clarity, we have used a solid line for the stable regions of the curve and a dashed line for the unstable regions.

Now we see why the numerical constants in (11.5-10) were so chosen: so that the phase transition occurs at U=1, which is convenient for the calculations, and since (11.5-14) allows s to take on all the values in its possible range, $-\frac{1}{2} \leq s \leq 1$, as $U \rightarrow \infty$. The reader may refer to [Edwards, 1991, §5.2] to see how this curve changes under the imposition of a steady shear flow.

As in the previous cases, the functional derivatives are required to belong to the same operating space as the descriptive variables of the system. Hence, $\delta F/\delta \mathbf{v}$ is again required to be divergence free and to vanish on the boundary $\partial\Omega$, and $\delta F/\delta \mathbf{m}$ is required to be symmetric and traceless, as is $\partial \mathbf{m}/\partial t$. Therefore, we define the functional derivatives appearing in (11.5-5) as

$$\frac{\delta F}{\delta v_\alpha} \equiv I_\alpha(\frac{\partial f}{\partial \mathbf{v}}) \,, \tag{11.5-15}$$

$$\begin{aligned}\frac{\delta F}{\delta m_{\alpha\beta}} \equiv \frac{1}{2}\left(\frac{\partial f}{\partial m_{\alpha\beta}} + \frac{\partial f}{\partial m_{\beta\alpha}}\right) - \frac{1}{3}\delta_{\alpha\beta}\frac{\partial f}{\partial m_{\gamma\gamma}} \\ - \frac{1}{2}\left(\nabla_\gamma \frac{\partial f}{\partial m_{\alpha\beta,\gamma}} + \nabla_\gamma \frac{\partial f}{\partial m_{\beta\alpha,\gamma}}\right) + \frac{1}{3}\delta_{\alpha\beta}\nabla_\gamma \frac{\partial f}{\partial m_{\epsilon\epsilon,\gamma}} \,,\end{aligned} \tag{11.5-16}$$

$$\frac{\delta F}{\delta w_{\alpha\beta}} \equiv \frac{\partial f}{\partial w_{\alpha\beta}} \,. \tag{11.5-17}$$

Note that no constraint is imposed *a priori* on the tensorial parameter $\mathbf{w}$.

Now we need to define the brackets which appear in the dynamical expression, (11.5-5). For the Poisson bracket, the obvious extension of (11.4-2) for arbitrary functionals $F[\mathbf{v},\mathbf{c},\mathbf{w}]$ and $G[\mathbf{v},\mathbf{c},\mathbf{w}]$ (where $\mathbf{c}$ is the unconstrained conformation tensor) is

$$\begin{aligned}\{F,G\} = &- \int_\Omega v_\alpha \left[\frac{\delta F}{\delta v_\gamma}\nabla_\gamma \frac{\delta G}{\delta v_\alpha} - \frac{\delta G}{\delta v_\gamma}\nabla_\gamma \frac{\delta F}{\delta v_\alpha}\right] d^3x \\ &- \int_\Omega c_{\alpha\beta}\left[\frac{\delta F}{\delta v_\gamma}\nabla_\gamma \frac{\delta G}{\delta c_{\alpha\beta}} - \frac{\delta G}{\delta v_\gamma}\nabla_\gamma \frac{\delta F}{\delta c_{\alpha\beta}}\right] d^3x\end{aligned}$$

$$
\begin{aligned}
&- \int_\Omega \left[\frac{\delta F}{\delta w_{\alpha\beta}} \frac{\delta G}{\delta c_{\alpha\beta}} - \frac{\delta G}{\delta w_{\alpha\beta}} \frac{\delta F}{\delta c_{\alpha\beta}} \right] \mathrm{d}^3x \\
&- \int_\Omega w_{\alpha\beta} \left[\frac{\delta F}{\delta v_\gamma} \nabla_\gamma \frac{\delta G}{\delta w_{\alpha\beta}} - \frac{\delta G}{\delta v_\gamma} \nabla_\gamma \frac{\delta F}{\delta w_{\alpha\beta}} \right] \mathrm{d}^3x \ .
\end{aligned}
\tag{11.5-18}
$$

It can be shown that the above expression satisfies both the antisymmetry property and the Jacobi identity for a Poisson bracket.

In order to obtain the appropriate Poisson bracket that satisfies both of the above properties under the constraint imposed by (11.5-2), we need to modify the bracket (11.5-18). For this purpose, we once again introduce the projection mapping of (5.5-16a), P'_m:

$$
P'_m[c_{\alpha\beta}] \rightarrow \frac{c_{\alpha\beta}}{\mathrm{tr}\,\mathbf{c}} \equiv m_{\alpha\beta} \ , \tag{11.5-19}
$$

which projects an arbitrary tensor to a tensor with unit trace. As proven in §5.5.3, we note the following theorem which is needed to make the transformation.

Theorem 11.2:

$$
\frac{\delta F}{\delta c_{\alpha\beta}} = \frac{\delta F}{\delta m_{\gamma\varepsilon}} \frac{1}{\mathrm{tr}\,\mathbf{c}} (\delta_{\alpha\gamma}\delta_{\beta\varepsilon} - m_{\gamma\varepsilon}\delta_{\alpha\beta}) \ . \tag{11.5-20}
$$

Substitution of Theorem 11.2 into the bracket (11.5-18) yields the Poisson bracket for the constrained tensor, $\mathbf{m}$:

$$
\begin{aligned}
\{F,G\} = &- \int_\Omega v_\alpha \left[\frac{\delta F}{\delta v_\gamma} \nabla_\gamma \frac{\delta G}{\delta v_\alpha} - \frac{\delta G}{\delta v_\gamma} \nabla_\gamma \frac{\delta F}{\delta v_\alpha} \right] \mathrm{d}^3x \\
&- \int_\Omega m_{\alpha\beta} \left[\frac{\delta F}{\delta v_\gamma} \nabla_\gamma \frac{\delta G}{\delta m_{\alpha\beta}} - \frac{\delta G}{\delta v_\gamma} \nabla_\gamma \frac{\delta F}{\delta m_{\alpha\beta}} \right] \mathrm{d}^3x \\
&- \int_\Omega \left[\frac{\delta F}{\delta w_{\alpha\beta}} \frac{\delta G}{\delta m_{\alpha\beta}} - \frac{\delta G}{\delta w_{\alpha\beta}} \frac{\delta F}{\delta m_{\alpha\beta}} \right] \mathrm{d}^3x \\
&+ \int_\Omega m_{\gamma\varepsilon} \left[\frac{\delta F}{\delta w_{\alpha\alpha}} \frac{\delta G}{\delta m_{\gamma\varepsilon}} - \frac{\delta G}{\delta w_{\alpha\alpha}} \frac{\delta F}{\delta m_{\gamma\varepsilon}} \right] \mathrm{d}^3x \\
&- \int_\Omega w_{\alpha\beta} \left[\frac{\delta F}{\delta v_\gamma} \nabla_\gamma \frac{\delta G}{\delta w_{\alpha\beta}} - \frac{\delta G}{\delta v_\gamma} \nabla_\gamma \frac{\delta F}{\delta w_{\alpha\beta}} \right] \mathrm{d}^3x \ ,
\end{aligned}
\tag{11.5-21}
$$

where F and G are now functionals of **v**, **m**, and **w**.

Likewise, a dissipation bracket may be defined in analogy with (11.4-16) as

$$[F,G] = -\int_\Omega R_{\alpha\beta\gamma\varepsilon} \nabla_\alpha\left(\frac{\delta F}{\delta v_\beta}\right) \nabla_\gamma\left(\frac{\delta G}{\delta v_\varepsilon}\right) d^3x$$

$$+ \int_\Omega \alpha_2^m \left[\frac{\delta F}{\delta w_{\alpha\beta}} - m_{\alpha\gamma}\nabla_\gamma \frac{\delta F}{\delta v_\beta} - m_{\beta\gamma}\nabla_\gamma \frac{\delta F}{\delta v_\alpha}\right]$$

$$\times \left[\frac{\delta G}{\delta w_{\alpha\beta}} - m_{\alpha\gamma}\nabla_\gamma \frac{\delta G}{\delta v_\beta} - m_{\beta\gamma}\nabla_\gamma \frac{\delta G}{\delta v_\alpha}\right] d^3x \qquad (11.5\text{-}22)$$

$$- \int_\Omega \alpha_3^m \left[\frac{\delta F}{\delta w_{\alpha\beta}} + m_{\alpha\gamma}\nabla_\beta \frac{\delta F}{\delta v_\gamma} + m_{\beta\gamma}\nabla_\alpha \frac{\delta F}{\delta v_\gamma}\right]$$

$$\times \left[\frac{\delta G}{\delta w_{\alpha\beta}} + m_{\alpha\gamma}\nabla_\beta \frac{\delta G}{\delta v_\gamma} + m_{\beta\gamma}\nabla_\alpha \frac{\delta G}{\delta v_\gamma}\right] d^3x \,,$$

where

$$R_{\alpha\beta\gamma\varepsilon} \equiv \tfrac{1}{2}\alpha_1^m(m_{\alpha\gamma}m_{\beta\varepsilon} + m_{\alpha\varepsilon}m_{\beta\gamma}) + \tfrac{1}{2}\alpha_4^m(\delta_{\alpha\gamma}\delta_{\beta\varepsilon} + \delta_{\alpha\varepsilon}\delta_{\beta\gamma})$$

$$+ \tfrac{1}{2}\alpha_5^m(m_{\alpha\gamma}\delta_{\beta\varepsilon} + m_{\alpha\varepsilon}\delta_{\beta\gamma} + \delta_{\alpha\gamma}m_{\beta\varepsilon} + \delta_{\alpha\varepsilon}m_{\beta\gamma})$$

$$+ \tfrac{1}{2}\alpha_6^m(m_{\alpha\zeta}m_{\zeta\gamma}\delta_{\beta\varepsilon} + m_{\alpha\zeta}m_{\zeta\varepsilon}\delta_{\beta\gamma} + \delta_{\alpha\gamma}m_{\beta\zeta}m_{\zeta\varepsilon} + \delta_{\alpha\varepsilon}m_{\beta\zeta}m_{\zeta\gamma})$$

$$+ \tfrac{1}{2}\alpha_7^m(m_{\alpha\zeta}m_{\zeta\gamma}m_{\beta\varepsilon} + m_{\alpha\zeta}m_{\zeta\varepsilon}m_{\beta\gamma} + m_{\alpha\gamma}m_{\beta\zeta}m_{\zeta\varepsilon} + m_{\alpha\varepsilon}m_{\beta\zeta}m_{\zeta\gamma})$$

$$+ \tfrac{1}{2}\alpha_8^m(m_{\alpha\zeta}m_{\zeta\gamma}m_{\beta\eta}m_{\eta\varepsilon} + m_{\alpha\zeta}m_{\zeta\varepsilon}m_{\beta\eta}m_{\eta\gamma}) \,. \qquad (11.5\text{-}23)$$

The parameters α_i^m in Eqs. (11.5-22,23) are new phenomenological viscosity coefficients which take the place of the α_i of the LE theory. There are now eight of these parameters, rather than the five independent ones of LE theory, since more information is contained in the conformation tensor theory. The expression (11.5-23) is a general form of the phenomenological tensor **R** (for an incompressible fluid), given the unit-trace constraint on **m** and the Cayley/Hamilton theorem. The eight coefficients α_i^m may, in general, be functions of the second and third invariants of **m**.

Using the definitions provided by (11.5-15,16,17), we may next evaluate the functional derivatives of the Hamiltonian, (11.5-6), as

$$\frac{\delta H_m}{\delta v_\alpha} = v_\alpha \,, \tag{11.5-24}$$

$$\frac{\delta H_m}{\delta m_{\alpha\beta}} = \frac{\delta H_m}{\delta S_{\alpha\beta}} = -\tfrac{1}{2}\chi_a(H_\alpha H_\beta - \tfrac{1}{3}\delta_{\alpha\beta}H_\gamma H_\gamma) - b_1 S_{\alpha\beta,\gamma,\gamma}$$
$$- b_2(\tfrac{1}{2}S_{\beta\gamma,\alpha} + \tfrac{1}{2}S_{\alpha\gamma,\beta} - \tfrac{1}{3}\delta_{\alpha\beta}S_{\gamma\varepsilon,\varepsilon})_{,\gamma} + 2a_1 S_{\alpha\beta}$$
$$+ 3a_2(S_{\beta\gamma}S_{\gamma\alpha} - \tfrac{1}{3}S_{\mu\nu}S_{\mu\nu}\delta_{\alpha\beta}) + 4a_3(S_{\beta\gamma}S_{\gamma\varepsilon}S_{\varepsilon\alpha} - \tfrac{1}{3}S_{\mu\gamma}S_{\gamma\varepsilon}S_{\varepsilon\mu}\delta_{\alpha\beta}) \,, \tag{11.5-25}$$

$$\frac{\delta H_m}{\delta w_{\alpha\beta}} = \frac{w_{\alpha\beta}}{\sigma^m} \,. \tag{11.5-26}$$

In the above, we have made implicit use of the following theorem.

Theorem 11.3:

$$\frac{\delta F}{\delta S_{\alpha\beta}} = \frac{\delta F}{\delta m_{\alpha\beta}} \,. \tag{11.5-27}$$

(Theorem 11.3 is proven in Example 11.2, which follows.) Hence we are able to calculate the functional derivative $\delta F/\delta \mathbf{m}$ without the necessity of rewriting H_b, (11.5-9), in terms of **m** via (11.3-7).

With the above derivatives of the Hamiltonian, it is now an easy task to arrive at the evolution equations for the variables **v**, **m**, and **w** simply by comparing the two forms of the dynamical equation for F, (11.5-5). These are

$$\dot{v}_\alpha = F^m_\alpha - p_{,\alpha} - \left(m_{\gamma\varepsilon,\alpha} \frac{\partial W}{\partial m_{\gamma\varepsilon,\beta}} \right)_{,\beta} + T_{\alpha\beta,\beta} \,, \tag{11.5-28}$$

$$\sigma^m \dot{m}_{\alpha\beta} = w_{\alpha\beta} - w_{\gamma\gamma} m_{\alpha\beta} \quad \rightarrow \quad w_{\alpha\beta} = \sigma^m \dot{m}_{\alpha\beta} + \varepsilon m_{\alpha\beta} \,, \tag{11.5-29}$$

$$\dot{w}_{\alpha\beta} = -\frac{\delta H_m}{\delta m_{\alpha\beta}} + \delta_{\alpha\beta} m_{\gamma\varepsilon} \frac{\delta H_m}{\delta m_{\gamma\varepsilon}} + \alpha_2^m \left[\frac{1}{\sigma^m} w_{\alpha\beta} - m_{\alpha\gamma} v_{\beta,\gamma} - m_{\beta\gamma} v_{\alpha,\gamma}\right]$$
$$- \alpha_3^m \left[\frac{1}{\sigma^m} w_{\alpha\beta} + m_{\alpha\gamma} v_{\gamma,\beta} + m_{\beta\gamma} v_{\gamma,\alpha}\right] \,, \tag{11.5-30}$$

with the force vector

$$F^m_\alpha = \chi_a H_\beta H_{\gamma,\alpha} m_{\beta\gamma} \,, \tag{11.5-31}$$

and the extra stress

$$T_{\alpha\beta} = R_{\beta\alpha\gamma\varepsilon}A_{\gamma\varepsilon} + 2\alpha_2^m m_{\beta\gamma}[\frac{1}{\sigma^m}w_{\alpha\gamma} - m_{\alpha\varepsilon}v_{\gamma,\varepsilon} - m_{\varepsilon\gamma}v_{\alpha,\varepsilon}]$$

$$+ 2\alpha_3^m m_{\alpha\gamma}[\frac{1}{\sigma^m}w_{\beta\gamma} + m_{\gamma\varepsilon}v_{\varepsilon,\beta} + m_{\beta\varepsilon}v_{\varepsilon,\gamma}] . \qquad (11.5\text{-}32)$$

Yet, from (11.5-29,30) we obtain

$$\sigma^m \dot{m}_{\alpha\beta} = \dot{w}_{\alpha\beta} - \dot{w}_{\gamma\gamma}m_{\alpha\beta} - w_{\gamma\gamma}\dot{m}_{\alpha\beta}$$

$$= -\frac{\delta H_m}{\delta m_{\alpha\beta}} - 3m_{\gamma\varepsilon}\frac{\delta H_m}{\delta m_{\gamma\varepsilon}}(m_{\alpha\beta} - ⅓\delta_{\alpha\beta}) + (\alpha_2^m - \alpha_m^m)\dot{m}_{\alpha\beta}$$

$$- \alpha_2^m[m_{\alpha\gamma}v_{\beta,\gamma} + m_{\beta\gamma}v_{\alpha,\gamma}] - \alpha_3^m[m_{\alpha\gamma}v_{\gamma,\beta} + m_{\beta\gamma}v_{\gamma,\alpha}]$$

$$+ 2(\alpha_2^m + \alpha_3^m)m_{\alpha\beta}m_{\gamma\varepsilon}v_{\gamma,\varepsilon} - \varepsilon\dot{m}_{\alpha\beta} . \qquad (11.5\text{-}33)$$

Setting $\varepsilon=0$, as it must from (11.5-29) in order for $\dot{\mathbf{m}}=\delta H_m/\delta\mathbf{w}=\mathbf{w}/\sigma^m$, we obtain the generalized theory in terms of tensorial parameters in the inertial form.

EXAMPLE 11.2: Since we wish to work with a Hamiltonian which is in terms of **S** and $\nabla\mathbf{S}$ rather than **m** and $\nabla\mathbf{m}$, it is necessary to use $\delta F/\delta\mathbf{S}$ rather than $\delta F/\delta\mathbf{m}$. Therefore, we require Theorem 11.3, which shows that these two quantities are equivalent due to the linear, one-to-one relationship, (11.3-30), between the two terms. Prove Theorem 11.3.

Proof: **S** is required to be symmetric and traceless, and, consequently, so is $\delta F/\delta\mathbf{S}$. Hence, we define $\delta F/\delta\mathbf{S}$ via (11.5-16) as

$$\frac{\delta F}{\delta S_{\alpha\beta}} \equiv \frac{1}{2}(\frac{\partial f}{\partial S_{\alpha\beta}} + \frac{\partial f}{\partial S_{\beta\alpha}}) - \frac{1}{3}\delta_{\alpha\beta}\frac{\partial f}{\partial S_{\gamma\gamma}}$$

$$- \frac{1}{2}(\nabla_\gamma\frac{\partial f}{\partial S_{\alpha\beta,\gamma}} + \nabla_\gamma\frac{\partial f}{\partial S_{\beta\alpha,\gamma}}) + \frac{1}{3}\delta_{\alpha\beta}\nabla_\gamma\frac{\partial f}{\partial S_{\varepsilon\varepsilon,\gamma}} .$$

By considering the functions f which depend on **S** only through their dependence on **m**, we can use (11.3-30), its gradient $\nabla_\eta m_{\gamma\varepsilon}=\nabla_\eta S_{\gamma\varepsilon}$,

$$\frac{\partial f}{\partial S_{\alpha\beta}} = \frac{\partial f}{\partial m_{\gamma\varepsilon}}\frac{\partial m_{\gamma\varepsilon}}{\partial S_{\alpha\beta}} + \frac{\partial f}{\partial(\nabla_\eta m_{\gamma\varepsilon})}\frac{\partial(\nabla_\eta m_{\gamma\varepsilon})}{\partial S_{\alpha\beta}} ,$$

$$\frac{\partial f}{\partial(\nabla_\eta S_{\alpha\beta})} = \frac{\partial f}{\partial m_{\gamma\varepsilon}}\frac{\partial m_{\gamma\varepsilon}}{\partial(\nabla_\eta S_{\alpha\beta})} + \frac{\partial f}{\partial(\nabla_\zeta m_{\gamma\varepsilon})}\frac{\partial(\nabla_\zeta m_{\gamma\varepsilon})}{\partial(\nabla_\eta S_{\alpha\beta})} ,$$

to substitute into the above equation and to lead us directly to Theorem 11.3. ■

11.5.2 The non-inertial form

Now we may apply the procedure of §11.4.2 to the generalized tensorial theory to obtain the non-inertial equations. Again, we consider that the parameter $\sigma^m \to 0$, so that the functional F is now written as

$$F[\mathbf{v},\mathbf{m}] \equiv \int_\Omega f(\mathbf{v},\mathbf{m},\nabla\mathbf{m})\, d^3x \; , \tag{11.5-34}$$

along with the corresponding Hamiltonian

$$H_m[\mathbf{v},\mathbf{m}] = \int_\Omega \left(\frac{1}{2}\mathbf{v}\cdot\mathbf{v} + W + \psi_m\right) d^3x + H_b \; . \tag{11.5-35}$$

The non-inertial Poisson bracket for the unconstrained tensor **c** is known from §5.5.2 to be

$$\begin{aligned}\{F,G\} = &- \int_\Omega v_\alpha \left[\frac{\delta F}{\delta v_\gamma}\nabla_\gamma \frac{\delta G}{\delta v_\alpha} - \frac{\delta G}{\delta v_\gamma}\nabla_\gamma \frac{\delta F}{\delta v_\alpha}\right] d^3x \\ &- \int_\Omega c_{\alpha\beta} \left[\frac{\delta F}{\delta v_\gamma}\nabla_\gamma \frac{\delta G}{\delta c_{\alpha\beta}} - \frac{\delta G}{\delta v_\gamma}\nabla_\gamma \frac{\delta F}{\delta c_{\alpha\beta}}\right] d^3x \\ &- \int_\Omega c_{\alpha\gamma} \left[\nabla_\gamma\left(\frac{\delta F}{\delta v_\beta}\right)\frac{\delta G}{\delta c_{\alpha\beta}} - \nabla_\gamma\left(\frac{\delta G}{\delta v_\beta}\right)\frac{\delta F}{\delta c_{\alpha\beta}}\right] d^3x \\ &- \int_\Omega c_{\beta\gamma} \left[\nabla_\gamma\left(\frac{\delta F}{\delta v_\alpha}\right)\frac{\delta G}{\delta c_{\alpha\beta}} - \nabla_\gamma\left(\frac{\delta G}{\delta v_\alpha}\right)\frac{\delta F}{\delta c_{\alpha\beta}}\right] d^3x \; ;\end{aligned} \tag{11.5-36}$$

however, under the transformation of Theorem 11.2, this bracket becomes

$$\begin{aligned}\{F,G\} = &- \int_\Omega v_\alpha \left[\frac{\delta F}{\delta v_\gamma}\nabla_\gamma \frac{\delta G}{\delta v_\alpha} - \frac{\delta G}{\delta v_\gamma}\nabla_\gamma \frac{\delta F}{\delta v_\alpha}\right] d^3x \\ &- \int_\Omega m_{\alpha\beta} \left[\frac{\delta F}{\delta v_\gamma}\nabla_\gamma \frac{\delta G}{\delta m_{\alpha\beta}} - \frac{\delta G}{\delta v_\gamma}\nabla_\gamma \frac{\delta F}{\delta m_{\alpha\beta}}\right] d^3x\end{aligned}$$

$$- \int_\Omega m_{\alpha\gamma} \left[\nabla_\gamma \left(\frac{\delta F}{\delta v_\beta}\right) \frac{\delta G}{\delta m_{\alpha\beta}} - \nabla_\gamma \left(\frac{\delta G}{\delta v_\beta}\right) \frac{\delta F}{\delta m_{\alpha\beta}} \right] d^3x$$

$$- \int_\Omega m_{\beta\gamma} \left[\nabla_\gamma \left(\frac{\delta F}{\delta v_\alpha}\right) \frac{\delta G}{\delta m_{\alpha\beta}} - \nabla_\gamma \left(\frac{\delta G}{\delta v_\alpha}\right) \frac{\delta F}{\delta m_{\alpha\beta}} \right] d^3x$$

$$+ 2 \int_\Omega m_{\gamma\varepsilon} m_{\alpha\beta} \left[\nabla_\beta \left(\frac{\delta F}{\delta v_\alpha}\right) \frac{\delta G}{\delta m_{\gamma\varepsilon}} - \nabla_\beta \left(\frac{\delta G}{\delta v_\alpha}\right) \frac{\delta F}{\delta m_{\gamma\varepsilon}} \right] d^3x \; . \tag{11.5-37}$$

The dissipation bracket corresponding to (11.4-29) is

$$[F,G] = - \int_\Omega R^m_{\alpha\beta\gamma\varepsilon} \nabla_\alpha \left(\frac{\delta F}{\delta v_\beta}\right) \nabla_\gamma \left(\frac{\delta G}{\delta v_\varepsilon}\right) d^3x$$

$$- \int_\Omega P^m_{\alpha\beta\gamma\varepsilon} \frac{\delta F}{\delta m_{\alpha\beta}} \frac{\delta G}{\delta m_{\gamma\varepsilon}} d^3x$$

$$- \int_\Omega L^m_{\alpha\beta\gamma\varepsilon} \left[\nabla_\alpha \left(\frac{\delta F}{\delta v_\beta}\right) \frac{\delta G}{\delta m_{\gamma\varepsilon}} - \nabla_\alpha \left(\frac{\delta G}{\delta v_\beta}\right) \frac{\delta F}{\delta m_{\gamma\varepsilon}} \right] d^3x$$

$$- \int_\Omega L^m_{\eta\zeta\gamma\gamma} m_{\alpha\beta} \left[\nabla_\eta \left(\frac{\delta F}{\delta v_\zeta}\right) \frac{\delta G}{\delta m_{\alpha\beta}} - \nabla_\eta \left(\frac{\delta G}{\delta v_\zeta}\right) \frac{\delta F}{\delta m_{\alpha\beta}} \right] d^3x \; , \tag{11.4-38}$$

where

$$R^m_{\alpha\beta\gamma\varepsilon} \equiv \tfrac{1}{2} \beta^m_1 (m_{\alpha\gamma} m_{\beta\varepsilon} + m_{\alpha\varepsilon} m_{\beta\gamma}) + \tfrac{1}{2} \beta^m_4 (\delta_{\alpha\gamma} \delta_{\beta\varepsilon} + \delta_{\alpha\varepsilon} \delta_{\beta\gamma})$$

$$+ \tfrac{1}{2} \beta^m_2 (m_{\alpha\gamma} \delta_{\beta\varepsilon} + m_{\alpha\varepsilon} \delta_{\beta\gamma} + \delta_{\alpha\gamma} m_{\beta\varepsilon} + \delta_{\alpha\varepsilon} m_{\beta\gamma})$$

$$+ \tfrac{1}{2} \beta^m_3 (m_{\alpha\zeta} m_{\zeta\gamma} \delta_{\beta\varepsilon} + m_{\alpha\zeta} m_{\zeta\varepsilon} \delta_{\beta\gamma} + \delta_{\alpha\gamma} m_{\beta\zeta} m_{\zeta\varepsilon} + \delta_{\alpha\varepsilon} m_{\beta\zeta} m_{\zeta\gamma})$$

$$+ \tfrac{1}{2} \beta^m_5 (m_{\alpha\zeta} m_{\zeta\gamma} m_{\beta\varepsilon} + m_{\alpha\zeta} m_{\zeta\varepsilon} m_{\beta\gamma} + m_{\alpha\gamma} m_{\beta\zeta} m_{\zeta\varepsilon} + m_{\alpha\varepsilon} m_{\beta\zeta} m_{\zeta\gamma})$$

$$+ \tfrac{1}{2} \beta^m_6 (m_{\alpha\zeta} m_{\zeta\gamma} m_{\beta\eta} m_{\eta\varepsilon} + m_{\alpha\zeta} m_{\zeta\varepsilon} m_{\beta\eta} m_{\eta\gamma}) \; , \tag{11.5-39}$$

$$P^m_{\alpha\beta\gamma\varepsilon} \equiv \frac{1}{\beta^m_7}[\tfrac{1}{2}(\delta_{\alpha\varepsilon}\delta_{\beta\gamma} + \delta_{\beta\varepsilon}\delta_{\alpha\gamma}) + 3m_{\alpha\beta}m_{\gamma\varepsilon}] , \qquad (11.5\text{-}40)$$

$$L^m_{\alpha\beta\gamma\varepsilon} \equiv \tfrac{1}{2}(\beta^m_8 - 1)(\delta_{\alpha\varepsilon}m_{\beta\gamma} + \delta_{\beta\varepsilon}m_{\alpha\gamma} + \delta_{\alpha\gamma}m_{\beta\varepsilon} + \delta_{\beta\gamma}m_{\alpha\varepsilon}) . \qquad (11.5\text{-}41)$$

All of the phenomenological matrices have been chosen in analogy with §11.4.2; however, $\mathbf{P}^m$ has an additional term (i.e., $3m_{\alpha\beta}m_{\gamma\varepsilon}$) for reasons to be explained later. Note that the fourth integral in the above dissipation bracket was obtained by applying Theorem 11.2 to the equivalent unconstrained dissipation bracket (not shown) in a similar fashion to the preceding cases. It is now evident from the form of the dissipation that β^m_7 plays the role of the system relaxation time and that β^m_8 expresses the degree of non-affine motion.

Once again using the dynamical equation for F, (11.5-5), via the brackets of (11.5-37,38), we may arrive at the evolution equations for the variables $\mathbf{v}$ and $\mathbf{m}$. These are

$$\dot{v}_\alpha = F^m_\alpha - p_{,\alpha} - \left(m_{\gamma\varepsilon,\alpha}\frac{\partial W}{\partial m_{\gamma\varepsilon,\beta}}\right)_{,\beta} + T^m_{\alpha\beta,\beta} , \qquad (11.5\text{-}42)$$

$$\dot{m}_{\alpha\beta} = -\frac{1}{\beta^m_7}\left[\frac{\delta H_m}{\delta m_{\alpha\beta}} + 3m_{\gamma\varepsilon}\frac{\delta H_m}{\delta m_{\gamma\varepsilon}}(m_{\alpha\beta} - \frac{1}{3}\delta_{\alpha\beta})\right] + \frac{1+\beta^m_8}{2}[m_{\alpha\gamma}v_{\beta,\gamma} + m_{\beta\gamma}v_{\alpha,\gamma}] - \frac{1-\beta^m_8}{2}[m_{\alpha\gamma}v_{\gamma,\beta} + m_{\beta\gamma}v_{\gamma,\alpha}] - 2\beta^m_8 m_{\alpha\beta}m_{\gamma\varepsilon}v_{\gamma,\varepsilon} , \qquad (11.5\text{-}43)$$

where the extra stress is defined as

$$T^m_{\alpha\beta} \equiv R^m_{\beta\alpha\gamma\varepsilon}A_{\gamma\varepsilon} + (1+\beta^m_8)m_{\beta\gamma}\frac{\delta H_m}{\delta m_{\gamma\alpha}} + (\beta^m_8 - 1)m_{\alpha\gamma}\frac{\delta H_m}{\delta m_{\gamma\beta}} - 2\beta^m_8 m_{\alpha\beta}m_{\gamma\varepsilon}\frac{\delta H_m}{\delta m_{\gamma\varepsilon}} . \qquad (11.5\text{-}44)$$

The above set of equations, (11.5-42,43,44), now describes completely the liquid-crystalline system in the non-inertial form. In defining the brackets in analogy with the non-inertial LE theory, we may feel confident that we have correctly incorporated most of the essential physics into the problem formulation. To be sure, however, we can check this by restricting the inertial form of the conformation tensor theory to the limit $\sigma^m \to 0$, as in §11.3.1.B. To do this, one must realize that the $\dot{\mathbf{w}}$ in (11.5-30) vanishes in this limit, and then solve this equation for $\mathbf{w}/\sigma^m$. Substituting the resulting expression into (11.5-29) then yields the evolution equation for $\dot{\mathbf{m}}$, (11.5-43). Substituting $\mathbf{w}/\sigma^m$ into (11.5-32) then gives the extra-stress expression, (11.5-44), provided that we set

$$\beta_1^m = \alpha_1^m + \frac{8\alpha_2^m\alpha_3^m}{(\alpha_2^m - \alpha_3^m)} , \quad \beta_4^m = \alpha_4^m , \quad \beta_5^m = \alpha_7^m ,$$

$$\beta_6^m = \alpha_8^m , \quad \beta_2^m = \alpha_5^m , \quad \beta_3^m = \frac{4\alpha_2^m\alpha_3^m}{(\alpha_2^m - \alpha_3^m)} + \alpha_6^m , \qquad (11.5\text{-}45)$$

$$\beta_7^m = \alpha_3^m - \alpha_2^m , \quad \beta_8^m = \frac{(\alpha_2^m + \alpha_3^m)}{(\alpha_2^m - \alpha_3^m)} .$$

Hence, the system of equations in the non-inertial form checks with the inertial version, and we have completed our modeling for the time being. Note, however, that the additional term $3m_{\alpha\beta}m_{\gamma\varepsilon}$ which was included in $\mathbf{P}^m$, (11.5-40), is necessary for the two forms to be equivalent in the limit $\sigma^m \to 0$.

As a last point in this section, let us develop thermodynamic relationships among the parameters β_i^m, similar to those of Example 11.1. To do this, we follow the same procedure, setting $[H_m, H_m] \geq 0$, except that we use $\mathbf{m}$ instead of $\mathbf{n}$ and require that $m_{11}=1$ and that all other components vanish with no loss of generality (which is strictly valid only for a uniaxial LC). In this case, we arrive at the following conditions on the β_i^m:

$$\beta_4^m \geq 0 , \quad \beta_7^m \geq 0 , \quad 2\beta_4^m + K \geq 0 ,$$

$$J + 6\beta_4^m + K \geq 0 , \quad 4J + 6\beta_4^m + 4K \geq 0 ,$$

$$(J + 6\beta_4^m + K)^2 - (2J - 6\beta_4^m + 2K)^2 \geq 0 ,$$

$$(4J + 6\beta_4^m + 4K)(J + 6\beta_4^m + K) - (-4J - 6\beta_4^m - 4K)^2 \geq 0 , \qquad (11.5\text{-}46)$$

where

$$J = \beta_1^m + \beta_5^m + \tfrac{1}{2}\beta_6^m \quad \text{and} \quad K = 2\beta_2^m + 2\beta_3^m . \qquad (11.5\text{-}47)$$

One can obtain conditions on the α_i^m of the inertial theory simply by using the equalities of (11.5-45).

11.6 Comparison of the Conformation Tensor Theory to Previous Theories

Now that we have completed the modeling stage of this chapter, we need to compare the results of this exercise with the previous theories described in §11.3. Consequently, we shall consider each of the three previous models in turn, and, as we shall see, all of these theories arise as a special case of the conformation tensor theory, developed in the preceding section.

At this point, let us interject a comment concerning comparison practices in the literature and their differences and similarities with the comparisons discussed in the present section. A number of authors have tried to compare the Doi theory with the LE theory by finding expressions for the Leslie viscosity coefficients (the α_i) in terms of the Doi parameters [Doi, 1981; Marrucci, 1982; Semenov, 1983; Kuzuu and Doi, 1983; 1984; Lee, 1988; Edwards *et al.* , 1990a]. Despite the fact, as seen more clearly below, that the LE and Doi theories really involve different components of the same general theory (the conformation tensor theory), such a comparison still makes sense from a phenomenological viewpoint when one attempts to match the predictions of the microscopic (Doi) theory—seen as a given reality—against the predictions of the phenomenological theory (LE theory). By necessity, such a comparison can only be done in the common domain of validity of the theories involved (i.e., in the nematic state and for vanishingly small shear rates), and it is valid in an asymptotic sense; i.e., the two theories provide similar results up to first order in the small parameter involved, in this case, the shear rate. This is also obvious from the fact that the LE theory is a linear theory with respect to the rate of deformation tensor, whereas the Doi theory, in any of its approximations, is not. The value of the comparison relies on the transfer of information concerning the predicted dynamic behavior according to one of the theories with that predicted by the other—see the pertinent discussion in §11.3.4.A.

In contrast, comparisons which involve a full mathematical equivalence between the two theories, under a certain number of assumptions (primarily related to the structure of the Hamiltonian), involve the identification of the values of the model parameters for which the two theories are completely *identical*, in all their predictions and for the entire range of their applicability. This involves a *reduction* of the more general to the more specific theory, and constitutes the subject of the first four subsections, §11.6.1 through §11.6.4, where the conformation theory is reduced, under certain simplifications, to the LE theory, the scalar/vector theory, the original Doi theory, and the extended Doi theory. This exercise demonstrates that all of these previous theories are particular cases of the most general conformation tensor theory, developed in the present chapter using the bracket formalism. In addition, the conformation theory can be used to generalize the restricted (continuum) version of the Doi theory, with the advantage that, at the expense of introducing one more additional parameter (the dimensionless number β_8^m), a broader range of dynamic behavior can be predicted. In particular, director tumbling phenomena, as also allowed by the kinetic theory, can be described using values for β_8^m smaller than unity. This is demonstrated in the last subsection, §11.6.5, where the Leslie coefficients corresponding to the general conformation theory are obtained for comparison with previous results.

11.6.1 Reduction to the Leslie/Ericksen theory

In order to obtain the LE theory from the conformation tensor theory, we require that **m** take the form corresponding to a uniaxial nematic LC,

$$m_{\alpha\beta} = n_\alpha n_\beta \;, \tag{11.6-1}$$

which obviously implies that

$$\dot{m}_{\alpha\beta} = \dot{n}_\alpha n_\beta + n_\alpha \dot{n}_\beta \;, \tag{11.6-2}$$

$$\ddot{m}_{\alpha\beta} = \ddot{n}_\alpha n_\beta + n_\alpha \ddot{n}_\beta + 2\dot{n}_\alpha \dot{n}_\beta \;. \tag{11.6-3}$$

Substitution of the restriction (11.6-1) into the Hamiltonian, (11.5-6), yields some useful information. For instance, we see that H_b reduces to a constant which can be neglected (or, alternatively, one can consider that a_i=0, i=1,2,3). Furthermore, we have already stated that the W of (11.5-6) reduces to the W of (11.4-2) provided that k_1=k_3=$2b_1$+b_2 and k_2=$2b_1$. Consequently, it only remains to correlate σ and σ^m, which, if we put (11.6-2) into **w**:**w** and compare with (11.4-2), we see immediately that σ=$2\sigma^m$. Hence the Hamiltonian of (11.5-6) reduces to the Hamiltonian of the LE theory under the restriction of (11.6-1).

Now we wish to compare the inertial form of the conformation tensor theory under the restriction (11.6-1) with the inertial form of the LE theory. First, it is necessary to determine the viscosity coefficients in the tensor theory in terms of those of the LE theory. Thus we define

$$\begin{gathered} \alpha_1 + \alpha_2 - \alpha_3 \equiv \alpha_1^m + \alpha_8^m + 2\alpha_7^m \;, \\ \alpha_2 + \alpha_5 = \alpha_6 - \alpha_3 \equiv \alpha_5^m + \alpha_6^m \;, \\ \alpha_4 \equiv \alpha_4^m \;, \quad ½\,\alpha_2 \equiv \alpha_2^m \;, \quad ½\,\alpha_3 \equiv \alpha_3^m \;. \end{gathered} \tag{11.6-4}$$

The only tricky part of this procedure is in restricting the functional derivative of the Hamiltonian, $\delta H_m/\delta\mathbf{m}$, to $\delta H_m/\delta\mathbf{n}$. Since the tensorial theory contains additional information not present in the vectorial theory, we must neglect a certain portion of that information in order to accomplish the reduction. Therefore, since we also have a difference in constraints between the vector and tensor cases, we need to express the functional derivative $\delta H_m/\delta \mathbf{n}$ in terms of the functional derivatives $\delta H_m/\delta \mathbf{m}$ through the relation

$$\frac{\delta H_m}{\delta n_\alpha} = 2\frac{\delta H_m}{\delta m_{\gamma\alpha}} n_\gamma - 2\frac{\delta H_m}{\delta m_{\beta\gamma}} n_\beta n_\gamma n_\alpha \,, \tag{11.6-5}$$

obtained by differentiation by parts and the use of (11.6-1), along with the corresponding projection operations (similar to the procedure used in Example 11.2). Also, due to the extra information contained in **m**, if we take $n_\alpha n_\beta \ddot{m}_{\alpha\beta}$ with $\ddot{m}_{\alpha\beta}$ given by (11.6-3), we see that

$$n_\alpha n_\beta \ddot{m}_{\alpha\beta} = 2\ddot{n}_\alpha n_\alpha \,; \tag{11.6-6}$$

yet we know that for a unit vector that

$$n_\alpha \ddot{n}_\alpha + \dot{n}_\alpha \dot{n}_\alpha = 0 \,, \tag{11.6-7}$$

so that

$$n_\alpha n_\beta \ddot{m}_{\alpha\beta} = -\, 2\dot{n}_\alpha \dot{n}_\alpha \,. \tag{11.6-8}$$

Multiplying (11.5-33) by **nn**, we find that

$$\sigma^m n_\alpha n_\beta \ddot{m}_{\alpha\beta} = -\, 3 n_\alpha n_\beta \frac{\delta H_m}{\delta m_{\alpha\beta}} \,, \tag{11.6-9}$$

which implies that in order to make the transformation from **m** to **n**, we must realize that

$$-\, 3 n_\alpha n_\beta \frac{\delta H_m}{\delta m_{\alpha\beta}} = -\, 2\sigma^m \dot{n}_\alpha \dot{n}_\alpha \,. \tag{11.6-10}$$

Under the above assertions, the inertial form of the tensorial theory reduces exactly to the inertial LE equations, (11.3-2,3,4). We can just as easily show the equivalence of the non-inertial forms, but we shall wait until the next subsection to do so (for the scalar-vector theory).

11.6.2 *Reduction to the scalar-vector theory*

In this subsection, we wish to show the equivalence of the conformation tensor theory and the scalar-vector theory under the restriction implied by (11.5-11):

$$m_{\alpha\beta} = s n_\alpha n_\beta + \tfrac{1}{3}(1-s)\delta_{\alpha\beta} . \tag{11.6-11}$$

Here we study the non-inertial form of this theory, since we compared the inertial forms in the preceding subsection. Notice that when $s=1$ (i.e., perfect alignment) we return to the restriction (11.6-1). On account of (11.6-11), we also have the relations

$$m_{\alpha\beta,\gamma} = s_{,\gamma}(n_\alpha n_\beta - \tfrac{1}{3}\delta_{\alpha\beta}) + s(n_{\alpha,\gamma} n_\beta + n_\alpha n_{\beta,\gamma}) , \tag{11.6-12}$$

$$\dot{m}_{\alpha\beta} = \dot{s}(n_\alpha n_\beta - \tfrac{1}{3}\delta_{\alpha\beta}) + s(\dot{n}_\alpha n_\beta + n_\alpha \dot{n}_\beta) . \tag{11.6-13}$$

As in the preceding subsection, we may use differentiation by parts to relate the functional derivatives involved in the transformation. It is quite easy to show that

$$\frac{\delta H_m}{\delta n_\alpha} = 2s\left(\frac{\delta H_m}{\delta m_{\alpha\gamma}} n_\gamma - \frac{\delta H_m}{\delta m_{\gamma\beta}} n_\beta n_\gamma n_\alpha\right) , \tag{11.6-14}$$

$$\frac{\delta H_m}{\delta s} = \frac{\delta H_m}{\delta m_{\alpha\beta}} n_\alpha n_\beta , \tag{11.6-15}$$

$$\frac{\partial W}{\partial n_{\gamma,\beta}} = 2s \frac{\partial W}{\partial m_{\gamma\varepsilon,\beta}} n_\varepsilon , \quad \text{and} \quad \frac{\partial W}{\partial s_{,\beta}} = \frac{\partial W}{\partial m_{\gamma\varepsilon,\beta}} n_\varepsilon n_\gamma . \tag{11.6-16}$$

Substitution of the above identities into (11.5-43) and double-dotting with **nn** yields the evolution equation for the scalar order parameter, using the equalities of (11.5-45):

$$\dot{s} = \frac{3}{2} \frac{1+2s^2}{(\alpha_2^m - \alpha_3^m)} \frac{\delta H_m}{\delta s} + \frac{(\alpha_2^m + \alpha_3^m)}{(\alpha_2^m - \alpha_3^m)} (2s+1)(1-s) n_\beta n_\gamma A_{\beta\gamma} . \tag{11.6-17}$$

Again substituting the identities into (11.5-43), but this time dotting with **n** and then substituting (11.6-17), gives

$$\dot{\mathbf{n}} = \boldsymbol{\Omega}\cdot\mathbf{n} + \frac{1}{(\alpha_2^m - \alpha_3^m)} \frac{1}{2s^2} \frac{\delta H_m}{\delta n_\alpha} - \frac{(\alpha_2^m + \alpha_3^m)}{(\alpha_2^m - \alpha_3^m)} \frac{2+s}{3s} \left[\mathbf{n}(\mathbf{n}^{\mathrm{T}}\cdot\mathbf{A}\cdot\mathbf{n}) - \mathbf{A}\cdot\mathbf{n}\right] . \tag{11.6-18}$$

Using the identities in Eqs. (11.6-42,44) produces the constitutive equation for the stress:

$$\Pi_{\alpha\beta} = \beta_1^m s^2 n_\alpha n_\beta n_\gamma n_\varepsilon A_{\gamma\varepsilon} + (\beta_1^m s\frac{1-s}{3} + \beta_2^m s + \beta_3^m s\frac{s+2}{3})(n_\gamma n_\alpha A_{\gamma\beta} + n_\beta n_\gamma A_{\alpha\gamma})$$

$$+ [\beta_4^m + \frac{2}{3}\beta_2^m(1-s) + \frac{2}{9}\beta_3^m(1-s)^2 + \frac{1}{9}\beta_1^m(1-s)^2]A_{\alpha\beta} - s_{,\alpha}\frac{\partial W}{\partial s_{,\beta}}$$

$$- n_{\gamma,\alpha}\frac{\partial W}{\partial n_{\gamma,\beta}} + \frac{(\alpha_2^m + \alpha_3^m)}{(\alpha_2^m - \alpha_3^m)}(1-s)(2s+1)\frac{\delta H_m}{\delta s}n_\alpha n_\beta + K'\delta_{\alpha\beta}$$

$$+ \frac{1}{2}\frac{2+s}{3s}\frac{(\alpha_2^m + \alpha_3^m)}{(\alpha_2^m - \alpha_3^m)}[n_\alpha\frac{\delta H_m}{\delta n_\beta} + n_\beta\frac{\delta H_m}{\delta n_\alpha}] - \frac{1}{2}[n_\alpha\frac{\delta H_m}{\delta n_\beta} - n_\beta\frac{\delta H_m}{\delta n_\alpha}]\,, \tag{11.6-19}$$

provided that one uses the constraint

$$\frac{2}{3}\frac{\delta H_m}{\delta m_{\alpha\beta}} = (n_\alpha n_\beta - \tfrac{1}{3}\delta_{\alpha\beta})\frac{\delta H_m}{\delta s} + \frac{1}{3s}[n_\alpha\frac{\delta H_m}{\delta n_\beta} + n_\beta\frac{\delta H_m}{\delta n_\alpha}]\,, \tag{11.6-20}$$

to restrict the excess information contained in the tensorial theory. Note that this constraint is entirely consistent with the definitions (11.6-14,15). In (11.6-19), K' is an isotropic function of H_m which can be incorporated into the pressure, and $\beta_5^m=\beta_6^m=0$ (solely for computational convenience).

In order to see how the constraint (11.6-20) arises, we must consider the lowest-order, weighted-residual approximation to $\delta H_m/\delta \mathbf{m}$. Starting with

$$\frac{\delta' m_{\alpha\beta}}{\delta' m_{\gamma\delta}} = \frac{\partial m_{\alpha\beta}}{\partial s}\frac{\delta' s}{\delta' m_{\gamma\delta}} + \frac{\partial m_{\alpha\beta}}{\partial n_\varepsilon}\frac{\delta' n_\varepsilon}{\delta' m_{\gamma\delta}} \tag{11.6-21}$$

$$= \tfrac{1}{2}(\delta_{\alpha\gamma}\delta_{\beta\delta} + \delta_{\alpha\delta}\delta_{\beta\gamma}) - \tfrac{1}{3}\delta_{\delta\gamma}\delta_{\alpha\beta}\,,$$

we realize that, in attempting to evaluate $\delta' s/\delta' \mathbf{m}$ and $\delta' \mathbf{n}/\delta' \mathbf{m}$, we have more equations than unknowns, and thus an over-determinate system.[20] We therefore seek the best approximation, using the weighted-residuals method, by taking weighted averages of the above equation provided by the double-dot product of Eq. (11.6-21) with $\boldsymbol{\delta}$ and dotting with $\mathbf{n}$. We obtain, respectively,

[20]The derivative $\delta' a/\delta' b$ denotes a partial derivative which is constrained to belong to the operating space, P.

$$\frac{\partial m_{\alpha\alpha}}{\partial s}\frac{\delta' s}{\delta' m_{\gamma\delta}} + \frac{\partial m_{\alpha\alpha}}{\partial n_\varepsilon}\frac{\delta' n_\varepsilon}{\delta' m_{\gamma\delta}} = 0 \; , \tag{11.6-22a}$$

$$n_\beta \frac{\partial m_{\alpha\beta}}{\partial s}\frac{\delta' s}{\delta' m_{\gamma\delta}} + n_\beta \frac{\partial m_{\alpha\beta}}{\partial n_\varepsilon}\frac{\delta' n_\varepsilon}{\delta' m_{\gamma\delta}} = \tfrac{1}{2}(\delta_{\alpha\gamma}n_\delta + \delta_{\alpha\delta}n_\gamma) - \tfrac{1}{3}\delta_{\gamma\delta}n_\alpha \; . \tag{11.6-22b}$$

Now we have a consistent set of four equations for four unknowns, $\delta' n_\alpha/\delta' m_{\gamma\delta}$, i=1,2,3, and $\delta' s/\delta' m_{\gamma\delta}$ for every partial derivative with respect to $m_{\gamma\delta}$. Using Eq. (11.6-11), we can calculate $\partial m_{\alpha\beta}/\partial s$ and $\partial m_{\alpha\beta}/\partial n_\varepsilon$, so that we may solve (11.6-22) to obtain

$$\frac{\delta' n_\alpha}{\delta' m_{\gamma\delta}} = \frac{1}{2s}(\delta_{\alpha\gamma}n_\delta + \delta_{\alpha\delta}n_\gamma) - \frac{1}{s}n_\alpha n_\gamma n_\delta \; , \tag{11.6-23}$$

$$\frac{\delta' s}{\delta' m_{\gamma\delta}} = \frac{3}{2}(n_\gamma n_\delta - \frac{1}{3}\delta_{\gamma\delta}) \; , \tag{11.6-24}$$

realizing, of course, that these values are the best estimates based on the weighted-residuals procedure. Finally, using differentiation by parts,

$$\frac{\delta H_m}{\delta m_{\alpha\beta}} = \frac{\delta H_m}{\delta s}\frac{\delta' s}{\delta' m_{\alpha\beta}} + \frac{\delta H_m}{\delta n_\gamma}\frac{\delta' n_\gamma}{\delta' m_{\alpha\beta}} \; , \tag{11.6-25}$$

we can arrive directly at the constraint (11.6-20).

By comparing Eqs. (11.6-17,18,19) to (11.3-20,21,23), we see that the two theories are equivalent provided the following relationships hold:

$$\hat\beta_2 = -2\frac{(\alpha_2^m - \alpha_3^m)}{3+6s^2} \; , \quad \frac{\hat\beta_1}{\hat\beta_2} = -\frac{(\alpha_2^m + \alpha_3^m)}{(\alpha_2^m - \alpha_3^m)}(2s+1)(1-s) \; ,$$

$$\hat\alpha_4 = \beta_4^m + \frac{2}{3}\beta_2^m(1-s) + \frac{2}{9}\beta_3^m(1-s)^2 + \frac{1}{9}\beta_1^m(1-s)^2 \; ,$$

$$\hat\alpha_1 + \frac{\hat\gamma_2^2}{\hat\gamma_1} - \frac{\hat\beta_1^2}{\hat\beta_2} = \beta_1^m s^2 \; , \quad \frac{\hat\gamma_2}{\hat\gamma_1} = -\frac{(\alpha_2^m + \alpha_3^m)}{(\alpha_2^m - \alpha_3^m)}\frac{2+s}{3s} \; ,$$

$$\hat\alpha_5 - \hat\alpha_2\frac{\hat\gamma_2}{\hat\gamma_1} = \beta_1^m s\frac{1-s}{3} + \beta_2^m s + \beta_3^m s\frac{s+2}{3} \; , \quad \hat\gamma_1 = 2s^2(\alpha_3^m - \alpha_2^m) \; . \tag{11.6-26}$$

Inclusion of non-zero β_5^m and/or β_6^m modifies trivially the right-hand side of the third, fourth, and sixth of the above equations. The important observation, however, is that by dividing $\hat{\beta}_1/\hat{\beta}_2$ by $\hat{\gamma}_2/\hat{\gamma}_1$, we always get the constraint

$$\frac{\hat{\beta}_1}{\hat{\beta}_2}\frac{\hat{\gamma}_1}{\hat{\gamma}_2} = 3s\,\frac{(2s+1)(1-s)}{2+s}\ , \tag{11.6-27}$$

which reduces the number of independent parameters in the scalar-vector theory from seven to six. The underlying physical meaning behind this constraint is that the dependencies of the time variations of s and $\mathbf{n}$ on $\mathbf{A}$ are correlated since they arise from a single evolution equation for $\mathbf{m}$.

11.6.3 *Reduction to the original Doi theory*

We now turn to the original molecular theory of Doi described in §11.3.3.A, as expressed by Eqs. (11.3-39,40). This theory only exists in the non-inertial form, so we start from the set of equations of (11.5-42,43,44). Many special assumptions are needed to get started, as the equations of §11.5.2 are much more general than the Doi equations. First, the Doi theory ignores the presence of an external field, so we set $\mathbf{F}=\mathbf{0}$ in the linear momentum balance, (11.5-42). Also, since Doi considers in the original formulation only the elastic contribution to the stress, (11.3-40), ignoring the viscous and solvent stresses, (11.3-36,37), we do not need to worry about the matrix $\mathbf{R}^m$ either. Furthermore, the Doi theory does not allow for non-affine motion, so we must set the phenomenological matrix $\mathbf{L}^m=\mathbf{0}$ (or, alternatively, set $\beta_8^m=1$).[21] Last, the Doi theory does not allow

[21]Note that even when the contribution of the system Hamiltonian is negligible, conditions under which, formally, the conformation theory should reduce to the LE theory, a negligible matrix $\mathbf{R}^m$ does not automatically imply that $\beta_8^m=1$. As (11.4-45) shows, in the limit of $\alpha_i\to 0$, $i=2,3$, β_8^m becomes undetermined and so it can have, in principle, any value (constrained only by the thermodynamic inequalities). This subtle observation is of importance in order to restore in the Doi theory the capability to predict tumbling phenomena while preserving its consistency with the inertial theory even for a negligible viscous matrix $\mathbf{R}^m$ (see also the discussion in the next paragraph). Mathematically, β_8^m controls the character of the objective time derivative that enters the evolution equation for the conformation tensor. The value $\beta_8^m=1$ corresponds to an upper-convected derivative, which is the one used in the restricted Doi theory, whereas values of $\beta_8^m<1$ introduce a lower-convected component. Note that not all conformation tensor theories by necessity require an upper-convected derivative. In particular, one of the earliest (if not the first) conformation tensor models proposed by de Gennes [1969b] made use of a corotational derivative defined as an average of an upper- and a lower-convected derivative which is therefore obtained for $\beta_8^m=0$. The corotational (or Jaumann) derivative expresses physically the relative rate of change in a coordinate system which corotates with the fluid—see Bird *et al.* [1987a, p. 507] and references therein.

spatial inhomogeneities in the orientation, and we therefore set $W=0$ as well.

One last assumption needs to be made for this subsection. As the Doi theory was not derived from the inertial form, we do not have the matching term appearing in the phenomenological matrix $\mathbf{P}^m$, (11.5-40). Hence, we let

$$P^m_{\alpha\beta\gamma\varepsilon} = \tfrac{1}{2}\Lambda(\delta_{\alpha\gamma}\delta_{\beta\varepsilon} + \delta_{\alpha\varepsilon}\delta_{\beta\gamma}) , \tag{11.6-28}$$

which, in actuality, amounts to neglecting one term from the general theory since it is of the same order of magnitude as the other terms. This is one of the defects of the Doi theory, developed in a non-inertial frame. Λ in the above expression is related to the rotational diffusivity of Doi (inverse relaxation time), defined later.

In view of the above assumptions, the evolution equation for the order parameter tensor ($\mathbf{S}$ rather than $\mathbf{m}$) and the extra stress of the general theory, (11.5-43,44), become

$$\frac{\partial S_{\alpha\beta}}{\partial t} = -\Lambda\frac{\delta H_m}{\delta S_{\alpha\beta}} + \frac{1}{3}(\nabla_\alpha v_\beta + \nabla_\beta v_\alpha) + S_{\mu\beta}\nabla_\mu v_\alpha$$
$$+ S_{\mu\alpha}\nabla_\mu v_\beta - \frac{2}{3}\delta_{\alpha\beta}S_{\mu\nu}\nabla_\nu v_\mu - 2S_{\alpha\beta}S_{\mu\nu}\nabla_\nu v_\mu , \tag{11.6-29}$$

$$T^m_{\alpha\beta} = 2(S_{\beta\gamma} + \frac{1}{3}\delta_{\beta\gamma})\frac{\delta H_m}{\delta S_{\alpha\gamma}} - 2(S_{\alpha\beta} + \frac{1}{3}\delta_{\alpha\beta})S_{\mu\nu}\frac{\delta H_m}{\delta S_{\mu\nu}} . \tag{11.6-30}$$

Now the terms involving $\delta H_m/\delta\mathbf{S}$ are picked up from (11.5-25), realizing that $\chi_a=b_1=b_2=0$ from the above assumptions. The a_i, however, are not those of (11.5-10), which were chosen for general conditions, but must be picked up especially to match the Doi equations. To find out what these a_i must be, we look at the Doi evolution equation, (11.3-39), under static (no-flow) conditions, whereupon it becomes

$$0 = (1-\tfrac{1}{3}U)S_{\alpha\beta} - U(S_{\alpha\mu}S_{\beta\mu} - \tfrac{1}{3}\delta_{\alpha\beta}S_{\mu\nu}S_{\mu\nu}) + US_{\alpha\beta}S_{\mu\nu}S_{\mu\nu} . \tag{11.6-31}$$

Substituting the uniaxial approximation, (11.5-11), into this expression then gives

$$0 = (1-\tfrac{1}{3}U)s - \tfrac{1}{3}Us^2 + \tfrac{2}{3}Us^3 \equiv \frac{\delta H_b'}{\delta s}\frac{1}{nk_BT} , \tag{11.6-32}$$

which, upon integration, yields the bulk free energy (to within a constant)

$$\frac{H_b'}{nk_BT} = \frac{1}{2}(1-\frac{U}{3})s^2 - \frac{U}{9}s^3 + \frac{U}{6}s^4 . \tag{11.6-33}$$

In order to continue the comparison, it is necessary to rewrite the bulk free energy of (11.5-9) using no higher than the third power of **S**. Therefore we make use of the following theorem, which is proven in Example 11.3.

Theorem 11.4:

$$\mathrm{tr}\,(\mathbf{S}\cdot\mathbf{S}\cdot\mathbf{S}\cdot\mathbf{S}) = \tfrac{1}{2}[\mathrm{tr}\,(\mathbf{S}\cdot\mathbf{S})]^2 \,. \tag{11.6-34}$$

Using Theorem 11.4, we can rewrite H_b of (11.5-9) as

$$H_b = \int_\Omega [a_1 S_{\alpha\beta}S_{\alpha\beta} + a_2 S_{\alpha\beta}S_{\beta\gamma}S_{\gamma\alpha} + a_3'(S_{\alpha\beta}S_{\beta\alpha})^2]\, d^3x \,, \tag{11.6-35}$$

which upon substitution of (11.5-11) becomes

$$H_b = \frac{2}{3}a_1 s^2 + \frac{2}{9}a_2 s^3 + \frac{2}{9}a_3' s^4 \,. \tag{11.6-36}$$

Comparison of Eqs. (11.6-33,36) then yields the a_i as

$$a_1 = \frac{3}{4}(1-\frac{U}{3})nk_BT \,, \quad a_2 = -\frac{U}{2}nk_BT \,, \quad a_3' = \frac{3}{4}Unk_BT \,. \tag{11.6-37}$$

For the new bulk free energy, (11.6-35), the functional derivative is evaluated as

$$\frac{\delta H_m}{\delta S_{\alpha\beta}} = \frac{\delta H_b}{\delta S_{\alpha\beta}} = \frac{3}{2}(1-\frac{U}{3})nk_BTS_{\alpha\beta} - \frac{3}{2}Unk_BT(S_{\beta\gamma}S_{\alpha\gamma} - \frac{1}{3}S_{\mu\nu}S_{\mu\nu}\delta_{\alpha\beta}) + \frac{3}{2}Unk_BTS_{\alpha\beta}S_{\mu\nu}S_{\mu\nu} \,, \tag{11.6-38}$$

so that substitution of this expression into (11.6-29) gives

$$\begin{aligned}\frac{\partial S_{\alpha\beta}}{\partial t} &= \tfrac{1}{3}(\nabla_\alpha v_\beta + \nabla_\beta v_\alpha) + S_{\mu\beta}\nabla_\mu v_\alpha + S_{\mu\alpha}\nabla_\mu v_\beta \\ &\quad - \tfrac{2}{3}\delta_{\alpha\beta}S_{\mu\nu}\nabla_\nu v_\mu - 2S_{\alpha\beta}S_{\mu\nu}\nabla_\nu v_\mu - \tfrac{3}{2}\Lambda\Big[(1-\frac{U}{3})S_{\alpha\beta} \\ &\quad - U(S_{\beta\gamma}S_{\alpha\gamma} - \tfrac{1}{3}S_{\mu\nu}S_{\mu\nu}\delta_{\alpha\beta}) + S_{\alpha\beta}S_{\mu\nu}S_{\mu\nu}\Big] \,.\end{aligned} \tag{11.6-39}$$

This is easily seen to match (11.3-39) provided that we set

$$\Lambda = 4D_r/nk_BT \ . \tag{11.6-40}$$

Using (11.6-30) and (11.6-38), we may also calculate the extra stress as

$$\frac{T^m_{\alpha\beta}}{3nk_BT} = (1-\frac{U}{3})S_{\alpha\beta} + (\frac{2}{3}U-1)\delta_{\alpha\beta}S_{\mu\nu}S_{\mu\nu} + (3-2U)S_{\beta\gamma}S_{\gamma\alpha}$$
$$+ (3U-3)S_{\alpha\beta}S_{\mu\nu}S_{\mu\nu} - 3US_{\beta\gamma}S_{\gamma\varepsilon}S_{\varepsilon\alpha} + US_{\gamma\varepsilon}S_{\varepsilon\nu}S_{\nu\gamma}\delta_{\alpha\beta}$$
$$+ 3US_{\alpha\beta}S_{\gamma\varepsilon}S_{\varepsilon\nu}S_{\nu\gamma} - US_{\mu\nu}S_{\mu\nu}S_{\gamma\varepsilon}S_{\gamma\varepsilon}\delta_{\alpha\beta}$$
$$+ 3US_{\beta\gamma}S_{\gamma\alpha}S_{\mu\nu}S_{\mu\nu} - 3US_{\alpha\beta}(S_{\mu\nu}S_{\mu\nu})^2 \ . \tag{11.6-41}$$

Although we were able to match exactly the evolution equations for the order parameter tensor of the general and Doi theories, comparing the stress expressions (11.3-40) and (11.6-41) reveals that there is a discrepancy in the two relations. This discrepancy is manifested in the higher-order terms; but near the isotropic-to-liquid-crystalline phase transition (U=3 for the values of the a_i in 11.6-37), the expressions match up to and including second-order terms. This disagreement between the two relations arises due to the lower-order calculation of the stress by Doi, as well as his neglect of the traceless constraint upon the derivative of the bulk free energy. However, the predicted stress behavior is qualitatively similar for two simple flow situations: steady shear and steady extensional flows [Edwards, 1991, pp. 114-18].

EXAMPLE 11.3: Prove Theorem 11.4.

For any 3×3 matrix **c**, the Cayley/Hamilton theorem states that

$$\mathbf{c}\cdot\mathbf{c}\cdot\mathbf{c} - I_1\mathbf{c}\cdot\mathbf{c} + I_2\mathbf{c} - I_3\boldsymbol{\delta} = 0 \ , \tag{11.6-42}$$

where the I_i are the invariants of **c** defined by Eqs. (8.1-66b,c,d). Multiplying this expression by **c**, then taking the trace yields

$$\mathrm{tr}(\mathbf{c}\cdot\mathbf{c}\cdot\mathbf{c}\cdot\mathbf{c}) - I_1\,\mathrm{tr}(\mathbf{c}\cdot\mathbf{c}\cdot\mathbf{c}) + I_2\,\mathrm{tr}(\mathbf{c}\cdot\mathbf{c}) - I_3\,\mathrm{tr}\,\mathbf{c} = 0 \ . \tag{11.6-43}$$

Substitution of **S** for **c**, knowing that tr **S**=0, immediately yields Theorem 11.4. ■

11.6.4 Reduction to the extended Doi theory

Lastly, we turn to the extended Doi theory of §11.3.3.B, and wish to express the remaining parameters of the general theory in terms of the parameters of the non-inertial molecular theory. As mentioned in §11.4.2, the extended theory tries to incorporate the effects of spatial inhomogeneities into the system, thereby forcing the addition of translational diffusivities into the system of equations. Since we did not include such effects in our modeling in §11.5, we shall find that we must add an additional term to the dissipation bracket of (11.5-38) to account for them. This being done, we present a simple calculation of the initial stages of the spinodal decomposition from the isotropic to the liquid-crystalline state and compare to a similar (but much more complicated) calculation by Shimada *et al.* [1988, §III] in order to match the b_i (the elastic constants) to molecular parameters.

From §11.4.2.B, we recall that the diffusion equation for the orientational distribution function is expressed as (keeping only the term due to the translational diffusivity)

$$\frac{\partial f}{\partial t} = \nabla_\gamma\Big[[(D_{\parallel}-D_{\perp})q_\gamma q_\varepsilon + D_\perp \delta_{\gamma\varepsilon}]\, f\, \nabla_\varepsilon(\ln f + W)\Big] , \qquad (11.6\text{-}44)$$

with

$$W(\mathbf{x},\mathbf{q},t) = \int_\Omega \int V^*(\mathbf{x}-\mathbf{x}',\mathbf{q},\mathbf{q}')\, f(\mathbf{x}',\mathbf{q}',t)\; d^2q'\, d^3x \ . \qquad (11.6\text{-}45)$$

By denoting a free energy, A, in terms of the free energy density, a, as

$$\begin{aligned} A &= \int_\Omega \int a(\mathbf{x},\mathbf{q},t)\; d^2q\, d^3x \\ &= nk_BT \int_\Omega \int (\ln f + W)\, f\; d^2q\, d^3x \ , \end{aligned} \qquad (11.6\text{-}46)$$

we can rewrite (11.6-44) as

$$\frac{\partial f}{\partial t} = \frac{1}{nk_BT}\nabla_\gamma\Big[[(D_{\parallel}-D_\perp)q_\gamma q_\varepsilon + D_\perp\delta_{\gamma\varepsilon}]\, f\, \nabla_\varepsilon a\Big] . \qquad (11.6\text{-}47)$$

Now in order to accomplish our objective, we need to express the free energy A in terms of $\mathbf{S}$ and the functional derivatives of the Hamiltonian, $\delta H_m/\delta \mathbf{S}$. Indeed, for the Hamiltonian of the general theory, (11.5-35), we write the free energy as

$$A'(\mathbf{S}) = H_b + \int_\Omega W\; d^3x \ , \qquad (11.6\text{-}48)$$

which includes all of the contributions (excluding external fields) which depend on the order parameter tensor. This expression we approximate as

$$A' \approx A'_0 + \frac{1}{2}\frac{dH_m}{d\mathbf{S}}:\mathbf{S} \;, \tag{11.6-49}$$

which becomes exact for small **S** (since then H_m can be replaced by a quadratic function of **S**). A'_0 is a constant with respect to **S** and need not concern us any further. Since **S** is given by (11.3-30,31), we are able to rewrite (11.6-49) as

$$A' \approx \int_\Omega \int \tfrac{1}{2}(q_\alpha q_\beta - \tfrac{1}{3}\delta_{\alpha\beta})\frac{\delta H_m}{\delta S_{\alpha\beta}} f(\mathbf{x},\mathbf{q},t)\; d^2q\; d^3x \;, \tag{11.6-50}$$

so that by comparison with (11.6-46) we find that

$$a \approx \frac{1}{2}(q_\alpha q_\beta - \frac{1}{3}\delta_{\alpha\beta})\frac{\delta H_m}{\delta S_{\alpha\beta}} \;. \tag{11.6-51}$$

Upon substitution of (11.6-51) into (11.6-47), multiplying by **qq**, and subsequently integrating over d^2q, we obtain

$$\frac{\partial m_{\alpha\beta}}{\partial t} = \frac{1}{2nk_BT}\nabla_\gamma\Bigg[[(D_\parallel - D_\perp)q_\gamma q_\varepsilon + D_\perp\delta_{\gamma\varepsilon}] \times [m_{\alpha\eta}m_{\beta\zeta} - m_{\alpha\beta}m_{\mu\eta}m_{\mu\zeta}]\nabla_\varepsilon(\frac{\delta H_m}{\delta S_{\eta\zeta}})\Bigg] \;, \tag{11.6-52}$$

where we have made use of the decoupling approximation

$$\langle q_\alpha q_\beta q_\zeta q_\eta q_\gamma q_\varepsilon\rangle \approx \langle q_\alpha q_\beta\rangle\langle q_\zeta q_\eta\rangle\langle q_\gamma q_\varepsilon\rangle \;. \tag{11.6-53}$$

We chose this decoupling approximation based on arguments which follow for the matrix **B**.[22]

Thus far, we have taken the extended Doi equations, which were originally written in terms of the distribution function, and, using a couple of presumptuous approximations, we have arrived at a serviceable equation in terms of the conformation tensor **m**. Now we wish to show how this equation arises through the bracket procedure, and in so doing, justify the assumptions that we have used in the above development relative to any others which we might have used; i.e., we wish to show that the

[22]Alas, for we have no proof whatsoever as to its validity.

bracket formulation leads to the equation (11.6-52) and thus validates our choices for the approximations.

As already mentioned, we have neglected one possible coupling from the dissipation bracket of (11.4-38). Since we let the Hamiltonian depend not only on **m**, but on **∇m** as well, we must allow for the possibility of a coupling between the gradients of the functional derivatives of the Hamiltonian. This coupling corresponds to the diffusive flux of chapter 8, and may be thought of as a diffusive transport due to a driving force which arises due to inhomogeneities in the structure of the medium. This effect does not appear in the LE theory. By incorporating in the dissipation bracket of (11.5-38) the integral

$$-\int_{\Omega} B_{\mu\eta\gamma\epsilon\nu\zeta}\,(\delta_{\mu\alpha}\delta_{\nu\beta}-\delta_{\mu\nu}m_{\alpha\beta})\,\nabla_{\gamma}(\frac{\delta F}{\delta m_{\alpha\beta}})\,\nabla_{\epsilon}(\frac{\delta F}{\delta m_{\eta\zeta}})\;d^3x\;, \quad (11.6\text{-}54)$$

we can incorporate these effects into the system of equations directly. Note that the second term in this integral insures that the result is traceless, as arising from Theorem 11.2.

Now we must choose the best expression for the phenomenological matrix **B**, or rather the best compromise between complexity and usefulness. Specifically, we could write out the most general possible expression for **B**, as we did for $\mathbf{R}^m$, but this would result in a ridiculously complex equation. Rather, we choose a simple form for this matrix which satisfies certain symmetry requirements. We require that **B** ensures that the evolution equation for **m** is symmetric, and we require, as before, that $B_{\alpha\eta\gamma\epsilon\beta\zeta}=B_{\eta\alpha\gamma\epsilon\beta\zeta}$=etc. We also employ two orientational diffusivities, one parallel and one perpendicular to the preferred direction of orientation, $\Lambda_{\parallel}$ and $\Lambda_{\perp}$, respectively. Hence we choose as our matrix

$$B_{\alpha\eta\gamma\epsilon\beta\zeta} = m_{\alpha\eta}\,[(\Lambda_{\parallel}-\Lambda_{\perp})\,m_{\gamma\epsilon} + \Lambda_{\perp}\delta_{\gamma\epsilon}]\,m_{\beta\zeta}\;, \quad (11.6\text{-}55)$$

which for $\Lambda_{\parallel}=D_{\parallel}/2nk_BT$ and $\Lambda_{\perp}=D_{\perp}/2nk_BT$ gives Eq. (11.6-52) directly. Granted, the choice of **B** is still somewhat arbitrary, but we justify it simply enough from the form of the integration which gave us (11.6-52). Hence our reasoning has been rather circular: we chose a decoupling approximation based upon symmetry requirements on the matrix **B**, and chose **B** based on the form of the integration. Yet it is a closed circle, and thus should allow us to do some simple calculations and arrive at realistic results.

As an example of the applicability of the present model to the study of non-homogeneous systems, the spinodal decomposition is examined as an LC in the isotropic state (**S**=**0**) is brought into a parameter region where the isotropic state is unstable. Recently, Shimada *et al.* [1988, §III,IV] have investigated the same problem by examining the initial behavior of the fluctuation of the concentration and orientation in the isotropic phase. At this point, we wish to show that the same results can be produced using the present model with much less effort, and to show that the remaining parameters of the general theory (b_1 and b_2) can also be matched with the extended Doi theory. Here, we perform a linear stability analysis of the kinetic equation governing the spinodal decomposition of the isotropic state to the liquid-crystalline state. Thus, we only concern ourselves here with the initial stages of this decomposition, where the physics of the problem is more likely to be represented by the linearized equations. For an initially isotropic fluid with **S=0**, we wish to see for what parameter (concentration) range arbitrary fluctuations (perturbations) may induce a decomposition into the liquid-crystalline phase.

In the initial stages of the disturbance, it is reasonable to assume no flow (**v**=**0**) and to assume that the parallel translational diffusion is the only relaxation mechanism for the molecules (Λ=0, $\Lambda_\perp$=0) [Shimada *et al.*, 1988, p. 7183]. Under these assumptions, and retaining only terms linear with respect to the components of the order parameter tensor, (11.6-52) becomes

$$\begin{aligned}\frac{\partial S_{\alpha\beta}}{\partial t} &= \nabla_\gamma\Big(\frac{D_{\parallel}}{27}\delta_{\alpha\eta}\delta_{\gamma\varepsilon}\delta_{\beta\zeta}\nabla_\varepsilon[a_1 S_{\zeta\eta} - b_1 S_{\zeta\eta,\mu,\mu} \\ &\quad - b_2\big(\frac{1}{2}S_{\mu\zeta,\mu,\eta} + \frac{1}{2}S_{\mu\eta,\mu,\zeta} - \frac{1}{3}\delta_{\eta\zeta}S_{\mu\nu,\mu,\nu}\big)]\Big) \\ &= \frac{D_{\parallel}}{27}[a_1 S_{\alpha\beta} - b_1 S_{\alpha\beta,\mu,\mu} - b_2\big(\frac{1}{2}S_{\mu\alpha,\mu,\beta} \\ &\quad + \frac{1}{2}S_{\mu\beta,\mu,\alpha} - \frac{1}{3}\delta_{\beta\alpha}S_{\mu\nu,\mu,\nu}\big)]_{,\gamma,\gamma}\ ,\end{aligned} \tag{11.6-56}$$

where we have used

$$B_{\alpha\eta\gamma\varepsilon\beta\zeta} = \frac{1}{27nk_BT}D_{\parallel}\delta_{\alpha\eta}\delta_{\gamma\varepsilon}\delta_{\beta\zeta}\ , \tag{11.6-57}$$

which is the limiting (**S**=**0**) expression for **B** arising from (11.6-55).

In order to investigate the time evolution of various modes, let $\mathbf{S}_k$ be the k-th Fourier component of the order parameter tensor

$$\mathbf{S}_k \equiv \Re\,[\mathbf{A}_k(t)\exp(ikx_3)]\ , \tag{11.6-58}$$

in general complex, where $\mathbf{A}_k(t)$ is a traceless, symmetric tensor. Substitution of (11.6-58) into (11.6-56) leads to a system of five independent, ordinary differential equations coupling the five independent components of $\mathbf{A}_k(t)$. As already observed by Shimada *et al.* [1988, pp. 7183, 7184]), these equations can be separated into five independent sets of equations, each one governing the (initial) evolution in time of five orientational modes. These are equations involving A_{k12}, $A_{k11}-A_{k22}$, A_{k13}, A_{k23}, and A_{k33}.

In particular, the equations have exactly the same form discovered by Shimada *et al.* [1988], separating the fluctuation modes into three types.

1. The "twist" mode, with similar equations followed by A_{k12} and $A_{k11}-A_{k22}$, is

$$\frac{\partial A_{k12}}{\partial t} = -\frac{2D_{\parallel}}{3L^2}\left[(1-\frac{U}{3})K^2 + \frac{8\overline{b}_1}{9L^2}K^4\right]A_{k12}\,, \tag{11.6-59}$$

where $K \equiv kL/2$ and $\overline{b}_1 \equiv b_1/(nk_BT)$. Eq. (11.6-59) is the same as the corresponding equation of Shimada *et al.* [1988, Eq. (3.2)], with the only minor difference being that the numerical factor 2/3 appears instead of 4/7, provided that the coefficient $\overline{b}_1$ is defined as

$$\overline{b}_1 \equiv \frac{UL^2}{32}\,. \tag{11.6-60}$$

Note that this expression is consistent with the one obtained from Onsager's rigid-rod theory, (11.2-44), to which it becomes identical (in the fully oriented nematic limit, s=1) for $\nu_2 \approx 0.32$. This condition is obtained using the relationship of $\overline{b}_1$ with the Frank moduli, $\overline{b}_1 = k_{22}/(2nk_BT)$, from (11.2-30b), and expression (11.3-29) for U. For $U<3$, the coefficient of A_{k12} in (11.6-59) is negative, corresponding to a negative eigenvalue, which implies a decaying fluctuation for every wavelength k. For $U>3$ however, the fluctuation will grow for small enough wavelengths with the maximum growth rate λ_m,

$$\lambda_m = 6D_{\parallel}\frac{(1-\frac{U}{3})^2}{UL^2}\,, \tag{11.6-61}$$

attained for the "most unstable" wavenumber, k_m, given as

$$k_m = \frac{6}{L}\left(\frac{-2(1-\frac{U}{3})}{U}\right)^{\frac{1}{2}}. \tag{11.6-62}$$

Thus U=3 corresponds to the critical concentration beyond which the isotropic state becomes unstable to infinitesimal perturbations, in agreement with the free energy analysis of the static system.

2. The "bend" mode, with similar equations followed by A_{k13} and A_{k23}, is

$$\frac{\partial A_{k13}}{\partial t} = -\frac{2D_{\parallel}}{3L^2}\left[(1-\frac{U}{3})K^2 + \frac{4(2\bar{b}_1+\bar{b}_2)}{9L^2}K^4\right]A_{k13}\,, \tag{11.6-63}$$

where $\bar{b}_2 \equiv b_2/(nk_BT)$. Eq. (11.6-63) is the same as the corresponding equation, (Eq. 3.3), of Shimada *et al.* [1988], with the only minor difference that the numerical factor 2/3 appears instead of 12/7, provided that the coefficient b_2 is defined as

$$\bar{b}_2 \equiv \frac{UL^2}{9}. \tag{11.6-64}$$

Note that a value for $\bar{b}_2 = UL^2/8$ is consistent with the value for $\bar{b}_1$ according to (11.6-60) and the Onsager rigid-rod theory, (11.2-44). The small difference between this value and that of (11.6-64) is surprising given the fact that the expression for the Frank elasticity utilized here imposes the restriction $k_{11}=k_{33}$ which, at least for high S values, is violated according to Onsager's theory—see the discussion following Eq. (11.2-44).

3. The "splay" mode, with the following coupled equations for A_{k33}, A_{k11}, and A_{k22}, is

$$\frac{\partial A_{k33}}{\partial t} = -\frac{2D_{\parallel}}{3L^2}\left[(1-\frac{U}{3})K^2 + \frac{8(3\bar{b}_1+2\bar{b}_2)}{27L^2}K^4\right]A_{k33}\,, \tag{11.6-65}$$

$$\frac{\partial A_{k11}}{\partial t} = -\frac{2D_{\parallel}}{3L^2}\left[(1-\frac{U}{3})K^2 + \frac{8\bar{b}_1}{9L^2}K^4\right]A_{k11} - \frac{2D_{\parallel}}{3L^2}\frac{8\bar{b}_2}{27L^2}K^4A_{k33}\,, \tag{11.6-66}$$

$$\frac{\partial A_{k22}}{\partial t} = -\frac{2D_{\parallel}}{3L^2}\left[(1-\frac{U}{3})K^2 + \frac{8\bar{b}_1}{9L^2}K^4\right]A_{k22} - \frac{2D_{\parallel}}{3L^2}\frac{8\bar{b}_2}{27L^2}K^4A_{k33}\,. \tag{11.6-67}$$

This mode has two independent equations (if the traceless condition is satisfied at all times). There are two eigenvalues, both of them remaining negative as long as U is less than the critical value of 3. If $U>3$, however, these eigenvalues can become positive for small wavenumbers. These characteristics are also exhibited by the corresponding equations for the "splay" mode of Shimada *et al.* [1988, p. 7184].

All the growth rates of the above modes have the same dependence on the wavenumber:

$$\lambda = \lambda_m [1 - (\frac{k^2}{k_m^2} - 1)^2] . \qquad (11.6\text{-}68)$$

In fact, based upon the maximum wavenumber, one can make an order-of-magnitude estimate of the "domain" size into which an isotropic material will decay during a phase transition from the isotropic to the liquid-crystalline state.

With the consistency checking for the b_i according to (11.6-60,64), we have completed the investigation for the relationship between the general theory and the molecular theories. We have shown that most, if not all, of the important physics occurring on the molecular level can be described by a much simpler set of equations on the level of a second-rank tensor.

11.6.5 Equivalent Leslie coefficients for the general conformation tensor theory

The LE coefficients which correspond to the general conformation theory, as obtained by a similar fitting of the original Doi theory (see pertinent discussion in §11.6.3) with the addition of the β_8^m as an additional parameter, are presented in Table 11.2. The coefficients are obtained by following the approach used before in Edwards *et al.* [1991a, §5]. Briefly, assuming a homogeneous steady flow, the evolution equation for **m**, (11.5-43), is used in order to solve for $\delta H/\delta \mathbf{m}$ in terms of the rate of deformation tensor, **A**, and the antisymmetric velocity gradient, $\boldsymbol{\Omega}$. Then this expression is substituted into the stress constitutive equation, (11.5-44), with the tensor **m** substituted by its uniaxial approximation, (11.6-11), valid at slow flows. The final expression is compared term by term with the equivalent expression of the LE theory, Eqs. (11.3-9) and (11.3-11), evaluated at steady state (although the steady-state assumption is not necessary). The procedure is lengthy, but straightforward. Moreover, any explicit introduction of an external field—in contrast to Marrucci's approach [Marrucci, 1982]—is avoided (although some type of unspecified symmetry-breaking effect must occur within the system).

In interpreting the results, it is useful to remind the reader of the information conveyed by the Leslie coefficients. As far as the expected dynamic behavior is concerned, this is primarily conveyed through the coefficient λ [Marrucci, 1982, p. 160]:

$$\lambda \equiv \frac{\alpha_5 - \alpha_6}{\alpha_3 - \alpha_2} = \frac{\alpha_2 + \alpha_3}{\alpha_2 - \alpha_3} = \frac{1 + \varepsilon}{1 - \varepsilon} , \qquad (11.6\text{-}69)$$

Parameter†	Predictions from the conformation tensor theory with arbitrary β_8^m
α_1	$-\frac{2}{3}S^2\frac{(5+8S-4S^2)}{1+S^2}(\beta_8^m)^2$
α_2	$-S\frac{(2+S)}{3}\beta_8^m - S^2$
α_3	$-S\frac{(2+S)}{3}\beta_8^m + S^2$
α_4	$\frac{4}{9}(1-S)^2(\beta_8^m)^2$
α_5	$S\frac{(4-S)}{3}(\beta_8^m)^2 + S\frac{(S+2)}{3}\beta_8^m$
α_6	$S\frac{(4-S)}{3}(\beta_8^m)^2 - S\frac{(S+2)}{3}\beta_8^m$

†The parameters are scaled with respect to $\beta_7^m \equiv \frac{1}{2} n k_B T D_r$.

Table 11.2: The Leslie coefficients, α_i, for a nematic LC of rigid rods as predicted from the general conformation theory.

where [Semenov, 1983, p. 323]

$$\varepsilon \equiv \frac{\alpha_3}{\alpha_2}\,, \tag{11.6-70}$$

the second equality in (11.6-69) being valid because of the use of the Parodi relation. The parameter λ (equivalently, ε) determines the existence of a steady-state solution (if $\lambda>1$) or a tumbling behavior for the director (if $|\lambda| < 1$) in simple shear flows, in the limit of vanishingly small elasticity, according to the LE theory (see §11.3.4.A).

As can be easily calculated from Table 11.2, the corresponding coefficient λ for the conformation tensor theory is

$$\lambda = \beta_8^m \frac{(2+S)}{S} . \tag{11.6-71}$$

Notice that it reduces to the previously reported value for the restricted Doi model [Marrucci, 1982; Edwards *et al.*, 1990, Eq. (5.13)] for β_8^m=1 (see also Table 11.1). For this value of β_8^m, as noted before [Edwards *et. al.*, 1990, p. 69], the parameter λ is always larger than one, thus implying always an aligning behavior. This is in contrast with the predictions of the molecular theory (see also Table 11.1). The parameter β_8^m allows us to restore the agreement between the two theories. A glance at Table 11.1 shows that it suffices for that purpose to choose β_8^m as

$$\beta_8^m = \Gamma \frac{3S}{S+2} , \tag{11.6-72}$$

where, of course, Γ should also be considered as a function of S (or concentration) [Kuzuu and Doi, 1984, Fig. 1]. It is interesting to note that given the dependence of Γ on S, treating β_8^m as a constant is a very good approximation. In terms of the physical interpretation of (11.6-71), we can venture to say that the parameter β_8^m communicates information pertaining to the tumbling behavior of the molecules corresponding to a perfect molecular alignment (S=1). The S-dependent correction reflects the stabilizing effect at high concentrations of a non-uniform (S<1) distribution. Similar behavior is predicted from the molecular theory. Thus, with the addition of one single extra parameter (β_8^m), we see that the consistency between the molecular and tensorial theories is restored.

The implications from this new flexibility offered to the tensorial model are numerous. For one thing, the tensor theory can now be used for the description of domain phenomena associated with director tumbling. Indeed, some preliminary calculations for a planar flow have already been performed [Edwards, 1991, ch. 5], and the results exhibit the same richness in phenomena already found through the solution of the time-dependent diffusion equation for the distribution function [Larson, 1990]. In the absence of Frank elasticity and for small enough values of the β_8^m parameter (β_8^m=0.86), the transients observed during the start-up of shear flow, as shown in Figure 11.9, are astonishingly similar to the sustained oscillations (wagging) obtained by the solution of the diffusion equation for the distribution function [Larson, 1990, Fig. 7]. Edwards [1991, §5.3] also reported the results of a two-dimensional calculation including a non-zero Oseen/Frank elasticity for planar shear flow between two parallel plates at a distance d_0 apart with the upper plate

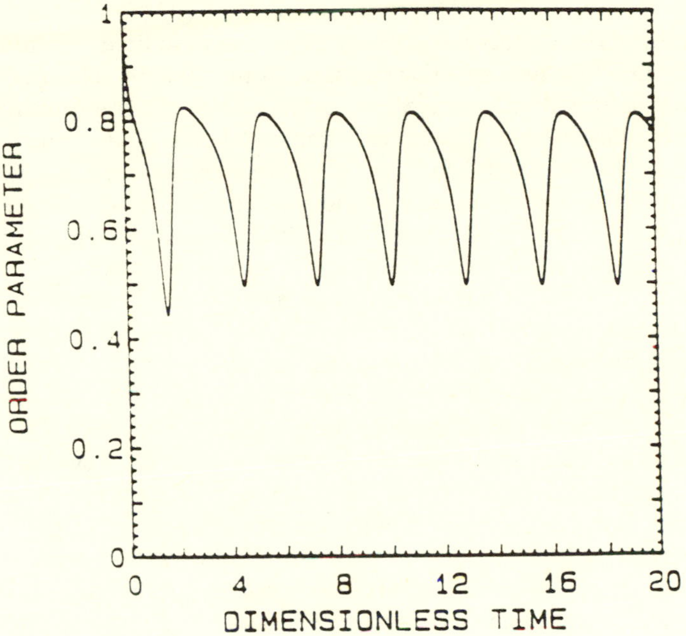

Figure 11.9: Order parameter versus dimensionless time, $\tilde{t}\equiv t\nu/\beta_7^m$ for U=2.0, β_8^m=0.86, and $\tilde{\dot{\gamma}}\equiv\dot{\gamma}\beta_7^m/\nu$. (Reprinted with permission from [Edwards, 1991, p. 264].)

moving with velocity V parallel to the bottom. These calculations showed that incorporation of Frank elasticity results in an immediate damping of the oscillations, very similar to the one observed in actual experiments [Yang, 1992]. Of course, a much more detailed (computational) study is necessary in order to appropriately extract all the wealth of information which we believe is incorporated into the model in addressing issues of importance, e.g., domain/texture formation and deformation under flow.

In addition, note that the expressions provided for the Leslie coefficients in Table 11.2 satisfy the Parodi relationship, (11.3-10), and the Leslie inequalities (see Example 11.1) for reasonable values of the parameters S and β_8^m, $0\leq S$, $\beta_8^m\leq 1$. Finally, although in calculating the results reported in Table 11.2 no viscous terms were included in the conformation tensor theory ($\mathbf{R}^m$=**0**), addition of these terms can be done in a straightforward fashion: this would only affect α_4 and α_1.

11.7 Concluding remarks

In conclusion, in this chapter we have systematically investigated the inherent structure corresponding to the dynamics of liquid-crystalline materials. We showed the connection between the existing dynamical LC theories by reformulating them through the generalized bracket. This led to a conformation tensor theory which nicely reduces to all of the previous LC theories in characteristic limits, while simultaneously providing their natural generalization into a single theory combining liquid-crystalline thermodynamics, Frank elasticity, and viscous fluid dynamics. The wonderful features of this generalized theory, however, are that it leads to a system of continuum equations which, in principle, is no more difficult to solve than the continuum theories which preceded it, while simultaneously incorporating the essential physics.

Clearly, in view of the inherent complexity and diversity of liquid-crystalline materials, an overview of which was presented in the first two sections of this chapter, much more still remains to be done in incorporating more of the physical features encountered in the studies of liquid-crystalline systems at equilibrium in theories of LC dynamics. Hopefully, we have managed to show that the bracket formalism provides the natural medium for this transition of information to be accomplished. Thus, we believe that the exercise of improving the existing models involves merely repeating the same type of procedure followed here using a different (enlarged) set of internal variables—for example, addition of the lyotropic LC concentration as a dynamic variable—in order to more faithfully represent the system's internal microstructure. The success of a research program in this direction will make the present efforts worthwhile.

12
Multi-Fluid Transport/Reaction Models with Application in the Modeling of Weakly Ionized Plasma Dynamics

We must therefore discover some method of investigation which allows the mind at every step to lay hold of a clear physical conception, without being committed to any theory founded on the physical science from which that conception is borrowed, so that it is neither drawn aside from the subject in pursuit of analytical subtleties, nor carried beyond the truth by a favourite hypothesis.

—*James Clerk Maxwell*

The industrial use of low-ambient-temperature, weakly ionized plasmas as a reaction environment is growing rapidly.[1] This is primarily evident in the manufacturing technologies of advanced materials, such as the ones used in micro-electronic devices [Jensen, 1987]. The advantages of the plasma environment are due primarily to the presence of high energy electrons which allow high energy chemistry to take place at low ambient temperatures. An example is the successful plasma-enhanced chemical vapor deposition of silicon nitride at temperatures as low as 250-350°C versus temperatures in the range of 700-900°C required for thermal deposition [Reif, 1984]. Thus emerges a need for modeling of the reaction chemistry and the transport phenomena within complex, multicomponent, charged-particle systems, under the influence of externally-imposed electric and magnetic fields. The present chapter addresses this need within the framework of a multi-fluid reactive continuum [Woods, 1975, ch. 9].

Multi-fluid continuum descriptions have arisen as a natural generalization of multicomponent systems in order to account for the absence of momentum and/or energy equilibria between different species populations within the same system [Enz, 1974; Woods, 1975, ch. 9]. The key underlying assumption is that of interpenetrating continua: each one of the mutually interacting, constituent subsystems is characterized as a separate continuum with its own (macroscopic) state variables. Hidden within this assumption is the local equilibrium hypothesis, not between

[1]This chapter is written in collaboration with J. B. Gustafson and closely follows ch. 4 of his Ph.D. thesis [Gustafson, 1991].

different subsystems—that would have resulted in the more traditional multicomponent description—but *within each subsystem* in order for the description of each subsystem using (equilibrium) state variables to be meaningful. This is both an asset and a liability of the multi-fluid approach: an asset, because the whole framework of equilibrium thermodynamics is still applicable at the subsystem level, resulting, among other things, in a description requiring only a few well-defined macroscopic state variables; a liability, because it places very stringent requirements on the type of systems to which this theory can be applied.

The multi-fluid approach is valid only for phenomena with characteristic time scales much larger than the time scale for each subsystem to reach *internal* (local) thermodynamic equilibrium. In turn, in order for this approach to provide useful results, the experimental time scale needs to be smaller, or at least of the same order of magnitude, as the time scale required for equilibrium *between* the different subsystems to be established. Obviously, these requirements are easier to satisfy when a partial equilibrium between the subsystems is allowed. This was the case examined in §9.2.2 for the modeling of shear-induced concentration effects in dilute polymer solutions. There the two subsystems (the macromolecules and the solvent) were assumed to be in thermal but not in hydrodynamic equilibrium. Alternatively, the multi-fluid (two-fluid) description is appropriate for describing the interaction between *thermal excitations* ("first fluid") governed by traditional local conservation equations for mass, momentum, and energy, and a *condensed phase* ("second fluid") defined by the microscopic dynamics of long-wavelength dynamical modes [Enz, 1974]. An example of this type of application is the description of second sound in dielectric crystals, as referred to in §10.1.2.

In the present chapter, we are concerned with the thermodynamics of a multi-fluid system, the subsystems of which are physically identified with its chemical constituents (species). These are modeled as thermodynamic open systems exchanging, in general, mass, momentum, and energy through chemical reactions and internal friction. They are assumed to be in local thermodynamic and hydrodynamic equilibrium characterized by different (in general) temperature and velocity variables, respectively. Since the best application for such a description is in dilute plasma systems, we, by necessity, focus on the form of equations that are suitable to describe such media. Multi-fluid descriptions of reactive plasma systems, based on a consistent application of thermodynamic principles, are already available in the literature [Woods, 1975, chs. 9 and 10]. Thus, the benefits from the utilization of the bracket formalism to develop such a description are more modest than those realized in the applications discussed in the previous chapters.

The benefits are mostly realized in elucidating the thermodynamics of chemical reactions. Previous descriptions of reactive multi-fluid systems have either not incorporated the reaction kinetics in any detail [Woods, 1975, ch. 9; Bataille and Kestin, 1977] or, more recently, have based the modeling on phenomenology [Morro and Romeo, 1988; Keck, 1990]. On the other hand, considerable work has been devoted to extending linear, non-equilibrium thermodynamic principles to incorporate nonlinear reaction kinetics in single-fluid, multicomponent descriptions [Wei, 1962; Aris, 1965; 1968; Krambeck, 1970; Feinberg, 1972; Lengyel, 1989a,b]—see also § 7.1.1 and references therein. Here, a generic, Marcelin-De Donder [Lengyel, 1989b, p. 577], representation of the nonlinear kinetics is provided for use within the multi-fluid formalism. Moreover, the application of the bracket formalism to a general reactive multi-fluid medium clarifies the description of entropy exchange between the reactive subsystems following the mass, momentum, and energy exchanges accompanying a general reaction network.

The addition of the reactions into the modeling, under the most general theoretical framework which can be accommodated using continuum concepts, completes the analysis of complex systems with an internal microstructure by demonstrating the capability of the formalism to accommodate phenomena characterized by nonlinear dissipation. The analysis of nonlinear dissipation processes is still, as far as we know, limited to chemical kinetics. As such, the incorporation of chemical reaction into a theory of nonequilibrium thermodynamics has consistently been used in the past in order to test that theory's suitability to handle nonlinear phenomena—see §7.1.1. Thus, the results presented in this chapter can also be considered as yet another (successful) test of the bracket theory. We certainly hope to see additional contributions of this nature, leading to the understanding of new concepts.

The introduction, in §12.1, offers background information on the modeling of weakly ionized, low-temperature plasmas and multi-fluid reactive media. In order to introduce the formalism for a multi-fluid system as smoothly as possible, as well as to familiarize the reader with the notation, the evolution equations for a non-dissipative, multi-fluid system are derived first via the bracket formalism in §12.2. The proper form of the dissipation bracket is addressed in §12.3, which extends the previous development to the non-reacting dissipative system of weakly interacting fluids. Section 12.4 introduces the bracket for reactions in single-fluid systems. The resulting equations are consistent with those commonly derived under the framework of irreversible thermodynamics. In §12.5, these equations are extended to the multi-fluid case, where deviation from thermal and mechanical equilibrium must be considered. There the formalism plays a significant role in identifying the form of the reaction driving force. In the last section of the chapter, §12.6, the continuum-level two-fluid model for weakly ionized plasmas commonly used in the simulation of glow discharges [Graves, 1986] is shown to be a limiting case of the more general equations.

12.1 Introduction

12.1.1 *Elements from the modeling of weakly ionized plasma dynamics*

The underlying physics at the microscopic level of plasmas is well understood (classical mechanics and electrodynamics [Nicholson, 1983]). However, descriptions of collective phenomena, occurring at a larger scale, still pose a challenge because of the large range of the electromagnetic interactions [Nicholson, 1983]. An exact description of the plasma dynamics is given by the well-known Boltzmann kinetic equation which governs the time evolution of the species energy distribution function, $f_i(\rho\mathbf{v}, \mathbf{r}, t)$, in the six-dimensional phase space and time [Chung *et al.*, 1975]. However, the solution of this equation is not generally tractable for other than a few simplified cases because of the high dimensionality and the need for information on the discrete particle interactions [Kline and Kushner, 1989, p. 2]. It appears that a Monte-Carlo particle simulation [Kushner, 1987] is computationally more efficient than a direct solution to the continuum Boltzmann equations, although even this method is still computationally limited to relatively simple plasma reaction environments.

In parallel, there has been considerable progress made toward simulating plasma-enhanced chemical processing through simplified continuum models. The first successful approximations involved continuum models with their transport coefficients evaluated from the solution of the Boltzmann kinetic equation under simplified assumptions. An example is the continuum approximation [Graves, 1986], which is derived by taking the moments of the Boltzmann kinetic equation (not unlike the approach used in the application of kinetic theory to viscoelastic modeling—see ch. 8) assuming a Maxwell/Boltzmann (thermal equilibrium) form for the electron energy distribution function (EEDF) [Chung *et al.*, 1975]. Alternatively, in continuum models one can assume that the EEDF is in equilibrium with the local electric field (the local field approximation) [Boeuf, 1988]. A similar approach was also followed by Gustafson [1991, ch. 5] in order to represent better the effects of a changing electric field on the EEDF and, correspondingly, the reaction rates in a weakly ionized, glow discharge plasma. The need to accommodate further changes in the EEDF has been demonstrated, for example, by Capitelli *et al.* [1987; 1988], who studied the effects of the various collision processes on the electron energy distribution. More recently, hybrid techniques have been used in non-homogeneous, multi-dimensional plasma simulations as, for example, the "beam-bulk" method for electron transport. According to this approach, the EEDF is represented as the superposition of two components: the beam containing energetic electrons which may have ballistic trajectories, and the bulk representing lower-energy electrons which can be described through a continuum fluid model [Pak and Kushner, 1990].

Thus, in a manner similar to the viscoelastic flow modeling, the most promising approach appears again to be one which attempts to capture the relevant microscopic phenomena in a continuum model. The latter can be used subsequently for the simulation of the macroscopic phenomena of interest at reasonable computational costs. We do not attempt to cover here either the microscopic foundations or the macroscopic simulation and analysis of plasma hydrodynamics, which would have taken us far away from the scope of this monograph. We rather refer the interested reader to the accumulated (already extensive) literature on these subjects [Woods, 1987; 1993; Kline and Kushner, 1989; Graves, 1989]. Instead, in the present chapter we want to examine the implications to the form of the continuum models that arise from the use of the bracket formalism. An additional objective is to demonstrate the flexibility of the latter formalism in describing the time evolution of yet another complex, structured medium.

Several continuum representations are, of course, available for the description of a multicomponent and reactive plasma, depending on the level of the microscopic information which is taken into account. In keeping with the didactic spirit of the presentation of the applications of the bracket formalism, we shall limit ourselves here to only one of the simpler continuum descriptions, that of a multi-fluid reactive continuum. Under this description, each one of the various components (i.e., electrons, ions, neutral species) in a low-density plasma is assumed to have reached, individually, mechanical and thermal equilibrium, although this might not necessarily be the case for the system as a whole. Thus, to each species it corresponds a separate set of dynamic and thermal variables—velocity and temperature, respectively—which might have different values between different species even at the same physical location. This approach has been proven suitable for the description of weakly ionized plasmas in the past [Woods, ch. 10; Graves, 1986]. In addition, we also allow for a complex system of chemical reactions.

A multi-fluid model implies that the components in a mixture interact only weakly [Bataille and Kestin, 1977]. Specifically, we are interested in the case when the constituent fluids need not be in mechanical and thermal equilibrium (i.e., the fluids have different momenta and temperatures). This state can be achieved when the characteristic times of momentum and energy transfer are on the same order or longer than the characteristic time of the macroscopic process of interest. Low-temperature, weakly ionized plasmas, under conditions when the Lorentz approximation is valid (i.e., when the elastic collisions between electrons and neutrals dominate), are the primary example of such a system. Indeed, the fractional energy transfer from species i to species j during a binary elastic collision with the velocities of the colliding particles at angle θ is given by [Chapman, 1980, pp. 11-13]

$$\frac{E_j}{E_i} = \frac{4M_iM_j}{\left(M_i+M_j\right)^2}\cos^2\theta\,, \qquad (12.1\text{-}1)$$

where M_i represents the atomic mass of species i. For an ion-electron plasma where $M_{ion}/M_e > 2000$, the maximum fractional energy transfer is less than 10^{-3}. Thus, many collisions are required to transfer a significant fraction of the energy from the high-energy electrons to the heavy ions. In contrast, inter-species collisions (the masses between the colliding particles being equal) are very efficient in establishing thermal and momentum equilibration between the populations for each species. These arguments can be further substantiated from a more detailed kinetic theory analysis [Woods, 1993]. For example, for a hydrogen plasma, Woods [1993, p. 230] has shown that thermal equlibrium between the electrons is established 43 times faster than between the ions and 918 times faster than between the ions and the electrons. Therefore, both of the assumptions for the validity of the multi-fluid approach are satisfied.

12.1.2 *Multi-fluid descriptions of nonlinear kinetics*

Continuum models of plasma systems are a special case of the more general multi-fluid models. The functional form of the governing equations for multi-fluid reacting systems has been well established by phenomenological development under the framework of irreversible thermodynamics [Dunwoody and Müller, 1968; Bataille and Kestin, 1977; Woods, 1987]—see also [Bedford and Drumheller, 1983, §1a] for a historical presentation of the subject. However, special issues, such as the effects of deviations from thermal and mechanical equilibrium on the driving force for chemical reactions, continue to attract interest in the literature [Bataille *et al.*, 1978; Morro and Romeo, 1988; Lengyel, 1989b; Kurzyński, 1990].

For homogeneous, single-fluid systems with no flow, irreversible thermodynamics yields a functional form for the driving force for chemical reactions which reduces to a linear form near equilibrium [Bataille *et al.*, 1978]—see also pertinent discussion on the extensions of the Onsager reciprocal relations to nonlinear dissipation expressions in §7.1.1. The thermodynamic driving force for chemical reaction has also been shown to be valid far from chemical equilibrium [Kurzyński, 1990]. More recently, work has been pursued to extend this development to systems in which the species exhibit different temperatures and momenta [Morro and Romeo, 1988; Keck, 1990]; however, the resulting equations are phenomenologically based. The methods are typically based on proposing forms of the mass, momentum, and total energy conservation equations, and then, afterwards, imposing the requirement on the rate of entropy production. The derivation of multi-fluid conservation equations with chemical reactions represents one of the outstanding areas of research being pursued in the field of irreversible thermodynamics.

The description of the model equations is presented here in a general form for an arbitrary, multi-fluid reactive system, thus generalizing, in a thermodynamically consistent fashion, the individual component transport equations—presented originally in §7.3 for a homogeneous, multicomponent system—to account for chemical reactions and individual species' thermal and mechanical equilibria. One of the outcomes of the use of the bracket formalism is that the assumptions necessary to arrive at the multi-fluid model equations are made evident. The corresponding bracket explicitly accounts for entropy production and entropy exchange due to chemical reactions between the various subsystems. The formalism also provides the natural definition of the thermodynamic driving forces, such as the rate of chemical reactions, in systems which are not in thermal or mechanical equilibrium. The occurrence of multi-particle reactions is accounted for through a sequence of nonlinear dissipation terms corresponding to binary, ternary, quaternary, etc., collision processes. Each one of these effects is explicitly associated with a corresponding exchange between the participating species' populations of mass, momentum, and energy.

12.2 The Non-Dissipative Multi-Fluid System

The Hamiltonian structure of conservative particle systems such as collisionless plasmas is well established. This structure can be exploited by applying a canonical Poisson bracket to derive the evolution equations for the canonical variables [Marsden and Weinstein, 1982, §4,5]. However, for many systems of interest, the canonical variables are difficult to identify and are not the variables of physical interest. The resolution of this conflict was realized about ten years ago when Morrison and Greene [1980, Eq. (9)] introduced a non-canonical Poisson bracket in terms of the physical variables of the equations of hydrodynamics and ideal magnetohydrodynamics. This method was later applied to other non-dissipative plasma systems. The Maxwell/Vlasov and Maxwell/Poisson equations were considered by Morrison [1980;[2] and

[2]The expression for the Poisson bracket originally developed in that paper has been shown not to satisfy the Jacobi identity [Weinstein and Morrison, 1981] and has been corrected by evaluating it using reduction methods from its canonical form [Marsden and Weinstein, 1982, Eq. (7.1)].

1982], Weinstein and Morrison [1981], and Marsden and Weinstein [1982]. The reduced magnetohydrodynamics equations were developed within the Poisson bracket formalism by Morrison and Hazeltine [1984]. The Hamiltonian structure of the non-dissipative, two-fluid approximation was considered by Spencer and Kaufman [1982] and Spencer [1982; 1984]. The Hamiltonian formulation has also found considerable use in non-linear stability analyses of fluid and plasma systems [Holm *et al.*, 1985].

The bracket approach will be illustrated through the development of the model equations for non-dissipative, multi-fluid systems in the presence of electromagnetic fields. These systems have been studied previously [Spencer, 1984] and have a known Hamiltonian structure. The current derivation differs from that of Spencer [1984] in that, following Morrison and Greene [1980, p. 792] and the general development of the bracket in this book, the entropy density is used in the formulation instead of the specific entropy (entropy per unit mass).

The assumption of interpenetrating, continuous fluids is evident in the choice of primary variables. The primary system variables are the momentum, mass, and entropy densities of each species, ρ_i, $\mathbf{m}_i$, and s_i respectively, and the electric and magnetic fields, **E** and **B**, respectively. The operating space, P, for these primary variables is thus given by

$$P \equiv \left\{ \begin{array}{l} \mathbf{m}^i(\mathbf{x},t) : \mathbf{m}^i \in \mathbb{R}^3;\ \mathbf{m}^i = \mathbf{0} \text{ on } \partial\Omega,\ i=1,2,\ldots,n;\ \sum_{i=1}^{n} \mathbf{m}^i = \mathbf{m} \\ \rho_i(\mathbf{x},t) : \rho_i \in \mathbb{R}^+;\ i=1,2,\ldots,n;\ \sum_{i=1}^{n} \rho_i = \rho \\ s_i(\mathbf{x},t) : s_i \in \mathbb{R};\ i=1,2,\ldots,n;\ \sum_{i=1}^{n} s_i = s \\ \mathbf{E}(\mathbf{x},t) : \mathbf{E} \in \mathbb{R}^3 \\ \mathbf{B}(\mathbf{x},t) : \mathbf{B} \in \mathbb{R}^3 \end{array} \right\}, \tag{12.2-1}$$

where, ρ, $\mathbf{m}$, and s are the mass, momentum, and entropy density of the total system. In contrast, in the governing equations for the single-fluid multicomponent model, previously derived using the bracket formulation (see §7.3), the primary system variables are the individual species mass densities and the total fluid momentum and entropy densities.

The Hamiltonian for the multi-fluid system (i.e., the total energy of the system in terms of the primary variables) is given by[3]

[3]Where, for simplicity, suitable units for **E** and **B** have been employed—see also Marsden and Weinstein [1982, Eq. (1.7)].

$$H[\mathbf{m}^s,\rho_s,s_s,\mathbf{E},\mathbf{B}] = \int_\Omega \left[\sum_{i=1}^{n} \frac{1}{2}\frac{|\mathbf{m}^i|^2}{\rho_i} + \sum_{i=1}^{n} \rho_i e_p^i + u + \frac{1}{2}\left(|\mathbf{E}|^2 + |\mathbf{B}|^2\right) \right] \mathrm{d}^3x \ . \tag{12.2-2}$$

The subscript/superscript s on the left-hand side implies the vector array i=1,2,...,n, where n is the number of fluids. The first term under the volume integral is the kinetic energy of each fluid. The second term is the potential energy of each fluid, where e_p^i is the potential acting on each species as a result of body forces other than the electric and magnetic fields. The third term is the total system internal energy, which is a function of the individual fluid mass and entropy densities:

$$u = u(\rho_s, s_s) = \sum_{i=1}^{n} u_i(\rho_i, s_i) \ . \tag{12.2-3}$$

The final term in the Hamiltonian is the energy of the electric and magnetic fields.

The Poisson bracket, equivalent to that of Spencer and Kaufman [1982], for this non-dissipative system in terms of the arbitrary functionals F and G is given by

$$\begin{aligned}
\{F,G\} = &-\sum_{i=1}^{n} \int_\Omega \left[\frac{\delta F}{\delta \rho_i} \nabla_\beta \left(\frac{\delta G}{\delta m_\beta^i} \rho_i \right) - \frac{\delta G}{\delta \rho_i} \nabla_\beta \left(\frac{\delta F}{\delta m_\beta^i} \rho_i \right) \right] \mathrm{d}^3x \\
&-\sum_{i=1}^{n} \int_\Omega \left[\frac{\delta F}{\delta m_\alpha^i} \nabla_\beta \left(\frac{\delta g}{\delta m_\beta^i} m_{\alpha i} \right) - \frac{\delta G}{\delta m_\alpha^i} \nabla_\beta \left(\frac{\delta F}{\delta m_\beta^i} m_\alpha^i \right) \right] \mathrm{d}^3x \\
&-\sum_{i=1}^{n} \int_\Omega \left[\frac{\delta F}{\delta s_i} \nabla_\beta \left(\frac{\delta G}{\delta m_\beta^i} s_i \right) - \frac{\delta G}{\delta s_i} \nabla_\beta \left(\frac{\delta F}{\delta m_\beta^i} s_i \right) \right] \mathrm{d}^3x \\
&+\int_\Omega \left[\varepsilon_{\alpha\beta\gamma} \frac{\delta F}{\delta E_\alpha} \nabla_\beta \frac{\delta G}{\delta B_\gamma} - \varepsilon_{\alpha\beta\gamma} \frac{\delta G}{\delta E_\alpha} \nabla_\beta \frac{\delta F}{\delta B_\gamma} \right] \mathrm{d}^3x \\
&+\sum_{i=1}^{n} \int_\Omega a_i \rho_i \left(\frac{\delta F}{\delta m_\alpha^i} \frac{\delta G}{\delta E_\alpha} - \frac{\delta G}{\delta m_\alpha^i} \frac{\delta f}{\delta E_\alpha} \right) \mathrm{d}^3x \\
&+\sum_{i=1}^{n} \int_\Omega a_i \rho_i \varepsilon_{\alpha\beta\gamma} B_\alpha \frac{\delta F}{\delta m_\beta^i} \frac{\delta G}{m_\gamma^i} \mathrm{d}^3x \ ,
\end{aligned} \tag{12.2-4}$$

where the Einstein summation convention is used over the Cartesian spatial coordinates indicated by Greek subscripts. Each term in the bracket describes a coupling between primary variables. The first three terms give the non-dissipative coupling of the species mass, momentum, and entropy densities with the species momentum. The fourth term represents the coupling between the electric and magnetic fields and results in the approximate Maxwell equations valid for non-relativistic velocities. In general, the Maxwell equations defined in the (relativistic) four-dimensional space (i.e., three spatial dimensions and time) the dielectric properties of the vacuum determine the conformal properties of space-time whereas the electromagnetic field determines the symplectic structure of the eight-dimensional space time for a relativistic particle [Guillemin and Sternberg, 1984, §20]—see also §2.6. The fifth and sixth terms represent the coupling between the particle motions and the electric and magnetic fields, respectively. The coefficient, a_i, is the charge of the species in fluid i.

For the non-dissipative system, the dynamical equation for the arbitrary functional F is given by

$$\frac{\mathrm{d}F}{\mathrm{d}t} = \{F,H\} , \tag{12.2-5}$$

where the Hamiltonian, H, and the Poisson bracket, { , }, are as defined above. Following the usual procedure of substituting the Hamiltonian, Eq. (12.2-2), into Eq. (12.2-5) and matching terms with those resulting from a direct differentiation in time of the functional expression, the evolution equations for the primary system variables are recovered:

$$\frac{\partial \rho_i}{\partial t} + \nabla_\alpha(\rho_i \upsilon^i_\alpha) = 0 , \tag{12.2-6}$$

$$\frac{\partial s_i}{\partial t} + \nabla_\alpha(s_i \upsilon^i_\alpha) = 0 , \tag{12.2-7}$$

$$\rho_i \frac{\partial \upsilon^i_\alpha}{\partial t} + \rho_i \upsilon^i_\beta \nabla_\beta \upsilon^i_\alpha = a_i \rho_i \left(E_\alpha + \varepsilon_{\alpha\beta\gamma}\delta_\alpha \upsilon^i_\beta B_\gamma\right) - \nabla_\alpha p_i , \tag{12.2-8}$$

$$\begin{aligned} \nabla_\alpha E_\alpha &= \sum_{i=1}^{n} a_s \rho_s , \\ \varepsilon_{\alpha\beta\gamma}\delta_\alpha \nabla_\beta B_\gamma &= \sum_{i=1}^{n} a_i \rho_i \upsilon^i_\alpha + \frac{\partial E_\alpha}{\partial t} , \\ \varepsilon_{\alpha\beta\gamma}\delta_\alpha \nabla_\beta B_\gamma &= -\frac{\partial B_\alpha}{\partial t} , \\ \nabla_\alpha B_\alpha &= 0 . \end{aligned} \tag{12.2-9}$$

Eqs. (12.2-6,7) are species mass and entropy continuity, Eq. (12.2-8) is species momentum conservation, where all body forces other than the electric and magnetic fields have been neglected, and the four relations in Eq. (12.2-9) are the Maxwell equations for the evolution of the fields, the first and fourth of which are constants of the system [Morrison, 1980] and therefore result from the gauge invariance of electromagnetism [Marsden and Weinstein, 1982] and not the generalized bracket. The partial pressures, p_i, are given by

$$p_i = \rho_i \frac{\partial u_i}{\partial \rho_i} + s_i \frac{\partial u_i}{\partial s_i} - u_i \ . \tag{12.2-10}$$

Note that Eqs. (12.2-6,7,8) are defined for each fluid (i.e., species) and are easily identified as the evolution equations for a single ideal fluid applied to each of the fluids. Since there is no dissipation, the fluids can only interact through the fields, and thus each fluid must obey the single-fluid equations.

The above analysis simplifies significantly for a Coulomb plasma, when the magnetic field is negligible (**B**=0) and the electric field can be defined in terms of the gradient of a scalar potential, ϕ:

$$E_\alpha = -\nabla_\alpha \phi \ . \tag{12.2-11}$$

The Maxwell equations reduce to the Poisson equation for the scalar electric potential,

$$\nabla_\alpha \nabla_\alpha \phi = -\sum_{i=1}^{n} a_i \rho_i \ , \tag{12.2-12}$$

and the operating space no longer includes the fields. The Hamiltonian reduces to

$$H[\mathbf{m}^s, \rho_s, s_s] = \int_\Omega \left[\sum_{i=1}^{n} \frac{1}{2} \frac{|\mathbf{m}^i|^2}{\rho_i} + \sum_{i=1}^{n} \rho_i e_p^i + u \right] d^3x \ , \tag{12.2-13}$$

where the potential, e_p^i, acting on each fluid is given by

$$e_p^i = -g_\alpha x_\alpha + a_i \phi_i \ , \tag{12.2-14}$$

when gravity is the only body force other than the electric field. Finally, since the scalar electric potential is included in the general potential and is defined by the local charge density, the coupling of the fields to the other primary variable is not included in the Poisson bracket:

$$\{F,G\} = -\sum_{i=1}^{n} \int_\Omega \left[\frac{\delta F}{\delta \rho_i} \nabla_\beta \left(\frac{\delta G}{\delta m_\beta^i} \rho_i \right) - \frac{\delta G}{\delta \rho_i} \nabla_\beta \left(\frac{\delta F}{\delta m_\beta^i} \rho_i \right) \right] d^3x$$

$$-\sum_{i=1}^{n}\int_{\Omega}\left[\frac{\delta F}{\delta m_{\alpha}^{i}}\nabla_{\beta}\left(\frac{\delta g}{\delta m_{\beta}^{i}}m_{\alpha}^{i}\right)-\frac{\delta G}{\delta m_{\alpha}^{i}}\nabla_{\beta}\left(\frac{\delta F}{\delta m_{\beta}^{i}}m_{\alpha}^{i}\right)\right]d^{3}x$$

$$-\sum_{i=1}^{n}\int_{\Omega}\left[\frac{\delta F}{\delta s_{i}}\nabla_{\beta}\left(\frac{\delta G}{\delta m_{\beta}^{i}}s_{i}\right)-\frac{\delta G}{\delta s_{i}}\nabla_{\beta}\left(\frac{\delta F}{\delta m_{\beta}^{i}}s_{i}\right)\right]d^{3}x\ . \tag{12.2-15}$$

The form of this bracket is characteristic of the classical Poisson structure.

The only resulting changes to the evolution equations for this reduced case are the elimination of the magnetic field terms and the substitution of Eq. (12.2-11) for the electric field in terms of the scalar electric potential. For simplicity, only the reduced case in terms of the electric potential (or a general scalar potential) will be considered further. This should not restrict the generality of the results since dissipation is imposed on the particle motions and not on the fields or the coupling of the fields to the particle motion directly.

The time evolution of the internal energy, u_i, of each fluid can be found by its total derivative,

$$\frac{du_i}{dt}=\frac{\partial u_i}{\partial \rho_i}\frac{d\rho_i}{dt}+\frac{\partial u_i}{\partial s_i}\frac{ds_i}{dt}\ , \tag{12.2-16}$$

the definition of the partial pressure, Eq. (12.2-10), and Eqs. (12.2-6,7) for the time evolution of the fluid mass and entropy densities:

$$\frac{\partial u_i}{\partial t}=-\nabla_{\alpha}(\upsilon_{\alpha}^{i}u_i)-p_i\nabla_{\alpha}\upsilon_{\alpha}^{i}\ . \tag{12.2-17}$$

The evolution equation for the kinetic energy of each fluid can be found by dotting the momentum equation, Eq. (12.2-8), with the species velocity, $\boldsymbol{\upsilon}^i$:

$$\rho_i\frac{\partial(\frac{1}{2}|\boldsymbol{\upsilon}^i|^2)}{\partial t}=-\rho_i\upsilon_{\beta}^{i}\nabla_{\beta}(\tfrac{1}{2}|\boldsymbol{\upsilon}^i|^2)-\nabla_{\alpha}(p_i\upsilon_{\alpha}^{i})+p_i\nabla_{\alpha}\upsilon_{\alpha}^{i}-a_i\rho_i\upsilon_{\alpha}^{i}\nabla_{\alpha}e_p^{i}\ . \tag{12.2-18}$$

Finally, by summing the two energy equations, we arrive at the evolution equation for the total energy, $e_i = u_i + ½\rho_i|\boldsymbol{\upsilon}^i|^2$:

$$\frac{\partial e_i}{\partial t}=-\nabla_{\alpha}(\upsilon_{\alpha}^{i}e_i)-\nabla_{\alpha}(p_i\upsilon_{\alpha}^{i})-a_i\rho_i\upsilon_{\alpha}^{i}\nabla_{\alpha}e_p^{i}\ . \tag{12.2-19}$$

These results again match the results of a single ideal fluid applied to each fluid individually, §7.2, and the results of Spencer and Kaufman [1982] if the entropy density is replaced by $s_i = \rho_i\hat{s}_i$, where $\hat{s}_i$ is the specific entropy.

12.3 The Dissipative Multi-Fluid System

We are now at a point to consider dissipation. For single-fluid systems, the most general form of the dissipative bracket is given by Eq. (7.1-19)

$$[F,G] = \int_\Omega \Xi\left(L\left[\frac{\delta F}{\delta \boldsymbol{\omega}}, \boldsymbol{\nabla}\frac{\delta F}{\delta \mathbf{w}}\right]; \frac{\delta G}{\delta \mathbf{w}}, \boldsymbol{\nabla}\frac{\delta G}{\delta \mathbf{w}}\right) \mathrm{d}^3x$$

$$- \int_\Omega \frac{\frac{\delta F}{\delta s}}{\frac{\delta G}{\delta s}} \Xi\left(L\left[\frac{\delta G}{\delta \boldsymbol{\omega}}, \boldsymbol{\nabla}\frac{\delta G}{\delta \mathbf{w}}\right]; \frac{\delta G}{\delta \mathbf{w}}, \boldsymbol{\nabla}\frac{\delta G}{\delta \mathbf{w}}\right) \mathrm{d}^3x \;, \tag{12.3-1}$$

where the vectors of the variables $\mathbf{w}$ and $\boldsymbol{\omega}$ are defined after Eq. (7.1-15). In order to generalize this expression to multi-fluid systems, we must first define the assumptions that are implied herein.

A weakly interacting multi-fluid system refers to the case where the characteristic times of momentum and energy transfer are on the same order or longer than the characteristic time of the macroscopic process of interest. Under these conditions, the mixture may also be assumed to be simple [Bataille and Kestin, 1977], which implies that the partial quantities sum to the whole fluid values,

$$u = \sum_{i=1}^{n} u_i(\rho_i, s_i), \quad s = \sum_{i=1}^{n} s_i \;, \tag{12.3-2}$$

and that the individual fluid chemical potential and partial pressure are well defined,

$$\frac{\partial u_i}{\partial \rho_i} = \mu_i \;, \quad \mathbf{P}_{i_{eq}} = p_i \mathbf{I} \;. \tag{12.3-3}$$

These conditions are necessary in order to write partial energy and entropy balances for each of the fluids. Note that the assumption of weak interaction is exact in the case of a non-dissipative system, as presented in the first part of the chapter, Eq. (12.2-2).

The generalized bracket can be extended to multi-fluid systems by first writing a bracket which partitions the dissipation into component dissipation within each fluid, $[F,G]_i$, binary dissipation which results from binary interactions between fluids, $[F,G]_{ij}$ and $[F,G]_{ji}$, ternary dissipation which results from three-particle interactions, $[F,G]_{ijk}$ and permutations, etc.:

$$[F,G] = \sum_{i=1}^{n} [F,G]_i + \sum_{i=1}^{n}\sum_{j=1}^{n} [F,G]_{ij} + \sum_{i=1}^{n}\sum_{j=1}^{n}\sum_{k=1}^{n} [F,G]_{ijk} + \ldots \tag{12.3-4}$$

This dissipative bracket must satisfy requisite properties similar to the single-fluid case; $[H,H]=0$, $[S,H]\geq 0$, $[\wp,H]=0$, as outlined in §7.1.1.

Additionally, the multi-fluid bracket must satisfy constraints to ensure that energy and mass are conserved and that entropy is not decreased during inter-fluid interactions. Thus, for a multi-fluid system with binary interactions, the constraints $[H_i+H_j,H_i+H_j]_{ij}=0$, $[S_i+S_j,H_i+H_j]_{ij}\geq 0$ and $[\wp_i+\wp_j,H_i+H_j]_{ij}=0$ must also be imposed. Notice that the constraints $[H_i, H]=0$ and $[\wp_i,H]=0$ are not imposed since energy and mass exchange must be allowed between the fluids. The multi-fluid bracket must also reduce to the single-fluid bracket in the absence of other fluids, consistent with the principle that in the absence of other fluids each fluid must be governed by the single-fluid equations [Müller, 1968].

The component dissipations within each fluid, $[F,G]_i$, are described by dissipation functionals which are equivalent to the single-fluid case given by Eq. (7.1-19) applied to each fluid individually, such that $\mathbf{w}^i = (a,b,c,\ldots,\rho_i,\mathbf{M}^i,s_i)$:

$$[F,G]_i = \int_\Omega \Xi_i\left(L\left[\frac{\delta F}{\delta \boldsymbol{\omega}^i},\boldsymbol{\nabla}\frac{\delta F}{\delta \mathbf{w}^i}\right];\frac{\delta G}{\delta \mathbf{w}^i},\boldsymbol{\nabla}\frac{\delta G}{\delta \mathbf{w}^i}\right)\mathrm{d}^3x$$
$$-\int_\Omega \frac{\dfrac{\delta F}{\delta s_i}}{\dfrac{\delta G}{\delta s_i}}\,\Xi_i\left(L\left[\frac{\delta G}{\delta \boldsymbol{\omega}^i},\boldsymbol{\nabla}\frac{\delta G}{\delta \mathbf{w}^i}\right];\frac{\delta G}{\delta \mathbf{w}^i},\boldsymbol{\nabla}\frac{\delta G}{\delta \mathbf{w}^i}\right)\mathrm{d}^3x\ . \tag{12.3-5}$$

For binary and higher-order interactions, we must specifically account for the distribution of the entropy production resulting from fluid-fluid interactions among the fluids involved. This is in contrast to the single-fluid case, in which only a total system entropy is accounted for, and thus it is the only one affected by the entropy production associated with any one of the dissipation phenomena taken into account. Thus, when separate fluid entropies are taken into account, the dissipation bracket resulting from a direct generalization of the single-fluid dissipation bracket, Eq. (12.3-1), is too restrictive for our purposes. We need to use the most general form

$$[F,G]_{ij} = \int_\Omega \Xi_{ij}\left(L\left[\frac{\delta F}{\delta \mathbf{w}^i},\frac{\delta F}{\delta \mathbf{w}^j},\boldsymbol{\nabla}\frac{\delta F}{\delta \mathbf{w}^i},\boldsymbol{\nabla}\frac{\delta F}{\delta \mathbf{w}^j}\right];\frac{\delta G}{\delta \mathbf{w}},\boldsymbol{\nabla}\frac{\delta G}{\delta \mathbf{w}}\right)\mathrm{d}^3x\ , \tag{12.3-6}$$

where $\mathbf{w}$ represents the vector of all variables, in contrast to $\mathbf{w}^i$ which involves only the variables associated with fluid i. In Eq. (12.3-6), the contribution accounting for the Volterra derivatives with respect to the fluid entropies, s_i and s_j, is included with that of the rest of the variables in $\mathbf{w}^i$, since there are no separate entropy terms. This bracket is easily generalized to higher-order interactions by extending the linear dependence of the functional, Ξ_{ij}, to include properties of additional fluids, e.g., Ξ_{ijk}. Moreover, without loss of generality, we can assume that when two of the indices are the same, the bracket is identically equal to zero, i.e., $[F,G]_{ii}=0$, since that would have corresponded to lower-order interactions, which can be included in the lower-order terms. The resulting dissipation bracket, $[F,G]$, must of course satisfy the constraints ensuring mass and energy conservation and entropy production given in the preceding paragraph.

For the case of binary interaction, a form of the dissipation bracket can be written explicitly, showing the entropy production terms, by directly exploiting the physics of the binary interactions. Notice that the conservation constraints on the complete dissipation bracket imply that changes in mass and energy in fluid i must be balanced by changes in the rest of the fluids. Thus, if we assume that each interaction is independent, it is reasonable to seek a form for the component dissipation which will satisfy the mass, energy, and entropy constraints by itself. Thus, we can propose a particular form of (12.3-6) which involves a linear dependence on the difference between the functional derivative of fluids i and j plus an (asymmetric) entropy correction:

$$[F,G]_{ij} = \int_\Omega \Psi_{ij}\left(L\left[\frac{\delta F}{\delta \mathbf{w}^i} - \frac{\delta F}{\delta \mathbf{w}^j}, \boldsymbol{\nabla}\frac{\delta F}{\delta \mathbf{w}^i} - \boldsymbol{\nabla}\frac{\delta F}{\delta \mathbf{w}^j}\right]; \frac{\delta G}{\delta \mathbf{w}}, \boldsymbol{\nabla}\frac{\delta G}{\delta \mathbf{w}}\right) d^3x$$

$$+ \int_\Omega \frac{\dfrac{\delta F}{\delta s_i}}{\dfrac{\delta G}{\delta s_i}} \Psi_{ij}\left(L\left[\frac{\delta G}{\delta \mathbf{w}^i} - \frac{\delta G}{\delta \mathbf{w}^j}, \boldsymbol{\nabla}\frac{\delta G}{\delta \mathbf{w}^i} - \boldsymbol{\nabla}\frac{\delta G}{\delta \mathbf{w}^j}\right]; \frac{\delta G}{\delta \mathbf{w}}, \boldsymbol{\nabla}\frac{\delta G}{\delta \mathbf{w}}\right) d^3x \,. \tag{12.3-7}$$

A dependence on the full set of system variables, $\mathbf{w}^i$, is retained inside the linear operator in Eq. (12.3-7), in order to preserve the symmetry of the bracket. Note that the bracket is by necessity asymmetric with respect to i,j since an entropy correction proportional to $(\delta F/\delta s_i - \delta F/\delta s_j)$ would have had no effect on the total entropy, i.e., for $F=S$. The entropy production in each fluid consists of a contribution from convection of the primary variables from one fluid to the other (calculated from the first term of the right-hand side of 12.3-7) and a contribution from the distribution of the entropy production among the fluids (calculated from the second term of the right-hand side of 12.3-7). As a result of the spatial structure of Eq. (12.3-7), the second contribution to the entropy production from $[F,G]_{ij}$ is non-zero only for fluid i. However, consideration of both terms $[F,G]_{ij}$ and $[F,G]_{ji}$ allows for the description of the most general distribution of the entropy production among fluids i,j arising as a result of a binary interaction. The asymmetric character of the interaction bracket becomes handy in the description of chemical reactions, as seen below.

In all cases considered, except that of chemical reactions, the system will be close enough to equilibrium such that the functionals, Ξ_{ij} and Ψ_{ij}, may be considered to be bilinear with respect to G and F. In particular, if Ψ_{ij} is taken to be bilinear with respect to $(\delta G/\delta \mathbf{w}^i - \delta G/\delta \mathbf{w}^j)$ and $(\delta F/\delta \mathbf{w}^i - \delta F/\delta \mathbf{w}^j)$, it is easy to show that Onsager's reciprocal relations are automatically obeyed. Furthermore, note that the form of Eq. (12.3-7) explicitly accounts for the distribution of the dissipation among the fluids.

For the remainder of this chapter, we shall neglect the higher-order dissipations caused by interaction between more than two fluids and by differences in the gradients of the system variables of the interacting fluids. The bracket defined by Eq. (12.3-7) will be used for the dissipation from binary fluid interactions. The dissipative bracket for a multi-fluid system will now be considered term by term, each of which accounts for a different dissipative mechanism. In the current formulation, each fluid is considered to be a single species. Thus, the mass densities of the fluids are non-dissipative variables (excluding reactions). Only dissipation of the momentum and the entropy of each species, which will be split into internal dissipation, due to gradients in the primary variables in each fluid, and external dissipation, due to the fluids having different thermodynamic states, will be considered.

The internal momentum dissipation, $[F,G]_i^m$, for each fluid is analogous to single-fluid viscous losses and results from gradients in the velocity of each fluid. Thus, this dissipation can be represented for each fluid by terms similar to those previously derived for viscous loss in a single fluid. Hence,

$$[F,G]_i^m = -\int_\Omega \Phi^i_{\alpha\beta\gamma\varepsilon} \nabla_\alpha \frac{\delta F}{\delta m_\beta^i} \nabla_\gamma \frac{\delta G}{\delta m_\varepsilon^i} d^3x + \int_\Omega \Phi^i_{\alpha\beta\gamma\varepsilon} \frac{\dfrac{\delta F}{\delta s_i}}{\dfrac{\delta G}{\delta s_i}} \nabla_\alpha \frac{\delta G}{\delta m_\beta^i} \nabla_\gamma \frac{\delta G}{\delta m_\varepsilon^i} d^3x \;, \tag{12.3-8}$$

where the phenomenological coefficients, $\Phi^i_{\alpha\beta\gamma\varepsilon}$, are defined for an isotropic fluid as

$$\Phi^i_{\alpha\beta\gamma\varepsilon} = \eta_i \delta_{\alpha\gamma}\delta_{\beta\varepsilon} + \eta_i \delta_{\alpha\varepsilon}\delta_{\beta\gamma} + (\kappa_i - \tfrac{2}{3}\eta_i)\delta_{\alpha\beta}\delta_{\gamma\varepsilon} \;, \tag{12.3-9}$$

where η_i and κ_i are the shear and bulk viscosities of fluid i, respectively—see §7.2.2.

The external momentum dissipation, $[F,G]_{ij}^m$, accounts for the relaxation that must occur when the particles of two gradient-free fluids of differing momentum density are interacting with each other. This term involves the coupling between the momentum of the different fluids and is of the form given by equation (12.3-7), which explicitly takes the conservation of the primary variables into account:

$$[F,G]^m_{ij} = -\int_\Omega \Lambda^{ij}_{\gamma\beta}\left(\frac{\delta F}{\delta m^i_\gamma} - \frac{\delta F}{\delta m^j_\gamma}\right)\left(\frac{\delta G}{\delta m^i_\beta} - \frac{\delta G}{\delta m^j_\beta}\right)d^3x$$

$$+\int_\Omega \Lambda^{ij}_{\gamma\beta}\frac{\dfrac{\delta F}{\delta s_i}}{\dfrac{\delta G}{\delta s_i}}\left(\frac{\delta G}{\delta m^i_\gamma} - \frac{\delta G}{\delta m^j_\gamma}\right)\left(\frac{\delta G}{\delta m^i_\beta} - \frac{\delta G}{\delta m^j_\beta}\right)d^3x \; . \tag{12.3-10}$$

Note that this is a particular form of dissipation due to internal relaxation within the system, similar to that postulated for the conformation tensor of viscoelastic fluids by Grmela [1988, p. 85] and Beris and Edwards [1990b, p. 69]—see also ch. 8. In Eq. (12.3-10), the phenomenological coefficients, $\Lambda^{ij}_{\gamma\beta}$, are related to the collision cross section for momentum transfer between fluids i and j by the relationship

$$\Lambda^{ij}_{\alpha\beta} + \Lambda^{ji}_{\alpha\beta} = n_i n_j m_{ij} \alpha_{ij} \delta_{\alpha\beta} \; , \tag{12.3-11}$$

where α_{ij} is the effective momentum transfer cross section, n_i and n_j are the number densities of species i and j, and m_{ij} is the reduced mass [Woods, 1987]. The effective momentum transfer cross section can be found by the relation

$$\alpha_{ij} = \frac{m_{ij}}{3T_{ij}} \int \upsilon^3 \sigma_{ij} f(\upsilon) d\upsilon \; , \tag{12.3-12}$$

where υ is the magnitude of the collision velocity, m_{ij} and T_{ij} are the reduced mass and temperature, respectively, and σ_{ij} is the integral cross section for momentum transfer [Braginskii, 1965]. The distribution of frictional dissipation between the fluids determines the relationship between $\Lambda^{ij}_{\alpha\beta}$ and $\Lambda^{ji}_{\alpha\beta}$. For example, the coefficients are related by the expression,

$$\Lambda^{ij}_{\alpha\beta} = \left(\Lambda^{ij}_{\alpha\beta} + \Lambda^{ji}_{\alpha\beta}\right)\left(\frac{m_j}{m_i + m_j}\right) , \tag{12.3-13}$$

if the thermal energy is distributed inversely proportionally to the particle mass (e.g., elastic collisions).

There is also internal and external thermal dissipation accounted for by single-species and binary-component entropy dissipation terms. In a similar manner to the momentum case, the single-species (internal) dissipation of the species entropy, $[F,G]^s_i$, arises from gradients within each fluid, and it is expressed in the same form as for the single-fluid case:

$$[F,G]_i^s = -\int_\Omega \Theta^i_{\gamma\beta} \nabla_\gamma \left(\frac{\delta F}{\delta s_i}\right) \nabla_\beta \left(\frac{\delta G}{\delta s_i}\right) d^3x$$

$$+ \int_\Omega \Theta^i_{\gamma\beta} \frac{\dfrac{\delta F}{\delta s_i}}{\dfrac{\delta G}{\delta s_i}} \nabla_\gamma \left(\frac{\delta G}{\delta s_i}\right) \nabla_\beta \left(\frac{\delta G}{\delta s_i}\right) d^3x \ . \tag{12.3-14}$$

The phenomenological coefficients, $\Theta^i_{\gamma\beta}$, can be identified as proportional to the species thermal conductivity divided by the species temperature (see §7.2.2):

$$\Theta^i_{\gamma\beta} = \frac{k_i}{T_i} \delta_{\gamma\beta} \ . \tag{12.3-15}$$

A term in the dissipation bracket is also necessary to account for the exchange of thermal energy between two gradient-free fluids, with different temperatures, that are in contact with each other. This dissipation is given by a term similar to that for external momentum dissipation,

$$[F,G]_{ij}^s = -\int_\Omega \Pi_{ij} \left(\frac{\delta F}{\delta s_i} - \frac{\delta F}{\delta s_j}\right)\left(\frac{\delta G}{\delta s_i} - \frac{\delta G}{\delta s_j}\right) d^3x$$

$$+ \int_\Omega \Pi_{ij} \frac{\dfrac{\delta F}{\delta s_i}}{\dfrac{\delta G}{\delta s_i}} \left(\frac{\delta G}{\delta s_i} - \frac{\delta G}{\delta s_j}\right)\left(\frac{\delta G}{\delta s_i} - \frac{\delta G}{\delta s_j}\right) d^3x \ , \tag{12.3-16}$$

where the phenomenological coefficients, Π_{ij}, are related to the effectiveness of heat transfer between fluids i and j. They can be defined in terms of physical quantities when electrons and ions are species i and j, respectively, by

$$\left(\Pi_{ij} T_j + \Pi_{ji} T_i\right) = \frac{3m_i}{m_j} \frac{k_B n_i}{\tau_i} \ , \tag{12.3-17}$$

where τ_i is the characteristic collision time for the electrons (species i) [Woods, 1987, Eq. (12.113)].

By summing the above dissipative terms and adding them to the Poisson bracket, (12.2-15), and following the usual procedure, the (dissipative) evolution equations for the species mass, momentum, and entropy densities can be derived:

$$\frac{\partial \rho_i}{\partial t} = - \nabla_\beta(\rho_i \upsilon^i_\beta) \; , \tag{12.3-18}$$

$$\begin{aligned}\rho_i \frac{\partial \upsilon^i_\alpha}{\partial t} + \rho_i \upsilon^i_\beta \nabla_\beta \upsilon^i_\alpha &= -\rho_i \nabla_\alpha e_p{}^i + \nabla_\beta\left[\Phi^i_{\alpha\beta\gamma\varepsilon} \nabla_\gamma(\upsilon^i_\varepsilon)\right] \\ &\quad - \sum_{j=1}^{n} \left(\Lambda^{ij}_{\alpha\beta} + \Lambda^{ji}_{\alpha\beta}\right)(\upsilon^i_\beta - \upsilon^j_\beta) - \nabla_\alpha p_i \; ,\end{aligned} \tag{12.3-19}$$

$$\begin{aligned}\frac{\partial s_i}{\partial t} &= -\nabla_\beta(\upsilon^i_\beta s_i) + \Phi^i_{\alpha\beta\gamma\varepsilon} \frac{1}{T_i} \nabla_\alpha \upsilon^i_\beta \nabla_\gamma \upsilon^i_\varepsilon + \nabla_\gamma\left(\Theta^i_{\gamma\beta} \nabla_\beta T_i\right) \\ &\quad + \Theta^i_{\gamma\beta} \frac{1}{T_i} \nabla_\gamma T_i \nabla_\beta T_i + \sum_{j=1}^{n} \Lambda^{ij}_{\alpha\beta} \frac{1}{T_i}\left(\upsilon^i_\alpha - \upsilon^j_\alpha\right)(\upsilon^i_\beta - \upsilon^j_\beta) \\ &\quad - \sum_{j=1}^{n} \left(\Pi_{ij} + \Pi_{ji}\right)\left(T_i - T_j\right) + \sum_{j=1}^{n} \Pi_{ij} \frac{1}{T_i}\left(T_i - T_j\right)^2 .\end{aligned} \tag{12.3-20}$$

The partial pressures, p_i, are again given by equation (12.2-10), and with the Poisson equation, (12.2-12), and the definitions of the phenomenological coefficients, this is a closed system of equations.

Following the same procedure as for the non-dissipative case, we can find the evolution equations for the internal energy,

$$\begin{aligned}\frac{\partial u_i}{\partial t} &= -\nabla_\alpha\left(\upsilon^i_\alpha u_i\right) - p_i \nabla_\alpha \upsilon^i_\alpha + \phi^i_{\alpha\beta\gamma\varepsilon} \nabla_\alpha \upsilon^i_\beta \nabla_\gamma \upsilon^i_\varepsilon \\ &\quad + T_i \nabla_\gamma\left(\Theta^i_{\gamma\beta} \nabla_\beta T_i\right) + \Theta^i_{\gamma\beta} \nabla_\gamma T_i \nabla_\beta T_i \\ &\quad + \sum_{j=1}^{n} \Lambda^{ij}_{\alpha\beta}\left(\upsilon^i_\alpha - \upsilon^j_\alpha\right)\left(\upsilon^i_\beta - \upsilon^j_\beta\right) \\ &\quad - \sum_{j=1}^{n} \left(\Pi_{ij} + \Pi_{ji}\right) T_i \left(T_i - T_j\right) + \sum_{j=1}^{n} \Pi_{ij}\left(T_i - T_j\right)^2 \; ,\end{aligned} \tag{12.3-21}$$

and the kinetic energy,

$$\begin{aligned}\rho_i \frac{\partial(\frac{1}{2}|\boldsymbol{\upsilon}^i|^2)}{\partial t} &= -\rho_i \upsilon^i_\beta \nabla_\beta(\tfrac{1}{2}|\boldsymbol{\upsilon}^i|^2) - \nabla_\alpha(p_i \upsilon^i_\alpha) + p_i \nabla_\alpha \upsilon^i_\alpha - \rho_i \upsilon^i_\alpha \nabla_\alpha e_p{}^i \\ &\quad + \nabla_\beta\left(\phi^i_{\beta\alpha\gamma\varepsilon} \upsilon^i_\alpha \nabla_\gamma \upsilon^i_\varepsilon\right) - \phi_{\beta\alpha\gamma\varepsilon} \nabla_\beta \upsilon^i_\alpha \nabla_\gamma \upsilon^i_\varepsilon\end{aligned}$$

$$-\sum_{j=1}^{n} \upsilon_\alpha^i \left(\Lambda_{\alpha\beta}^{ij} + \Lambda_{\alpha\beta}^{ji}\right)\left(\upsilon_\beta^i - \upsilon_\beta^j\right) . \tag{12.3-22}$$

Finally, summing the two energy equations, we arrive at the evolution equation for the total energy of each fluid in a non-reacting dissipative system:

$$\begin{aligned}\frac{\partial e_i}{\partial t} &= -\nabla_\alpha(\upsilon_\alpha^i e_i) - \nabla_\alpha(p_i \upsilon_\alpha^i) - \rho_i \upsilon_\alpha^i \nabla_\alpha e_p^i \\ &+ \nabla_\beta\left(\phi_{\beta\alpha\gamma\varepsilon}^i \upsilon_\alpha^i \nabla_\gamma \upsilon_\varepsilon^i\right) + \nabla_\gamma\left(\Theta_{\gamma\beta}^i T_i \nabla_\beta T_i\right) \\ &+ \sum_{j=1}^{n} \Lambda_{\alpha\beta}^{ij}\left(\upsilon_\alpha^i - \upsilon_\alpha^j\right)\left(\upsilon_\beta^i - \upsilon_\beta^j\right) - \sum_{j=1}^{n} \upsilon_\alpha^i\left(\Lambda_{\alpha\beta}^{ij} + \Lambda_{\alpha\beta}^{ji}\right)\left(\upsilon_\beta^i - \upsilon_\beta^j\right) \\ &- \sum_{j=1}^{n}\left(\Pi_{ij} T_j + \Pi_{ji} T_i\right)\left(T_i - T_j\right) .\end{aligned} \tag{12.3-23}$$

The first three terms in equation (12.3-23) are the non-dissipative convection, expansion, and potential contributions to the rate of change of the total energy of each fluid. The fourth term accounts for the internal viscous dissipation. The fifth term is the dissipation from temperature gradients within each fluid. The fraction of the friction loss dissipated in the i-th fluid is given by the sixth term. The seventh term accounts for the energy exchange due to momentum transfer from the other fluids to the i-th fluid. Finally, the last term is the transfer of energy to the i-th fluid from the other fluids, which are at temperatures different than that of the i-th fluid.

The final set of equations, (12.3-18), (12.3-19), and (12.3-23), are equivalent to those derived via irreversible thermodynamics [Woods, 1987; Bataille and Kestin, 1977] for a thermodynamically consistent, weakly interacting, multi-fluid model, neglecting chemical reactions. In addition, the formulation exhibits the same phenomenology as the macroscopic moment equations derived from the Boltzmann kinetic equation under the assumptions of no chemical reactions, a Maxwellian distribution for each species energy, and under conditions for which the self-collisions dominate as compared to collisions among different species [Goldman and Sirovich, 1967; Struminskii, 1974]. Thus, as indicated above, the phenomenological coefficients may be estimated by comparison with the kinetic theory results. We note that higher-order terms resulting from extending the standard perturbation solution of the Boltzmann kinetic equation do not appear in our formulation [Struminskii and Shavaliyev, 1987]. This is consistent with our assumptions of weakly interacting fluids and that the higher-order dissipation caused by differences in the gradients of the system variables of the interacting fluids are negligible.

12.4 Chemical Reactions in a Multicomponent Single-Fluid System

The net rate of production of species i by r independent, reversible chemical reactions is given by

$$\frac{\partial \rho_i}{\partial t}_{RXN} = \sum_{I=1}^{r} \left(\gamma_{Ii} M_i J_I\right) , \tag{12.4-1}$$

where γ_{Ii} is the stoichiometric coefficient for species i in reaction I, J_I is the flux or rate of reaction I, and M_i is the molecular weight of species i. The stoichiometric coefficients for each reaction, I, are constrained by mass conservation in each reaction,

$$\sum_{k=1}^{n} \gamma_{Ik} M_k = 0, \quad I = 1,2,\ldots,r , \tag{12.4-2}$$

such that the overall mass conservation,

$$\sum_{k=1}^{n} \frac{\partial \rho_k}{\partial t}_{RXN} = \sum_{k=1}^{n} \sum_{I=1}^{r} \gamma_{Ik} M_k J_I = 0 , \tag{12.4-3}$$

is trivially satisfied.

The flux of reaction I can be derived from irreversible thermodynamics for a multicomponent, single-fluid system [Bataille, *et al.*, 1978]:

$$J_I = k_I(P,T) \left\{ \exp\left(\frac{-A_I^-}{RT} \right) - \exp\left(\frac{-A_I^+}{RT} \right) \right\} . \tag{12.4-4}$$

The net reaction rate coefficient, k_I, is a function of the system pressure and temperature and is related to the forward and reverse reaction rate constants by

$$\begin{aligned} k_I(P,T) &= k_I^-(P,T) \exp\left(\sum_{k=1}^{n} \frac{\gamma_{Ik}^- M_i g_k(T)}{RT} \right) \\ &= k_I^+(P,T) \exp\left(\sum_{k=1}^{n} \frac{\gamma_{Ik}^+ M_i g_k(T)}{RT} \right) . \end{aligned} \tag{12.4-5}$$

In the above equation, $\gamma_{Ik}^{\pm}$, are the forward and reverse stoichiometric coefficients for species k in reaction I defined by

$$\gamma_{Ik}^{\pm} = \frac{1}{2}\left(\left|\gamma_{Ik}\right| \pm \gamma_{Ik}\right) \geq 0 \;, \tag{12.4-6}$$

and g_k is the pure species chemical potential at temperature T. The forward and reverse affinities, A_I^- and A_I^+, are defined by

$$A_I^{\pm} = -\sum_{k=1}^{n} \gamma_{Ik}^{\pm} M_k \mu_k \;, \tag{12.4-7}$$

where μ_k is the chemical potential of species k defined by

$$\mu_k = g_k(T) + RT\log(f_k) \;, \tag{12.4-8}$$

and f_k is the species fugacity. The reaction flux can be rewritten in terms of the total affinity, which is given by $A_I = A_I^+ - A_I^-$ as

$$J_I = k_I(P,T)\exp\left(\frac{-A_I^-}{RT}\right)\left\{1 - \exp\left(\frac{-A_I}{RT}\right)\right\} . \tag{12.4-9}$$

Close to equilibrium, the reaction flux reduces to a linear form,

$$J_I \approx \frac{k_I(P,T)}{RT}\exp\left(\frac{-A_I^-(eq)}{RT}\right)A_I \approx k_I' A_I \;, \tag{12.4-10}$$

since the total affinity, A_I, is approximately equal to zero. In addition, the nonlinear form of the reaction driving force, equation (12.4-9), has been shown to be valid far from chemical equilibrium [Kurzyński, 1990].

Except for the special case close to equilibrium, the driving force for chemical reaction is a nonlinear functional of the system variables. Thus, the nonlinear form of the dissipation bracket given earlier must be used to derive the dissipation terms for the r independent, reversible chemical reactions. The case of a multicomponent, single-fluid system, where all species are assumed to be in thermal and mechanical equilibrium (e.g., $T_i = T_j$ for all i and j) will be used as a starting point. The dissipation bracket can be defined as

$$[F,H]_i = -\sum_{I=1}^{r}\int_{\Omega} k_I \Gamma_{Ii} \frac{\dfrac{\delta F}{\delta \rho_i}}{R\dfrac{\delta H}{\delta s}} \exp\left(-\sum_{k=1}^{n} \frac{\gamma_{Ik}^- M_k}{R\dfrac{\delta H}{\delta s}}\frac{\delta H}{\delta \rho_k}\right)\left\{1 - \exp\left(-\sum_{k=1}^{n}\frac{\gamma_{Ik} M_k}{R\dfrac{\delta H}{\delta s}}\frac{\delta H}{\delta \rho_k}\right)\right\} d^3x$$

$$+\sum_{l=1}^{r}\int_{\Omega}k_l\Gamma_{li}\frac{\dfrac{\delta H}{\delta\rho_i}\dfrac{\delta F}{\delta s}}{R\dfrac{\delta H}{\delta s}\dfrac{\delta H}{\delta s}}\exp\left(-\sum_{k=1}^{n}\frac{\gamma_{lk}^{-}M_k}{R\dfrac{\delta H}{\delta s}}\frac{\delta H}{\delta\rho_k}\right)\left\{1-\exp\left(-\sum_{k=1}^{n}\frac{\gamma_{lk}M_k}{R\dfrac{\delta H}{\delta s}}\frac{\delta H}{\delta\rho_k}\right)\right\}d^3x\ ,$$

(12.4-11)

where s is the total system entropy density. The net reaction rate coefficient, k_l, is defined by equation (12.4-5). Near equilibrium, the derivatives of the Hamiltonian approach zero, $\delta H/\delta \mathbf{w}_i \to \mathbf{0}$, and thus, the bracket reduces to

$$[F,H]_i=-\sum_{l=1}^{r}\sum_{k=1}^{n}\int_{\Omega}k_l'\Gamma_{li}\frac{\dfrac{\delta F}{\delta\rho_i}}{R\dfrac{\delta H}{\delta s}}\Gamma_{lk}\frac{\dfrac{\delta H}{\delta\rho_k}}{R\dfrac{\delta H}{\delta s}}d^3x$$

(12.4-12)

$$+\sum_{l=1}^{r}\sum_{k=1}^{n}\int_{\Omega}k_l'\Gamma_{li}\frac{\dfrac{\delta F}{\delta s}}{\dfrac{\delta H}{\delta s}}\frac{\dfrac{\delta H}{\delta\rho_i}}{R\dfrac{\delta H}{\delta s}}\Gamma_{lk}\frac{\dfrac{\delta H}{\delta\rho_i}}{R\dfrac{\delta H}{\delta s}}d^3x\ ,$$

which is symmetric, quadratic, and bilinear as expected.

The time derivatives of the species densities and the total system entropy can be found by substituting the Hamiltonian for a multi-component single-fluid system, Eq. (7.3-4), into Eq. (12.4-11) and matching terms with the corresponding Eq. (7.1-20):

$$\left.\frac{\partial\rho_i}{\partial t}\right|_{RXT}=-\sum_{l=1}^{r}k_l\Gamma_{li}\exp\left(-\sum_{k=1}^{n}\frac{\gamma_{lk}^{-}M_k}{RT}\mu_k^*\right)\left\{1-\exp\left(-\sum_{k=1}^{n}\frac{\gamma_{lk}M_k}{RT}\mu_k^*\right)\right\}$$

$$=-\sum_{l=1}^{r}k_l\Gamma_{li}\exp\left(\frac{-A_l^{*-}}{RT}\right)\left\{1-\exp\left(\frac{-A_l^{*}}{RT}\right)\right\}\ , \qquad (12.4\text{-}13)$$

$$\left.\frac{\partial s}{\partial t}\right|_{RXN}=+\sum_{i=1}^{n}\sum_{l=1}^{r}k_l\Gamma_{li}\frac{\mu_i^*}{T}\exp\left(-\sum_{k=1}^{n}\frac{\gamma_{lk}^{-}M_k}{RT}\mu_k^*\right)\left\{1-\exp\left(-\sum_{k=1}^{n}\frac{\gamma_{lk}M_k}{RT}\mu_k^*\right)\right\}$$

$$= + \sum_{i=1}^{n}\sum_{I=1}^{r} k_I \Gamma_{Ii} \frac{\mu_i^*}{T} \exp\left(\frac{-A_I^{*-}}{RT}\right)\left\{1-\exp\left(\frac{-A_I^*}{RT}\right)\right\} = -\sum_{i=1}^{n} \frac{\mu_i^*}{T}\frac{\partial \rho_i}{\partial t}_{RXN} \; . \tag{12.4-14}$$

The generalized chemical potential, μ_k^*, is defined by Eq. (7.3-5d),

$$\mu_k^* = \frac{\delta H}{\delta \rho_k} = \frac{\partial u}{\partial \rho_k} - \tfrac{1}{2}\upsilon^2 + g_\alpha x_\alpha \, , \tag{12.4-15}$$

and the generalized affinities, $A_I^{*\pm}$, are defined by Eq. (12.4-7) above, in terms of the generalized chemical potentials. By comparing Eq. (12.4-13) to Eqs. (12.4-1) and (12.4-9), the empirical coefficient, $\Gamma_{\alpha i}$, can now be identified as the product of the species stoichiometric coefficient and the species molecular weight:

$$\Gamma_{Ii} = -\gamma_{Ii} M_i \; . \tag{12.4-16}$$

The right-hand side terms in Eqs. (12.4-13) and (12.4-14) can be added to the right-hand sides in Eqs. (7.3-11a) and (7.3-11c) to form the complete dissipative multicomponent transport equations for a reactive system. The momentum equation, (7.3-11b), remains unchanged since internal momentum cannot be created [Bataille and Kestin, 1977]. The chemical reaction dissipation is also found not to contribute to the internal and total energy balances. This is expected since kinetic energy and potential energy are conserved in a chemical reaction. Thus, in the absence of fluid motion and potential forces, the Hamiltonian reduces to the internal energy, which must be conserved.

12.5 Chemical Reactions in Multi-Fluid Systems

For a multi-fluid system where the system variables are the individual fluid mass, momentum, and entropy densities, the dissipation bracket for chemical reactions can be given by

$$[F,H]_{ij} = -\sum_{I=1}^{r}\int_\Omega \left[\left(\frac{\delta F}{\delta \rho_i} - \frac{\delta F}{\delta \rho_j}\right) + \frac{m_\alpha^j}{\rho_j}\left(\frac{\delta F}{\delta m_\alpha^i} - \frac{\delta F}{\delta m_\alpha^j}\right) + \frac{s_j}{\rho_j}\left(\frac{\delta F}{\delta s_i} - \frac{\delta F}{\delta s_j}\right)\right]\left[\Gamma_{Iij}^+ J_I^+ + \Gamma_{Iij}^- J_I^-\right] d^3x \tag{12.5-1}$$

$$+\sum_{I=1}^{r}\int_{\Omega}\frac{\dfrac{\delta F}{\delta s_i}}{\dfrac{\delta H}{\delta s_i}}\left[\left(\frac{\delta H}{\delta \rho_i}-\frac{\delta H}{\delta \rho_j}\right)+\frac{m_\alpha^j}{\rho_j}\left(\frac{\delta H}{\delta m_\alpha^i}-\frac{\delta H}{\delta m_\alpha^j}\right)\right. \tag{12.5-1}$$

$$\left.+\frac{s_j}{\rho_j}\left(\frac{\delta H}{\delta s_i}-\frac{\delta H}{\delta s_j}\right)\right]\left[\Gamma_{Iij}^{+}J_I^{\pm}+\Gamma_{Iij}^{-}J_I^{-}\right]\mathrm{d}^3x \ .$$

The phenomenological coefficient, $\Gamma_{Iij}^{\pm}$, is the stoichiometric coefficient for the production of species i in reaction I from species j. As before, the plus/minus notation denotes the forward (+) and reverse (–) reactions. The coefficient can be defined by the relation

$$\Gamma_{Iij}^{\pm} = -\gamma_{Ij}^{\mp}M_j\left[\frac{\gamma_{Ii}^{\pm}M_i}{\sum_{k=1}^{n}\gamma_{Ik}^{\pm}M_k}\right] . \tag{12.5-2}$$

The form of the dissipation bracket ensures that mass, momentum, and energy are conserved in each chemical reaction. Thus, the fluxes of the primary variables of species j must be equal to (but opposite in sign to) the fluxes of species i caused by the reaction driving forces, $J_I^{\pm}$. The reaction driving forces, $J_I^{\pm}$, are now defined for the forward and reverse reactions of the r independent, reversible reactions given above. This is necessary since, even at chemical equilibrium when the forward and reverse reactions rates are equal, momentum and energy will be transferred from one species to the other.

The most general form of the reaction driving force, $J_I^{\pm}$, is a nonlinear function of the system primary variables:

$$J_I^{\pm} = k_I\left\{1-\exp\left[-\sum_{l=1}^{n}\sum_{m=1}^{n}\left[\left(\Gamma_{Ilm}^{\pm}-\Gamma_{Iml}^{\pm}\right)\frac{\delta H}{\delta \rho_l}\right.\right.\right. \tag{12.5-3}$$

$$\left.\left.\left.+\left(\Gamma_{Ilm}^{\pm}\frac{m_\alpha^m}{\rho_m}-\Gamma_{Iml}^{\pm}\frac{m_\alpha^l}{\rho_l}\right)\frac{\delta H}{\delta m_\alpha^l}+\left(\Gamma_{Ilm}^{\pm}\frac{s_m}{\rho_m}-\Gamma_{Iml}^{\pm}\frac{s_l}{\rho_l}\right)\frac{\delta H}{\delta s_l}\right]\right]\right\} .$$

Note that the rate coefficient, k_I, is the same for the forward and reverse driving forces. This is consistent with Onsager's reciprocal relations. As in the single fluid case, this expression becomes linear near equilibrium, when $\delta H/\delta \mathbf{w}_i \to 0$,

$$J_I^{\pm}\Big|_{Eq} = k_I \sum_{l=1}^{n}\sum_{m=1}^{n} \left\{ \left(\Gamma_{Ilm}^{\pm} - \Gamma_{Iml}^{\pm}\right)\frac{\delta H}{\delta \rho_l} + \left(\Gamma_{Ilm}^{\pm}\frac{m_\alpha^m}{\rho_m} - \Gamma_{Iml}^{\pm}\frac{m_\alpha^l}{\rho_l}\right)\frac{\delta H}{\delta m_\alpha^l} \right.$$
$$\left. + \left(\Gamma_{Ilm}^{\pm}\frac{s_m}{\rho_m} - \Gamma_{Iml}^{\pm}\frac{s_l}{\rho_l}\right)\frac{\delta H}{\delta s_l}\right\} , \tag{12.5-4}$$

which leads to the expected symmetric, bilinear, and quadratic form for the dissipation bracket, $[F,H]$, given by Eq. (12.4-15). Also, when $T_i = T_j$ and $\mathbf{m}^i = \mathbf{m}^j$ for all i and j, i.e., when the system is in thermal and mechanical equilibrium, the reaction driving force and the reaction bracket for the multi-fluid system reduce to the single-fluid case given in the previous section.

Following the usual procedure, the time derivatives of the fluid mass, momentum, and entropy density as a result of chemical reactions can be found for the multi-fluid system:

$$\frac{\partial \rho_i}{\partial t}\bigg|_{RXN} = -\sum_{I=1}^{r}\sum_{j=1}^{n}\left[\left(\Gamma_{Iij}^{+} - \Gamma_{Iji}^{+}\right)J_I^{+} + \left(\Gamma_{Iij}^{-} - \Gamma_{Iji}^{-}\right)J_I^{-}\right] , \tag{12.5-5}$$

$$\frac{\partial m_\alpha^i}{\partial t}\bigg|_{RXN} = -\sum_{I=1}^{r}\sum_{j=1}^{n}\left[\left(\Gamma_{Iij}^{+}\upsilon_\alpha^j - \Gamma_{Iji}^{+}\upsilon_\alpha^i\right)J_I^{+} + \left(\Gamma_{Iij}^{-}\upsilon_\alpha^j - \Gamma_{Iji}^{-}\upsilon_\alpha^i\right)J_I^{-}\right] , \tag{12.5-6}$$

$$\frac{\partial s_i}{\partial t}\bigg|_{RXN} = -\sum_{I=1}^{r}\sum_{j=1}^{n}\left[\left\{\frac{\Gamma_{Iij}^{+}}{T_i}\left(\left(\mu_i^* - \mu_j^*\right) + \left(\upsilon_\alpha^i - \upsilon_\alpha^j\right)\upsilon_\alpha^j + \frac{s_j}{\rho_j}T_j\right) - \Gamma_{Iji}^{+}\frac{s_i}{\rho_i}\right\}J_I^{+}\right.$$
$$\left. + \left\{\frac{\Gamma_{Iij}^{-}}{T_i}\left(\left(\mu_i^* - \mu_j^*\right) + \left(\upsilon_\alpha^i - \upsilon_\alpha^j\right)\upsilon_\alpha^j + \frac{s_j}{\rho_j}T_j\right) - \Gamma_{Iji}^{-}\frac{s_i}{\rho_i}\right\}J_I^{-}\right] . \tag{12.5-7}$$

Now the generalized chemical potential is defined by

$$\mu_i^* = \frac{\delta H}{\delta \rho_i} = -\tfrac{1}{2}|\boldsymbol{\upsilon}^i|^2 + e_p^i + \frac{\partial u_i}{\partial \rho_i} . \tag{12.5-8}$$

The right-hand sides of Eqs. (12.5-5,6,7) can be added to those of Eqs. (12.3-18,19,20) to arrive at the governing equations for a dissipative, reactive, multi-fluid system:

$$\frac{\partial \rho_i}{\partial t} = -\nabla_\beta(\rho_i \upsilon_\beta^i) - \sum_{I=1}^{r}\sum_{j=1}^{n}\left[\left(\Gamma_{Iij}^{+} - \Gamma_{Iji}^{+}\right)J_I^{+} + \left(\Gamma_{Iij}^{-} - \Gamma_{Iji}^{-}\right)J_I^{-}\right] , \tag{12.5-9}$$

$$\rho_i \frac{\partial \upsilon^i_\alpha}{\partial t} = -\rho_i \upsilon^i_\beta \nabla_\beta \upsilon^i_\alpha - \rho_i \nabla_\alpha e_p{}^i + \nabla_\beta \left[\Phi^i_{\alpha\beta\gamma\varepsilon} \nabla_\gamma (\upsilon^i_\varepsilon)\right]$$
$$- \sum_{j=1}^{n} \left(\Lambda^{ij}_{\alpha\beta} + \Lambda^{ji}_{\alpha\beta}\right)(\upsilon^i_\beta - \upsilon^j_\beta) - \nabla_\alpha p_i \qquad (12.5\text{-}10)$$
$$- \sum_{I=1}^{r} \sum_{j=1}^{n} \left[\left\{ \left(\Gamma^+_{Iij} \upsilon^j_\alpha - \Gamma^+_{Iji} \upsilon^i_\alpha\right) - \left(\Gamma^+_{Iij} + \Gamma^+_{Iji}\right) \upsilon^i_\alpha \right\} J^+_I \right.$$
$$\left. + \left\{ \left(\Gamma^-_{Iij} \upsilon^j_\alpha - \Gamma^-_{Iji} \upsilon^i_\alpha\right) - \left(\Gamma^-_{Iij} + \Gamma^-_{Iji}\right) \upsilon^i_\alpha \right\} J^-_I \right],$$

$$\frac{\partial s_i}{\partial t} = -\nabla_\beta (\upsilon^i_\beta s_i) + \Phi^i_{\alpha\beta\gamma\varepsilon} \frac{1}{T_i} \nabla_\alpha \upsilon^i_\beta \nabla_\gamma \upsilon^i_\varepsilon + \nabla_\gamma \left(\Theta^i_{\gamma\beta} \nabla_\beta T_i\right)$$
$$+ \Theta^i_{\gamma\beta} \frac{1}{T_i} \nabla_\gamma T_i \nabla_\beta T_i + \sum_{j=1}^{n} \Lambda^{ij}_{\alpha\beta} \frac{1}{T_i} \left(\upsilon^i_\alpha - \upsilon^j_\alpha\right)(\upsilon^i_\beta - \upsilon^j_\beta)$$
$$- \sum_{j=1}^{n} \left(\Pi_{ij} + \Pi_{ji}\right)\left(T_i - T_j\right) + \sum_{j=1}^{n} \Pi_{ij} \frac{1}{T_i} \left(T_i - T_j\right)^2$$
$$- \sum_{I=1}^{r} \sum_{j=1}^{n} \left[\left\{ \frac{\Gamma^+_{Iij}}{T_i} \left(\left(\mu^*_i - \mu^*_j\right) - \left(\upsilon^i_\alpha - \upsilon^j_\alpha\right) \upsilon^j_\alpha + \frac{s_j}{\rho_j} T_j \right) - \Gamma^+_{Iji} \frac{s_i}{\rho_i} \right\} J^+_I \right.$$
$$\left. + \left\{ \frac{\Gamma^-_{Iij}}{T_i} \left(\left(\mu^*_i - \mu^*_j\right) - \left(\upsilon^i_\alpha - \upsilon^j_\alpha\right) \upsilon^j_\alpha + \frac{s_j}{\rho_j} T_j \right) - \Gamma^-_{Iji} \frac{s_i}{\rho_i} \right\} J^-_I \right].$$
$$(12.5\text{-}11)$$

The partial pressures, p_i, are given again by Eq. (12.2-13). The Poisson equation, (12.2-15), along with the definition of the reaction fluxes, (12.5-3) or (12.5-4), and of the phenomenological coefficients, represents a closed system of equations.

Following the usual procedure, we can find the evolution equations for the internal energy,

$$\frac{\partial u_i}{\partial t} = -\nabla_\alpha \left(\upsilon^i_\alpha u_i\right) - p_i \nabla_\alpha \upsilon^i_\alpha + \phi^i_{\alpha\beta\gamma\varepsilon} \nabla_\alpha \upsilon^i_\beta \nabla_\gamma \upsilon^i_\varepsilon$$
$$+ T_i \nabla_\gamma \left(\Theta^i_{\gamma\beta} \nabla_\beta T_i\right) + \Theta^i_{\gamma\beta} \nabla_\gamma T_i \nabla_\beta T_i + \sum_{j=1}^{n} \Lambda^{ij}_{\alpha\beta} \left(\upsilon^i_\alpha - \upsilon^j_\alpha\right)\left(\upsilon^i_\beta - \upsilon^j_\beta\right)$$
$$- \sum_{j=1}^{n} \left(\Pi_{ij} + \Pi_{ji}\right) T_i \left(T_i - T_j\right) + \sum_{j=1}^{n} \Pi_{ij} \left(T_i - T_j\right)^2$$

$$-\sum_{I=1}^{r}\sum_{j=1}^{n}\left[\left\{\left(\mu_j + \frac{s_j}{\rho_j}T_j\right)\Gamma^{+}_{Iij} - \left(\mu_i + \frac{s_i}{\rho_i}T_i\right)\Gamma^{+}_{Iji} + \tfrac{1}{2}\left(\upsilon^i_\alpha - \upsilon^j_\alpha\right)\left(\upsilon^i_\alpha - \upsilon^j_\alpha\right)\Gamma^{+}_{Iij}\right\}J^{+}_I\right.$$

$$\left.+\left\{\left(\mu_j + \frac{s_j}{\rho_j}T_j\right)\Gamma^{-}_{Iij} - \left(\mu_i + \frac{s_i}{\rho_i}T_i\right)\Gamma^{-}_{Iji} + \tfrac{1}{2}\left(\upsilon^i_\alpha - \upsilon^j_\alpha\right)\left(\upsilon^i_\alpha - \upsilon^j_\alpha\right)\Gamma^{-}_{Iij}\right\}J^{-}_I\right] ,$$

(12.5-12)

and the kinetic energy,

$$\rho_i\frac{\partial\left(\frac{1}{2}|\boldsymbol{\upsilon}^i|^2\right)}{\partial t} = -\rho_i\upsilon^i_\beta\nabla_\beta\tfrac{1}{2}|\boldsymbol{\upsilon}^i|^2 - \nabla_\alpha(p_i\upsilon^i_\alpha) + p_i\nabla_\alpha\upsilon^i_\alpha - \rho_i\upsilon^i_\alpha\nabla_\alpha e_p{}^i$$

$$+ \nabla_\beta\left(\phi^i_{\beta\alpha\gamma\varepsilon}\upsilon^i_\alpha\nabla_\gamma\upsilon^i_\varepsilon\right) - \phi_{\beta\alpha\gamma\varepsilon}\nabla_\beta\upsilon^i_\alpha\nabla_\gamma\upsilon^i_\varepsilon - \sum_{j=1}^{n}\upsilon^i_\alpha\left(\Lambda^{ij}_{\alpha\beta} + \Lambda^{ji}_{\alpha\beta}\right)\left(\upsilon^i_\beta - \upsilon^j_\beta\right)$$

$$-\sum_{I=1}^{r}\sum_{j=1}^{n}\left[\left\{\frac{1}{2}|\boldsymbol{\upsilon}^j|^2\Gamma^{+}_{Iij} - \frac{1}{2}|\boldsymbol{\upsilon}^i|^2\Gamma^{+}_{Iji} - \frac{1}{2}\left(\upsilon^i_\alpha - \upsilon^j_\alpha\right)\left(\upsilon^i_\alpha - \upsilon^j_\alpha\right)\Gamma^{+}_{Iij}\right\}J^{+}_I\right.$$

$$\left.+\left\{\frac{1}{2}|\boldsymbol{\upsilon}^j|^2\Gamma^{-}_{Iij} - \frac{1}{2}|\boldsymbol{\upsilon}^i|^2\Gamma^{-}_{Iji} - \frac{1}{2}\left(\upsilon^i_\alpha - \upsilon^j_\alpha\right)\left(\upsilon^i_\alpha - \upsilon^j_\alpha\right)\Gamma^{-}_{Iij}\right\}J^{-}_I\right] .$$

(12.5-13)

Finally, summing the two energy equations, the evolution equation for the total energy of each fluid in a dissipative system is obtained:

$$\frac{\partial e_i}{\partial t} = -\nabla_\alpha(\upsilon^i_\alpha e_i) - \nabla_\alpha(p_i\upsilon^i_\alpha) - \rho_i\upsilon^i_\alpha\nabla_\alpha e_p{}^i$$

$$+ \nabla_\beta\left(\phi^i_{\beta\alpha\gamma\varepsilon}\upsilon^i_\alpha\nabla_\gamma\upsilon^i_\varepsilon\right) + \nabla_\gamma\left(\Theta^i_{\gamma\beta}T_i\nabla_\beta T_i\right)$$

$$+ \sum_{j=1}^{n}\Lambda^{ij}_{\alpha\beta}\left(\upsilon^i_\alpha - \upsilon^j_\alpha\right)\left(\upsilon^i_\beta - \upsilon^j_\beta\right) - \sum_{j=1}^{n}\upsilon^i_\alpha\left(\Lambda^{ij}_{\alpha\beta} + \Lambda^{ji}_{\alpha\beta}\right)\left(\upsilon^i_\beta - \upsilon^j_\beta\right)$$

$$-\sum_{j=1}^{n}\left(\Pi_{ij}T_j + \Pi_{ji}T_i\right)\left(T_i - T_j\right)$$

$$-\sum_{I=1}^{r}\sum_{j=1}^{n}\left[\left\{\left(\frac{1}{2}|\boldsymbol{v}^j|^2 + h_j\right)\Gamma^+_{Iij} - \left(\frac{1}{2}|\boldsymbol{v}^i|^2 + h_i\right)\Gamma^+_{Iji}\right\}J^+_I\right.$$

$$\left.+\left\{\left(\frac{1}{2}|\boldsymbol{v}^j|^2 + h_j\right)\Gamma^-_{Iij} - \left(\frac{1}{2}|\boldsymbol{v}^i|^2 + h_i\right)\Gamma^-_{Iji}\right\}J^-_I\right] , \qquad (12.5\text{-}14)$$

where, $h_j = \mu_j + s_j T_j / \rho_j$, is the specific enthalpy for fluid j. These results are consistent with those of Woods [1987, pp. 262, 308] and Bataille and Kestin [1977, §4] as derived via irreversible thermodynamics for a system of weakly interacting, reacting fluids.

12.6 Weakly Ionized Plasma Model

The multi-fluid equations can be rearranged to arrive, via the bracket formalism, at the continuum level equations commonly used for the simple case of a weakly ionized, three-component plasma consisting of electrons, ions, and neutrals [Graves, 1986]. The system of equations consists of four equations: electron conservation, ion conservation, Poisson's equation, and electron energy conservation. The parameters in the model are the species diffusivities, mobilities, and charges, the electron thermal conductivity, and the ionization rate constant. The problem variables are the species mass densities and temperatures.

The first step is to change from the stationary frame of reference to one which moves with the total mass average velocity, $\mathbf{v}$,

$$\mathbf{v} = \frac{1}{\rho}\sum_{i=1}^{n} \rho_i \boldsymbol{v}^i . \qquad (12.6\text{-}1)$$

The continuity equation for the total mass of the system can be found by summing Eq. (12.5-9) over all species,

$$\frac{\partial \rho}{\partial t} + \nabla_\beta\left(\rho v_\beta\right) = 0 . \qquad (12.6\text{-}2)$$

Subtracting Eq. (12.6-2) from the species continuity equation, (12.5-9), results in the species continuity equation in terms of the species flux, $\mathbf{j}_i$, where

$$\mathbf{j}^i = \rho_i\left(\boldsymbol{v}^i - \mathbf{v}\right) \equiv \rho_i \mathbf{v}^i . \qquad (12.6\text{-}3)$$

and $\mathbf{v}^i$ is the velocity of the i-th species relative to the mass average velocity:

$$\rho\frac{\partial c_i}{\partial t} + \rho v_\beta \nabla_\beta c_i + \nabla_\beta j^i_\beta = r_i \ . \tag{12.6-4}$$

The species concentration is given by $c_i = \rho_i / \rho$ and the term r_i is the volumetric rate of production of species i by reaction given by equation (12.5-5) above. To solve this equation, an equation for the species flux is needed.

Following a procedure similar to that used for the species continuity equations, an equation for the species flux can be derived from the species momentum equation:

$$\rho_i\frac{\partial v^i_\alpha}{\partial t} = -\rho_i v^i_\beta \nabla_\beta v^i_\alpha - \rho_i \frac{\partial v_\alpha}{\partial t} - \rho_i v_\beta \nabla_\beta v_\alpha - \rho_i v^i_\beta \nabla_\beta v_\alpha - \rho_i v_\beta \nabla_\beta v^i_\alpha$$

$$- a_i \rho_i \nabla_\alpha e_p{}^i + \nabla_\beta\left(\phi^i_{\alpha\beta\gamma\varepsilon}\nabla_\gamma\left(v^i_\varepsilon - v_\varepsilon\right)\right) - \nabla_\alpha p_i - \sum_{j=1}^{n}\left(\Lambda^{ij}_{\alpha\beta} + \Lambda^{ji}_{\alpha\beta}\right)\left(v^i_\beta - v^j_\beta\right)$$

$$+ \sum_{I=1}^{r}\sum_{j=1}^{n}\left[\left\{\Gamma^+_{Iij}\left(v^i_\alpha - v^j_\alpha\right) - 2\Gamma^+_{Iji}\left(v^i_\alpha + v_\alpha\right)\right\}J^+_I\right.$$

$$\left. + \left\{\Gamma^-_{Iij}\left(v^i_\alpha - v^j_\alpha\right) - 2\Gamma^-_{Iji}\left(v^i_\alpha + v_\alpha\right)\right\}J^-_I\right] . \tag{12.6-5}$$

This equation is a general result and can be used under any condition when the multi-fluid model is valid. For the special case of a weakly ionized, three-component (electron, ion, and neutral) plasma, only collisions with neutral species are significant [Braginskii, 1965]. Thus, the following assumptions can be made: only collision terms involving the neutral species will be significant, viscous effects are negligible for the ions and electrons, and inertial effects are negligible:

$$\sum_{j=1}^{n}\left(\Lambda^{ij}_{\alpha\beta} + \Lambda^{ji}_{\alpha\beta}\right)\left(v^i_\beta - v^j_\beta\right) \approx \left(\Lambda^{iN} + \Lambda^{Ni}\right)\left(v^i_\alpha - v^N_\alpha\right) ,$$

$$\rho_N \gg \rho_i > \rho_e \Rightarrow \mathbf{v} \approx \boldsymbol{v}^N - \mathbf{v}^N = \mathbf{0} \ . \tag{12.6-6}$$

Thus, when the momentum transfer due to reaction is neglected, the fluxes of the electrons and ions can be approximated by

$$\mathbf{j}^i = \rho_i\left(-a_i \mu_{iN} \boldsymbol{\nabla} V_i - D_{iN} \boldsymbol{\nabla} \ln p_i\right) , \tag{12.6-7}$$

where μ_{iN} and D_{iN} are the mobility and diffusivity of species i in the neutral species defined by

$$\mu_{iN} \equiv \frac{\rho_i}{\Lambda^{iN} + \Lambda^{Ni}} \; ; \quad D_{iN} \equiv \mu_{iN} T_i \; . \tag{12.6-8}$$

Finally, under the similar assumptions, neglecting viscous dissipation within each fluid and convection due to the total fluid velocity (i.e., $\mathbf{v} \approx \mathbf{0}$), the energy equations reduce to

$$\frac{\partial e_i}{\partial t} = -\nabla_\beta q^i_\beta - j^i_\beta \nabla_\beta e^i_p - \nabla_\beta \left(p_i v^i_\beta \right) + r_i H_i \; , \tag{12.6-9}$$

where the heat flux is given by

$$\mathbf{q}^i = \mathbf{j}^i \frac{e_i}{\rho_i} - k_i \boldsymbol{\nabla} T_i \; , \tag{12.6-10}$$

and $r_i H_i$ is the enthalpy change of species i due to interactions with other fluids. It is defined by

$$\begin{aligned} r_i H_i \equiv & -\sum_{j=1}^{n} \left(\Lambda^{ij}_{\alpha\beta} \upsilon^i_\alpha + \Lambda^{ji}_{\alpha\beta} \upsilon^j_\alpha \right) \left(\upsilon^i_\beta - \upsilon^j_\beta \right) \\ & -\sum_{j=1}^{n} \left(\Pi_{ij} T_i + \Pi_{ji} T_j \right) \left(T_i - T_j \right) \\ & -\sum_{l=1}^{r} \sum_{j=1}^{n} \left[\left(\tfrac{1}{2} |\boldsymbol{v}^j|^2 + h_j \right) \Gamma^{\pm}_{lij} - \left(\tfrac{1}{2} |\boldsymbol{v}^i|^2 + h_i \right) \Gamma^{\pm}_{lji} \right] J^{\pm}_l \; . \end{aligned} \tag{12.6-11}$$

The first term accounts for energy transferred from fluid j to fluid i by momentum transfer interactions; the second term accounts for the thermal energy transferred from fluid j to fluid i in non-reacting interactions; and the last term is the energy transferred during chemical reactions.

12.7 Conclusions

In this chapter, the bracket formalism has been extended from the previous development for single-fluid systems to multi-fluid and chemically reacting systems. The advantage of the bracket formulation over classical irreversible thermodynamic approaches is twofold. First, the primary physical assumptions are elucidated from an analysis of the constituent terms of the bracket. Second, by realizing that the dissipative bracket essentially involves the products of the fluxes in the dynamic variables with their associated driving forces, once the bracket has been formulated, the formalism results in a natural definition of the driving forces associated with each flux. Thus, the derivation of a thermodynamically consistent driving force for chemical reactions in multi-fluid systems which are not in thermal or mechanical equilibrium was possible. Near equilibrium, the reaction driving force reduces to a symmetric, quadratic, and bilinear form. In addition, when the system is in thermal and mechanical equilibrium, the reaction driving force for the multicomponent system reduces to the standard form derived through irreversible thermodynamics for a multicomponent, single-fluid system.

However, the multi-fluid model as derived here is limited to the case where the electrons are in thermal equilibrium, although not necessarily at the same temperature as the other species. This limitation may possibly be relieved by the inclusion of internal parameters to account for the deviation of the energy distribution of the electrons from the Maxwellian. A similar approach has been used in the development of models of liquid-crystalline systems, as seen in chapter 11, and of the behavior of polymeric liquids near solid surfaces, as described in chapter 9. Thus the analysis offered in this chapter can be used as the basis for the future development of a model which would allow for deviation from thermal equilibrium of the electrons. Another limitation of the model developed here is the involvement of phenomenological coefficients, such as $\Lambda_{\alpha\beta}^{ij}$ in Eq. (12.3-10), which may be complex functions of the dynamic variables. For the model to be useful, these coefficients need to be evaluated through some microscopic theory, such as kinetic theory, which will reduce the adjustable parameters to a manageable number to be finally determined through a comparison of the model predictions with experimental results.

Epilogue

Καὶ τὰ μέν γε ἄλλα οὐκ ἂν πάνυ ὑπὲρ τοῦ λόγου
διισχυρισαίμην, ὅτι δ' οἰόμενοι δεῖν ζητεῖν ἃ μή τις
οἶδε, βελτίους ἂν εἶμεν καὶ ἀνδρικώτεροι καὶ ἧττον
ἀργοὶ ἢ εἰ οἰοίμεθα ἃ μὴ ἐπιστάμεθα, μηδὲ δυνατὸν
εἶναι εὑρεῖν μηδὲ δεῖν ζητεῖν, περὶ τούτου πάνυ ἂν
διαμαχοίμην, εἰ οἷός τε εἴην, καὶ λόγῳ καὶ ἔργῳ.

Plato—words from the mouth of Socrates, Menon 86C[1]

At this point, after all these elaborate discussions, we consider it useful
to summarize the main conclusions. The topic which the present
monograph has addressed is the study of dynamic phenomena within
complex systems, from a theoretical modeler's perspective. Given the
complexity of the subject, we are the first to admit that the study of non-
equilibrium phenomena is far from being completed. Even the question
of whether these phenomena can be adequately described using only a
few variables is, for the most general case, as yet unanswered. However,
we consider it worthwhile to try. We believe that what we have offered
here, based heavily on the work of many previous researchers, is a more
systematic approach to the modeling of the many phenomena
encountered within this vast subject. We have shown that, if used by
itself (in isolation from any presumed understanding of the microscopic
physics), this approach leads to general continuum models for the
dynamic behavior of complex systems with an internal microstructure.
It consistently combines the continuum modeling of dissipative
phenomena offered by irreversible thermodynamics and that of
conservative processes described by a (non-canonical) Poisson bracket
formulation.

Moreover, we believe that greater profit can be realized from the
combination of both microscopic and macroscopic descriptions: the major
role that we see for the bracket formulation is in facilitating the transfer
of information between different scales of length and time. Although sev-

---

[1]English translation: And with respect the other points I would not dare to adhere to
my view with great insistence, but one thing I would fight for to the end, both in word and
deed if I were able—that if we believed that we must try to find out what is not known, we
should be better and braver and less idle than if we believed that what we do not know it
is impossible to find out and that we need not even try [Rouse, 1956, p. 51].

eral examples have been discussed here, there are many more which have not. We see little additional value in our describing additional applications. We feel that a lot more can be gained by letting the experts in each particular area work on further applications. Moreover, if, in the process of illustrating the flexibility of the bracket formulation, we left out important contributions in the applications we discussed here, we duly apologize. We anticipate considerable gains to be realized in future work by focused investigations on specific topics with specific experimental data and microscopic pictures at hand. We should be very happy if the researcher of the future who endeavors to undertake this type of modeling were to find this work of some usefulness.

A
Introduction to Differential Manifolds

A.1 Differential Manifolds

An n-dimensional differential manifold, $\Im$ (or $\Im_n$), is a set of points a_i, i=1,2,..., together with a set of smooth (infinitely differentiable) maps $\wp_1, \wp_2, \ldots$, so that for every point a_i in the manifold there exists a one-to-one map $\wp_i$ which maps an open neighborhood A_i of that point to an open subset S_i of $\mathbb{R}^n$:

$$\begin{aligned} &\forall a \in \Im_n \ \exists\ \wp_i : \Im_n \supseteq A_i \to S_i \subseteq \mathbb{R}^n : a \in A_i \ , \\ &\bigcup A_i = \Im_n \ , \\ &\forall i,j : A_i \cap A_j \neq \varnothing \ \Rightarrow \ \wp_j \circ \wp_i^{-1} \in C^\infty \ . \end{aligned} \tag{A-1}$$

Because of this direct correspondence between the points of an n-dimensional manifold and the points in an n-dimensional Euclidean space, we can always identify (at least locally) $\Im_n$ with $\mathbb{R}^n$ without loss of generality.

EXAMPLE A.1: The n-dimensional sphere can be identified as the collection of points

$$(x_1, x_2, \ldots, x_n)^{\mathrm{T}} \in \mathbb{R}^n : \sum_{i=1}^{n} x_i^2 \equiv R^2 \ . \tag{A-2}$$

This is an $(n-1)$-dimensional manifold since we can map these points to $\mathbb{R}^{n-1}$ through the collection of two partially-overlapping, differentiable, one-to-one maps

$$\wp_1 : \mathbf{x} \in \Im \setminus \{(0,0,\ldots,R)^{\mathrm{T}}\} \to \mathbf{u} \in \mathbb{R}^{n-1} : u_i = \frac{R}{R - x_n} x_i \ , \quad i \neq n \ , \tag{A-3}$$

$$\wp_2 : \mathbf{x} \in \Im \setminus \{(0,0,\ldots,-R)^{\mathrm{T}}\} \to \mathbf{u} \in \mathbb{R}^{n-1} : u_i = \frac{R}{R + x_n} x_i \ , \quad i \neq n \ . \tag{A-4}$$

These two maps are the n-dimensional extension of the traditional spherical projection onto a plane (see Figure A.1).

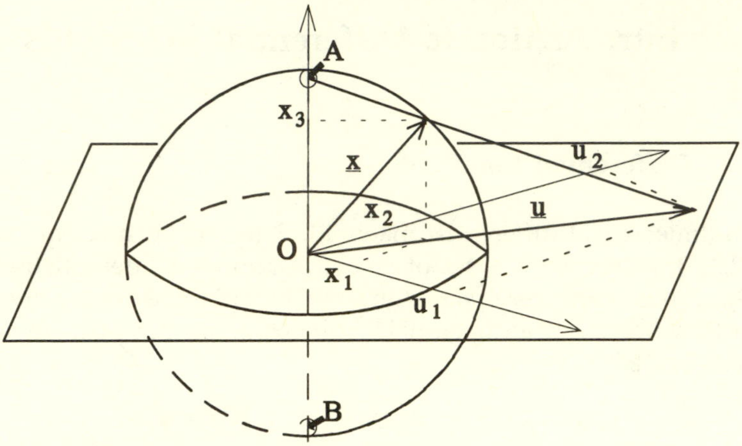

Figure A.1: $\wp_1$ projection of a sphere into a plane.

Note that the inverses of both maps exist. For example,

$$\wp_1^{-1}: \mathbf{u} \in \mathbb{R}^{n-1} \to \mathbf{x} \in \Im \setminus \{(0,0,\ldots,R)^T\}:$$

$$x_n = R\,\frac{r^2 - R^2}{r^2 + R^2}\ ; \quad x_i = \frac{2R^2}{r^2 + R^2}\,u_i\ , \quad i \neq n\ , \tag{A-5}$$

where $$r^2 \equiv \sum_{i=1}^{n-1} u_i^2\ ,$$

with a similar expression for $\wp_2^{-1}$. ■

A.2 Curves

A curve f in a differentiable manifold $\Im_n$ is a set of points within $\Im$ corresponding to the image of a (differentiable) map from $\mathbb{R}$ to $\Im_n$. Using any local map of the manifold $\Im_n$ to $\mathbb{R}^n$, a (differentiable) curve can be (locally) represented by a continuously differentiable mapping from $\mathbb{R}$ into $\mathbb{R}^n$:

$$f: \gamma \in \mathbb{R} \rightarrow \mathbf{f}(\gamma) \in \mathbb{R}^n \; . \tag{A-6}$$

The curve f is then said to be *parameterized* by γ. In the following, a curve f on a manifold will be loosely represented by the same symbol either in the manifold itself or, after mapping, in its image on $\mathbb{R}^n$ with the understanding of the equivalence between the two through a smooth mapping.

EXAMPLE A.2: Let the curve f be defined as $\mathbf{f}(\gamma)=(\gamma,0,0,\ldots,0)^{\mathrm{T}} \in \mathbb{R}^{n-1}$, $\gamma \in \mathbb{R}$, using the first of the two maps defined in Example 2.1. Then the curve f, being a straight line in the mapped space $\mathbb{R}^{n-1}$, corresponds in the original space, $\mathbb{R}^n$, where the manifold $\Im$ (the n-th dimensional hypersphere) is defined, to a circle in the $x_1 - x_n$ plane, of radius R with its center at the origin, (0,0):

$$\wp_1^{-1}\big(\mathbf{f}(\gamma)\big) = \left(\frac{2\gamma R^2}{\gamma^2 + R^2}, 0, 0, \ldots, R\frac{\gamma^2 - R^2}{\gamma^2 + R^2}\right)^{\mathrm{T}} . \quad \blacksquare$$

A.3 Tangent Spaces

At a given point a of an n-dimensional manifold, $\Im_n$, different curves are considered equivalent (they belong to the same equivalent class) if they are asymptotically (in an infinitesimal neighborhood of a) the same. Since the curve asymptotically close to every differential curve f is the same as its linearization, $a + (\mathrm{d}\mathbf{f}/\mathrm{d}\gamma)(\gamma - \gamma_a)$, there is a one-to-one correspondence between every equivalent class and the tangent $\mathrm{d}f/\mathrm{d}\gamma\,(a)$. Using any local map of the manifold $\Im_n$ to $\mathbb{R}^n$, the tangent of a curve at a point a can be represented by an n-dimensional vector, $\mathrm{d}\mathbf{f}/\mathrm{d}\gamma\,(a) = (\mathrm{d}f_1/\mathrm{d}\gamma\,(a), \mathrm{d}f_2/\mathrm{d}\gamma\,(a), \ldots, \mathrm{d}f_n/\mathrm{d}\gamma\,(a))^{\mathrm{T}}$, or by its corresponding first-order differential operator

$$L(f,a) \equiv \frac{\mathrm{d}\mathbf{f}}{\mathrm{d}\gamma}(a)\cdot\frac{\partial}{\partial \mathbf{x}} = \sum_{i=1}^{n} \frac{\mathrm{d}f_i}{\mathrm{d}\gamma}(a)\,\frac{\partial}{\partial x_i} \; . \tag{A-7}$$

Therefore, it follows that all possible tangents at a point a of an n-dimensional manifold $\Im_n$ form a vector space, called the tangent space, $T\Im_n$, of the same dimensionality n. All points other than a have their own tangent spaces, defined analogously.

The collection of all tangent spaces of a given manifold $\Im$ (i.e., the set composed of the tangent spaces from all of the points in the manifold) also forms a space called the tangent bundle, denoted as $T\Im \times \Im$; its elements correspond to the realization of vector fields $\mathbf{v}(\mathbf{x})$. A fundamental theorem from the theory of ordinary differential equations

[Atkinson, 1978, p. 292] ascertains that, provided $\mathbf{v}(\mathbf{x})$ is a differentiable function, it can be integrated as $\mathrm{d}\mathbf{x}/\mathrm{d}t = \mathbf{v}(\mathbf{x})$, subject to certain initial conditions, $\mathbf{x}(0) = \mathbf{x}_0$, to yield a family of solutions, $\mathbf{x} = \mathbf{x}(\mathbf{x}_0, t)$, which for given $\mathbf{x}_0$ represents a curve parameterized by t (*integral curve*). Alternatively, for given t, $\mathbf{x} = \mathbf{x}(\mathbf{x}_0, t)$ can be considered as a mapping between $\mathbf{x}_0$ and $\mathbf{x}$.

A.4 Differential Forms

For every linear vector space, V, there corresponds another vector space, called the *dual vector space*, formed by all possible linear mappings of the original vector space V into the real space $\mathbb{R}$ (functionals). For any vector space of finite dimensions, V, the dual vector space is isomorphic to V. It can be shown [Greenberg, 1978, p. 325] that in a finite-dimensional Hilbert space (i.e., a vector space where an inner product is defined) every functional can be expressed as the inner product with a given element of V, $\mathbf{a}_f$, i.e., $f \equiv \langle \mathbf{a}_f, \cdot \rangle$. The dual space of the tangent space, $T\Im$, is the cotangent space, $T^*\Im$, formed by the functionals of the tangents, which are called *first-order differential forms* or 1-forms.

The total derivative of a function g, $g{:}\Im \rightarrow \mathbb{R}$, along a curve f, $f{:}\mathbb{R} \rightarrow \Im$, at a point a of the manifold $\Im$ is defined as $\mathrm{D}g/\mathrm{D}t(a) \equiv \partial g/\partial \mathbf{f}(a) \cdot \mathrm{d}\mathbf{f}/\mathrm{d}t(a)$. Thus, the result is a real number which, for a given point a, depends linearly on the tangent of the curve, $\mathrm{d}\mathbf{f}/\mathrm{d}t$ and on the gradient of g, $\partial g/\partial \mathbf{x}$, evaluated at the same point a. Therefore, the cotangent space at a given point can be identified as the vector space generated by the functional gradients at that point and as such it has the same dimensionality, n, as the tangent space and the original manifold. Thus, a given 1-form ω may be represented as a *differential*:

$$\omega(a) \equiv \sum_{i=1}^{n} \omega_i(a)\, \mathrm{d}x_i = \sum_{i=1}^{n} \frac{\partial g}{\partial x_i}(a)\, \mathrm{d}x_i \ ,$$

$$\text{where} \qquad g = \sum_{i=1}^{n} \omega_i(a)\, x_i \ , \tag{A-8}$$

$$\text{and} \qquad \omega(a) \circ \left(\sum_{i=1}^{n} v_i(a) \frac{\partial}{\partial x_i} \right) \equiv \sum_{i=1}^{n} \omega_i(a)\, v_i(a) \ .$$

In this expression, $\mathrm{d}x_i$, i=1,..,n, are a set of basis vectors which represent the functionals defined by the projection of the tangent vectors on the i-th coordinate axis.

The collection of the cotangent spaces at all points also forms a space, that of cotangent bundles with the 1-form fields as elements defined over the entire manifold. The gradient of a real function g defined over $\mathfrak{S}$ is, by definition, a 1-form

$$\mathrm{d}g \equiv \sum_{i=1}^{n} \frac{\partial g}{\partial x_i} \mathrm{d}x_i \ , \tag{A-9}$$

in which case its corresponding differential, dg, is called *exact*. However, not all 1-form fields ω correspond to the gradient of a function. In order for ω to be an exact differential, i.e., ω=df for some function f on $\mathfrak{S}$, the compatibility conditions

$$\frac{\partial \omega_i}{\partial x_j} = \frac{\partial \omega_j}{\partial x_i} \ , \quad i,j = 1,\ldots,n \ , \tag{A-10}$$

need to be satisfied between the components $\omega_1,\omega_2,\ldots,\omega_n$ (analogous to Maxwell relations in thermodynamics—see also chapter 4). Note that exact differentials are linear functionals (1-forms) with the property (directly verified through a Taylor expansion around the point α)

$$\mathrm{d}f \circ \Delta\mathbf{v} = \lim_{\Delta\mathbf{v}\to 0} [f(\alpha+\Delta\mathbf{v}) - f(\alpha)] \equiv \int_{\alpha}^{\alpha+\Delta\mathbf{v}} \mathrm{d}f \ . \tag{A-11}$$

Note that the above property can be used in order to define the differential operator, d, when applied to real functions. Similarly, from a given 1-form field ω we can construct a *2-form* field dω, called the *exterior derivative* of ω, as

$$\mathrm{d}\omega \circ \mathbf{uv} \equiv \lim_{\mathbf{u},\mathbf{v}\to 0} [\omega(\alpha)\circ\mathbf{u} + \omega(\alpha+\mathbf{u})\circ\mathbf{v} - \omega(\alpha+\mathbf{v})\circ\mathbf{u} - \omega(\alpha)\circ\mathbf{v}] \ , \tag{A-12}$$

or, if the application of the 1-form ω on an infinitesimal tangent **v**, $\omega\circ\mathbf{v}$ is interpreted as the integral along the infinitesimal segment **v** (see the definition of Eq. (A-11)), we can interpret (A-12) as defining dω to be the integrand of a double integral along the area of a rectangle defined by the vectors **u** and **v** so that $\iint \mathrm{d}\omega$ is equal to the closed integral $\oint \mathrm{d}\omega$ evaluated along the rectangle's periphery (see also Figure A.2).

An arbitrary *2-form* is constructed as a linear superposition of exact differentials of 1-forms. From this definition of the 2-forms, it is easy to see that they form a linear vector space representing the space of the bilinear, antisymmetric functionals mapping $T\mathfrak{S}\times T\mathfrak{S}$ into $\mathbb{R}$. Alternatively, each 2-form can be expressed as a linear superposition of the *exterior* or *wedge* products $\mathrm{d}x_i\wedge\mathrm{d}x_j$, $i,j=1,\ldots,n$, $i<j$. In particular, in a two-dimensional space (x,y), 2-forms can be expressed solely in terms of the wedge product $\mathrm{d}x\wedge\mathrm{d}y$. In general, the exterior or wedge product of two 1-forms ω_1 and ω_2, $\omega_1\wedge\omega_2$, is defined to be an antisymmetric, bilinear functional such as

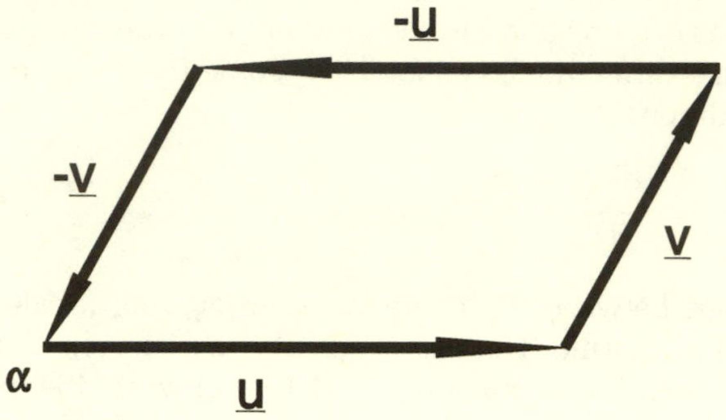

Figure A.2: Elementary integration area defined by vectors **u** and **v**.

$$\omega_1 \wedge \omega_2 \circ \mathbf{uv} \equiv [\omega_1 \circ \mathbf{u}]\,[\omega_2 \circ \mathbf{v}] - [\omega_1 \circ \mathbf{v}][\omega_2 \circ \mathbf{u}] \;, \quad \forall \mathbf{u},\mathbf{v} \in T\mathfrak{I} \;. \tag{A-13}$$

Then, through Eqs. (A-12) and (A-13), and a Taylor expansion of ω around the point α, one can deduce that the exterior derivative dω of an arbitrary 1-form $\omega = \Sigma_i \omega_i \mathrm{d}x_i$ is

$$\mathrm{d}\omega \equiv \sum_{i=1}^{n} \sum_{j=1}^{n} \frac{\partial \omega_i}{\partial x_j} \mathrm{d}x_j \wedge \mathrm{d}x_i \;. \tag{A-14}$$

EXAMPLE A.3: In a two-dimensional space, the exterior derivative dω of the 1-form $\omega = f\mathrm{d}x + g\mathrm{d}y$ is $\mathrm{d}w = (\partial g/\partial x - \partial f/\partial y)\,\mathrm{d}x \wedge \mathrm{d}y$. ■

Formulae (A-13) and (A-14) can be generalized for arbitrary order *k*-forms

$$\omega_1 \wedge \omega_2 \wedge \ldots \wedge \omega_k \circ \mathbf{u}_1 \mathbf{u}_2 \ldots \mathbf{u}_k \equiv \mathrm{Det} \begin{pmatrix} \omega_1 \circ \mathbf{u}_1 & \omega_1 \circ \mathbf{u}_2 & \ldots & \omega_1 \circ \mathbf{u}_k \\ \omega_2 \circ \mathbf{u}_1 & \omega_2 \circ \mathbf{u}_2 & \ldots & \omega_2 \circ \mathbf{u}_k \\ \ldots & \ldots & \ldots & \ldots \\ \omega_k \circ \mathbf{u}_1 & \omega_k \circ \mathbf{u}_2 & \ldots & \omega_k \circ \mathbf{u}_k \end{pmatrix} , \tag{A-15}$$

$$d\omega \equiv \sum_{i_1=1}^{n} \dots \sum_{i_k=1}^{n} \sum_{j=1}^{n} \frac{\partial \omega_{i_1 \dots i_k}}{\partial x_j} dx_j \wedge dx_{i_1} \wedge \dots \wedge dx_{i_k} , \tag{A-16}$$

where $\omega = \sum_{i_1=1}^{n} \dots \sum_{i_k=k}^{n} \omega_{i_1 \dots i_k} dx_{i_1} \wedge \dots \wedge dx_{i_k}$.

Notice that by using the bilinearity of the wedge product, one can extend the definition (A-15) to the wedge product between arbitrary k- and 1-forms (see also Bamberg and Sternberg [1988, p. 538] for more details).

Finally, using the definition (A-16) for the differential, it is easy to show that for an arbitrary k-form ω (including a function which can be considered as a 0-form)

$$d(d\omega) = 0 , \tag{A-17}$$

i.e., the exterior derivative of an exact form is zero or every exact form is *closed*. It can be shown that the inverse is also true, i.e., every closed form is exact.

A.5 Symplectic Forms and Symplectic Transformations

A symplectic form is a non-degenerate 2-form ω. For every symplectic form there corresponds a symplectic product (with properties as defined by Eq. (2.3-1)) and vice versa. For example, the corresponding symplectic form (canonical form) to the symplectic product in $\mathbb{R}^{2n}$, defined by Eq. (2.3-3), is

$$\omega = \sum_{i=1}^{n} dp_i \wedge dq_i , \tag{A-18}$$

where the manifold $\Im_{2n}$, and thus also the tangent space $T\Im_{2n}$, are isomorphic to $\mathbb{R}^{2n}=\{(\mathbf{q},\mathbf{p}),\ \mathbf{q},\mathbf{p} \in \mathbb{R}^n\}$. A manifold is called "symplectic" if a symplectic form is defined at every point (symplectic form field). The concept of symplectic manifolds parallels that of Hilbert spaces with the symplectic product «·,·» defined in lieu of the inner product <·,·>. Both products are scalar functions, bilinear on $T\Im_{2n} \times T\Im_{2n}$, corresponding to a scalar number, «**a**,**b**» and <**a**,**b**>, respectively, for every pair of vectors, **a**, **b** $\in T\Im_{2n}$. However, the symplectic product is *skew-symmetric* (*antisymmetric*) whereas the inner product is *symmetric*.

The above example of the canonical form is more generic in that it is always possible by a suitable choice of basis vectors to transform a symplectic form (and correspondingly, its associated symplectic product), at least locally in a symplectic manifold, into the form indicated by (A-18) (correspondingly, Eq. (2.3-3)—the Darboux theorem [Fomenko, 1988, p. 87]). Alternatively phrased, there is always a symplectic basis for every symplectic space $\mathbb{R}^{2n}$. Moreover, we can take any non-zero vector $\mathbf{v}\in\mathbb{R}^{2n}$ as the first vector of the symplectic basis. The construction of the symplectic basis from a complete set of independent vectors parallels the Gram/Schmidt orthonormalization procedure in the construction of an orthonormal basis in a Hilbert space, with the concept of orthogonality replaced by that of *skew-orthogonality*: two vectors **u** and **v** are skew-orthogonal if their symplectic product is zero, i.e., «**u**,**v**»=0.

By virtue of the above-mentioned correspondence between a symplectic form and a symplectic product, a linear mapping (isomorphism) between $2n$-dimensional tangent spaces of a symplectic manifold $\mathfrak{I}$, $dg:V\to V'$, is symplectic if and only if it preserves the corresponding symplectic form; i.e., the transformed 2-form $\hat{\omega}$

$$\hat{\omega} \equiv \sum_{i,j=1}^{n} \omega_{ij}\, dg_i \otimes dg_j = \sum_{i,j=1}^{n} \sum_{k,l=1}^{n} \omega_{ij} \frac{\partial g_i}{\partial x_k} \frac{\partial g_j}{\partial x_l}\, dx_k \otimes dx_l \,, \tag{A-19}$$

which is the 2-form induced to the transformed tangent space V' by the mapping g from the original 2-form $\omega\equiv\sum_{i,j}\omega_{ij}dx_i\otimes dx_j$ on V, is the same as the ω field on V (defined in the symplectic manifold in a suitable coordinate frame so that the coefficients ω_{ij} are constants). This requirement, using (A-19), implies

$$\omega_{kl} = \hat{\omega}_{kl} \equiv \omega_{ij} \frac{\partial g_i}{\partial x_k} \frac{\partial g_j}{\partial x_l} \,, \tag{A-20}$$

which together with the expression for the symplectic form, as, for example, provided by (A-18), provides the necessary and sufficient conditions for a transformation to be symplectic. In (A-19) the symbol $\otimes$ has been used to denote the tensor product between two 1-forms used to define the basis functions for arbitrary bilinear functionals on $T\mathfrak{I}\times T\mathfrak{I}$. Note that the notion of a symplectic transformation parallels that of a unitary transformation in a Hilbert space with the requirement of invariance of the inner product replacing that of the symplectic product.

By taking the determinant of both sides of (A-20), we get as a necessary condition (in accordance to the discussion in Section 2.3.2)

$$\mathrm{Det}\left(\frac{\partial g_i}{\partial x_j}\right) = 1 \,, \tag{A-21}$$

which also guarantees that the transformation is non-singular (one-to-one) and hence an isomorphism. A further consequence of the non-singular character of the symplectic transformations is that they form a group (as is also recognized in §2.3.2). The physical significance of a symplectic transformation is that it transforms canonical coordinates into canonical coordinates.

A.6 The Lagrangian Manifold

A vector subspace TL of a symplectic tangent space $T\Im_{2n}$ is *isotropic* if it is skew-orthogonal to itself; i.e., every pair of vectors of that subspace—including the trivial case when they are the same—are skew-orthogonal. Note that in any symplectic space, there are non-trivial (i.e., with dimensionality different than zero) isotropic subspaces; i.e., subspaces which are skew-orthogonal to themselves. For example, since every vector is skew-orthogonal to itself, the subspace generated by any vector is isotropic. In contrast, in a Hilbert space the only subspace orthogonal to itself is the trivial one which contains only the zero vector, **0**. In consequence, in Hilbert spaces it only makes sense to discuss the orthogonality between two different subspaces.

If the dimension of an isotropic subspace is n (half of the space $T\Im$), then the isotropic subspace is called *Lagrangian*. Thus, as a result of the properties of the symplectic transformation mentioned in the previous section, a symplectic transformation maps isotropic subspaces into isotropic subspaces and, in particular, Lagrangian subspaces into Lagrangian ones, with respect to the same symplectic product.

A submanifold L of a symplectic $2n$-dimensional manifold $\Im$ is called a *Lagrangian manifold* if all its tangent subspaces are Lagrangian subspaces in $\mathbb{R}^{2n}=T\Im$. Furthermore, its dimension is n. A necessary and sufficient condition for an n-dimensional submanifold of a $2n$-dimensional manifold $\Im$ to be Lagrangian is the restriction of its symplectic form field ω to be identically equal to zero. An important example of a Lagrangian submanifold is the one generated from the graph of the gradient of a smooth function $S(q_1,...,q_n)$, i.e., where the **p** coordinates are related to the **q** as

$$p_i = \frac{\partial S(q_1,...,q_n)}{\partial q_i} , \quad i=1,...,n . \tag{A-22}$$

Indeed, in this case, the standard (canonical) symplectic form ω, restricted to the manifold generated by the graph of the gradient of $S(q_1,...,q_n)$, becomes

$$\begin{aligned}\omega &= \sum_{i=1}^{n} \mathrm{d}p_i \wedge \mathrm{d}q_i = \sum_{i=1}^{n} \mathrm{d}(\frac{\partial S}{\partial q_i}) \wedge \mathrm{d}q_i \\ &= \sum_{i=1}^{n} \sum_{j=1}^{n} \frac{\partial^2 S}{\partial q_i \partial q_j} \mathrm{d}q_j \wedge \mathrm{d}q_i = 0 ,\end{aligned} \tag{A-23}$$

as a sum of antisymmetric 2-forms with symmetric coefficients. Inversely, it can be proven [Fomenko, 1988, p. 53] that any Lagrangian submanifold can be locally represented as the graph of the gradient of a smooth function S.

EXAMPLE A.4: A great (equatorial) circle on a sphere in a three-dimensional space (n=3 in Examples A.1 and A.2) is always a Lagrangian manifold. This is easily seen from the fact that a great circle, as a one-parameter curve (apply Example A.2 for n=3), is a one-dimensional submanifold of the original two-dimensional manifold (sphere) and the fact that the tangent space at any point of a great circle, such as a one-dimensional subspace of a symplectic space, is (as explained above) always isotropic. ■

A.7 The Hamiltonian Vector Field

A critical mathematical differential quantity, derived from a function f defined within a symplectic manifold $\mathfrak{I}$, is the *skew-symmetric gradient*, **sgrad**(f). This is the corresponding quantity in a symplectic space to grad(f) in a Euclidean (metric) space. It is defined as the vector field such that

$$\omega \circ \mathbf{u}\, \mathbf{sgrad}(f) = L_{\mathbf{u}}(f) \,, \tag{A-24}$$

where $L_{\mathbf{u}} f$ is the linear differential operator corresponding to **u** applied on a smooth function f, and **u** is any smooth vector field on $\mathfrak{I}$. This definition leads to a set of linear equations for the components of **sgrad**(f), the solution of which, for an arbitrary symplectic form ω, $\omega \equiv \sum_{i,j} \omega_{ij} dx_i \otimes dx_j$, is

$$[\mathbf{sgrad}(f)]_i = \sum_{j=1}^{n} \omega^{ij} \frac{\partial f}{\partial x_j} \,, \tag{A-25}$$

where ω^{ij} represents the (i,j) component of the inverse of the anti-symmetric matrix Ω constructed from the components ω_{ij} of the symplectic form ω. A particularly easy expression for **sgrad**(f) is obtained when the canonical form, (A-18), is used for ω

$$\mathbf{sgrad}(f) = \left(\frac{\partial f}{\partial p_1}, \ldots, \frac{\partial f}{\partial p_n}, -\frac{\partial f}{\partial q_1}, \ldots, -\frac{\partial f}{\partial q_n} \right)^{\mathrm{T}} \,, \tag{A-26}$$

where the canonical coordinates (**q**,**p**) are used.

EXAMPLE A.5: Evaluate **sgrad**(f) in the two-dimensional canonical (symplectic) space $(q,p)^{\mathrm{T}} \in \mathbb{R}^2$, using the definition provided by (A-25).

Using (A-18) for the definition of the (canonical) symplectic form ω, we can evaluate its components ω_{ij}, $\omega \equiv \sum_{i,j} \omega_{ij} dx_i \otimes dx_j$ as

$$\boldsymbol{\omega} = \begin{pmatrix} 0 & -1 \\ 1 & 0 \end{pmatrix}, \tag{A-27}$$

from which the components of the inverse matrix ω^{ij}, $\omega^{-1} \equiv \sum_{i,j} \omega^{ij} dx_i \otimes dx_j$ can be easily evaluated as

$$\boldsymbol{\omega}^{-1} = \begin{pmatrix} 0 & 1 \\ -1 & 0 \end{pmatrix}. \tag{A-28}$$

Using (A-25), **sgrad**(f) can now be directly evaluated through a matrix-vector multiplication as

$$\mathbf{sgrad}(f) \equiv \boldsymbol{\omega}^{-1} \cdot \begin{pmatrix} \dfrac{\partial f}{\partial q} \\ \dfrac{\partial f}{\partial p} \end{pmatrix} = \begin{pmatrix} 0 & 1 \\ -1 & 0 \end{pmatrix} \cdot \begin{pmatrix} \dfrac{\partial f}{\partial q} \\ \dfrac{\partial f}{\partial p} \end{pmatrix} = \begin{pmatrix} \dfrac{\partial f}{\partial p} \\ -\dfrac{\partial f}{\partial q} \end{pmatrix}, \tag{A-29}$$

which is the same as the expression provided by (A-26) for n=1, as it should be. ■

A *Hamiltonian vector field* is a vector field **u** that can be written as the skew-symmetric gradient of a function f

$$\mathbf{u} = \mathbf{sgrad}(f) \ . \tag{A-30}$$

In this respect there is a close connection between a Hamiltonian vector field in a symplectic space and a potential field in a Hilbert space, the skew-symmetric gradient in the definition provided by (A-30) being replaced by the gradient in the latter case.

THEOREM A.1: [Fomenko, 1988, p. 78] A Hamiltonian vector field $\mathbf{u} \equiv (\mathbf{Y},\mathbf{X})^T$, $\mathbf{Y},\mathbf{X} \in \mathbb{R}^n$, preserves the symplectic form ω, i.e., $d\omega/dt=0$ along the integral curve generated by **u** and parametrized by t, and vice versa. In the following we will prove the opposite, i.e., if a vector field **u** preserves the symplectic form ω, then there is a function f such that (A-30) is true, i.e., the vector field is Hamiltonian.

PROOF:

$$\frac{\mathrm{d}}{\mathrm{d}t}\omega(t) = \frac{\mathrm{d}}{\mathrm{d}t}\sum_{i=1}^{n} \mathrm{d}p_i(t) \wedge \mathrm{d}q_i(t)$$

$$= \sum_{i=1}^{n} \left\{\left(\frac{\mathrm{d}}{\mathrm{d}t}\mathrm{d}p_i(t)\right) \wedge \mathrm{d}q_i(t) + \mathrm{d}p_i(t) \wedge \left(\frac{\mathrm{d}}{\mathrm{d}t}q_i(t)\right)\right\}$$

$$= \sum_{i=1}^{n} \left\{\mathrm{d}\left(\frac{\mathrm{d}p_i}{\mathrm{d}t}\right) \wedge \mathrm{d}q_i + \mathrm{d}p_i \wedge \mathrm{d}\left(\frac{\mathrm{d}q_i}{\mathrm{d}t}\right)\right\} = \sum_{i=1}^{n} \left\{\mathrm{d}X_i \wedge \mathrm{d}q_i + \mathrm{d}p_i \wedge \mathrm{d}Y_i\right\}$$

$$= \sum_{i=1}^{n}\sum_{k=1}^{n} \left\{\left(\frac{\partial X_i}{\partial p_k}\mathrm{d}p_k + \frac{\partial X_i}{\partial q_k}\mathrm{d}q_k\right) \wedge \mathrm{d}q_i + \mathrm{d}p_i \wedge \left(\frac{\partial Y_i}{\partial p_k}\mathrm{d}p_k + \frac{\partial Y_i}{\partial q_k}\mathrm{d}q_k\right)\right\}$$

$$= \sum_{i,k}^{n} \left(\frac{\partial X_i}{\partial p_k} + \frac{\partial Y_i}{\partial q_i}\right)\mathrm{d}p_k \wedge \mathrm{d}q_i + \sum_{k<i}^{n} \left(\frac{\partial X_i}{\partial q_k} - \frac{\partial X_k}{\partial q_i}\right)\mathrm{d}q_k \wedge \mathrm{d}q_i$$

$$+ \sum_{i<k}^{n} \left(\frac{\partial Y_i}{\partial p_k} - \frac{\partial Y_k}{\partial p_i}\right)\mathrm{d}p_k \wedge \mathrm{d}p_i \equiv 0 \,. \tag{A-31}$$

By equating to zero the coefficients of the independent 2-forms in the last equality of the previous equation, (A-31), the necessary and sufficient conditions for the 1-form

$$\alpha = \sum_{i=1}^{n} \left[Y_i\,\mathrm{d}p_i - X_i\,\mathrm{d}q_i\right] , \tag{A-32}$$

to be closed (i.e., $\mathrm{d}\alpha=0$) are established. But every closed form is exact, i.e., it can be written as the differential of a smooth function H, $\alpha=\mathrm{d}H$, which implies, based on the definition of α through (A-32), that

$$Y_i = \frac{\partial H}{\partial p_i} , \quad X_i = -\frac{\partial H}{\partial q_i} , \tag{A-33}$$

or $\mathbf{u}=\mathbf{sgrad}(H)$ as required. ■

The significance of the previous theorem stems from the fact that ω preservation results in the invariance of ω along an integral curve which, based on the interpretation of the integral curve as a collection of mappings from the initial point of integration $\mathbf{x}_0$ to $\mathbf{x}(t)$, implies that the mapping from $\mathbf{x}_0$ to any $\mathbf{x}(t)$ is symplectic. Thus, if a system evolves through a sequence of symplectic maps, then the previous theorem guarantees the existence of a Hamiltonian vector field, the integral curves of which provide the evolution of the system under consideration.

Again, Theorem A.1 is very closely related to the theorem from differential calculus in Hilbert spaces which states that if the curl of a vector field is zero then this is a potential field and there is a potential function f so that the potential field can be expressed as the gradient of f. Once more we see the close relationship between the calculus in symplectic manifolds and the (more familiar) one in Hilbert (inner-product) spaces.

A.8 The Poisson Bracket

The Poisson bracket is defined in a symplectic manifold as the mapping of two smooth functions, f and g to another smooth function, $\{f,g\}$, specified as

$$\begin{aligned}\{f,g\} &\equiv \omega \circ \mathbf{sgrad}(f)\,\mathbf{sgrad}(g) \\ &= \langle\!\langle \mathbf{sgrad}(f), \mathbf{sgrad}(g) \rangle\!\rangle = L_{\mathbf{sgrad}(f)}(g)\ ,\end{aligned} \tag{A-34}$$

the third equality being valid by the definition of the skew-symmetric gradient, (A-24). Furthermore, if (A-25) is used for the evaluation of $\mathbf{sgrad}(f)$, then we arrive at the expression for the Poisson bracket as

$$\{f,g\} = \sum_{i,j=1}^{n} \omega^{ij} \frac{\partial f}{\partial x_j} \frac{\partial g}{\partial x_i}\ . \tag{A-35}$$

Starting from this expression, it is easy to verify the following three fundamental properties of the Poisson bracket:

1. the mapping $\{f,g\}$ is *bilinear*,

$$\{\alpha f + \beta g, h\} = \alpha\{f,h\} + \beta\{g,h\}\ ,\quad \alpha,\beta \in \mathbb{R} \text{ or } \mathbb{C}\ , \tag{A-36}$$

2. the mapping $\{f,g\}$ is *antisymmetric*,

$$\{f,g\} = -\{g,f\}\ , \tag{A-37}$$

3. the *Jacobi identity*,

$$\{f,\{g,h\}\} + \{g,\{h,f\}\} + \{h,\{f,g\}\} = 0\ , \tag{A-38}$$

is satisfied for any selection of three smooth functions, f, g, and h. Thus the function space on a symplectic manifold is endowed with a Lie algebra structure; the Lie derivative corresponding to the Poisson bracket.

In addition, a *Lie isomorphism* (i.e., a one-to-one correspondence preserving the Lie derivative) can be constructed between the function space $F(\mathfrak{I})$ and the subspace $H(\mathfrak{I})$ of all the Hamiltonian vector fields on the manifold, using the mapping

$$\alpha : f \in F(\mathfrak{I}) \to \boldsymbol{\alpha}(f) \equiv \mathbf{sgrad}(f) \in H(\mathfrak{I}) \,, \tag{A-39}$$

where the Lie derivative $[\![\mathbf{u},\mathbf{v}]\!]$ is defined within the space of smooth vector fields on the manifold, $\mathbf{u},\mathbf{v} \in V(\mathfrak{I})$ $[H(\mathfrak{I}) \subseteq V(\mathfrak{I})]$ as

$$\begin{aligned} [\![\mathbf{u},\mathbf{v}]\!] : L_{[\![\mathbf{u},\mathbf{v}]\!]} \circ f &\equiv L_{\mathbf{u}} \circ (L_{\mathbf{v}} \circ f) - L_{\mathbf{v}} \circ (L_{\mathbf{u}} \circ f) \\ \Rightarrow \quad [\![\mathbf{u},\mathbf{v}]\!]_i &= \sum_{i=1}^{n} \left(u_j \frac{\partial v_i}{\partial x_j} - v_j \frac{\partial u_i}{\partial x_j} \right) . \end{aligned} \tag{A-40}$$

Indeed, the mapping of the Poisson bracket of two smooth functions f and g, $\boldsymbol{\alpha}(\{f,g\})$, is expressed as

$$\begin{aligned} \boldsymbol{\alpha}(\{f,g\}) &= \mathbf{sgrad}(\{f,g\}) = \mathbf{sgrad}(L_{\mathbf{sgrad}(f)}\, g) \\ &= \mathbf{sgrad}\left(\sum_{k=1}^{n} \sum_{l=1}^{n} \omega^{kl} \frac{\partial f}{\partial x_l} \frac{\partial g}{\partial x_k} \right) \\ &= \sum_{i,j,k,l=1}^{n} \omega^{ij} \omega^{kl} \left(\frac{\partial^2 f}{\partial x_l \partial x_j} \frac{\partial g}{\partial x_k} + \frac{\partial f}{\partial x_l} \frac{\partial^2 g}{\partial x_k \partial x_j} \right) \mathbf{e}_i \,, \end{aligned} \tag{A-41}$$

where $\mathbf{e}_i$, $i=1,\ldots,n$, are the unit vectors in $T\mathfrak{I}$. This is the same as the Lie bracket of the mappings of the two functions individually, $[\![\boldsymbol{\alpha}(f),\boldsymbol{\alpha}(g)]\!]$:

$$\begin{aligned} [\![\boldsymbol{\alpha}(f),\boldsymbol{\alpha}(g)]\!] &= [\![\mathbf{sgrad}(f),\mathbf{sgrad}(g)]\!] \\ &= \left[\!\!\left[\sum_{i,j=1}^{n} \omega^{ij} \frac{\partial f}{\partial x_j} \mathbf{e}_i \,, \sum_{k,l=1}^{n} \omega^{kl} \frac{\partial g}{\partial x_l} \mathbf{e}_k \right]\!\!\right] \\ &= \sum_{i,j,k,l=1}^{n} \left(\omega^{il} \omega^{jk} \left(\frac{\partial f}{\partial x_k} \frac{\partial^2 g}{\partial x_l \partial x_j} \right) - \omega^{il} \omega^{jk} \left(\frac{\partial^2 f}{\partial x_l \partial x_j} \frac{\partial g}{\partial x_k} \right) \right) \mathbf{e}_i \\ &= \sum_{i,j,k,l=1}^{n} \omega^{ij} \omega^{kl} \left(\frac{\partial^2 f}{\partial x_l \partial x_j} \frac{\partial g}{\partial x_k} + \frac{\partial f}{\partial x_l} \frac{\partial^2 g}{\partial x_k \partial x_j} \right) \mathbf{e}_i \,. \end{aligned} \tag{A-42}$$

Mathematical entity	Hilbert space	Symplectic space
Quadratic form $\boldsymbol{\omega} \cdot \mathbf{ab} = \Sigma_{ij} \omega_{ij} a_i b_j$	Inner product $\boldsymbol{\omega}$: symmetric, positive-definite	Symplectic product $\boldsymbol{\omega}$: skew-symmetric, non-degenerate
Invariant transformations	Unitary transformations	Symplectic transformations
Differential operator on a function	Gradient	Skew-gradient
Vector field	Potential field	Hamiltonian field

Table A.1: Analogies between a Hilbert vector space and a symplectic vector space.

Finally, note that in the study of symplectic manifolds we could have used the properties of the Poisson bracket rather than those of the symplectic form ω as our starting point. In fact, quoting Fomenko [1988, p. 83], "The concept of a Poisson bracket is in a certain sense more flexible than that of the canonical symplectic structure generated by the nondegenerate form ω. The fact is that the degenerate Poisson bracket can be successfully used in many problems." We summarize the main analogies between a symplectic vector space and a Hilbert space in Table A.1.

p. 619: in Table A.1: $\boldsymbol{\omega} \cdot \mathbf{ab}$ should be $\boldsymbol{\omega} : \mathbf{ab}$

B
The Legendre Dual Transformation

Consider a function, f, which depends continuously on the N independent variables $a_1,a_2,...,a_N$: $f=f(a_1,a_2,...,a_N)$. Now we wish to introduce a new set of variables, $b_1,b_2,...,b_N$, which derive from the transformations

$$b_i = \frac{\partial f}{\partial a_i} , \quad i=1,2,...,N , \tag{B-1}$$

such that the N variables b_i are independent. Now we can define a new function, $g=g(b_1,b_2,...,b_N)$, in terms of the previous function, f, as

$$g \equiv \sum_{i=1}^{N} a_i b_i - f , \tag{B-2}$$

where the a_i are written in terms of the b_i. Taking the exact differential of the function g, we know that

$$\mathrm{d}g = \sum_{i=1}^{N} \frac{\partial g}{\partial b_i} \mathrm{d}b_i , \tag{B-3}$$

but, in terms of (B-2), we also have

$$\mathrm{d}g = \sum_{i=1}^{N} (a_i \mathrm{d}b_i + b_i \mathrm{d}a_i) - \mathrm{d}f . \tag{B-4}$$

Using the exact differential of f, as in (B-3), Eq. (B-4) can be rewritten as

$$\mathrm{d}g = \sum_{i=1}^{N} (a_i \mathrm{d}b_i + [b_i - \frac{\partial f}{\partial a_i}] \mathrm{d}a_i) , \tag{B-5}$$

from which, via (B-1), the second term on the right-hand side is seen to vanish. Therefore, by comparison of Eqs. (B-3) and (B-5), we recognize immediately that

$$\frac{\partial g}{\partial b_i} = a_i , \tag{B-6}$$

and thus arrive at the duality of this special transformation, i.e., the new variables, b_i, are the partial derivatives of the old function, f, and the old variables, a_i, are the partial derivatives of the new function, g. In fact, the concept of old and new is completely meaningless, as one can start with either system and arrive at the other. Parenthetically, the Legendre transformation arises from a basic theorem in line geometry—see Modell and Reid [1983, §5.4] and references therein.

Now let f be a function of M additional independent variables, $c_1, c_2, \ldots, c_M$: $f = f(a_1, \ldots, a_N, c_1, \ldots c_M)$; however, these new variables do not participate in the transformation. Therefore, the new function g will also depend upon these M variables: $g = g(b_1, \ldots, b_N, c_1, \ldots c_M)$. The transformation relation (B-1) still holds, as well as (B-2), so that

$$dg = \sum_{i=1}^{N} \frac{\partial g}{\partial b_i} db_i + \sum_{i=1}^{M} \frac{\partial g}{\partial c_i} dc_i \ . \tag{B-7}$$

Equation (B-5) now becomes

$$dg = \sum_{i=1}^{N} (a_i \, db_i + [b_i - \frac{\partial f}{\partial a_i}] \, da_i) - \sum_{i=1}^{M} \frac{\partial f}{\partial c_i} dc_i \ , \tag{B-8}$$

so that we immediately see that

$$\frac{\partial g}{\partial c_i} = - \frac{\partial f}{\partial c_i} \ , \quad i = 1,2,\ldots,M \ . \tag{B-9}$$

The a_i are generally termed the *active transformation variables* and the c_i are called the *passive transformation variables*. The connection of the above development with the Lagrangian/Hamiltonian transformation appearing throughout the book is obvious.

C
Poisson Brackets for Arbitrary Second-Rank Deformation Tensors

In §5.5.2, we saw how an absolute, second-rank, contravariant deformation tensor transformed the material Poisson bracket into its spatial counterpart. This choice was completely arbitrary, however, and we could just as easily have quantified the deformation of the medium with a second-rank, covariant deformation tensor,

$$c_{\alpha\beta}(\mathbf{F}) = \frac{\partial r_\gamma}{\partial Y_\alpha}\frac{\partial r_\gamma}{\partial Y_\beta} ; \tag{C-1}$$

i.e., specifying the new coordinate function $\overline{\mathbf{Y}}=\overline{\mathbf{Y}}(\mathbf{Y})$ and its inverse, we find that

$$\overline{c}_{\alpha\beta}(\mathbf{F}) = \frac{\partial r_\gamma}{\partial \overline{Y}_\alpha}\frac{\partial r_\gamma}{\partial \overline{Y}_\beta} = \frac{\partial r_\gamma}{\partial Y_\varepsilon}\frac{\partial Y_\varepsilon}{\partial \overline{Y}_\alpha}\frac{\partial r_\gamma}{\partial Y_\eta}\frac{\partial Y_\eta}{\partial \overline{Y}_\beta} = \frac{\partial Y_\varepsilon}{\partial \overline{Y}_\alpha}\frac{\partial Y_\eta}{\partial \overline{Y}_\beta} c_{\varepsilon\eta} , \tag{C-2}$$

which is indeed a covariant transformation.

In order to derive the Poisson bracket in terms of this covariant tensor, we replace the material Poisson bracket of (5.5-10) with

$$\{C_{\alpha\beta}(\mathbf{x},t), M_\gamma(\mathbf{z},t)\}_L = \int_\Omega \left[\rho_o c_{\alpha\beta}\frac{\partial \delta^3[\mathbf{Y}-\mathbf{x}]}{\partial Y_\gamma} - \frac{\partial}{\partial r_\eta}(\rho_o \delta^3[\mathbf{Y}-\mathbf{x}]\frac{\partial c_{\alpha\beta}}{\partial F_{\gamma\eta}})\right] \tag{C-3}$$
$$\times\ \delta^3[\mathbf{Y}-\mathbf{z}]\ d^3r ,$$

and by using the identity obtained from differentiating (see also Truesdell [1966, p. 18]),

$$\frac{\partial(\frac{\partial r_\varepsilon}{\partial Y_\alpha})}{\partial(\frac{\partial Y_\gamma}{\partial r_\eta})} = -\frac{\partial r_\varepsilon}{\partial Y_\gamma}\frac{\partial r_\eta}{\partial Y_\alpha} , \tag{C-4}$$

we find that

$$\{C_{\alpha\beta}(\mathbf{x},t), M_\gamma(\mathbf{z},t)\}_L = \int_{\Omega'} \left[\rho c_{\alpha\beta} \frac{\partial \delta^3[\mathbf{Y}-\mathbf{x}]}{\partial Y_\gamma(\mathbf{r},t)} + \frac{1}{J}\frac{\partial}{\partial r_\eta}(\rho J \delta^3[\mathbf{Y}-\mathbf{x}] \right.$$

$$\left. \times \{c_{\beta\gamma}\frac{\partial r_\eta}{\partial Y_\alpha} + c_{\alpha\gamma}\frac{\partial r_\eta}{\partial Y_\beta}\}) \right] \delta^3[\mathbf{Y}-\mathbf{z}]\, \mathrm{d}^3 Y$$

$$= \int_{\Omega'} \left[\rho c_{\alpha\beta} \frac{\partial \delta^3[\mathbf{Y}-\mathbf{x}]}{\partial Y_\gamma} + \frac{1}{J}\frac{\partial Y_\varepsilon}{\partial r_\eta}\frac{\partial}{\partial Y_\varepsilon}(\rho J \delta^3[\mathbf{Y}-\mathbf{x}] \right.$$

$$\left. \times \{c_{\beta\gamma}\frac{\partial r_\eta}{\partial Y_\alpha} + c_{\alpha\gamma}\frac{\partial r_\eta}{\partial Y_\beta}\}) \right] \delta^3[\mathbf{Y}-\mathbf{z}]\, \mathrm{d}^3 Y \qquad \text{(C-5)}$$

$$= \int_{\Omega'} \left[\rho c_{\alpha\beta} \frac{\partial \delta^3[\mathbf{Y}-\mathbf{x}]}{\partial Y_\gamma} + \frac{\partial}{\partial Y_\alpha}(\rho c_{\beta\gamma}\delta^3[\mathbf{Y}-\mathbf{x}] \right.$$

$$\left. + \frac{\partial}{\partial Y_\beta}(\rho c_{\alpha\gamma}\delta^3[\mathbf{Y}-\mathbf{x}]) \right] \delta^3[\mathbf{Y}-\mathbf{z}]\, \mathrm{d}^3 Y$$

$$= C_{\alpha\beta}(\mathbf{z},t)\frac{\partial \delta^3[\mathbf{z}-\mathbf{x}]}{\partial z_\gamma}$$

$$+ \frac{\partial}{\partial z_\alpha}(\delta^3[\mathbf{z}-\mathbf{x}]C_{\beta\gamma}) + \frac{\partial}{\partial z_\beta}(\delta^3[\mathbf{z}-\mathbf{x}]C_{\alpha\gamma}) \; .$$

Substituting this expression into the Poisson sub-bracket of (5.5-9) then yields the spatial Poisson bracket in terms of an absolute, second-rank, covariant tensor as

$$\{F,G\}_E^C = -\int_{\Omega'} \left[\frac{\delta F}{\delta C_{\alpha\beta}} \nabla_\gamma \left(C_{\alpha\beta}\frac{\delta G}{\delta M_\gamma} \right) - \frac{\delta G}{\delta C_{\alpha\beta}} \nabla_\gamma \left(C_{\alpha\beta}\frac{\delta F}{\delta M_\gamma} \right) \right] \mathrm{d}^3 x$$

$$+ \int_{\Omega'} C_{\gamma\alpha} \left[\frac{\delta G}{\delta C_{\alpha\beta}} \nabla_\beta \frac{\delta F}{\delta M_\gamma} - \frac{\delta F}{\delta C_{\alpha\beta}} \nabla_\beta \frac{\delta G}{\delta M_\gamma} \right] \mathrm{d}^3 x \qquad \text{(C-6)}$$

$$+ \int_{\Omega'} C_{\gamma\beta} \left[\frac{\delta G}{\delta C_{\alpha\beta}} \nabla_\alpha \frac{\delta F}{\delta M_\gamma} - \frac{\delta F}{\delta C_{\alpha\beta}} \nabla_\alpha \frac{\delta G}{\delta M_\gamma} \right] \mathrm{d}^3 x \; .$$

Hence we can derive the evolution equation for **C** as

$$\frac{\partial C_{\alpha\beta}}{\partial t} = -\nabla_\gamma (v_\gamma C_{\alpha\beta}) - C_{\gamma\alpha} \nabla_\beta v_\gamma - C_{\gamma\beta} \nabla_\alpha v_\gamma , \tag{C-7}$$

with the extra stress defined as

$$\sigma_{\alpha\beta} \equiv -2C_{\gamma\alpha} \frac{\delta H}{\delta C_{\beta\gamma}} . \tag{C-8}$$

Similarly, we could work in terms of a second-rank, mixed tensor as well, for example,

$$c_{\alpha\beta}(\mathbf{F}) = \frac{\partial r_\gamma}{\partial Y_\alpha} \frac{\partial Y_\beta}{\partial r_\gamma} , \tag{C-9}$$

which does indeed follow a mixed transformation law:

$$\bar{c}_{\alpha\beta}(\mathbf{F}) = \frac{\partial r_\gamma}{\partial \bar{Y}_\alpha} \frac{\partial \bar{Y}_\beta}{\partial r_\gamma} = \frac{\partial r_\gamma}{\partial Y_\varepsilon} \frac{\partial Y_\varepsilon}{\partial \bar{Y}_\alpha} \frac{\partial \bar{Y}_\beta}{\partial Y_\eta} \frac{\partial Y_\eta}{\partial r_\gamma} = \frac{\partial Y_\varepsilon}{\partial \bar{Y}_\alpha} \frac{\partial \bar{Y}_\beta}{\partial Y_\eta} c_{\varepsilon\eta} . \tag{C-10}$$

(Actually, this case is rather banal since, via (C-9), **c**=$\boldsymbol{\delta}$.) Using a combination of the arguments for the contravariant and covariant tensors, we can derive

$$\begin{aligned} \{F,G\}_E^C = &-\int_{\Omega'} \left[\frac{\delta F}{\delta C_{\alpha\beta}} \nabla_\gamma \left(C_{\alpha\beta} \frac{\delta G}{\delta M_\gamma} \right) - \frac{\delta G}{\delta C_{\alpha\beta}} \nabla_\gamma \left(C_{\alpha\beta} \frac{\delta F}{\delta M_\gamma} \right) \right] d^3x \\ &- \int_{\Omega'} C_{\gamma\alpha} \left[\frac{\delta G}{\delta C_{\alpha\beta}} \nabla_\gamma \frac{\delta F}{\delta M_\beta} - \frac{\delta F}{\delta C_{\alpha\beta}} \nabla_\gamma \frac{\delta G}{\delta M_\beta} \right] d^3x \\ &+ \int_{\Omega'} C_{\gamma\beta} \left[\frac{\delta G}{\delta C_{\alpha\beta}} \nabla_\alpha \frac{\delta F}{\delta M_\gamma} - \frac{\delta F}{\delta C_{\alpha\beta}} \nabla_\alpha \frac{\delta G}{\delta M_\gamma} \right] d^3x , \end{aligned} \tag{C-11}$$

from which we obtain

$$\frac{\partial C_{\alpha\beta}}{\partial t} = -\nabla_\gamma (v_\gamma C_{\alpha\beta}) - C_{\gamma\alpha} \nabla_\beta v_\gamma + C_{\gamma\beta} \nabla_\gamma v_\alpha , \tag{C-12}$$

$$\sigma_{\alpha\beta} \equiv C_{\beta\gamma} \frac{\delta H}{\delta C_{\gamma\alpha}} - C_{\gamma\alpha} \frac{\delta H}{\delta C_{\beta\gamma}} . \tag{C-13}$$

It is now obvious that the choice of the deformation tensor affects not only the evolution equation for **C**($\mathbf{x}$,t), but the form of the extra stress expression as well. Hence one must be very careful in defining this quantity, as this definition will have drastic implications concerning the predicted behavior of the resulting system of equations. Unfortunately, there is no universally accepted method for determining which definition of the deformation tensor (if any) corresponds best to the physics of the material in a particular instance. As shown by Oldroyd [1950, p. 541], however, the contravariant derivative does indeed predict the experimentally observed phenomenon of rod climbing.

D
Calculations of the Random-Flight Model

In order to preserve the flow of the development in §9.1.3, a couple of minor calculations were omitted. These are described in this appendix.

D.1 Calculation of the End-to-End Distribution Function near a Solid Surface

The problem of wall-induced conformational loss can be attacked in a number of different ways. For instance, we might use the method of Wiener (path) integrals [Wiegel, 1986, §1.4]; however, we choose here to make the passage of the distribution function to a differential equation. In this respect, the approach followed here is closely related to those of Edwards [1965], Freed [1972; 1987, ch. 3], and de Gennes [1969a; 1979, ch. 9] for the study of excluded-volume effects and adsorption phenomena. The Gaussian distribution function, (9.3-13), has been shown to be the solution to the partial differential equation [Chandrasekhar, 1943, §1.5]

$$\frac{\partial W}{\partial t} = D_1 \frac{\partial^2 W}{\partial \xi^2} + D_2 \frac{\partial^2 W}{\partial \eta^2} + D_3 \frac{\partial^2 W}{\partial \zeta^2} , \tag{D.1-1}$$

where

$$D_1 = \frac{1}{2}\nu<\xi>^2 , \quad D_2 = \frac{1}{2}\nu<\eta>^2 , \quad \text{and} \quad D_3 = \frac{1}{2}\nu<\zeta>^2 , \tag{D.1-2}$$

and ν is the number of displacements per unit time, $\nu=N/t$. The differential equation (D.1-1,2) is subject to the homogeneous boundary conditions

$$\begin{gathered} W = \delta(\xi-\xi_0)\,\delta(\eta-\eta_0)\,\delta(\zeta-\zeta_0) \quad \text{at} \quad t=0 , \\ W = 0 \quad \text{at} \quad \xi = \pm\infty , \quad \text{at} \quad \eta = \pm\infty , \quad \text{and at} \quad \zeta = \pm\infty , \end{gathered} \tag{D.1-3}$$

so that

$$c_{\xi\xi} = 2D_1 t , \quad c_{\eta\eta} = 2D_2 t , \quad c_{\zeta\zeta} = 2D_3 t . \tag{D.1-4}$$

In this case, (ξ_0,η_0,ζ_0) is the point of origin of the chain and (ξ,η,ζ) is its endpoint.

Since we are interested only in two-dimensional flows at this point, we shall consider the problem in the (x,y) coordinate system, where the x-axis is parallel to the wall and the y-axis is perpendicular to it. Let us also denote (ξ,η) as the orthogonal coordinate system defined by the eigenvectors of the matrix (9.3-12) (restricted to two dimensions); i.e., ξ and η represent the principal axes of the ellipse. If $\hat{a}$ is the rotation angle of the ellipse with respect to the solid surface, then the condition that $y \geq 0$ is translated into $\eta \geq -\xi \tan(\hat{a})$.

For the particular problem at hand, the distribution function for wall-induced conformational reduction is given by the solution to the boundary value problem

$$\frac{\partial W}{\partial t} = D_1 \frac{\partial^2 W}{\partial \xi^2} + D_2 \frac{\partial^2 W}{\partial \eta^2} , \tag{D.1-5}$$

with the boundary conditions

$$\begin{gathered} W = \delta(\xi - \xi_0)\, \delta(\eta - \eta_0) \quad \text{at} \quad t = 0 , \\ W = 0 \quad \text{at} \quad \xi = \pm\infty , \quad \text{at} \quad \eta = \infty , \quad \text{and at} \quad \eta = -\xi \tan(\hat{a}) . \end{gathered} \tag{D.1-6}$$

In this system of coordinates, (ξ_0,η_0) is the point of origin of the chain. An analytic solution for this problem may be obtained by incorporating D_1 and D_2 into generalized coordinates and subsequently applying the method of images [Feller, 1968, p. 369]. The solution is given by (9.3-16,17).

D.2 Calculation of the Partition Function

Rewriting (9.3-14) in the (x,y) coordinate system, we obtain

$$\begin{aligned} W = (4\pi^2 \det \mathbf{c})^{-1/2} \Big\{ & \exp[-\tfrac{1}{2}\mathbf{c}^{-1} : (\mathbf{x}-\mathbf{x}_0)(\mathbf{x}-\mathbf{x}_0)] \\ & - \exp[-\tfrac{1}{2}\mathbf{c}^{-1} : (\mathbf{x}-\mathbf{x}_2)(\mathbf{x}-\mathbf{x}_2)] \Big\} , \end{aligned} \tag{D.2-1}$$

where $\mathbf{x}_0 \equiv (0,y_0)^T$ is the location of the chain origin and

$$\mathbf{x}_2 \equiv y_0 \left(\frac{2(1 - \frac{D_1}{D_2}) \tan\hat{a}}{1 + \frac{D_1}{D_2} \tan^2\hat{a}} , -1 \right)^T . \tag{D.2-2}$$

Then, by inserting (D.2-1) into (9.3-18) and rearranging the integration limits $-\infty<x<\infty$ and $0\leq y<\infty$ for the new variables $\mathbf{q}=\mathbf{x}-\mathbf{x}_0$ and $\mathbf{p}=\mathbf{x}-\mathbf{x}_2$, we find that

$$\Theta = (4\pi^2 \det \mathbf{c})^{-1/2} \left[\int_{-y_o}^{\infty} \int_{-\infty}^{\infty} \exp[-\tfrac{1}{2}\mathbf{c}^{-1} : \mathbf{qq}]\, dq_1\, dq_2 - \int_{y_o}^{\infty} \int_{-\infty}^{\infty} \exp[-\tfrac{1}{2}\mathbf{c}^{-1} : \mathbf{pp}]\, dp_1\, dp_2 \right] . \tag{D.2-3}$$

In this case, $\mathbf{c}$ is the conformation tensor and $\mathbf{c}^{-1}$ is its inverse in the (x,y) coordinate system. However,

$$\int_{-\infty}^{\infty} \exp[-\tfrac{1}{2}\mathbf{c}^{-1} : \mathbf{qq}]\, dq_1 = (2\pi / c_{11}^{-1})^{1/2} \exp\left[\tfrac{1}{2} q_2^2 \{ \frac{(c_{12}^{-1})^2}{c_{11}^{-1}} - c_{22}^{-1} \} \right] . \tag{D.2-4}$$

A similar relation holds for the inner integral of the second term in (D.2-3) over p_1. Using these identities in (D.2-3), integrating over q_2 and p_2, and properly rearranging the terms, we obtain the final result, (9.3-19).

Bibliography

[Newton wrote to Halley, who was supervising the publication of the great Principia, *that he would not give Hooke any credit.] That, alas, is vanity. You find it in so many scientists. You know, it has always hurt me to think that Galileo did not acknowledge the work of Kepler.*

—Albert Einstein [Cohen, 1979, p.41]

Abarbanel, H.D.I., Brown, R., and Yang, Y.M. (1988). Hamiltonian formulation of inviscid flows with free boundaries. *Phys. Fluids* 31: 2802-09.

Abe, A., and Ballauff, M. (1991). The Flory lattice model. In: *Liquid Crystallinity in Polymers: Principles and Fundamental Properties*, Cifferi, A. (ed.). VCH Publishers: New York, pp. 131-67.

Abraham, R., Marsden, J.E., and Ratiu, T. (1988). *Manifolds, Tensor Analysis, and Applications*, 2nd ed. Springer-Verlag: New York.

Abramowitz, M., and Stegun, I. (1964). *Handbook of Mathematical Functions with Formulas, Graphs, and Mathematical Tables*. National Bureau of Standards: Washington.

Acierno, D., La Mantia, F.P., Marrucci, G., and Titomanlio, G. (1976a). A nonlinear viscoelastic model with structure-dependent relaxation times. I. Basic formulation. *J. Non-Newtonian Fluid Mech.* 1: 125-46.

Acierno, D., La Mantia, F.P., Marrucci, G., Rizzo, G., and Titomanlio, G. (1976b). A non-linear viscoelastic model with structure-dependent relaxation times. II. Comparison with L.D. Polyethylene transient stress results. *J. Non-Newtonian Fluid Mech.* 1: 147-57.

Aero, E.L., Bulygin, A.N., and Kuvshinskii, E.V. (1965). Asymmetric hydromechanics. *J. Appl. Math. Mech.* 29: 333-46.

Aifantis, E.C. (1980). On the problem of diffusion in solids. *Acta Mech.* 37: 265-96.

Akyildiz, F., Jones, R.S., and Walters, K. (1990). On the spring-dashpot representation of linear viscoelastic behaviour. *Rheol. Acta* 29: 482-84.

Allender, D.W., and Lee, M.A. (1984). Landau theory of biaxial nematic liquid crystals. *Mol. Cryst. Liq. Cryst.* 110: 331-39.

Altmann, S.L. (1992). *Icons and Symmetries*. Clarendon Press: Oxford.

Apelian, M.R., Armstrong, R.C., and Brown, R.A. (1988). Impact of the constitutive equation and singularity on the calculation of stick-slip flow: The modified upper-convected Maxwell model (MUCM). *J. Non-Newtonian Fluid Mech.* 27: 299-321.

Aris, R. (1962). *Vectors, Tensors, and the Basic Equations of Fluid Mechanics*. Prentice-Hall, Inc.: New Jersey.

Aris, R. (1965). Prolegomena to the rational analysis of systems of chemical reactions. *Arch. Rational Mech. Anal. 19,* 81-99.

Aris, R. (1968). Prolegomena to the rational analysis of systems of chemical reactions II. Some addenda. *Arch. Rational Mech. Anal.* 22: 356-64.

Arnold, V.I. (1965). Variational principle for three-dimensional steady-state flows of an ideal fluid. *J. Appl. Math. Mech. 29,* 1002-8.

Arnold, V.I. (1966a). Sur la géométrie différentielle des groupes de Lie de dimension infinie et ses applications a l'hydrodynamique des fluides parfaits. *Ann. Inst. Fourier* 16: 319-61.

Arnold, V.I. (1966b). Sur un principe variationnel pour les écoulements stationnaires des liquides parfaits et ses applications aux problèmes de stabilité non linéaires. *J. Mécanique* 5: 29-43.

Arnold, V.I. (1978). *Mathematical Methods of Classical Mechanics*. Springer-Verlag: New York.

Ashtekar, A. (1991). *Lectures on Non-Perturbative Canonical Gravity*. World Scientific: Singapore.

Atkinson, K.E. (1978). *An Introduction to Numerical Analysis*. John Wiley and Sons: New York.

Aubert, J.H., and Tirrell, M. (1980). Macromolecules in non-homogeneous velocity gradient fields. *J. Chem. Phys.* 72: 2694-701.

Aubert, J.H., and Tirrell, M. (1982). Effective viscosity of dilute polymer solutions near confining boundaries. *J. Chem. Phys.* 77: 553-61.

Ausserré, D., Hervet, H., and Rondelez, F. (1986). Concentration profile of polymer solutions near a solid wall. *Phys. Rev. Lett.* 54: 1948-51.

Ausserré, D., Edwards, J., Lecourtier, J., Hervet, H., and Rondelez, F. (1991). Hydrodynamic thickening of depletion layers in colloidal solutions. *Europhys. Lett.* 14: 33-8.

Bagley, E., and Long, F.A. (1955). Two-stage sorption and desorption of organic vapors in cellulose acetate. *J. Amer. Chem. Soc.* 77: 2172-78.

Balescu, R. (1975). *Equilibrium and Nonequilibrium Statistical Mechanics*. John Wiley & Sons: New York.

Ballauff, M. (1989). The Flory lattice model of nematic fluids. *Mol. Cryst. Liq. Cryst.* 168: 209-28.

Bamberg, P.G., and Sternberg, S. (1988). *A Course in Mathematics for Students of Physics*, vol. 1. Cambridge University Press: Cambridge.

Barnes, H.A., Hutton, J.F., and Walters, K. (1989). *An Introduction to Rheology*. Elsevier: Amsterdam.

Barut, A.O. (1986). From kinematical groups to dynamical groups. Originally appeared in: *Lecture Notes in Physics*, vol. 261. Springer-Verlag: New York; pp. 3-21. Also reprinted as Appendix D in: Barut, A.O. (1989). *Geometry and Physics, Non-Newtonian Forms of Dynamics*. Bibliopolis: Napoli.

Bataille, J., and Kestin, J. (1977). Thermodynamics of mixtures. *J. Non-Equilib. Thermodyn.* 2: 49-65.

Bataille, J., Edelen, D.G., Kestin, J. (1978). Non-equilibrium thermodynamics of the nonlinear equations of chemical kinetics. *J. Non-Equilib. Thermodyn.* 3: 153-68.

Batchelor, G.K. (1967). *An Introduction to Fluid Dynamics*. Cambridge University Press: Cambridge.

Bearman, R.J., and Kirkwood, J.G. (1958). Statistical mechanics of transport processes. XI. Equations of transport in multicomponent systems. *J. Chem. Phys.* 28: 136-45.

Bedford, A. (1985). *Hamilton's Principle in Continuum Mechanics*. Pitman Publishing: Boston.

Bedford, A., and Drumheller, D.S. (1983). Theories of immiscible and structured mixtures. *Int. J. Engng. Sci.* 21: 863-960.

Berens, A.R., and Huvard, G.S. (1987). Interaction of polymers with near-critical carbon dioxide. Manuscript prepared for presentation at *1987 Annual AIChE Meeting*, New York City, November 15-20, paper No. 37e.

Beris, A.N., and Edwards, B.J. (1990a). Poisson bracket formulation of incompressible flow equations in continuum mechanics. *J.Rheol.* 34: 55-78.

Beris, A.N., and Edwards, B.J. (1990b). Poisson bracket formulation of viscoelastic flow equations of differential type: A unified approach. *J. Rheol.* 34: 503-38.

Beris, A.N., and Edwards, B.J. (1993). On the admissibility criteria for linear viscoelasticity kernels. *Rheol. Acta* 32: 505-10.

Beris, A.N., Mavrantzas, V.G., and Edwards, B.J. (1994). Investigation of the properties of a coupled two-mode Maxwell model: Slow, slowly-varying flows. *J. Non-Newtonian Fluid Mech.*, submitted for publication.

Bernstein, B., Kearsley, A.E., and Zapas, L. (1963). A study of stress relaxation with finite strain. *Trans. Soc. Rheol.* 7: 391-410.

Bernstein, S. (1928). Sur les fonctions absolument monotones. *Acta Mathematica* 52: 1-66.

Berreman, D.W. (1973). Alignment of liquid crystals by grooved surfaces. *Mol. Cryst. Liq. Cryst.* 23: 215-31.

Berry, G.C. (1988). Rheological properties of nematic solutions of rodlike polymers. *Mol. Cryst. Liq. Cryst.* 165: 333-60.

Berry, G.C. (1991). Bingham award lecture—1990: Rheological and rheo-optical studies on nematic solutions of a rodlike polymer. *J. Rheol.* 35: 943-83.

Bhave, A.V., Armstrong, R.C., and Brown, R.A. (1991). Kinetic theory and rheology of dilute, nonhomogeneous polymer solutions. *J. Chem. Phys.* 95: 2988-3000.

Biller, P., Öttinger, H.C., and Petruccione, F. (1986). Consistently averaged hydrodynamic interaction for dumbbell models in elongational flow. *J. Chem. Phys.* 85: 1672-5.

Billmeyer, F.W. (1984). *Textbook of Polymer Science,* 3rd ed. John Wiley and Sons: New York.

Bird, R.B., and Curtiss, C.F. (1985). Molecular theory expressions for the stress tensor in flowing polymeric liquids. *J. Polym. Sci.: Polym. Symp.* 73: 187-99.

Bird, R.B., and DeAguiar, J.R. (1983). An encapsulated dumbbell model for concentrated polymer solutions and melts I. Theoretical development and constitutive equation. *J. Non-Newtonian Fluid Mech.* 13: 149-60.

Bird, R.B., and Öttinger, H.C. (1992). Transport properties of polymeric liquids. *Annual Review of Physical Chemistry* 43: 371-406.

Bird, R.B., and Wiest, J.M. (1985). Anisotropic effects in dumbbell kinetic theory. *J. Rheol.* 29: 519-32.

Bird, R.B., Armstrong, R.C., and Hassager, O. (1987a). *Dynamics of Polymeric Fluids,* vol. 1, 2nd ed. John Wiley and Sons: New York.

Bird, R.B., Curtiss, C.F., Armstrong, R.C., and Hassager, O. (1987b). *Dynamics of Polymeric Fluids,* vol. 2, 2nd ed. John Wiley and Sons: New York.

Bitsanis, I., and Hadziioannou, G. (1990). Molecular dynamics simulations of the structure and dynamics of confined polymer melts. *J. Chem. Phys.* 92: 3827-47.

Bloore, F.J. (1973). Identity of commutator and Poisson bracket. *J. Phys. A: Math. Nucl. Gen.* 6: L7-8.

Boeuf, J.-P. (1988). A two-dimensional model of dc glow discharges. *J. Appl. Phys.* 63: 1342-9.

Booij, H.C. (1984). The energy storage in the Rouse model in an arbitrary flow field. *J. Chem. Phys.* 80: 4571-2.

Braginskii, S.I. (1965). Transport processes in a plasma. *Reviews of Plasma Physics* 1: 205-311.

Brennan, M., Jones, R.S., and Walters, K. (1988). Linear viscoelasticity revisited. In: *Proc. Xth Intern. Congr. Rheology,* vol. 1. Sydney, Australia, Aug. 14-19, Uhlherr, P.H.T. (ed.). Australian Society of Rheology, pp. 207-9.

Breuer, S., and Onat, E.T. (1962). On uniqueness in linear viscoelasticity. *Q. Appl. Math.* 19: 355-9.

Brochard-Wyart, F., de Gennes, P.G., and Pincus, P., (1992). Suppression du glissement à l' interface de deux polymères fondus incompatibles. *C.R. Acad. Sci. Paris* Ser. 2 314: 873-8.

Bronshtein, I.N., and Semendyayev, K.A. (1985). *Handbook of Mathematics.* Van Nostrand Reinhold Co.: New Yrok.

Brown, R.A. (1992). Stretching viscoelastic fluid mechanics: The importance of corners and elongational flows. In: *Fundamental*

Research in Fluid Mechanics: Invited Lecture. AIChE 1992 annual meeting, November 1-6, 1992, Miami Beach, Florida.

Brunn, P.O. (1976). The effect of a solid wall for the flow of dilute macro-molecular solutions. *Rheol. Acta* 15: 23-9.

Brunn, P.O. (1984). Non-uniform concentration profiles of dilute macromolecular solutions in rotational viscometric flows. *J. Chem. Phys.* 80: 3420-6.

Brunn, P.O. (1985). Wall effects for dilute polymer solutions in arbitrary unidirectional flows. *J. Rheol.* 29: 859-86.

Brunn, P.O., and Grisafi, S. (1987). Pressure driven flow of dilute polymer solutions in a channel: Analytic results for very small and very large channels. *Rheol. Acta* 26: 211-6.

Burghardt, W.R., and Fuller, G.G. (1990). Transient shear flow of nematic liquid crystals: Manifestations of director tumbling. *J. Rheol.* 34: 959-92.

Burghardt, W.R., and Fuller, G.G. (1991). Role of director tumbling in the rheology of polymer liquid crystal solutions. *Macromolecules* 24: 2546-55.

Burke, W.L. (1985). *Applied Differential Geometry*. Cambridge University Press: Cambridge.

Cahn, J.W. (1961). On spinodal decomposition. *Acta Metall.* 9: 795-801.

Cahn, J.W., and Hilliard, J.E. (1958). Free energy of a nonuniform system. I. Interfacial free energy. *J. Chem. Phys.* 28: 258-67.

Calderer, M.C., Cook, L.P., and Schleiniger, G. (1989). An analysis of the Bird-DeAguiar model for polymer melts. *J. Non-Newtonian Fluid Mech.* 31: 209-25.

Callen, H.B. (1960). *Thermodynamics*. John Wiley and Sons: New York.

Camera-Roda, G., and Sarti, G.C. (1990). Mass transport with relaxation in polymers. *A.I.Ch.E. J.* 36: 851-60.

Capitelli, M., Celiberto, R., Gorse, C., Winkler, R., and Wilhelm, J. (1987). Electron energy distribution function in He-CO radio frequency plasmas: The role of vibrational and electronic superelastic collisions. *J. Appl. Phys.* 62: 4398-403.

Capitelli, M., Gorse, C., Winkler, R., and Wilhelm, J. (1988). On the modulation of the electron energy distribution function in radio frequency SiH_4, SiH_4-H_2 bulk plasmas. *Plasma Chem. Plasma Proc.* 8: 399-424.

Carlsson, T. (1984). Theoretical investigation of the shear flow of nematic liquid crystals with Leslie viscosity $\alpha_3>0$: Hydrodynamic analogue of first order phase transitions. *Mol. Cryst. Liq. Cryst.* 104: 307-34.

Casimir, H.B.G. (1945). On Onsager's principle of microscopic reversibility. *Rev. Mod. Phys.* 17: 343-50.

Casti, J.L. (1989). *Alternate Realities*. John Wiley and Sons: New York.

Cattaneo, C. (1948). Sulla conduzione del calore. *Atti Sem. Mat. Fis. Univ. Modena* 3: 3-21.

Chandler, D. (1987). *Introduction to Modern Statistical Mechanics*. Oxford University Press: New York.

Chandrasekhar, S. (India) (1977). *Liquid Crystals*. Cambridge University Press: Cambridge.

Chandrasekhar, S. (USA) (1943). Stochastic problems in physics and astronomy. *Rev. Mod. Phys.* 15: 1-89.

Chapman, B. (1980). *Glow Discharge Processes, Sputtering and Plasma Etching*. John Wiley and Sons: New York.

Chauveteau, G. (1982). Rodlike polymer solution flow through fine pores: Influence of pore size on rheological behavior. *J. Rheol.* 26: 111-42.

Chilcott, M.D., and Rallison, J.M. (1988). Creeping flow of dilute polymer solutions past cylinders and spheres. *J. Non-Newtonian Fluid Mech.* 29: 381-432.

Chu, K.C., and McMillan, W.L. (1977). Unified Landau theory for the nematic, smectic A, and smectic C phases of liquid crystals. *Phys. Rev.* A 15: 1181-7.

Chung, P.H., Talbot, L., and Touryon, K.J. (1975). *Electric Probes in Stationary and Flowing Plasmas: Theory and Application*. Springer-Verlag: New York.

Ciferri, A. (ed.) (1991a). *Liquid Crystallinity in Polymers: Principles and Fundamental Properties*. VCH Publishers: New York.

Ciferri, A. (1991b). Phase behavior of rigid and semirigid mesogens, in: *Liquid Crystallinity in Polymers: Principles and Fundamental Properties*, Ciferri, A. (ed.). VCH Publishers: New York; pp. 209-59.

Ciferri, A., Krigbaum, W.R., and Meyer, R.B. (eds.) (1982). *Polymer Liquid Crystals*. Academic Press: New York.

Cladis, P.E., and Torza, S. (1975). Stability of nematic liquid crystals in Couette flow. *Phys. Rev. Lett.* 19: 1283-6.

Clark, M.G., Saunders, F.C., Shanks, I.A., and Leslie, F.M. (1981). A study of alignment instability during rectilinear oscillatory shear of nematics. *Mol. Cryst. Liq. Cryst.* 70: 195-222.

Cohen, A. (1991). A Padé approximant to the inverse Langevin function. *Rheologica Acta* 30: 270-273.

Cohen, I.B. (1979). Einstein and Newton. In: *Einstein: A Centenary Volume,* A.P. French (ed.). Harvard University Press: Cambridge, Massachusetts; pp. 40-2.

Cohen, R., Hardt, R., Kinderlehrer, D., Lin, S.-Y., and Lyskin, M. (1987). Minimum energy configurations for liquid crystals: Computational results. In: *Theory and Applications of Liquid Crystals*, Ericksen, J.L., and Kinderlehrer, D., (eds.). IMA vol. 5. Springer-Verlag: New York; pp. 99-121.

Cohen, Y., and Metzner, A.B. (1985). Apparent slip flow of polymer solutions. *J. Rheol.* 29: 67-102.

Cohen-Tannoudji, C., Dupont-Roc, J., and Grynberg, G. (1989). *Photons and Atoms*. John Wiley and Sons: New York.

Coleman, B.D., and Truesdell, C. (1960). On the reciprocal relations of Onsager. *J. Chem. Phys.* 33: 28-31.

Coleman, B.D., Markovitz, H., and Noll, W. (1966). *Viscometric Flows of Non-Newtonian Fluids*. Springer: New York.

Conlon, L. (1993). *Differentiable manifolds. A First Course*. Birkhäuser: Boston.

Cook, L.P., and Schleiniger, G. (1991). Inlet layer in the flow of viscoelastic fluids. *J. Non-Newtonian Fluid Mech.* 40: 307-21.

Courant, R., and Hilbert, D. (1962). *Methods of Mathematical Physics. Vol. II. Partial Differential Equations*. John Wiley and Sons: New York.

Cowin, S.C. (1968). Polar fluids. *Phys.* Fluids *11*: 1919-68.

Cowin, S.C. (1974). The Theory of Polar Fluids, in: *Advances in Applied Mechanics*, vol. 14. Yih, C.-S. (ed.). Academic Press: New York; pp. 279-347.

Crank, J. (1953). A theoretical investigation of the influence of molecular relaxation and internal stress on diffusion in polymers. *J. Polym. Sci.* 11: 151-68.

Crochet, M.J. (1975). A non-isothermal theory of viscoelastic materials. In: *Theoretical Rheology*. Hutton, J.F., Pearson, J.R.A., and Walters, K. (eds.). Halsted Press: New York; pp. 111-122.

Curie, P. (1894). Sur la symétrie dans les phénomènes physiques, symétrie d'un champ électrique et d'un champ magnétique. *J. Physique, 3e serie,* III: 393-415. Also in: (1908). *Oeuvres de Pierre Curie*. Gauthier-Villars: Paris; pp. 118-41.

Currie, P.K., and Macsithigh, G.P. (1979). The stability and dissipation of solutions for shearing flow of nematic liquid crystals. *Q. J. Mech. Appl. Math.* 32: 499-511.

Curtiss, C.F., and Bird, R.B. (1981a). A kinetic theory for polymer melts. I. The equation for the single-link orientational distribution function. *J. Chem. Phys.* 74: 2016-25.

Curtiss, C.F., and Bird, R.B. (1981b). A kinetic theory for polymer melts. II. The stress tensor and the rheological equation of state. *J. Chem. Phys.* 74: 2026-33.

Cussler, E.L. (1984). *Diffusion: Mass Transfer in Fluid Systems*. Cambridge University Press: Cambridge.

DeAguiar, J.R. (1983). An encapsulated dumbbell model for concentrated polymer solutions and melts II. Calculation of material functions and experimental comparisons. *J. Non-Newtonian Fluid Mech.* 13: 161-79.

Dean, W.R., and Montagnon, P.E. (1949). On the steady motion of a viscous liquid in a corner. *Proc. Camb. Phil. Soc.* 45: 389-94.

De Cleyn, G., and Mewis, J. (1981). A constitutive equation for polymeric liquids: Application to shear flow. *J. Non-Newtonian Fluid Mech.* 9: 91-105.

de Gennes, P.G. (1969a). Some conformation problems for long macromolecules. *Rep. Progr. Phys.* 32: 187-205.

de Gennes, P.G. (1969b). Phenomenology of short-range-order effects in the isotropic phase of nematic materials. *Phys. Lett.* 30A: 454-5.

de Gennes, P.G. (1971). Short range order effects in the isotropic phase of nematics and cholesterics. *Mol. Cryst. Liq. Cryst.* 12: 193-214.

de Gennes, P.G. (1972). Some remarks on the polymorphism of smectics. *Mol. Cryst. Liq. Cryst.* 21: 49-76.

de Gennes, P.G. (1974). *The Physics of Liquid Crystals*. Clarendon Press: Oxford.

de Gennes, P.G. (1977). Polymeric liquid crystals: Frank elasticity and light scattering. *Mol. Cryst. Liq. Cryst.* 34 (Letters): 177-82.

de Gennes, P.G. (1979). *Scaling Concepts in Polymer Physics*. Cornell University Press: Ithaca, New York.

de Gennes, P.G. (1981). Polymer solutions near an interface. 1. Adsorption and depletion layers. *Macromolecules* 14: 1637-44.

de Groot, S.R. (1974). The Onsager relations: Theoretical basis. In: *Foundations of Continuum Thermodynamics*, Domingos, J.J.D., Nina, M.N.R., and Whitelaw, J.H. (eds.). Macmillan Press: London; pp. 159-72.

de Groot, S.R., and Mazur, P. (1962). *Non-Equilibrium Thermodynamics*. North-Holland Publishing Company: Amsterdam. (Also available in paperback, (1984). Dover: New York.)

de Haro, M.L., and Rubi, J.M. (1988). Consistently averaged hydrodynamic interaction beyond the Oseen approximation for Rouse-Zimm-Bueche dumbbells in steady shear flow. *J. Chem. Phys.* 88: 1248-52.

Dekker, H. (1981). Classical and quantum mechanics of the damped harmonic oscillator. *Phys. Rep.* 80: 1-112.

de Pablo, J.J., Öttinger, H.C., and Rabin, Y. (1992). Hydrodynamic changes of the depletion layer of dilute polymer solutions near a wall. *A.I.Ch.E. J.* 38: 273-83.

des Cloizeaux, J., and Jannink, G. (1989). *Polymers in Solution: Their Modelling and Structure*. Oxford University Press: Oxford (English translation from the French edition of 1987).

Diaz, F.G., de la Torre, J.G., and Freire, J.J. (1989). Hydrodynamic interaction effects in the rheological properties of Hookean dumbbells in steady shear flow: A Brownian dynamics simulation study. *Polymer* 30: 259-64.

Dill, K.A. and Zimm, B.H. (1979). A rheological separator for very large DNA molecules. *Nucleic Acids Research* 7: 735-49.

Dirac, P.A.M. (1925). The fundamental equations of quantum mechanics. *Proc. Roy. Soc. A* 109: 642-53.

Dobb, M.G. and McIntyre, J.E. (1984). Properties and applications of liquid-crystalline main-chain polymers. *Adv. Polym. Sci.* 60/61: 61-98.

Doi, M. (1981). Molecular dynamics and rheological properties of concentrated solutions of rodlike polymers in isotropic and liquid crystalline phases. *J. Polym. Sci.: Polym. Phys. Ed.* 19: 229-43.

Doi, M. (1983). Viscoelasticity of concentrated solutions of stiff polymers. *Faraday Symp. Chem. Soc.* 18: 49-56.

Doi, M. (1990). Effects of viscoelasticity on polymer diffusion, in: *Dynamics and Patterns in Complex Fluids: New Aspects of Physics and Chemistry of Interfaces*, Onuki, A., and Kawasaki, K. (eds.). Springer: New York.

Doi, M., and Edwards, S.F. (1978). Dynamics of concentrated polymer systems. Part 1. Brownian motion in the equilibrium state, *J. Chem. Soc. Faraday Trans. II 74*, 1789-801; Part 2. Molecular motion under flow. *ibid.:* 1802-1817; Part 3. The constitutive equation. *ibid.:* 1818-32.

Doi, M., and Edwards, S.F. (1979). Dynamics of concentrated polymer sys-tems. Part 4. Rheological properties, *J. Chem. Soc. Faraday Trans. II 75*, 38-54.

Doi, M., and Edwards, S.F. (1986). *The Theory of Polymer Dynamics*. Clarendon Press: Oxford.

Doi, M., Shimada, T. and Okano, K. (1988). Concentration fluctuation of stiff polymers. II. Dynamical structure factor of rod-like polymers in the isotropic phase. *J. Chem. Phys.* 88: 4070-5.

Domingos, J.J.D. (1974). From thermostatics to thermodynamics. In: *Foundations of Continuum Thermodynamics*, Domingos, J.J.D., Nina, M.N.R., and Whitelaw, J.H. (eds.). Macmillan Press: London; pp. 1-21.

Duering, E., and Rabin, Y. (1990). Polymers in shear flow near repulsive boundaries. *Macromolecules* 23: 2232-7.

Dugas, R. (1988). *A History of Mechanics*. Dover Publications: New York.[1]

Dunwoody, N.T., and Müller, I. (1968). A thermodynamic theory of two chemically reacting ideal gases with different temperatures. *Arch. Rat. Mech. Anal.* 29: 344-69.

DuPré, D.B., and Yang, S.-J. (1991). Liquid crystalline properties of solutions of persistent polymer chains. *J. Chem. Phys.* 94: 7466-77.

Dupret, F., and Marchal, J.M. (1986a). Sur le signe des valeurs propres du tenseur des extra-contraintes dans un écoulement de fluide de Maxwell. *J. Méc. Théor. Appli.* 5: 403-27.

[1]This is a recent English translation of the original work *Histoire de la Mécanique*: Éditions du Griffon, Neuchâtel, Switzerland, 1955.

Dupret, F., and Marchal, J.M. (1986b). Loss of evolution in the flow of viscoelastic fluids. *J. Non-Newtonian Fluid Mech.* 20: 143-71.

Durning, C.J., Spencer, J.L., and Tabor, M. (1985). Differential sorption and permeation in viscous media. *J. Polym. Sci.: Polym. Lett. Ed.* 23: 171-81.

Durning, C.J., and Tabor, M. (1986). Mutual diffusion in concentrated polymer solutions under a small driving force. *Macromolecules* 19: 2220-32.

Dutta, A., and Mashelkar, R.A. (1984). Hydrodynamics in media with migrating macromolecules: Development of FDCF asymptote. *J. Non-Newtonian Fluid Mech.* 16: 279-302.

Dutta, A., and Mashelkar, R.A. (1987). Thermal conduction in polymeric liquids. In: *Transport Phenomena in Polymeric Systems, Part 1,* Mashelkar, R.A., Mujumdar, A.S., and Kamal, R. (eds.). Ellis Horwood: Chichester; pp. 285-338.

Dzyaloshinskii, I.E., and Volovick, G.E. (1980). Poisson brackets in condensed matter physics. *Ann. Phys.* 125: 67-97.

Ebin, D.G., and Marsden, J. (1970). Groups of diffeomorphisms and the motion of an incompressible fluid. *Ann. Math.* 92: 102-63.

Eckart, C. (1948a). The thermodynamics of irreversible processes. IV. The theory of elasticity and anelasticity. *Phys. Rev.* 73: 373-82.

Eckart, C. (1948b). The theory of the anelastic fluid. *Rev. Mod. Phys.* 20: 232-5.

Edelen, D.G.B. (1972). A nonlinear Onsager theory of irreversibility. *Int. J. Eng. Sci.* 10: 481-90.

Edelen, D.G.B., and McLennan, J.A. (1973). Material indifference: A principle or a convenience. *Int. J. Eng. Sci.* 11: 813-7.

Edwards, B.J. (1991). *The Dynamical Continuum Theory of Liquid Crystals.* Ph.D. Dissertation: University of Delaware.

Edwards, B.J., and Beris, A.N. (1989a). Order parameter representation of spatial inhomogeneities in polymeric liquid crystals. *J. Rheol.* 33: 1189-93.

Edwards, B.J., and Beris, A.N. (1989b). Flow-induced orientation in monodomain systems of polymeric liquid crystals. *J. Rheol.* 33: 537-57.

Edwards, B.J., and Beris, A.N. (1990). Remarks concerning compressible viscoelastic fluid models. *J. Non-Newtonian Fluid Mech.* 36: 411-7.

Edwards, B.J., and Beris, A.N. (1991a). Unified view of transport phenomena based on the generalized bracket formulation. *Ind. Eng. Chem. Res.* 30: 873-81.

Edwards, B.J., and Beris, A.N. (1991b). Noncanonical Poisson bracket for nonlinear elasticity with extensions to viscoelasticity. *J. Phys. A: Math. Gen.* 24: 2461-80.

Edwards, B.J., and Beris, A.N. (1992). The dynamics of a thermotropic liquid crystal. *Eur. J. Mech. B/Fluids* 11: 121-42.

Edwards, B.J., Beris, A.N., and Grmela, M. (1990a). Generalized constitutive equation for polymeric liquid crystals. Part 1. Model formulation using the Hamiltonian (Poisson bracket) formulation. *J. Non-Newtonian Fluid Mech.* 35: 51-72.

Edwards, B.J., Beris, A.N., Grmela, M., and Larson, R.G. (1990b). Generalized constitutive equation for polymeric liquid crystals. Part 2. Non-homogeneous systems. *J. Non-Newtonian Fluid Mech.* 36: 243-54.

Edwards, B.J., Beris, A.N., and Grmela, M. (1991). The dynamical behavior of liquid crystals: A continuum description through generalized brackets. *Mol. Cryst. Liq. Cryst.* 201: 51-86.

Edwards, B.J., and McHugh, A.J. (1994). Hamiltonian mechanics and bracket structures for polar fluids. *Q. J. Appl. Mech.* Math.: submitted for publication.

Edwards, D.A., Brenner, H., and Wasan, D.T. (1991). *Interfacial Transport Processes and Rheology*. Butterworth-Heinemann: Boston.

Edwards, S.F. (1965). The statistical mechanics of polymers with excluded volume. *Proc. Phys. Soc.* 85: 613-24.

Einaga, Y., Berry, G.C., and Chu, S.G. (1985). Rheological properties of rodlike polymers in solution. III. Transient and steady-state studies on nematic solutions. *Polymer J.* 17: 239-51.

Einstein, A. (1905). On the movement of small particles suspended in a stationary liquid demanded by the molecular theory of heat. *Ann. d. Physik* 17: 549-59. For an English translation, see: Einstein, A. (1926). *Investigations on the Theory of the Brownian Movement*. Methuen and Co, also republshed (1956). Dover: New York.

Einstein, A. (1906). *Ann. Physik* 19, 289; corrected (1911). 34, 591. See republication cited above for complete dissertation.

Ekwall, P. (1975). Composition, properties and structures of liquid crystalline phases in systems of amphiphilic compounds. In: *Advances in Liquid Crystals,* vol. 1. Brown, G.H. (ed.). Academic Press: New York; pp. 1-142.

El-Kareh, A.W., and Leal, G.L. (1989). Existence of solutions for all Deborah numbers for a non-Newtonian model modified to include diffusion. *J. Non-Newtonian Fluid Mech.* 33: 257-87.

Enz, C.P. (1974). Two-fluid hydrodynamic description of ordered systems. *Rev. Mod. Phys.* 46: 705-53.

Ericksen, J.L. (1960a). Anisotropic fluids. *Arch. Rat. Mech. Anal.* 4: 231-7.

Ericksen, J.L. (1960b). Transversely isotropic fluids. *Kolloid Z.* 173: 117-22.

Ericksen, J.L. (1961). Conservation laws for liquid crystals. *Trans. Soc. Rheol.* 5: 23-34.

p. 641: 15 lines down, *Q. J. Appl. Mech.* Math. should be *Q. J. Mech. Appl. Math.*

Ericksen, J.L. (1962). Nilpotent energies in liquid crystal theory. *Arch. Rat. Mech. Anal.* 10: 189-96.

Ericksen, J.L. (1966). Inequalities in liquid crystal theory. *Phys. Fluids* 9: 1205-7.

Ericksen, J.L. (1976). Equilibrium theory of liquid crystals In: *Advances in Liquid Crystals,* vol. 2. Brown, G.H. (ed.). Academic Press: New York; pp. 233-98.

Ericksen, J.L. (1991). Liquid crystals with variable degree of orientation. *Arch. Rat. Mech. Anal.* 113: 97-120.

Eringen, A.C. (1962). *Non-Linear Theory of Continuous Media.* McGraw Hill: New York.

Eringen, A.C. (1966). Theory of micropolar fluids. *J. Math. Mech. 16*: 1-18.

Fan, C.-P., and Stephen, M.J. (1970). Isotropic-nematic transition in liquid crystals. *Phys. Rev. Lett.* 25: 500-3.

Fan, X.-J. (1985a). Viscosity, first normal-stress coefficient, and molecular stretching in dilute polymer solutions. *J. Non-Newtonian Fluid Mech.* 17: 125-44.

Fan, X.-J. (1985b). Viscosity, first and second normal-stress coefficients, and molecular stretching in concentrated solutions and melts. *J. Non-Newtonian Fluid Mech.* 17: 251-65.

Fan, X.-J. (1986). A numerical investigation of the hydrodynamic interaction for Hookean dumbbell suspensions in steady state shear flow. *J. Chem. Phys.* 85: 6237-8.

Fan, X.-J. (1987). The effect of the hydrodynamic interaction on the rheological properties of Hookean dumbbell suspensions in steady state shear flow, *Applied Mathematics and Mechanics* (Published by SUT, Shanghai, China-English Ed.) 8, 829-38.

Faulkner, D.L., Hopfenberg, H.B., and Stannett, V.T. (1977). Residual solvent removal and n-hexane sorption in blends of atactic and isotactic polystyrene. *Polymer* 18: 1130-6.

Feinberg, M. (1972). On chemical kinetics of a certain class. *Arch. Rational Mech. Anal.* 46: 1-41.

Feller, W. (1968). *An Introduction to Probability Theory and Its Applications,* vol. 1, 3rd ed. John Wiley and Sons: New York.

Ferry, J.D. (1980). *Viscoelastic Properties of Polymers.* 3rd ed. John Wiley and Sons: New York.

Feynman: R.P. (1950). The mathematical formulation of the quantum theory of electromagnetic interaction. *Phys. Rev.* 80: 440-57.

Feynman, R.P. (1965). *The Character of Physical Law.* MIT Press: Cambridge.

Feynman, R.P. (1985). *QED, The Strange Theory of Light and Matter.* Princeton University Press: Princeton, New Jersey.

Feynman, R.P., and Hibbs, A.R. (1965). *Quantum Mechanics and Path Integrals*. McGraw-Hill: New York.

Fitzgibbon, D.R., and McCullough, R.L. (1989). Influences of a neutral surface on polymer molecules in the vicinity of the surface. *J. Polym. Sci.: Polym. Phys. Ed.* 27: 655-71.

Fixman, M. (1981). Inclusion of hydrodynamic interaction in polymer dynamical simulations. *Macromolecules* 14: 1710-7.

Flory, P.J. (1953). *Principles of Polymer Chemistry*. Cornell University Press: Ithaca.

Flory, P.J. (1956). Phase equilibria in solutions of rod-like particles. *Proc. Roy. Soc. London, Ser. A* 234: 73-89.

Flory, P.J. (1969). *Statistical Mechanics of Chain Molecules*. Hansen Publishers (1988 edition): Munich.

Flory, P.J. (1982). Molecular theories of liquid crystals, in: *Polymer Liquid Crystals*, Ciferri, A., Krigbaum, W.R., and Meyer, R.B. (eds.). Academic Press: New York; pp. 103-13.

Flory, P.J. (1989). Conformational rearrangement in the nematic phase of a polymer comprising rigid and flexible sequences in alternating succession, orientation-dependent interactions included. *Mat. Res. Soc. Symp. Proc.* 134: 3-9.

Flory, P.J., and Ronca, G. (1979a). Theory of systems of rodlike particles. I. Athermal systems. *Mol. Cryst. Liq. Cryst.* 54: 289-310.

Flory, P.J., and Ronca, G. (1979b). Theory of systems of rodlike particles. II. Thermotropic systems with orientation-dependent interactions. *Mol. Cryst. Liq. Cryst.* 54: 311-30.

Fomenko, A.T. (1988). *Symplectic Geometry*. Gordon and Breach Science Publishers: New York.

Førland, K.S., Førland, T., and Ratkje, S.K. (1988). *Irreversible Thermodynamics: Theory and Applcations*. John Wiley and Sons: New York.[2]

Frank, F.C. (1958). On the theory of liquid crystals. *Disc. Faraday Soc.* 25: 19-28.

Freed, K.F. (1972). Functional integrals and polymer statistics. *Adv. Chem. Phys.* 22: 1-128.

Freed, K.F. (1987). *Renormalization Group Theory of Macromolecules*. John Wiley and Sons: New York.

Freidzon, Ya.S., and Shibaev, V.P. (1993). Liquid-crystal polymers of the cholesteric type, in: *Liquid-Crystal Polymers*, Platé, N.A. (ed.), Plenum Press: New York; pp. 251-302.

Friedlander, F.G. (1982). *Introduction to the Theory of Distributions*. Cambridge University Press: Cambridge.

[2]This text is rather unique among the books on traditional irreversible thermodynamics in that it is largely concerned with biological applications.

Gabriele, M.C. (1990). A Challenge for the 90's: How to process LCP's. *Plastics Technology, April* 1990: 92-8.

Garner, F.H., and Nissan, A.H. (1946). Rheological properties of high-viscosity solutions of long molecules. *Nature* 158: 634-5.

Geigenmüller, U., Titulaer, U.M., and Felderhof, B.U. (1983a). Systematic elimination of fast variables in linear systems. *Physica* 119A: 41-52.

Geigenmüller, U., Titulaer, U.M., and Felderhof, B.U. (1983b). The approximate nature of the Onsager-Casimir reciprocal relations. *Physica* 119A: 53-66.

Giacomin, A.J., Jeyaseelan, R.S., and Oakley, J.G. (1993). Technical note: Structure dependent moduli in the contravariant derivative of structural network theories for melts. *J. Rheol.* 37: 127-132.

Giesekus, H. (1982). A simple constitutive equation for polymer fluidsbased on the concept of deformation-dependent tensorial mobility. *J. Non-Newtonian Fluid Mech.* 11: 69-109.

Giesekus, H. (1985a). Constitutive equations for polymer fluids based on the concept of configuration-dependent molecular mobility: A generalized mean-configuration model. *J. Non-Newtonian Fluid Mech.* 17: 349-72.

Giesekus, H. (1985b). A comparison of molecular and network-constitutive theories for polymer fluids, in: *Viscoelasticity and Rheology*, Lodge, A.S., Renardy, M., and Nohel, J.A. (eds.). Academic Press: Orlando, Florida; pp. 157-80.

Giesekus, H. (1988). Flow hardening in polymer solutions and its modelling with a modified mean-configuration constitutive equation. In: *Proc. Xth Intern. Congr. Rheology*, vol. 1. Sydney, Australia, Aug. 14-19, Uhlherr, P.H.T. (ed.). Australian Society of Rheology, pp. 341-43.

Giesekus, H. (1990). Carried along on a pathline in modelling constitutive equations of viscoelastic fluids. *Rheol. Acta* 29: 500-11.

Goldman, E., and Sirovich, L. (1967). Equations of gas mixtures. *Phys. Fluids* 10: 1928-40.

Goldstein, H. (1980). *Classical Mechanics*, 2nd ed. Addison-Wesley: Reading, Massachusetts.

Goossens, W.J.A. (1978). Electrodynamic instabilities in nematic liquid crystals. In: *Advances in Liquid Crystals*, vol. 3. Brown, G.H. (ed.). Academic Press: New York; pp. 1-39.

Gordon, R.J., and Schowalter, W.R. (1972). Anisotropic fluid theory: A different approach to the dumbbell theory of dilute polymer solutions. *Trans. Soc. Rheol.* 16: 79-97.

Grad, H. (1952). Statistical mechanics, thermodynamics and fluid dynamics of systems with an arbitrary number of integrals. *Commun. Pure Appl. Math.* 5: 455-94.

Grad, H. (1958). Principles of the kinetic theory of gases. In: *Thermodynamics of Gases*, vol. 12. Flügge, S. (ed.), Encyclopedia of Physics. Springer-Verlag: Berlin; pp. 205-94.

Gramsbergen, E.F., Longa, L., and de Jeu, W.H. (1986). Landau theory of the nematic-isotropic phase transition. *Phys. Rep.* 135: 195-257.

Graves, D.B. (1986). *A Continuum Model of Low Pressure Gas Discharges*. Ph.D. Dissertation: Univ. of Minnesota.

Graves, D.B. (1989). Plasma processing in microelectronics manufacture. *AIChE J.* 35: 1-29.

Gray, G.W. (1976). Molecular geometry and the properties of nonamphiphilic liquid crystals. In: *Advances in Liquid Crystals*, vol. 2. Brown, G.H. (ed.). Academic Press: New York; pp. 1-72.

Green, M.S., and Tobolsky, A.V. (1946). A new approach to the theory of relaxing polymeric media. *J. Chem. Phys.* 14: 80-92.

Greenberg, M.D. (1978). *Foundations of Applied Mathematics*. Prentice-Hall: Englewood Cliffs, New Jersey.

Grisafi, S., and Brunn, P.O. (1989). Wall effects in the flow of a dilute polymer solution: Numerical results for intermediate channel sizes. *J. Rheol.* 33: 47-67.

Grmela, M. (1984). Bracket formulation of dissipative fluid mechanics equations. *Phys. Lett.* 102A: 355-58.

Grmela, M. (1985). Bracket formulation of dissipative time evolution equations. *Phys. Lett.* 111A: 36-40.

Grmela, M. (1986). Bracket formulation of diffusion-convection equations. *Physica* 21D: 179-212.

Grmela, M. (1988). Hamiltonian dynamics of incompressible elastic fluids. *Phys. Lett.* A130: 81-86.

Grmela, M. (1989a). Hamiltonian mechanics of complex fluids. *J. Phys. A: Math. Gen.* 22: 4375-4394.

Grmela, M. (1989b). Dependence of the stress tensor on the intramolecular viscosity. *J. Rheol.* 33: 207-31.

Grmela, M., and Carreau, P.J. (1987). Conformation tensor rheological models. *J. Non-Newtonian Fluid Mech.* 23: 271-94.

Grosberg, A.Y., and Zhestkov, A.V. (1986). The dependence of the elasticity coefficients of a nematic liquid crystal polymer on macromolecular rigidity. *Polymer Science U.S.S.R.* 28: 97-104.

Guillemin, V., and Sternberg, S. (1984). *Symplectic Techniques in Physics*. Cambridge University Press: Cambridge.

Guminski, K. (1980). On the Natanson principle of irreversible processes. *Acta Phys. Polonica* A58: 501-07.

Gurtin, M.E., and Sternberg, E. (1962). On the linear theory of viscoelasticity. *Arch. Rat. Mech. Anal.* 11: 291-356.

Gustafson, J.B. (1991). *Reaction and Transport Phenomena in Non-Thermal Equilibrium Plasmas*. Ph.D. Dissertation: University of Delaware.

Guyer, R.A., and Krumhansl, J.A. (1966). Solution of the linearized phonon Boltzmann equation. *Phys. Rev.* 148: 766-78.

Gyarmati, I. (1970). *Non-Equilibrium Thermodynamics: Field Theory and Variational Principles*. Springer-Verlag: New York.

Hajicek, P. (1991). Comment on "Time in quantum gravity: An hypothesis," *Phys. Rev.* D 44: 1337-8.

Halperin, B.I., Hohenberg, P.C., and Ma, S. (1974). Renormalization-group methods for critical dynamics: I. Recursion relations and effects of energy conservation. *Phys. Rev.* B 10: 139-53.

Hand, G.L. (1962). A theory of anisotropic fluids. *J. Fluid Mech.* 13: 33-46.

Hankins, T.L. (1980). *Sir William Rowan Hamilton*. The John Hopkins University Press: Baltimore.

Hatzikiriakos, S.C., and Dealy, J.M. (1991). Wall slip of molten high-density polyethylenes. I. Sliding plate rheometer studies. *J. Rheol.* 35: 497-523.

Hatzikiriakos, S.C., and Dealy, J.M. (1992). II. Capillary rheometer studies. *J. Rheol.* 36: 703-41.

Helfand, E., and Fredrickson, G.H. (1989). Large fluctuations in polymer solutions under shear, *Phys. Rev. Lett.* 62: 2468-71.

Helfrich, W. (1969). Conduction-induced alignment of nematic liquid crystals: Basic model and stability considerations. *J. Chem. Phys.* 51: 4092-105.

Henin, F. (1974). Entropy, dynamics and molecular chaos: The McKean's model. *Physica* 76: 201-23.

Herivel, J.W. (1955). The derivation of the equations of motion of an ideal fluid by Hamilton's principle. *Proc. Camb. Phil. Soc.* 51: 344-9.

Hermans, Jr., J. (1962). The viscosity of concentrated solutions of rigid rodlike molecules (poly-γ-benzyl-L-glutamate in m-Cresol). *J. Colloid Sci.* 17: 638-648.

Hermans, J.J. (ed.) (1978). *Polymer Solutions Properties Part II: Hydrodynamics and Light Scattering*. Dowden, Hutchinson and Ross. Inc.: Stroudburg, Pennsylvania.

Hess, S. (1974). Birefringence caused by the diffusion of macromolecules or colloidal particles. *Physica* 74: 277-93.

Hess, S. (1976a). Fokker-Planck-equation approach to flow alignment in liquid crystals. *Z. Naturforsch.* 31a: 1034-7.

Hess, S. (1976b). Pre- and post-transitional behavior of the flow alignment and flow-induced phase transition in liquid Crystals. *Z. Naturforsch.* 31a: 1507-13.

Hess, S. and Pardowitz, I (1981). On the unified theory for non-equilibrium phenomena in the isotropic and nematic phases of a liquid crystal; Spatial inhomogeneous alignment *Z. Naturforsch. 36a*, 554-8.

Hesselink, F.T. (1969). On the density distribution of segments of a terminally adsorbed macromolecule. *J. Phys. Chem.* 73: 3488-90.

Hesselink, F.T. (1971). On the theory of the stabilization of dispersions by adsorbed macromolecules. I. Statistics of the change of some configurational properties of adsorbed macromolecules on the approach of an impenetrable interface. *J. Phys. Chem.* 75: 65-71.

Hinch, E.J. (1977). Mechanical models of dilute polymer solutions in strong flows. *Phys. Fluids* 20: S22-30.

Hirschfelder, J.O., Curtiss, C.F., and Bird, R.B. (1954). *Molecular Theory of Gases and Liquids*. John Wiley and Sons: New York.

Ho, B.P., and Leal, L.G. (1974). Inertial migration of rigid spheres in two-dimensional unidirectional flows. *J. Fluid Mech.* 65: 365-400.

Hohenberg, P.C., and Halperin, B.I. (1977). Theory of dynamic critical phenomena. *Rev. Mod. Phys.* 49: 435-79.

Hojman, S.A., and Shepley, L.C. (1991). No Lagrangian? No quantization!. *J. Math. Phys.* 32: 142-6.

Holm, D.D., and Kupershmidt, B.A. (1983). Poisson brackets and Clebsch representations for magnetohydrodynamics, multifluid plasmas, and elasticity. *Physica* 6D: 347-63.

Holm, D.D., Marsden, J.E., Ratiu, T., and Weinstein, A. (1985). Nonlinear stability of fluid and plasma equilibria. *Phys. Rep.* 123: 1-116.

Honerkamp, J., and Öttinger, H.C. (1986). Nonlinear force and tensorial mobility in a kinetic theory for polymer liquids. *J. Chem. Phys.* 84: 7028-35.

Horwitz, L.P., Arshansky, R.I., and Elitzur, A.C. (1988). On the two aspects of time: The distinction and its implications. *Found. Phys.* 18: 1159-93.

Hudson, S.D., Thomas, E.L., and Lenz, R.W. (1987). Imaging of textures and defects of thermotropic liquid crystalline polyesters by electron microscopy. *Mol. Cryst. Liq. Cryst.* 153: 63-72.

Huilgol, R. (1975). *Continuum Mechanics of Viscoelastic Liquids*. John Wiley and Sons: New York.

Hulsen, M.A. (1988). Some properties and analytical expressions for plane flow of Leonov and Giesekus models. *J. Non-Newtonian Fluid Mech.* 30: 85-92.

Hulsen, M.A. (1990). A sufficient condition for a positive definite configuration tensor in differential models. *J. Non-Newtonian Fluid Mech.* 38: 93-100.

Husain, V. (1993). Ashtekar variables, self-dual metrics, and w_∞. *Class. Quant. Grav.* 10: 543-50.

Ide, Y., and White, J.L. (1977). Investigation of failure during elongational flow of polymer melts. *J. Non-Newtonian Fluid Mech.* 2: 281-98.

Ikenberry, E., and Truesdell, C. (1956). On the pressures and the flux of energy in a gas according to Maxwell's kinetic theory. *J. Rat. Mech. Anal.* 5: 1-54.

Isihara, A. (1951). Theory of anisotropic colloidal solutions. *J. Chem. Phys.* 19: 1142-7.

Jackson, C.L., and Shaw, M.T. (1991). Polymer liquid crystalline materials. *Int. Mater. Rev.* 36: 165-86.

Jähnig, F. (1979). Molecular theory of lipid membrane order. *J. Chem. Phys.* 70: 3279-90.

James, H.M., and Guth, E. (1941). Theoretical stress-strain curve for rubberlike materials. *Phys. Rev.* 59: 111.

Jeffery, G.B. (1922). The motion of ellipsoidal particles immersed in a viscous fluid. *Proc. Roy. Soc. London A* 102: 161-79.

Jeffrey, D.J., and Acrivos, A. (1976). The rheological properties of suspensions of rigid particles. *A.I.Ch.E. J.* 22: 417-32.

Jeffreys, H. (1924). *The Earth: Its Origin, History and Physical Constitution.* Cambridge University Press: Cambridge.

Jensen, K.F. (1987). Micro-reaction engineering applications of reaction engineering to processing of electronic and photonic materials. *Chem. Eng. Sci.* 42: 923-58.

Joanny, J.F., Leibler, L., and de Gennes, P.G. (1979). Effects of polymer solutions on colloid stability. *J. Polym. Sci.: Polym. Phys. Ed.* 17: 1073-84.

Johnson, M.W., and Segalman, D. (1977). A model for viscoelastic fluid behavior which allows non-affine deformation. *J. Non-Newtonian Fluid Mech.* 2: 225-70.

Jongschaap, R.J.J. (1981). Derivation of the Marrucci model from transient-network theory. *J. Non-Newtonian Fluid Mech.* 8: 183-90.

Joseph, D.D. (1990). *Fluid Dynamics of Viscoelastic Liquids*. Springer-Verlag: New York.

Joseph, D.D., and Preziosi, L. (1989). Heat waves. *Rev. Mod. Phys.* 61: 41-73.

Joseph, D.D., and Preziosi, L. (1990). Addendum to the paper "Heat waves," *Rev. Mod. Phys. 62*: 375-91.

Joseph, D.D., Renardy, M., and Saut, J.C. (1985). Hyperbolicity and change of type in the flow of viscoelastic fluids. *Arch. Ration. Mech. Anal.* 87: 213-51.

Joseph, D.D., and Saut, J.C. (1986). Change of type and loss of evolution in the flow of viscoelastic fluids. *J. Non-Newtonian Fluid Mech.* 20: 117-41.

Jou, D., Camacho, J., and Grmela, M. (1991). On the nonequilibrium thermodynamics of non-Fickian diffusion. *Macromolecules* 24: 3597-3602.

Jou, D., Casas-Vázquez, J., and Lebon, G. (1988). Extended irreversible thermodynamics. *Rep. Prog. Phys.* 51: 1105-79.

Kalika, D.S., and Denn, M.M. (1987). Wall slip and extrudate distortion in linear low-density polyethylene. *J. Rheol.* 31: 815-34.

Kalospiros, N.S., Astarita, G., and Paulaitis, M.E. (1993a). Coupled diffusion and morphological change in solid polymers. *Chem. Eng. Sci.* 48: 23-40.

Kalospiros, N.S., Edwards, B.J., and Beris, A.N. (1993b). Internal variables for relaxation phenomena in heat and mass transfer. *Int. J. Heat Mass Trans.* 36: 1191-200.

Kalospiros, N.S., Ocone, R., Astarita, G., and Meldon, J.H. (1991). Analysis of anomalous diffusion behavior and relaxation in polymers. *Ind. Eng. Chem. Res.* 30: 851-64.

Kamiya, Y., Mizoguchi, K., Hirose, T., and Naito, Y. (1989). Sorption and dilation in poly(ethyl methacrylate)-carbon dioxide system. *J. Polym. Sci.: Polym. Phys. Ed.* 27: 879-92.

Kaufman, A.N. (1984). Dissipative Hamiltonian systems: A unifying principle. *Phys. Lett.* 100A: 419-22.

Kawaguchi, M., and Takahashi, A. (1980). Effect of solvent power on the adsorption of polystyrene onto a metal surface. *J. Polym. Sci.: Polym. Phys. Ed.* 18: 2069-76.

Kaye, A. (1966). An equation of state for non-Newtonian fluids. *Brit. J. Appl. Phys.* 17: 803-6.

Kayser, Jr., R.F., and Raveché, H.J. (1978). Bifurcation in Onsager's model of the isotropic-nematic transition. *Phys. Rev.* A 17: 2067-72.

Keck, J.C. (1990). Rate-controlled constrained-equilibrium theory of chemical reactions in complex systems. *Prog. Energy Combust. Sci.* 16: 125-54.

Kestin, J. (1974). Entropy and entropy production: Discussion paper. In: *Foundations of Continuum Thermodynamics*, Domingos, J.J.D., Nina, M.N.R., and Whitelaw, J.H. (eds.). Macmillan Press: London; pp. 143-51.

Khan, S.A., and Armstrong, R.C. (1986). Rheology of foams: I. Theory for dry foams. *J. Non-Newtonian Fluid Mech.* 22: 1-22.

Khan, S.A., and Armstrong, R.C. (1987). Rheology of foams: II. Effects of polydispersity and liquid viscosity for foams having gas fraction approaching unity. *J. Non-Newtonian Fluid Mech.* 25: 61-92.

Khan, S.A., Schnepper, C.A., and Armstrong, R.A. (1988). Foam rheology: III. Measurement of shear flow properties. *J. Rheol.* 32: 69-92.

Khokhlov, A.R. (1978). Liquid-crystalline ordering in the solution of semi-flexible macromolecules. *Phys. Lett.* 68A: 135-6.

Khokhlov, A.R. (1991). Theories based on the Onsager approach. In: *Liquid Crystallinity in Polymers: Principles and Fundamental Properties*, Cifferi, A. (ed.). VCH Publishers: New York; pp. 97-129.

Khokhlov, A.R., and Semenov, A.N. (1981). Liquid-crystalline ordering in the solution of long persistent chains. *Physica* 108A: 546-56.

Khokhlov, A.R., and Semenov, A.N. (1982). Liquid-crystalline ordering in the solution of partially flexible macromolecules. *Physica* 112A: 605-14.

Khokhlov, A.R., and Semenov, A.N. (1985). On the theory of liquid-crystalline ordering of polymer chains with limited flexibility. *J. Stat. Phys.* 38: 161-82.

Kirkwood, J.G. (1967). *Macromolecules.* Gordon and Breach: New York.

Kiss, G., and Porter, R.S. (1978). Rheology of concentrated solutions of poly(γ-benzyl-glutamate). *J. Polym. Sc.: Polym. Symp.* 65: 193-211.

Kléman, M. (1975). Defects in liquid crystals. In: *Advances in Liquid Crystals,* vol. 1. Brown, G.H. (ed.). Academic Press: New York; pp. 266-311.

Kléman, M. (1983). *Points, Lines and Walls.* John Wiley and Sons: New York.

Kléman, M. (1991). Defects and textures in liquid crystalline polymers. In: *Liquid Crystallinity in Polymers: Principles and Fundamental Properties.* Cifferi, A. (ed.). VCH Publishers: New York; pp. 365-394.

Kline, L.E., and Kushner, M.J. (1989). Computer simulation of materials processing plasma discharges. *Critical Reviews in Solid State and Material Sciences* 16: 1-35.

Kolkka, R.W., Malkus, D.S., Hansen, M.G., Ierley, G.R., and Worthing, R.A. (1988). Spurt phenomena of the Johnson-Segalman fluid and related models. *J. Non-Newtonian Fluid Mech.* 29: 303-35.

Korn, G.A., and Korn, T.M. (1968). *Mathematical Handbook for Scientists and Engineers.* McGraw-Hill: New York.

Krambeck, F.J. (1970). The mathematical structure of chemical kinetics in homogeneous single-phase systems. *Arch. Rational Mech. Anal.* 38: 317-47.

Kraynik, A.M., Reinelt, D.A., and Princen, H.M. (1991). The nonlinear elastic behavior of polydisperse hexagonal foams and concentrated emulsions. *J. Rheol.* 35: 1235-53

Kremer, G.M. (1987). Extended thermodynamics of mixtures of ideal gases. *Int. J. Engng. Sci.* 25: 95-115.

Kuhn, W. (1934). The shape of fibrous molecules in solutions. *Kolloid Z.* 68: 2-15.

Kuhn, W. (1936). The relation between molecular size, statistical molecular shape and elastic properties of highly polymerized substances. *Kolloid Z.* 76: 258-71.

Kuhn, W., and Grün, F. (1942). Relation between the elasticity constant and extension double diffraction of highly elastic substances. *Kolloid Z.* 101: 248-71.

Kuhn, W., and Grün, F. (1946). Statistical behavior of the single chain molecule and its relation to the statistical behavior of assemblies consisting of many chain molecules. *J. Polym. Sci.* 1: 183-99.

Kurzyński, M. (1990). Chemical reactions from the point of view of statistical thermodynamics. *J. Chem. Phys.* 93: 6793-9.

Kushner, M.J. (1987). Application of a particle simulation to modeling commutation in a linear thyratron. *J. Appl. Phys.* 61: 2784-94.

Kuznetsov, E.A., and Mikhailov, A.V. (1980). On the topological meaning of Clebsch variables. *Phys. Lett.* 77A: 37-8.

Kuzuu, N., and Doi, M. (1983). Constitutive equation for nematic liquid crystals under weak velocity gradient derived from a molecular kinetic equation. *J. Phys. Soc. (Japan)* 52: 3486-94.

Kuzuu, N., and Doi, M. (1984). Constitutive equation for nematic liquid crystals under weak velocity gradient derived from a molecular kinetic equation. II. Leslie coefficients for rodlike polymers. *J. Phys. Soc. (Japan)* 53: 1031-40.

Kwolek, S.L., Morgan, P.W., Schaefgen, J.R., and Gulrich, L.W. (1977). Synthesis, anisotropic solutions and fibers of poly(1,4-benzamide). *Macromolecules* 10: 1390-6.

Lamb, H. (1932). *Hydrodynamics*, 6th ed. Dover (1945 republication): New York.

Lanczos, C. (1970). *The Variational Principles of Mechanics, ed. 4.* University of Toronto Press: Toronto.[3] Also available in paperback. (1986). Dover: New York.

Lanczos, C. (1972). The Poisson bracket. In: *Aspects of Quantum Theory*, Salam, A., and Wigner, E.P. (eds.). Cambridge University Press: Cambridge; pp. 169-78.

Landau, L.D. (1937). On the theory of phase transitions. Part 1. *Soviet Phys.* 11: 26; Part 2. *Soviet Phys.* 11: 545. Reprinted (English translation) in: *Collected Papers of L. D. Landau*, Ter Haar, D. (ed.). Gordon and Breach: New York; pp. 193-216.

Landau, L.D., and Lifshitz, E.M. (1958). *Statistical Physics (Landau and Lifshitz Course of Theoretical Physics*, vol. 5*)*. Pergamon Press: London.

Landau, L.D., and Lifshitz, E.M. (1959). *Fluid Mechanics (Landau and Lifshitz Course of Theoretical Physics*, vol. 6*)*. Pergamon Press: Oxford.

Langer, J.S. (1986). Models of pattern formation in first-order phase transitions. In: *Directions in Condensed Matter Physics*, Grinstein, G., and Mazenko, G. (eds.). World Scientific Publishing: Singapore; pp. 165-86.

[3]This is probably the most accessible account of introductory classical mechanics. The author claims that the book is written [p. viii] "in a humble spirit and is written for humble people."

Larson, R.G. (1983). Convection and diffusion of polymer network strands. *J. Non-Newtonian Fluid Mech.* 13: 279-308.

Larson, R.G. (1988). *Constitutive Equations for Polymer Melts and Solutions*. Butterworths: Boston.

Larson, R.G. (1990). Arrested tumbling in shearing flows of liquid crystal polymers. *Macromolecules* 23: 3983-92.

Larson, R.G. (1992). Flow-induced mixing, demixing, and phase transitions in polymeric fluids. *Rheol. Acta* 31: 497-520.

Larson, R.G., and Doi, M. (1991). Mesoscopic domain theory for textured liquid crystalline polymers. *J. Rheol.* 35: 539-63.

Larson, R.G., and Mead, D.W. (1989). Time and shear-rate scaling laws for liquid crystal polymers. *J. Rheol.* 33: 1251-81.

Lasher, G. (1970). Nematic ordering of hard rods derived from a scaled particle treatment. *J. Chem. Phys.* 53: 4141-6.

Lau, H.C., and Schowalter, W.R. (1986). A model for adhesive failure of viscoelastic fluids during flow. *J. Rheol.* 30: 193-206.

Lauprêtre, N., and Noël, C. (1991). Conformational analysis of mesogenic polymers. In: *Liquid Crystallinity in Polymers: Principles and Funda-mental Properties*. Ciferri, A. (ed.). VCH Publishers: New York; pp. 3-60.

Lavenda, B.H. (1978). *Thermodynamics of Irreversible Processes*. John Wiley and Sons: New York.

Lee, S.-D. (1988). The Leslie coefficients for a polymer nematic liquid crystal. *J. Chem. Phys.* 88: 5196-201.

Lee, S.-D., and Meyer, R.B. (1986). Computations of the phase equilibrium, elastic constants, and viscosities of a hard-rod nematic liquid crystal. *J. Chem. Phys.* 84: 3443-8.

Lee, S.-D., and Meyer, R.B. (1991). Elastic and viscous properties of lyotropic polymer nematics. In: *Liquid Crystallinity in Polymers: Principles and Fundamental Properties*. Cifferi, A. (ed.). VCH Publishers: New York; pp. 343-64.

Leighton, D., and Acrivos, A. (1987). The shear-induced migration of particles in concentrated suspensions. *J. Fluid Mech.* 181: 415-39.

Lekkerkerker, H.N.V., Coulon, P., Van Der Haegen, R., and Deblieck, R. (1984). On the isotropic-liquid crystal phase separation in a solution of rodlike particles of different lengths. *J. Chem. Phys.* 80: 3427-33.

Lengyel, S. (1989a). Chemical kinetics and thermodynamics. A history of their relationship. *Computers Math. Applic.* 17: 443-55.

Lengyel, S. (1989b). On the relationship between thermodynamics and chemical kinetics. *Z. Phys. Chemie Leipzig* 270: 577-89.

Lengyel, S., and Gyarmati, I. (1981). Nonlinear thermodynamic studies of homogeneous chemical kinetic systems. *Period. Polytechn. Chem. Eng.* 25: 63-99.

Leonov, A.I. (1976). Nonequilibrium thermodynamics and rheology of viscoelastic polymer media. *Rheol. Acta* 15: 85-98.

Leonov, A.I. (1987). On a class of constitutive equations for viscoelastic fluids. *J. Non-Newtonian Fluid Mech.* 25: 1-59.

Leslie, F.M. (1966). Some constitutive equations for anisotropic fluids. *Quart. J. Mech. Appl. Math.* 19: 357-70.

Leslie, F.M. (1968). Some constitutive equations for liquid crystals. *Arch. Rat. Mech. Anal.* 28: 265-83.

Leslie, F.M. (1976). An analysis of the flow instability in nematic liquid crystals. *J. Phys.* D 9: 925-37.

Leslie, F.M. (1979). Theory of flow phenomena in liquid crystals. In: *Advances in Liquid Crystals,* vol. 4. Brown, G.H. (ed). Academic Press: New York; pp. 1-81.

Leslie, F.M. (1981). Viscometry of nematic liquid crystals. *Mol. Cryst. Liq. Cryst.* 63: 111-28.

Leslie, F.M. (1987). Theory of flow phenomena in nematic liquid crystals. In: *Theory and Applications of Liquid Crystals*. Ericksen, J.L., and Kinderlehrer, D., (eds.). Institute of Mathematics and Its Applications (IMA) vol. 7. Springer-Verlag: New York; pp. 235-54.

Lewis, D., Marsden, J., Montgomery, R., and Ratui, T. (1986). The Hamiltonian structure for dynamic free boundary problems. *Physica* 18D: 391-404.

Lifshitz, I.M. (1969). Some problems of the statistical theory of biopolymers. *Soviet Phys. JETP* 28: 1280-1286. Translation from: (1968). *Zh. Eksp. Teor. Fiz.* 55: 2408-22.

Lim, F.R., and Schowalter, W.R. (1989). Wall slip of narrow molecular weight distribution polybutadienes. *J. Rheol.* 33: 1359-82.

Lin, C.C. (1963). Hydrodynamics of Helium II. In: *Liquid Helium.* Careri, G., (ed.). Academic Press: New York; pp. 93-146.

Lin, F.-H. (1989). Nonlinear theory of defects in nematic liquid crystals; phase transition and flow phenomena. *Comm. Pure Appl. Math.* 42: 789-814.

Lindner, P., and Oberthür, R. C. (1988). Shear induced deformation of polystyrene coils in dilute solution from small angle neutron scattering. 2. Variation of shear gradient, molecular mass and solvent viscosity. *Colloid Polym.* Sci. 266: 886-97.

Lighthill, M.J. (1958). *Introduction to Fourier Analysis and Generalized Functions*. Cambridge University Press: Cambridge.

Lighthill, M.J. (1986). *An Informal Introduction to Theoretical Fluid Mechanics*. Clarendon Press: Oxford.

Lo, W.-S., and Pelcovits, R.A. (1990). Nonlocal elasticity theory of polymeric liquid crystals. *Phys. Rev.* A 42: 4756-63.

Lodge, A.S. (1956). A network theory of flow birefringence and stress in concentrated polymer solutions. *Trans. Faraday* Soc. 52: 120-30.

Lodge, A.S. (1964). *Elastic Liquids*. Academic Press: New York.

Lodge, A.S. (1974). *Body Tensor Fields in Continuum Mechanics*. Academic Press: New York.

Lodge, A.S., and Wu, Y. (1971). Constitutive equations for polymer solutions derived from the bead/spring model of Rouse and Zimm. *Rheol. Acta* 10: 539-53.

Lonberg, F., Fraden, S., Hurd, A.J., and Meyer, R.E. (1984). Field-induced transient periodic structures in nematic liquid crystals: The twist-Fréedericksz transition. *Phys. Rev. Lett.* 52: 1903-6.

Machlup, S., and Onsager, L. (1953). Fluctuations and irreversible processes. II. Systems with kinetic energy. *Phys. Rev.* 91: 1512-15.

Mackay, A.L. (1991). *A Dictionary of Scientific Quotations*. Adam Hilger: Bristol.

Mada, H. (1979). Study on the surface alignment of nematic liquid crystals: Temperature dependence of pretilt angles. *Mol. Cryst. Liq. Cryst.* 51: 43-56.

Mada, H. (1984). A unified phenomenology of nematic liquid crystals. *Mol. Cryst. Liq. Cryst.* 108: 317-32.

Magda, J.J., Baek, S.-G., DeVries, K.L., and Larson, R.G. (1991). Shear flows of liquid crystal polymers: Measurements of the second normal stress difference and the Doi molecular theory. *Macromolecules* 24: 4460-68.

Maier, W., von, and Saupe, A. (1958). Eine einfache molekulare Theorie des nematischen kristallinflüssigen Zustandes. *Z. Naturforschg.* 13a: 564-66.

Malkus, D.S., Nohel, J.A., and Plohr, B.J. (1990). Dynamics of shear flow of a non-Newtonian fluid. *J. Comput. Phys.* 87: 464-87.

Malkus, D.S., Nohel, J.A., and Plohr, B.J. (1991). Analysis of new phenomena in shear flow of non-Newtonian fluids. *SIAM J. Appl. Math.* 51: 899-929.

Maluf, J.W. (1993). Degenerate triads and reality conditions in canonical gravity. *Class. Quant. Grav.* 10: 805-9.

Manneville, P. (1981). The transition to turbulence in nematic liquid crystals. *Mol. Cryst. Liq. Cryst.* 70: 223-50.

Manneville, P., and Dubois-Violette, E. (1976a). Shear flow instability in nematic liquids: Theory of steady simple shear flows. *J. Physique* 37: 285-96.

Manneville, P., and Dubois-Violette, E. (1976b). Steady Poiseuille flow in nematics: Theory of uniform instability. *J. Physique* 37: 1115-24.

Mansfield, K.F., and Theodorou, D.N. (1989). Interfacial structure and dynamics of macromolecular liquids: A Monte Carlo simulation approach. *Macromolecules* 22: 3143-52.

Marrucci, G. (1972). The free energy constitutive equation for polymer solutions from the dumbbell model. *Trans. Soc. Rheol.* 16: 321-30.

Marrucci, G. (1982). Prediction of Leslie coefficients for rodlike polymer nematics. *Mol. Cryst. Liq. Cryst.* 72 (Letters): 153-61.

Marrucci, G. (1985). Rheology of liquid crystalline polymers. *Pure & Appl. Chem.* 57: 1545-52.

Marrucci, G. (1991). Tumbling regime of liquid-crystalline polymers. *Macromolecules* 24: 4176-82.

Marrucci, G. and Greco, F. (1991). The elastic constants of Maier-Saupe rodlike molecule nematics. *Mol. Cryst. Liq. Crysr.* 206: 17-30.

Marrucci, G., and Grizzuti, N. (1984). Predicted effect of polydispersity on rodlike polymer behaviour in concentrated solutions. *J. Non-Newtonian Fluid Mech.* 14: 103-19.

Marrucci, G., and Maffettone, P.L. (1989). Description of the liquid-crystalline phase of rodlike polymers at high shear rates. *Macromolecules* 22: 4076-82.

Marrucci, G., and Maffettone, P.L. (1990a). Nematic phase of rodlike polymers. I. Prediction of transient behavior at high shear rates. *J. Rheol.* 34: 1217-1230.

Marrucci, G., and Maffettone, P.L. (1990b). Nematic phase of rodlike polymers. II. Polydomain predictions in the tumbling regime. *J. Rheol.* 34: 1231-44.

Marrucci, G., Titomanlio, G., and Sarti, G.C. (1973). Testing of a constitutive equation for entangled networks by elongational and shear data of polymer melts. *Rheol. Acta* 12: 269-75.

Marsden, J.E., Ratiu, T., and Weinstein, A. (1984). Semidirect products and reduction in mechanics. *Trans. Amer. Math. Soc.* 281: 147-77.

Marsden, J.E., and Weinstein, A. (1982). The Hamiltonian structure of the Maxwell-Vlasov equations. *Physica* 4D: 394-406.

Marsden, J., and Weinstein, A. (1983). Coadjoint orbits, vortices, and Clebsch variables for incompressible fluids. *Physica* 7D: 305-23.

Martin, P.G., Moore, J.S., and Stupp, S.I. (1986). Surface-enhanced rate of molecular alignment in a liquid crystal polymer. *Macromolecules* 19: 2459-61.

Martin, P.G., and Stupp, S.I. (1987). Solidification of a main-chain liquid-crystal polymer: Effects of electric fields, surfaces and mesophase aging. *Polymer* 28: 897-906.

Maugin, G.A., and Drouot, R. (1983). Internal variables and the thermodynamics of macromolecule solutions. *Int. J. Engng. Sci.* 21: 705-24.

Mavrantzas, V.G., and Beris, A.N. (1992a). Theoretical study of wall effects on the rheology of dilute polymer solutions. *J. Rheol.* 36: 175-213.

Mavrantzas, V.G., and Beris, A.N. (1992b). Modelling of the rheology and flow-induced concentration changes in polymer solutions. *Phys. Rev. Lett.* 69: 273-6; errata. *ibid.* 70: 2659 (1993).

Mavrantzas, V.G., and Beris, A.N. (1993). Pseudospectral calculations of stress-induced concentration changes in viscometric flows of polymer solutions. *Theoret. Computat. Fluid Dynamics* 5: 3-33.

Mavrantzas, V.G., and Beris, A.N. (1994a). A conformation tensor, self-consistent field theory-based study of surface effects on polymer solutions: A) General formulation for equilibrium and non-equilibrium (flow) conditions, with application to a neutral surface. *Macromolecules*, in preparation.

Mavrantzas, V.G., and Beris, A.N. (1994b). A conformation tensor, self-consistent field theory-based study of surface effects on polymer solutions: B) Adsorption under static conditions. *Macromolecules*, in preparation.

Mavrantzas, V.G., and Beris, A.N. (1994c). A conformation tensor, self-consistent field theory-based study of surface effects on polymer solutions: C) Polymer adsorption under flow conditions. *Macromolecules*, in preparation.

Maxwell, J.C. (1867). On the dynamical theory of gases. *Phil. Trans. Roy. Soc.* A157: 49-88.

McMillan, W.L. (1975). Phase transitions in liquid crystals. *J. Physique Colloq.* C1 36: 103-5.

Meixner, J. (1974). Entropy and entropy production. In: *Foundations of Continuum Thermodynamics*. Domingos, J.J.D., Nina, M.N.R., and Whitelaw, J.H. (eds.). Macmillan Press: London; pp. 129-42.

Metzner, A.B. (1985). Rheology of suspensions in polymeric liquids. *J. Rheol.* 29: 739-75.

Metzner, A.B., Cohen, Y., and Rangel-Nafaile, C. (1979). Inhomogeneous flows of non-Newtonian fluids: Generation of spatial concentration gradients. *J. Non-Newtonian Fluid Mech.* 5: 449-62.

Metzner, A.B., and Prilutski, G.M. (1986). Rheological properties of polymeric liquid crystals. *J. Rheol.* 30: 661-91.

Mewis, J., and Denn, M.M. (1983). Constitutive equation based on the transient network concept. *J. Non-Newtonian Fluid Mech.* 12: 69-83.

Meyer, R.B. (1982). Macroscopic phenomena in nematic polymers. In: *Polymer Liquid Crystals*, Ciferri, A., Krigbaum, W.R., and Meyer, R.B. (eds.). Academic Press: New York; pp. 133-63.

Mezlov-Deglin, L.P. (1964). Thermal conductivity of solid He^4. *Sov. Phys. JETP* 19: 1297-8.

Mezlov-Deglin, L.P. (1966). Measurement of the thermal conductivity of crystalline He^4. *Sov. Phys. JETP* 22: 47-56.

Miesowicz, M. (1935). Influence of a magnetic field on the viscosity of para-azoxyanisol. *Nature* 136: 261.

Miesowicz, M. (1946). The three coefficients of viscosity of anisotropic liquids. *Nature* 158: 27.

Miesowicz, M. (1983). Liquid crystals in my memories and now—the role of anisotropic viscosity in liquid crystals research. *Mol. Cryst. Liq. Cryst.* 97: 1-11.

Migler, K.B., Hervet, H. and Leger, L. (1993). Slip transition of a polymer melt under shear stress. *Phys. Rev. Lett.* 70: 287-90.

Miller, D.G. (1974). The Onsager relations; experimental evidence. In: *Foundations of Continuum Thermodynamics*. Domingos, J.J.D., Nina, M.N.R., and Whitelaw, J.H. (eds.). Macmillan Press: London; pp. 185-214.

Milner, S.T. (1991). Hydrodynamics of semidilute polymer solutions, *Phys. Rev. Lett. 66*, 1477-80.

Modell, M., and Reid, R.C. (1983). *Thermodynamics and its Applications*. Prentice Hall: Englewood Cliffs, New Jersey.

Moffatt, H.K. (1964). Viscous and resistive eddies near a sharp corner. *J. Fluid Mech.* 18: 1-18.

Moldenaers, P., Fuller, G., and Mewis, J. (1989). Mechanical and optical rheometry of polymer liquid-crystal domain structure. *Macromolecules* 22: 960-5.

Moldenaers, P., and Mewis, J. (1986). Transient behavior of liquid crystalline solutions of poly(benzylglutamate). *J. Rheol.* 26: 567-84.

Moldenaers, P., and Mewis, J. (1990). Relaxational phenomena and anisotropy in lyotropic polymeric liquid crystals. *J. Non-Newtonian Fluid Mech.* 34: 359-74.

Moore, J.S., and Stupp, S.I. (1987). Orientation dynamics of main-chain liquid crystal polymers. 2. Structure and kinetics in a magnetic field. *Macromolecules* 20: 282-93.

Morrison, P.J. (1980). The Maxwell-Vlasov equations as a continuous Hamiltonian system. *Phys. Lett.* 80A: 383-6.

Morrison, P.J. (1982). Poisson brackets for fluids and plasmas. In: *Mathematical Methods in Hydrodynamics and Integrability in Dynamical Systems*, Tabor, M., and Treve, Y.M. (eds.). AIP Press: New York; pp.13-46.

Morrison, P.J. (1984). Bracket formulation for irreversible classical fields. *Phys. Lett.* 100A: 423-7.

Morrison, P.J. (1986). A paradigm for joined Hamiltonian and dissipative systems. *Physica* 18D: 410-9.

Morrison, P.J., and Greene, J.M. (1980). Noncanonical Hamiltonian density formulation of hydrodynamics and ideal magnetohydrodynamics. *Phys. Rev. Lett.* 45: 790-4; errata. *ibid.* 48: 569 (1982).

Morrison, P.J., and Hazeltine, R.D. (1984). Hamiltonian formulation of reduced magnetohydrodynamics. *Phys. Fluids* 27: 886-97.

Morro, A., and Romeo, M. (1988). The law of mass action for fluid mixtures with several temperatures and velocities. *J. Non-Equil. Thermodyn.* 13: 339-53.

Muir, M.C., and Porter, R.S. (1989). Processing rheology of liquid crystalline polymers: A review. *Mol. Cryst. Liq. Cryst.* 169: 83-95.

Müller, I. (1968). A thermodynamic theory of mixtures of fluids. *Arch. Rat. Mech. Anal.* 28: 1-39.

Müller, I. (1972). On the frame dependence of stress and heat flux. *Arch. Rat. Mech. Anal.* 45: 241-50.

Müller, I., and Villaggio, P. (1976). Condition of stability and wave speeds of fluid mixtures. *Meccanica* 11: 191-5.

Müller-Mohnssen, H., Weiss, D., and Tippe, A. (1990). Concentration dependent changes of apparent slip in polymer solution flow. *J. Rheol.* 34: 223-44.

Murdoch, A.I. (1983). On material frame-indifference. Intrinsic spin, and certain constitutive relations motivated by the kinetic theory of gases. *Arch. Rat. Mech. Anal.* 83: 185-94.

Nakahara, M. (1990). *Geometry, Topology and Physics.* Adam Hilger: Bristol.

Natanson, L. (1896). On the laws of irreversible phenomena. *Phil. Mag.* 41: 385-406.

Neogi, P. (1983a). Anomalous diffusion of vapors through solid polymers. Part I: Irreversible thermodynamics of diffusion and solution processes. *A.I.Ch.E. J.* 29: 829-33.

Neogi, P. (1983b). Anomalous diffusion of vapors through solid polymers. Part II: Anomalous sorption. *A.I.Ch.E. J.* 29: 833-9.

Nicholson, D.R. (1983). *Introduction to Plasma Theory.* John Wiley and Sons: New York.

Nutting, P.G. (1921). A new general law of deformation. *J. Franklin Inst.* 191: 679-85.

Odijk, T. (1986). Theory of lyotropic liquid crystals. *Macromolecules* 19: 2313-29.

Odijk, T., and Lekkerkerker, H.N.W. (1985). Theory of the isotropic-liquid crystal phase separation for a solution of bidisperse rodlike macromolecules. *J. Phys. Chem.* 89: 2090-6.

Oláh, K. (1987). Thermokinetics. An introduction. *Period. Polytechn. Chem. Eng.* 31: 19-27.

Oldroyd, J.G. (1950). On the formulation of rheological equations of state. *Proc. Roy. Soc., London* A200: 523-41.

Olver, P.J. (1986). *Applications of Lie Groups to Differential Equations.* Springer-Verlag: New York.

Omari, A., Moan, M., and Chauveteau, G. (1989a). Hydrodynamic behavior of semirigid polymer at a solid-liquid interface. *J. Rheol.* 33: 1-13.

Omari, A., Moan, M., and Chauveteau, G. (1989b). Wall effects in the flow of flexible polymer solutions through small pores. *Rheol. Acta* 28: 520-6.

Onogi, S., and Asada, T. (1980). Rheology and rheo-optics of polymer liquid crystals. In: *Rheology*, vol. 1. Astarita, G., Marrucci, G., and Nicolais, L., (eds.). (Papers presented at the Eighth International Congress on Rheology, Naples, 1980.) Plenum Press: New York; pp. 127-47.

Onsager, L. (1931a). Reciprocal relations in irreversible processes. Part 1. *Phys. Rev.* 37: 405-26.

Onsager, L. (1931b). Reciprocal relations in irreversible processes. Part 2. *Phys. Rev.* 38: 2265-79.

Onsager, L. (1949). The effects of shape on the interaction of colloidal particles. *Ann. N.Y. Acad. Sci.* 51: 627-59.

Onuki, A. (1989). Elastic effects in the phase transitions of polymer solutions under shear flow. *Phys. Rev. Lett.* 62: 2472-5.

Onuki, A. (1990). Dynamic equations of polymers with deformations in semidilute regions. *J. Phys. Soc. (Japan)* 59: 3423-6.

Osaki, K., and Doi, M. (1991). On the concentration gradient of polymer solution in a rotating rheometer. *J. Rheol.* 35: 89-92.

Oseen, C.W. (1933). The theory of liquid crystals. *Trans. Faraday Soc.* 29: 883-99.

Öttinger, H.C. (1985). Consistently averaged hydrodynamic interaction for Rouse dumbbells in steady shear flow. *J. Chem. Phys.* 83: 6535-6.

Öttinger, H.C. (1986a). Consistently averaged hydrodynamic interaction for Rouse dumbbells. Series expansions. *J. Chem. Phys.* 84: 4068-73.

Öttinger, H.C. (1986b). Consistently averaged hydrodynamic interaction for Rouse dumbbells: The rheological equation of state. *J. Chem. Phys.* 85: 1669-71.

Öttinger, H.C. (1987). Generalized Zimm model for dilute polymer solutions under theta conditions, *J. Chem. Phys.* 86: 3731-49; erratum. *ibid.* 87: 1460.

Öttinger, H.C. (1989a). Diffusivity of polymers in dilute solutions undergoing homogeneous flows. *A.I.Ch.E. J.* 35: 279-86.

Öttinger, H.C. (1989b). Gaussian approximation for Rouse chains with hydrodynamic interactions. *J. Chem. Phys.* 90: 463-73.

Öttinger, H.C. (1992). Incorporation of polymer diffusivity and migration into constitutive equations. *Rheol. Acta* 31: 14-21.

Öttinger, H.C., and Rabin, Y. (1989). Renormalization-group calculation of viscometric functions based on conventional polymer kinetic theory. *J. Non-Newtonian Fluid Mech.* 33: 53-93.

Pak, H., and Kushner, M.J. (1990). Multi-beam-bulk model for electron transport during commutation in an optically triggered pseudospark thyratron. *Appl. Phys. Lett.* 57: 1619-21.

Pangelinan, A. (1991). *Surface-Induced Molecular Weight Segregation in Thermoplastic Composites*. Ph.D. Dissertation: University of Delaware.

Pao, Y.-H., and Banerjee, D.K. (1973). Thermal pulses in dielectric crystals. *Lett. Appl. Eng. Sci.* 1: 3-41.

Papkov, S.P., Kulichikhin, U.G., Kalmykova, V.D., and Malkin, A.Y. (1974). Rheological properties of anisotropic poly(para-benzamide) solutions. *J. Polym. Sci.: Polym. Phys. Ed.* 12: 1753-70.

Papoulis, A. (1984). *Probability, Random Variables, and Stochastic Processes*, 2nd ed. McGraw-Hill: New York.

Pardowitz, I., and Hess, S. (1982). Elasticity coefficients of nematic liquid crystals. *J. Chem. Phys.* 76: 1485-9.

Parodi, O. (1970). Stress tensor for a nematic liquid crystal. *J. Phys. (Paris)* 31: 581-4.

Parsons, J.D. (1979). Nematic ordering in a system of rods. *Phys. Rev.* A 19: 1225-30.

Pearson, J.R.A. (1979). Non-isothermal rheology of polymers and its significance in polymer processing. *J. Non-Newtonian Fluid Mech.* 6: 81-95.

Pennington, G.J., and Cowin, S.C. (1969). Couette flow of a polar fluid. *Trans. Soc. Rheo*l. 13: 387-403.

Penrose, O., and Fife, P.C. (1990). Thermodynamically consistent models of phase-field type for the kinetics of phase transitions. *Physica* 43D: 44-62.

Penrose, R. (1989). *The Emperor's New Mind.* Oxford University Press: Oxford.

Peppas, N.A., and Urdahl, K.G. (1988). Anomalous penetrant transport in glassy polymers—VIII. Solvent-induced cracking in polystyrene. *Eur. Polym. J.* 24: 13-20.

Pérez-González, J., de Vargas, L., and Tejero, J. (1992). Flow development of xanthan solutions in capillary rheometers. *Rheol. Acta* 31: 83-93.

Peterlin, A. (1961a). Einfluss der endlichen Moleküllänge auf die Gradientenabhängigkeit des Staudinger-Index. *Makromol. Chem.* 43-46: 338-46.

Peterlin, A. (1961b). Streaming birefringence of soft linear macromolecules with finite chain length. *Polymer* 2: 257-64.

Peterlin, A. (1966). Hydrodynamics of macromolecules in a velocity field with longitudinal gradient, *J. Polym. Sci.: Polym. Lett.* B4: 287-91.

Phan-Thien, N. (1978). A nonlinear network viscoelastic model. *J. Rheol.* 22: 259-83.

Phan-Thien, N., and Atkinson, J.D. (1985). A note on the encapsulated dumbbell model. *J. Non-Newtonian Fluid Mech.* 17: 111-6.

Phan-Thien, N., and Tanner, R.I. (1977). A new constitutive equation derived from network theory. *J. Non-Newtonian Fluid Mech.* 2: 353-65.

Phelan, F.R., Malone, M.F., and Winter, H.H. (1989). A purely hyperbolic model for unsteady viscoelastic flow. *J. Non-Newtonian Fluid Mech.* 32: 197-224.

Phillips, R.J., Armstrong, R.C., Brown, R.A., Graham, A.L., and Abbott, J.R. (1992). A constitutive equation for concentrated suspensions that accounts for shear-induced particle migration. *Phys. Fluids* A 4: 30-40.

Pieranski, P., and Guyon, E. (1973). Shear-flow-induced transition in nematics. *Solid State Comm.* 13: 435-7.

Pieranski, P., and Guyon, E. (1974). Two shear-flow regimes in nematic p-n-Hexyloxybenzilidene-p'-aminobensonitrile. *Phys. Rev. Lett.* 32: 924-6.

Pieranski, P., Guyon, E., and Pikin, S.A. (1976). Nouvelles instabilités de cisaillement dans les nématiques. *J. Physique Colloque* 37: C1-3-6.

Pikin, S.A. (1974). Couette flow of a nematic liquid. *Sov. Phys. JETP* 38: 1246-50.

Pipkin, A.C. (1986). *Lectures on Viscoelasticity Theory,* 2nd ed. Springer-Verlag: New York.

Ploehn, H.J., and Russel, W.B. (1989). Self-consistent field model of polymer adsorption: Matched asymptotic expansion describing tails. *Macromolecules* 22: 266-76.

Ploehn, H.J., Russel, W.B., and Hall, C.K. (1988). Self-consistent field model of polymer adsorption: Generalized formulation and ground-state solution. *Macromolecules* 21: 1075-85.

Poisson, S.D. (1809). Sur la variation des constantes arbitraires dans les questions de mécanique. *J. de l'École Polytechnique* 8: 266-344.

Pokrovskii, V.N. (1972). Stresses, viscosity, and optical anisotropy of a moving suspension of rigid ellipsoids. *Soviet Phys. USPEKHI* 14: 737-46.

Poniewierski, A., and Stecki, J. (1979). Statistical theory of the elastic constants of nematic liquid crystals. *Molecular Physics* 38: 1931-40.

Prakash, J.R., and Mashelkar, R.A. (1991). The diffusion tensor for a flowing dilute solution of Hookean dumbbells: Anisotropy and flow rate dependence. *J. Chem. Phys.* 95: 3743-8.

Priest, R.G. (1972). A calculation of the elastic constants of a nematic liquid crystal. *Mol. Cryst. Liq. Cryst.* 17: 129-37.

Priest, R.G. (1973). Theory of the Frank elastic constants of nematic liquid crystals. *Phys. Rev. A* 7: 720-9.

Prigogine, I. (1955). *Thermodynamics of Irreversible Processes.* Interscience Publishers: New York.

Prigogine, I. (1967). *Introduction to Thermodynamics of Irreversible Processes, 3rd Ed.* John Wiley and Sons: New York.

Prigogine, I. (1973). Irreversibility as a symmetry-breaking process. *Nature* 246 (November): 67-71.

Prigogine, I., George, C., Henin, F., and Rosenfeld, L. (1973). A unified formulation of dynamics and thermodynamics. *Chemica Scripta* 4: 5-32.

Princen, H.M. (1983). Rheology of foams and highly concentrated emulsions: I. Elastic properties and yield stress of a cylindrical model system. *J. Colloid Interface Sci.* 91: 160-75.

Pulffy-Muhoray, P., de Bruyn, J.J., and Dunmur, D.A. (1985). Mean field theory of binary mixtures of nematic liquid crystals. *Mol. Cryst. Liq. Cryst.* 127: 301-19.

Rallison, J.M., and Hinch, E.J. (1988). Do we understand the physics of the constitutive equation?. *J. Non-Newtonian Fluid Mech.* 29: 37-55.

Ramamurthy, A.V. (1986a). Wall slip in viscous fluids and the influence of materials of construction. *J. Rheol.* 30: 337-57.

Ramamurthy, A.V. (1986b). LLDPE rheology and blown film fabrication. *Adv. Polym. Techn.* 6: 489-99.

Rangel-Nafaile, R., Metzner, A.B., and Wissbrun, K.F. (1984). Analysis of stress-induced phase separations in solutions. *Macromolecules* 17: 1187-95.

Reichl, L.E. (1980). *A Modern Course in Statistical Physics.* University of Texas Press: Austin.

Reif, R. (1984). Plasma enhanced chemical vapor deposition of thin crystalline semiconductor and conductor films. *J. Vac. Sci.* 2: 429-35.

Renardy, M. (1988a). Inflow boundary conditions for steady flows of viscoelastic fluids with differential constitutive laws. *Rocky Mountain J. Math.* 18: 445-53; erratum. *ibid.* 19: 561.

Renardy, M. (1988b). Recent advances in the mathematical theory of steady flows of viscoelastic fluids. *J. Non-Newtonian Fluid Mech.* 29: 11-24.

Renardy, M., Hrusa, W.J., and Nohel, J.A. (1987). *Mathematical Problems in Viscoelasticity.* Pitman Monographs and Surveys in Pure and Applied Mathematics 35, Longman Scientific and Technical: Essex.

Rey, A.D. (1991). Periodic textures of nematic polymers and orientational slip. *Macromolecules* 24: 4450-6.

Richards, J.I., and Youn, H.K. (1990). *Theory of Distributions: A Non-Technical Introduction.* Cambridge University Press: Cambridge.

Rivlin, R.S. (1980). Material symmetry and constitutive equations. *Ingenieur-Archiv* 49: 325-36.

Rivlin, R.S., and Ericksen, J.L. (1955). Stress-deformation relations for isotropic materials, *J. Rat. Mech. Anal.* 4: 323-425.

Rondelez, F., Ausserré, D., and Hervet, H. (1987). Experimental studies of polymer concentration profiles at solid-liquid and liquid-gas interfaces by optical and x-ray evanescent wave techniques. *Ann. Rev. Phys. Chem.* 38: 317-47.

Rotne, J., and Prager, S. (1969). Variational treatment of hydrodynamic interaction in polymers. *J. Chem. Phys.* 50: 4831-7.

Rouse, P.E. (1953). A theory of the linear viscoelastic properties of dilute solutions of coiling polymers. *J. Chem. Phys.* 21: 1272-80.

Rouse, W.H.S. (1956). *Great Dialogues of Plato (transl.)*. A Mentor book: New York.

Rovelli, C. (1991). Time in quantum gravity: An hypothesis. *Phys. Rev. D* 43: 442-56, and 44: 1339-41.

Rowlinson, J.S., and Widom, B. (1989). *Molecular Theory of Capillarity*, (pbk ed.). Oxford University Press: Oxford.

Rusakov, V.V., and Shliomis, M.I. (1985). Landau-de Gennes free energy expression for nematic polymers. *J. Physique Lett.* 46: L935-43.

Rutkevich, I.M. (1969). Some general properties of the equations of viscoelastic incompressible fluid dynamics. *J. Applied Math. Mech. (PMM, English ed.)* 33: 30-9.

Rutkevich, I.M. (1970). The propagation of small perturbations in a viscoelastic fluid. *J. Applied Math. Mech. (PMM, English ed.)* 34: 35-50.

Rutkevich, I.M. (1972). On the thermodynamic interpretation of the evolutionary conditions of the equations of the mechanics of finitely deformable viscoelastic media of Maxwell type. *J. Applied Math. Mech. (PMM, English ed.)* 36: 283-95.

Ryskin, G. (1985). Misconception which led to the "material frame-indifference" controversy. *Phys. Rev.* A 32: 1239-40.

Ryskin, G. (1987). Reply to "Comments on the 'material frame-indifference' controversy," *Phys. Rev.* A 36: 4526.

Samulski, E.T., and DuPré, D.B. (1979). Polymeric liquid crystals. In: *Advances in Liquid Crystals*, vol. 4. Brown, G.H. (ed.). Academic Press: New York; pp. 121-45.

Sandler, S.I., and Dahler, J.S. (1964). Nonstationary diffusion. *Phys. Fluids* 7: 1743-6.

Sarti, G.C., and Marrucci, G. (1973). Thermomechanics of dilute polymer solutions: Multiple bead-spring model. *Chem. Eng. Sci.* 28: 1053-9.

Scheutjens, J.M.H.M., and Fleer, G.J. (1979). Statistical theory of the adsorption of interacting chain molecules. 1. Partition function, segment density distribution, and adsorption isotherms. *J. Phys. Chem.* 83: 1619-35.

Scheutjens, J.M.H.M., and Fleer, G.J. (1980). Statistical theory of the adsorption of interacting chain molecules. 2. Train, loop, and tail size distribution. *J. Phys. Chem.* 84: 178-90.

Schowalter, W.R. (1978). *Mechanics of Non-Newtonian Fluids*. Pergamon Press: New York.

Schümmer, P., and Otten, B. (1984). A compliment to the bead/spring theory of polymer solutions: Inclusion of the relationship between microscopic and macroscopic velocity gradients. In: *Advances in Rheology, vol 1: Theory*. Mena, B., García-Réjon, A., and Rangel-Nafaile, C., (eds.). Universidad Nacional Autonoma de México, México; pp. 399-403.

Sekiguchi, T., and Wolf, K.B. (1987). The Hamiltonian formulation of optics. *Amer. J. Phys.* 55:830-5.

Seliger, R.L., and Whitham, G.B. (1968). Variational principles in continuum mechanics. *Proc. Roy. Soc.* 305A: 1-25.

Semenov, A.N. (1983). Rheological properties of a liquid-crystal solution of rodlike molecules. *Sov. Phys. JETP* 58: 321-26.

Semenov, A.N. (1987). Rheological properties of a nematic solution of semiflexible macromolecules. *Sov. Phys. JETP* 66: 712-6.

Semenov, A.N., and Kokhlov, A.R. (1988). Statistical physics of liquid-crystalline polymers. *Sov. Phys. Usp.* 31: 988-1014.

Shafer, R.M., Laiken, N., and Zimm, B.H. (1974). Radial migration of DNA molecules in cylindrical flow. I. Theory of the free-draining model. *Biophys. Chem.* 2: 180-4.

Sheng, P., and Priestley, E.B. (1974). The Landau-de Gennes theory of liquid crystal phase transitions. In: *Introduction to Liquid Crystals*, Priestley, E.B., Wojtowicz, P.J., and Sheng, P. (eds.). Plenum Press: New York; pp. 143-201.

Sherry, G.C. (1990). A generalized Hamiltonian formalism unifying classical and quantum mechanics. *Found. Phys. Lett.* 3: 255-65.

Shimada, T., Doi, M. and Okano, K. (1988). Concentration fluctuation of stiff polymers. III. Spinodal decomposition. *J. Chem. Phys.* 88: 7181-6.

Sieniutycz, S. (1981). Thermodynamics of coupled heat, mass and momentum transport with finite wave speed. II. Examples of transformations of fluxes and forces. *Int. J. Heat and Mass Transfer* 24: 1759-69.

Simo, J.C., Marsden, J.E., and Krishnaprasad, P.S. (1989). The Hamiltonian structure of nonlinear elasticity: The material and convective repre-sentations of solids, rods, and plates. *Arch. Rat. Mech. Anal.* 104: 125-83.

Skarp, K., Carlsson, T., Lagerwall, S.T., and Stebler, B. (1981). Flow properties of nematic 8CB: An example of diverging and vanishing α_3. *Mol. Cryst. Liq. Cryst.* 66: 199-208.

Slattery, J.C. (1990). *Interfacial Transport Phenomena.* Springer-Verlag: New York.

Smith, M.J., and Peppas, N.A. (1985). Effect of the degree of crosslinking of penetrant transport in polystyrene. *Polymer* 26: 569-74.

Sobotka, Z. (1984). *Rheology of Materials and Engineering Structures. Rheology Series*, 2. Elsevier: Amsterdam.

Souvaliotis: A., and Beris, A.N. (1992). An extended White-Metzner viscoelastic fluid model based on an internal structural parameter. *J. Rheol.* 36: 241-71.

Spencer, R.G. (1982). The Hamiltonian structure of multi-species fluid electrodynamics. In: *Mathematical Methods in Hydrodynamics and Integrability in Dynamical Systems*, Tabor, M., and Treve, Y.M. (Eds.). La Jolla Institute 1981, APS: New York; pp. 121-6.

Spencer, R.G. (1984). Poisson structure of the equations of ideal multispecies fluid electrodynamics. *J. Math. Phys.* 25: 2390-6.

Spencer, R.G., and Kaufman, A.N. (1982). Hamiltonian structure of two-fluid plasma dynamics. *Phys. Rev.* A 25: 2437-9.

Speziale, C.G. (1987). Comments on the "material frame-indifference" controversy. *Phys. Rev.* A 36: 4522-5.

Speziale, C.G. (1988). The Einstein equivalence principle. Intrinsic spin and the invariance of constitutive equations in continuum mechanics. *Int. J. Eng. Sci.* 26: 211-20.

Srinivasarao, M., and Berry, G.C. (1991). Rheo-optical studies of aligned nematic solutions of a rodlike polymer. *J. Rheol.* 35: 379-97.

Stephen, M.J., and Straley, J.P. (1974). Physics of liquid crystals. *Rev. Mod. Phys.* 46: 617-704.

Straley, J.P. (1973a). Frank elastic constants of the hard-rod liquid crystal. *Phys. Rev.* A 8: 2181-3.

Straley, J.P. (1973b). The gas of long rods as a model for lyotropic liquid crystals. *Mol. Cryst. Liq. Cryst.* 22: 333-57.

Straley, J.P. (1973c). Third virial coefficient for the gas of long rods. *Mol. Cryst. Liq. Cryst.* 24: 7-20.

Struminskii, V.V. (1974). The effect of diffusion rate on the flow of gas mixtures. *Appl. Math. Mech.* 34: 181-8.

Struminskii, V.V., and Shavaliyev, M.S. (1987). Transport phenomena in multivelocity, multitemperature gas mixtures. *Appl. Math. Mech.* 50: 59-64.

Stueckelberg, E.C.G. (1942). La mécanique du point matériel en théorie de relativité et en théorie des quanta. *Helv. Phys. Acta* 15: 23-37.

Tanner, R.I. (1975). Stresses in dilute solutions of bead-nonlinear-spring macromolecules. II. Unsteady flows and approximate constitutive relations. *Trans. Soc. Rheol.* 19: 37-65.

Tanner, R.I. (1985). *Engineering Rheology*. Oxford University Press: Oxford.

Tanner, R.I., and Stehrenberger, W. (1971). Stresses in dilute solutions of bead-nonlinear-spring macromolecules. I. Steady potential and plane flows, *J. Chem. Physics* 55: 1958-64.

ten Bosch, A., Maïssa, P., and Sixou, P. (1983). A Landau-de Gennes theory of nematic polymers. *J. Physique—Lett.* 44: L-105-11.

ten Bosch, A., and Sixou, P. (1987). Elastic constants of semirigid liquid-crystal polymers. *J. Chem. Phys.* 86: 6556-9.

Ter Haar, D. (1969). *Men of Physics: L.D. Landau,* vol. 2: *Thermodynamics, Plasma Physics and Quantum Mechanics*. Pergamon Press: Oxford.

Theodorou, D.N. (1988a). Microscopic structure and thermodynamic properties of bulk copolymers and surface-active polymers at interfaces. 1. Theory. *Macromolecules* 21: 1411-21.

Theodorou, D.N. (1988b). Microscopic structure and thermodynamic properties of bulk copolymers and surface-active polymers at interfaces. 2. Results for some representative chain architectures. *Macromolecules* 21: 1422-36.

Thomas, N., and Windle, A.H. (1978). Transport of methanol in poly-(methyl methacrylate). *Polymer* 19: 255-65.

Thomas, N.L., and Windle, A.H. (1982). A theory of Case II diffusion. *Polymer* 23: 529-42.

Thomson, J.J. (1931). *James Clerk Maxwell: A Commemoration Volume*. Cambridge University Press: Cambridge.

Threefoot, S.A. (1991). *Relaxation Modes in Liquid Crystalline Polymers*. Ph.D. Dissertation: University of Delaware.

Tirrell, M., and Malone, M.F. (1977). Stress-induced diffusion of macromolecules, *J. Polym. Sci.: Polym. Phys. Ed.* 15: 1569-83.

Titow, W.V., Braden, M., Currell, B.R., and Loneragan, R.J. (1974). Diffusion and some structural effects of two chlorinated hydrocarbon solvents in bisphenol A polycarbonate. *J. Appl. Polym. Sci.* 18: 867-86.

Tolédano, J.-C., and Tolédano, P. (1987). *The Landau theory of phase transitions*. World Scientific: Singapore.

Treloar, L.R.G. (1942-43). The structure and elasticity of rubber. *Rept. Progr. Phys.* 9: 113-36.

Treloar, L.R.G. (1975). *The Physics of Rubber Elasticity,* 3rd Ed. Oxford University Press: London.

Truesdell, C. (1962). Mechanical basis for diffusion. *J. Chem. Phys.* 37: 2336-44.

Truesdell, C. (1966). *The Mechanical Foundations of Elasticity and Fluid Dynamics*. Gordon and Breach: New York.

Truesdell, C. (1976). Correction of two errors in the kinetic theory of gases which have been used to cast unfounded doubt upon the principle of material indifference. *Meccanica* 11: 196-9.

Truesdell, C. (1984). *Rational Thermodynamics,* 2nd ed. Springer-Verlag: New York.

Truesdell, C., and Muncaster, R.G. (1980). *Fundamentals of Maxwell's Kinetic Theory of a Simple Monatomic Gas*. Academic Press: New York.

Truesdell, C., and Noll, W. (1965). The non-linear field theories of mechanics. In: *Handbuch der Physic,* vol. III/3. Flügge, S. (ed.). Springer-Verlag: Berlin.

Tsang, W.K.-W., and Dealy, J.M. (1981). The use of larde transient deformations to evaluate rheological models for molten polymers. *J. Non-Newtonian Fluid Mech.* 9: 203-22.

Tschoegl, N.W. (1989). *The Phenomenological Theory of Linear Viscoelastic Behavior: An Introduction*. Springer-Verlag: Berlin.

VanArsdale, W.E. (1985). Fractional rates of deformation. *J. Rheol.* 29: 851-7.

Van der Brule, B.H.A.A. (1990). The non-isothermal elastic dumbbell: A model for the thermal conductivity of a polymer solution. *Rheol. Acta* 29: 416-22.

Van der Zanden, J., and Hulsen, M. (1988). Mathematical and physical requirements for successful computations with viscoelastic fluid models. *J. Non-Newtonian Fluid Mech.* 29: 93-117.

Van Kampen, N.G. (1973). Nonlinear irreversible processes. *Physica* 67: 1-22.

Van Rysselberghe, P. (1962). General reciprocity relation between the rates and affinities of simultaneous chemical reactions. *J. Chem. Phys.* 36: 1329-30.

Van Rysselberghe, P. (1965). On a reciprocity relation of Péneloux and Van Rysselberghe. *J. Chem. Phys.* 43: 3422.

Varshney, S.K. (1986). Liquid crystalline polymers: A novel state of material. *Rev. Macromol. Chem. Phys.* C26: 551-650.

Verdier, C., and Joseph, D.D. (1989). Change of type and loss of evolution of the White-Metzner model. *J. Non-Newtonian Fluid Mech.* 31: 325-43.

Vertogen, G. (1982). The theory of elastic constants of nematics. *Phys. Lett.* 89A: 448-50.

Vertogen, G., and de Jeu, W.H. (1988). *Thermotropic Liquid Crystals, Fundamentals*. Springer-Verlag: Berlin.

Vinogradov, G., Malkin, A., Yanovskii, Y., Borisenkova, E., Yarlykov, B., and Berezhnaya, G. (1972). Viscoelastic properties and flow of narrow distribution polybutadienes and polyisoprenes. *J. Polymer Sci. Part A-2* 10: 1061-84.

Vrentas, J.S., Duda, J.L., and Hou, A.-C. (1984). Anomalous sorption in poly(ethyl methacrylate). *J. Appl. Polym. Sci.* 29: 399-406.

Vroege, G.J., and Odijk, T. (1988). Induced chain rigidity, splay modulus, and other properties of nematic polymer liquid crystals. *Macromolecules* 21: 2848-58.

Wall, F.T. (1942a). Statistical thermodynamics of rubber I. *J. Chem. Phys.* 10: 132-4.

Wall, F.T. (1942b). Statistical thermodynamics of rubber II. *J. Chem. Phys.* 10: 485-8.

Wagner, M.H. (1979). Zur Ntzwerktheorie von Polymer-Schmelzen. *Rheol. Acta* 18: 33-50.

Wang, C.C. (1975). On the concept of frame indifference in continuum mechanics and the kinetic theory of gases. *Arch. Rat. Mech. Anal.* 58: 381-93.

Ware, R.A., Tirtowidjojo, S., and Cohen, C. (1981). Diffusion and induced crystallization in polycarbonate. *J. Appl. Polym. Sci.* 29: 2975-88.

Warner, Jr., H.R. (1972). Kinetic theory and rheology of dilute suspensions of finitely extendible dumbbells. *Ind. Eng. Chem. Fundam.* 11: 379-87.

Warner, M., Gunn, J.M.F., and Baumgärtner, A.B. (1985). Rod to coil transitions in nematic polymers. *J. Phys. A: Math. Gen.* 18: 3007-26.

Wedgewood, L.E. (1988). A constitutive equation for a dilute solution of Hookean dumbbells with approximate hydrodynamic interaction. *J. Non-Newtonian Fluid Mech.* 30: 21-35.

Wedgewood, L.E. (1989). A Gaussian closure of the second-moment equation for a Hookean dumbbell with hydrodynamic interaction. *J. Non-Newtonian Fluid Mech.* 31: 127-42.

Wedgewood, L.E., and Bird, R.B. (1988). From molecular models to the solution of flow problems. *Ind. Eng. Chem. Res.* 27: 1313-20.

Wedgewood, L.E., Ostrov, D.N., and Bird, R.B. (1991). A finitely extensible bead-spring chain model for dilute polymer solutions. *J. Non-Newtonian Fluid Mech.* 40: 119-39.

Wei, J. (1962). Axiomatic treatment of chemical reaction systems. *J. Chem. Phys.* 36: 1578-84.

Wei, J., and Pratter, C.D. (1962). The structure and analysis of complex reaction systems. In: *Advances in Catalysis,* vol. 13. Eley, D.D., Selwood, P.W., and Weisz, P.B. (eds.). Academic Press: New York; pp. 203-392.

Wei, J., and Zahner, J.C. (1965). Comment on the general reciprocity relation of Van Rysselberghe. *J. Chem. Phys.* 43: 3421.

Weinstein, A., and Morrison, P.J. (1981). Comments on: The Maxwell-Vlasov equations as a continuous Hamiltonian system. *Phys. Lett.* 86A: 235-6.

White, J.L. (1990). *Principles of Polymer Engineering Rheology.* John Wiley and Sons: New York.

White, J.L., and Metzner, A.B. (1963). Development of constitutive equations for polymeric melts and solutions. *J. Appl. Polym. Sci.* 7: 1867-89.

Wiegel, F.W. (1969a). A network model for viscoelastic fluids. *Physica* 42: 156-64.

Wiegel, F.W. (1969b). Rheological properties of a network model for macromolecular fluids. *Physica* 43: 33-44.

Wiegel, F.W. (1986). *Introduction to Path-Integral Methods in Physics and Polymer Science.* World Scientific: Singapore.

Wiest, J.M. (1989). A differential constitutive equation for polymer melts. *Rheol. Acta* 28: 4-12.

Williams, M.C. (1965). Normal stresses in polymer solutions with remarks on the Zimm treatment. *J. Chem. Phys.* 42: 2988-9; erratum. *J. Chem. Phys.* 43: 4542.

Winter, H.H., and Chambon, F. (1986). Analysis of linear viscoelasticity of a crosslinking polymer at the gel point. *J. Rheol.* 30: 367-82.

Wissbrun, K.F. (1981). Rheology of rod-like polymers in the liquid crystalline state. *J. Rheol.* 25: 619-62.

Wissbrun, K.F. (1985). A model for domain flow of liquid-crystal polymers. *Faraday Discuss. Chem. Soc.* 79: 161-73.

Wolff, U., Greubel, W., and Krüger, H. (1973). The homogeneous alignment of liquid crystal layers. *Mol. Cryst. Liq. Cryst.* 23: 187-96.

Woods, L.C. (1975). *The Thermodynamics of Fluid Systems*. Oxford University Press: Oxford.[4]

Woods, L.C. (1987). *Principles of Magnetoplasma Dynamics*. Oxford University Press: Oxford.

Woods, L.C. (1993). *An Introduction to the Kinetic Theory of Gases and Magnetoplasmas*. Oxford University Press: Oxford.

Wright, D.C., and Mermin, N.D. (1989). Crystalline liquids: The blue phases. *Rev. Mod. Phys.* 61: 385-432.

Wu, X.L., Pine, D.J., and Dixon, P.K. (1991). Enhanced concentration fluctuations in polymer solutions under shear flow. *Phys. Rev. Lett.* 66: 2408-11.

Yamakawa, H. (1970). Transport properties of polymer chains in dilute solution: Hydrodynamic interaction. *J. Chem. Phys.* 53: 436-43.

Yanase, H., and Asada, T. (1987). Rheology of a lyotropic polymer liquid crystal of rodlike polymers and main-chain thermotropic liquid-crystalline polymers. *Mol. Cryst. Liq. Cryst.* 153: 281-90.

Yang, I.-K. (1992). *The Rheology of an Electrically Oriented Liquid Crystalline Polymer*. Ph.D. Dissertation: University of Delaware.

Yoon, D.Y., and Flory, P.J. (1989). Conformational rearrangement in the nematic phase of a polymer comprising rigid and flexible sequences in alternating succession. II. Theoretical results and comparison with experiments. *Mat. Res. Soc. Symp. Proc.* 134: 11-9.

Yourgrau, W., van der Merwe, A., and Raw, C. (1966). *Treatise on Irreversible and Statistical Thermophysics*. MacMillan Co.: New York. Also available in paperback, (1982). Dover: New York.

Ziegler, H. (1958). An attempt to generalize Onsager's principle, and its significance for rheological problems. *ZAMP* 9: 748-63.

Zimm, B.H. (1956). Dynamics of polymer molecules in dilute solution: Viscoelasticity, flow birefringence and dielectric loss. *J. Chem. Phys.* 24: 269-78.

[4]In the humble opinion of the authors, this is the most decisive description of the traditional formulation of irreversible thermodynamics written to date.

Zocher, H. (1933). The effect of a magnetic field on the nematic state. *Trans. Faraday Soc.* 29: 945-57.

Zúñiga, I., and Leslie, F.M. (1989). Shear flow instabilities in nonaligning nematic liquid crystals. *Europhys. Lett.* 9: 689-93.

Zylka, W. (1991). Gaussian approximation and Brownian dynamics simulations for Rouse chains with hydrodynamic interaction undergoing simple shear flow. *J. Chem. Phys.* 94: 4628-36.

Zylka, W., and Öttinger, H.C. (1989). A comparison between simulations and various approximations for Hookean dumbbells with hydrodynamic interaction. *J. Chem. Phys.* 90: 474-80.

Zylka, W., and Öttinger, H.C. (1991). Calculation of various universal properties for dilute polymer solutions undergoing shear flow. *Macromolecules* 24: 484-94.

Author Index

Subject Index